# THE YEASTS
Volume 2

# THE YEASTS

*Edited by*

ANTHONY H. ROSE
*School of Biological Sciences, Bath University,*
*Claverton Down, Bath, England.*

AND

J. S. HARRISON
*The Distillers Co. (Yeast) Ltd., Great Burgh, Epsom,*
*Surrey, England.*

**Volume 2**

**PHYSIOLOGY AND BIOCHEMISTRY OF YEASTS**

1971

ACADEMIC PRESS·LONDON and NEW YORK

ACADEMIC PRESS INC. (LONDON) LTD.
BERKELEY SQUARE HOUSE
BERKELEY SQUARE
LONDON, WIX 6BA

*U.S. Edition published by*
ACADEMIC PRESS INC.
111 FIFTH AVENUE
NEW YORK, NEW YORK 10003



*Library of Congress Catalog Card Number:* 72–85465

SBN: 12–596402–1



Printed in Great Britain by
William Clowes and Sons Limited, London, Colchester and Beccles

# Contributors

BARTNICKI-GARCIA, S., *Department of Plant Pathology, University of California, Riverside, California, U.S.A.*

CHICHESTER, C. O., *Department of Food and Resource Chemistry, University of Rhode Island, Kingston, Rhode Island, U.S.A.*

DELAFUENTE, G., *Consejo Superior de Investigaciones Científicas, Madrid, Spain.*

GANCEDO, C., *Consejo Superior de Investigaciones Científicas, Madrid, Spain.*

HARRISON, J. S., *The Distillers Co. (Yeast) Ltd., Epsom, Surrey, England.*

HUNTER, K., *School of Biological Sciences, Bath University, Bath, Somerset, England.*

MANNERS, D. J., *Department of Brewing & Biochemistry, Heriot-Watt University, Edinburgh, Scotland.*

MCMURROUGH, I., *Department of Plant Pathology, University of California, Riverside, California, U.S.A.*

MOUNOLOU, J.-C., *Institut de Biologie Moleculaire, Faculté des Sciences, Paris, France.*

OURA, E., *State Alcohol Monopoly, Helsinki, Finland.*

PHAFF, H. J., *Department of Food Science & Technology, University of California, Davis, California, U.S.A.*

ROBICHON-SZULMAJSTER, H. DE, *Laboratoire d'Enzymologie, CNRS, Gif-sur-Yvette, Seine-et-Oise, France.*

ROSE, A. H., *School of Biological Sciences, Bath University, Bath, Somerset, England.*

SIMPSON, K. L., *Department of Food and Resource Chemistry, University of Rhode Island, Kingston, Rhode Island, U.S.A.*

SOLS, A., *Consejo Superior de Investigaciones Científicas, Madrid, Spain.*

STOKES, J. L., *Department of Bacteriology & Public Health, Washington State University, Pullman, Washington, U.S.A.*

SUOMALAINEN, H., *State Alcohol Monopoly, Helsinki, Finland.*

SURDIN-KERJAN, Y., *Laboratoire d'Enzymologie, CNRS, Gif-sur-Yvette, Seine-et-Oise, France.*

VAN UDEN, N., *Laboratory of Microbiology, Gulbenkian Institute of Science, Oeiras, Portugal.*

# Preface to Volume I

It is often said that yeasts have been more intimately associated than any other group of micro-organisms with the progress and well-being of mankind. Many would base this claim largely on the capacity of certain yeasts to bring about a rapid and effective alcoholic fermentation, an activity which has been exploited for centuries in the manufacture of alcoholic beverages and in the leavening of bread. Others may prefer to acknowledge the contribution which yeasts have made in the elucidation of the basic metabolic processes of living cells. Few, however, would deny that yeast species, although less numerous than those of other major groups of micro-organisms, are sufficiently important that information on them needs to be fully documented so that scholars can continue to exploit their characteristic qualities. Over the years, many excellent texts have been published that deal comprehensively with yeasts and their activities. The present set of three volumes aims to bring the literature on this subject up to date.

So much has been written on yeasts that it is now impossible to encompass our knowledge of their microbiology in one volume. Moreover, it is beyond the capacity of any one person to write such a review, and so recourse must be made to a team of specialists each reviewing his or her own field of interest.

The first of the present volumes deals with the biology of yeasts, the second with yeast physiology and biochemistry, and the third with yeast technology. Together, these volumes present an up-to-date account of knowledge on yeasts and their activities. At the same time, each volume is sufficiently complete in itself to be perused separately.

We are extremely grateful to the authors of the chapters, whose labours have made this effort possible. Their forbearance, often under conditions of pressure and always with a deadline looming ahead, is very greatly appreciated. Our hope is that readers of these volumes will consider the venture to have been worthwhile.

*July 1969*

Anthony H. Rose
J. S. Harrison

# Contents

Contributors v

Preface to Volume I vii

Contents of Volumes I and 3 xii & xiii

Abbreviations xiv

1. Introduction . . . . . . . . . . . . 1
   A. H. Rose and J. S. Harrison

2. Yeast Nutrition and Solute Uptake . . . . . . 3
   H. Suomalainen and E. Oura
   - I. Nutrition . . . . . . . . . . . 3
   - II. Solute Uptake . . . . . . . . . 36
   - III. Outflow of Compounds . . . . . . . 57
   - IV. Acknowledgement . . . . . . . . 59
   - References . . . . . . . . . . 60

3. Kinetics and Energetics of Yeast Growth . . . . . 75
   N. van Uden
   - I. Introduction . . . . . . . . . . 75
   - II. Notation . . . . . . . . . . . 76
   - III. Exponential Growth . . . . . . . . 79
   - IV. Dependence of the Specific Growth Rate on the Concentration of a Limiting Nutrient . . . . . . . . 82
   - V. Dependence of the Specific Growth Rate on Temperature . 91
   - VI. Yield and Maintenance Relations . . . . . . 100
   - VII. A Model . . . . . . . . . . . 109
   - References . . . . . . . . . . 116

4. Influence of Temperature on the Growth and Metabolism of Yeasts . . . . . . . . . . . . 119
   J. L. Stokes
   - I. Introduction . . . . . . . . . . 119
   - II. Growth . . . . . . . . . . . 120
   - III. Sporulation . . . . . . . . . . 127

IV. Survival . . . . . . . . . . 127
V. Other Aspects . . . . . . . . . 131
VI. Acknowledgements . . . . . . . . 132
References . . . . . . . . . . 132

5. Structure and Biosynthesis of the Yeast Cell Envelope . . 135

H. J. Phaff

I. Introduction . . . . . . . . . 135
II. Chemistry and Biochemistry of Yeast Cell Walls . . 137
III. Chemistry and Biochemistry of Capsular Polysaccharides . 174
IV. Enzymic Lysis of Yeast Cell Walls . . . . . . 186
V. Regeneration of Cell Walls from Sphaeroplasts . . . 193
VI. Morphology . . . . . . . . . . 198
VII. Cell Wall Composition as a Basis for Differentiating Yeasts . 200
References . . . . . . . . . . 202

6. Yeast Lipids and Membranes . . . . . . . 211

K. Hunter and A. H. Rose

I. Introduction . . . . . . . . . . 211
II. Yeast Lipids . . . . . . . . . . 212
III. Lipid Metabolism . . . . . . . . . 234
IV. Composition, Structure and Function of Yeast Membranes . 254
V. Acknowledgements . . . . . . . . 264
References . . . . . . . . . . 264

7. Energy-Yielding Metabolism in Yeasts . . . . . . 271

A. Sols, C. Gancedo and G. Delafuente

I. Introduction . . . . . . . . . . 271
II. Anaerobic Metabolism . . . . . . . . 273
III. Aerobic Metabolism . . . . . . . . 286
IV. Energy Sources other than Glucose . . . . . . 290
V. Energy Reserves and their Mobilization . . . . . 292
VI. Regulation of Energy Metabolism . . . . . . 297
VII. Acknowledgements . . . . . . . . 303
References . . . . . . . . . . 303

8. The Properties and Composition of Yeast Nucleic Acids . 309

J.-C. Mounolou

I. Introduction . . . . . . . . . . 309
II. Deoxyribonucleic Acids . . . . . . . . 310
III. Ribonucleic Acids . . . . . . . . . 316
IV. Mitochondrial Nucleic Acids . . . . . . . 324
V. Concluding Remarks . . . . . . . . 326
References . . . . . . . . . . 327

**9. Nucleic Acid and Protein Synthesis in Yeasts: Regulation of Synthesis and Activity** . . . . . . . . . 335

H. DE ROBICHON-SZULMAJSTER and Y. SURDIN-KERJAN

I. Introduction . . . . . . . . . . . 336
II. Biosynthesis of Purine and Pyrimidine Derivatives and its Regulation . . . . . . . . . . . 338
III. Biosynthesis of Amino Acids and its Regulation . . . . 348
IV. Protein Biosynthesis . . . . . . . . . 388
V. Nucleic Acid Biosynthesis . . . . . . . 402
References . . . . . . . . . . . 409

**10. The Structure and Biosynthesis of Storage Carbohydrates in Yeasts** . . . . . . . . . . . 419

D. J. MANNERS

I. Introduction . . . . . . . . . . . 419
II. Trehalose . . . . . . . . . . . 420
III. Glycogen . . . . . . . . . . . 423
References . . . . . . . . . . . 437

**11. Biochemistry of Morphogenesis in Yeasts** . . . . . . 441

S. BARTNICKI-GARCIA and IAN MCMURROUGH

I. Introduction . . . . . . . . . . . 441
II. Cell Wall Biochemistry: Morphogenetic Aspects . . . . 444
III. Vegetative Development . . . . . . . . 464
IV. Sexual Sporulation . . . . . . . . . 483
V. Acknowledgements . . . . . . . . . 487
References . . . . . . . . . . . 487

**12. Carotenoid Pigments of Yeasts** . . . . . . . . 493

K. L. SIMPSON, C. O. CHICHESTER and H. J. PHAFF

I. Introduction . . . . . . . . . . . 493
II. Cultural Conditions for Pigment Production . . . . 494
III. Extraction of Pigments . . . . . . . . 499
IV. Chromatography and Identification . . . . . . 501
V. Biosynthesis of Carotenoids . . . . . . . 504
VI. Conclusion . . . . . . . . . . . 513
References . . . . . . . . . . . 513

**Author Index** . . . . . . . . . . . 517

**Subject Index** . . . . . . . . . . . 545

# Contents of Volume 1

## Biology of Yeasts

1. Introduction
   by A. H. Rose and J. S. Harrison
2. Taxonomy and Systematics of Yeasts
   by N. J. W. Kreger-van Rij
3. Distribution of Yeasts in Nature
   by Lídia do Carmo-Sousa
4. Yeasts as Human and Animal Pathogens
   by J. C. Gentles and C. J. La Touche
5. Yeasts Associated with Living Plants and their Environs
   by F. T. Last and D. Price
6. Yeast Cytology
   by Ph. Matile, H. Moor and C. F. Robinow
7. Sporulation and Hybridization of Yeasts
   by R. R. Fowell
8. Yeast Genetics
   by Robert K. Mortimer and Donald C. Hawthorne
9. Life Cycles in Yeasts
   by R. R. Fowell

## Contents of Volume 3

### Yeast Technology

1. Introduction
   by J. S. Harrison and A. H. Rose
2. Yeasts in Wine-making
   by R. E. Kunkee and M. A. Amerine
3. The Role of Yeasts in Cider-making
   by F. W. Beech and R. R. Davenport
4. Brewer's Yeasts
   by C. Rainbow
5. Saké Yeast
   by K. Kodama
6. Yeasts in Distillery Practice
   by J. S. Harrison and J. C. J. Graham
7. Baker's Yeast
   by S. Burrows
8. Food Yeasts
   by H. J. Peppler
9. Yeasts as Spoilage Organisms
   by H. W. Walker and J. C. Ayres
10. Miscellaneous Products from Yeast
    by J. S. Harrison

# Abbreviations

The following abbreviations are used for names of yeast genera:

| | | | |
|---|---|---|---|
| *Aureobasidium* | *A.* | *Lipomyces* | *L.* |
| *Ashbya* | *Ash.* | *Metschnikowia* | *M.* |
| *Bullera* | *B.* | *Nematospora* | *N.* |
| *Brettanomyces* | *Br.* | *Oospora* | *O.* |
| *Candida* | *C.* | *Pichia* | *Pi.* |
| *Citeromyces* | *Cit.* | *Pityrosporum* | *Pit.* |
| *Coccidiascus* | *Co.* | *Pullularia* | *Pull.* |
| *Cryptococcus* | *Cr.* | *Rhodotorula* | *Rh.* |
| *Debaryomyces* | *D.* | *Saccharomyces* | *Sacch.* |
| *Endomycopsis* | *E.* | *Saccharomycodes* | *S.* |
| *Eremascus* | *Erem.* | *Schizosaccharomyces* | *Schizosacch.* |
| *Geotrichum* | *G.* | *Schwanniomyces* | *Schw.* |
| *Hansenula* | *H.* | *Sporobolomyces* | *Sp.* |
| *Hanseniaspora* | *Ha.* | *Torulopsis* | *T.* |
| *Itersonilia* | *I.* | *Trichosporon* | *Trich.* |
| *Kloeckera* | *Kl.* | *Trigonopsis* | *Trig.* |
| *Kluyveromyces* | *Kluyv.* | *Zygosaccharomyces* | *Zygosacch.* |

The abbreviations used for chemical compounds are those recommended by the *Biochemical Journal* (1967; **102,** 1). Enzymes are referred to by the trivial names recommended by the "Report of the Commission on Enzymes of the International Union of Biochemistry" (1961, Pergamon Press, Oxford). All temperatures recorded in this book are in degrees Centigrade.

## *Chapter 1*

# Introduction

Anthony H. Rose
*School of Biological Sciences, Bath University,*
*Claverton Down, Bath, England*

AND

J. S. Harrison
*Research Department, The Distillers Co. (Yeast) Ltd.,*
*Great Burgh, Epsom, Surrey, England*

The association between yeasts and the science of biochemistry is an old and well-established one. Indeed, biochemistry was itself born out of yeast technology. The brothers Hans and Eduard Buchner, working in Germany in 1897, were interested in preparing extracts of yeast for medicinal purposes. To do this, they ground brewer's yeast with sand, mixed it with kieselguhr, and squeezed out the juice. In an attempt to find some way of preserving their extracts, they tried adding large amounts of cane sugar, since it was known that solutions which contain high concentrations of sugar are less prone to microbial infection. To their surprise, they discovered that the cane sugar was rapidly fermented by the yeast juice. For the first time, therefore, a means had been found of studying alcoholic fermentation in a test tube in the absence of living cells. This discovery sparked off a whole series of studies in which chemists—or rather biochemists as they came to be known—set out to discover the nature of the intermediate steps in the fermentation of glucose to ethanol and carbon dioxide. This work spanned several years, and was associated with the names of many famous biochemists including Embden, Meyerhof, Parnas, Harden and Young.

This charting of the first metabolic pathway set a pattern for all subsequent investigations in intermediary metabolism of living organisms. The ease with which brewer's and baker's yeast can be obtained in bulk quantities has been the major factor in ensuring the popularity of *Saccharomyces cerevisiae* as an experimental organism in biochemical studies. This yeast species has another advantage in that it has a sexual cycle and so is amenable to genetical analysis. The main competitor to *Sacch. cerevisiae* as the biochemist's "work horse" has been *Escherichia coli*, a bacterium which also can easily be grown in the laboratory and which can be analysed genetically. Indeed, during the past decade, many of the major advances in molecular biology have come from studies on *E. coli* rather than *Sacch. cerevisiae*, so much so that it has been proposed by Francis H. C. Crick that one of the major themes to be explored in a projected European molecular biology programme should be the "complete solution" of *Escherichia coli* K-12; Crick has christened this Project K. Although there are some dissenting voices, it cannot be denied that there are compelling reasons for mounting a concentrated programme of this type. It must be remembered, however, that Project K is concerned with a procaryotic microbe, in which there is a minimum of intracellular differentiation into membrane-bound structures and organelles, and that the data obtained will not be entirely applicable to cells of higher organisms.

It is clear, nevertheless, that a "complete solution" of one or more types of unicellular higher organism, that is of eucaryotic cells, which have membrane-bound nuclei and a more or less extensive system within intracellular membranes, is equally desirable. When this task is contemplated, it will be difficult to overlook the candidature of *Saccharomyces cerevisiae*. Such an effort—Project Y—would nicely complement Crick's proposed Project K. It cannot be denied that the amount of published material on the physiology and biochemistry of *Sacch. cerevisiae* is very extensive indeed. If this volume, the second in the three-volume treatise on *The Yeasts*, in any way helps to foster Project Y, then that alone will be a justification for its presentation.

## *Chapter 2*

# Yeast Nutrition and Solute Uptake

HEIKKI SUOMALAINEN AND ERKKI OURA

*Research Laboratories of the State Alcohol Monopoly (Alko), Helsinki, Finland*

I. NUTRITION . . . . . 3
- A. Introduction . . . . . 3
- B. Carbon Sources . . . . . 4
- C. Sources of Nitrogen . . . . . 10
- D. Sources of Phosphorus . . . . . 15
- E. Sources of Sulphur . . . . . 16
- F. Minerals . . . . . 17
- G. Growth Factors . . . . . 20
- H. Reserve Carbohydrates, Fat and Phosphate . . . . . 26
- I. Natural and Industrial Sources of Yeast Nutrients . . . . . 28

II. SOLUTE UPTAKE . . . . . 36
- A. Introduction . . . . . 36
- B. Uptake of Solutes . . . . . 39
- C. The Roles of Enzymes Located on the Cell Surface . . . . . 55

III. OUTFLOW OF COMPOUNDS . . . . . 57

IV. ACKNOWLEDGEMENT . . . . . 59

REFERENCES . . . . . 60

## I. Nutrition

### A. INTRODUCTION

Yeasts are facultative organisms which have the ability to produce energy for their own use from suitable organic compounds under both anaerobic and aerobic conditions. In the yeast cells, the sugar is metabolized to carbon dioxide and water, for energy production, under aerobic conditions; the sugar molecule has thus provided the cell with the maximum amount of energy. If no oxygen is available, the energy production per sugar molecule is poor, as under anaerobic conditions the fermentation of sugar primarily leads to the formation of ethanol. This may also occur in an aerated medium if a large excess of sugar is present.

As a result of the extensive surface area in relation to cell mass, the uptake of nutrients by yeast takes place rapidly. One gram of pressed brewer's yeast, with a dry matter content of 25%, has an area of 3·9 square metres (Just, 1940); one cell of brewer's yeast is able to transform about $10^7$ molecules of maltose per second to carbon dioxide and alcohol (Just, 1940). Correspondingly, the rate of glucose uptake is about $2 \times 10^7$ molecules per second (Musfeld, 1942). Carbohydrates, only poorly soluble in fat, penetrate into the cell so rapidly that the transport mechanism cannot be explained merely by diffusion (cf. Ørskov, 1945); the transfer into the cell occurs by the agency of facilitated diffusion or active uptake. The affinity of the glucose uptake system is lower under aerobic than it is under anaerobic conditions, and this seems to be one of the factors involved in the Pasteur effect (Kotyk and Kleinzeller, 1967). Nevertheless, varying opinions are expressed on whether or not the uptake of sugar is a rate-limiting stage of metabolism (Scharff and Kremer, 1962; Kotyk and Kleinzeller, 1967).

## B. CARBON SOURCES

### 1. *Sugar Fermentation*

*a. Hexoses.* The yeasts are able to ferment anaerobically the hexoses, D-glucose, D-fructose and D-mannose; the term "zymohexoses" was originally applied to these sugars (Fischer and Thierfelder, 1894). Baker's and brewer's yeasts ferment glucose and fructose equally rapidly (Slator, 1906, 1908), although as far as brewer's yeast is concerned this applies only in glucose concentrations of 1–10%, and in fructose concentrations of 2–8%; in lower concentrations fructose is fermented more slowly than glucose (Hopkins and Roberts, 1935; Menzinsky, 1943). In low concentrations, the rate of fermentation of mannose is 20–25% less than that of glucose and, in higher concentrations, the difference is 10–15%. Under cold conditions, the rate of fermentation of mannose is barely half that of glucose. Baker's and brewer's yeasts behave in a similar way (Menzinsky, 1943; Gottschalk, 1947a, 1949).

The rate of fermentation of $\alpha$- and $\beta$-glucose by brewer's yeast is the same, provided that the original concentration of glucose exceeds 1%; if the glucose concentration is less than 1%, the rate of fermentation of $\alpha$-glucose by brewer's and baker's yeast is higher than that of $\beta$-glucose (Hopkins and Roberts, 1936; Heredia *et al.*, 1968). *Alpha*-mannose is fermented somewhat more rapidly than the $\beta$-form, both by baker's and brewer's yeast (Gottschalk, 1947a).

Glucose and mannose are present in aqueous solutions as pyranoses (Isbell and Pigman, 1938). Fructose appears in solution in water mainly as fructopyranose, although it rapidly mutarotates to the fermentable furanose form. The fructose residue in sucrose is a fructofuranose, which is split off by invertase and is directly fermentable (Suomalainen and Toivonen, 1948; Hopkins and Horwood, 1950).

Brewer's and baker's yeasts are capable of fermenting D-galactose after adaptation (Stone and Tollens, 1888). The adaptation to galactose is genetically controlled (Lindegren, 1945; Suomalainen, 1948a). During galactose induction, several enzymes are formed; the *de novo* synthesis of protein which participates in the sugar transport occurs before galactokinase activity has appeared (Kotyk and Haškovec, 1968).

*b. Unfermentable sugars.* All L-sugars and other monosaccharides, apart from those mentioned above, are unfermentable (Fischer and Thierfelder, 1894). A substance originally referred to as glutose, which occurs in the unfermentable part of cane molasses, has proved to be a mixture of several components; in the main, it contains anhydrides of fructose and condensation products of sugars, in particular of glucose with amino acids (Sattler, 1948). D-Allulose is an unfermentable reducing sugar present in cane molasses (Zerban and Sattler, 1942).

*c. Oligosaccharides.* (i) *Sucrose.* Sucrose, common in plants and vegetables, and the most important disaccharide in beet and cane molasses, can be fermented by most yeasts. Invertase (saccharase, sucrose, β-D-fructofuranosidase), which hydrolyses sucrose to glucose and fructose, is located at the cell surface (cf. Suomalainen and Oura, 1958b; Lampen, 1968). In 1932, Wilkes and Palmer found that a cell-free preparation of invertase, and intact yeast cells, gave identical pH curves for the enzyme activity (Fig. 1). They came to the conclusion that invertase is located close to the outer surface of the cell. The results of experiments with uranyl salts made by Demis *et al.* (1954) confirmed this. The uranyl ion non-competitively inhibits invertase activity in intact yeast cells. Uranyl inhibition could be reversed by orthophosphate addition corresponding to one-hundredth of the intracellular amount of orthophosphate. As early as 1923, von Euler and Myrbäck reported the possibility of regenerating, by means of hydrogen sulphide, invertase which had been inactivated by silver nitrate; this is true both in an enzyme preparation and in intact yeast cells. Mercuric chloride is also capable of inhibiting invertase in yeast; the inhibition can be reversed quantitatively, for example by hydrogen sulphide, provided that enzyme and inhibitor have not been incubated together for any appreciable time (Myrbäck, 1926, 1957).

The rate of sucrose hydrolysis rises as high as 300 times greater than the rate of fermentation by yeast (Demis *et al.*, 1954). The invertase activity of baker's yeast increases during the industrial production process on gradual transfer from anaerobic to aerobic growth conditions (Suomallainen and Oura, 1957). During this process, the invertase activity of baker's yeast follows the change in the mannose content of the cells (Suomalainen *et al.*, 1960b). Invertase is a polysaccharide-protein

(Sumner and O'Kane, 1948); the polysaccharide part is predominantly mannan, with a small percentage of glucosamine (Neumann and Lampen, 1967). When mannose is used by baker's yeast as a carbon source, the mannose content and the invertase activity of the cells are significantly higher than when the cells have been grown on a glucose medium (Suomalainen *et al.*, 1967a).

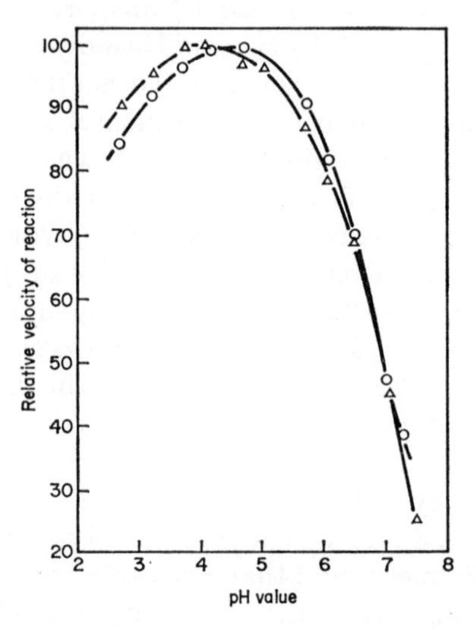


FIG. 1. Effect of pH value on inversion of sucrose by intact yeast cells (o—o) and a cell-free invertase preparation (△—△). From Wilkes and Palmer (1932).

(ii) *Maltose and other α-glucosides*. Brewer's yeasts are known to ferment maltose quite rapidly under normal growth conditions in wort, in which maltose is the most important sugar. Fermentation of maltose by intact baker's and brewer's yeasts may even proceed more rapidly than that of glucose (Willstätter and Steibelt, 1921; Willstätter and Bamann, 1926a, b; Leibowitz and Hestrin, 1942; DelaFuente and Sols, 1962). Maltose fermentation by an unadapted yeast requires an induction time (Willstätter and Steibelt, 1921; Willstätter and Rohdewald, 1937; Rhoades, 1941; Myrbäck and Renvall, 1946), which may be decreased by the addition of a small amount of glucose (Leibowitz and Hestrin, 1939), or by the presence of oxygen (Schultz *et al.*, 1940a). Hexoses activate maltose fermentation by baker's yeast in the following order of efficiency: glucose > fructose > mannose (Suomalainen *et al.*, 1956).

According to Leibowitz and Hestrin (1942, 1945), maltose fermentation by baker's and brewer's yeasts is in general "direct"; only brewer's bottom yeast is capable of fermenting maltose "indirectly" under neutral conditions involving hydrolysis by maltase as the initial step.

Gottschalk (1949) has pointed out that the permeability factor should be taken into consideration and, with this in mind, there existed insufficient evidence for "direct" maltose fermentation. The problem of "direct/indirect" fermentation has become a question of whether maltose hydrolysis takes place inside the cell or on its surface (Suomalainen, 1948b; DelaFuente and Sols, 1962). The studies made by DelaFuente and Sols (1962) firmly support the view that a specific transport system is the first step in the fermentation of maltose, followed by splitting by intracellular enzymes.

*Alpha*-methylglucoside is vigorously fermentable by intact bottom-fermenting brewer's yeast at neutral pH values in the range which is optimal for maltase activity, but not below pH 4·0 (Leibowitz and Hestrin, 1942). In the case of intact baker's yeast, $\alpha$-methylglucoside is a very poor maltose substitute (DelaFuente and Sols, 1962), and a powerful retardant of maltose fermentation (Leibowitz and Hestrin, 1942); a similar retardation is caused by trehalose (Suomalainen and Oura, 1956). Maltase also hydrolyses maltotriose, a fermentable trisaccharide (Cook and Phillips, 1957); however, the rate of hydrolysis of maltotriose is low in comparison with that of maltose (Myrbäck and Ahlborg, 1942; Myrbäck and Leissner, 1943; Myrbäck, 1947; DelaFuente and Sols, 1962). Brewer's yeast maltase is also capable of hydrolysing the $\alpha$-glucosidic glucose component of sucrose (Weidenhagen, 1928; Winge and Roberts, 1958) as much as twice as rapidly as that of maltose (Suomalainen *et al.*, 1961).

It has been reported that *Saccharomyces* species contain six polymeric genes $M_1$–$M_6$, which influence the ability of yeast to ferment maltose. Of these genes, $M_1$ and $M_4$ direct production of maltases, which catalyse hydrolysis of maltose and sucrose, and $M_2$, $M_5$ and $M_6$ direct production of maltases which catalyse the hydrolysis of only maltose; $M_3$ may be adapted also to sucrose (Winge and Roberts, 1958). While it was found in this way that differences existed in specificity between maltases produced by different genes in intact yeasts, these same differences were not observable in maltase preparations that had been partly purified (Halvorson *et al.*, 1963). However, the capacity of different genes to induce $\alpha$-glucosidic activity is variable (Rudert and Halvorson, 1963).

Baker's yeast may contain trehalose to the extent of 14% of its dry weight (Myrbäck, 1936; Myrbäck and Örtenblad, 1936), while brewer's yeast contains only small amounts of trehalose (Stewart *et al.*, 1950). The content of trehalose in baker's yeast cells is dependent upon the aeration; the more vigorous this is the higher is the trehalose content (Brandt, 1941; Suomalainen and Oura, 1956; Suomalainen and Pfäffli, 1961). When acetone-dried baker's yeast is used in glucose fermentation, 20–30% of the added glucose is transformed into trehalose (Elander,

1956). In addition to trehalose, its monophosphate is formed during glucose fermentation with dried baker's or brewer's yeast in the presence of toluene (Elander, 1959). Trehalose 6-phosphate is an intermediate in trehalose synthesis by baker's yeast (Elander, 1963).

The non-reducing disaccharide, 2,2′-dideoxy-$\alpha$,$\alpha'$-trehalose, accumulates in baker's yeast cells under anaerobic conditions, when the cells are incubated in a medium containing 2-deoxy-D-glucose (Farkaš *et al.*, 1969). Further, uridine diphosphate-2-deoxy-D-glucose (Biely and Bauer, 1966) and guanosine diphosphate-2-deoxy-D-glucose (Biely and Bauer, 1969) are formed; the first-mentioned of these is an intermediate in the formation of dideoxytrehalose (Farkaš *et al.*, 1969).

Intact baker's and brewer's yeast ferment extracellular trehalose but at a poor rate (Myrbäck and Örtenblad, 1936; Suomalainen, 1948b; Myrbäck, 1949). At the beginning of each growth phase in the industrial production of baker's yeast, trehalose is rapidly and almost completely metabolized, but is resynthesized during the later part of the growth phase (Suomalainen and Pfäffli, 1961). This rapidly mobilizing reserve carbohydrate may accumulate as an energy reserve to be used during the lag phase when cells prepare for division (Panek, 1963). Trehalase is an $\alpha$-glucosidase which specifically hydrolyses trehalose (Myrbäck, 1949).

(iii) *$\beta$-Glucosides.* $\beta$-Glucosides, such as cellobiose (Skraup and König, 1901; Myrbäck, 1940) and gentiobiose (Zemplén, 1915), are not fermented by brewer's or baker's yeast.

(iv) *$\alpha$- and $\beta$-Galactosides.* Baker's yeast, along with top- and bottom-fermenting brewer's yeasts, is able to ferment the fructose moiety of raffinose. After the splitting off of fructose, the remaining disaccharide is the $\alpha$-galactoside melibiose. Melibiose can be fermented by brewer's bottom yeast (*Saccharomyces carlsbergensis*), not by top-fermenting brewer's yeast or by baker's yeast (*Saccharomyces cerevisiae*) (Rischbiet and Tollens, 1886; Berthelot, 1889). The first step in the fermentation of melibiose is hydrolysis outside the plasma membrane (Friis and Ottolenghi, 1959; DelaFuente and Sols, 1962).

Lactose is not fermented by bottom- or top-fermenting yeast (Fischer and Thierfelder, 1894; Kluyver and Custers, 1940). Lactose-fermenting yeasts adapted to glucose, galactose, or lactose ferment lactose faster than an equivalent mixture of glucose and galactose (Rogosa, 1948).

*d. Selective fermentation.* In a mixture of glucose and mannose, brewer's yeast initially ferments the glucose (Slator, 1908; Gottschalk, 1947b). According to Gottschalk (1947b), this is explained by the slow penetration of mannose through the membrane of the yeast cell, and most probably mannose has a lower affinity for hexokinase than glucose. When sucrose is fermented, it is hydrolysed to glucose and fructose; both baker's and brewer's yeast ferment glucose more rapidly than fructose

(Gottschalk, 1947b; Suomalainen and Toivonen, 1948), while Sauterne yeasts ferment fructose more rapidly than glucose (Sobotka and Reiner, 1930). Pentoses, particularly xylose, exert a retarding effect upon the rate of fermentation of hexoses by brewer's yeast; this pentose effect could not be observed with baker's yeast (Sobotka *et al.*, 1936). The pentoses themselves are not fermented (Dickens, 1938).

### 2. *Oxidative Metabolism*

*a. Aerobic metabolism.* Yeasts have two pathways of glycolytic metabolism. Under anaerobic conditions, the extent to which the Embden-Meyerhof pathway is used is about 90%, both with baker's yeast and *Candida utilis* (Blumenthal *et al.*, 1954; Simon and Medina, 1968). Under aerobic conditions, the hexose monophosphate pathway is responsible for 30–50% of glycolysis in *C. utilis* (Blumenthal *et al.*, 1954), and for 6–30% in *Saccharomyces cerevisiae* (Wang *et al.*, 1956, 1958; Chen, 1959). Under aerobic conditions, where fermentation occurs in part, the rate of aeration does not exercise any influence upon the activity of the hexose monophosphate pathway in *Sacch. cerevisiae* (Chen, 1959). *Rhodotorula gracilis*, which lacks metabolic activity under anaerobic conditions, metabolizes 60–80% of its glucose through the hexose monophosphate pathway (Höfer, 1968; Nakagawa and Tatsumi, 1968).

Absence of air is not the only condition conducive to alcoholic fermentation; a high sugar concentration in the presence of oxygen may result in metabolism occurring by fermentation (Tustanoff and Bartley, 1964b). A glucose concentration in excess of 5% totally inhibits the synthesis of respiratory enzymes (Singer *et al.*, 1966). The disappearance of sugars from the growth medium has been found to induce an increase in the activities of the enzymes of the citric acid cycle, and in the emergence of considerable activity on the part of the enzymes of the glyoxylate cycle (Polakis and Bartley, 1965). The influence of the diminished glucose concentration on the activity of the electron-transport enzymes is also obvious (Oura, 1969; Suomalainen, 1969). The presence of sugar inhibits the formation of mitochondria (Polakis *et al.*, 1964, 1965; Nurminen and Suomalainen, 1968). Under strictly anaerobic conditions, an inhibition of ethanol fermentation appears as compared with that which occurs in oxygen, air, or in a mixture of oxygen and nitrogen. This inhibition is preventable by the supply of a small amount of oxygen before the addition of the sugar to be fermented (Wikén and Richard, 1953, 1954a, b).

*b. Substrates used under aerobic conditions.* Yeasts metabolize the same sugars aerobically as they ferment. Moreover, they may respire some disaccharides, such as maltose, which *Candida utilis*, for example, is unable to ferment under anaerobic conditions (Kluyver and Custers, 1940).

Of the pentoses, xylose is a good carbon source for *C. utilis*, and arabinose is inferior in this respect (Lechner, 1938, 1940). Ethanol improves the growth of *C. utilis* to an even greater extent than glucose. Although acetic acid also induces some growth of *C. utilis*, the yields are lower than with sugars (Fink and Krebs, 1938). If, during the growth of baker's yeast, the medium contains not only sugar but also about 4% ethanol, yeast growth is considerable; ethanol alone induces a lower yield than sugar (White, 1954). Baker's yeast also respires acetic acid (Krebs, 1935), which appears to be readily incorporated by brewer's bottom yeast into the carbohydrate-metabolizing system of the cell (White and Werkman, 1947). Other simple carbon sources for the growth of *Saccharomyces cerevisiae* include dihydroxyacetone, α-ketoglutaric acid, and oxaloacetic acid (Jacob and Abadie, 1967).

Furthermore, the nature of the carbon source markedly influences the carbohydrate metabolism of yeast. Cells from a young galactose-grown culture have a greater oxidative activity than cells from a young glucose-grown culture, and also contain mitochondria (Polakis and Bartley, 1965; Polakis *et al.*, 1965). Yeast grown anaerobically on galactose can respire immediately with glucose, fructose, galactose or pyruvate (Tustanoff and Bartley, 1964a). Maltose as a carbon source exerts an effect on the oxidative activity which is the same as that of galactose. This indicates the importance of adaptation to a carbon source as a regulatory factor in the development of oxidative carbohydrate metabolism (Görts, 1967).

Yeasts which are capable of using hydrocarbons are to be found in the genera *Candida*, *Hansenula*, *Torula* and *Torulopsis* (Fuhs, 1961; Johnson, 1964; Quayle, 1968).

## C. SOURCES OF NITROGEN

### 1. *Inorganic Nitrogen*

All yeasts are able to utilize ammonium sulphate as a source of nitrogen (Wickerham, 1946). Comparing the influence of different ammonium salts on the growth of yeast, Pirschle (1930) found that diammonium phosphate, when used as nitrogen source, promoted growth of baker's yeast most efficiently. Sources found to be almost as good include mono- and tri-ammonium phosphate, ammonium sulphate, bicarbonate, carbonate, acetate, lactate and tartrate, while ammonium chloride has proved inferior as a source of nitrogen. The ability to assimilate nitrate-nitrogen has been employed as a criterion in the classification of yeast. Baker's and brewer's yeasts are unable to assimilate nitrates, while *Candida utilis* can do so: low-nitrogen *C. utilis* also takes up nitrate from an aerated solution with comparative rapidity (Virtanen *et al.*, 1949). It has been suggested that all yeasts would utilize nitrate-nitrogen in the presence of sufficient growth factors (Ingram, 1955). With regard to low-nitrogen *C. utilis*, which is highly suitable for the study of nitrogen uptake, assimilation of ammonia-nitrogen is dependent on the aeration of the culture (Roine, 1947).

According to Metcalfe and Chayen (1954), Roberts and Wilson (1954) and Schanderl (1955), at least some *Rhodotorula* yeasts can assimilate molecular nitrogen.

2. *Amino Acids*

Thorne (1933, 1941) investigated growth of both bottom- and top-fermenting brewer's yeast with an amino acid as the sole source of nitrogen. He observed (Thorne, 1941) that, when sufficient growth factors had been added to the medium, aspartic acid, asparagine and glutamic acid produced a considerable growth, followed in order by $\alpha$-alanine, $\alpha$-amino-butyric acid, valine, leucine, isoleucine, serine, ornithine, arginine, phenylalanine, tyrosine and proline, which induced reasonable growth, and with tryptophan and oxyproline displaying a distinctly poorer nutrient value; finally, histidine, glycine, cystine and lysine can scarcely be regarded as nutrients. Schultz and Pomper (1948), with several *Saccharomyces* species, strains and variants, studied the amino acids which could be utilized as a source of nitrogen. They achieved results similar to those reported by Thorne (1941), and also found that there are particular amino acids that are taken up well by some *Saccharomyces* species, and less well by others; these amino acids are isoleucine, methionine, phenylalanine, proline, serine, tryptophan and tyrosine. According to Abadie (1967), *Sacch. cerevisiae* exhibited inferior growth when histidine, lysine or proline was used as nitrogen source, an observation in agreement with the results obtained by Thorne (1941) and Schultz and Pomper (1948); however, Abadie (1967) obtained a divergent result with glycine.

Asparagine and glutamine, abundantly present in germinated malt, are well known as excellent sources of nitrogen (Euler and Lindner, 1915). Aspartic acid (Massart and Horens, 1953) and asparagine (Mothes, 1953) are assimilated rapidly when sugar is present; at the same time, consumption of oxygen is increased and, moreover, it has been found that ammonium sulphate has to be present in the medium. In the presence of ammonium sulphate, the yeast assimilates amino- and amido-nitrogen from aspartic acid and asparagine at twice the rate at which ammonia-nitrogen is taken up from ammonium salts. Nonetheless, although glutamic acid is well assimilated if no ammonium salt is present, it remains unassimilated if the medium contains ammonium sulphate (Hartelius, 1938, 1939).

According to Hartelius (1938, 1939), the length of the carbon chain of the amino acid influences the assimilation of amino acids by yeast (*Sacch. cerevisiae*). He found that glycine was assimilated very poorly; the higher the number of carbon atoms in the chain, the more easily are the following four homologous amino acids taken up, while aminocaprylic acid is again assimilated as poorly as glycine. Furthermore, branched-chain amino acids are not taken up as well as those with a straight chain.

*Saccharomyces cerevisiae* does not assimilate $\beta$-amino acids such as $\beta$-alanine and $\beta$-aminobutyric acid (Nielsen, 1943). $\beta$-Alanine, although not utilizable as a source of nitrogen, is a growth factor and, when added in small amounts, has been shown to increase the growth of yeast in the presence of a source of nitrogen (Nielsen and Hartelius, 1938). $\gamma$- and $\delta$-Amino acids are assimilated, although slowly, whereas neither $\epsilon$-amino acids (Nielsen, 1943) nor $\alpha$-substituted synthetic amino acids are assimilated (Suomalainen and Kahanpää, 1963).

With regard to the optical isomers, in general yeast takes up only the L-form (Nielsen, 1936b). Both D- and L-asparagine and aspartic acid are assimilated by *Sacch. cerevisiae*; the two isomers of asparagine are taken up at the same speed, while the D-isomer of aspartic acid is assimilated more slowly (Nielsen, 1943; cf. also Harris, 1958). LaRue and Spencer (1967a) examined the growth of 91 yeast strains from 19 genera, with D-amino acids as the source of nitrogen. They found that, although the yeast grew on almost all the L-amino acids, it could use no more than one or just a few of the D-isomers. In some cases, the yeast was able to metabolize the D-amino acid, but not the corresponding L-acid. Of the D- and L-isomers, the two $\alpha$-alanines were most readily assimilated.

### 3. *Mixtures of Amino Acids*

Growth of yeast on wort is more rapid than in a medium where an ammonium salt is used as a source of nitrogen, although an ammonium salt, again, is a better source than any amino acid alone. In a synthetic mixture of amino acids which contains an ammonium salt in addition to the amino acids, the consumption of amino nitrogen is about one and a half times as rapid as that of ammonia nitrogen (Thorne, 1949). Yeast can assimilate about 40–50% of the total nitrogen in wort (Nielsen, 1935), and this is mostly amino nitrogen (Nielsen and Lund, 1936). Barton-Wright (1949) has applied chromatographic and microbiological methods in studies of the changes occurring in the contents of the following amino acids during fermentation of wort: arginine, asparagine, cysteine, glutamine, histidine, isoleucine, leucine, lysine, methionine, phenylalanine, proline, serine, threonine, tyrosine, tryptophan and valine. He observed the following order of assimilation: the straight-chain or aliphatic amino acids are utilized first (methionine, lysine, leucine, aspartic acid, isoleucine); then follow the cyclic amino acids phenylalanine, tyrosine, tryptophan and histidine, with the exception of proline, which remains almost untouched. It was found that brewer's bottom yeast differs from top yeast in that methionine is utilized more slowly and not completely, and that glutamic acid, initially used almost completely, is at a later stage excreted to a greater extent than any other amino acid. In general, most amino acids of wort are not assimilated as well by bottom yeasts as by top yeasts.

On investigation of the amino-acid assimilation of brewer's bottom yeasts, Sandegren *et al.* (1954) obtained results which differ in part from those mentioned above. According to them, proline is clearly assimilated. This subject has been studied further in top fermentations by Pierce (1966a, b). He divided the amino acids which could be absorbed from wort into four groups by virtue of their rate of absorption. The amino acids of Group 1 were those most rapidly absorbed, namely glutamic acid, aspartic acid, asparagine, glutamine, serine, threonine, lysine, arginine; next in absorption came the amino acids of Group 2, namely valine, methionine, leucine, isoleucine, histidine; then Group 3 which includes glycine, phenylalanine, tyrosine, tryptophan, alanine, ammonia, and finally proline, falling in Group 4. Palmqvist and Äyräpää (1969) have reported the same grouping in bottom fermentations, but they further observed that the order is not quite constant; uptake of arginine and of ammonia are particularly subject to variations. Thorne (1949) and Barton-Wright and Thorne (1949) have suggested that utilization of amino acids takes place, at least to some extent, by the assimilation mechanism used by yeast for building proteins direct from the amino acids taken up. The results obtained by Jones *et al.* (1969) demonstrate that the amino acids are not incorporated into yeast protein intact, but participate in reactions which include transaminations prior to their involvement in protein synthesis. Glutamic acid plays a central role in the transamination system (Jones *et al.*, 1969; cf. Roine, 1947).

4. *Peptides*

Hartelius (1938, 1939) observed that yeast takes up the dipeptides leucylglycine and glycylleucine from the medium. Nielsen (1943) investigated the assimilation of 15 different peptides. Not only dipeptides but also the polypeptides are assimilable, for instance DL-leucyllysylglycylglycine. It was found that the structure of peptides clearly influences the assimilation; thus, glycyl-L-leucine is assimilated to the extent of 92–93% of the original nitrogen content, whereas the corresponding proportion for glycyl-D-leucine does not exceed 2–4%. This work proved that growth of yeast is inferior with amino acids when peptides are used as a source of nitrogen. Polypeptides are assimilated also from wort (Jones *et al.*, 1965; Pierce *et al.*, 1969).

5. *Other Sources of Nitrogen*

For *Saccharomyces* species, urea provides a source of nitrogen which is just as good as ammonium sulphate (Schultz and Pomper, 1948; DiCarlo *et al.*, 1953; Abadie, 1968). However, if the yeast is to grow as well with urea as with ammonium sulphate, the medium must contain abundant amounts of biotin (Schultz *et al.*, 1940b; DiCarlo *et al.*, 1953). Among *Saccharomyces* species, a large variation is apparent in the ability to grow on purine and pyrimidine bases. LaRue and Spencer (1968) studied the ability of 123 species of yeast to grow on different purines or pyrimidines as sources of nitrogen. *Saccharomyces cerevisiae* grows well on allantoin or allantoic acid, and to some extent on adenine, guanine or cytosine (DiCarlo *et al.*, 1951; LaRue and Spencer, 1968). *Candida utilis* can utilize a greater number of nitrogen-containing compounds than *Sacch. cerevisiae* (Nielsen, 1943; LaRue and Spencer, 1968). *Saccharomyces cerevisiae* grows only slowly on histidine, and is completely unable to utilize other imidazole compounds (LaRue and Spencer, 1967b). *Candida utilis* grows well in the presence of L-histidine, D-histidine or histamine (LaRue and Spencer, 1967b).

6. *Excretion of Nitrogen-Containing Compounds*

During growth and fermentation, baker's and brewer's yeasts, as well as *Candida utilis*, excrete compounds containing nitrogen (Nielsen, 1936a; Nielsen and Hartelius, 1937; Fink and Krebs, 1939; Drews *et al.*, 1953). Apart from amino acids, oligopeptides are excreted (Drews *et al.*, 1953). The amino acids found in the medium are the same as those in the acid-soluble nitrogen fraction of baker's yeast, but in amounts which are considerably less than those in the yeast cell; on comparison of the amount of amino acids with the amount of keto acids excreted into the medium, it was found that the content of amino acids was much smaller (Suomalainen and Keränen, 1967). During the early stages of fermentation in hopped worts, the amino acids glycine, alanine and proline, together with ammonia, are excreted into the medium (Jones *et al.*, 1965). The results arrived at by Jones *et al.* (1965) with a synthetic medium proved that alanine, glycine, proline and amides are not excreted in the medium until after 38 hours of growth and that, with the exception of proline, excretion then continues during the fermentation. Both the excretion and the absorption of amino acids are dependent upon the energy source present (Jones *et al.*, 1965).

It has been reported that nucleotides are excreted from the cell when yeast is suspended in water (Delisle and Phaff, 1961; Lewis and Phaff,

1963). In these experiments, the amino acids excreted in glucose solutions were reabsorbed, while the nucleotides were not; moreover, the nucleotides were excreted in larger amounts in a glucose solution than in water (Delisle and Phaff, 1961). Lee and Lewis (1968a) identified the following nucleotides excreted by *Saccharomyces carlsbergensis*: nicotinamide adenine dinucleotide, cytidine 5′-monophosphate, adenosine 5′-monophosphate, guanosine 5′-monophosphate, uridine 5′-monophosphate, adenosine diphosphate, cytidine diphosphate, guanosine diphosphate, uridine diphosphate; and the following five free bases: adenine, guanine, uracil, hypoxanthine, cytosine; and four nucleosides: adenosine, cytidine, uridine and guanosine. They further identified the same components in the intracellular pool.

### D. SOURCES OF PHOSPHORUS

Phosphorus is necessary for the growth of yeasts and for fermentation. Baker's and brewer's yeasts are capable of growing briefly on a medium without phosphate, but in that case the phosphate reserves of cells are used for the growth (Rosenbaum, 1931, 1935; Markham *et al.*, 1966). The yeast cell assimilates orthophosphate from the medium against a concentration gradient (Goodman and Rothstein, 1957). If potassium dihydrogen phosphate is used as a source of phosphate, considerably more phosphate can be taken up than with disodium hydrogen phosphate. Yeast takes up phosphate as a monovalent anion, $H_2PO_4^-$, and the bivalent anion is not absorbable (Rothstein, 1961). *Candida utilis*, poor in phosphorus, rapidly assimilates phosphate from the medium, and to an extent such that it may contain twice its normal amount (Rautanen and Miikkulainen, 1951). $^{32}$P-Orthophosphate is taken up so rapidly that, in at least ten organic phosphate esters, labelling is found after 0·1 sec (Miettinen, 1964). Savioja and Miettinen (1966a) have identified a number of different components in the acid-soluble phosphorus fraction. When $^{32}$P-orthophosphate was assimilated by *C. utilis* at a temperature above normal, the distribution and the amounts of radioactivity differed from those at normal temperature (Savioja and Miettinen, 1966b). At 38–48°, trehalose 6-phosphate is accumulated in the food yeast, *C. utilis*, and in baker's yeast, *Saccharomyces cerevisiae*, under both aerobic and anaerobic conditions; at 30°, traces of this compound are detectable only if the cell suspension is agitated by pure oxygen (Savioja and Miettinen, 1966c).

The inorganic phosphate taken up by yeast cells is stored as metaphosphate in volutin granules (Wiame, 1946, 1947). Metaphosphate is present in yeast as two different fractions, both of which give a metachromatic reaction with toluidine blue. One metaphosphate fraction is soluble, and the other insoluble; metabolically the insoluble metaphosphate fraction is more active than the soluble one, being rapidly and reversibly transformed into orthophosphate in the cell (Wiame, 1949). Kornberg (1956) has isolated trimeta- and tripolyphosphate from baker's yeast but, as opposed to what Wiame (1949) had found with metaphosphate polymers, none of these displayed the properties of metachromasy.

Metaphosphate synthesis in the cells is a requisite for the rapid growth of yeast (Lindegren, 1947, 1951). Yoshida (1953) and Yoshida and Yamataka (1953) have established that metaphosphate could well be the store of the phosphate energy in yeast cells. Nevertheless, apparently the polyphosphates do not constitute any appreciable energy reserve in the cells of baker's yeast (Suomalainen and Pfäffli, 1961).

Growth of top-fermenting brewer's yeast increases when the phosphate concentration is raised to a definite limit (Markham and Byrne, 1967). A lack of phosphate induces weakening of the growth, and the formation of cells with a fat content above the normal (Schulze, 1950; Nielsen and Nilsson, 1953; Maas-Förster, 1955). When baker's yeast or *C. utilis* is cultured in a phosphate-poor medium, the activity of acid phosphatase, located on the cell surface, is found to increase (Rautanen and Kärkkäinen, 1951; Rautanen and Kylä-Siurola, 1954; Suomalainen *et al.*, 1960a). With the aid of acid phosphatase, the yeast in a phosphate-free medium can take up the necessary phosphate from phosphate esters (Günther and Kattner, 1968).

### E. SOURCES OF SULPHUR

Almost all yeasts will take up the sulphur they need from inorganic sulphate (Maw, 1965). Sulphate may be replaced partially or wholly by other inorganic or organic sulphur compounds (Schultz and McManus, 1950; Maw, 1960). Most *Saccharomyces* species display good growth when sulphate is replaced by sulphite or thiosulphate (Schultz and McManus, 1950). *Saccharomyces* species are unable to use amino acids containing sulphur, such as cysteine or cystine, as the source of sulphur (Schultz and McManus, 1950; Maw, 1960, 1961, 1963b); some *Saccharomyces* species are able to replace sulphite by methionine to varying extents (Schultz and McManus, 1950; Maw, 1960, 1963b). In general, utilization of the sulphur of glutathione is inferior to that of methionine (Maw, 1965). Growth factors which contain sulphur, such as biotin and thiamin, cannot be used by brewer's top yeast as sources of sulphur (Maw, 1960). Not only sulphate, but also methionine and glutathione, when present in the medium, can increase the growth rate (Maw, 1963c).

*Candida utilis* can use a larger number of sulphur sources than *Saccharomyces cerevisiae*; *C. utilis* grows well if the sulphate is replaced by sulphite, thiosulphate, sulphide, glutathione, methionine, cystine or cysteine, cysteic acid or taurine (Schultz and McManus, 1950; Maw, 1965).

Uptake of sulphate requires energy, and both glucose and nitrogen-containing nutrients need to be present in the medium (Kotyk, 1959; Crocomo and Menard, 1962; Maw, 1963a). The rate of aeration in the propagation of *Sacch. cerevisiae* influences its ability to utilize different sources of sulphur (Maw, 1960). Increased aeration and poorer nutrition during the industrial propagation of baker's yeast lead to a diminution in the total content of sulphur in the cells (Suomalainen and Keränen, 1955). Nonetheless, Kotyk (1959) has proved that the uptake of sulphate proceeds at the same rate both aerobically and anaerobically.

Thiosulphate, sulphite and selenate inhibit the utilization of sulphate (Fels and Cheldelin, 1949; Kleinzeller *et al.*, 1959; Kotyk, 1959; Maw, 1963a). Some closely related compounds, such as ethionine and methionine sulphone, also inhibit the uptake of amino acids containing sulphur (Maw, 1963c).

## F. MINERALS

### 1. *Influence of Elements*

Yeasts need certain mineral compounds which act as functional components of proteins, as activators for enzymes or as stabilizers for proteins. Some mineral compounds, although unnecessary for propagation, may stimulate the growth of yeast (Morris, 1958).

Spectrographic analyses have shown that yeasts contain various trace elements, e.g. silver, barium, cobalt, chromium, copper, molybdenum, nickel, lead, tin, vanadium and zinc (O. Erämetsä, H. Suomalainen and A. J. A. Keränen, unpublished observations). White and Munns (1951) divided about 50 common elements which they had studied into four groups, designated very poisonous, moderately toxic, slightly toxic, and very slightly toxic or non-toxic. In the very poisonous group, cadmium, copper, silver, osmium, mercury and palladium completely inhibit growth in concentrations of 0·4–10·0 p.p.m. The elements in the moderately toxic group completely inhibit growth in concentrations of 115–400 p.p.m.; this group contains cobalt, lithium fluorine and tin. Slightly poisonous elements, including selenium and thallium, completely inhibit growth in concentrations of 500–600 p.p.m. Very slightly poisonous, or non-poisonous elements, include lead, iron, the halogens in the form of their potassium salts, antinomy, strontium and barium. This group further contains magnesium, potassium, sodium, calcium, sulphur and phosphorus, which are normally required for yeast growth, and have no toxic properties. Growth media which contain organic compounds, such as molasses and wort, may possess the ability to protect the yeast from the influence of some inhibitors by precipitating them or by forming complexes (White and Munns, 1951).

2. *Alkali Metals*

Potassium is necessary for the growth of yeast and for fermentation (Lasnitzki and Szörényi, 1934, 1935; Atkin *et al.*, 1949b; Meyerhof and Kaplan, 1951). The absorption of $K^+$ ions is facilitated by the uptake of glucose; when this is consumed, the $K^+$ ions are retransported into the medium (Pulver and Verzár, 1940a, b). When $K^+$ is absent from the medium, phosphate cannot be absorbed (Schmidt *et al.*, 1949). Fermenting yeast absorbs twice as much $K^+$ as respiring yeast (Horschak and Wartenberg, 1967). "Potassium yeast" has been prepared from normal commercial baker's yeast by allowing the yeast to ferment for different periods in a fresh medium containing both potassium citrate and glucose. Between the periods of fermentation, the cells were centrifuged and washed. The resting oxygen-consumption rate of potassium yeast is about twice as great as that of normal yeast, and the rate of carbon dioxide production about half that of normal yeast. The growth rates of potassium yeast and of normal yeast in a low-potassium medium are the same (Conway and Moore, 1954).

It has been shown that other alkali ions can partly replace the growth-promoting and fermenting effect of the $K^+$ ion (Lasnitzki and Szörényi, 1934). With respect to this, the alkali ions may be set in the following order: $Rb^+ > Na^+ = Li^+ = Cs^+$ (Rothstein and Demis, 1953). From baker's yeast, Conway and Moore (1954) have prepared yeast rich in sodium, in which the potassium of the cell has been replaced by sodium. The "sodium yeast" and the reference potassium yeast exhibited similar fermentation rates, which were less than that of commercial baker's yeast. The growth rate, and the resting oxygen-consumption rate, were less with sodium yeast than with potassium yeast and commercial baker's yeast.

Under certain conditions, the potassium of intact baker's yeast is entirely replaceable by ammonia (Conway and Breen, 1945). "Ammonia yeast" ferments glucose to the extent of about 40% of the rate of potassium yeast. It can grow in the absence of potassium, but the initial rate of growth is much slower. The oxygen consumption of ammonia yeast is higher than that of potassium yeast. The $NH_4^+$ ion replaces the $K^+$ as poorly as does lithium or caesium (Rothstein and Demis, 1953).

3. *Magnesium and Calcium*

Magnesium is a necessary growth factor for yeasts (Fulmer *et al.*, 1921, 1936; Morris, 1958). Magnesium is the commonest enzyme activator, and is of particular importance in activating the large class of phosphate transferases, and a number of decarboxylases (Bowen, 1966). However, when potassium is replaced by magnesium, as was effected by Conway and Beary (1962) in "magnesium yeast", growth is inhibited and both the oxygen uptake and the fermentation rate are low.

Although calcium is apparently not essential for growth of yeast cells (Morris, 1958), it stimulates growth and fermentation (Fulmer *et al.*, 1921; Richards, 1925). Atkin and Gray (1947) reported that, in the absence of $Mg^{2+}$, $Ca^{2+}$ did not affect the fermentation rate at low concentrations, and was somewhat inhibitory at higher concentrations. They further observed a likelihood of some interrelationships existing between calcium and magnesium. According to Thorne (1954) $Ca^{2+}$ exerts a small depressant effect on the rate of fermentation.

4. *Heavy Metals*

Copper, in low concentrations, is necessary for the growth of yeasts (Elvehjem, 1931; McHargue and Calfee, 1931). Hanson and Baldwin (1941) found that an addition of copper, as sulphate, to the wort promotes the growth of baker's yeast by affecting the redox potential. Furthermore, the presence of small amounts of iron is necessary for growth of yeast (Olson and Johnson, 1949). Copper and iron induce increased activity in the cytochromes and in the citric acid cycle (Elvehjem, 1931; Kauppinen, 1963). With *Candida guilliermondii*, iron deficiency in the medium results in an accumulation of ethanol, pyruvate, and acyl phosphate, together with an increased excretion of riboflavin (Enari and Kauppinen, 1961; Kauppinen, 1963). Addition of zinc to the medium augments the growth of yeast (McHargue and Calfee, 1931), and when such an addition is made to the wort before yeast inoculation the rate of fermentation rises (Densky *et al.*, 1966). Zinc uptake is influenced by the temperature of the medium and the concentration of $Cu^{2+}$ (Frey *et al.*, 1967; Gesswagner and Altmann, 1967). Autoradiography has proved that incorporated $^{55}Fe$, $^{63}Ni$ and $^{65}Zn$ are located in outer parts of the cell (van Kleeff *et al.*, 1969). Manganese, in small concentrations, promotes the growth of yeast, but excessive quantities are toxic (McHargue and Calfee, 1931). As opposed to this, Olson and Johnson (1949) have reported that manganese does not exercise any influence on growth of yeast.

Lead, even in quite large concentrations, does not completely inhibit the ability of yeast to ferment (Prinsen Geerligs, 1940; Suomalainen and Kuronen, 1944; White and Munns, 1951). It has been found that the addition of boron, cobalt, iodine or tin does not affect the yield of baker's yeast or *Candida utilis* (Olson and Johnson, 1949). Cobalt in concentrations of $10^{-3}$–$10^{-4}$ *M* is toxic to the growth of the yeasts *Sacch. cerevisiae*, *C. utilis*, *C. albicans* and *C. guilliermondii* (Nickerson and Zerahn, 1949; Erkama and Enari, 1956; Enari, 1958). Cobalt considerably enhances the production of riboflavin by *C. guilliermondii* (Enari, 1958). Greaves *et al.*

(1928) found that iodine or iodide increased the growth of yeast, but Olson and Johnson (1949) have subsequently reported findings contradictory to this. Chlorine appears to be essential for growth, and is readily available in most media (Morris, 1958).

## G. GROWTH FACTORS

### 1. *Requirements for Growth Factors*

Different yeasts have extremely divergent demands with respect to their growth factors. The most common growth factors, taken up alone or together with others, are biotin, pantothenic acid, inositol, thiamin, nicotinic acid and pyridoxine.

### 2. *Biotin*

Kögl and Tönnis (1936) isolated and identified a growth factor from eggs and yeast, which they named biotin. They discovered that it promotes the growth of baker's and distiller's yeast; subsequently it has been found that all top and bottom yeasts, and a large number of other yeasts, need D-biotin, either alone or with other growth factors (Williams *et al.*, 1940; Leonian and Lilly, 1942; Atkin *et al.*, 1949a; Weinfurtner *et al.*, 1959). Examples of yeasts which do not need biotin are some species of *Hansenula*, *Mycoderma* and *Brettanomyces* (Burkholder *et al.*, 1944). *Candida utilis* and *Hansenula anomala* do not take up any noticeable amounts of biotin, even if biotin is present in the medium (Chang and Peterson, 1949). However, Keränen (1969) has shown that the addition of biotin induces inhibition of the biosynthesis of biotin in fodder yeast, *C. utilis*. With regard to *Saccharomyces cerevisiae*, which needs biotin, the biotin content of the medium influences the uptake of biotin (Winzler *et al.*, 1944; Chang and Peterson, 1949; White and Munns, 1953). Addition of diaminopelargonic acid to the medium results in an accumulation of desthiobiotin in baker's yeast (Keränen, 1969).

Of the biotin analogues, D-desthiobiotin (Melville *et al.*, 1943), biocytin (Wright *et al.*, 1951) and biotin-D-sulphoxide (Melville *et al.*, 1954) can replace biotin in *Sacch. cerevisiae*. Oxybiotin only partly replaces biotin, and is not transformed into biotin itself (Hofmann and Winnick, 1945; Rubin *et al.*, 1945). Biotin-L-sulphoxide is scarcely active (Melville *et al.*, 1954). The data in Table I illustrates the fact that the ability of biotin analogues and derivatives to replace biotin as a growth factor varies in different micro-organisms. Consequently, in examination of the influence exercised by growth factors, it is necessary that the organism used should be the same as that for which the growth factor is intended.

Long chain $C_{16}$ and $C_{18}$ fatty acids, both unsaturated and saturated, and their ethyl esters, when added together with aspartic acid, are

TABLE I. *The Ability of Biotin Analogues to Replace Biotin as a Growth Factor for Different Micro-organisms*

| Compound | *Saccharomyces cerevisiae* | *Lactobacillus arabinosus* | *Lactobacillus casei* | References |
|---|---|---|---|---|
| D-Biotin[a] | 100 | 100 | 100 | |
| D-Biotin methyl ester | 100 | | 0 | Kögl and Tönnis (1936) |
| | | | | Shull *et al.* (1942) |
| Biocytin | 100 | 0 | 100 | Lichstein *et al.* (1950) |
| | | | | Wright *et al.* (1951) |
| Biotin D-sulphoxide | 100 | 100 | 0·00001 | Melville *et al.* (1954) |
| Biotin L-sulphoxide | 0·002 | 5 | 5 | Melville *et al.* (1954) |
| D-Desthiobiotin | 100 | 0 | 0 | Melville *et al.* (1943) |
| | | | | Lilly and Leonian (1944) |
| | | | | Dittmer *et al.* (1944) |
| DL-Oxybiotin | 25 | 50 | 40 | Pilgrim *et al.* (1945) |
| | | | | Rubin *et al.* (1945) |

[a] Growth with biotin taken as 100.

capable of replacing biotin in the growth of baker's yeast under aerobic conditions (Suomalainen and Keränen, 1963a, 1968).

Biotin participates in yeast anabolism in many ways: in the carboxylation of pyruvic acid (Lichstein, 1951), in the synthesis of pyridine nucleotides (Rose, 1960), in nucleic acid synthesis (Ahmad *et al.*, 1961), in the formation of both purine and pyrimidine bases (A. J. A. Keränen and H. Suomalainen, unpublished observations), in protein synthesis (Suomalainen *et al.*, 1958; Ahmad *et al.*, 1961), in the synthesis of polysaccharides (Dunwell *et al.*, 1961), and in the synthesis of fatty acids (Lynen, 1959, 1961; Wakil, 1961; Wakil and Waite, 1962; Suomalainen and Keränen, 1963a, b). Its deficiency also leads to damage of the plasma membrane (Dixon and Rose, 1964).

### 3. *Pantothenic Acid*

Pantothenic acid was first isolated from liver (Williams *et al.*, 1938); it was found to be present in a variety of plant and animal tissues (Williams *et al.*, 1933), and to be capable of stimulating the growth of *Saccharomyces cerevisiae*. Most *Saccharomyces* species need pantothenic acid for growth (Williams *et al.*, 1940; Leonian and Lilly, 1942; Burkholder *et al.*, 1944; Weinfurtner *et al.*, 1959).

In small amounts, a moiety of pantothenic acid, β-alanine, can replace the pantothenic acid requirement of yeasts (Nielsen and Hartelius, 1938; Sarett and Cheldelin, 1945). The growth-promoting effect of β-alanine is apparent only in the absence of asparagine, although asparagine does not influence the activity of pantothenic acid (Williams and Rohrman, 1936; Atkin *et al.*, 1944). Taurine and pantoyltaurine are inhibitors of pantothenic acid, but do not inhibit the effect of β-alanine (Sarett and Cheldelin, 1945).

Panthothenic acid influences the metabolism of yeast under both anaerobic and aerobic conditions (Williams *et al.*, 1936). Pantothenic acid participates in the transfer of the acyl group, as a component of coenzyme A, in carbohydrate and fatty acid metabolism (Lipmann and Kaplan, 1946; Lynen and Reichert, 1951; Jaenicke and Lynen, 1960).

### 4. *Inositol*

Eastcott (1928) isolated *myo*-inositol from tea dust, and demonstrated that it augments the growth of yeast. *myo*-Inositol (*meso*-inositol) is an optically inactive stereo-isomer of hexahydroxycyclohexane. The concentrations of inositol necessary to produce maximum growth of yeasts are relatively high (Williams and Saunders, 1934). To some extent, baker's and brewer's bottom yeast are able to synthesize inositol, or an inositol active substance, in the absence of added inositol (White and Munns, 1950; Smith, 1951). However, in the cultivation of baker's yeast, inositol must be added if optimum yields are to be produced, although beet molasses contains it in sufficient amounts (White and Munns, 1950). Some strains among the brewer's top and bottom yeasts need inositol, and others do not (Williams *et al.*, 1940; Leonian and Lilly, 1942; Weinfurtner *et al.*, 1959).

D-Inositol, L-inositol, pinitol (the monomethyl ether of D-inositol), quebrachitol (L-inositol monomethyl ether) and inositol hexa-acetate are unable to act as growth factors (Woolley, 1941). Oxidation products of *myo*-inositol, including rhodizonic acid, tetrahydroxy-*p*-quinone and triquinonyl, partly replace inositol as promotors of the growth of *Saccharomyces carlsbergensis* (Johnston *et al.*, 1962). However, this effect cannot be explained by the assumption that the compounds named undergo reduction to *myo*-inositol in the cell, since their action is completely independent of that of *myo*-inositol and may result from the influence they exercise upon the redox conditions of the medium (Dworsky and Hoffmann-Ostenhof, 1967).

*Saccharomyces carlsbergensis* rapidly assimilates inositol from the medium. A lack of inositol leads to a less efficient cell division (Smith, 1951), and to morphological changes (Challinor and Daniels, 1955; Lewin, 1965), particularly with regard to the cytoplasmic granules and to the cell wall (Challinor *et al.*, 1964; Power and Challinor, 1969). The metabolic fate of inositol in *Sacch. carlsbergensis* is not known (Dworsky and Hoffmann-Ostenhof, 1967), but there are yeasts such as *Schwanniomyces occidentalis* which synthesize an enzyme that can decompose *myo*-inositol; as a consequence, yeasts of this type are able to grow with *myo*-inositol as the source of carbon (Sivak and Hoffmann-Ostenhof, 1962; Dworsky and Hoffmann-Ostenhof, 1965). Inositol deficiency produces a weakened glucose metabolism under both anaerobic and aerobic conditions (Ridgway and Douglas, 1958b; Lewin and Shafai, 1967). Inositol, mainly attached to lipids, acts as a structural component (Ridgway and Douglas, 1958a). Although it is unknown whether inositol possesses coenzymic properties, Ghosh and Bhattacharyya (1967) are of the opinion that phosphofructokinase activity is affected by inositol deficiency.

### 5. *Thiamin*

Williams and Roehm (1930) found that thiamin stimulates the growth of baker's yeast. However, only a few strains of *Saccharomyces cerevisiae* require thiamin for growth (Williams *et al.*, 1940; Leonian and Lilly, 1942). It is typical of brewer's top yeasts that they need either thiamin or pyridoxine; in this respect they differ from brewer's bottom yeasts which grow well on a medium which contains neither thiamin nor pyridoxine (Atkin *et al.*, 1949a, b). Both brewer's bottom yeast and baker's yeast rapidly take up thiamin from the medium (Fink and Just, 1941;

Sperber and Renvall, 1941; Sperber, 1942; Weinfurtner *et al.*, 1964). Uptake of thiamin by baker's yeast takes place in the same way as a typical active transport (Suzuoki, 1955). A requirement for thiamin is characteristic of yeasts which ferment lactose (Rogosa, 1944), and of *Rhodotorula* yeasts (Ahearn *et al.*, 1962; Mackenzie and Auret, 1963).

Brewer's bottom yeast is unable to synthesize thiamin from the individual moieties of thiamin, thiazole and pyrimidine, even if both are present. In baker's yeast and brewer's top yeast, synthesis of thiamin occurs if both components are present (Fink and Just, 1942c), although the ability to synthesize one component when the other is absent is extremely poor or completely lacking (Schultz *et al.*, 1938; Fink and Just, 1942b; Moses and Joslyn, 1953).

Westenbrink *et al.* (1940) described a phosphatase which hydrolyses thiamin phosphate. Sperber (1942) demonstrated that uptake of thiamin, when diphosphothiamin was used as substrate, has the same optimum pH value as the phosphatase described by Westenbrink *et al.* (1940). Furthermore, he noted that both are inhibited by 2-methyl-4-amino-5-methylaminopyrimidine. Sperber (1942) took this as an indication that the phosphatase of Westenbrink *et al.* (1940), by dephosphorylating the thiamin diphosphate, participates in the uptake of thiamin. Thiamin mono- and diphosphates, when incubated with *Saccharomyces carlsbergensis*, are completely absorbed by the yeast, and appear as free thiamin in the cell. A cell-free extract of *Sacch. carlsbergensis* has been shown to have phosphatase activity and to catalyse the hydrolysis of thiamin phosphates (Kawasaki *et al.*, 1968).

Thiamin diphosphate is the coenzyme for cocarboxylase. Forsander (1956) has shown that thiamin phosphorylation occurs in one step as a direct transfer of a pyrophosphate group from ATP to thiamin. The purified thiaminokinase from baker's yeast is specific for ATP and other nucleoside triphosphates as phosphate donor, and active with thiamin as phosphate acceptor, whereas ADP or thiamin monophosphate is inactive as substrate (Kaziro, 1959).

### 6. *Nicotinic Acid*

Yeasts which require nicotinic acid include some *Kloeckera*, *Candida*, *Zygosaccharomyces*, *Saccharomyces* and *Schizosaccharomyces* species (Burkholder *et al.*, 1944). Such a requirement is not shared by *Sacch. cerevisiae* and *Sacch. carlsbergensis* strains (Wiles, 1953; Weinfurtner *et al.*, 1959). Nevertheless, the ability of baker's yeast to synthesize nicotinic acid under anaerobic conditions is limited, and nicotinic acid is thus a necessary growth factor (Suomalainen *et al.*, 1965b). Although the addition of nicotinic acid under aerobic conditions promotes the synthesis of NAD (Suomalainen, 1963; Suomalainen *et al.*, 1965a), it inhibits synthesis of nicotinic acid by the yeast itself, and may stop it completely (Suomalainen *et al.*, 1965a). Nicotinic acid is necessary for yeasts which ferment lactose (Rogosa, 1943a, b, 1944).

Nicotinic acid and its amide seem to be equally active growth factors for yeast (Fries, 1965). Both compounds penetrate into the cell in much the same way, and uptake is dependent upon the metabolic activity of the yeast (Oura and Suomalainen, 1969). Nicotinamide is a part of the hydrogen-transferring coenzymes NAD and NADP.

### 7. *Pyridoxine*

Schultz *et al.* (1939b) have reported that vitamin $B_6$ promotes the growth of *Saccharomyces cerevisiae*, although most strains of *Sacch. cerevisiae* and *Sacch. carlsbergensis* do not require pyridoxine for growth (Leonian and Lilly, 1942; Weinfurtner *et al.*, 1959). The yeasts which require pyridoxine include not only some *Saccharomyces* strains, but also species of *Kloeckera, Torulopsis, Pichia* and *Brettanomyces* (Burkholder *et al.*, 1944).

The $B_6$ group of vitamins includes pyridoxine, pyridoxal and pyridoxamine. In those *Sacch. carlsbergensis* strains which need pyridoxine, this is replaceable by pyridoxal or pyridoxamine (Melnick *et al.*, 1945) although the influence of pyridoxamine is somewhat weaker (Morris *et al.*, 1959). Some strains of *Sacch. cerevisiae* require pyridoxine, which can be replaced only partly by pyridoxal or pyridoxamine (Melnick *et al.*, 1945). Although *Sacch. carlsbergensis* rapidly takes up pyridoxal, pyridoxal phosphate is not absorbed, and a cell-free extract of this yeast is unable to hydrolyse pyridoxal phosphate (Kawasaki *et al.*, 1968). Pyridoxal phosphate acts as a coenzyme of enzymes which participate in amino-acid metabolism, such as the aminotransferases (Snell, 1944; Lichstein *et al.*, 1945).

### 8. *Riboflavin*

All yeasts are capable of synthesizing riboflavin (Burkholder *et al.*, 1944; Ingram, 1955; Weinfurtner *et al.*, 1959). When baker's yeast cells are transferred from anaerobic to aerobic culture conditions during industrial propagation, the riboflavin content rises and reaches its maximum in the semi-aerobic growth phase (Oura and Suomalainen, 1961).

### 9. para-*Aminobenzoic Acid*

Only a few yeast strains need *p*-aminobenzoic acid for growth; they include *Rhodotorula* yeasts (Ahearn *et al.*, 1962; Nyman and Fries, 1962) and some strains of brewer's top yeast (Rainbow, 1948; Cutts and Rainbow, 1949, 1950). In high concentrations, it inhibits the invertase activity

(Myrbäck, 1926, 1941), and growth of yeast when sucrose is used as substrate (H. Suomalainen, unpublished observations).

### 10. *Folic Acid*

Folic acid is synthesized by all yeasts. It is not assimilated from wort (Weinfurtner *et al.*, 1964).

### 11. *Factors Required Under Anaerobic Conditions*

In 1953, Andreasen and Stier found that, under completely anaerobic conditions, ergosterol is a necessary growth factor for distiller's yeast. Synthesis of ergosterol is inhibited by components which inhibit respiration, such as sodium cyanide and dinitrophenol (Parks and Starr, 1963). The entire molecular structure of ergosterol is not required for the growth-promoting effect, since epicholestanol (5-$\alpha$-cholestane-3-$\alpha$-ol) is just as efficient as ergosterol (Proudlock *et al.*, 1968). Andreasen and Stier (1953) added ergosterol to the medium dissolved in Tween 80 (polyoxyethylene sorbitan mono-oleate). Moreover, Andreasen and Stier (1954) found that Tween 80, by virtue of its oleic acid content, is a growth-promoting factor under anaerobic conditions; oleic and linoleic acid proved to be more active than linolenic acid. Bloch and his colleagues (Lennarz and Bloch, 1960; Bloch *et al.*, 1961) made the observation that the 9-(and 10-)hydroxystearic acids normalize the growth of yeast under anaerobic conditions. A mixture of 9-(and 10-)hydroxystearic acid changes in the yeast into 9-(and 10-)acetoxystearic acid (Light *et al.*, 1962). 9-(and 10-)Acetoxystearate, 11-*cis*-octadecenoic acid (*cis*-vaccenic acid)—although not *trans*-vaccenic acid (Axelrod *et al.*, 1947, 1948)—and 9-octadecynoic acid (stearolic acid) can each replace oleic acid as a growth factor for yeast under anaerobic conditions (Light *et al.*, 1962).

## H. RESERVE CARBOHYDRATES, FAT AND PHOSPHATE

The glycogen content of baker's yeast increases when glucose is fermented under culture conditions deficient in nitrogen (Brandt, 1942; Trevelyan and Harrison, 1956; see also Chapter 10, p. 419). During the industrial production of baker's yeast, its glycogen content increases on transfer from anaerobic to aerobic conditions. At the beginning of each growth period, the glycogen content diminishes but the polysaccharide is finally resynthesized; the glycogen content of commercial baker's yeast is about 12% (Suomalainen and Pfäffli, 1961). A high carbohydrate content is positively correlated with the resistance of baker's yeast to autolysis (Suomalainen and Pfäffli, 1961; Suomalainen *et al.*, 1965d). The iodine absorption spectrum of glycogen, separated from baker's yeast, differs from the spectrum of glycogen of animal origin (H. Suomalainen and A. J. A. Keränen, unpublished observations). Contrary to the situation with baker's yeast, the content of glycogen in bottom-fermenting brewer's yeast first diminishes in a medium deficient in nitrogen and later increases only slowly (H. Suomalainen and A. J. A. Keränen, unpublished observations).

Baker's yeast may contain trehalose up to 14% of its dry weight (Myrbäck and Örtenblad, 1936). Even brewer's top and bottom yeasts, grown anaerobically, contain trehalose, but considerably less than baker's yeast (Stewart *et al.*, 1950). During the industrial production of baker's yeast, the trehalose content of the cells increases on transfer from anaerobic to aerobic conditions (Brandt, 1941; Suomalainen and Pfäffli, 1961). At the beginning of each growth period, the trehalose almost completely disappears from the cells, but is again resynthesized at the end of the period (Suomalainen and Pfäffli, 1961).

Fink and Just (1938) proved that *Candida utilis* grown on wood hydrolysate contained dulcitol up to 1% of the dry matter. They stated that galactose was the probable origin of dulcitol. During the aerobic dissimilation of galactose, some yeasts excrete dulcitol into the medium; thus *Pichia farinosa* produced dulcitol to the extent of 28–43%, and *Torulopsis versatilis* 3–25%, of the galactose utilized, depending on the original concentration of galactose. Whether dulcitol is an intracellular minor reserve substance, or an extracellular major metabolic product of galactose, is not settled (Onishi and Suzuki, 1968).

Under certain conditions, some yeasts are able to produce fat from carbohydrates, and to store it in their cells; such organisms include *Candida pulcherrima*, *C. utilis*, *Torula alba* and *Hansenula anomala* (Bernhauer, 1943). A suitable organism for industrial fat production is *Rhodotorula gracilis*, with a fat content rising to 60%. Such a fat content is reached by aerating a medium deficient in nitrogen and phosphorus. For the production of 1 g of fat, the yeast uses about 4·5 g of glucose (Enebo *et al.*, 1946). *Rhodotorula gracilis* also produces fat when xylose is used as the source of carbon (Nielsen and Nilsson, 1950).

Phosphate taken up from the medium by baker's yeast is stored as metaphosphate in volutin granules (Wiame, 1946, 1947). This polyphosphate appears as either a soluble or insoluble form; the insoluble form is rapidly and reversibly transformed into orthophosphate in the cell (Wiame, 1949). Polyphosphates are energy reserves (Hoffmann-Ostenhof and Weigert, 1952; Yoshida, 1953), and are thought to be a prerequisite for rapid growth of the yeast (Lindegren, 1947, 1951), but they are not considered to form any noteworthy reserve of energy (Suomalainen and Pfäffli, 1961). A high content of polyphosphates tends to decrease the stability of baker's yeast cells (Suomalainen and Pfäffli, 1961).

## I. NATURAL AND INDUSTRIAL SOURCES OF YEAST NUTRIENTS

### 1. *Molasses*

Cane or beet molasses, or a mixture of both, is the most important raw material in the industrial production of baker's yeast. The types of molasses which have been utilized for yeast propagation are beet, high test cane, refinery cane and blackstrap. Beet molasses contains about 50% sugar, practically all of it sucrose (Olbrich, 1956); in addition, it contains raffinose (about 2%) and small amounts of invert sugar and kestose (de Whalley, 1952; Olbrich, 1956). Baker's yeast is able to utilize only a third of the hexose residues in raffinose. Nitrogen-free non-sugar compounds in beet molasses include pectins and organic acids, such as oxalic, formic, acetic, butyric, ketoglutaric, lactic, saccharinic, succinic, propionic, huminic and arabic (Vogel, 1949).

The sugar content of both cane and beet molasses is about 50%, but cane molasses has a lower content of sucrose in proportion to the total sugar content and contains more invert sugar than does beet molasses (Olbrich, 1956). The presence of aconitic acid is characteristic of cane molasses (Binkley and Wolfrom, 1953; Olbrich, 1956).

The total nitrogen content of beet molasses is 1–2%, the bulk of which is represented by amino acids (Olbrich, 1956). Alanine is present to the largest extent; other amino acids found are $\gamma$-aminobutyric acid, glutamic acid, leucine, aspartic acid, glycine, serine, valine, tyrosine, and threonine (Mariani and Torraca, 1951). For the most part, glutamic acid appears as pyrrolidone carboxylic acid (Olbrich, 1956); glutamic acid and pyrrolidone carboxylic acid together form the bulk of the amino nitrogen of beet molasses, rising to a proportion of approximately 55% (Brukner, 1955; Olbrich, 1956). Betaine is a nitrogenous compound typical of beet molasses, and does not appear in cane molasses (Vogel, 1949; Olbrich, 1956). The betaine-nitrogen content of beet molasses varies from 0·08% to 0·7% of the total nitrogen (Olbrich, 1956); it is unassimilable and remains in fermented molasses (White, 1954). Other organic-nitrogen compounds in beet molasses are purines, such as adenine, guanine, xanthine and hypoxanthine (Schneider and Hoffmann-Walbeck, 1955). The nitrogen content of blackstrap molasses varies from 0·4% to 1·4%; this form of molasses has been found to contain alanine, $\gamma$-aminobutyric acid, asparagine, aspartic acid, glutamic acid, glycine, leucine (or isoleucine), lysine and valine. Cane molasses contains guanine, xanthine and hypoxanthine, and in addition the pyrimidine 5-methylcytosine (Binkley and Wolfrom, 1953).

The biotin content differs in cane and beet molasses, varying in beet molasses from 0·04 to 0·13 p.p.m. and in blackstrap molasses from 2·7 to 3·2 p.p.m. The lack of biotin limits the growth of baker's yeast in industrial production (Dawson and Harrison, 1949; Munns and White, 1949; Menzinsky, 1950). If biotin is added to biotin-poor beet molasses, the yield of yeast may be increased up to 20% (Suomalainen *et al.*, 1958) and in extreme cases even up to 40% (Suomalainen, 1963). In general, the content of calcium D-pantothenate is higher in beet molasses (50–110 p.p.m.) than in blackstrap (54 p.p.m.). For this reason, a mixture of the two molasses gives a better result in the industrial production of baker's yeast than does each separately (White and Munns, 1950). According to Rogers and Mickelson (1948), the content of pantothenic acid is 1·3 p.p.m. in beet molasses and 21·4 p.p.m. in blackstrap molasses. The content of *myo*-inositol is of the same magnitude in the two kinds of molasses; White and Munns (1950) have reported 6,000 p.p.m. in blackstrap molasses, and 5,700–8,000 p.p.m. in beet molasses. Blackstrap molasses contains a considerably larger number of growth factors than does refinery cane molasses (White and Munns, 1950). A larger content of nicotinic acid is found in beet molasses than in blackstrap molasses, while a larger content of riboflavin occurs in blackstrap molasses (Rogers and Mickelson, 1948; Binkley and Wolfrom, 1953). The nicotinic acid content in beet molasses is, however, not sufficient for the needs of baker's yeast, although the yeast is able to synthesize the deficient amount of nicotinic acid under aerobic conditions (Suomalainen *et al.*, 1965a, b). It has been reported that both beet and blackstrap molasses additionally contain thiamin, pyridoxine and folic acid (Rogers and Mickelson, 1948; Binkley and Wolfrom, 1953).

The ash content of beet molasses is about 10%, while that of cane molasses is lower (Olbrich, 1956). The amount and the composition of ash in beet molasses are influenced by the quality of the soil, the fertilizer treatment, and the species of beet. The raw ash contains approximately 80% of potassium and sodium carbonates. Considerable variation exists in the content of magnesium and calcium, dependent upon the quality of the soil. The phosphorus content of ash in beet molasses is low, since the phosphate in the beets does not remain in the molasses. Chloride, sulphate, silicate and phosphate constitute about 20% of the raw ash of beet molasses. Nitrate-nitrogen and ammonium salts appear in beet molasses in very small amounts, 0·04–0·064% of nitrate-nitrogen, and 0·011–0·023% of ammonium-nitrogen (Olbrich, 1956). Beet molasses normally contains 100 μg magnesium/g, 30 μg zinc/g, 150 μg iron/g and 20 μg copper/g. Refinery cane molasses contains more magnesium than does beet molasses or blackstrap molasses (Pyke, 1958). The variety of trace elements in molasses is considerable: Co, B, Fe, Cu, Mn, Mo, Zn, I (Olbrich, 1956), and also Ag, Ga, Ni, Pb, Ti, V, Ba, Be, Bi, Cd, Sn, Y have been found (O. Erämetsä, H. Suomalainen and A. J. A. Keränen, unpublished observations).

### 2. *Dough*

Wheat flour contains starch, small amounts of sugars, gluten-forming proteins, and small amounts of soluble proteins, fats, mineral salts, along with traces of vitamins and other substances (Tipper, 1954). Wheat varies widely in chemical composition; the percentages of protein, minerals, vitamins, pigments and enzymes exhibit up to a five-fold range among different cargoes of wheat. The bread-making potentialities of a wheat flour depend upon the chemical composition, which is again influenced by the wheat variety, the environmental and soil conditions under which the wheat is grown, the process used to mill it into flour, and the extraction of the flour (Pomeranz, 1968). The sugars in dough consist of a mixture of maltose, glucose, fructose, sucrose and levosin (a fructose oligosaccharide complex) all of which are preformed before the flour is made into dough (Koch *et al.*, 1951; Tipper, 1954). When yeast is added to the dough, it initially ferments glucose, sucrose and, to a varying extent, levosin (Pelshenke *et al.*, 1940; White, 1954). At the beginning the amount of maltose in the dough increases and baker's yeast utilizes it only in the later phase of panary fermentation (Pelshenke *et al.*, 1940; Atkin *et al.*, 1946). The pentosans in flours, composed of D-xylose and L-arabinose residues, affect the loaf volume (Pomeranz, 1968).

The nitrogen sources of the dough are the soluble nitrogen compounds of the flour, amino acids liberated by proteolytic enzymes in the flour and, thirdly, dough ingredients with a nitrogen content, other than flour (Atkin *et al.*, 1946). Initially, the amino-acid content rises in the dough, but subsequently diminishes to an almost constant level (Samuel, 1935).

In general, flours contain an abundance of phosphate, sulphur, magnesium and potassium for dough fermentation (Atkin *et al.*, 1946). Addition of salt to the dough, apart from a direct effect exerted on the taste of the baked product, further influences the ability of the yeast to ferment by preventing it from working too rapidly in long-process dough (White, 1954). Of the vitamins, flour contains at least thiamin and pyridoxine (Schultz *et al.*, 1939a; Atkin *et al.*, 1946). The relationship between yeast growth in dough and the rate of fermentation is not clear. The rate of gas evolution from a dough characteristically shows an increase with time (Atkin *et al.*, 1946).

### 3. *Malt Worts*

When paper chromatography became available, glucose, fructose, sucrose, maltose, maltotriose and maltotetraose were identified in malt wort (Barton-Wright and Harris, 1951; Harris *et al.*, 1951), as were small amounts of the pentoses arabinose, xylose and ribose, along with galactose (Green and Stone, 1952; Montreuil *et al.*, 1961). Maltotriose and maltotetraose are not the only oligosaccharides of wort.

Montreuil *et al.* (1961), in a study of the dialysable fraction, found in all 22 oligo- and polysaccharides; these oligo- and polysaccharides are all polymers of glucose. The carbohydrate composition of ale worts and lager worts differs with regard to the sucrose content which is, in ale wort, significantly higher than that in lager wort (McFarlane and Held, 1953).

Both top- and bottom-fermenting brewer's yeasts at first utilize the glucose of the wort, and then the fructose and sucrose (Barton-Wright and Harris, 1951; Harris *et al.*, 1951; Barton-Wright, 1953). Fermentation of maltose does not begin until the glucose concentration is decreased to a negligible level (Montreuil *et al.*, 1961). The utilization of sugars in wort occurs more slowly in bottom fermentation than it does in top fermentation (Harris *et al.*, 1951). According to McFarlane and Held (1953), beer does not, after lager fermentation, contain maltose, whereas small amounts of maltose are found in ales. Barton-Wright (1953) has observed that in pale-ale fermentation the maltose is consumed in 48 hours, whereas it lasts 120 hours in bottom fermentation processes. Maltotriose is not fermented until most of the maltose has been utilized and it can, in general, be found in the finished beer (Barton-Wright, 1953; McFarlane and Held, 1953; Montreuil *et al.*, 1961). Glucose polymers of higher molecular weight, together with pentoses and galactose, are not fermented (Phillips, 1955; Montreuil *et al.*, 1961). Yeasts of the so-called "super-attenuating type", *Saccharomyces diastaticus*, *Schizosaccharomyces pombe* and *Brettanomyces bruxellensis*, vigorously attack maltotetraose and other more complex carbohydrates (Phillips, 1955).

The nitrogenous compounds of wort vary with the quality of barley and the treatment (Jones and Pierce, 1963). The amounts of low molecular-weight nitrogenous compounds increase by about five times during the malting process; during the mashing, further formation of low molecular-weight nitrogenous compounds takes place. The wort contains ammonium nitrogen to the extent of 2·18–2·44 mg per 100 ml wort (Ljungdahl and Sandegren, 1953). The content of $\alpha$-amino nitrogen in wort varies between 99·5 and 162·2 $\mu$g/g dry grain, depending on the total nitrogen (Jones and Pierce, 1963). Amino acids are removed from the wort in an orderly manner, as discussed above in connection with sources of nitrogen (Barton-Wright, 1949; Sandegren *et al.*, 1954; Pierce, 1966a, b). The finished beer contains 6–7% nitrogenous compounds (Schuster, 1968).

As a rule, malt wort is poorer in biotin, pantothenate and inositol than beet molasses (White and Munns, 1950). Hopped wort contains biotin (0·0176–0·0230 $\mu$g/ml) and sweet wort slightly less. Yeast does not use all of the biotin during the course of fermentation; bottled ale still contains biotin (Lynes and Norris, 1948; Schuster, 1968). Malt wort contains sufficient pantothenic acid and riboflavin for yeast growth, and these compounds may be found even more abundantly in beer (Hopkins, 1945;

Weinfurtner *et al.*, 1964). Absorption of thiamin from wort occurs rapidly and nearly quantitatively (Fink and Just, 1942a; Hopkins, 1945; Just, 1951; Weinfurtner *et al.*, 1964). Bottom yeast takes up about 46% of the inositol and 44% of the pyridoxine in wort (Weinfurtner *et al.*, 1964). The contents of nicotinic acid in wort and in beer are almost the same after top fermentation (Hopkins, 1945), while bottom-fermenting yeast takes up approximately 40% of the nicotinamide of wort (Weinfurtner *et al.*, 1964).

The minerals in wort originate from the malt and from the water used for brewing. In addition to mineral substances necessary for the yeast, malt contains trace elements, but the amounts of these elements in beer are very small (Hudson, 1955). The filtered solids of finished beer contain 3–4% of mineral substances, and one-third of the cations are accounted for by sodium and potassium (Schuster, 1968).

### 4. *Grape Must*

The sugar content of grape must varies; on the average it is 100–250 g sugar per litre, with the bulk consisting of glucose and fructose. Completely fermented wine does not contain any, or only small amounts of, fructose and glucose. On the other hand, L-arabinose, rhamnose and xylose, present in grape must, are found in wine. The most important acids of grape must are malic and tartaric, and in addition citric, succinic, glycolic, oxalic and tannic acids are found. The acid content in grape must varies, depending on the crop and, in bad vintage years, it rises to many times the normal amount. During fermentation, a change in the acid composition occurs; malic acid is broken down to lactic and succinic acids, and small amounts of volatile acids are formed. During maturation, a diminution occurs in the acid content, through the precipitation of tartar and breakdown of acids. Tartaric, malic and lactic acids along with citric, acetic and succinic are the most important acids of wine. The most important nitrogenous compounds of grape must are ammonia, some amines, amino acids, polypeptides and albumin. The bulk of nitrogenous compounds of grape must is utilized by the yeast during fermentation. The most important minerals in both grape must and wine are potassium, calcium and magnesium phosphates. The content of minerals in wine varies from 1·5 to 3·0 g/l, being somewhat higher in grape must (Ribéreau-Gayon and Peynaud, 1958; Vogt, 1968).

### 5. *Sulphite Waste Liquor*

Sulphite waste liquor contains fermentable carbohydrates, which are used as raw material in both ethanol and food yeast production (Hägglund, 1915; Krohn, 1924). In experiments on *Candida utilis*, Fink and

Lechner (1936) observed that the yeast grows well following the addition of ammonia as a source of nitrogen. To improve the growth, diammonium phosphate and potassium chloride are generally added to the sulphite waste liquor (Inskeep *et al.*, 1951; Forss, 1961). Sulphite waste liquor is also used as raw material in the production of baker's yeast (Suomalainen, 1964). The composition of sulphite waste liquor is presented in Table II. It is subject to some variation with the kind of wood used and the method and degree of cooking in the pulping process (Inskeep *et al.*, 1951). Sundman (1947) has reported that the hexoses of the strong-pulp waste

TABLE II. *Quantitative Distribution of the Organic Dry Matter in Spruce Sulphite Waste Liquor* (*Forss, 1961*)

| Component | Amount (%) |
|---|---|
| Lignosulphonic acids and related compounds | 57·0 |
| Aldonic acids | 4·4 |
| Uronic acids and derivatives | 3·5 |
| Polysaccharides | 3·2 |
| Monosaccharides | |
| D-Glucose | 2·6 |
| D-Mannose | 11·0 |
| D-Galactose | 2·6 |
| L-Arabinose | 0·9 |
| D-Xylose | 4·6 |
| Low molecular-weight compounds, mainly acetic acid | 5·7 |
| Substances not investigated | 3·0 |
| Total | 98·5 |

liquors from spruce consist of 90% mannose and 10% galactose, while the hexoses of rayon-pulp waste liquors consist of 10% glucose and 90% mannose. Finnish sulphite waste liquor contains on the average 2–4% sugar, of which about 65% is fermentable (Sundman, 1949).

### 6. *Hydrocarbons*

Just *et al.* (1951) have reported that the yeasts *Candida lipolytica*, *C. tropicalis* and *Torulopsis colliculosa* grow well on paraffin oil and crude paraffin wax. The aqueous growth medium was aerated during growth and contained not only the source of carbon, but also mineral substances and ammonium sulphate as the source of nitrogen. *Candida lipolytica* was able to use synthetic *n*-alkanes ($C_{16}H_{34}$ and $C_{18}H_{38}$) and alkenes ($C_{16}H_{32}$ and $C_{18}H_{36}$) but not the alkane $C_{12}H_{26}$. When urea was added to a medium containing minerals and an ammonium salt, growth increased considerably (exceeding 200%; Just *et al.*, 1951). Wawzonek *et al.* (1960), as opposed to Just *et al.* (1951), observed that *C. lipolytica* also grows well when the alkane $C_{12}H_{26}$ is used as the source of carbon. It is unnecessary to separate the *n*-alkanes from the petroleum fraction which contains them; during yeast growth, the petroleum fraction is dewaxed, and in some cases its value is increased. One ton of *n*-alkane yielded a ton of protein-vitamin concentrates with yeasts which had been adapted to grow on petroleum (Champagnat *et al.*, 1963). The production costs for the manufacture of single-cell protein on an industrial scale are comparable with the costs for the production of a corresponding feedstuff (Evans, 1968). Industrially produced single-cell protein contains considerable amounts of lysine, but has a low content of sulphur-containing amino acids (Evans, 1968).

When *n*-alkanes with an even carbon number from $C_{12}$ to $C_{18}$ are used, the generation time of *C. intermedia* decreases with increased chain length. The cell yield on substrates between $C_{14}$ and $C_{18}$ is almost the same (83·0%, 81·7%, 82·5%); on the first hydrocarbon of the series it is only 59·8%. Of the odd number *n*-alkanes, $C_{17}H_{36}$ gives a yield inferior to that of the even number *n*-alkanes. The hydrocarbons $C_{20}H_{42}$ and $C_{22}H_{46}$ are solid at the temperature (30°) generally employed, and yeast growth is poor on these hydrocarbons, as an even dispersion in an aqueous solution is unobtainable. When $C_{20}H_{42}$ and $C_{22}H_{46}$ were dissolved in 2,6,10,14-tetramethyl pentadecane, which is not attacked by *C. intermedia*, rapid growth of the yeast and a rich yield were obtained (Johnson, 1964). The length of the carbon chain markedly influences the content of the crude fat, but only slightly changes the content of the crude protein of the cell (Yamada *et al.*, 1968). Table III indicates the assimilation of alkanes and alkenes by some yeasts.

In the oxidation of *n*-alkanes, the terminal methyl group is oxidized to give an alcohol, aldehyde or carboxylic acid; the penultimate methylene group is oxidized and the hydrocarbon transformed into a methyl ketone; or a diterminal oxidation takes place, in which an $\alpha$-dicarboxylic acid is formed (McKenna and Kallio, 1965; Quayle, 1968). The ability of yeasts to oxidize aromatic hydrocarbons is not as common as their ability to assimilate alkanes and alkenes (Fuhs, 1961). Breakdown of the aromatic ring occurs *via* a hydroxylated derivative (McKenna and Kallio, 1965; Quayle, 1968). The breakdown of benzene takes place *via* catechol (Van der Linden and Thijsse, 1965). Yeasts which grow on catechol are found among the genera *Oospora, Candida, Debaryomyces, Pichia* and *Saccharomyces* (Evans, 1963).

TABLE III. *Assimilation of Hydrocarbons by Different Yeasts* (*Markovitz and Kallio, 1964*)

[+ indicates growth; – no growth; ± questionable growth]

| Yeast | Substrate | | | | | | | | | |
|---|---|---|---|---|---|---|---|---|---|---|
| | Decane | Do-decane | Tetra-decane | Hexa-decane | Octa-decane | 1-Decene | 1-Do-decene | 1-Tetra-decene | 1-Hexa-decene | 1-Octa-decene |
| *Candida lipolytica* | + | + | + | + | + | + | + | + | + | + |
| *Candida pulcherrima* | + | + | + | + | + | + | + | + | + | + |
| *Candida reukaufii* | + | + | + | ± | ± | – | – | ± | – | – |
| *Cryptococcus laurentii* | – | – | – | – | – | – | – | ± | – | – |
| *Debaryomyces kloeckeri* | – | ± | + | ± | – | – | – | + | – | – |
| *Debaryomyces membranaefaciens* | – | – | ± | – | – | – | – | ± | – | – |
| *Hansenula anomala* | – | – | + | – | – | – | – | + | – | – |
| *Hansenula saturnus* | – | – | ± | – | – | – | – | ± | – | – |
| *Rhodotorula glutinis* | – | ± | + | ± | – | – | – | + | ± | – |
| *Rhodotorula gracilis* | + | + | + | + | + | – | + | + | + | ± |
| *Rhodotorula mucilaginosa* | – | ± | + | ± | ± | – | ± | + | ± | – |
| *Schizoblastosporion starkeyi-henricii* | – | – | – | – | – | – | – | ± | – | – |
| *Trichosporon capitatum* | – | – | + | – | – | – | – | + | – | – |

## II. Solute Uptake

### A. INTRODUCTION

The wall which protects the yeast cell is composed mainly of glucans and mannans, but also contains proteins, lipids and mineral substances (Northcote and Horne, 1952; Suomalainen *et al.*, 1967c). The space occupied by the cell wall, which Conway and Downey (1950) have named the outer metabolic region of the yeast cell, forms about one-tenth of the cell volume. The cell wall is impermeable to large molecules, such as inulin and peptone (Conway and Downey, 1950). Some enzymes, such as the glycoproteins, invertase and acid phosphatase (Boer and Steyn-Parvé, 1966), are located in the cell wall.

The cell wall may be removed from the yeast cell enzymically, while the released protoplast remains intact (Eddy and Williamson, 1957). However, Suomalainen *et al.* (1967b) have reported that the protoplast certainly cannot be considered simply as a yeast cell from which the cell wall has been removed by digestion, as many cofactors and some enzymes, normally located in the cell, are released into the medium during protoplast formation.

The plasma membrane surrounding the protoplast and limited by the cell wall in the intact cell, is selectively permeable. The most important components of the protoplast membrane of *Saccharomyces cerevisiae* are lipids and proteins. The lipids, of which, on an average, one-fifth consists of phospholipids, represent about 40%, and the proteins about 40–50% (Boulton and Eddy, 1962; Boulton, 1965; Longley *et al.*, 1968; Nurminen and Suomalainen, 1969; Suomalainen, 1969; Suomalainen and Nurminen, 1969). Several models of molecular organization of the membrane have been presented and, until recently, the Davson-Danielli bilayer model has been the one generally accepted. Today it appears more likely that different portions of the same membrane may have different structures, or that the same section of membrane may exist in different states at different times (Korn, 1968; Pardee, 1968; Rothfield and Finkelstein, 1968). It has been proved with certain bacteria that the enzymes which take part in the transport of nutrients are located on or near the cell membrane (Heppel, 1967; Pardee, 1967, 1968), but the question is unsolved as far as yeast is concerned.

By reason of the non-polar structure of the lipid molecules of the membrane, these molecules limit penetration into the cell by mere diffusion of the ionizable compounds dissolved in the medium. However, lipid solubility alone is inadequate to explain the permeation of compounds into the yeast cell. Äyräpää (1950) determined the permeability of baker's yeast to 19 different weak bases and observed that some low molecular-weight molecules, such as ammonia, methylamine and hydrazine, penetrate more rapidly than somewhat larger molecules of corresponding lipid solubility. Moreover, the molecular volume does not determine the rate of penetration. Likewise, glucose and xylose penetrate into the yeast cell more rapidly than would be presumed on the basis of the sizes of the molecules, or their lipid solubility (Ørskov, 1945). In general, however, the more lipid-soluble the compounds are, the faster they penetrate into the cell (Collander *et al.*, 1931; Collander and Bärlund, 1933; Äyräpää, 1950; Oura *et al.*, 1959). Thus fatty acids penetrate into

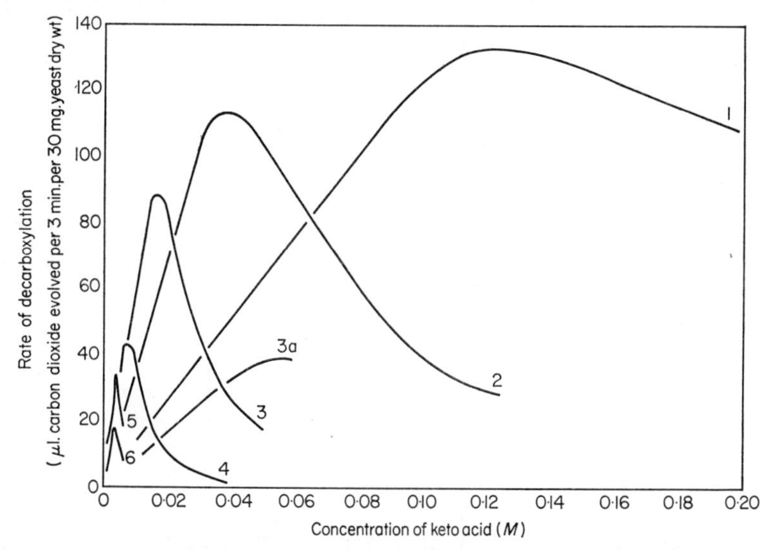


FIG. 2. Decarboxylation of α-keto acids by intact baker's yeast at pH 2 as a function of substrate concentration. Curve 1 shows data obtained with pyruvic acid, curve 2 with α-ketobutyric acid, curve 3 with α-ketovaleric acid, curve 3a with α-ketoisovaleric acid, curve 4 with α-ketocaproic acid, curve 5 with α-ketoenanthic acid, and curve 6 with α-ketocaprylic acid. The slopes of the ascending parts of the curves are a measure of the rate of penetration of the keto acid. At the maximum, the rate of penetration equals the rate of breakdown of the acid by decarboxylation. At higher concentrations, the acids accumulate in the cells and cause a lowering of the rate of decarboxylation. From Suomalainen and Oura (1958a).

the cells only in an undissociated form (Collander *et al.*, 1931). Penetration of some fatty acids, such as caproic and valeric, into the baker's yeast cell is so rapid that equilibrium is attained within a few seconds (Oura *et al.*, 1959). The influence exercised by lipid solubility upon the permeation rate of keto acids is clearly apparent from the observations made by Suomalainen and Oura (1958a, 1959), in which the rate of decarboxylation reflects the rate of permeation (Fig. 2). It has further been proved that, although the membrane of the baker's yeast is impermeable to α-ketoglutaric acid, this acid is transformed into an easily permeable form when made lipid-soluble by esterification of one or both of the carboxyl groups (Suomalainen *et al.*, 1969; Fig. 3).

It has been confirmed that the plasma membrane of yeast has small invaginations, which suggests that macromolecules may be transferred into the cell by the agency of pinocytosis or another similar mechanism (Agar and Douglas, 1957; Bakerspigel, 1964; Holter, 1965). *Candida utilis*

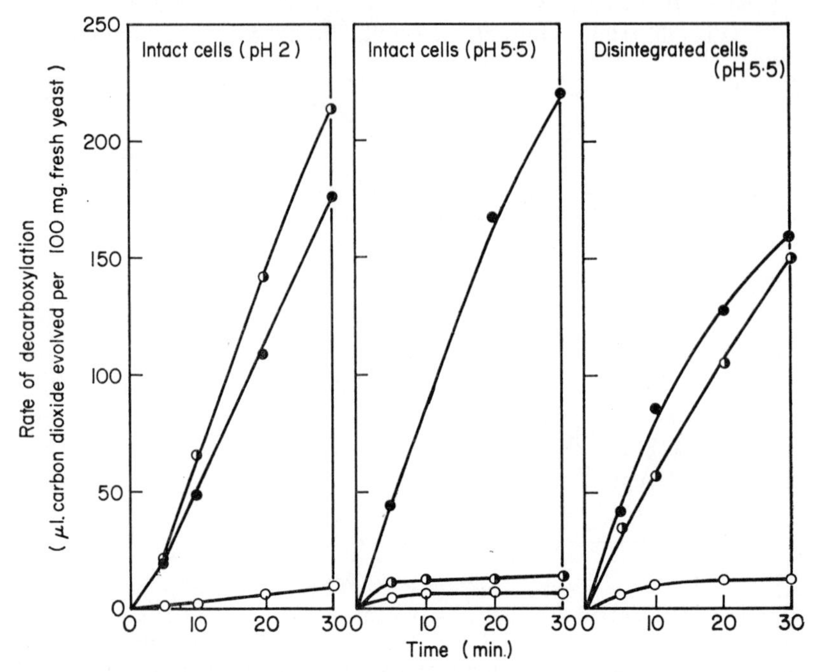


FIG. 3. Decarboxylation of α-ketoglutaric acid (o—o), α-ketoglutaric acid γ-monomethyl ester (◑—◑), and α-ketoglutaric acid diethyl ester (●—●) by intact baker's yeast at pH 2 and pH 5·5, and by disintegrated yeast at pH 5·5, as a function of time. *Alpha*-ketoglutaric acid diethyl ester penetrates into the intact cell at both pH values; α-ketoglutaric acid γ-monoethyl ester penetrates only at pH 2 where it exists in an undissociated form. After de-esterification, α-ketoglutaric acid diethyl ester is decarboxylated as the γ-ester. From Suomalainen *et al.* (1969).

rapidly takes up proteins of moderate size from the medium, such as ribonuclease, protamine, lysozyme, bovine serum albumin, cytochrome *c* and myoglobin (Yphantis *et al.*, 1967). Further, with regard to *Saccharomyces* strains in a common laboratory breeding stock, about half of the strains used have been found capable of retaining bovine serum albumin (Ottolenghi, 1967). In consequence of the influence of these proteins, certain ultraviolet-absorbing compounds and some of the amino-acid pool are released into the medium, and the viability of the cell ceases rapidly. A process of penetration, for example by pinocytosis, leaving the membrane intact or followed by repair, appears incompatible with the observations of Yphantis *et al.* (1967).

## B. UPTAKE OF SOLUTES

### 1. *Uptake of Sugars*

*a. General features of sugar uptake.* The plasma membrane of the yeast cell is impermeable to sugars (Wertheimer, 1934; Rothstein, 1954; Heredia *et al.*, 1968). The sugar is transported into the cell either by a passive transport system, or by an active mechanism. For the passive transport, no energy other than thermal motion is needed, whereas the active transport mechanism demands energy originating from metabolism, and may lead to transport of sugar against the concentration gradient. For the main part, the transport of sugar has been investigated with non-metabolizable sugars or, in the study of glucose transport, by preventing metabolism of glucose in the cell. To explain the transport process, a "carrier" concept has been employed (Cirillo, 1961a).

It is characteristic of this hypothetical carrier that it is presumed to act similarly to enzyme systems. Thus, the carrier is stereospecific, and carriers combine with nutrients by undefined bonds, although to an extent which is dependent upon the nature of the carrier and nutrient; the carrier-nutrient complex can cross the membrane, and must be physically movable to ensure its availability to compounds first from one side of the membrane and then from the other. Moreover, the rates of transit of free carrier and carrier–nutrient complex can differ (Cirillo, 1961a; Stein, 1967).

The rate of sugar transport in yeast follows Michaelis-Menten kinetics; it approaches a degree of saturation when the external sugar concentration is increased. The Michaelis constant ($K_m$) is higher for the uptake of glucose under aerobic than it is under anaerobic conditions (Kotyk and Kleinzeller, 1967). Heredia *et al.* (1968) have observed that the $K_m$ values for uptake of D-glucose, D-fructose and D-mannose are of the same order of magnitude, approximately 1 m*M* for glucose and 5 m*M* for fructose and mannose.

In baker's yeast, fermentable and non-fermentable sugars are transported into the cell through the agency of a common membrane mechanism (Cirillo, 1961b), and the stereospecific structure of the sugar, not its fermentability, is the controlling factor (Cirillo, 1968a). Kotyk (1967) divided sugars, on the basis of their stereospecificity, into two groups each with its own carrier. The first group includes the monosaccharides, with an equatorial hydroxyl group in positions 1 and 4 of the C1 chair conformation, and those with an equatorial hydroxyl group in position 2 and an equatorial $-CH_2OH$ group in position 5 of the C1 chair conformation.

In the first group are D-glucose, 2-deoxy-D-glucose, D-mannose, D-xylose, D-arabinose, D-lyxose and L-glucose; sugars of the second group include D-gulose, D-galactose, D-fucose, L-rhamnose, D-ribose, L-arabinose and L-xylose.

Sugars with the same stereospecific grouping compete for the same carrier. Moreover, experiments have shown that the carrier for the second group is probably shared by all the sugars, and that by reason of great differences in affinity, a competitive effect is observable in one direction but not in the other. 2-Deoxy-D-glucose, D-mannose, D-xylose, D-arabinose, D-fructose and L-sorbose, and in part L-lyxose, inhibit the uptake of D-glucose (Kotyk, 1967). With regard to their stereospecific structure, these sugars fall within the same group. Sorbitol, L-arabinose, L-xylose, turanose and 1,4-anhydro-D-glucitol do not inhibit the uptake of D-mannose (Heredia *et al.*, 1968). Uptake of D-arabinose is inhibited by D-glucose, 2-deoxy-D-glucose, D-mannose, D-xylose, D-fructose, L-sorbose, D-galactose, D-lyxose and N-acetyl-D-glucosamine (Kotyk, 1967; Heredia *et al.*, 1968). In experiments on inhibition of transport of L-sorbose and D-xylose, Cirillo (1968a) obtained results similar to those of Kotyk (1967). It is evident that the ketoses, L-sorbose and D-fructose, compete for the same carrier as glucose (Kotyk, 1967; Cirillo, 1968a). The difference in rates of diffusion of L- and D-arabinose, demonstrated by Sobotka *et al.* (1936), is explained by Kotyk (1967) as being attributable to the diverging stereospecific structure of the sugars. According to Van Steveninck (1968), the competition of two sugars, when the transport of at least one sugar is active, is much more complicated than a mere competitive inhibition.

The counter-transport concept has been introduced to describe the phenomenon of sugar being transported, under certain conditions, against a concentration gradient. In this case, the cell is initially in equilibrium with relation to some sugar; then another sugar is added to the medium, and transported by the agency of the carrier of the sugar first mentioned. Thus, the added sugar is transferred into the cell, and the first sugar (in equilibrium) out of the cell against the concentration gradient. Cirillo (1961b) has found that in baker's yeast glucose inhibits the uptake of sorbose, and induces a transference of sorbose out of the cell against a concentration gradient. This release of sorbose may be inhibited by uranyl ions. Wilkins (1967) has studied the counter-transport of D-arabinose, induced by glucose, and thus determined the maximum rate of glucose transport. He found the maximum transport to be constant when he varied the extracellular glucose concentration from 2 to 100 m$M$. According to Wilkins (1967), this proves that the glucose transport through the cell membrane of yeast is a symmetrical system, and acts independently of metabolism.

*b. Inhibitors of sugar uptake.* Yeast cells take up $UO_2^{2+}$ ions from the medium almost quantitatively if the concentration of $UO_2^{2+}$ in the medium is low; for example, with a concentration of 1 $\mu M$, 90% is removed. Uranyl ions are bound to the anionic groups of the cell surface; of these, the most important from a chemical aspect resemble polyphosphate. The ions inhibit the uptake of sugar, but not the endogenous respiration of the cell (Rothstein, 1955). The inhibition induced by uranyl ions is reversible, and consequently provides a means of interrupting sugar uptake at a desired stage (Cirillo and Wilkins, 1964). Nickel ions partly inhibit uptake of glucose (Van Steveninck and Rothstein, 1965). The $Ni^{2+}$ inhibition depends upon temperature; the inhibition of glucose transport, for example, increases from 9% to 87% on a rise in temperature from 8° to 35° (Van Steveninck, 1966). The $Ni^{2+}$ inhibition is not dependent upon the concentration of $Ni^{2+}$ ions, provided the concentration exceeds a stated minimum. Nickel ions are bound reversibly on the cell surface to polyphosphate groups (Van Steveninck and Booij, 1964; Van Steveninck and Rothstein, 1965). Furthermore, $Co^{2+}$ may be bound to polyphosphate groups on the surface of the cell, but does not inhibit the uptake of glucose; this may be the consequence of induced changes in the spatial arrangement of polyphosphate chains (Van Steveninck and Booij, 1964).

*c. The transport reaction.* Opinions vary as to whether the uptake of sugar by yeast is active or not, and some confusion is apparent with regard to what is meant by "active uptake" (Mitchell, 1961, 1967; Van Steveninck and Rothstein, 1965; Heredia *et al.*, 1968). Heredia *et al.* (1968) have stated that no reason exists for the assumption that there is more than one mechanism for transport of the constitutive, fermentable sugars, and that this transport is a typical facilitated diffusion. Kotyk and Michaljaničová (1968) reported that D-glucose, D-arabinose, D-xylose, D-ribose and D-galactose, at low external concentrations (0·02–2·0 m$M$), are transferred into the cell against a concentration gradient, even by as much as 85 times, in comparison with the reverse transfer into the medium. Although it is generally considered that transfer against a concentration gradient is an active transport (Rothstein, 1954; Cirillo, 1961a), the transport observed by Kotyk and Michaljaničová (1968) is not completely explicable by an active transport mechanism.

Van Steveninck and Rothstein (1965) have divided processes for the uptake of sugars into two categories: a facilitated diffusion which occurs through a rather unspecific carrier system, and an active transport which is specific and connected with metabolism. The first category includes the uptake of sorbose and of galactose by non-induced cells, and the uptake of glucose in cells induced with iodoacetate. This facilitated diffusion is not affected by the presence of $Ni^{2+}$ ions on the cell surface, and is not inhibited until higher $UO_2^{2+}$ concentrations are reached; the value for the Michaelis constant is high, and the maximum diffusion rate is low. Active transport includes the uptake of glucose and galactose in induced cells. Characteristic features here are the partially diminished rate of uptake due to the influence of $Ni^{2+}$ ions, inhibition by $UO_2^{2+}$ ions even at low concentrations, and low $K_m$ and high $V_{max.}$ values. In the case of galactose, a change from facilitated diffusion to active transport occurs during induction. The active transport mechanism is dependent on, and the passive is independent of, the pH value (Van Steveninck and Dawson, 1968).

Van Steveninck (1963) and Van Steveninck and Booij (1964) have observed that active uptake of glucose by baker's yeast is connected with enzymic phosphorylation, with the polyphosphate acting as phosphate donor. The inhibiting effect of $Ni^{2+}$ ions results from the effect exerted by the ions on the polyphosphate; polyphosphate does not take part in facilitated diffusion (Van Steveninck, 1966). According to Van Steveninck (1968), facilitated diffusion occurs by the following steps:

$$S + C \rightleftharpoons (SC),$$

and active transport by:

$$R + E \rightleftharpoons (RE)$$

$$(RE) + C + (KPO_3)_n \rightleftharpoons (R\text{–}E\text{–phosphate–}C) + (KPO_3)_{n-1}$$

$$(R\text{–}E\text{–phosphate–}C) \rightleftharpoons (R\text{–phosphate–}C) + E$$

where S is a passively transported substrate; R, an actively transported substrate; C, a carrier; E, a permease; and $(KPO_3)_n$, the polyphosphate. The next step is transfer of the (SC)- and (R–phosphate–C)-complex through the membrane. Both mechanisms use the same carrier.

Deierkauf and Booij (1968) have used a *Saccharomyces cerevisiae* mutant for examination of the changes in phosphatide components on active transport. From their results, they proposed the following series of reactions:

phosphatidylglycerol + phosphate → phosphatidylglycerol phosphate

phosphatidylglycerol phosphate + carrier + substrate →
carrier–phosphate–substrate + phosphatidylglycerol

carrier–phosphate–substrate → transmembrane transport

The reaction scheme is illustrated in Fig. 4.

In baker's yeast, actively transported glucose could not be recovered from the cells as free glucose; it was found to be phosphorylated. The intracellular concentration of ATP appeared to be far too low to account for sugar phosphorylation *via* the hexokinase reaction, and probably the phosphorylation occurs with polyphosphate as phosphate donor (Van Steveninck, 1969). The transport capacity of baker's yeast varies inversely with the intracellular concentration of glucose 6-phosphate. Like glucose, glucose 6-phosphate is bound to the yeast cell membrane, but its binding site appears to differ from the principal binding site of glucose (Azam and Kotyk, 1969).

Belaich *et al.* (1968) have employed micro-calorimetry for preparing thermograms of a mutant strain of *Sacch. cerevisiae* under anaerobic conditions. They found that the glucose concentration influences the $K_m$ value for glucose uptake. This observation explains the assumption that

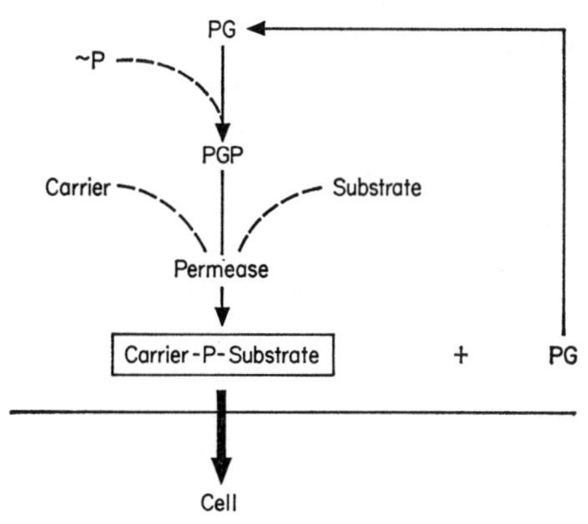


FIG. 4. Generalized hypothesis for the role in active transport of the reaction: phosphatidylglycerol phosphate (PGP) → phosphatidylglycerol (PG). From Deierkauf and Booij (1968).

the sugar can be transported and utilized in two ways: one limits the catabolism at low external concentration, and the other at high external concentration. It could be assumed that these two systems involve two different transport systems, were it not for the observations that the $V_{max.}$ value obtained is the same in both concentration ranges. Another possibility is that glucose transport is effected by only one system, with an affinity which changes when the sugar concentration falls below a given value (1·33 m*M*).

Pre-incubation of yeast with D-galactose results in the formation of a transport-system; following this, galactose, and also D-ribose, L-arabinose and L-xylose, achieve a state of equilibrium with the water content of the cell (Kotyk, 1967). According to Kotyk (1967), baker's yeast has two carriers, as mentioned above in connection with stereospecificity; in addition to these, a third carrier is formed for cells adapted to galactose. By the agency of all three of these carriers, the uptake of sugar takes place as a facilitated diffusion (Kotyk and Haškovec, 1968). Cirillo (1967, 1968b) has reported the rate of uptake of L-arabinose in baker's yeast cells grown on galactose to be more than 100 times as great as that on cells grown on glucose.

As a result of D-galactose induction, the transferred monosaccharides belong to the same stereospecific group as D-galactose, e.g. D-fucose and L-arabinose (Cirillo, 1966, 1968b; Kotyk, 1967; Kotyk and Haškovec, 1968). Induction of the system required for the transport of D-fucose and L-arabinose is regulated by the same gene as the induction of galactose transport (Cirillo, 1968b). It seems possible that a stereospecific carrier is formed as a result of the induction and, as a separate occurrence, a permease synthesis (Cirillo, 1968b). D-Galactose induces the formation of a number of galactose-utilizing enzymes; the monosaccharide carrier, apparently a protein, is synthesized *de novo* as much as two hours before galactokinase (Kotyk and Haškovec, 1968). An attempt was made to isolate the protein associated with the formation of the inducible D-galactose-transporting system in baker's yeast. Only the fraction sedimenting at 4000 *g* and containing a predominance of membranes was found to bind D-galactose (Haškovec and Kotyk, 1969).

*d. Uptake of sugars by species of* Rhodotorula *and* Candida. Kotyk and Höfer (1965) have found an active transport of sugars in the lipid-forming yeast *Rhodotorula gracilis*, by reason of the accumulation of both metabolizable and non-metabolizable sugars in the cells against concentration gradients. The process requires metabolic energy, but operates also to a limited extent even anaerobically when no gas exchange and substrate utilization are detectable in the cells. In the transport system there is a carrier which transfers one sugar at a time, and has a different affinity for the substrate inside to that outside the membrane (Kotyk and Höfer, 1965). Höfer and Kotyk (1968) could not separate the active transport of monosaccharides and a facilitated diffusion in *Rh. gracilis*. Active transport of monosaccharides (accumulation against a concentration gradient) occurred also in *Candida beverwijkii*. As compared with *C. beverwijkii*, uptake of D-ribose and L-rhamnose differed in *Rh. gracilis* in that they never reached a diffusion equilibrium with the water content of the cell.

*e. Uptake of disaccharides.* Uptake of maltose by *Saccharomyces cerevisiae* takes place by virtue of an uptake mechanism specific for maltose (Robertson and Halvorson, 1957). It has been observed that the cells of top-fermenting brewer's yeast can accumulate maltose up to a 15-fold concentration when compared with the concentration in the medium (Harris and Thompson, 1961). The accumulation mechanism is inducible and specific for maltose; only methyl-$\alpha$-glucoside and trehalose, in rather high concentrations, are able to inhibit maltose uptake (Leibowitz and

Hestrin, 1942; Suomalainen and Oura, 1956; Harris and Thompson, 1961). Baker's yeast does not utilize maltose in an alkaline medium where glucose is readily fermented (Genevois and Pavloff, 1935; Genevois, 1937; Somogyi, 1937; Schultz and Atkin, 1939; Stark and Somogyi, 1942), although it penetrates into the yeast (Harris and Thompson, 1961). A protein, which Harris and Thompson (1961) call a maltose permease, participates in the uptake of maltose.

During maltose induction, there occurs a successive induction of the transport system and of the hydrolysing enzyme (Harris and Millin, 1963). Sols and DelaFuente (1961) have reported that the first stage of fermentation of maltose and α-glucosides is transport through the membrane. This stage is stereospecific. The transport is followed by intracellular hydrolysis, and phosphorylation of the liberated hexoses. By experiments using $^{14}C$-labelled substrate, a specific permease system has also been found for maltotriose (Harris and Thompson, 1960).

2. *Uptake of Amino Acids*

Two functionally distinct amino-acid pools are found in *Candida utilis*. The first of these, the expandable pool, concentrating amino acids supplied exogenously, is variable in size depending on exogenous amino-acid concentration, and is readily exchangeable with external amino acids. The second pool, the internal one, interconverts and selects amino acids for protein incorporation, and is constant in size (Cowie and McClure, 1959). In general, exogenous amino acids are first transported to the expandable pool, and from there to the internal pool, although at high concentrations of exogenous amino acids they may be used for protein synthesis without equilibration with the expandable pool (Halvorson and Cohen, 1958; Halvorson and Cowie, 1961; Fig. 5). Accumulation of amino acids in the expandable pool is a process which demands energy; accumulation occurs against a concentration gradient, and is inhibited by azide and 2,4-dinitrophenol, but not by arsenate (Halvorson *et al.*, 1955; Halvorson and Cowie, 1961).

Permeases are responsible for the transport of amino acids into the cell (Halvorson and Cowie, 1961; Surdin *et al.*, 1965). The process for transporting amino acids from the medium to the expandable pool has a low specificity (Halvorson and Cowie, 1961). A number of α-amino acids, although dissimilar in structure, compete for the same transport mechanism in *Saccharomyces cerevisiae*; thus the accumulation of valine is inhibited by methionine, isoleucine, phenylalanine and *p*-fluorophenylalanine. Substitution of the amino or carboxyl groups destroys their inhibitory capacity, and consequently DL-valinamide and DL-N-monomethylvaline do not inhibit the accumulation of valine (Halvorson and Cohen, 1958; Halvorson and Cowie, 1961). Of the amino acids studied, the D-forms were also concentrated, but had affinities which were much lower than those of the L-forms (Surdin *et al.*, 1965).

The number of the amino-acid permeases in yeast is unknown. In *Sacch. cerevisiae*, Grenson *et al.* (1966) found a specific permease for L-arginine, which was competitively inhibited by some basic amino acids. *Saccharomyces cerevisiae* also possesses a very specific lysine permease which transports lysine into the cell; moreover, lysine may enter the cell by the agency of the arginine permease already mentioned

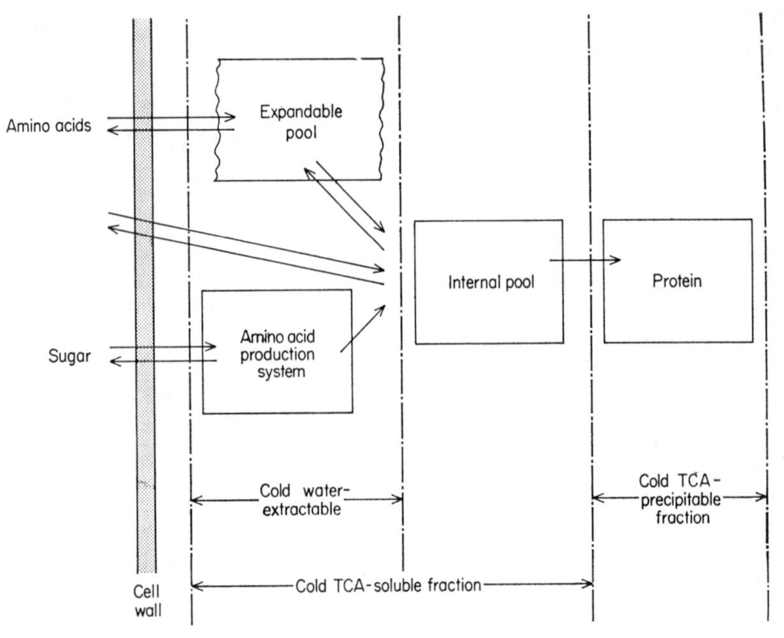


FIG. 5. Flow diagram of carbon compounds during yeast metabolism. From Halvorson and Cowie (1961).

(Grenson, 1966). Methionine uptake similarly occurs by the agency of at least two different systems, one of which has a high affinity for methionine, and the other, with a low affinity for methionine, corresponds to some other amino-acid permease (Gits and Grenson, 1967). Maw (1963b), with top-fermenting brewer's yeast, has observed that the uptake of sulphur-containing amino acids is inhibited by some other amino acids closely related structurally to them, when they are present in a high concentration. Mutants of *Saccharomyces cerevisiae* have been used for the study of amino-acid permeases (Surdin *et al.*, 1965; Grenson, 1966; Gits and Grenson, 1967). In these mutants, the ability of the yeast to produce the permease needed for the transport of a certain amino acid is affected.

For example, the mutant *arg-p* lacks an arginine permease; it is at the same time resistant to canavanine, and does not have an unspecific system for the permeation of lysine (Grenson, 1966). In *Saccharomyces chevalieri*, biosynthesis of the specific proline-transport system is repressed during growth in complete medium, or in mineral medium containing ammonium sulphate as the sole nitrogen source, and derepression occurs when cells are subjected to nitrogen starvation (Magaña-Schwencke and Schwencke, 1969; Schwencke and Magaña-Schwencke, 1969). It has been proved that the transport system is metabolically controlled also in the case of certain other amino acids (Gits and Grenson, 1969; Joiris and Grenson, 1969).

Uptake of some amino acids by yeast rich in sodium leads to excretion of sodium. Sodium excretion was measured at pH 6, at which value the highest excretion was caused by basic amino acids with an isoelectric point within the range 7·6–10·8. Dicarboxylic amino acids and alanine induce only a little, or no, excretion of sodium (Conway and Duggan, 1958). Davies *et al.* (1953) have also shown that when lysine, a basic amino acid, is accumulated losses of sodium and potassium occur from cells of *Saccharomyces fragilis*; this is not the case when glutamic acid, an acidic amino acid, is accumulated. Accumulation of amino acids is largely dependent upon the pH value. The pH optimum differs for different amino acids (Surdin *et al.*, 1965; Eddy, 1969.

As compared with the sodium ion, potassium and ammonium ions inhibit the uptake of amino acids (Surdin *et al.*, 1965; Eddy, 1969). Eddy has proposed two hypotheses to explain the transport mechanism for glycine: there exist either two different systems, which by some means compete with each other, or else the $K^+$ and $H^+$ ions directly influence a carrier system which facilitates the uptake of glycine.

### 3. *Uptake of Organic Acids*

Organic acids, in undissociated form, penetrate by diffusion into an intact cell (Collander *et al.*, 1931). Thus, the irreversible toxic effect of iodoacetic acid is proportional to the concentration of the undissociated acid (Aldous, 1948; Suomalainen and Konttinen, 1969). The commonly used buffers include systems either able to buffer solely the solution surrounding the cells and the surfaces of the cells, or to affect, as does the acetic acid-acetate buffer, the acidity inside the cells as well (Suomalainen and Oura, 1955). The significance of the permeability of the plasma membrane as a factor influencing the metabolism of organic acids by yeast cells was first indicated in Hägglund's investigations with pyruvic acid (Hägglund and Augustsson, 1926). Lynen successfully continued studies on the penetration of keto acids, and in connection with these, the intermediates of the citric acid cycle (Lynen, 1939; Lynen and Neciullah, 1939; Lynen and Kalb, 1955). The permeation constants of the straight-chain fatty acids are approximately proportional to their relative lipid solubilities, except that smaller molecules have somewhat faster permeation than would be inferred from their solubility properties alone (Oura *et al.*, 1959). With regard to keto acids, permeation of pyruvic acid is slow in comparison with that of acetic acid and propionic acid, and the permeability of α-keto acids increases with the length of the carbon chain (Suomalainen and Oura, 1958a, 1959; Fig. 2, p. 37).

Di- and tricarboxylic acids, which are highly dissociated and have poor lipid solubility, permeate into the intact yeast cell with difficulty or not at all (Lynen, 1939; Lynen and Neciullah, 1939; Lynen and Kalb, 1955; Conway and Downey, 1950; Malm, 1950; Oura *et al.*, 1959). Neglect of this fact resulted in an erroneous explanation of yeast metabolism until Lynen and Kalb (1955) pointed out that attention should be paid to the permeability of the substrate concerned. The poor permeation of di- and tricarboxylic acids makes them usable in the preparation of buffer solutions (Runnström *et al.*, 1939; Suomalainen and Oura, 1955).

Not only the chain length, but also the substituents in the acid molecule, influence the permeation characteristics of the acid. Thus the lipophobic hydroxyl group, for example in the acid pairs butyric and β-oxobutyric acid or pyruvic and hydroxypyruvic acid, considerably retards the permeation (Oura *et al.*, 1959; Suomalainen *et al.*, 1969). Since a prerequisite for easy permeability is lipid solubility, esterification of an acid should greatly affect its permeability. This has been proved with respect to α-ketoglutaric acid, where the esterification of one or both carboxylic groups has been found greatly to facilitate its permeation into baker's yeast cells (Suomalainen *et al.*, 1969; Fig. 3, p. 38).

At low pH values, acetic acid inhibits accumulation of phosphate and fermentation of glucose (Samson *et al.*, 1955). According to Fencl (1961), the inhibiting effect of acetic acid in an acid medium is directed towards the uptake of polyatomic anions such as phosphate, but not the uptake of monoatomic anions such as $I^-$ or $Cl^-$. Furthermore, higher fatty acids, butyric acid (Hägglund, 1914), and even $C_8$–$C_{11}$ acids (Suomalainen and Keränen, 1968) exert an inhibitory effect on yeast growth.

In addition to the diffusion described above, an active uptake of anions, for example of pyruvate, may take place under aerobic conditions (Smythe, 1938; Foulkes, 1955).

### 4. *Uptake of Monovalent Cations*

*a. Uptake of potassium ions in connection with metabolism.* The influx of $K^+$ is an active transport process in yeast. Commercial baker's yeast takes up potassium ions from the medium even against a concentration gradient in excess of 1000 : 1. Uptake of potassium ions is connected with the transfer of the $H^+$ ion from the cell, which may occur against a gradient of 50 : 1 (Rothstein and Bruce, 1958a). Uptake of $K^+$ follows Michaelis-Menten kinetics. The enzyme-substrate model explains the occurrence of the uptake of $K^+$ as a combination of the $K^+$ ion with the membrane receptor, the carrier (Rothstein, 1961).

Uptake of $K^+$ is connected with the assimilation of the metabolizable substrate. In 1940, Pulver and Verzár (1940b) proved that, while baker's yeast assimilates glucose from the medium, it simultaneously takes up $K^+$ ions, and the $Na^+$ content in the medium remains unchanged. At the beginning of the test, the $K^+$ ions were rapidly assimilated; during the course of fermentation their retransfer to the medium began. Finally the concentration of $K^+$ ions was even higher than that at the beginning of the experiment. In Rothstein and Enns' (1946) experiments, the uptake and utilization of substrate (glucose or ethanol) are combined with an exchange of $H^+$ ions from the cells for $K^+$ ions from the medium. When the substrate has been completely utilized, some of the $K^+$ ions may diffuse back into the medium, although a proportion is retained in the cell. Glucose is metabolized to a greater extent under aerobic than under anaerobic conditions, and correspondingly more $K^+$ ions remain in the cell under aerobic than under anaerobic conditions. During the dissimilation of cellular carbohydrate reserves, $K^+$ ions diffuse from the cells in exchange for $H^+$ ions from the medium.

In the presence of glucose, cation exchange occurs more rapidly under aerobic than under anaerobic conditions. The cation exchange is very slow even under aerobic conditions, and is extremely slow under anaerobic conditions when no exogenous substrate is present. An exchange of ions, which is independent of metabolism, that is ion diffusion, hardly occurs at all (Rothstein, 1961). Riemersma (1964) has reported that the uptake of potassium ions is more probably connected with some metabolic process other than the uptake of glucose. Reilly (1967b) found, when the sodium and potassium transport was inhibited by *o*-phenanthroline in a wild-type and a mutant strain of *Saccharomyces cerevisiae*, that apparently there exist two different mechanisms for the transport of $Na^+$ and $K^+$ ions. Efflux of $Na^+$ is less dependent upon the metabolic activity of yeast than is $K^+$ uptake.

*b. Uptake of potassium ions in connection with the diffusion of organic anions.* Uptake of the $K^+$ ion is connected with $H^+$ ion exchange, and simultaneously transfer of the organic anion to the medium occurs. However, diffusion of the organic anion is not a premise for the uptake of potassium ions. If the $K^+$ ion is replaced by $Na^+$ ion, or if the medium contains no salt, diffusion of the $H^+$ ion together with the organic anion still takes place (Rothstein and Enns, 1946).

Conway and Brady (1950) have carried out fermentations with baker's yeast in an unbuffered solution of glucose (5%) with a volume of cells approximately equal to that of the sugar solution, and found that during a period of 80 min 40–50 m-equiv. of organic acid per litre were excreted; this acid was mainly succinic acid. When the glucose solution was made 0·1–0·2 *M* in respect of potassium chloride, only a small increase in excretion occurred as compared with the total acid excretion, although the proportion of succinic acid diminished. When the cells were oxygenated 24–28 hours before fermentation, excretion of succinic acid diminished considerably or ceased completely. However, in the presence of potassium chloride, the concentration of acid-labile carbon dioxide ($HCO_3^-$) is increased in the cells as compared with the control not containing potassium chloride. The succinic acid excreted from the cells during fermentation remains unchanged in the medium, but the $H^+$ ions disappear within 3–4 hours.

*c.* $K^+$–$H^+$ *exchange.* Conway *et al.* (1954), with sodium-rich yeast prepared from commercial baker's yeast, have arrived at results which indicate the presence of two different carriers in yeast; one transfers $K^+$ ions into the cell, and the other $Na^+$ ions from the cell. In the absence of $K^+$ ions, the $Na^+$ ions can be transferred into the cell by a carrier which normally conveys $K^+$ ions. The $Na^+$ carrier is inhibited by cyanide (2 m*M*) and by anoxia, but not by azide (2 m*M*) or by 2,4-dinitrophenol (2 m*M*). The $K^+$ carrier is inhibited by anoxia, cyanide, azide and 2,4-dinitrophenol. Foulkes (1956) obtained, with inhibition experiments, results which differed from those of Conway *et al.* (1954) and concluded that only one carrier is involved in the transport of both $K^+$ and $Na^+$ ions.

Rothstein (1961) has stated that the transfer of the $K^+$ ion (or any other monovalent ion) into the cell by the $K^+$ carrier is connected with the transfer of a $Na^+$, $H^+$ or $K^+$ ion from the cell by the $Na^+$ or $H^+$ carrier. This exchange of one ion for another may be a direct combination of two systems (Rothstein, 1961). Apart from the net uptake of the $K^+$ ion, an efflux of the $K^+$ ion occurs in accordance with the concentration gradient. The $K^+$ concentration regulates the influx and efflux rates (Rothstein and Bruce, 1958a).

Efflux of $K^+$ is primarily a $K^+$–$H^+$ exchange, and thus the efflux rate increases at low pH values. At higher pH values, some monovalent cations (such as $Na^+$ and the triethylamine ion) are capable of increasing the efflux (Rothstein and Bruce, 1958b).

*d. Specificity of the carrier.* In a commercial baker's yeast, the pH value affects the ability of the yeast cell to discriminate between $Na^+$ and $K^+$ ions. Potassium ions are favoured considerably at low pH values. The results of kinetic studies concerned with different alkali-metal ions in fermenting yeasts have indicated three different pH ranges. Below pH 4, $H^+$ ions competitively inhibit the transport of all cations. Within the pH range 4–6, $H^+$ ions are a non-competitive inhibitor, and at pH 6–8, $H^+$ ions do not exercise any influence. Within the pH range 6–8, uptake of the alkali metal ion occurs in accordance with the Michaelis-Menten equation. The order of specificity, according to the affinity, is $K^+ > Rb^+ > Cs^+ > Na^+ > Li^+$ (Armstrong and Rothstein, 1964). For the relative affinities of the ions potassium, rubidium, caesium, sodium, lithium and magnesium, Conway and Duggan (1958) have derived the following values: 100, 42·7, 3·8, 0·5, and 0·5. The same values were obtained under two different sets of conditions, during fermentation, and with non-fermenting yeast rich in $Na^+$.

The $K^+$ carrier also transports $NH_4^+$ ions into the cell; the affinity of the transport system for $NH_4^+$ is only one-fifth of that for $K^+$. The affinity of the $K^+$ transport system for ethylamine is quite appreciable, amounting to about one-tenth of that of $K^+$. Competition experiments have shown that $Na^+$ and $K^+$ compete for the same active group; the $K^+$ carrier attained its maximum rate of action at a $K^+$ concentration of 1·6 m*M*. The maximum transport speed of other ions is approximately the same as that of $K^+$, but it is reached under very different external concentrations; with $Li^+$, the concentration was 300 m*M* (Conway and Duggan, 1958).

The specificity of the transport mechanism differs inside and outside the cell (Conway *et al.*, 1954). Rothstein (1965) has stated that the membrane is asymmetrical; it thus chooses $K^+$ on the outer surface, and discriminates against it on the inner surface. The maintenance of such an asymmetry is a metabolic function, which may be attributable to the cyclic change of the carrier from a $K^+$-selective to a $H^+$- or $Na^+$-selective form.

*e. The uptake mechanism.* Influx and efflux of monovalent cations depend upon metabolism (Conway *et al.*, 1954; Rothstein and Bruce, 1958a). The ion pump in animal cells derives its energy from ATP, whereas with yeast ouabain, which prevents the availability of energy from ATP, does not influence the $K^+/Na^+$ rate (Jennings and Souter, 1963; Dee and Conway, 1968). According to Conway (1951, 1953, 1955), active ion transport by yeast is not dependent upon ATP, but is effected by means of the redox pump. The redox dyes, which raise or lower the potential during fermentation, also raise or lower the excretion of $H^+$ or the active absorption of $K^+$. For instance, $K^+$ absorption or $H^+$ excretion are decreased to about 30% or less of the control values when the potential is lowered by about 60 mV. Inorganic redox systems have a similar but less pronounced effect (Conway and Kernan, 1955). The redox pump theory states that transport is effected by the cycle which contains the metal ion and enzyme (Conway, 1951; Reilly, 1967a). In this cycle, the redox systems act as carriers (Conway and Kernan, 1955). Riemersma (1964), however, does not consider that the redox carrier hypothesis provides a satisfactory explanation, at least as far as $H^+$ excretion is concerned.

Permeation of ammonium hydroxide into an intact cell is, according to Collander (Collander *et al.*, 1931; Collander, 1957), a diffusion of undissociated ammonia through the cell membrane. The momentary penetration of ammonium hydroxide in comparison with that of sodium hydroxide or potassium hydroxide can be clearly demonstrated with baker's yeast cells pretreated with neutral red (Collander and Äyräpää, 1947). Ammonia–ammonium chloride buffer extends its effect into the cell, inside the plasma membrane (Suomalainen and Oura, 1955).

5. *Uptake of Bivalent Cations*

Yeast cells are relatively impermeable to bivalent cations. If cells are suspended in distilled water, scarcely any leakage of bivalent cations takes place (Rothstein, 1955). Moreover, resting yeast cells are impermeable to bivalent cations such as $Ca^{2+}$ and $Mn^{2+}$. Although no exchange occurs with bivalent cations in the cytoplasm in resting cells, the extracellular $Mn^{2+}$ and $Ca^{2+}$ rapidly equilibrate with the cell surface (Rothstein and Hayes, 1956). The surface layer of the cell contains fixed anionic groups of two species, polyphosphate and carboxyl (Rothstein, 1955). Of the bivalent cations, $UO_2^{2+}$ has the greatest affinity for yeast, while other bivalent cations including $Ba^{2+}$, $Zn^{2+}$, $Mg^{2+}$, $Ca^{2+}$, $Sr^{2+}$, $Mn^{2+}$, $Cu^{2+}$ and $Hg^{2+}$ have the same affinity (Rothstein and Hayes, 1956). Similarly, the $Ni^{2+}$, $Co^{2+}$ and $Fe^{2+}$ ions are bound to the same sites as uranyl ions on the outer surface of the cell (Van Steveninck and Booij, 1964). The binding of exogenous bivalent cations by the yeast cell is rapid and reversible (Rothstein and Hayes, 1956).

Besides this rather unspecific attachment to the cell surface, another specific system occurs in baker's yeast; this transports the bivalent cations into the cell, into a virtually non-exchangeable pool. It is characteristic of this transport system that surface binding by fixed negative groups on the cell is not involved. The reaction is effectively irreversible (Rothstein *et al.*, 1958). The uptake is the same under anaerobic and aerobic conditions, which suggests that fermentative reactions can supply the energy for transport (Rothstein *et al.*, 1958; Fuhrmann and Rothstein, 1968). Synthesis of a carrier for the transport of bivalent cations involves a phosphorylation step that is closely coupled with reactions involved in the absorption of phosphate (Jennings *et al.*, 1958). Yeast cells can take up $Ni^{2+}$, $Co^{2+}$ and $Zn^{2+}$ into a non-exchangeable pool by a system that also transports $Mg^{2+}$ and $Mn^{2+}$ (Fuhrmann and Rothstein, 1968). The order of affinity is $Mg^{2+} > Co^{2+} > Zn^{2+} > Mn^{2+} > Ni^{2+} > Ca^{2+} > Sr^{2+}$ (Rothstein *et al.*, 1958; Fuhrmann and Rothstein, 1968). Uptake is decreased at low pH values (below 5·0), but an $H^+$ exchange system is not involved. Instead, two $K^+$ ions are secreted for each bivalent cation absorbed (Fuhrmann and Rothstein, 1968). Potassium ions stimulate uptake of bivalent cations at low concentrations, but inhibit it at a $K^+$ concentration in excess of 20 m*M* (Rothstein *et al.*, 1958).

Conway and his colleagues (Conway and Beary, 1956, 1958; Conway and Duggan, 1956, 1958) have described another kind of transport system for $Mg^{2+}$ ions. In this system, magnesium ions are transferred into the cell by the same carrier as that for $K^+$ ions. This cation carrier has a low affinity for the $Mg^{2+}$ ion (Conway and Duggan, 1958). Potassium ions inhibit the uptake of magnesium ions even at low concentrations (Conway and Duggan, 1956), and the uptake is balanced by $H^+$ secretion (Conway and Beary, 1958). As regards $Mg^{2+}$ transport, the activity of this carrier is dependent on the presence of oxygen (Conway and Beary, 1956, 1958). This carrier system functions in the same way as that for monovalent cations, by means of the redox pump system (Conway and Gaffney, 1966).

6. *Uptake of Inorganic Anions*

*a. Phosphate.* Penetration of the orthophosphate ion into the yeast cell does not take place by mere diffusion; under aerobic and anaerobic conditions, energy is also needed (Holzer, 1953; Goodman and Rothstein, 1957; Schönherr and Borst Pauwels, 1967). Moreover, despite the rather high intracellular concentration of orthophosphate and phosphate esters, only traces of these compounds leak from cells suspended in distilled water (Rothstein, 1961). During the uptake of phosphate, almost no efflux of the ion occurs (Goodman and Rothstein, 1957; Rothstein, 1963). Consequently, uptake of phosphate is an active process which derives its energy from metabolism. Orthophosphate is taken up against a concentration gradient as high as 100:1. Phosphate uptake follows saturation kinetics, suggesting that the entry of phosphate involves its combination with a receptor or carrier (Goodman and Rothstein, 1957). Only the monovalent ion $H_2PO_4^-$ is transported. Uptake of $H_2PO_4^-$ is balanced electrically by the appearance of an equivalent amount of hydroxyl ion in the medium (Rothstein, 1961, 1963). Arsenate competes with phosphate for transport into the yeast cell (Rothstein, 1963), and at the same time arsenate continuously inactivates the transport system in an irreversible manner, so that the uptake eventually ceases (Jung and Rothstein, 1965). $^{74}$As-Arsenate was bound exclusively to the phosphoinositide fraction of the cellular lipids in *Saccharomyces carlsbergensis*. High affinity sites for arsenate transport are formed during arsenate adaptation, and phosphate does not exercise any influence upon these (Cerbón, 1969).

The presence of the $K^+$ ion is necessary for uptake of phosphate, although its influence is not associated directly with this process; this is indicated by the occurrence of uptake even if the cells are preloaded with

$K^+$ ions (Schmidt *et al.*, 1949; Rothstein, 1961). Further, in a medium containing a potassium salt and phosphate, uptake of $K^+$ occurs rather rapidly, and uptake of phosphate continues even when the $K^+$ has been completely absorbed (Rothstein *et al.*, 1958; Rothstein, 1961). The presence of the $K^+$ ion augments the ability to take up phosphate by increasing the buffering capacity of the cell. A cell which contains only a few $K^+$ ions is capable of absorbing only a few phosphate ions, as the phosphate uptake leads to an acidification of the cell, which again retards the phosphate-transporting system (Rothstein, 1961). The effect of potassium ions is a specific one, since sodium ions are unable to stimulate phosphate uptake (Goodman and Rothstein, 1957). The active uptake of bivalent cations involves a phosphorylation step, closely coupled with the reaction involved in the absorption of phosphate (Jennings *et al.*, 1958; Rothstein *et al.*, 1958).

When cells of *Saccharomyces carlsbergensis* are suspended in a solution of fermentable sugar, phosphate and also $K^+$ and $Mg^{2+}$ are rapidly released into the medium; on further incubation, the phosphate is reabsorbed: leakage is dependent upon an increase in the permeability of the cytoplasmic membrane during sugar utilization. This release does not occur when the cells are suspended in water, neither is it attributable to osmotic pressure, as the release of phosphate does not take place in a galactose solution (Stephanopoulos and Lewis, 1968).

*b. Sulphate.* Uptake of sulphate depends on the availability of metabolic energy in baker's and brewer's yeast and in *Candida utilis* (Kotyk, 1959; Crocomo and Menard, 1962; Maw, 1963a). For optimum uptake, a further requirement is the presence of citrate in the medium (Maw, 1963a). Sulphate uptake follows saturation kinetics (Maw, 1965). In baker's yeast, uptake of sulphate proceeds at the same rate anaerobically and aerobically (Kotyk, 1959). Sulphite, thiosulphate and selenate competitively inhibit uptake of sulphate (Kotyk, 1959; Maw, 1963a). The uptake is partly suppressed by sulphur-containing amino acids, such as L-methionine and L-cysteine (Kleinzeller *et al.*, 1959; Maw, 1963a). The mechanism of sulphate uptake by yeast is unknown but, according to Kleinzeller *et al.* (1959) and Maw (1965), the sulphate ion may, during the course of transport, be bound to a carrier as 3′-phosphoadenosine 5′-phosphosulphate and as adenosine 5′-phosphosulphate.

*c. Fluoride.* Fluorine permeates into the yeast cell as undissociated fluoride, and thus the permeation is dependent upon pH value (Runnström and Sperber, 1938; Malm, 1940, 1947). The inhibiting effect of fluoride on respiration and fermentation depends on temperature, and the temperature coefficient of fluoride permeation corresponds with the coefficient characteristic of the diffusion process. The transfer of fluoride into the cell induces leakage of potassium (Malm, 1947).

7. *Uptake of Nucleotides*

*Candida utilis*, by active transport, takes up adenine, hypoxanthine, guanine, xanthine, uric acid, 2,6-diaminopurine and isoguanine (Roush *et al.*, 1959). Also through an active transport mechanism, *Torulopsis candida* accumulates guanine, xanthine and uric acid, but not adenine and hypoxanthine; the last of these compounds does not support growth and flavinogenesis in this yeast (Roush and Shieh, 1962). Harris and Thompson (1962) could not prove in their studies of top-fermenting brewer's yeast that any specific concentration mechanism takes part in the uptake of adenine.

*Saccharomyces carlsbergensis* takes up high molecular-weight RNA from the medium, although to date no knowledge is available on the mechanism (Bloemers and Koningsberger, 1967).

C. THE ROLES OF ENZYMES LOCATED ON THE CELL SURFACE

The enzymes that are located either in the cell wall or on the outer surface of the plasma membrane, and which are active outside the plasma membrane, are hydrolases, with the exception of catalase. Several of these enzymes, such as invertase and melibiase, hydrolyse impermeable di- and trisaccharides. Invertase is located in the cell wall, and closer to the external than to the internal surface (Demis *et al.*, 1954; Myrbäck and Willstaedt, 1955; Preiss, 1958; Lampen, 1968). In brewer's bottom yeast adapted to raffinose fermentation, a melibiase is also located in the cell wall; melibiase, like invertase, is released into the medium if the cell wall is removed (Friis and Ottolenghi, 1959). Hydrolysis is the first stage in melibiose fermentation, and is followed by transfer of the released hexoses into the cell (DelaFuente and Sols, 1962).

With regard to the α-glucosidase, Suomalainen (1948b) has suggested that baker's yeast contains two different α-glucosidases, one corresponding to yeast maltase and located within the interior of the cell, and the other active in acid media and located on the surface of the cell. However, in brewer's yeast lacking α-glucosidase activity in acid media, maltase seems to be located on the cell surface. Sutton and Lampen (1962) state that the maltase of baker's yeast is obviously only intracellular. Brewer's bottom and top yeasts decompose cellobiose, but do so extremely slowly (Weidenhagen, 1930). β-Glucosidase activity has been reported to be located at the surface of some strains of *Saccharomyces cerevisiae* (Kaplan and Tacreiter, 1966).

*Saccharomyces diastaticus* has been found capable of fermenting starch and dextrin (Andrews and Gilliland, 1952; Gilliland, 1953); the fermentation is attributable to the activity of glucamylase, which is able to diffuse out of the cells (Hopkins, 1955, 1958). *Saccharomyces fragilis* produces, extracellularly and intracellularly, inulinase, a β-fructosidase, which hydrolyses inulin, bacterial levans and the fructose portion of raffinose (Snyder and Phaff, 1960). $\beta$-Glucanase also belongs to the hydrolysing enzymes released into the medium, and is involved in the budding process. The $\beta$-1,3-glucanase activity found in a culture medium of *Sacch. fragilis* is about seven times as great as that of *Sacch. cerevisiae* (Abd-El Al and Phaff, 1968). The glucanase isolated from *Sacch. cerevisiae* hydrolyses the $\beta$-1,3- and $\beta$-1,6-bonds of glucan (Abd-El Al and Phaff, 1967).

The acid phosphatase is entirely, or at least principally, external to the plasma membrane (Suomalainen *et al.*, 1960a; Tonino and Steyn-Parvé, 1963). Protoplasts synthesize the acid phosphatase, and release it into the medium (McLellan and Lampen, 1963), and the acid phosphatase molecule contains a carbohydrate moiety (Boer and Steyn-Parvé, 1966). Following phosphate starvation of *Candida utilis* and baker's yeast, the activity of the acid phosphatase increases (Rautanen and Kärkkäinen, 1951; Rautanen and Kylä-Siurola, 1954; Suomalainen *et al.*, 1960a). Phosphate-starved cells, with a high acid phosphatase activity, hydrolyse fructose 1,6-diphosphate and ferment it about three times more rapidly than normal baker's yeast (Suomalainen *et al.*, 1960a). In intact top yeast, phosphatase with a pH optimum of 3·7 catalyses hydrolysis of the pyrophosphate group of thiamin pyrophosphate, with thiamin monophosphate appearing as an intermediate (Westenbrink *et al.*, 1940; Steyn-Parvé, 1962). The cell surface of baker's yeast contains an easily extractable phospholipase (Kokke *et al.*, 1963; Kokke, 1966; Nurminen and Suomalainen, 1969; Suomalainen and Nurminen, 1969).

Ahearn *et al.* (1968) have observed extracellular proteolytic activity in a number of yeasts and yeast-like fungi; the most active organisms were strains of *Candida lipolytica*, *C. punicea*, *Aureobasidium pullulans* and species of *Cephalosporium*. Activity was not found in *Sacch. cerevisiae* and *Sacch. carlsbergensis*. Chen and Miller (1968) have reported proteolytic activity by intact cells of three *Sacch. cerevisiae* strains during sporulation; according to them, it is unlikely that the substrate, haemoglobin, could penetrate the yeast cell wall and, since they were unable to detect proteolytic activity in sporulation medium, it appears that the haemoglobin degradation occurred at the surface of the sporulating cells. The proteolytic enzymes formed during autolysis of brewer's yeast cells are glycoprotein in nature, and contain glucan and mannan, whereas the proteases of the protoplasts are free from mannan and glucan. The glucans and mannans present in the enzymic preparations seemed to reflect the composition of yeast cell walls. Protease is secreted from cells of intact brewer's yeast into a medium which contains protein, but the glycoprotein nature of this protease is unknown (Maddox and Hough, 1969). Aminopeptidase has been found outside the membrane of *Sacch. cere-*

*visiae*. Most of the peptidase has been removed together with the cell walls during the preparation of protoplasts. However, in the intact cells, only part of the aminopeptidase is accessible to substrate (Matile, 1969).

Although some yeast catalase appears in intact cells, the rest is found only after treatment of the cells with heat or solvents (von Euler and Blix, 1919; von Euler *et al.*, 1927). Baker's yeast grown anaerobically has a very small content of catalase, but the enzyme is formed when the yeast suspension is aerated (Chantrenne and Courtois, 1954; Kaplan, 1963). During anaerobic growth, when inhibition of catalase synthesis becomes apparent, the "cryptic catalase", i.e. that hidden within the cell, is progressively converted into the "patent form", active *in vivo* against external substrate. Cryptic enzyme is inactivated more easily than the patent form (Kaplan, 1962). Uranyl and mercuric ions inhibit the patent catalase activity (Kaplan, 1965). It is concluded that the patent catalase activity of aerobic yeast results from enzyme molecules held at the cell membrane, or external to it in the cell wall (Kaplan, 1965). The plasma membranes separated from anaerobically-grown *Sacch. cerevisiae* contain lipids, polysaccharides and protein. The activity of an ATPase is detectable in isolated plasma membrane (Boulton and Eddy, 1962; Boulton, 1965; Matile *et al.*, 1967; Suomalainen *et al.*, 1967c; Nurminen *et al.*, 1970).

## III. Outflow of Compounds

The plasma membrane exerts a limiting and selective effect upon the penetration of compounds from the medium into the cell. Correspondingly, it influences the release of metabolites from the cell into the medium. It is well known that fermentation products, such as ethanol and glycerol, are excreted into the fermentation solution, where additional intermediates and by-products of sugar fermentation, such as acetaldehyde and fusel alcohols, are to be found. In a study of the appearance of keto acids in yeast cells and fermentation solutions, Suomalainen and Ronkainen (1963) observed that all of the keto acids present in the cell were also present, and in even larger amounts, in the fermentation medium. The amounts correlated with the enzyme composition of the yeast. Thus, the fermented wash of decarboxylase-poor commercial-stage baker's yeast contained twice as much pyruvic acid, but only one-fifth of the amount of $\alpha$-ketoglutaric acid, as was present in the wash using decarboxylase-rich baker's yeast (Suomalainen and Ronkainen, 1963; Trevelyan and Harrison, 1954). Apart from pyruvic acid and $\alpha$-ketoglutaric acid, the following keto acids have been found in the medium: *p*-hydroxyphenylpyruvic acid, $\alpha$-keto-$\beta$-methylvaleric acid, $\alpha$-keto-isovaleric acid, $\alpha$-keto-isocaproic acid, $\beta$-phenylpyruvic acid,

$\alpha$-ketobutyric acid, oxaloacetic acid and $\alpha$-keto-$\gamma$-methylbutyric acid (Suomalainen and Ronkainen, 1963; Suomalainen and Keränen, 1967).

Some amino acids also appear in the medium during yeast growth, but in amounts which are very small in comparison with those of keto acids (Suomalainen and Keränen, 1967). This again points to the selective function of the plasma membrane in the release of metabolites from the cell. It seems that the uptake of glucose in particular induces an outflow of amino acids from the cells of *Saccharomyces carlsbergensis*, by effecting a change in the permeability of the plasma membrane. A rapid release of amino acids occurs when the yeast cells are suspended in a solution of glucose; later, the amino acids are re-absorbed (Lewis and Stephanopoulos, 1967).

Acetaldehyde is the most important of the carbonyl compounds formed during alcoholic fermentation; other aldehydes found in beer and/or the fermented medium are propionaldehyde, butyraldehyde, isobutyraldehyde, valeraldehyde, isovaleraldehyde, optically active valeraldehyde, hexanal, heptanal and glycolaldehyde (Suomalainen *et al.*, 1968; Suomalainen and Ronkainen, 1968b). The formation of diacetyl in yeast fermentations has long been known, and recent investigations have proved that it is formed from the $\alpha$-acetolactic acid synthesized inside the cell, which is transferred from the cell into the fermenting medium and there transformed to diacetyl by spontaneous decarboxylation (Suomalainen and Ronkainen, 1968a; Inoue *et al.*, 1968). The bulk of the carbonyl compounds in the medium are formed by the decarboxylation of keto acids (Suomalainen and Linnahalme, 1966).

During the course of alcoholic fermentation, higher alcohols (fusel alcohols) are released into the medium in addition to ethanol. The fusel oil contains fusel alcohols, and other compounds, depending upon the raw material utilized. The main components of fusel oil are isoamyl alcohol, optically active amyl alcohol and isobutyl alcohol. Several other alcohols also appear; their number and quantity vary according to the raw material (Webb and Ingraham, 1963; Suomalainen *et al.*, 1968). Other aroma components formed during the fermentation are esters and free fatty acids (see Vol. 3, Chapter 6). The predominant role played by yeast in the formation of aroma compounds has been demonstrated by the results reported by Suomalainen and Nykänen (1964, 1966a; Suomalainen, 1968a, b). They have further stated (Suomalainen and Nykänen, 1966b) that sherry yeast produces the same aroma components, irrespective of whether grape or berry wine is used as raw material.

During the fermentation, baker's and brewer's yeasts release nucleotides (Delisle and Phaff, 1961; Lewis and Phaff, 1963). Release of the nucleotides and other ultraviolet-absorbing compounds depends upon the pH value of the medium, the temperature, the concentration of fermentable sugars, and also on the presence of membrane-protecting ($Ca^{2+}$ or $Mg^{2+}$) or membrane-damaging (butanol, detergent) reagents. Little leakage occurs when the yeast is suspended in water, but substantial leakage is observable in the presence of glucose (Lee and Lewis, 1968b).

A dried yeast cell has lost the selectivity of its plasma membrane, and consequently it becomes permeable to compounds which do not enter an intact yeast cell. In an aqueous suspension, a considerable quantity of constituents of low molecular-weight is released from active dry yeast (Herrera *et al.*, 1956; Harrison and Trevelyan, 1963; Suomalainen *et al.*, 1965c). Harrison and Trevelyan (1963) attributed the high leakage of dried yeast to the action of a lecithinase on the phospholipids of the plasma membrane. The loss is largely dependent upon the mechanism of rehydration following admixture with water, and on the temperature at which the reconstitution takes place (Herrera *et al.*, 1956; Suomalainen *et al.*, 1965c). Rehydration of the semipermeable membrane under cold conditions is slow, and allows the soluble constituents to escape (Herrera *et al.*, 1956). When dried baker's yeast is rehydrated at 5–10°, practically all the thiamin and riboflavin remain in the cell, but almost all of the nicotinic acid is released into the medium. Moreover, the bulk of the NAD and related compounds in dried baker's yeast, to the extent of 65–80%, enters the water phase after extraction at 10°. In addition to NAD, a substantial part of the adenosine nucleotides is also extracted from air-dried baker's yeast, rising to 80% at temperatures between 5° and 10° (Suomalainen *et al.*, 1965c). In 1923, von Euler and Myrbäck observed that NAD was readily extractable from bottom yeast.

In the preparation of protoplasts, nicotinic acid, riboflavin and thiamin are released into the medium; furthermore, proteins, such as glutamic-oxaloacetic transaminase and alkaline phosphatase, are to a small extent released from the protoplasts (Suomalainen *et al.*, 1967b). The membranes of cells of *Candida utilis* and *Saccharomyces cerevisiae* are destroyed by ribonuclease, and proteins such as cytochrome *c* and bovine serum albumin, and proteins are then able to penetrate into the cytoplasm of the cells. This results in a release of ultraviolet-absorbing compounds, mostly nucleotides and coenzymes, as well as amino acids into the medium (Schlenk and Dainko, 1965, 1966; Ottolenghi, 1967; Yphantis *et al.*, 1967).

## IV. Acknowledgement

The authors wish to express their gratitude to Miss Hertta Kyyhkynen, M.Sc. for her efficient assistance in the preparation of the manuscript.

## References


Abadie, F. (1967). *Annls Inst. Pasteur, Paris* **113**, 81–95.

Abadie, F. (1968). *Annls Inst. Pasteur, Paris* **115**, 197–211.

Abd-El Al, A. T. H. and Phaff, H. J. (1967). *Bact. Proc.* 32–33.

Abd-El Al, A. T. H. and Phaff, H. J. (1968). *Biochem. J.* **109**, 347–360.

Agar, H. D. and Douglas, H. C. (1957). *J. Bact.* **73**, 365–375.

Ahearn, D. G., Meyers, S. P. and Nichols, R. A. (1968). *Appl. Microbiol.* **16**, 1370–1374.

Ahearn, D. G., Roth, F. J., Jr. and Meyers, S. P. (1962). *Can. J. Microbiol.* **8**, 121–132.

Ahmad, F., Rose, A. H. and Garg, N. K. (1961). *J. gen. Microbiol.* **24**, 69–80.

Aldous, J. G. (1948). *J. biol. Chem.* **176**, 83–88.

Andreasen, A. A. and Stier, T. J. B. (1953). *J. cell. comp. Physiol.* **41**, 23–36.

Andreasen, A. A. and Stier, T. J. B. (1954). *J. cell. comp. Physiol.* **43**, 271–281.

Andrews, J. and Gilliland, R. B. (1952). *J. Inst. Brew.* **58**, 189–196.

Armstrong, W. McD. and Rothstein, A. (1964). *J. gen. Physiol.* **48**, 61–71.

Atkin, L. and Gray, P. P. (1947). *Archs Biochem.* **15**, 305–322.

Atkin, L., Gray, P. P., Moses, W. and Feinstein, M. (1949a). *Biochim. biophys. Acta* **3**, 692–708.

Atkin, L., Gray, P. P., Moses, W. and Feinstein, M. (1949b). *Wallerstein Labs Commun.* **12**, 153–170.

Atkin, L., Schultz, A. S. and Frey, C. N. (1946). *In* "Enzymes and Their Role in Wheat Technology", (J. A. Anderson, ed.), pp. 321–352. Interscience Publishers, New York.

Atkin, L., Williams, W. L., Schultz, A. S. and Frey, C. N. (1944). *Ind. Engng Chem. analyt. Edn* **16**, 67–71.

Axelrod, A. E., Hofmann, K. and Daubert, B. F. (1947). *J. biol. Chem.* **169**, 761–762.

Axelrod, A. E., Mitz, M. and Hofmann, K. (1948). *J. biol. Chem.* **175**, 265–274.

Äyräpää, T. (1950). *Physiologia Pl.* **3**, 402–429.

Azam, F. and Kotyk, A. (1969). *FEBS Lett.* **2**, 333–335.

Bakerspigel, A. (1964). *J. Bact.* **87**, 228–230.

Barton-Wright, E. C. (1949). *European Brewery Convention, Congr. Lucerne* 19–31.

Barton-Wright, E. C. (1953). *Wallerstein Labs Commun.* **16**, 5–28.

Barton-Wright, E. C. and Harris, G. (1951). *Nature, Lond.* **167**, 560–561.

Barton-Wright, E. C. and Thorne, R. S. W. (1949). *J. Inst. Brew.* **55**, 383–386.

Belaich, J.-P., Senez, J. C. and Murgier, M. (1968). *J. Bact.* **95**, 1750–1757.

Bernhauer, K. (1943). *Ergebn. Enzymforsch.* **9**, 297–360.

Berthelot, M. (1889). *C.r. hebd. Séanc. Acad. Sci., Paris* **109**, 548–550.

Biely, P. and Bauer, Š. (1966). *Biochim. biophys. Acta* **121**, 213–214.

Biely, P. and Bauer, Š. (1968). *Biochim. biophys. Acta* **156**, 432–434.

Binkley, W. W. and Wolfrom, M. L. (1953). *Adv. carb. Chem.* **8**, 291–314.

Bloch, K., Baronowsky, P., Goldfine, H., Lennarz, W. J., Light, R., Norris, A. T. and Scheverbrandt, G. (1961). *Fedn Proc. Fedn Am. Socs exp. Biol.* **20**, 921–927.

Bloemers, H. P. J. and Koningsberger, V. V. (1967). *Nature, Lond.* **214**, 487–488.

Blumenthal, H. J., Lewis, K. F. and Weinhouse, S. (1954). *J. Am. chem. Soc.* **76**, 6093–6097.

Boer, P. and Steyn-Parvé, E. P. (1966). *Biochim. biophys. Acta* **128**, 400–402.

Boulton, A. A. (1965). *Expl Cell Res.* **37**, 343–359.

Boulton, A. A. and Eddy, A. A. (1962). *Biochem. J.* **82**, 16P–17P.

Bowen, H. J. M. (1966). "Trace Elements in Biochemistry", p. 123. Academic Press, London.
Brandt, K. M. (1941). *Biochem. Z.* **309**, 190–201.
Brandt, K. M. (1942). *Protoplasma* **36**, 77–119.
Brukner, B. (1955). *In* "Technologie des Zuckers" (Verein der Zuckerindustrie, ed.), pp. 501–508. Verlag M. & H. Schaper, Hanover.
Burkholder, P. R., McVeigh, I. and Moyer, D. (1944). *J. Bact.* **48**, 385–391.
Cerbón, J. (1969). *J. Bact.* **97**, 658–662.
Challinor, S. W. and Daniels, N. W. R. (1955). *Nature, Lond.* **176**, 1267–1268.
Challinor, S. W., Power, D. M. and Tonge, R. J. (1964). *Nature, Lond.* **203**, 250–251.
Champagnat, A., Vernet, C., Lainé, B. and Filosa, J. (1963). *Nature, Lond.* **197**, 13–14.
Chang, W.-S. and Peterson, W. H. (1949). *J. Bact.* **58**, 33–44.
Chantrenne, H. and Courtois, C. (1954). *Biochim. biophys. Acta* **14**, 397–400.
Chen, A. W.-C. and Miller, J. J. (1968). *Can. J. Microbiol.* **14**, 957–963.
Chen, S. L. (1959). *Biochim. biophys. Acta* **32**, 470–479.
Cirillo, V. P. (1961a). *A. Rev. Microbiol.* **15**, 197–218.
Cirillo, V. P. (1961b). *In* "Membrane Transport and Metabolism", (A. Kleinzeller and A. Kotyk, eds.), pp. 343–351. Academic Press, London–Czechoslovak Academy of Sciences, Prague.
Cirillo, V. P. (1966). *Bact. Proc.* 105–106.
Cirillo, V. P. (1967). *Bact. Proc.* 106.
Cirillo, V. P. (1968a). *J. Bact.* **95**, 603–611.
Cirillo, V. P. (1968b). *J. Bact.* **95**, 1727–1731.
Cirillo, V. P. and Wilkins, P. O. (1964). *J. Bact.* **87**, 232–233.
Collander, R. (1957). *Societas Scientiarum Fennica, Commentationes Biologicae* **16**, No. 15.
Collander, R. and Äyräpää, T. (1947). *Acta physiol. scand.* **14**, 171–173.
Collander, R. and Bärlund, H. (1933). *Acta bot. Fennica* **11**, 1–114.
Collander, R., Turpeinen, O. and Fabritius, E. (1931). *Protoplasma* **13**, 348–362.
Conway, E. J. (1951). *Science, N.Y.* **113**, 270–273.
Conway, E. J. (1953). *Int. Rev. Cytol.* **2**, 419–445.
Conway, E. J. (1955). *Int. Rev. Cytol.* **4**, 377–396.
Conway, E. J. and Beary, M. E. (1956). *Nature, Lond.* **178**, 1044.
Conway, E. J. and Beary, M. E. (1958). *Biochem. J.* **69**, 275–280.
Conway, E. J. and Beary, M. E. (1962). *Biochem. J.* **84**, 328–333.
Conway, E. J. and Brady, T. G. (1950). *Biochem. J.* **47**, 360–369.
Conway, E. J. and Breen, J. (1945). *Biochem. J.* **39**, 368–371.
Conway, E. J. and Downey, M. (1950). *Biochem. J.* **47**, 347–355.
Conway, E. J. and Duggan, P. F. (1956). *Nature, Lond.* **178**, 1043–1044.
Conway, E. J. and Duggan, P. F. (1958). *Biochem. J.* **69**, 265–274.
Conway, E. J. and Gaffney, H. M. (1966). *Biochem. J.* **101**, 385–391.
Conway, E. J. and Kernan, R. P. (1955). *Biochem. J.* **61**, 32–36.
Conway, E. J. and Moore, D. T. (1954). *Biochem. J.* **57**, 523–528.
Conway, E. J., Ryan, H. and Carton, E. (1954). *Biochem. J.* **58**, 158–167.
Cook, A. H. and Phillips, A. W. (1957). *Archs Biochem. Biophys.* **69**, 1–9.
Cowie, D. B. and McClure, F. T. (1959). *Biochim. biophys. Acta* **31**, 236–245.
Crocomo, O. J. and Menard, L. N. (1962). *Nature, Lond.* **193**, 502.
Cutts, N. S. and Rainbow, C. (1949). *Nature, Lond.* **164**, 234–235.
Cutts, N. S. and Rainbow, C. (1950). *J. gen. Microbiol.* **4**, 150–155.
Davies, R., Folkes, J. P., Gale, E. F. and Bigger, L. C. (1953). *Biochem. J.* **54**, 430–437.

Dawson, E. R. and Harrison, J. S. (1949). *1st Intern. Congr. Biochem.*, Cambridge. Abstr. Communs., pp. 546–547.
Dee, E. and Conway, E. J. (1968). *Biochem. J.* **107**, 265–271.
Deierkauf, F. A. and Booij, H. L. (1968). *Biochim. biophys. Acta* **150**, 214–225.
DelaFuente, G. and Sols, A. (1962). *Biochim. biophys. Acta* **56**, 49–62.
Delisle, A. L. and Phaff, H. J. (1961). *Proc. Am. Soc. Brew. Chem.* 103–118.
Demis, D. J., Rothstein, A. and Meier, R. (1954). *Archs Biochem. Biophys.* **48**, 55–62.
Densky, H., Gray, P. J. and Buday, A. (1966). *Proc. Am. Soc. Brew. Chem.* 93–100.
DiCarlo, F. J., Schultz, A. S. and Kent, A. M. (1953). *Archs Biochem. Biophys.* **44**, 468–474.
DiCarlo, F. J., Schultz, A. S. and McManus, D. K. (1951). *J. biol. Chem.* **189**. 151–175.
Dickens, F. (1938). *Biochem. J.* **32**, 1645–1653.
Dittmer, K., Melville, D. B. and du Vigneaud, V. (1944). *Science, N.Y.* **99**, 203–205.
Dixon, B. and Rose, A. H. (1964). *J. gen. Microbiol.* **35**, 411–419.
Drews, B., Just, F. and Weinhold, G. (1953). *Brauerei, Wiss. Beil.* **6**, 73–76.
Dunwell, J. L., Ahmad, F. and Rose, A. H. (1961). *Biochim. biophys. Acta* **51**, 604–607.
Dworsky, P. and Hoffmann-Ostenhof, O. (1965). *Biochem. Z.* **343**, 394–398.
Dworsky, P. and Hoffmann-Ostenhof, O. (1967). *Z. allg. Mikrobiol.* **7**, 1–6.
Eastcott, E. V. (1928). *J. Phys. Chem.* **32**, 1094–1111.
Eddy, A. A. (1969). Proc. 2nd Symposium on Yeasts, Bratislava, 1966. (A. Kochová-Kratochvílová, ed.), pp. 457–462. Vydavatel'stvo Slovenskej Akadémie Vied, Bratislava.
Eddy, A. A. and Williamson, D. H. (1957). *Nature, Lond.* **179**, 1252–1253.
Elander, M. (1956). *Ark. Kemi* **9**, 191–224.
Elander, M. (1959). *Ark. Kemi* **13**, 457–474.
Elander, M. (1963). *Ark. Kemi* **21**, 317–333.
Elvehjem, C. A. (1931). *J. biol. Chem.* **90**, 111–132.
Enari, T.-M. (1958). Dissertation: University of Helsinki.
Enari, T.-M. and Kauppinen, V. (1961). *Acta chem. scand.* **15**, 1513–1516.
Enebo, L., Anderson, L. G. and Lundin, H. (1946). *Archs Biochem.* **11**, 383–395.
Erkama, J. and Enari, T.-M. (1956). *Suom. Kemistilehti* **29B**, 176–178.
Euler, H. von and Blix, R. (1919). *Hoppe-Seyler's Z. physiol. Chem.* **105**, 83–114.
Euler, H. von, Fink, H. and Hellström, H. (1927). *Hoppe-Seyler's Z. physiol. Chem.* **169**, 10–51.
Euler, H. and Lindner, P. (1915). "Chemie der Hefe und der alkoholischen Gärung", p. 234. Akademische Verlagsgesellschaft, Leipzig.
Euler, H. von and Myrbäck, K. (1923). *Hoppe-Seyler's Z. physiol. Chem.* **131**, 179–203.
Evans, W. C. (1963). *J. gen. Microbiol.* **32**, 177–184.
Evans, G. H. (1968). *In* "Single-Cell Protein", (R. I. Mateles and S. R. Tannenbaum, eds.), pp. 243–254. The M.I.T. Press, Cambridge, Massachusetts.
Farkaš, V., Bauer, Š. and Zemek, J. (1969). *Biochim. biophys. Acta* **184**, 77–82.
Fels, G. and Cheldelin, V. H. (1949). *Archs Biochem. Biophys.* **22**, 402–405.
Fencl, Z. (1961). *In* "Membrane Transport and Metabolism", (A. Kleinzeller and A. Kotyk, eds.), pp. 296–304. Academic Press, London–Czechoslovak Academy of Sciences, Prague.
Fink, H. and Just, F. (1938). *Biochem. Z.* **296**, 306–314.
Fink, H. and Just, F. (1941). *Biochem. Z.* **309**, 212–218.
Fink, H. and Just, F. (1942a). *Biochem. Z.* **311**, 61–72.
Fink, H. and Just, F. (1942b). *Biochem. Z.* **311**, 287–306.

Fink, H. and Just, F. (1942c). *Biochem. Z.* **313**, 39–47.

Fink, H. and Krebs, Jos. (1938). *Biochem. Z.* **300**, 59–77.

Fink, H. and Krebs, Jos. (1939). *Biochem. Z.* **301**, 13–14.

Fink, H. and Lechner, R. (1936). *Angew. Chem.* **49**, 775–777.

Fischer, E. and Thierfelder, H. (1894). *Ber. dt. chem. Ges.* **27**, 2031–2037.

Forsander, O. (1956). Dissertation: University of Helsinki.

Forss, K. (1961). Dissertation: Åbo Akademi, Turku.

Foulkes, E. C. (1955). *J. gen. Physiol.* **38**, 425–430.

Foulkes, E. C. (1956). *J. gen. Physiol.* **39**, 687–704.

Frey, S. W., DeWitt, W. G. and Bellomy, B. R. (1967). *Proc. Am. Soc. Brew. Chem.* 199–205.

Fries, N. (1965). *In* "The Fungi", (G. C. Ainsworth and A. S. Sussman, eds.). Vol. 1, pp. 491–523. Academic Press, New York.

Friis, J. and Ottolenghi, P. (1959). *C.r. Trav. Lab. Carlsberg* **31**, 272–281.

Fuhrmann, G.-F. and Rothstein, A. (1968). *Biochim. biophys. Acta* **163**, 325–330.

Fuhs, G. W. (1961). *Arch. Mikrobiol.* **39**, 374–422.

Fulmer, E. I., Nelson, V. E. and Sherwood, F. F. (1921). *J. Am. chem. Soc.* **43**, 191–199.

Fulmer, E. I., Underkofler, L. A. and Lesh, J. B. (1936). *J. Am. chem. Soc.* **58**, 1356–1358.

Genevois, L. (1937). *Ann. Fermentations* **13**, 600–615.

Genevois, L. and Pavloff, M. (1935). *Bull. Soc. Chim. biol.* **17**, 991–1021.

Gesswagner, D. and Altmann, H. (1967). *Naturwissenschaften* **54**, 147.

Ghosh, A. and Bhattacharyya, S. N. (1967). *Biochim. biophys. Acta* **136**, 19–26.

Gilliland, R. B. (1953). *Proc. European Brewery Convention*, 121–134.

Gits, J. J. and Grenson, M. (1967). *Biochim. biophys. Acta* **135**, 507–516.

Gits, J. J. and Grenson, M. (1969). *Archs int. Physiol. Biochim.* **77**, 153–154.

Goodman, J. and Rothstein, A. (1957). *J. gen. Physiol.* **40**, 915–923.

Görts, C. P. M. (1967). *Antonie van Leeuwenhoek* **33**, 451–463.

Gottschalk, A. (1947a). *Biochem. J.* **41**, 276–280.

Gottschalk, A. (1947b). *Wallerstein Labs Commun.* **10**, 109–117.

Gottschalk, A. (1949). *Wallerstein Labs Commun.* **12**, 55–67.

Greaves, J. E., Zobell, C. E. and Greaves, J. D. (1928). *J. Bact.* **16**, 409–430.

Green, S. R. and Stone, I. (1952). *Wallerstein Labs Commun.* **15**, 347–361.

Grenson, M. (1966). *Biochim. biophys. Acta* **127**, 339–346.

Grenson, M., Mousset, M., Wiame, J. M. and Bechet, J. (1966). *Biochim. biophys. Acta* **127**, 325–338.

Günther, Th. and Kattner, W. (1968). *Z. Naturf.* **23b**, 77–80.

Hägglund, E. (1914). "Hefe und Gärung in ihrer Abhängigkeit von Wasserstoff- und Hydroxylionen". Akademische Abhandlung, Univ. Stockholm.

Hägglund, E. (1915). "Die Sulfitablauge und ihre Verarbeitung auf Alkohol". Friedr. Vieweg & Sohn, Braunschweig.

Hägglund, E. and Augustsson, A. M. (1926). *Biochem. Z.* **170**, 102–125.

Halvorson, H. O. and Cohen, G. N. (1958). *Annls Inst. Pasteur, Paris* **95**, 73–87.

Halvorson, H. O. and Cowie, D. B. (1961). *In* "Membrane Transport and Metabolism", (A. Kleinzeller and A. Kotyk, eds.), pp. 479–487. Academic Press, London–Czechoslovak Academy of Sciences, Prague.

Halvorson, H. O., Fry, W. and Schwemmin, D. (1955). *J. gen. Physiol.* **38**, 549–573.

Halvorson, H. O., Winderman, S. and Gorman, J. (1963). *Biochim. biophys. Acta* **67**, 42–53.

Hanson, A. M. and Baldwin, I. L. (1941). *J. Bact.* **41**, 94.

Harris, G. (1958). *In* "The Chemistry and Biology of Yeasts", (A. H. Cook, ed.), pp. 437–533. Academic Press, New York.
Harris, G., Barton-Wright, E. C. and Curtis, N. S. (1951). *J. Inst. Brew.* **57**, 264–280.
Harris, G. and Millin, D. J. (1963). *Biochem. J.* **88**, 89–95.
Harris, G. and Thompson, C. C. (1960). *J. Inst. Brew.* **66**, 293–297.
Harris, G. and Thompson, C. C. (1961). *Biochim. biophys. Acta* **52**, 176–183.
Harris, G. and Thompson, C. C. (1962). *Biochim. biophys. Acta* **56**, 293–302.
Harrison, J. S. and Trevelyan, W. E. (1963). *Nature, Lond.* **200**, 1189–1190.
Hartelius, V. (1938). *Biochem. Z.* **299**, 317–333.
Hartelius, V. (1939). *C.r. Trav. Lab. Carlsberg, Sér. Physiol.* **22**, No. 19, 303–322.
Haškovec, C. and Kotyk, A. (1969). *Eur. J. Biochem.* **9**, 343–347.
Heppel, L. A. (1967). *Science, N.Y.* **156**, 1451–1455.
Heredia, C. F., Sols, A. and DelaFuente, G. (1968). *Eur. J. Biochem.* **5**, 321–329.
Herrera, T., Peterson, W. H., Cooper, E. J. and Peppler, H. J. (1956). *Archs Biochem. Biophys.* **63**, 131–143.
Höfer, M. (1968). *Folio microbiol., Praha* **13**, 373–378.
Höfer, M. and Kotyk, A. (1968). *Folio microbiol., Praha* **13**, 197–204.
Hoffmann-Ostenhof, O. and Weigert, W. (1952). *Naturwissenschaften* **39**, 303–304.
Hofmann, K. and Winnick, T. (1945). *J. biol. Chem.* **160**, 449–453.
Holter, H. (1965). *Symp. Soc. gen. Microbiol.* **15**, 89–114.
Holzer, H. (1953). *Biochem. Z.* **324**, 144–155.
Hopkins, R. H. (1945). *Wallerstein Labs Commun.* **8**, 110–117.
Hopkins, R. H. (1955). *Proc. European Brewery Convention* 52–64.
Hopkins, R. H. (1958). *Wallerstein Labs Commun.* **21**, 309–322.
Hopkins, R. H. and Horwood, M. (1950). *Biochem. J.* **47**, 95–97.
Hopkins, R. H. and Roberts, R. H. (1935). *Biochem. J.* **29**, 931–936.
Hopkins, R. H. and Roberts, R. H. (1936). *Biochem. J.* **30**, 76–83.
Horschak, R. and Wartenberg, H. (1967). *Ber. dt. bot. Ges.* **80**, 345–355.
Hudson, J. R. (1955). *J. Inst. Brew.* **61**, 127–133.
Ingram, M. (1955). "An Introduction to the Biology of Yeasts". Isaac Pitman & Sons, London.
Inoue, T., Masuyma, K., Yamamoto, Y., Okada, K. and Kuroiwa, Y. (1968). *Proc. Am. Soc. Brew. Chem.* 158–165.
Inskeep, G. C., Wiley, A. J., Holderby, J. M. and Hughes, L. P. (1951). *Ind. Engng Chem. ind. Edn* **43**, 1702–1711.
Isbell, H. S. and Pigman, W. W. (1938). *J. Res. Natl Bur. Std.* **20**, 773–798.
Jacob, F. and Abadie, F. (1967). *Mycopath. Mycol. appl.* **33**, 113–124.
Jaenicke, L. and Lynen, F. (1960). *In* "The Enzymes", (P. D. Boyer, H. Lardy and K. Myrbäck, eds.). Vol. 3, p. 103. Academic Press, New York.
Jennings, D. H., Hooper, D. C. and Rothstein, A. (1958). *J. gen. Physiol.* **41**, 1019–1026.
Jennings, D. H. and Souter, G. A. (1963). Quoted in Jennings, D. H. "The Absorption of Solutes by Plant Cells", p. 103. Oliver & Boyd, Edinburgh.
Johnson, M. J. (1964). *Chemy Ind.* 1532–1537.
Johnston, J. A., Ghadially, R. C., Roberts, R. N. and Fuhr, B. W. (1962). *Archs Biochem. Biophys.* **99**, 537–538.
Joiris, C. R. and Grenson, M. (1969). *Archs int. Physiol. Biochim.* **77**, 154–156.
Jones, M. and Pierce, J. S. (1963). *Proc. European Brewery Convention* **1963**, 110–134.
Jones, M., Power, D. M. and Pierce, J. S. (1965). *Proc. European Brewery Convention* **1965**, 182–194.
Jones, M., Pragnell, M. J. and Pierce, J. S. (1969). *J. Inst. Brew.* **75**, 520–536.

Jung, C. and Rothstein, A. (1965). *Biochem. Pharmac.* **14**, 1093–1112.
Just, F. (1940). *Wochschr. Brau.* **57**, 262–264.
Just, F. (1951). *Brauerei, Wiss. Beil.* **4**, 17–20.
Just, F., Schnabel, W. and Ullmann, S. (1951). *Brauerei, Wiss. Beil.* **4**, 57–60, 71–75.
Kaplan, J. G. (1962). *Nature, Lond.* **196**, 950–952.
Kaplan, J. G. (1963). *Enzymologia* **25**, 359–366.
Kaplan, J. G. (1965). *Nature, Lond.* **205**, 76–77.
Kaplan, J. G. and Tacreiter, W. (1966). *J. gen. Physiol.* **50**, 9–24.
Kauppinen, V. (1963). Dissertation: University of Helsinki.
Kawasaki, C., Kishi, T. and Nishihara, T. (1968). *Bitamin* **38**, 202–206; quoted in *Chem. Abstr.* (1968) **69**, 84387s.
Kaziro, Y. (1959). *J. Biochem., Tokyo* **46**, 1523–1539.
Keränen, A. J. A. (1969). *Antonie van Leeuwenhoek* **35**, *Suppl., Third International Symposium on Yeasts, Delft—The Hague*, pp. H7–H8.
van Kleeff, B. H. A., Kokke, R. and Nieuwdorp, P. J. (1969). *Antonie van Leeuwenhoek* **35**, *Suppl., Third International Symposium on Yeasts, Delft—The Hague*, pp. G9–G10.
Kleinzeller, A., Kotyk, A. and Kováč, L. (1959). *Nature, Lond.* **183**, 1402–1403.
Kluyver, A. J. and Custers, M. Th. J. (1940). *Antonie van Leeuwenhoek* **6**, 121–162.
Koch, R. B., Geddes, W. F. and Smith, F. (1951). *Cereal Chem.* **28**, 727–730.
Kögl, F. and Tönnis, B. (1936). *Hoppe-Seyler's Z. physiol. Chem.* **242**, 43–73.
Kokke, R. (1966). The action of a baker's yeast supernatant on phospholipids. Proefschrift, Rijksuniv. Leiden.
Kokke, R., Hooghwinkel, G. J. M., Booij, H. L., Van den Bosch, H., Zelles, L., Mulder, E. and Van Deenen, L. L. M. (1963). *Biochim. biophys. Acta* **70**, 351–354.
Korn, E. D. (1968). *J. gen. Physiol.* **52**, No. 1, Part 2, 257S–274S.
Kornberg, S. R. (1956). *J. biol. Chem.* **218**, 23–31.
Kotyk, A. (1959). *Folio microbiol., Praha* **4**, 363–373.
Kotyk, A. (1967). *Folio microbiol., Praha* **12**, 121–131.
Kotyk, A. and Haškovec, C. (1968). *Folio microbiol., Praha* **13**, 12–19.
Kotyk, A. and Höfer, M. (1965). *Biochim. biophys. Acta* **102**, 410–422.
Kotyk, A. and Kleinzeller, A. (1967). *Biochim. biophys. Acta* **135**, 106–111.
Kotyk, A. and Michaljaničová, D. (1968). *Folio microbiol., Praha* **13**, 212–220.
Krebs, H. A. (1935). *Biochem. J.* **29**, 1620–1644.
Krohn, V. (1924). *Ann. Acad. Sci. Fennicae.* Ser. A, **23** No. 8.
Lampen, J. O. (1968). *Antonie van Leeuwenhoek* **34**, 1–18.
LaRue, T. A. and Spencer, J. F. T. (1967a). *Can. J. Microbiol.* **13**, 777–788.
LaRue, T. A. and Spencer, J. F. T. (1967b). *Can. J. Microbiol.* **13**, 789–794.
LaRue, T. A. and Spencer, J. F. T. (1968). *Can. J. Microbiol.* **14**, 79–86.
Lasnitzki, A. and Szörenyi, E. (1934). *Biochem. J.* **28**, 1678–1683.
Lasnitzki, A. and Szörenyi, E. (1935). *Biochem. J.* **29**, 580–587.
Lechner, R. (1938). *Biochem. Z.* **300**, 204–207.
Lechner, R. (1940). *Biochem. Z.* **306**, 218–223.
Lee, T. C. and Lewis, M. J. (1968a). *J. Fd Sci.* **33**, 119–123.
Lee, T. C. and Lewis, M. J. (1968b). *J. Fd Sci.* **33**, 124–128.
Leibowitz, J. and Hestrin, S. (1939). *Enzymologia* **6**, 15–26.
Leibowitz, J. and Hestrin, S. (1942). *Biochem. J.* **36**, 772–785.
Leibowitz, J. and Hestrin, S. (1945). *Adv. Enzymol.* **5**, 87–127.
Lennarz, W. J. and Bloch, K. (1960). *J. biol. Chem.* **235**, PC26.
Leonian, R. H. and Lilly, V. G. (1942). *Am. J. Bot.* **29**, 459–464.
Lewin, L. M. (1965). *J. gen. Microbiol.* **41**, 215–224.
Lewin, L. M. and Shafai, T. (1967). *Bact. Proc.* 33.

Lewis, M. J. and Phaff, H. J. (1963). *Proc. Am. Soc. Brew. Chem.* 114–123.

Lewis, M. J. and Stephanopoulos, D. (1967). *J. Bact.* **93**, 976–984.

Lichstein, H. C. (1951). *Vitamins and Hormones* **9**, 27–74.

Lichstein, H. C., Christman, J. F. and Boyd, W. L. (1950). *J. Bact.* **59**, 113–116.

Lichstein, H. C., Gunsalus, I. C. and Umbreit, W. W. (1945). *J. biol. Chem.* **161**, 311–320.

Light, R. J., Lennarz, W. J. and Bloch, K. (1962). *J. biol. Chem.* **237**, 1793–1800.

Lilly, V. G. and Leonian, L. H. (1944). *Science, N.Y.* **99**, 205–206.

Lindegren, C. C. (1945). *Bact. Rev.* **9**, 111–170.

Lindegren, C. C. (1947). *Nature, Lond.* **159**, 63–64.

Lindegren, C. C. (1951). *Expl Cell Res.* **2**, 275–278.

Lipmann, F. and Kaplan, N. O. (1946). *J. biol. Chem.* **162,** 743–744.

Ljungdahl, L. and Sandegren, E. (1953). *Proc. European Brewery Convention* 85–97.

Longley, R. P., Rose, A. H. and Knights, B. A. (1968). *Biochem. J.* **108**, 410–412.

Lynen, F. (1939). *Ann. Chem.* **539**, 1–39.

Lynen, F. (1959). *J. cell. comp. Physiol.* **54** Suppl. 1, 33–49.

Lynen, F. (1961). *Fedn Proc. Fedn Am. Socs. exp. Biol.* **20**, 941–951.

Lynen, F. and Kalb, H.-W. (1955). *Ann. Acad. Sci. Fennicae* Ser. A II No. **60**, 471–487.

Lynen, F. and Neciullah, N. (1939). *Ann. Chem.* **541**, 203–218.

Lynen, F. and Reichert, E. (1951). *Angew. Chem.* **63**, 47–48.

Lynes, K. J. and Norris, F. W. (1948). *J. Inst. Brew.* **54**, 150–157.

Maas-Förster, M. (1955). *Arch. Mikrobiol.* **22**, 115–144.

McFarlane, W. D. and Held, H. R. (1953). *Proc. European Brewery Convention* 110–120.

McHargue, J. S. and Calfee, R. K. (1931). *Pl. Physiol., Lancaster* **6**, 559–566.

McKenna, E. J. and Kallio, R. E. (1965). *A. Rev. Microbiol.* **19**, 183–208.

Mackenzie, D. W. R. and Auret, B. J. (1963). *J. gen. Microbiol.* **31**, 171–177.

McLellan, W. L., Jr. and Lampen, J. O. (1963). *Biochim. biophys. Acta* **67**, 324–326.

Maddox, I. S. and Hough, J. S. (1969). *Proc. European Brewery Convention* 315–325.

Magaña-Schwencke, N. and Schwencke, J. (1969). *Biochim. biophys. Acta* **173**, 313–323.

Malm, M. (1940). *Naturwissenschaften* **28**, 723–724.

Malm, M. (1947). *Ark. Kemi, Mineral. Geol.* **25A** No. 1.

Malm, M. (1950). *Physiologia Pl.* **3**, 376–401.

Mariani, E. and Torraca, G. (1951). *Nature, Lond.* **168**, 959.

Markham, E. and Byrne, W. J. (1967). *J. Inst. Brew.* **73**, 271–273.

Markham, E., Mills, A. K. and Byrne, W. J. (1966). *Proc. Am. Soc. Brew. Chem.* 76–85.

Markovitz, A. J. and Kallio, R. E. (1964). *J. Bact.* **87**, 968–969.

Massart, L. and Horens, J. (1953). *Enzymologia* **15**, 359–361.

Matile, Ph. (1969). *In* "Yeasts. The Proceedings of the 2nd Symposium on Yeasts, Bratislava 1966" (A. Kocková-Kratochvílová, ec.), pp. 503–508. Vydavatel'stvo Slovenskej Akadémie Vied, Bratislava.

Matile, Ph., Moor, H. and Mühlethaler, K. (1967). *Arch. Mikrobiol.* **58**, 201–211.

Maw, G. A. (1960). *J. Inst. Brew.* **66**, 162–167.

Maw, G. A. (1961). *J. Inst. Brew.* **67**, 57–63.

Maw, G. A. (1963a). *Folio microbiol., Praha* **8**, 325–332.

Maw, G. A. (1963b). *J. gen. Microbiol.* **31**, 247–259.

Maw, G. A. (1963c). *Pure Appl. Chem.* **7**, 655–668.

Maw, G. A. (1965). *Wallerstein Labs Commun.* **28**, 49–70.
Melnick, D., Hochberg, M., Himes, H. W. and Oser, B. L. (1945). *J. biol. Chem.* **160**, 1–14.
Melville, D. B., Dittmer, K., Brown, G. B. and du Vigneaud, V. (1943). *Science, N.Y.* **98**, 497–499.
Melville, D. B., Genghof, D. S. and Lee, J. M. (1954). *J. biol. Chem.* **208**, 503–512.
Menzinsky, G. (1943). *Biochem. Z.* **314**, 312–326.
Menzinsky, G. (1950). *Ark. Kemi* **2**, 1–94.
Metcalfe, G. and Chayen, S. (1954). *Nature, Lond.* **174**, 841–842.
Meyerhof, O. and Kaplan, A. (1951). *Archs Biochem. Biophys.* **33**, 282–298.
Miettinen, J. K. (1964). *In* "Rapid Mixing and Sampling Techniques in Biochemistry", (B. Change, R. H. Eisenhardt, Q. H. Gibson and K. K. Longberg-Holm, eds.), pp. 303–307. Academic Press, New York.
Mitchell, P. (1961). *In* "Membrane Transport and Metabolism", (A. Kleinzeller and A. Kotyk, eds.), p. 100. Academic Press, London–Czechoslovak Academy of Sciences, Prague.
Mitchell, P. (1967). *In* "Comprehensive Biochemistry", (M. Florkin and E. H. Stotz, eds.), Vol. 22, pp. 167–197. Elsevier, London.
Montreuil, J., Mullet, S. and Scriban, R. (1961). *Wallerstein Labs Commun.* **24**, 304–315.
Morris, E. O. (1958). *In* "The Chemistry and Biology of Yeasts", (A. H. Cook, ed.), pp. 251–321. Academic Press, New York.
Morris, J. G., Hughes, D. T. D. and Mulder, C. (1959). *J. gen. Microbiol.* **20**, 566–575.
Moses, W. and Joslyn, M. A. (1953). *J. Bact.* **66**, 197–203.
Mothes, K. (1953). *Planta* **42**, 64–80.
Munns, D. J. and White, J. (1949). *Abstr. Communs 1st Intern. Congr. Biochem.*, Cambridge, pp. 551–552.
Musfeld, W. (1942). *Ber. schweiz. Bot. Ges.* **52**, 583–620.
Myrbäck, K. (1926). *Hoppe-Seyler's Z. physiol. Chem.* **158**, 160–301.
Myrbäck, K. (1936). *Svensk kem. Tidskr.* **48**, 55–61.
Myrbäck, K. (1940). *Biochem. Z.* **304**, 147–159.
Myrbäck, K. (1941). *In* "Die Methoden der Fermentforschung", (E. Bamann and K. Myrbäck, eds.), Bd. 2, pp. 1494–1502. Georg Thieme, Leipzig.
Myrbäck, K. (1947). *Archs Biochem.* **14**, 53–56.
Myrbäck, K. (1949). *Ergebn. Enzymforsch.* **10**, 168–190.
Myrbäck, K. (1957). *Ark. Kemi* **11**, 471–479.
Myrbäck, K. and Ahlborg, K. (1942). *Biochem. Z.* **311**, 213–226.
Myrbäck, K. and Leissner, E. (1943). *Ark. Kemi, Mineral. Geol.* **17A**, No. 18.
Myrbäck, K. and Örtenblad, B. (1936). *Biochem. Z.* **288**, 329–337.
Myrbäck, K. and Renvall, S. (1946). *Ark. Kemi, Mineral. Geol.* **23A**, No. 6.
Myrbäck, K. and Willstaedt, E. (1955). *Ark. Kemi* **8**, 367–374.
Nakagawa, M. and Tatsumi, C. (1968). *Nippon Nogei Kagaku Kaishi* **42**, 330–336; quoted in *Chem. Abstr.* (1968) **69**, 74672s.
Neumann, N. D. P. and Lampen, J. O. (1967). *Biochemistry, N.Y.* **6**, 468–475.
Nickerson, W. J. and Zerahn, K. (1949). *Biochim. biophys. Acta* **3**, 476–483.
Nielsen, N. (1935). *C.r. Trav. Lab. Carlsberg, Sér. Physiol.* **21** No. 5, 113–138.
Nielsen, N. (1936a). *C.r. Trav. Lab. Carlsberg, Sér. Physiol.* **21** No. 10, 205–219.
Nielsen, N. (1936b). *C.r. Trav. Lab. Carlsberg, Sér. Physiol.* **21** No. 16, 395–425.
Nielsen, N. (1943). *Ergebn. Biol.* **19**, 375–408.
Nielsen, N. and Hartelius, V. (1937). *C.r. Trav. Lab. Carlsberg, Sér. Physiol.* **22**, No. 2, 23–47.
Nielsen, N. and Hartelius, V. (1938). *Biochem. Z.* **295**, 211–225.

Nielsen, N. and Lund, A. (1936). *C.r. Trav. Lab. Carlsberg, Sér. Physiol.* **21** No. 13, 239–246.
Nielsen, N. and Nilsson, N. G. (1950). *Archs Biochem.* **25**, 316–322.
Nielsen, N. and Nilsson, N. G. (1953). *Acta chem. scand.* **7**, 984–986.
Northcote, D. H. and Horne, R. W. (1952). *Biochem. J.* **51**, 232–236.
Nurminen, T., Oura, E. and Suomalainen, H. (1970). *Biochem. J.* **116**, 61–69.
Nurminen, T. and Suomalainen, H. (1968). *J. gen. Microbiol.* **53**, 275–285.
Nurminen, T. and Suomalainen, H. (1969). *Fedn European Biochem. Socs, 6th Symp.*, p. 66.
Nyman, B. and Fries, N. (1962). *Acta chem. scand.* **16**, 2306–2308.
Olbrich, H. (1956). "Die Melasse", Institut für Gärungsgewerbe, Berlin.
Olson, B. H. and Johnson, M. J. (1949). *J. Bact.* **57**, 235–246.
Onishi, H. and Suzuki, T. (1968). *J. Bact.* **95**, 1745–1749.
Ørskov, S. L. (1945). *Acta path. microbiol. scand.* **22**, 523–559.
Ottolenghi, P. (1967). *C.r. Trav. Lab. Carlsberg* **36**, 95–111.
Oura, E. (1969). *Antonie van Leeuwenhoek 35, Suppl., Third International Symposium on Yeasts, Delft—The Hague*, pp. G25–G26.
Oura, E. and Suomalainen, H. (1961). *Suom. Kemistilehti* **34B**, 138–139.
Oura, E. and Suomalainen, H. (1969). *In* "Yeasts. The Proceedings of the 2nd Symposium on Yeasts, Bratislava 1966", (A. Kocková-Kratochvílová, ed.), pp. 437–443. Vydavatel'stvo Slovenskej Akadémie Vied, Bratislava.
Oura, E., Suomalainen, H. and Collander, R. (1959). *Physiologia Pl.* **12**, 534–544.
Palmqvist, U. and Äyräpää, T. (1969). *J. Inst. Brew.* **75**, 181–190.
Panek, A. (1963). *Archs Biochem. Biophys.* **100**, 422–425.
Pardee, A. B. (1967). *Science, N.Y.* **156**, 1627–1628.
Pardee, A. B. (1968). *Science, N.Y.* **162**, 632–637.
Parks, L. W. and Starr, P. R. (1963). *J. cell. comp. Physiol.* **61**, 61–65.
Pelshenke, P., Rotsch, A. and Koeber, H.-J. (1940). *Biochem. Z.* **306**, 205–217.
Phillips, A. W. (1955). *J. Inst. Brew.* **61**, 122–126.
Pierce, J. S. (1966a). *Process Biochem.* **1**, 412–416.
Pierce, J. S. (1966b). *Tech. Quart. Master Brewers Assoc. Am.* **3**, 231–236.
Pierce, J. S., Jones, M. and Woof, J. B. (1969). *Tech. Quart. Master Brewers Assoc. Am.* **6**, 93–97.
Pilgrim, F. J., Axelrod, A. E., Winnick, T. and Hofmann, K. (1945). *Science, N.Y.* **102**, 35–36.
Pirschle, K. (1930). *Biochem. Z.* **218**, 412–444.
Polakis, E. S. and Bartley, W. (1965). *Biochem. J.* **97**, 284–297.
Polakis, E. S., Bartley, W. and Meek, G. A. (1964). *Biochem. J.* **90**, 369–374.
Polakis, E. S., Bartley, W. and Meek, G. A. (1965). *Biochem. J.* **97**, 298–302.
Pomeranz, Y. (1968). *Adv. Fd Res.* **16**, 335–455.
Power, D. M. and Challinor, S. W. (1969). *J. gen. Microbiol.* **55**, 169–176.
Preiss, J. W. (1958). *Archs Biochem. Biophys.* **75**, 186–195.
Prinsen Geerligs, H. C. (1940). *Rec. Trav. Chim.* **59**, 549–554.
Proudlock, J. W., Wheeldon, L. W., Jollow, D. J. and Linnane, A. W. (1968). *Biochim. biophys. Acta* **152**, 434–437.
Pulver, R. and Verzár, F. (1940a). *Helv. chim. Acta* **23**, 1087–1100.
Pulver, R. and Verzár, F. (1940b). *Nature, Lond.* **145**, 823–824.
Pyke, M. (1958). *In* "The Chemistry and Biology of Yeasts", (A. H. Cook, ed.), pp. 535–586. Academic Press, New York.
Quayle, J. R. (1968). *In* "Microbiology", (P. Hepple, ed.), pp. 21–34. The Institute of Petroleum, London–Elsevier, Amsterdam.

Rainbow, C. (1948). *Nature, Lond.* **162**, 572–573.
Rautanen, N. and Kärkkäinen, V. (1951). *Acta chem. scand.* **5**, 1216–1217.
Rautanen, N. and Kylä-Siurola, A.-E. (1954). *Acta chem. scand.* **8**, 106–111.
Rautanen, N. and Miikkulainen, P. (1951). *Acta chem. scand.* **5**, 89–96.
Reilly, C. (1967a). *Folio microbiol., Praha* **12**, 495–499.
Reilly, C. (1967b). *Nature, Lond.* **214**, 1330–1331.
Rhoades, H. E. (1941). *J. Bact.* **42**, 99–115.
Ribéreau-Gayon, J. and Peynaud, E. (1958). "Analyse et Contrôle des Vins", 2nd ed., Librairie Polytechnique Ch. Béranger, Paris et Liége.
Richards, O. W. (1925). *J. Am. chem. Soc.* **47**, 1671–1676.
Ridgway, G. J. and Douglas, H. C. (1958a). *J. Bact.* **75**, 85–88.
Ridgway, G. J. and Douglas, H. C. (1958b). *J. Bact.* **76**, 163–166.
Riemersma, J. C. (1964). Hydrogen ion transport during anaerobic fermentation by baker's yeast. Proefschrift, Rijksuniv. Leiden.
Rischbiet, P. and Tollens, B. (1886). *Justus Liebigs Annln Chem.* **232**, 172–201.
Roberts, E. R. and Wilson, T. G. G. (1954). *Nature, Lond.* **174**, 842.
Robertson, J. J. and Halvorson, H. O. (1957). *J. Bact.* **73**, 186–198.
Rogers, D. and Mickelson, M. N. (1948). *Ind. Engng Chem. ind. Edn* **40**, 527–529.
Rogosa, M. (1943a). *J. Bact.* **45**, 306–307.
Rogosa, M. (1943b). *J. Bact.* **46**, 435–440.
Rogosa, M. (1944). *J. Bact.* **47**, 159–170.
Rogosa, M. (1948). *J. biol. Chem.* **175**, 413–423.
Roine, P. (1947). Dissertation: University of Helsinki.
Rose, A. H. (1960). *J. gen. Microbiol.* **23**, 143–152.
Rosenbaum, E. (1931). *Z. Untersuch. Lebensm.* **61**, 80–84.
Rosenbaum, E. (1935). *Z. Untersuch. Lebensm.* **70**, 366–378.
Rothfield, L. and Finkelstein, A. (1968). *A. Rev. Biochem.* **37**, 463–496.
Rothstein, A. (1954). *In* "Protoplasmatologia. Handbuch der Protoplasmaforschung", (L. V. Heilbrunn and F. Weber, eds.), Bd. II E 4, Springer Verlag, Vienna.
Rothstein, A. (1955). *In* "Electrolytes in Biological Systems", (A. M. Shanes, ed.), pp. 65–100. American Physiological Society, Washington.
Rothstein, A. (1961). *In* "Membrane Transport and Metabolism", (A. Kleinzeller and A. Kotyk, eds.), pp. 270–284. Academic Press, London–Czechoslovak Academy of Sciences, Prague.
Rothstein, A. (1963). *J. gen. Physiol.* **46**, 1075–1085.
Rothstein, A. (1965). *In* "The Fungi", (G. C. Ainsworth and A. S. Sussman, eds.), Vol. 1, pp. 429–455. Academic Press, New York.
Rothstein, A. and Bruce, M. (1958a). *J. cell. comp. Physiol.* **51**, 145–159.
Rothstein, A. and Bruce, M. (1958b). *J. cell. comp. Physiol.* **51**, 439–455.
Rothstein, A. and Demis, C. (1953). *Archs Biochem. Biophys.* **44**, 18–29.
Rothstein, A. and Enns, L. H. (1946). *J. cell. comp. Physiol.* **28**, 231–252.
Rothstein, A. and Hayes, A. D. (1956). *Archs Biochem. Biophys.* **63**, 87–99.
Rothstein, A., Hayes, A. D., Jennings, D. H. and Hooper, D. C. (1958). *J. gen. Physiol.* **41**, 585–594.
Roush, A. H., Questiaux, L. M. and Domnas, A. J. (1959). *J. cell. comp. Physiol.* **54**, 275–286.
Roush, A. H. and Shieh, T. R. (1962). *Biochim. biophys. Acta* **61**, 255–264.
Rubin, S. H., Flower, D., Rosen, F. and Drekter, L. (1945). *Archs Biochem.* **8**, 79–90.
Rudert, F. and Halvorson, H. O. (1963). *Bull. Res. Coun. Israel Sect. A* **11**, 337–344.

Runnström, J., Borei, H. and Sperber, E. (1939). *Ark. Kemi, Mineral. Geol.* **13A**, No. 22.
Runnström, J. and Sperber, E. (1938). *Biochem. Z.* **298**, 340–367.
Samson, F. E., Katz, A. M. and Harris, D. L. (1955). *Archs Biochem. Biophys.* **54**, 406–423.
Samuel, L. W. (1935). *Biochem. J.* **29**, 2331–2333.
Sandegren, E., Enebo, L., Guthenberg, H. and Ljungdahl, L. (1954). *Proc. Am. Soc. Brew. Chem.* **1954**, 63–74.
Sarett, H. P. and Cheldelin, V. H. (1945). *J. Bact.* **49**, 31–39.
Sattler, L. (1948). *Adv. carb. Chem.* **3**, 113–128.
Savioja, T. and Miettinen, J. K. (1966a). *Acta chem. scand.* **20**, 2435–2443.
Savioja, T. and Miettinen, J. K. (1966b). *Acta chem. scand.* **20**, 2444–2450.
Savioja, T. and Miettinen, J. K. (1966c). *Acta chem. scand.* **20**, 2451–2455.
Schanderl, H. (1955). *Brauwissenschaft* **8**, 157–159.
Scharff, T. G. and Kremer, E. H., III (1962). *Archs Biochem. Biophys.* **97**, 192–198.
Schlenk, F. and Dainko, J. L. (1965). *J. Bact.* **89**, 428–436.
Schlenk, F. and Dainko, J. L. (1966). *Archs Biochem. Biophys.* **113**, 127–133.
Schmidt, G., Hecht, L. and Thannhauser, S. J. (1949). *J. biol. Chem.* **178**, 733–742.
Schneider, F. and Hoffmann-Walbeck, H. P. (1955). *In* "Technologie des Zuckers", (Verein der Zuckerindustrie, ed.), p. 53. Verlag M. & H. Schaper, Hanover.
Schönherr, O. Th. and Borst Pauwels, G. W. F. H. (1967). *Biochim. biophys. Acta* **135**, 787–790.
Schultz, A. S. and Atkin, L. (1939). *J. Am. chem. Soc.* **61**, 291–294.
Schultz, A. S., Atkin, L. and Frey, C. N. (1938). *J. Am. chem. Soc.* **60**, 490.
Schultz, A. S., Atkin, L. and Frey, C. N. (1939a). *Cereal Chem.* **16**, 648–651.
Schultz, A. S., Atkin, L. and Frey, C. N. (1939b). *J. Am. chem. Soc.* **61**, 1931.
Schultz, A. S., Atkin, L. and Frey, C. N. (1940a). *J. Am. chem. Soc.* **62**, 2271–2272.
Schultz, A. S., Atkin, L. and Frey, C. N. (1940b). *J. biol. Chem.* **135**, 267–271.
Schultz, A. S. and McManus, D. K. (1950). *Archs Biochem.* **25**, 401–409.
Schultz, A. S. and Pomper, S. (1948). *Archs Biochem.* **19**, 184–192.
Schulze, K. L. (1950). *Arch. Mikrobiol.* **15**, 315–351.
Schuster, K. (1968). *In* "Handbuch der Lebensmittelchemie", (J. Schormüller, ed.), Bd. 7, pp. 45–171. Springer-Verlag, Berlin.
Schwencke, J. and Magaña-Schwencke, N. (1969). *Biochim. biophys. Acta* **173**, 302–312.
Shull, G. M., Hutchings, B. L. and Peterson, W. H. (1942). *J. biol. Chem.* **142**, 913–920.
Simon, H. and Medina, R. (1968). *Z. Naturf.* **23b**, 326–329.
Singer, T. P., Rocca, E. and Kearney, E. B. (1966). *In* "Flavins and Flavoproteins", (E. C. Slater, ed.), pp. 391–426. Elsevier, Amsterdam.
Sivak, A. and Hoffmann-Ostenhof, O. (1962). *Biochem. Z.* **336**, 229–240.
Skraup, Z. H. and König, J. (1901). *Ber. dt. chem. Ges.* **34**, 1115–1118.
Slator, A. (1906). *J. chem. Soc.* **89**, 128–142.
Slator, A. (1908). *J. chem. Soc.* **93**, 217–242.
Smith, R. H. (1951). *J. gen. Microbiol.* **5**, 772–780.
Smythe, C. V. (1938). *J. biol. Chem.* **125**, 635–651.
Snell, E. E. (1944). *J. biol. Chem.* **154**, 313–314.
Snyder, H. E. and Phaff, H. J. (1960). *Antonie van Leeuwenhoek* **26**, 433–452.
Sobotka, H., Holzman, M. and Reiner, M. (1936). *Biochem. J.* **30**, 933–940.
Sobotka, H. and Reiner, M. (1930). *Biochem. J.* **24**, 1783–1786.

Sols, A. and DelaFuente, G. (1961). *In* "Membrane Transport and Metabolism", (A. Kleinzeller and A. Kotyk, eds.), pp. 361–377. Academic Press, London–Czechoslovak Academy of Sciences, Prague.
Somogyi, M. (1937). *J. biol. Chem.* **119**, 741–747.
Sperber, E. (1942). *Biochem. Z.* **313**, 62–74.
Sperber, E. and Renvall, S. (1941). *Biochem. Z.* **310**, 160–169.
Stark, I. E. and Somogyi, M. (1942). *J. biol. Chem.* **142**, 579–584.
Stein, W. D. (1967). "The Movement of Molecules across Cell Membranes", p. 157. Academic Press, New York.
Steyn-Parvé, E. P. (1962). *Biochim. biophys. Acta* **64**, 13–20.
Stephanopoulos, D. and Lewis, M. J. (1968). *J. Inst. Brew.* **74**, 378–383.
Stewart, L. C., Richtmyer, N. K. and Hudson, C. S. (1950). *J. Am. chem. Soc.* **72**, 2059–2061.
Stone, W. E. and Tollens, B. (1888). *Justus Liebigs Annln Chem.* **249**, 257–271.
Sumner, J. B. and O'Kane, D. J. (1948). *Enzymologia* **12**, 251–253.
Sundman, J. (1947). *Finnish Paper Timber J.* **29**, 113–114, 116–117.
Sundman, J. (1949). *Meddelanden från Industriens Centrallaboratorium* **71**, 1–143.
Suomalainen, H. (1948a). *Arch. Mikrobiol.* **14**, 154–156.
Suomalainen, H. (1948b). Dissertation: University of Helsinki.
Suomalainen, H. (1963). *Pure Appl. Chem.* **7**, 639–654.
Suomalainen, H. (1964). *Branntweinwirtschaft* **104**, 402, 404–406, 408.
Suomalainen, H. (1968a). *In* "Aspects of Yeast Metabolism. A Guinness Symposium, Dublin 1965", (A. K. Mills and H. Krebs, eds.), pp. 1–31. Blackwell Scientific Publications, Oxford.
Suomalainen, H. (1968b). *Suom. Kemistilehti* **41A**, 239–254.
Suomalainen, H. (1969). *Antonie van Leeuwenhoek.* In the Press.
Suomalainen, H., Axelson, E. and Oura, E. (1956). *Biochim. biophys. Acta* **20**, 319–322.
Suomalainen, H., Björklund, A., Vihervaara, K. and Oura, E. (1965a). *J. Inst. Brew.* **71**, 221–226.
Suomalainen, H., Christiansen, V. and Oura, E. (1967a). *Suom. Kemistilehti* **40B**, 286–287.
Suomalainen, H. and Kahanpää, H. (1963). *J. Inst. Brew.* **69**, 473–478.
Suomalainen, H., Kauppila, O., Nykänen, L. and Peltonen, R. J. (1968). *In* "Handbuch der Lebensmittechemie", (J. Schormüller, ed.), Bd. 7, pp. 496–653. Springer-Verlag, Berlin.
Suomalainen, H. and Keränen, A. J. A. (1955). *Z. analyt. Chem.* **148**, 81–91.
Suomalainen, H. and Keränen, A. J. A. (1963a). *Biochim. biophys. Acta* **70**, 493–503.
Suomalainen, H. and Keränen, A. J. A. (1963b). *Suom. Kemistilehti* **36B**, 88–90.
Suomalainen, H. and Keränen, A. J. A. (1967). *J. Inst. Brew.* **73**, 477–484.
Suomalainen, H. and Keränen, A. J. A. (1968). *Chem. Phys. Lipids* **2**, 296–315.
Suomalainen, H., Keränen, A. J. A. and Kitunen, M. (1958). *Proc. 7th Intern. Congr. Microbiol.*, Stockholm, p. 143.
Suomalainen, H. and Konttinen, K. (1969). *Suom. Kemistilehti,* **42B**, 255–256.
Suomalainen, H., Konttinen, K. and Oura, E. (1969). *Arch. Mikrobiol.* **64**, 251–261.
Suomalainen, H. and Kuronen, T. (1944). *Finnish Paper Timber J.* **26**, 145–146.
Suomalainen, H., Linko, M. and Oura, E. (1960a). *Biochim. biophys. Acta* **37**, 482–490.
Suomalainen, H. and Linnahalme, T. (1966). *Archs Biochem. Biophys.* **114**, 502–513.
Suomalainen, H. and Nurminen, T. (1969). *Fedn European Biochem. Socs., 6th Symp.*, Madrid, p. 65.

Suomalainen, H., Nurminen, T. and Oura, E. (1967b). *Archs Biochem. Biophys.* **118**, 219–223.
Suomalainen, H., Nurminen, T. and Oura, E. (1967c). *Suom. Kemistilehti* **40B**, 323–326.
Suomalainen, H., Nurminen, T., Vihervaara, K. and Oura, E. (1965b). *J. Inst. Brew.* **71**, 227–231.
Suomalainen, H. and Nykänen, L. (1964). *Suom. Kemistilehti* **37B**, 230–232.
Suomalainen, H. and Nykänen, L. (1966a). *J. Inst. Brew.* **72**, 469–474.
Suomalainen, H. and Nykänen, L. (1966b). *Suom. Kemistilehti* **39B**, 252–256.
Suomalainen, H. and Oura, E. (1955). *Expl Cell Res.* **9**, 355–359.
Suomalainen, H. and Oura, E. (1956). *Biochim. biophys. Acta* **20**, 538–542.
Suomalainen, H. and Oura, E. (1957). *Archs Biochem. Biophys.* **68**, 425–431.
Suomalainen, H. and Oura, E. (1958a). *Biochim. biophys. Acta* **28**, 120–127.
Suomalainen, H. and Oura, E. (1958b). *Suom. Kemistilehti* **31B**, 41–42.
Suomalainen, H. and Oura, E. (1959). *Biochim. biophys. Acta* **31**, 115–124.
Suomalainen, H., Oura, E. and Linko, M. (1961). *Biochim. biophys. Acta* **47**, 267–270.
Suomalainen, H., Oura, E. and Linnahalme, T. (1965c). *J. Inst. Brew.* **71**, 330–336.
Suomalainen, H., Oura, E. and Nevalainen, P. (1965d). *Fedn European Biochem. Socs., 2nd Symp.*, Abstr. Commun. p. 67.
Suomalainen, H. and Pfäffli, S. (1961). *J. Inst. Brew.* **67**, 249–254.
Suomalainen, H., Pfäffli, S. and Oura, E. (1960b). *Suom. Kemistilehti* **33B**, 205.
Suomalainen, H. and Ronkainen, P. (1963). *J. Inst. Brew.* **69**, 478–483.
Suomalainen, H. and Ronkainen, P. (1968a). *Nature, Lond.* **220**, 792–793.
Suomalainen, H. and Ronkainen, P. (1968b). *Tech. Quart., Master Brewers Assoc. Am.* **5**, 119–127.
Suomalainen, H. and Toivonen, T. (1948). *Archs Biochem.* **18**, 109–118.
Surdin, Y., Sly, W., Sire, J., Bordes, A. M. and de Robichon-Szulmajster, H. (1965). *Biochim. biophys. Acta* **107**, 546–566.
Sutton, D. D. and Lampen, J. O. (1962). *Biochim. biophys. Acta* **56**, 303–312.
Suzuoki, Z. (1955). *J. Biochem., Tokyo* **42**, 27–39.
Thorne, R. S. W. (1933). *J. Inst. Brew.* **39**, 608–621.
Thorne, R. S. W. (1941). *J. Inst. Brew.* **47**, 255–272.
Thorne, R. S. W. (1949). *J. Inst. Brew.* **55**, 201–222.
Thorne, R. S. W. (1954). *J. Inst. Brew.* **60**, 227–237.
Tipper, E. J. (1954). *In* White, J. "Yeast Technology", Chapter 26. Chapman & Hall, London.
Tonino, G. J. M. and Steyn-Parvé, E. P. (1963). *Biochim. biophys. Acta* **67**, 453–469.
Trevelyan, W. E. and Harrison, J. S. (1954). *Biochem. J.* **57**, 556–561.
Trevelyan, W. E. and Harrison, J. S. (1956). *Biochem. J.* **63**, 23–33.
Tustanoff, E. R. and Bartley, W. (1964a). *Biochem. J.* **91**, 595–600.
Tustanoff, E. R. and Bartley, W. (1964b). *Can. J. Biochem.* **42**, 651–665.
Van der Linden, A. C. and Thijsse, G. J. E. (1965). *Adv. Enzymol.* **27**, 469–546.
Van Steveninck, J. (1963). *Biochem. J.* **88**, 25P–26P.
Van Steveninck, J. (1966). *Biochim. biophys. Acta* **126**, 154–162.
Van Steveninck, J. (1968). *Biochim. biophys. Acta* **150**, 424–434.
Van Steveninck, J. (1969). *Archs Biochem. Biophys.* **130**, 244–252.
Van Steveninck, J. and Booij, H. L. (1964). *J. gen. Physiol.* **48**, 43–60.
Van Steveninck, J. and Dawson, E. C. (1968). *Biochim. biophys. Acta* **150**, 47–55.
Van Steveninck, J. and Rothstein, A. (1965). *J. gen. Physiol.* **49**, 235–246.
Virtanen, A. I., Csáky, T. Z. and Rautanen, N. (1949). *Biochim. biophys. Acta* **3**, 208–214.

Vogel, H. (1949). "Die Rohstoffe der Gärungsindustrie", pp. 17–20. Wepf & Co., Verlag, Basel.
Vogt, E. (1968). *In* "Handbuch der Lebensmittelchemie", (J. Schormüller, ed.), Bd. 7, pp. 172–310. Springer-Verlag, Berlin.
Wakil, S. J. (1961). *J. Lipid Res.* **2**, 1–24.
Wakil, S. J. and Waite, M. (1962). *Biochem. biophys. Res. Commun.* **9**, 18–24.
Wang, C. H., Gregg, C. T., Forbusch, I. A., Christensen, B. E. and Cheldelin, V. H. (1956). *J. Am. chem. Soc.* **78**, 1869–1872.
Wang, C. H., Stern, I., Gilmour, C. M., Klungsoyr, S., Reed, D. J., Bialy, J. J., Christensen, B. E. and Cheldelin, V. H. (1958). *J. Bact.* **76**, 207–216.
Wawzonek, S., Klimstra, P. D., Kallio, R. E. and Stewart, J. E. (1960). *J. Am. chem. Soc.* **82**, 1421–1424.
Webb, A. D. and Ingraham, J. L. (1963). *Adv. appl. Microbiol.* **5**, 317–353.
Weidenhagen, R. (1928). *Z. ver. Deut. Zucker-Ind.* **78**, 539–542.
Weidenhagen, R. (1930). *Z. ver. Deut. Zucker-Ind.* **80**, 11–24.
Weinfurtner, F., Eschenbecher, F. and Borges, W.-D. (1959). *Zentbl. Bakt. ParasitKde* (*Abt. II*) **113**, 134–162.
Weinfurtner, F., Eschenbecher, F. and Thoss, G. (1964). *Brauwissenschaft* **17**, 121–129.
Wertheimer, E. (1934). *Protoplasma* **21**, 522–560.
Westenbrink, H. G. K., van Dorp, D. A., Gruber, M. and Veldman, H. (1940). *Enzymologia* **9**, 73–89.
de Whalley, H. C. S. (1952). *Int. sugar J.* **54**, 127.
White, J. (1954). "Yeast Technology". Chapman & Hall, London.
White, J. and Munns, D. J. (1950). *J. Inst. Brew.* **56**, 194–202.
White, J. and Munns, D. J. (1951). *J. Inst. Brew.* **57**, 175–179.
White, J. and Munns, D. J. (1953). *Am. Brewer* **86**, 29–33, 83.
White, A. G. C. and Werkman, C. H. (1947). *Archs Biochem.* **13**, 27–32.
Wiame, J. M. (1946). *Bull. Soc. Chim. biol.* **28**, 552–556.
Wiame, J. M. (1947). *Biochim. biophys. Acta* **1**, 234–255.
Wiame, J. M. (1949). *J. biol. Chem.* **178**, 919–929.
Wickerham, L. J. (1946). *J. Bact.* **52**, 293–301.
Wikén, T. and Richard, O. (1953). *Experientia* **9**, 417–420.
Wikén, T. and Richard, O. (1954a). *Antonie van Leeuwenhoek* **20**, 385–405.
Wikén, T. and Richard, O. (1954b). *Schweiz. Z. allg. Path. Bakt.* **17**, 475–486.
Wiles, A. E. (1953). *J. Inst. Brew.* **59**, 265–284.
Wilkes, B. G. and Palmer, E. T. (1932). *J. gen. Physiol.* **16**, 233–242.
Wilkins, P. O. (1967). *J. Bact.* **93**, 1565–1570.
Williams, R. J., Eakin, R. E. and Snell, E. E. (1940). *J. Am. chem. Soc.* **62**, 1204–1207.
Williams, R. J., Lyman, C. M., Goodyear, G. H., Truesdail, J. H. and Holaday, D. (1933). *J. Am. chem. Soc.* **55**, 2912–2927.
Williams, R. J., Mosher, W. A. and Rohrman, E. (1936). *Biochem. J.* **30**, 2036–2039.
Williams, R. J. and Roehm, R. R. (1930). *J. biol. Chem.* **87**, 581–590.
Williams, R. J. and Rohrman, E. (1936). *J. Am. chem. Soc.* **58**, 695.
Williams, R. J. and Saunders, D. H. (1934). *Biochem. J.* **28**, 1887–1893.
Williams, R. J., Truesdail, J. H., Weinstock, H. H., Jr., Rohrman, E., Lyman, C. M. and McBurney, C. H. (1938). *J. Am. chem. Soc.* **60**, 2719–2723.
Willstätter, R. and Bamann, E. (1926a). *Hoppe-Seyler's Z. physiol. Chem.* **151**, 242–272.

Willstätter, R. and Bamann, E. (1926b). *Hoppe-Seyler's Z. physiol. Chem.* **152**, 202–214.
Willstätter, R. and Rohdewald, M. (1937). *Hoppe-Seyler's Z. physiol. Chem.* **247**, 269–280.
Willstätter, R. and Steibelt, W. (1921). *Hoppe-Seyler's Z. physiol. Chem.* **115**, 211–234.
Winge, Ö. and Roberts, C. (1958). *In* "The Chemistry and Biology of Yeasts", (A. H. Cook, ed.), pp. 123–156. Academic Press, New York.
Winzler, R. J., Burk, D. and du Vigneaud, V. (1944). *Archs Biochem.* **5**, 25–47.
Woolley, D. W. (1941). *J. biol. Chem.* **140**, 461–466.
Wright, L. D., Skeggs, H. R. and Cresson, E. L. (1951). *J. Am. chem. Soc.* **73**, 4144–4145.
Yamada, K., Takahashi, J., Kawabata, Y., Okada, T. and Onihara, T. (1968). *In* "Single-Cell Protein", (R. I. Mateles and S. R. Tannenbaum, eds.), pp. 192–207. The M.I.T. Press, Cambridge, Massachusetts.
Yoshida, A. (1953). *Sci. Papers Coll. Gen. Educ. Univ. Tokyo*, **3**, 151–168.
Yoshida, A. and Yamataka, A. (1953). *J. Biochem., Tokyo* **40**, 85–94.
Yphantis, D. A., Dainko, J. L. and Schlenk, F. (1967). *J. Bact.* **94**, 1509–1515.
Zemplén, G. (1915). *Ber. dt. chem. Ges.* **48**, 233–238.
Zerban, F. W. and Sattler, L. (1942). *Ind. Engng Chem. ind. Edn* **34**, 1180–1188.

# Chapter 3

# Kinetics and Energetics of Yeast Growth

N. VAN UDEN
*Laboratory of Microbiology, Gulbenkian Institute of Science, Oeiras, Portugal*

I. INTRODUCTION . . . . . 75
II. NOTATION . . . . . 76
III. EXPONENTIAL GROWTH . . . . . 79
IV. DEPENDENCE OF THE SPECIFIC GROWTH RATE ON THE CONCENTRATION OF A LIMITING NUTRIENT . . . . . 82
A. Early Equations . . . . . 82
B. Fundamental Relations . . . . . 83
V. DEPENDENCE OF THE SPECIFIC GROWTH RATE ON TEMPERATURE . . . . . 91
A. Arrhenius Relations . . . . . 91
B. Steady-State Concentrations of Native Enzymes . . . . . 94
C. Concurrent Exponential Death . . . . . 96
VI. YIELD AND MAINTENANCE RELATIONS . . . . . 100
A. Yield Factor and Specific Maintenance Rate . . . . . 100
B. Yield and Maintenance Analogues . . . . . 105
C. ATP Yields . . . . . 106
D. Yield and Maintenance Analysis . . . . . 108
VII. A MODEL . . . . . 109
A. Description . . . . . 109
B. The Sensitivity Coefficient . . . . . 114
REFERENCES . . . . . 116

## I. Introduction

For many purposes of growth analysis, yeasts have more similarities with heterotrophic bacteria than with the filamentous ascomycetes and basidiomycetes with which they are taxonomically related. If we exclude such genera of filamentous yeast-like fungi as *Endomycopsis* and *Trichosporon* and certain filamentous species of other genera (for instance as found in *Candida*), it may be said that yeasts growing in an agitated liquid environment are, in general, excellent approximations of a homogeneous population of unicellular heterotrophic organisms.

The present chapter is concerned with the kinetics and energetics of such a population. Though concrete examples have been chosen from the literature on yeasts, the treatment is often applicable to growth of populations of heterotrophic unicellular organisms in general. Therefore, the chapter may be of use in the teaching of the kinetics and energetics of microbial growth, and should be helpful as an introductory text to prospective workers in the field.

Since the references to the literature are selective, the chapter does not represent an effort towards a comprehensive review of the literature on yeast growth. For information on such aspects of growth as the cytology of budding, sporulation, nutritional requirements, regulation of mitosis, and morphogenesis, other chapters of this treatise may be consulted. Readers interested in synchronous growth of yeasts may consult Williamson (1964, 1966), Dawson (1969) and von Meyenburg (1969). Information on the distribution of generation times and cell ages in growing populations may be found in Painter and Marr (1968), Beran *et al.* (1967) and Beran (1969).

## II. Notation

The notations used in the description of the model for yeast growth are explained in the text of Section VII (p. 109). The following notations are used in other sections:

| | |
|---|---|
| $a$ | Specific maintenance rate. |
| $a^{gas}$ | Maintenance rate analogue with respect to a gaseous metabolite. |
| $a^{met}$ | Maintenance rate analogue with respect to a non-gaseous metabolite. |
| $A$ | Empirical constant in the Arrhenius equation. |
| $C$ and $C'$ | Various constants, as explained in the text. |
| $D$ | Dilution rate in open (continuous) culture systems. |
| $D_b$ | Diffusion coefficient in biomass. |
| $D_{sp}$ | Specific diffusion coefficient. |
| $D_w$ | Diffusion coefficient in water. |
| $E$ | Enzyme concentration; also "energy of activation" in the Arrhenius equation. |
| $E_D$ | Concentration of reversibly denaturated enzyme. |
| $E_N$ | Concentration of native enzyme. |
| $f$ | Flow rate in continuous culture systems; also other parameters as explained in the text. |
| $\Delta F^0$ | Standard free-energy change. |

| | |
|---|---|
| $\Delta F^*$ | Free energy of activation. |
| $G$ | Gradient across cell surface for unit molar concentration difference between the surface and the intracellular receptor site. |
| $h$ | Planck's constant; also other parameters as explained in the text. |
| $\Delta H^0$ | Standard heat change. |
| $\Delta H^*$ | Heat of activation. |
| $I$ | Concentration of competitive inhibitor. |
| $\boldsymbol{k}$ | Boltzmann constant. |
| $k_1, k_2$ | Rate constants. |
| $k$ | Specific transfer rate. |
| $k_g$ | Specific transfer rate for growth only. |
| $k_m$ | Specific transfer rate for maintenance only. |
| $k_m^{gas}$ | Specific output rate of a gaseous metabolite in connection with maintenance. |
| $k_m^{met}$ | Specific output rate of a non-gaseous metabolite in connection with maintenance. |
| $k_{max_1}$ | Maximum specific transport-rate in either direction. |
| $k_{max_2}$ | Maximum specific turnover-rate in forward direction of the first enzymic step. |
| $K$ | Equilibrium constant. |
| $K_i$ | Michaelis constant of competitive inhibitor. |
| $K_{m_1}$ | Michaelis constant of transport into the biomass. |
| $K_{m_2}$ | Michaelis constant of transport out of the biomass. |
| $K_{m_3}$ | Michaelis constant of first enzymic reaction in the forward direction. |
| $K_s$ | Substrate constant in the Monod equation. |
| $m$ | Number of thermosensitive sites per cell. |
| $M_w$ | Molecular weight. |
| $n$ | Parameters explained in text. |
| $N$ | Number of cells per unit of dry biomass. |
| $N_0, N_t$ | Number of cells per unit volume of medium after zero time and time $t$. |
| $N^{nv}$ | Number of non-viable cells per unit volume of medium. |
| $N^v$ | Number of viable cells per unit volume of medium. |
| $p$ | Probability. |
| $Q$ | Function in Powell equation (see p. 83). |
| $Q_{gas}$ | Volume of gas per unit of biomass per unit time. |

$Q^{ATP}$ Moles ATP per gram of dry biomass per hour.

$Q_g^{ATP}$ Moles ATP per gram of dry biomass per hour used for growth.

$Q_m^{ATP}$ Moles ATP per gram of dry biomass per hour used for maintenance.

$r$ Parameters explained in the text.

$R$ Gas constant.

$S$ Concentration of limiting nutrient; also function in Powell equation (see p. 83).

$S_1$ Concentration of limiting nutrient at the cell surface.

$S_2$ Concentration of limiting nutrient in the biomass.

$S_r$ Concentration of limiting nutrient in the medium reservoir of the chemostat.

$\Delta S^0$ Standard entropy change.

$\Delta S^*$ Entropy of activation.

$t$ Time.

$T$ Absolute temperature.

$T_{max}$ Maximum temperature for growth.

$T_{op}$ Optimum temperature for growth.

$v$ Rate or velocity.

$V$ Volume in fermenter of continuous culture systems.

$V_{max}$ Maximum velocity.

$x$ Population density (biomass per unit volume of medium).

$x_i$ Average concentration of any cell constituent or product (mass per unit volume of medium).

$Y$ Function in Powell equation (see p. 83).

$y$ Yield factor (biomass per unit mass).

$Y_{max}$ Maximum yield factor (biomass per unit mass).

$Y_{max}^{ATP}$ Maximum ATP yield (biomass per mole).

$Y_{max'}^{gas}$ and $Y_{max}^{gas}$ Maximum yield factor with respect to a gaseous metabolite (biomass per unit volume and per unit mass respectively).

$Y_{max}^{met}$ Maximum yield factor with respect to a non-gaseous metabolite (biomass per unit mass).

$\mu_d$ Specific death rate; also specific inactivation rate of thermosensitive sites.

$\mu_g$ Specific growth rate.

$\mu_{max}$ Maximum specific growth rate.

$\pi$ Probability.

## III. Exponential Growth

Let us consider a population of genetically identical cells, homogeneously dispersed in a liquid medium appropriate for growth. The cells take up nutrients, and process them through successive transport and enzymic reactions producing heat, waste products, carriers of chemical energy, micro- and macromolecular building blocks and cellular structures made thereof. This leads to cell growth, cell division and population increase.

The rate or velocity of this process may be evaluated by measuring the change of the population density with time:

$$v = \frac{dx}{dt} \tag{1}$$

If the instantaneous rate is divided by the instantaneous population density, we obtain the specific growth rate of that moment:

$$\mu_g = \frac{1}{x}\frac{dx}{dt} \tag{2}$$

If the medium has constant chemical and physical properties, the specific growth rate eventually becomes invariant with time and all constituents of the biomass must then increase with the same specific rate

$$\mu_g = \frac{1}{x}\frac{dx}{dt} = \frac{1}{x_1}\frac{dx_1}{dt} = \frac{1}{x_2}\frac{dx_2}{dt} = \ldots = \frac{1}{x_n}\frac{dx_n}{dt} \tag{3}$$

where $x_1, x_2 \ldots x_n$ are average concentrations of cell constituents at any given time.

Population growth, in which all extensive properties of the biomass increase with the same specific rate, has been called "balanced growth" (Campbell, 1957). If this specific rate is invariant with time, population growth is both balanced *and* exponential (Painter and Marr, 1968). Rewriting eq. (2):

$$\frac{dx}{dt} = \mu_g x \tag{4}$$

and integrating, we obtain the well-known expression for the population increase during exponential growth:

$$x_t = x_0 \cdot e^{\mu_g t} \tag{5}$$

Taking logarithms at both sides of eq. (5) produces a linear equation:

$$\log x_t = \log x_0 + \mu_g t \tag{6}$$

By plotting the logarithms of measured values of the population density against time, a series of points is obtained to which a straight line may be fitted. Its slope, after division by $\log_e$, is an estimate of the specific growth rate. The practical usefulness of the log form of eq. (5) probably is the reason why exponential growth is often (and erroneously) referred to as "logarithmic" growth.

In closed systems, as a batch culture for example, exponential growth is preceded and followed by non-exponential phases. The distinction and nomenclature of these phases go back to Monod (1942) and can now be found in many textbooks; they will not be analysed in the present treatment. Abrupt changes in the chemical or physical characteristics of the medium during exponential growth may unbalance growth and give rise to a so-called transient phase between two different states of balanced, exponential growth. Much information on the control of macromolecular synthesis has been obtained by analysing cell populations in experimental transient phases (Maaløe and Kjeldgaard, 1966). There is some evidence, obtained with *Candida utilis* in batch culture (Stanley, 1964) that, even in a constant environment, exponential growth may become unbalanced and also that the specific growth rate may attain different values dependent on unknown factors.

True steady states are more easily obtained when exponential growth occurs in open systems such as the chemostat and the turbidostat. For descriptions and mathematical treatments of the chemostat and turbidostat, see Monod (1950), Novick and Szilard (1950), Herbert *et al.* (1956), Pfennig and Jannasch (1962) and Fencl (1966).

Figure 1 shows the essential features of such systems. Fermenter B, containing a constant volume ($V$) of medium which supports an exponentially growing cell population, receives fresh medium with flow rate ($f$) from reservoir ($A$), while spent medium with cells leaves the fermenter through an overflow into receiving vessel (C). In the steady state, the dilution rate $D(=f/V)$ is balanced by the specific growth rate $\mu_g$. Though energy and matter are continuously transformed while flowing through the fermenter, all physical, chemical and mathematical parameters which define the contents of the fermenter have fixed values for each steady state, including energy and entropy content, concentrations of solutes in the medium and the pools of the biomass, composition and concentration of the biomass, number of cells, distributions of cell sizes, cell ages and generation times. The steady state may be disturbed by changing the composition of the inflowing medium, the dilution rate, or the temperature. If the new conditions allow a new steady state, the latter will be reached after a period of time during which the contents of the fermenter are in a so-called transient state.

An exponentially growing cell population in the steady state may be understood as a multistep system (van Uden, 1969b) with input (nutrients), throughput (turnover of pools through transfer and enzyme reactions) and output (new biomass and final metabolites).

The specific output rate of new biomass (i.e. the specific growth rate) as a function of nutrient concentration will be studied in Section IV (p. 82)

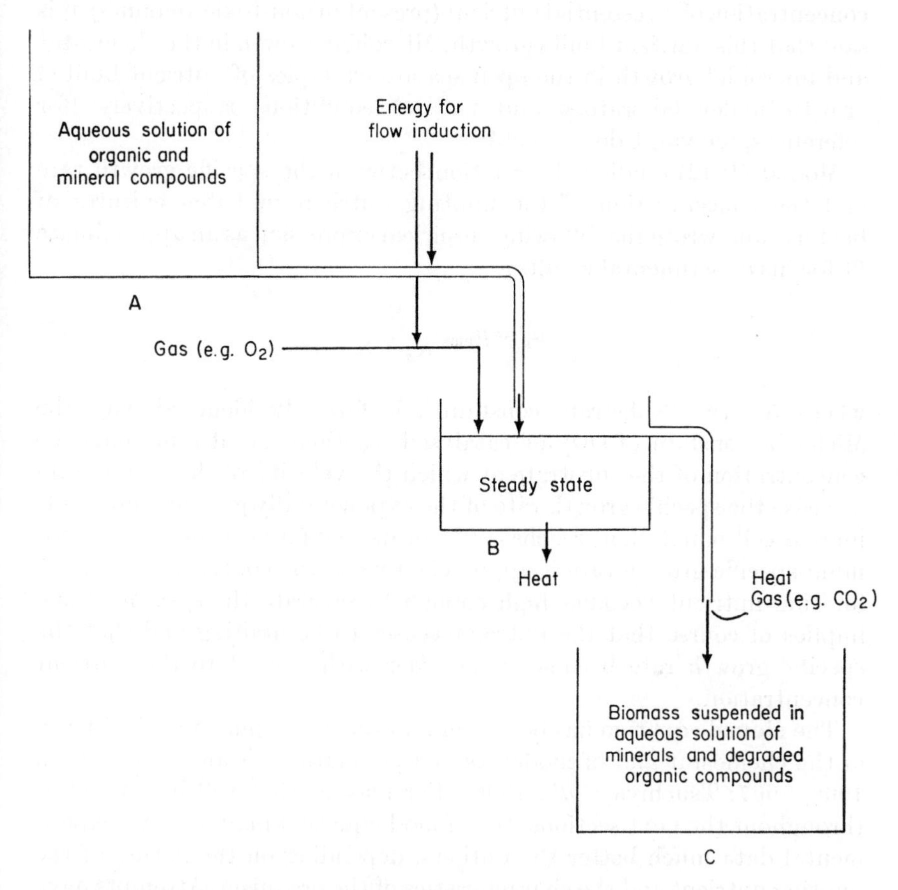


FIG. 1. Energy and mass relations in an open (continuous) culture system.

and as a function of the temperature in Section V (p. 91). In Section VI (p. 100) relations will be examined between the specific growth rate on the one hand and the specific input, output and turnover rates of nutrients and metabolites on the other hand. Finally, in Section VII (p. 109), a simple model will be presented which incorporates some of the essential features of the relations discussed and which is suitable for computer simulation using existing programmes.

## IV. Dependence of the Specific Growth Rate on the Concentration of a Limiting Nutrient

### A. Early Equations

When the specific growth rate responds measurably to variations in the concentration of an essential nutrient (present in non-toxic amounts), it is said that this nutrient limits growth. Microbial growth in the chemostat and microbial growth in the open sea are examples of nutrient-limited growth under laboratory and natural conditions respectively (for references, see van Uden, 1969b).

Monod (1942) studied the relation between the specific growth rate and the concentration of the limiting nutrient in batch cultures of bacteria and wrote the following empirical expression as an approximate fit for his experimental results:

$$\mu_g = \mu_{\max} \frac{S}{K_s + S} \tag{7}$$

where $K_s$, the "substrate constant", is formally identical with the Michaelis constant of enzyme-catalysed reactions, i.e. it represents the concentration of the substrate at which the velocity of the reaction (in our case the specific growth rate of the exponentially growing, nutrient-limited cell population) reaches 50% of its maximum value. The maximum specific growth rate is approached when the concentration of the limiting nutrient becomes high enough to saturate the system, which implies of course that the nutrient ceases to be limiting and that the specific growth rate becomes zero order with respect to the nutrient concentration.

The Monod equation has been useful in the development of the theory of the chemostat and of models of cell populations (Ramkrishna *et al.*, 1966, 1967; Tsuchiya *et al.*, 1966). For reasons that will become clear throughout the next sections, the Monod equation may fit some experimental data much better than others, depending on the nature of the limiting nutrient and the characteristics of the organism. Attempts have been made to improve the general applicability of the Monod equation by the introduction of additional constants (Moser, 1958; Contois, 1959; Ierusalimsky *et al.*, 1962; Ierusalimsky, 1967). Teissier (1936) proposed a different empirical equation which was applied to chemostat growth data by Schulze and Lipe (1964).

None of these equations and modifications thereof constitutes a theoretically sound starting point for the development of a kinetic theory of nutrient-limited cell population growth (Powell, 1967; van Uden, 1967a). Powell (1966, 1969) stated in a compact equation the complexity of the relation between the specific growth rate and the concentration of the limiting nutrient:

$$\mu_g = YQS \tag{8}$$

$Y$ represents the yield factor, i.e. the ratio between biomass formed and limiting nutrient consumed, $S$ is a function of the concentration of the limiting nutrient in the medium, and $Q$ stands for kinetic factors defined by the physiological state of the population.

A number of more explicit equations, sometimes applicable to the analysis of individual growth situations, have become available in recent years and will be introduced presently.

## B. FUNDAMENTAL RELATIONS

The growth rate is in general a function of the input rate of the limiting nutrient:

$$\frac{dx}{dt} = -y\frac{dS}{dt} \tag{9}$$

where $y$ is the yield factor (Monod, 1942) with respect to nutrient $S$. By dividing both sides by the population density, a relation between specific rates is obtained:

$$\frac{dx}{dt}\frac{1}{x} = -y\frac{dS}{dt}\frac{1}{x} \tag{10}$$

Writing this equation with appropriate symbols, this fundamental growth rate equation can be written in simple form (Powell, 1966; van Uden, 1967a):

$$\mu_g = yk \tag{11}$$

The specific transfer rate, $k$, of the limiting nutrient is defined as mass of nutrient transferred into one unit of biomass per unit time, and has thus the same dimension (reciprocal time) as the specific growth rate. It is a function of the nutrient concentration on both sides of the plasma membrane and sometimes of additional variables (such as cell size, inhibitors); the form of this function and the values of the parameters depend on many factors, including the nature of the nutrient and that of the organism. There is therefore no single generally valid explicit form of eq. (11), though many microbiologists (personal communications) still seem to think that such a form might be written if one were just clever enough.

Less general forms of eq. (11), applicable to distinct types of transfer kinetics, will now be presented.

### 1. *Michaelis-Menten Transfer*

The transfer of low molecular-weight nutrients into cells may obey either first-order kinetics with respect to the concentration gradient across the plasma membrane (true diffusion) or saturation kinetics of the

Michaelis-Menten type with respect to the extracellular and intracellular concentrations of the nutrient. The kinetics of uptake of sugars by yeasts for example is of the Michaelis-Menten type and may be represented as follows (Wilbrandt and Rosenberg, 1961):

$$v = V_{max_1} \frac{S_1}{K_{m_1} + S_1} - V_{max_2} \frac{S_2}{K_{m_2} + S_2} \quad (12)$$

It is useful to compare this Michaelis-Menten equation with the one that applies to a one-substrate one-product enzyme-catalysed chemical reaction:

$$v = V_{max_1} \frac{S_1}{K_{m_1} + \frac{K_{m_1}}{K_{m_2}} S_2 + S_1} - V_{max_2} \frac{S_2}{K_{m_2} + \frac{K_{m_2}}{K_{m_1}} S_1 + S_2} \quad (13)$$

which is often written as follows:

$$v = \frac{V_{max_1} \frac{S_1}{K_{m_1}} - V_{max_2} \frac{S_2}{K_{m_2}}}{1 + \frac{S_1}{K_{m_1}} + \frac{S_2}{K_{m_2}}} \quad (14)$$

The difference is that, in transfer, $S_1$ and $S_2$ are physically separated, i.e. they are in different compartments and do not compete. In other words, the kinetics of the unidirectional flux (in either direction) is dependent only on the concentration of $S$ in the compartment where the flux comes from. On the other hand, in enzyme reactions of the type represented by eqs. (13) and (14), the kinetics of the forward and the back reaction, considered separately, are both dependent on the concentrations of substrate and product.

Studies on sugar transfer in yeasts have suggested that two types of Michaelis-Menten transfer mechanisms occur in these organisms (Cirillo, 1961, 1962; Okada and Halvorson, 1964a, b; van Steveninck and Booij, 1964; van Steveninck and Rothstein, 1965; van Steveninck, 1966; Kotyk, 1967). They have been called "facilitated diffusion" and "active transport" respectively. The kinetic expression for "facilitated diffusion" is obtained by postulating that $V_{max_1} = V_{max_2}$ and that $K_{m_1} = K_{m_2}$ and we may write a simplified form of eq. (12):

$$v = V_{max} \left( \frac{S_1}{K_m + S_1} - \frac{S_2}{K_m + S_2} \right) \quad (15)$$

Net transfer will no longer occur when $S_1 = S_2$.

In "active transport", transfer may continue against a concentration gradient (i.e. $S_2 > S_1$) and metabolic energy is consumed. Kinetically this transfer mechanism requires that $V_{max_1} > V_{max_2}$ or that $K_{m_2} > K_{m_1}$, or both.

Experimental work has regularly revealed that, in active transport, the affinity of the nutrient for the transport system (as expressed by $K_m$) is different on the two sides of the plasma membrane, whereas differences in unidirectional transport capacity as expressed by $V_{max}$ are still controversial. In what follows we shall ignore possible differences between the unidirectional $V_{max}$ values. Under this condition, the introduction of eq. (12) into eq. (11) leads to the following expression:

$$\mu_g = yk_{max_1}\left(\frac{S_1}{K_{m_1}+S_1} - \frac{S_2}{K_{m_2}+S_2}\right) \tag{16}$$

The maximum specific transfer rate, $k_{max_1}$, represents the unidirectional specific transfer rate in either direction when the concentration of $S$ in the respective compartment saturates the transport mechanism. Under conditions of saturation in the external compartment (the culture medium in batch culture or the turbidostat), the specific growth rate attains its maximum value and eq. (16) reduces to:

$$\mu_{max} = yk_{max_1}\left(1 - \frac{S_2}{K_{m_2}+S_2}\right) \tag{17}$$

If a nutrient other than $S$ becomes limiting, eq. (17) is still valid with respect to $S$ but "$\mu_{max}$", $S_2$ and often "$y$" will have different values. Such variation is probably of great ecological significance, particularly with respect to microbial growth in the oceans.

Obviously an infinite number of "$\mu_{max}$" values can be conceived when the growth rate is limited by more than one nutrient. Imagine that $S$ and $S'$ are limiting nutrients; make $S$ saturating for different values of $S'$ and *vice versa*; in each case a different "$\mu_{max}$" value will appear. However, in ordinary microbiological jargon, $\mu_{max}$ refers to the specific growth rate in the exponential phase in a medium of given composition at a fixed temperature when all nutrients have saturating concentrations.

The maximum specific transfer rate is a function of the ratio between cell-surface area and biomass. Since it has been observed in a number of microbial species, including yeasts (Herbert, 1958, 1961; Button and Garver, 1966), that the average cell size varies with the specific growth rate, it would appear that $k_{max_1}$ must also vary with $\mu_g$.

The yield factor, more often than not, is a variable dependent on the specific growth rate, the range of the variation depending on the nature of the growth-limiting nutrient. For example, when the limiting nutrient is the energy source, eq. (16) becomes:

$$\mu_g + a = Y_{max}\, k_{max_1}\left(\frac{S_1}{K_{m_1}+S_1} - \frac{S_2}{K_{m_2}+S_2}\right) \tag{18}$$

where $a$ and $Y_{max}$ are constants (see Section VI, p. 100).

The Michaelis constants $K_{m_1}$ and $K_{m_2}$ may often be growth-rate independent. However, it has been suggested that glucose transport in baker's yeast may be regulated by feedback control (Sols, 1967; Kleinzeller and Kotyk, 1967) and recently it was found by Azam and Kotyk (1969) that glucose 6-phosphate decreases in a non-competitive way the affinity of baker's yeast plasma membranes for glucose. This suggests that the values of $K_{m_1}$ or $K_{m_2}$ or of both may be somewhat dependent on the concentration of glucose 6-phosphate in the biomass.

The intracellular steady-state concentration ($S_2$) is a function of the preceding, and most if not all following, steps (see Section VII, p. 109). In cases where $S_2$ is low enough to make back-transfer kinetically insignificant, eq. (16) reduces to:

$$\mu_g = yk_{\max_1} \frac{S_1}{K_{m_1} + S_1} \tag{19}$$

and it is said that growth is transport-limited (van Uden, 1967a, b).

Equation (19) has the same form as eq. (7), the Monod equation. From this, it follows that the Monod equation implies one special case, namely unidirectional Michaelis-Menten kinetics of the transfer of the limiting nutrient (i.e. transport-limited growth). In addition, it implies that $yk_{\max_1} = \mu_{\max}$, which at best is an approximation within experimental error. For example, if the limiting nutrient is the energy source and when its transfer is unidirectional over the whole range of specific growth rates, eq. (18) takes the following form:

$$\mu_g + a = Y_{\max} k_{\max_1} \frac{S_1}{K_{m_1} + S_1} \tag{20}$$

and

$$\mu_{\max} = Y_{\max} k_{\max_1} - a \tag{21}$$

Experimental results obtained with a population obeying eq. (20) may satisfy the Monod equation equally well when the value of "$a$" is very small and the experimental error of the $S_1$ values relatively great (as is often the case with data obtained with the chemostat).

When growth is studied in a chemostat, the following parameters are easily estimated: $\mu_g$ from the relation $D = \mu_g$; $y$ from the relation $x = y(S_r - S_1)$; and $S_1$ through direct quantitative estimations. If growth is transport-limited, the following linear double reciprocal plot (from eq. 16) may be obtained:

$$\frac{y}{D} = \frac{1}{k_{\max_1}} + \frac{K_{m_1}}{k_{\max_1}} \frac{1}{S_1} \tag{22}$$

and $k_{\max_1}$ and $K_m$ may be obtained therefrom. Transport-limited growth was observed in anaerobic glucose-limited cultures of *Saccharomyces cerevisiae* growing in a mineral vitamin-supplemented medium in the chemostat, and in a respiration-deficient mutant of this species growing in this medium with oxygen (van Uden, 1967b; Fig. 2).

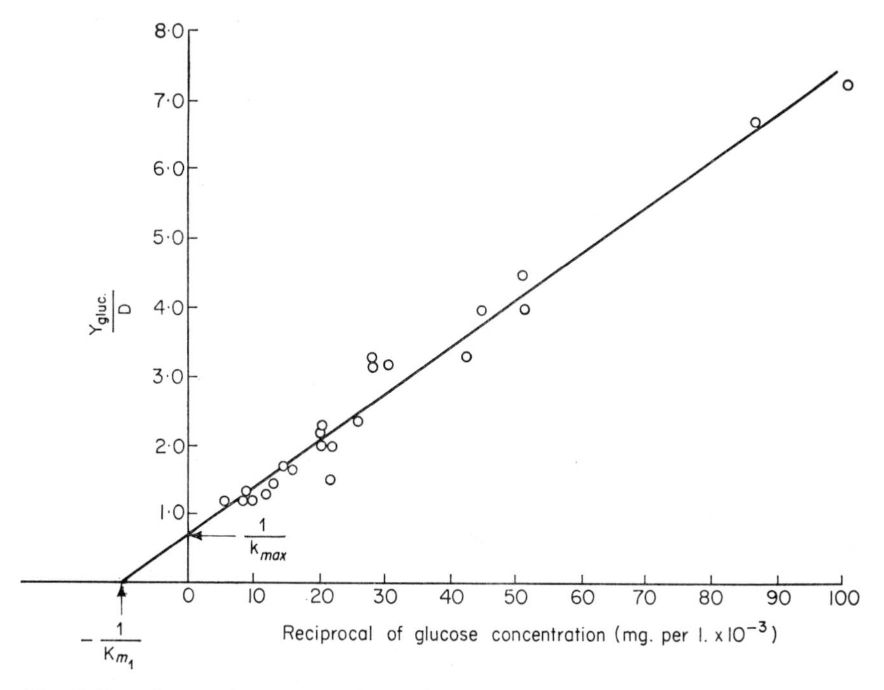


FIG. 2. Data from a chemostat culture of a respiration-deficient mutant of *Saccharomyces cerevisiae* in a glucose-limited mineral medium (van Uden, 1967b) at 20°, plotted according to eq. (22). Unpublished data of A. Madeira-Lopes and N. van Uden.

## 2. *Inhibited Michaelis-Menten Transfer*

Inhibitors present in the medium which affect the specific growth rate of nutrient-limited populations may be formally thought to change the yield factor or the specific transfer rate of the limiting nutrient or both. More often than not, the primary effect will be on the steady-state pool sizes and the turnover kinetics of one or more intracellular metabolites. At best, the effect on the specific growth can then be expressed by empirical equations (Ierusalimsky, 1967; Dean and Hinshelwood, 1966; Egambardiev and Ierusalimsky, 1969). When the primary effect is on the parameters of the transfer step only, explicit equations become possible. Some theoretical and experimental data on competitive inhibition of yeast growth limited by transport are available (van Uden, 1967a, b).

When a non-metabolizable competitive inhibitor of the transport of the limiting nutrient is present in concentrations sufficiently high to make transport growth-limiting, eq. (16) becomes:

$$\mu_g = yk_{\max_1} \frac{S_1}{K_{m_1} + \frac{K_{m_1}}{K_i} I + S_1} \tag{23}$$

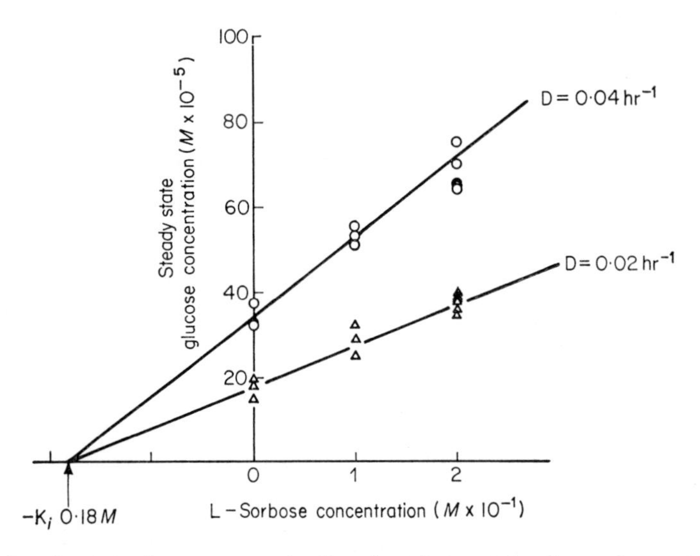


FIG. 3. Steady-state glucose concentrations in a chemostat culture of a respiration-deficient mutant of *Saccharomyces cerevisiae* at two dilution rates as affected by L-sorbose at two different concentrations. The data, which are from van Uden (1967b), are plotted according to eq. (24).

When the concentration of the competitive inhibitor is varied in the chemostat at a constant dilution rate, it can be shown easily that the following linear relation applies (van Uden, 1967a):

$$S = CK_{m_1} + C\frac{K_{m_1}}{K_i} I \tag{24}$$

where $C = \dfrac{D}{k_{\max_1} y - D}$

When eq. (24) is plotted for different values of the dilution rate (i.e. different values of $C$), the slopes of the straight lines vary accordingly. The lines converge to a point on the base line corresponding to $-K_i$ (Fig. 3).

### 3. *Diffusion Transfer*

When the growth-limiting nutrient is transferred to the biomass by simple diffusion, the specific transfer rate is a linear function of a concentration difference (van Uden, 1969a, b):

$$k = D_{sp}(S_1 - S_2) \tag{25}$$

The specific diffusion coefficient ($D_{sp}$) is equal to the mass of nutrient transferred into one unit of biomass per unit time under unit molar concentration difference between the cell surface and the intracellular receptor site. $S_1$ is the molar concentration of the nutrient in the medium (at the cell surface) and $S_2$ the molar concentration at the intracellular receptor site, i.e. at the site of the enzyme that catalyses the first metabolic step involving the nutrient. $D_{sp}$ has thus the dimensions of reciprocal time and of reciprocal molar concentrations (i.e. of volume).

The value of $D_{sp}$ depends on: (i) $D_b$, the diffusion coefficient of the nutrient in the biomass (nutrient transfer per unit time across unit area per unit concentration gradient). When $D_w$, the diffusion coefficient in water, is known, $D_b$ may either be predicted within reasonable limits or determined experimentally if the biomass is accessible to this approach (see Johnson, 1967, for some oxygen diffusion data); (ii) the total cell-surface per unit of biomass (dry weight). An estimate of this is $N$ (the number of cells per unit of biomass) times the average cell surface. More accurately, the surface area may be calculated from size distribution curves; (iii) $G$, the gradient across the cell surface for unit molar concentration difference between the surface and the intracellular receptor site. This leads to the following expression for the specific diffusion coefficient:

$$D_{sp} = D_b . N4\pi r^2 . G . 10^{-3} M_w \tag{26}$$

When $G$ is calculated from molar concentrations using centimetres as the linear unit, and $D_b$ is given as $cm^2 h^{-1}$, the factor $10^{-3} M_w$ (molecular weight of the nutrient) has to be introduced if $D_{sp}$ is to be obtained as grams (of nutrient) per gram of biomass (dry weight) per h under unit molar concentration difference between the surface and the intracellular receptor site.

The calculation of $G$ will now be discussed. Johnson (1967) derived a gradient expression for exponentially growing spherical cells in which the receptors of the diffusing growth-limiting nutrient are thought to be located on a shell concentric with the surface area; $r$ is the radius of the cells, $nr$ the radius of the shells, and $h$ is a constant (for fixed values of $\mu_g$, $y$, $r$ and $D_b$):

$$S_1 - S_2 = \frac{h}{r}\left(\frac{1-n}{n}\right) \tag{27}$$

The gradient in such cells at a distance $f$ from the cell centre, where $r > f > nr$, was written by Johnson (1967) as:

$$\frac{dS}{df} = \frac{h}{f^2} \tag{28}$$

From this it follows (Johnson, 1967) that the gradient at the surface of such cells is:

$$\frac{dS}{df} = \frac{h}{r^2} \tag{29}$$

For unit concentration difference between the cell surface and receptor site, eq. (27) yields the following expression for $h$:

$$h = \frac{rn}{1-n} \tag{30}$$

Substitution of eq. (30) in eq. (29) produces an expression for $G$, the concentration gradient across the cell surface for unit concentration difference between the medium and the receptor site, i.e. the gradient needed for the calculation of $D_{sp}$:

$$G = \frac{n}{(1-n)r} \tag{31}$$

If the funnelling of diffusion that occurs in spherical cells were not taken into account, $G$ would be written simply as:

$$G = \frac{1}{(1-n)r} \tag{32}$$

i.e. unit concentration difference between the surface and the receptor site divided by the distance from the surface to that site.

Calculating $G$ by the use of eq. (32) will give values $1/n$ times higher than values obtained with eq. (31). When $n$ approaches unity, i.e. when the receptor site is very near the cell membrane, the two equations yield closer values. By substituting eq. (31) in eq. (26), we obtain a practicable expression for the specific diffusion coefficient:

$$D_{sp} = 4.10^{-3}\, r \frac{n}{1-n} M_w N D_b \tag{33}$$

Now we return to eq. (25) and substitute in eq. (11):

$$\mu_g = y D_{sp}(S_1 - S_2) \tag{34}$$

Exponential growth of *Candida utilis*, limited by glycerol (Gancedo *et al.*, 1968) or oxygen (Button and Garver, 1966; Johnson, 1967) are instances of diffusion transfer of the limiting nutrient. In both cases, the receptor enzymes (glycerol kinase and cytochrome oxidase) catalyse virtually irreversible steps. If we assume that the irreversible step follows "first order" Michaelis-Menten kinetics with respect to the limiting nutrient, we may write:

$$\mu_g = yk_{\max_2} \frac{S_2}{K_{m_3} + S_2} \tag{35}$$

As was shown by van Uden (1969b), combination of eq. (35) with eq. (34) leads to:

$$\mu_g = yk_{\max_2} \frac{S_1}{K_{ap} + S_1} \tag{36}$$

where

$$K_{ap} = K_{m_3} + \frac{k_{\max_2}}{D_{sp}} \frac{K_{m_3}}{K_{m_3} + S_2} \tag{37}$$

Application (van Uden, 1969b) of eqs. (36) and (34) to experimental data on oxygen-limited growth of *Candida utilis* (Button and Garver, 1966; Johnson, 1967) revealed that $K_{m_3}$ and $K_{ap}$ were of the same order of magnitude and that the diffusion step had only very slight control over the specific growth rate.

## V. Dependence of the Specific Growth Rate on Temperature

### A. Arrhenius Relations

Up to the optimum temperature for growth ($T_{op}$), the specific growth rate is a more or less approximate Arrhenius function of the absolute temperature (Fig. 4).

$$\mu_g = A\mathrm{e}^{-E/(RT)} \tag{38}$$

This relationship implies that the dependence of the specific growth rate on the temperature is formally similar to that of rate constants of chemical reactions. Consequently, eq. (38) may be rewritten according to the theory of absolute reaction rates (Eyring, 1935; Stearn, 1949):

$$\mu_g = \frac{kT}{h} \mathrm{e}^{\Delta S^{\ddagger}/R} \mathrm{e}^{-\Delta H^{\ddagger}/RT} \tag{39}$$

and estimates of growth-rate parameters may be obtained which are, at least formally, analogues of the entropy ($\Delta S^{\ddagger}$) and the heat ($\Delta H^{\ddagger}$) of activation of a chemical reaction.

At temperatures above $T_{op}$, $\mu_g$ drops off with increasing sharpness till the maximum temperature ($T_{max}$) is reached, above which no growth occurs. The establishment of an optimum temperature for growth and the sharp decline of the $\mu_g$ values in the superoptimal temperature range require the simultaneous occurrence of *at least* two opposite processes: (i) a constructive one (i.e. biosynthesis leading to growth) with a relatively low heat of activation; and (ii) a destructive one with a relatively high heat of activation. Growth rate equations have been proposed based on different concepts of the destructive process. Hinshelwood (1946) derived an equation based on the assumption that the destructive process is irreversible, while an equation proposed by Johnson *et al.* (1954) implies that the specific growth rate in the superoptimal zone is limited by the concentration of the native (enzymically active) form of a key protein

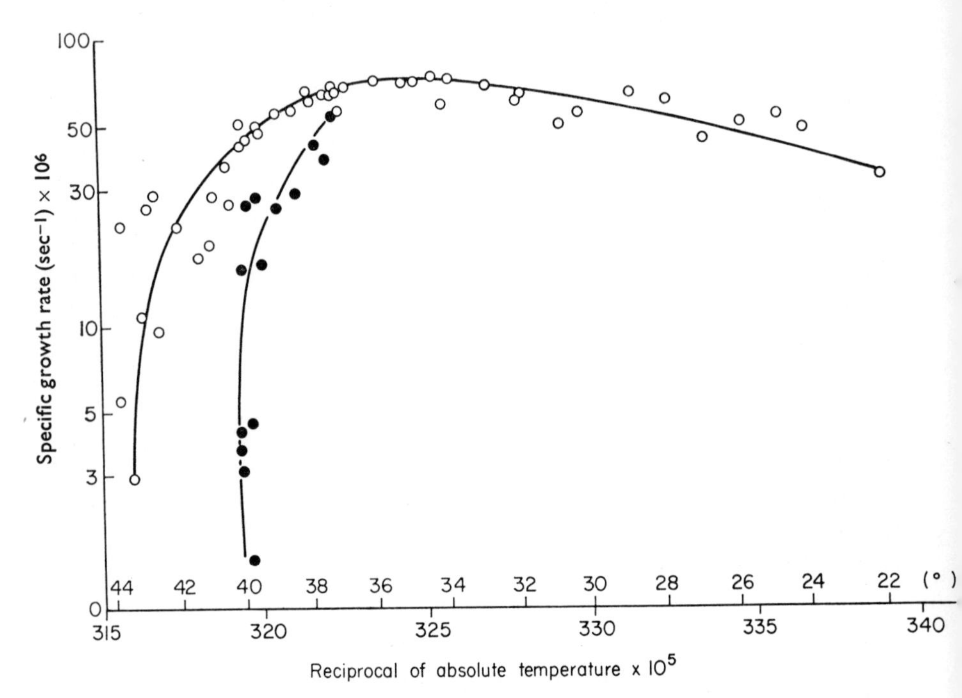


FIG. 4. Arrhenius plot of the maximum specific growth rates of a respiration-deficient mutant of *Saccharomyces cerevisiae*. The yeast was grown in batch cultures in glucose-mineral medium. Data of van Uden and Madeira-Lopes (1970).

which is in thermodynamic equilibrium with its reversibly denatured form. In the equilibrium state:

$$\frac{E_D}{E_N} = K \tag{40}$$

from which it follows that:

$$E_N = \frac{E}{1+K} \tag{41}$$

where $E = E_N + E_D$.

Following Johnson *et al.* (1954), the specific growth rate may now be expressed as follows:

$$\mu_g = C' \frac{E}{1 + K} k' \tag{42}$$

where $k'$ is the velocity constant of the rate-limiting step catalysed by the enzyme, and $C'$ is a proportionality constant.

When $k'$ and $K$ are written as their respective temperature functions and some of the constants are lumped together with $C'$ of eq. (42), the following equation is obtained (Johnson *et al.*, 1954):

$$\mu_g = \frac{CT\mathrm{e}^{-\Delta H^{\ddagger}/RT}}{1 + \mathrm{e}^{\Delta S^0/R}\,\mathrm{e}^{-\Delta H^0/RT}} \tag{43}$$

When suitable parameters are chosen, this equation fits some experimental results rather well, both in the suboptimal range (where the total key enzyme is virtually in the native state) and the superoptimal range (where, with increasing temperature, an increasing fraction of the enzyme is reversibly denatured). Only at temperatures very near $T_{max}$ do the observed $\mu_g$ values usually drop more precipitously than the theoretical curve.

There is mounting evidence that one of the principal postulates of the Johnson equation—the reversible heat denaturation of key enzymes as the dominant factor of the kinetics of growth in the superoptimal temperature range—is correct (Brandts, 1967). Nevertheless, the equation is qualitative and tentative, and is based on a number of highly simplified concepts most of which have been clearly acknowledged as such by its authors (Johnson *et al.*, 1954). Four of these concepts will be discussed presently:

1. The equilibrium constant of eqs. (40), (41) and (42) is substituted in eq. (43) in terms of its temperature function:

$$K = \mathrm{e}^{-\Delta F^0/RT} \tag{44}$$

where

$$\Delta F^0 = \Delta H^0 - T\Delta S^0 \tag{45}$$

In the case of reversible heat denaturation of proteins, however, the standard free energy change is *not* a linear function of the absolute temperature (see Brandts, 1967) and eq. (45) becomes a polynomial of a degree higher than one.

2. In a growing population, reversible chemical reactions—including reversible protein denaturation—do not normally attain thermodynamic equilibrium. This aspect will be discussed in Section V.B (p. 91).

3. The rate constant of eq. (42), which is substituted by its temperature function in eq. (43), is in reality composed of several constants and

variables. Some insight into this aspect may be obtained by studying the model developed in Section VII (p. 109).

4. The specific growth rate is only very rarely limited by a single step ("master reaction") catalysed by a "key enzyme". Johnson *et al.* (1954) were fully aware of the non-tenability of the concept of a "master" reaction. Recent work on this fundamental problem is discussed in Section VII.B (p. 114).

### B. STEADY-STATE CONCENTRATIONS OF NATIVE ENZYMES

Let us consider a steady-state cell population growing in a chemostat. In such a system, the concentration of all cell components and products is constant while the rate of their formation equals the rate of outflow:

$$\frac{\mathrm{d}x_i}{\mathrm{d}t} = \mu_g x_i \tag{46a}$$

$$-\frac{\mathrm{d}x_i}{\mathrm{d}t} = Dx_i \tag{46b}$$

$$\mu_g x_i = Dx_i \tag{46c}$$

where $\mu_g = D$, and $x_i$ (the concentration in the medium of any cell constituent or cell product) is constant.

The steady-state concentration of the native form, $E_N$, of any enzyme is established through the simultaneous occurrence of various processes: (1) The tendency toward thermodynamic equilibrium according to:

$$E_N \underset{k_2}{\overset{k_1}{\rightleftharpoons}} E_D \tag{47a}$$

from which it follows that:

$$\frac{\mathrm{d}E_N}{\mathrm{d}t} = k_2 E_D - k_1 E_N \tag{47b}$$

Since the steady-state concentration, $E$, of total enzyme is constant, we may rewrite eq. (47b):

$$\frac{\mathrm{d}E_N}{\mathrm{d}t} = k_2 E - (k_1 + k_2)E_N \tag{47c}$$

(2) *De novo* synthesis of enzymes on the ribosomes in agreement with eq. (46a):

$$\frac{\mathrm{d}E}{\mathrm{d}t} = \mu_g E \tag{48a}$$

If $p$ is the probability that a newly formed polypeptide with the primary structure of enzyme $E$ takes the configuration of the native form, any increase of the latter due to *de novo* synthesis may be expressed as follows:

$$\frac{dE_N}{dt} = p\mu_g E \tag{48b}$$

(3) Outflow in accordance with eq. (46b):

$$-\frac{dE_N}{dt} = DE_N \tag{49}$$

(4) Irreversible denaturation. We shall assume here that the contribution of this process is negligible.

Combining eqs. (47c), (48b) and (49) in such a way that eqs. (46a, b and c) are satisfied, we may write the following equality valid for the chemostat:

$$k_2 E - (k_1 + k_2)E_N + p\mu_g E = DE_N \tag{50a}$$

Since, in the steady state $D = \mu_g$ providing no cell death occurs, we obtain by reworking eq. (50a) the following expression for the steady-state fraction of the total enzyme present in the native form:

$$E_N = \frac{p\mu_g + k_2}{\mu_g + k_1 + k_2} E \tag{50b}$$

Similarly we obtain:

$$E_D = \frac{(1 - p)\mu_g + k_1}{\mu_g + k_1 + k_2} E \tag{50c}$$

From eqs. (50b) and (50c) follows the steady-state equation:

$$\frac{E_D}{E_N} = \frac{(1 - p)\mu_g + k_1}{p\mu_g + k_2} \tag{51}$$

Let us consider three situations: (1) When the specific growth rate, $\mu_g$, is zero (for example with resting cells or a cell-free enzyme solution), eq. (51) reduces to eq. (40), i.e. thermodynamic equilibrium. (2) For one value of $p$,

$$p = \frac{k_2}{k_1 + k_2} \tag{52}$$

the steady-state equilibrium becomes identical with the thermodynamic equilibrium, as can be seen by substituting this value of $p$ in eq. (51). (3) In all other cases (i.e. a growing cell population in the steady state with $p \neq k_2/(k_1 + k_2)$) the steady-state equation has a value different from the thermodynamic equilibrium constant.

It is not fruitful to try introducing eq. (50b) in the Johnson equation. It is perfectly feasible, however, to use the former equation in models of growing populations designed for computer simulation. The basic truth underlying this statement is that the dependence of such a complex variable as the specific growth rate on any independent variable (concentration of a nutrient, temperature, pH value) cannot normally be expressed in terms of a single explicit and workable equation.

### C. CONCURRENT EXPONENTIAL DEATH

It was found in a strain of *Saccharomyces cerevisiae* that, in the superoptimal temperature range, exponential population death concurs with

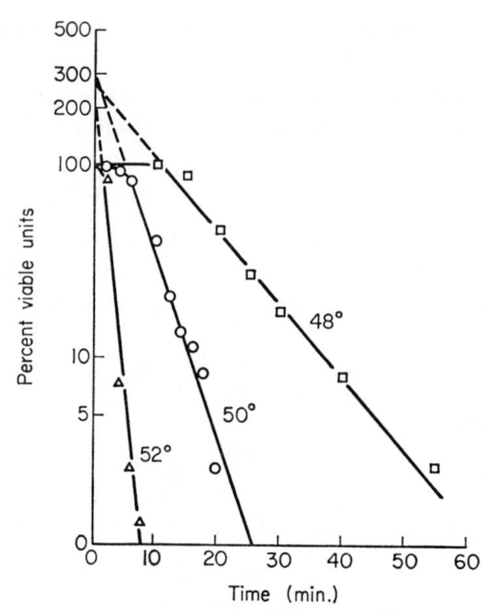


FIG. 5. Semilogarithmic survival curves of *Saccharomyces cerevisiae* exposed to various temperatures. After van Uden *et al.* (1968).

exponential population growth (van Uden and Madeira-Lopes, 1970). Subsequently the same mechanism was observed in a number of other yeast species (Oliveira-Baptista and van Uden, 1970). Exponential death sets in after an initial period of deathless growth, the duration of which is itself a function of temperature. During the second period, the observed specific growth rate is the difference between the specific growth rate of the initial period and the specific death rate. Thus two sets of superoptimal $\mu_g$ values can be obtained. The Arrhenius plots will then show two branches in the super-optimal temperature range and two values of $T_{max}$ (Fig. 4, p. 92). For a better understanding of this phenomenon, some aspects of death kinetics must now be discussed.

When a homogeneous cell suspension is submitted to a supermaximal temperature (i.e. a temperature above $T_{max}$) and the logarithm of the fraction of surviving cells is plotted against time, often a graph of the type depicted in Fig. 5 is obtained. For a discussion of other types of survival curves, see Chapter 4 (p. 126). A model fitting plots of the type shown in Fig. 6 was developed by Johnson *et al.* (1954) and Wood (1956). Since the underlying theory also satisfies the observations on concurrent exponential death in the superoptimal temperature range, it will be outlined below.

The following assumptions are made: (i) each cell contains $m$ thermosensitive sites; (ii) thermal inactivation of the sites follows first order

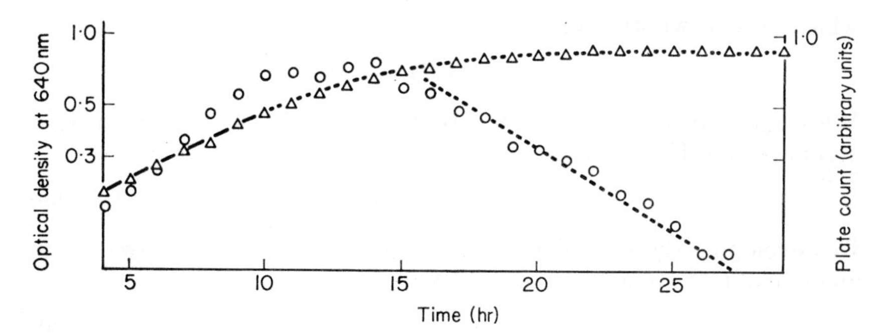


FIG. 6. A semilogarithmic plot of optical density (△) and viable count (○) for a respiration-deficient mutant of *Saccharomyces cerevisiae* growing in liquid medium at a superoptimal temperature. After van Uden and Madeira-Lopes (1970).

kinetics; (iii) death of a cell occurs when all of its $m$ sites have been inactivated.

In accordance with assumption (ii), we can write:

$$-\frac{dS}{dt} = \mu_d S \tag{53}$$

where $S$ is the number of native sites per unit volume of cell suspension. After integration we obtain an expression for the fraction of native (surviving) sites after time $t$:

$$\frac{S_t}{S_0} = e^{-\mu_d t} \tag{54}$$

The probability that a given site belongs to this fraction is given by the same expression, and it follows that the probability of a given site having been inactivated at time $t$ is:

$$p = 1 - e^{-\mu_d t} \tag{55}$$

In accordance with assumptions (i) and (iii), the cells surviving after any time $t$ belong to the cell classes that still have 1, 2, 3 ... $m$ native sites. Since each cell is supposed to have $m$ sites it follows, from eq. (55), that the probability of a cell belonging to any of these classes may be written as:

$$\pi = 1 - (1 - e^{-\mu_d t})^m \tag{56}$$

Since we are dealing with very large populations, the probability of an individual cell belonging to a class is practically identical with the frequency of the class. From this it follows that:

$$\frac{N_t}{N_0} = 1 - (1 - e^{-\mu_d t})^m \tag{57}$$

which may be written as:

$$N_t = N_0 - N_0(1 - e^{-\mu_d t})^m \tag{58}$$

When survival is measured in terms of viable units, cell grouping must be accounted for. If each group contains $n$ cells, eq. (58) becomes:

$$N_t = N_0 - N_0(1 - e^{-\mu_d t})^{nm} \tag{59}$$

By developing the binomial of eq. (59), it can easily be seen that, with increase of $t$, the equation approaches:

$$N_t = nmN_0 e^{-\mu_d t} \tag{60}$$

A semilogarithmic plot of eq. (59) produces a shoulder which falls off to a straight line that corresponds to the log form of eq. (60) (Fig. 6, p. 97).

The specific inactivation rate, $\mu_d$, of the thermosensitive sites is also the specific death rate of the cells once the exponential phase of death has been entered. As with the specific growth rate, the specific death rate is an Arrhenius function of the absolute temperature, and activation parameters may be calculated. Van Uden *et al.* (1968) determined parameters of thermal death in 10 strains of yeast species whose $T_{max}$ values ranged from 22° to 49°. The positional sequence of the Arrhenius plots correlated with the $T_{max}$ values, and $\mu_d$ values extrapolated to the superoptimal temperature range were of the same order of magnitude as $\mu_g$ values of that range.

Van Uden and Madeira Lopes (1970) extended the theory of thermal death outlined above to concurrent exponential death in the superoptimal temperature range. They assumed that the thermosensitive sites are inactivated exponentially at temperatures above $T_{op}$ with a specific inactivation rate $\mu_d$, the value of which decreases exponentially with the reciprocal of the absolute temperature (Arrhenius function). In the superoptimal range, two periods of growth may be distinguished. During the first period, each cell contains more than one active site and cell population growth is governed by the specific growth rate $\mu_g$, while the increase of the population of active sites is governed by $(\mu_g - \mu_d)$:

$$S_t = S_0 \, e^{(\mu_g - \mu_d)t} \tag{61}$$

where $\mu_d$ is again the specific inactivation rate of the thermosensitive sites. If we assume again that each cell contains $m$ sites, eq. (61) may be written as:

$$S_t = mN_0 \, e^{(\mu_g - \mu_d)t} \tag{62}$$

Since the specific growth rate $(\mu_g)$ of the cell population is greater than the specific growth rate $(\mu_g - \mu_d)$ of the population of active sites, the number of active sites per cell decreases with time. After a time $t_{pr}$ the number of active sites equals the number of cells (i.e. the *average* number of active sites per cell is unity) and we may write:

$$mN_0 \, e^{(\mu_g - \mu_d)t_{pr}} = N_0 \, e^{\mu_g t_{pr}} \tag{63}$$

from which it follows that:

$$t_{pr} = \frac{1}{\mu_d} \ln m \tag{64}$$

The second period of growth, governed by $(\mu_g - \mu_d)$, sets in when each cell contains only one active site. Under the simplifying assumption that the transition from the first to the second period occurs instantaneously when the average of one active site per cell is reached, eq. (64) expresses the duration of the first period.

During the second period of growth, death concurs with growth, i.e. the populations of both viable $(N^v)$ and non-viable $(N^{nv})$ cells change with time. The following equations may be derived (for a full discussion see van Uden and Madeira-Lopes, 1970).

The viable population changes exponentially with an apparent specific growth rate equal to $(\mu_g - \mu_d)$:

$$N_t^v = N_0^v \, e^{(\mu_g - \mu_d)t} \tag{65}$$

where $N_0^v$ is the viable population at the end of the first period of growth.

Three cases may be distinguished: (i) $\mu_g > \mu_d$, there is a net exponential increase of the viable population; (ii) $\mu_g = \mu_d$, growth balances death, consequently the viable population is in a steady state (with respect to its population density); (iii) $\mu_g < \mu_d$, there is a net exponential decrease of the viable population leading eventually to its extinction. In this case there may be an early stationary phase due to cell-group formation, the duration of which may be calculated (van Uden and Madeira-Lopes, 1970).

The non-viable population increase, through exponential death of the viable population, is given by:

$$\frac{dN^{nv}}{dt} = \mu_g N_0^v e^{(\mu_g - \mu_d)t} \tag{66}$$

which on integration leads to:

$$N_t^{nv} = \frac{\mu_d}{\mu_g - \mu_d} N_0^v e^{(\mu_g - \mu_d)t} - \frac{\mu_d}{\mu_g - \mu_d} N_0^v \tag{67}$$

By adding eqs. (65) and (67), an expression is obtained for the change of the total population ($N^T$) with time:

$$N_t^T = \frac{\mu_g}{\mu_g - \mu_d} N_0^v e^{(\mu_g - \mu_d)t} - \frac{\mu_d}{\mu_g - \mu_d} N_0^v \tag{68}$$

Curves corresponding to eq. (68) may be obtained by measuring the population density photometrically with time, and parameters may then be calculated therefrom (van Uden and Madeira-Lopes, 1970). Three characteristic shapes are obtained, depending on which of the three cases apply. Figure 6 (p. 97) depicts experimental results obtained with *Saccharomyces cerevisiae* at a temperature where $\mu_g < \mu_d$ (third case). The second period of growth may escape laboratory observation due to the duration of the first period. In nature, when temperatures are above $T_{op}$, the kinetics of the second period will govern more often than not the fitness of the cell population.

## VI. Yield and Maintenance Relations

### A. YIELD FACTOR AND SPECIFIC MAINTENANCE RATE

#### 1. *Yield Factor*

The specific growth rate is related through the appropriate yield factors to the specific input, turnover and output rates of nutrients and metabolites. The word "yield" has here an amplified sense in that it relates biomass production not only to nutrient consumption but also to metabolite production. The yield factor in the fundamental eq. (11) relates the specific growth rate to the specific input (transfer) rate of the limiting nutrient. Often this yield factor varies with the specific transfer rate, i.e. at different specific growth rates (or dilution rates in the chemostat) the yield factor takes different values (Postgate and Hunter, 1962; Postgate, 1965). Growth rate-dependent yield factor variation may be due to various mechanisms. Three such mechanisms will now be exemplified.

The composition of microbial biomass together with the cell size is generally a function of the specific growth rate (Herbert, 1961). With increasing specific growth rate the RNA content increases markedly, the protein content decreases somewhat, the cell size increases and, concomitantly, the DNA content decreases. There is an extensive literature on growth rate-dependent biomass variation in bacteria; yeasts appear to follow the same underlying laws (McMurrough and Rose, 1967; Fig. 7). Changes in the RNA content with specific growth rate can largely be accounted for in terms of cellular ribosome content (for references, see Tempest, 1969). On the other hand, there is a direct relationship between magnesium content and ribosome content in micro-organisms. In agreement with this relationship, it was found that the yield factor with respect to magnesium in magnesium-limited chemostat cultures of

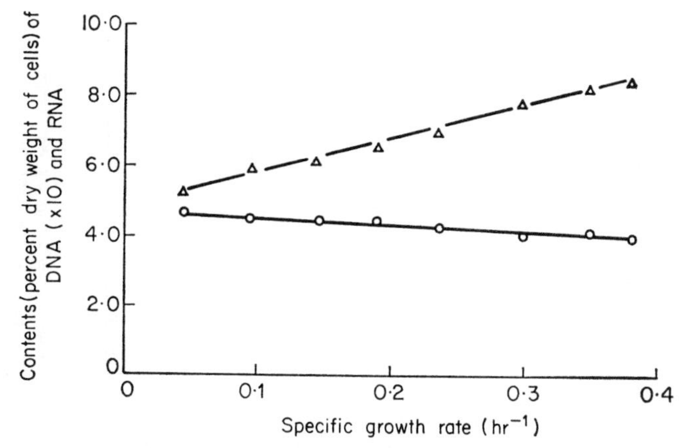


FIG. 7. RNA (△) and DNA (○) contents of the biomass as a function of specific growth rate in a chemostat culture of *Saccharomyces cerevisiae* in a glucose-limited medium. After McMurrough and Rose (1967).

*Candida utilis* and several bacterial species decreased with increasing specific growth rate (for references, see Tempest, 1967). Thus we have here an instance of yield factor variation due to growth rate-dependent variations in biomass composition.

A second type of yield-factor variation may arise when the stoichiometry of energy metabolism changes with the specific growth rate. A pertinent example is the changeable ratio between alcoholic fermentation and respiration as found in many yeast species. When such yeasts (*Saccharomyces cerevisiae* for example) grow under conditions of nutrient saturation with glucose (or another fermentable sugar) as the energy source, the ratio between the two pathways (as measured by the respiratory quotient) can be controlled externally through the oxygen tension in the medium (the Pasteur effect). Oxygen stimulates oxidative phosphorylation which leads to an increase in the size of the ATP pool and a decrease in the ADP and AMP pools. These compounds, together with citrate, regulate phosphofructokinase activity, the principal control point of the Pasteur effect. Chemostat studies have shown that a decrease in the external pool of fermentable sugar reinforces the Pasteur effect and, if the concentration in the pool is low enough, the Pasteur effect will be nearly 100% (i.e. alcoholic fermentation is virtually absent and the respiratory quotient will be unity). This finding is not surprising since the flux through the reaction catalysed by phosphofructokinase depends not only on the state of the enzyme but also on the input of metabolite. The shift from alcoholic fermentation to respiration leads to a yield-factor

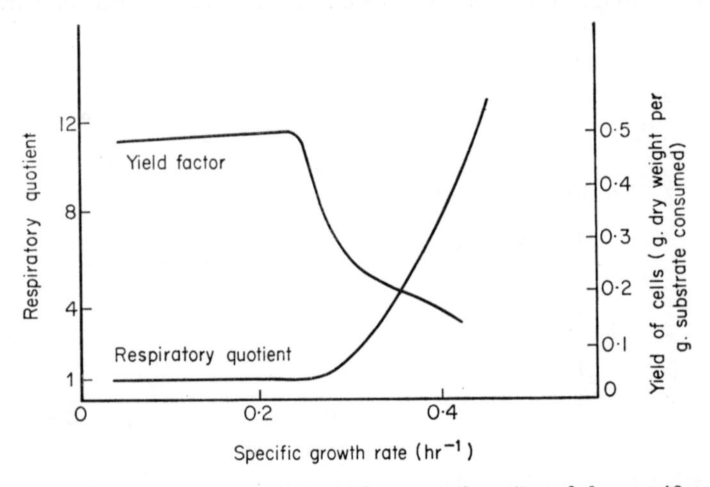


Fig. 8. Yield factor and respiration quotient as a function of the specific growth rate in a chemostat culture of *Saccharomyces cerevisiae* in a glucose-limited medium. After von Meyenburg (1969).

increase of about five-fold. For additional discussion on this subject, see Section VI.C (p. 106). Figure 8 shows the correlation between the yield-factor value and the ratio of the two pathways in a glucose-limited chemostat culture of *Sacch. cerevisiae* (von Meyenburg, 1969).

2. *Specific Maintenance Rate*

A third type of yield-factor variation, of general occurrence, is due to energy requirements for growth-independent maintenance, and may be observed in chemostat cultures when the limiting nutrient is the energy source. Maintenance energy is required for such processes as turnover of macromolecules, osmotic work, motion, and coping with adverse conditions, i.e. for functions other than the production of new biomass (Lamanna, 1963; Marr *et al.*, 1963; Schulze and Lipe, 1964; Wase and

Hough, 1966; van Uden, 1967a, 1969c). Pirt (1965) derived an equation for yield-factor variation due to maintenance. A simplified derivation (van Uden, 1969a) follows.

When the limiting nutrient is the energy source, the specific transfer rate ($k$) is the sum of two specific rates, one for growth only ($k_g$) and a second for maintenance only ($k_m$). Equation (11) may now be written as follows:

$$\mu_g = y(k_g + k_m) \tag{69}$$

If it is assumed that the specific growth rate is a linear function of the specific uptake rate of limiting nutrient used for growth only (i.e. if there are no yield-factor variations due to causes other than maintenance requirements), we may write:

$$\mu_g = Y_{\max} k_g \tag{70}$$

$Y_{\max}$, the "maximum yield factor", may also be defined as the limit of $y$ when $k_m$ tends to zero. Combination of eqs. (69) and (70) leads to:

$$y = Y_{\max} \frac{k_g}{k_m + k_g} \tag{71}$$

By multiplying the numerator and the denominator of the fraction in eq. (71) by $Y_{\max}$ we obtain:

$$y = Y_{\max} \frac{\mu_g}{a + \mu_g} \tag{72}$$

where by definition:

$$a = Y_{\max} k_m \tag{73}$$

Parameter "$a$" has thus the same dimension as the specific growth rate (reciprocal time), and may be thought of as the increase in specific growth rate that would occur if the energy required for maintenance could be channelled into growth.

Adding eqs. (70) and (73) we obtain:

$$\mu_g + a = Y_{\max} k \tag{74}$$

Equations (18), (20) and (21) are more explicit and less general forms of this fundamental relation, applicable to certain growth situations. Parameter "$a$" has been called the "specific maintenance rate" (Marr *et al.*, 1963; van Uden, 1967a) while $k_m$ is often referred to as the "maintenance coefficient" (Monod, 1950; Schulze and Lipe, 1964; Pirt, 1965).

In the chemostat, the following relation applies (van Uden, 1967a):

$$Dx + ax = Y_{\max}(S_r - S)D \tag{75}$$

from which it easily follows that:

$$y = Y_{\max} \frac{D}{D + a} \tag{76}$$

A double reciprocal plot of estimates of $y$ against experimental values of $D$:

$$\frac{1}{y} = \frac{1}{Y_{\max}} + \frac{a}{Y_{\max}} \frac{1}{D} \tag{77}$$

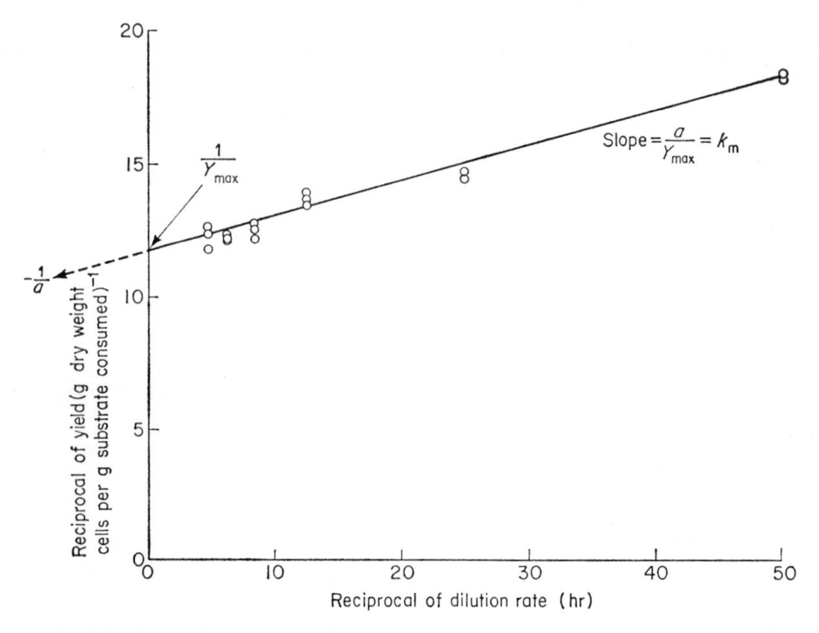


FIG. 9. Data from a chemostat culture of a respiration-deficient mutant of *Saccharomyces cerevisiae* in a glucose-limited mineral medium (van Uden, 1967b) at 30°, plotted according to eq. (77). Unpublished data of A. Madeira-Lopes and N. van Uden.

will be acceptably linear when $Y_{\max}$ and $a$ are invariant with respect to $D$, or nearly so. Other forms of eq. (77), applicable to the kinetics of product (final metabolite) formation in the chemostat, were derived by Pirt (1967).

Maintenance parameters have been estimated for a number of micro-organisms, mainly by applying eq. (77) or related forms to continuous culture data (Marr *et al.*, 1963; Schulze and Lipe, 1964; Pirt, 1965; Schuster and Schlegel, 1967; Hempfling and Vishniac, 1967). A. Madeira-Lopes and N. van Uden (unpublished observations) obtained a specific maintenance rate of 0·012 $h^{-1}$ for *Sacch. cerevisiae* in a glucose-limited chemostat culture (Fig. 9). Wase and Hough (1966), working with phenol-limited chemostat cultures of the yeast *Debaryomyces subglobosus*, obtained a much higher specific maintenance rate ($\pm 0{\cdot}08$ h$^{-1}$), presumably because phenol, though used as an energy source by that yeast, is toxic in the sense that it increases the energy-requiring turnover of cell constituents.

In the chemostat, when exponential death concurs with exponential growth and the energy source is limiting, the apparent specific maintenance rate may be high due to substrate consumption by the non-viable fraction of the population. If $f$ is the fraction of substrate consumed per unit of non-viable biomass as compared with substrate consumption per unit viable biomass, eq. (76) then takes the following form (for a derivation, see van Uden, 1969c):

$$y = Y_{\max} \frac{D(D+\mu_d)}{(D+\mu_d+a)(D+f\mu_d)} \tag{78}$$

When $f$ is zero, eq. (78) reduces to a form which permits a linear double reciprocal plot (if the specific death rate is known) from which the true specific maintenance rate may be estimated:

$$\frac{1}{y} = \frac{1}{Y_{\max}} + \frac{a}{Y_{\max}} \frac{1}{D+\mu_d} \tag{79}$$

At the other extreme, when $f$ is unity, the following linear double reciprocal plot applies:

$$\frac{1}{y} = \frac{1}{Y_{\max}} + \frac{a+\mu_d}{Y_{\max}} \frac{1}{D} \tag{80}$$

## B. YIELD AND MAINTENANCE ANALOGUES

The yield factors which relate the specific growth rate of a cell population limited by the energy source to the specific uptake, turnover and output rates of gaseous and non-gaseous metabolites pertaining to the pathways of energy metabolism, vary with $\mu_g$ (van Uden, 1968). Consider the following example:

$$\text{En} + \text{gas}_1 \rightarrow \text{Biomass} + \text{gas}_2 + \text{met} \tag{81}$$

Let this process refer to a steady-state cell population growing in a chemostat, and limited by the energy source En; $\text{gas}_1$ is taken up, while $\text{gas}_2$ and met, a non-gaseous metabolite, are produced in immediate connection with the energy metabolism. With respect to the non-gaseous metabolite, we may write in analogy to eq. (75):

$$\mu_g x + a^{\text{met}} x = Y_{\max}^{\text{met}} [\text{met}] D \tag{82}$$

where [met] is the concentration of the metabolite in the chemostat at dilution rate $D$. From this relation, a linear form is easily derived:

$$\frac{1}{y^{met}} = \frac{1}{Y_{max}^{met}} + \frac{a^{met}}{Y_{max}^{met}} \frac{1}{D} \tag{83}$$

The parameter $a^{met}$ has been called the "maintenance rate analogue" (van Uden, 1968) while $Y_{max}^{met}$ is the maximum yield-factor with respect to the metabolite. The relations of such parameters to "$a$" and $Y_{max}$ will be reviewed in Section VI.D (p. 108).

Maximum yield-factors and maintenance-rate analogues have also been defined with respect to gaseous metabolites consumed or produced in direct connection with the energy metabolism. In analogy to eq. (75), we can write:

$$\mu_g x + a^{gas} x = Y_{met}^{gas} \times \text{rate of gas output (uptake)} \tag{84}$$

By dividing both sides by $x$ and defining a maximum-yield factor, $Y_{met'}^{gas}$, with the dimensions of biomass per volume, we readily obtain a linear form applicable to the chemostat:

$$Q_{gas} = \frac{a^{gas}}{Y_{max'}^{gas}} + \frac{1}{Y_{max'}^{gas}} D \tag{85}$$

When $Q_{gas}$ data are available for a number of experimental $D$ values, estimates of $a^{gas}$ and $Y_{max'}^{gas}$ may be obtained by use of this equation; subsequently $Y_{max'}^{gas}$ can be converted into $Y_{max}^{gas}$.

### C. ATP YIELDS

Maximum-yield factors and maintenance-rate analogues may also be defined with respect to intracellular metabolites and relations established between the specific growth rate and the specific turnover rate of such metabolites. Interest in these relations has been mainly concerned with ATP (Bauchop and Elsden, 1960; Senez, 1962; Hadjipetrou *et al.*, 1964; Hernandez and Johnson, 1967; von Meyenburg, 1969). The ATP yield is usually defined as the number of grams of dry biomass produced per mole of (calculated) ATP, and the specific turnover rate as the number of moles of ATP produced per gram of dry biomass per hour.

Bauchop and Elsden (1960) grew several micro-organisms anaerobically in complex media with limited energy source. Under these conditions, virtually the total energy source was fermented (i.e. there was no assimilation and no respiration) and the ATP yield could easily be calculated based on a knowledge of the fermentative pathways and measured yields with respect to the energy source. Forrest (1969) has tabulated anaerobic ATP yields obtained by Bauchop and Elsden (1960) and by other authors. The results obtained with a wide variety of substrates with a number of bacteria and yeasts were similar (the maximum value of $Y_{ATP}$ was $10{\cdot}3 \pm 0{\cdot}3$ g per mole for 38 determinations). The following anaerobic ATP yields have been measured in yeasts with glucose as the energy source: *Saccharomyces cerevisiae*, $10{\cdot}0 \pm 0{\cdot}2$ (Bauchop and Elsden, 1960; Battley, 1960; Bulder, 1963); *Sacch. rosei*, 11·6 (Bulder, 1963, 1966).

The aerobic yield-factor of *Sacch. cerevisiae* with respect to glucose in a glucose–vitamins–mineral medium is about five times higher than the anaerobic yield-factor measured in a respiratory-deficient mutant of the same strain (van Uden, 1967b). This ratio is lower than would be expected from calculated ATP yields. Similar inconsistencies have been found in other micro-organisms (Forrest, 1969) and have been explained in terms of inefficient oxidative phosphorylation (i.e. P/O ratios lower than three; Chen, 1964; Hernandez and Johnson, 1967; Stouthamer, 1969).

Von Meyenburg (1969) approached the problem using gas exchange data from glucose-limited chemostat cultures of a strain of *Saccharomyces cerevisiae*; a short outline of the theory underlying his approach follows.

In analogy with eq. (70), we can write:

$$\mu_g = Y_{\max}^{\mathrm{ATP}} Q_g^{\mathrm{ATP}} \tag{86}$$

where $Y_{\max}^{\mathrm{ATP}}$ is the maximum ATP yield (the number of grams of dry biomass formed per mole of ATP used for growth) and $Q_g^{\mathrm{ATP}}$ is the molar ATP turnover rate for growth only (the number of moles of ATP utilized per gram of dry biomass per hour for growth). Using the following relation:

$$Q_g^{\mathrm{ATP}} = Q^{\mathrm{ATP}} - Q_m^{\mathrm{ATP}} \tag{87}$$

where $Q^{\mathrm{ATP}}$ is the total molar ATP turnover rate and $Q_m^{\mathrm{ATP}}$ the molar ATP turnover rate for maintenance, we arrive by substitution in eq. (86) at the following expression (von Meyenburg, 1969):

$$Y_{\max}^{\mathrm{ATP}} = \frac{D}{Q^{\mathrm{ATP}} - Q_m^{\mathrm{ATP}}} \tag{88}$$

where $D = \mu_g$ (in the steady state).

$Q^{\mathrm{ATP}}$ and $Q_m^{\mathrm{ATP}}$ may be estimated for each value of $D$ using the pertinent experimental oxygen and carbon dioxide exchange data and taking into account substrate-level phosphorylation and oxidative phosphorylation, the ratio of which changes with dilution rate (for details see von Meyenburg, 1969).

If it is assumed that $Y_{\max}^{\mathrm{ATP}}$ is invariant with respect to the specific growth rate, a plot of eq. (88) should give a straight line parallel to the base line provided that $Q^{\mathrm{ATP}}$ and $Q_m^{\mathrm{ATP}}$ are estimated correctly. Plotting eq. (88) for different assumed values of the P/O ratios in the calculation of the ATP formation rates, von Meyenburg (1969) obtained such a straight line for a P/O ratio of about 1·1.

This low efficiency of oxidative phosphorylation in growing yeast was also observed in resting preparations (Lynen and Koenigsberger, 1951; Chance, 1959a, b) and has been explained in terms of high rates of ATP wastage through dephosphorylation not linked to biosynthesis or work. Using his graphic method, von Meyenburg (1969) obtained an ATP yield of 12·0 for *Saccharomyces cerevisiae* growing aerobically in a chemostat with glucose as the limiting substrate. It should be understood that this yield corresponds to $Y_{max}^{ATP}$, i.e. the number of grams of biomass formed per mole ATP actually available for chemical and physical cell work (extrapolated to zero maintenance). It is only slightly higher than the ATP yield of about 10 calculated for anaerobic yeast. This may mean that wasteful dephosphorylation of ATP formed by substrate-level phosphorylation is of minor importance in the energy economy of the yeast cell.

### D. YIELD AND MAINTENANCE ANALYSIS

The chemostat is ideally suited for obtaining estimates of yield and maintenance parameters of cell populations limited by the energy source (Herbert, 1958; van Uden, 1968). By relating the parameters to each other, some information may be obtained on the stoichiometry of energy metabolism of growing cell populations (van Uden, 1968).

The biological significance of the yield and maintenance parameters and their interrelations can be readily appreciated by considering an example. Let the following equations refer to a steady-state cell population limited by the energy source and growing in a chemostat. Equation (89) refers to consumption of the energy source (En) for growth only; eq. (90) to consumption for maintenance only.

$$1\,\text{g En} + n_1\,\text{g gas}_1 \rightarrow m\,\text{g biomass} + n_2\,\text{g gas}_2 + n_3\,\text{g met} \tag{89}$$

$$1\,\text{g En} + r_1\,\text{g gas}_1 \rightarrow r_2\,\text{g gas}_2 + r_3\,\text{g met} \tag{90}$$

Since $Y_{max} = m$, $Y_{max}^{gas_1} = m/n_1$, $Y_{max}^{gas_2} = m/n_2$ and $Y_{max}^{met} = m/n_3$, it follows that:

$$\frac{Y_{max}}{Y_{max}^{gas_1}} = n_1, \quad \frac{Y_{max}}{Y_{max}^{gas_2}} = n_2 \quad \text{and} \quad \frac{Y_{max}}{Y_{max}^{met}} = n_3.$$

In general form these relations may be expressed as follows:

$$\frac{Y_{max}}{Y^{gas(met)}} = n \tag{91}$$

where $n$ indicates the mass of gaseous or non-gaseous metabolite consumed or produced per unit mass of energy source consumed for growth only.

Using the defining equations of the specific maintenance rate and of the maintenance-rate analogues (eq. 73), we can write: $a = Y_{max} k_m$, $a^{gas_1} = Y_{max}^{gas_1} k_m^{gas_1}$, $a^{gas_2} = Y_{max}^{gas_2} k_m^{gas_2}$ and $a^{met} = Y_{max}^{met} k_m^{met}$.

From eq. (90) it follows that:

$$\frac{k_m^{gas_1}}{k_m} = r_1, \quad \frac{k_m^{gas_2}}{k_m} = r_2 \quad \text{and} \quad \frac{k_m^{met}}{k_m} = r_3$$

and thus we have:

$$\frac{a^{gas_1}}{a} = \frac{r_1}{n_1}, \quad \frac{a^{gas_2}}{a} = \frac{r_2}{n_2} \quad \text{and} \quad \frac{a^{met}}{a} = \frac{r_3}{n_3}$$

More generally, these relations can be expressed as follows:

$$n \frac{a^{gas(met)}}{a} = r \qquad (92)$$

where $r$ indicates the mass of gaseous or non-gaseous metabolite consumed or produced per unit mass of energy source consumed for maintenance only.

In this way, values for the mass coefficients of the $n$ and $r$ series may be obtained. These coefficients can easily be transformed into molar coefficients and it becomes possible to write separate stoichiometric equations for growth and maintenance. The method has been applied to glycerol-limited growth of *Aerobacter aerogenes* (van Uden, 1968).

## VII. A Model

### A. DESCRIPTION

The model which I present in the following pages is concerned with a theoretical cell population limited by the energy source and growing in a chemostat. It is a highly simplified model designed for the following limited purposes: (i) to demonstrate that it is fruitful to understand a growing cell population as a multi-step system, particularly for the study of the specific growth rate as a function of the concentration of the limiting nutrient; (ii) to elucidate some basic aspects of yield and maintenance relations; (iii) to facilitate insight into the complexity of rate-limiting steps.

The more severe simplifications are: (i) the theoretical cell population is treated as a single homogeneous unit of biomass in balanced exponential growth; (ii) the composition of the biomass is held to be invariant with respect to the specific growth rate; (iii) the highly complex reaction system of real cells is reduced to a simple sequence of one transport step and six enzymic steps, having one biosynthetic branch, one energy-generating branch and one control mechanism, and leading from one external substrate to biomass and one metabolic product. Furthermore, there are provisions for pool leakage and maintenance.

The last of these simplifications implies that the model does not provide for biosynthetic steps in which at least one of the substrates and products is a macromolecular species or structure (such as synthesis of DNA, RNA, polysaccharide or protein, and organelle assembly). Maynard Smith (1969) recently discussed the possible rate-limiting weight of such steps with respect to the maximum specific growth rate. However, in heterotrophic organisms capable of growing in mineral media with an organic source for carbon and energy (and very few obligatory growth factors or none at all), the maximum specific growth rate increases in general when the medium is enriched with preformed organic

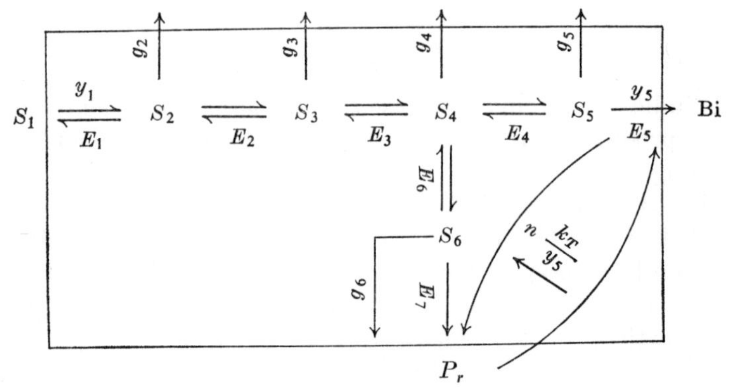


FIG. 10. Diagram of the model described in text (see p. 110).

nutrients (such as vitamins, amino acids). This implies that biosynthetic pathways involving only low-molecular weight compounds and which lead to such cell constituents may have considerable, if not total, rate-limiting weight when the population is growing in media lacking these constituents.

It is believed that the following description is detailed enough for mounting the model on one of the existing programmes for computer simulation of multi-enzyme systems (Garfinkel, 1969; Burns, 1969). The programme developed by Burns (1968, and personal communication) is particularly well suited for accommodating the model and extensions thereof. It calculates steady-state rates, pool sizes and sensitivity coefficients. The drawback is that the programme is rather specifically designed for the computer facilities of the University of Edinburgh.

In Fig. 10 the essential features of the model are depicted. The main features are: (1) the rectangle comprises one unit of biomass; (2) new biomass, Bi, is formed continuously by the unit and is excluded from it; (3) $S_1$ is the external steady-state concentration of the energy source (i.e. its concentration in the chemostat); (4) $P_r$ is a metabolic product which is excluded from the unit. Its concentration in the chemostat does not affect the kinetics of the model as long as the step leading to it is given unidirectional kinetics; (5) $S_2$, $S_3$, $S_4$, $S_5$ and $S_6$ represent kinetically effective concentrations of metabolites in the unit of biomass; (6) $E_1$ is the mass of the transport system in the unit of biomass; $E_2$, etc. have the same meaning with respect to the enzymes. $T$, the turnover number as used in the kinetic equations, has the meaning of mass of substrate processed unidirectionally per unit mass of enzyme (or transport system) per unit time under conditions of substrate saturation, and has therefore the dimensions of reciprocal time; (7) $y_1$ and $y_5$ are yield factors with respect to $S_1$ and $S_5$ respectively.

1. *Kinetics*

(1) All rates ($F$) are specific rates, i.e. they have the dimension of reciprocal time.

(2) The transport step follows compartmented Michaelis-Menten kinetics (compare with eqs. 12, 15, 16, 17):

$$F_{1,2}^{\text{net}} = F_{1,2}^{\max(\text{un})} \frac{S_1}{K_{1,2} + S_1} - F_{2,1}^{\max(\text{un})} \frac{S_2}{K_{2,1} + S_2} \tag{93}$$

Subscripts and symbols have the meaning as exemplified below:

$F_{1,2}^{\text{net}}$ = Net specific rate from pool 1 to pool 2

$F_{2,1}^{\max(\text{un})}$ = Maximum unidirectional specific rate from pool 2 to pool 1

$K_{1,2}$ = Michaelis constant of $S_1$ with respect to the system that mediates the step from $S_1$ to $S_2$

If it is assumed that $F_{1,2}^{\max(\text{un})} = F_{2,1}^{\max(\text{un})}$, we may write:

$$F_{1,2}^{\text{net}} = F_{1,2}^{\max(\text{un})} \left( \frac{S_1}{K_{1,2} + S_1} - \frac{S_2}{K_{2,1} + S_2} \right) \tag{94}$$

or when the programme of Burns (1968) is used for mounting the model:

$$F_{1,2}^{\text{net}} = E_1 T_1 \left( \frac{P_{1,2} S_1}{1 + P_{1,2} S_1} - \frac{P_{1,2} S_1}{1 + P_{2,1} S_1} \right) \tag{95}$$

where $T_1 = T_{1,2} = T_{2,1}$ and $P_{1,2} = 1/K_{1,2}$, etc.

The assumption implies that the Haldane relationship of the transport step reduces to:

$$K_{\text{eq}} = \frac{K_{2,1}}{K_{1,2}} \tag{96}$$

As long as $K_{1,2} = K_{2,1}$ the transport step does not require metabolic energy. Computer experiments for kinetic purposes only, where $K_{1,2} < K_{2,1}$ (active transport) may be performed without a special provision for metabolic transport energy.

(3) The steps catalysed by enzymes $E_2$, $E_3$, $E_4$ and $E_6$ are reversible one substrate-one product reactions with "first order" Michaelis-Menten kinetics (compare with eqs. 13, 14):

$$F_{i,j}^{\text{net}} = F_{i,j}^{\max(\text{un})} \frac{S_i}{K_{i,j} + \dfrac{K_{i,j}}{K_{j,i}} S_j + S_i} - F_{j,i}^{\max(\text{un})} \frac{S_j}{K_{j,i} + \dfrac{K_{j,i}}{K_{i,j}} S_i + S_j} \qquad (97)$$

If the programme of Burns (1969) is used this equation is conveniently written as:

$$F_{i,j}^{\text{net}} = \frac{P_{i,j} E_i T_{i,j} (S_i - S_j K_{\text{eq}})}{1 + P_{i,j} S_i + P_{j,i} S_j} \qquad (98)$$

where:

$$K_{\text{eq}} = \frac{T_{j,i} P_{j,i}}{T_{i,j} P_{i,j}} \quad \text{and} \quad P_{i,j} = \frac{1}{K_{i,j}} \quad \text{etc.}$$

(4) The steps catalysed by $E_5$ (leading to new biomass) and $E_7$ (leading to product) are conveniently written as unidirectional steps:

$$F_{5,\text{Bi}}^{\text{net}} = F_{5,\text{Bi}}^{\max(\text{un})} \frac{S_5}{K_{5,\text{Bi}} + S_5} \qquad (99)$$

(5) $\text{Step}_{6,P_r}$ generates metabolite energy; one arbitrary "unit" of energy for each mass unit of $S_6$ transformed into $P_r$. This energy is used in two processes: (a) for each mass unit of $S_5$ transformed into Bi, $n$ units of energy are used; (b) the permanent unit of biomass requires maintenance energy. Let us define $k_T$ as the specific turnover rate of biomass. The following assumptions are made: no material from metabolite pools is required for maintaining the biomass unit in good repair (i.e. it is continuously resynthesized from its own breakdown products); the amount of energy for resynthesizing the unit from its breakdown products is the same as the amount needed for synthesizing one unit of biomass from $S_5$ (and from the other building blocks not provided for in the model).

By definition we have:

$$\mu_g = y_5 F_{5,\text{Bi}}^{\text{net}} \qquad (100)$$

This implies that, for each unit of new biomass formed, $1/y_5$ units of $S_5$ are transformed and thus $n/y_5$ units of energy consumed. Consequently, with a specific turnover rate of $k_T$, the energy consumption for maintenance is $k_T n/y_5$. Since both energy requirements depend on energy generated by $\text{step}_{6,P_r}$ and since no energy is wasted in the model, this step must be controlled in such a way that the following relationship applies:

$$F_{6,P_r} = n\left(\frac{k_T}{y_5} + F_{5,\text{Bi}}\right) \tag{101}$$

The best way for building this control mechanism into the model is somewhat programme dependent. Figure 10 suggests a possible solution as may be used in the Burns programme (J. A. Burns, personal communication): the biosynthetic branch is linked to the energy-generating branch by an energy carrier cycle; for draining away maintenance energy, the cycle has an internal short-cut with zero order kinetics equal to $nk_T/y_5$.

(6) The symbol $g$ represents pool leakage into new biomass and equals $\mu S_i$. By substituting eq. (100) in this expression we obtain:

$$g = y_5 F^{\text{net}}_{5,\text{Bi}} S_i \tag{102}$$

where concentration $S_i$ is expressed as mass per unit biomass.

## 2. *Simulation Experiments*

Once the model is mounted, many different experiments may be executed. The following examples were chosen because they are relevant to some of the relations discussed in the present chapter; it is presumed that the model is mounted in the Burns (1968) programme or a modification thereof.

Give convenient values to the concentrations of the enzymes and the transport system, turnover numbers, Michaelis constants ($P$ values), equilibrium constants, $y_5$, $n$ and $k_T$. $S_1$ is an independent variable. $P_r$ and Bi can be ignored as long as the steps leading to them have unidirectional kinetics.

For different values of $S_1$, find steady-state pools, fluxes and sensitivity coefficients (see Section VII.B, p. 114). Calculate corresponding $\mu_g(=D)$ values from:

$$\mu_g = y_5 F_{5,\text{Bi}} \tag{103}$$

and yield factors with respect to $S_1$ from:

$$y_1 = y_5 \frac{F^{\text{net}}_{5,\text{Bi}}}{F^{\text{net}}_{1,2}} \tag{104}$$

Plot $\mu_g/y_1$ against $S_1$; a saturation curve should appear (see eq. 16).

Plot $y_1/\mu_g$ against $1/S_1$. By choosing convenient parameters two cases may be demonstrated:

(a) in general the plot is not a straight line (see eq. 16);

(b) when the backflux ($F_{2,1}$) is kinetically insignificant the double reciprocal plot (see eq. 22) will give a straight line:

$$\frac{y_1}{\mu_g} = \frac{1}{F^{\max}_{1,2}} + \frac{K_{1,2}}{F^{\max}_{1,2}} \frac{1}{S_1} \tag{105}$$

Compare these plots with the double reciprocal plot of the Monod equation ($1/\mu_g$, $1/S_1$).

Plot $1/y_1$ against $1/D(=1/\mu_g)$. This should produce a straight line (see eq. 77):

$$\frac{1}{y_1}=\frac{1}{Y_{\max}}+\frac{a}{Y_{\max}}\frac{1}{D} \tag{106}$$

Verify that in the model the following relation applies:

$$a=\frac{n}{n+1}k_T \tag{107}$$

B. THE SENSITIVITY COEFFICIENT

Using the symbolism of the model:

$$\mu_g=y_1(F_{1,2}-F_{2,1})=y_2(F_{2,3}-F_{3,2})=\ldots y_v(k_{vw}-k_{wv}) \tag{108}$$

Each of the specific rates is mediated either by a transport system or an enzyme and has corresponding kinetics.

Let us consider the following case:

$S_1$ (and all other) nutrients are present in saturating concentrations (see eq. 17):

$$\mu_g=y_1\left(F_{1,2}^{\max}-F_{1,2}^{\max}\frac{S_2}{K_{2,1}+S_2}\right) \tag{109}$$

Further increase of the concentration of $S_1$ will not increase the specific growth rate because $F_{1,2}^{\max}$, the specific influx rate of $S_1$, has already reached its maximum value. If the capacity of the transport system were increased (by increasing $E_1$ in the model) the maximum value of $\mu_g$ would also suffer some increase. However, transport of $S_1$ is *not* a rate-limiting step in the sense of a "master reaction", because many other changes can be made (for example increasing the concentration of one or more enzymes $E_2, E_3 \ldots E_v$) that may decrease the steady-state concentration of $S_2$ and thus increase $F_{1,2}^{\text{net}}$ and consequently the value of $\mu_g$. On the other hand an increase in the capacity of $E_1$ will tend to increase the size of pool $S_2$ and thus $F_{2,1}$ so that the positive effect on the specific growth rate will be partly buffered. However, the concept of the "master reaction" as it arose in connection with the effect of temperature on biological rates, including the specific growth rate of cell population, requires that the overall rate be a linear function of the concentration of the enzyme (in the native state) that catalyses the "master reaction".

Let us consider the case of transport-limited growth (see eq. 19):

$$\mu_g=y_1 F_{1,2}^{\max}\frac{S_1}{K_{1,2}+S_1} \tag{110}$$

For a given constant concentration of $S_1$ and under the assumption that $K_{1,2}$ is a constant (for example not subject to allosteric effects) we may write:

$$\mu_g = CF_{1,2}^{max} \tag{111}$$

where $C$ is a constant.

Assuming that the capacity of the transport system is a linear function of its concentration, we may now write:

$$\mu_g = CT_1 E_1 \tag{112}$$

and:

$$\frac{d\mu_g}{dE_1} = CT_1 = \text{constant} \tag{113}$$

Kacser and Burns (1968) expressed the rate-controlling weight of an enzyme by the sensitivity coefficient, $R$, which for the case of our model may be defined as:

$$\frac{d\mu_g}{dE}\frac{E}{\mu_g} = R$$

As can easily be seen, the sensitivity coefficient of the case expressed by eq. (113) is unity, and it may be said that a master reaction occurs when the control of its enzyme on the overall rate is characterized by a sensitivity coefficient of unity. Such reactions cannot exist in steady-state multi-step systems; near approximations (such as the transport step in transport-limited growth) of true master reactions occur but are rare. The Burns (1968) programme calculates sensitivity coefficients for each enzyme for any possible steady state. The reader who succeeds in mounting the model should try choosing parameters in such a way that one enzyme reaches a sensitivity coefficient of nearly unity. He will have the elucidating experience that this is only possible (except in the transport case) when assumptions are made which are rather unlikely from the biological point of view.

The sensitivity coefficients of the enzymes and transport mediators which compose a multi-step system are, as one would expect, interrelated. For example, in a linear system the sum of all sensitivity coefficients with respect to the overall rate is approximately unity. A better understanding of these relations is an essential for progress in the kinetics of growth. A much needed treatment of the properties of sensitivity coefficients was announced (Burns, 1968) but has not appeared in print at the time of writing.

## References

Azam, F. and Kotyk, A. (1969). *FEBS Lett.* **2**, 333–335.
Battley, E. H. (1960). *Physiologia Pl.* **13**, 192–203.
Bauchop, T. and Elsden, S. R. (1960). *J. gen. Microbiol.* **23**, 457–469.
Beran, K., Málek, I., Streiblová, E. and Lieblová, J. (1967). *In* "Microbial Physiology and Continuous Culture", (E. O. Powell, C. G. T. Evans, R. E. Strange and D. W. Tempest, eds.), pp. 57–67. H.M.S.O., London.
Beran, K. (1969). *In* "Continuous Cultivation of Microorganisms", (I. Málek, K. Beran, Z. Fencl, V. Murk, J. Řičica and H. Smrčková, eds.), pp. 87–103. Academia, Prague.
Brandts, J. F. (1967). *In* "Thermobiology", (A. H. Rose, ed.). Academic Press, London.
Bulder, C. J. E. A. (1963). Ph.D Thesis: Technical University, Delft.
Bulder, C. J. E. A. (1966). *Arch. Mikrobiol.* **53**, 189–194.
Burns, J. A. (1968). *In* "Quantitative Biology of Metabolism", (A. Locker, ed.), pp. 75–80. Springer-Verlag, Berlin.
Burns, J. A. (1969). *FEBS Lett.* **2** (Supplement), S30–S33.
Button, D. K. and Garver, J. C. (1966). *J. gen. Microbiol.* **45**, 194–204.
Campbell, A. (1957). *Bact. Rev.* **21**, 263–272.
Chance, B. (1959a). *J. biol. Chem.* **234**, 3036–3040.
Chance, B. (1959b). *J. biol. Chem.* **234**, 3041–3043.
Chen, S. L. (1964). *Nature, Lond.* **202**, 1135–1136.
Cirillo, V. P. (1961). *Trans. N.Y. Acad. Sci.* **23**, Series II, 725–734.
Cirillo, V. P. (1962). *J. Bact.* **84**, 485–491.
Contois, D. E. (1959). *J. gen. Microbiol.* **21**, 40–50.
Dawson, P. S. S. (1969). *In* "Continuous Cultivation of Microorganisms", (I. Málek, K. Beran, Z. Fencl, V. Murk, J. Řičica and H. Smrčková, eds.), pp. 71–85. Academia, Prague.
Dean, A. C. R. and Hinshelwood, C. N. (1966). "Growth Function and Regulation in Bacterial Cells". Clarendon Press, Oxford.
Egambardiev, N. B. and Ierusalimsky, D. D. (1969). *In* "Continuous Cultivation of Microorganisms", (I. Málek, K. Beran, Z. Fencl, V. Murk, J. Řičica and H. Smrčková, eds.), pp. 517–527. Academia, Prague.
Eyring, H. (1935). *Chem. Rev.* **17**, 65–77.
Fencl, Z. (1966). *In* "Theoretical and Methodological Basis of Continuous Culture of Microorganisms", (I. Málek and Z. Fencl, eds.), pp. 67–153. Academic Press, New York.
Forrest, W. W. (1969). *Symp. Soc. gen. Microbiol.* **19**, 65–86.
Gancedo, C., Gancedo, J. M. and Sols, A. (1968). *Eur. J. Biochem.* **5**, 165–172.
Garfinkel, D. (1969). *FEBS Lett.* **2** (Supplement), S9–S13.
Hadjipetrou, L. P., Gerrits, J. P., Teulings, F. A. G. and Stouthamer, A. H., (1964). *J. gen. Microbiol.* **36**, 139–150.
Hempfling, W. P. and Vishniac, W. J. (1967). *J. Bact.* **93**, 874–878.
Herbert, D. (1958). *Proc. Int. Congr. Microbiol.* (Stockholm) No. 7, p. 372.
Herbert, D. (1961). *Symp. Soc. gen. Microbiol.* **11**, 391–416.
Herbert, D., Elsworth, R. and Telling, R. C. (1956). *J. gen. Microbiol.* **14**, 601–622.
Hernandez, E. and Johnson, M. J. (1967). *J. Bact.* **94**, 996–1001.
Hinshelwood, C. N. (1946). "The Chemical Kinetics of the Bacterial Cell". Clarendon Press, Oxford.

Ierusalimsky, N. D. (1967). *In* "Microbial Physiology and Continuous Culture", (E. O. Powell, C. G. T. Evans, R. E. Strange and D. W. Tempest, eds.), pp. 23–33. H.M.S.O., London.

Ierusalimsky, N. D., Zaitseva, G. N. and Khmel, I. A. (1962). *Mikrobiologiya* **31**, 417–421.

Johnson, F. H., Eyring, H. and Polissar, M. J. (1954). "The Kinetic Basis of Molecular Biology". John Wiley and Sons, Inc., New York.

Johnson, M. J. (1967). *J. Bact.* **94**, 996–1001.

Kacser, H. and Burns, J. A. (1968). *In* "Quantitative Biology of Metabolism", (A. Locker, ed.), pp. 11–23. Springer-Verlag, Berlin.

Kleinzeller, A. and Kotyk, A. (1967). *In* "Symposium on Aspects of Yeast Metabolism", (A. K. Mills, ed.), pp. 33–45. Blackwell Scientific Publications, Oxford.

Kotyk, A. (1967). *Biochim. biophys. Acta* **135**, 112–119.

Lamanna, C. (1963). *Ann. N.Y. Acad. Sci.* **102**, 515.

Lynen, F. and Koenigsberger, R. (1951). *Justus Liebigs Annln Chem.* **573**, 60–84.

Maaløe, O. and Kjeldgaard, N. O. (1966). "Control of Macromolecular Synthesis". Benjamin, New York.

Marr, A. G., Nilson, E. H. and Clark, D. J. (1963). *Ann. N.Y. Acad. Sci.* **144**, 536–548.

Maynard Smith, J. (1967). *Symp. Soc. gen. Microbiol.* **19**, 1–13.

McMurrough, I. and Rose, A. H. (1967). *Biochem. J.* **105**, 189–203.

Meyenburg, H. K. von (1969). *Arch. Mikrobiol.* **66**, 289–303.

Monod, J. (1942). "Recherches sur la Croissance des Cultures Bactériennes". Herman, Paris.

Monod, J. (1950). *Annls Inst. Pasteur, Paris* **79**, 390–410.

Moser, H. (1958). "The Dynamics of Bacterial Populations Maintained in the Chemostat". Carnegie Inst., Wash. Publ.

Novick, A. and Szilard, L. (1950). *Science, N.Y.* **112**, 715–716.

Okada, H. and Halvorson, H. O. (1964a). *Biochim. biophys. Acta* **82**, 538–546.

Okada, H. and Halvorson, H. O. (1964b). *Biochim. biophys. Acta* **82**, 547–555.

Oliveira-Baptista, A. and van Uden, N. (1970). *Z. allg. Mikrobiol.* in press.

Painter, P. R. and Marr, A. G. (1968). *A. Rev. Microbiol.* **22**, 519–548.

Pfennig, N. and Jannasch, H. W. (1962). *Ergebn Biol.* **25**, 93–135.

Pirt, S. J. (1965). *Proc. R. Soc. B* **163**, 224–231.

Pirt, S. J. (1967). *In* "Microbial Physiology and Continuous Culture", (E. O. Powell, C. G. T. Evans, R. E. Strange and D. W. Tempest, eds.), pp. 162–172. H.M.S.O., London.

Postgate, J. R. (1965). *Lab. Pract.* **14**, 1140–1144.

Postgate, J. R. and Hunter, J. R. (1962). *J. gen. Microbiol.* **29**, 233–263.

Powell, E. O. (1966). *J. gen. Microbiol.* **45**, xi–xii.

Powell, E. O. (1967). *In* "Microbial Physiology and Continuous Culture", (E. O. Powell, C. G. T. Evans, R. E. Strange and D. W. Tempest, eds.), pp. 34–55. H.M.S.O., London.

Powell, E. O. (1969). *In* "Continuous Cultivation of Microorganisms", (I. Málek, K. Beran, Z. Fencl, V. Murk, J. Řičica and H. Smrčková, eds.), pp. 275–284. Academia, Prague.

Ramkrishna, D., Fredrickson, A. G. and Tsuchiya, H. M. (1966). *J. gen. appl. Microbiol., Tokyo* **12**, 311–327.

Ramkrishna, D., Fredrickson, A. G. and Tsuchiya, H. M. (1967). *Biotechnol. Bioengng* **9**, 129–170.

Schulze, K. L. and Lipe, R. S. (1964). *Arch. Mikrobiol.* **48**, 1–20.

Schuster, E. and Schlegel, H. G. (1967). *Arch. Mikrobiol.* **58**, 380–409.

Senez, J. C. (1962). *Bact. Rev.* **26**, 95–107.

Sols, A. (1967). *In* "Symposium on Aspects of Yeast Metabolism", (A. K. Mills, ed.), pp. 47–66. Blackwell Scientific Publications, Oxford.
Stanley, P. E. (1964). PhD Thesis: University of Bristol.
Stearn, A. E. (1949). *Adv. Enzymol.* **9**, 25–74.
Steveninck, J. van and Rothstein, A. (1965). *J. gen. Physiol.* **49**, 235–246.
Steveninck, J. van (1966). *Biochim. biophys. Acta* **126**, 154–162.
Steveninck, J. van and Booij, H. L. (1964). *J. gen. Physiol.* **48**, 43–60.
Stouthamer, A. H. (1969). *In* "Microbiological Methods", J. R. Norris and D. W. Ribbons, eds.). Vol. 1, pp. 629–663, Academic Press, London.
Teissier, G. (1936). *Ann. Physiol. physicochim. biol.* **12**, 527–586.
Tempest, D. W. (1969). *Symp. Soc. gen. Microbiol.* **19**, 87–111.
Tsuchiya, H. M., Fredrickson, A. G. and Aris, R. (1966). *Adv. Chem. Engng* **6**, 125–206.
Uden, N. van (1967a). *Arch. Mikrobiol.* **58**, 145–154.
Uden, N. van (1967b). *Arch. Mikrobiol.* **58**, 155–168.
Uden, N. van (1968). *Arch. Mikrobiol.* **62**, 34–40.
Uden, N. van (1969a). *Z. allg. Mikrobiol.* **9**, 385–396.
Uden, N. van (1969b). *A. Rev. Microbiol.* **23**, 473–486.
Uden, N. van (1969c). *In* "Continuous Cultivation of Microorganisms", (I. Málek, K. Beran, Z. Fencl, V. Murk, J. Řičica and H. Smrčková, eds.), pp. 49–58. Academia, Prague.
Uden, N. van, Abranches, P. and Cabeça-Silva, C. (1968). *Arch. Mikrobiol.* **61**, 381–393.
Uden, N. van and Madeira-Lopes, A. (1970). *Z. allg. Mikrobiol.* in press.
Wase, D. A. J. and Hough, J. S. (1966). *J. gen. Microbiol.* **42**, 13–23.
Wilbrandt, W. and Rosenberg, T. (1961). *Pharmac. Rev.* **13**, 109–183.
Williamson, D. H. (1964). *In* "Synchrony in Cell Division and Growth", (E. Zeuthen, ed.), pp. 351–379. Interscience Publishers, New York.
Williamson, D. H. (1966). *In* "Cell Synchrony", (I. L. Cameron and G. M. Padilla, eds.), pp. 81–101. Academic Press, New York.
Wood, T. H. (1956). *Adv. biol. med. Phys.* **4**, 119–165.

# Chapter 4

# Influence of Temperature on the Growth and Metabolism of Yeasts

J. L. Stokes

*Department of Bacteriology and Public Health, Washington State University, Pullman, Washington, U.S.A.*

I. INTRODUCTION . . . . . . . . . . . . 119
II. GROWTH . . . . . . . . . . . . 120
A. Cardinal Growth Temperatures . . . . . . . 120
B. Biochemistry . . . . . . . . . . . 124
III. SPORULATION . . . . . . . . . . . . 127
IV. SURVIVAL . . . . . . . . . . . . 127
A. Heat . . . . . . . . . . . . 127
B. Effect of Subzero Temperatures . . . . . . . 130
V. OTHER ASPECTS . . . . . . . . . . . 131
A. Fermentation . . . . . . . . . . . 131
B. Cell Composition . . . . . . . . . . 132
VI. ACKNOWLEDGEMENTS . . . . . . . . . . 132
REFERENCES . . . . . . . . . . . . . 132

## I. Introduction

Temperature can be expected to exert a profound effect on all aspects of the growth, metabolism and survival of yeasts. Although there are numerous incidental observations in the literature on the effects of temperature, there are surprisingly few systematic and comprehensive studies. This situation may be due, in part, to the unavailability in the past of refrigerated equipment necessary to maintain constant temperatures below ambient room temperatures. During the past decade, however, considerable interest has developed in the nature and ecology of psychrophilic yeasts, i.e. those which grow at low temperatures. This has led to many investigations on the influence of temperatures on yeasts. Moreover, for comparative purposes, the experiments have been extended to include mesophilic yeasts. As a result, there is now a considerable body of new information on the effects of temperature on the growth, chemical composition, substrate intake, and enzymic activities of a large variety of yeasts.

## II. Growth

### A. CARDINAL GROWTH TEMPERATURES

The cardinal growth temperatures for a variety of yeasts are given in Table I. Except for the obligately psychrophilic yeasts, most of the yeasts multiply at or close to 0°, although some have a minimum of 5°. Growth may be possible below 0° but usually this is not determined, probably because special incubators are required and also freezing of culture media becomes a problem (Larkin and Stokes, 1968). Growth is slow at these low temperatures. Most rapid growth occurs, usually, in the range of 20°–30°. The maximum growth temperatures for the common yeasts are, generally, in the range of 30°–40°, although an occasional yeast will grow at somewhat higher temperatures. There do not appear to be any thermophilic yeasts, i.e. those which can grow above 50°, although a few thermophilic filamentous fungi have been reported (Cooney and Emerson, 1964).

The obligately psychrophilic yeasts differ greatly from other yeasts with respect to growth temperatures. Many strains can develop below 0°. The *Candida* species of Lawrence *et al.* (1959) were not tested below 0°. All of the yeasts, however, grew optimally below 20° and usually at 15°. Moreover, virtually none of them can grow above 20°. All of their cardinal temperatures, therefore, are considerably lower than those for mesophilic yeasts. The growth curves of an obligately psychrophilic species of *Candida* at various temperatures are shown in Fig. 1.

Psychrophilic yeasts are common in Antarctic and Arctic soils and other materials (Straka and Stokes, 1960; di Menna, 1960, 1966a, b; Sinclair and Stokes, 1965). Recently Fell and Phaff (1967) reported the isolation of many psychrophilic yeasts from Antarctic water with maximum growth temperatures of 15°–20°. Yeasts capable of growth at 37° or higher temperatures were rare. Also, Stanley and Rose (1967) have isolated psychrophilic yeasts from Antarctica. The strains of *Cryptococcus* isolated from sea water by Norkans (1966) are probably psychrophiles since they grew well at 3°, which was the lowest temperature tested, compared to a minimum growth temperature of 10° for strains of *Saccharomyces cerevisiae* and *Sacch. pastorianus*.

Psychrophilic bacteria have been defined as those which make macroscopically visible growth at 0° in about a week (Ingraham and Stokes, 1959; Stokes, 1963). Moreover, it has been suggested that psychrophilic bacteria which grow most rapidly at or below 20° be named "obligate psychrophiles" and those which grow most rapidly above 20°, "facultative psychrophiles" (Stokes, 1963). These definitions for bacteria cannot be applied to yeasts without modification. It is evident from the data in Table I that many of the common yeasts can grow at or close to 0°. In

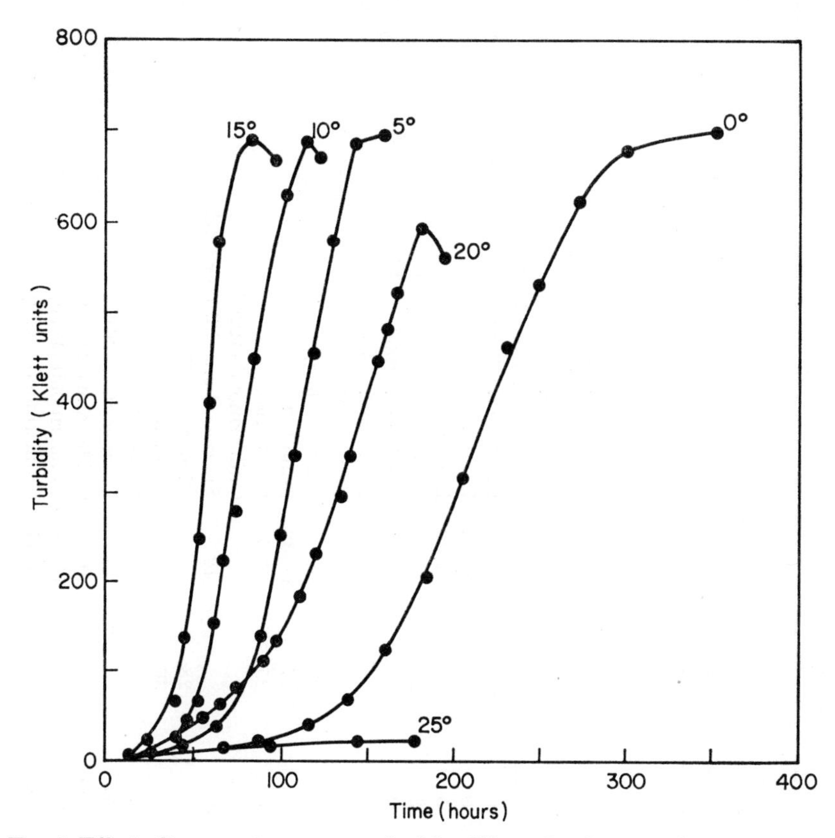


FIG. 1. Effect of temperature on growth of *Candida* sp. P25 in aerated tryptone-yeast extract-glucose medium. Redrawn from Sinclair and Stokes (1965).

general, yeasts appear to have a lower temperature range for growth than bacteria. Mesophilic bacteria tend to stop growing at about 10° and many of them have optimum growth temperatures of 30°–40° and maximum growth temperatures of 45°–50°. Only the obligately psychrophilic yeasts constitute a sharply defined group. Perhaps the term "psychrophilic" should be applied only to organisms in this group which grow at 0° or lower, most rapidly below 20° and usually fail to grow above 20°. In the future, when additional information becomes available on the ability of a wide variety of yeasts to grow below 0°, it may become possible to define

TABLE I. *Cardinal Growth Temperatures of Yeasts*

| Species | Cardinal growth temperatures (°) | | | Reference |
|---|---|---|---|---|
| | Minimum | Optimum | Maximum | |
| Mesophilic and Facultative Psychrophilic Yeasts | | | | |
| *Candida macedonensis* | 5 | | 45 | Phaff *et al.* (1966) |
| *Candida parapsilosis* | 0 | 20–25 | 30 | Lund (1958) |
| *Candida slooffii* | 27 | | 45 | Phaff *et al.* (1966) |
| *Candida utilis* | 5–10 | | | Rose and Evison (1965) |
| *Debaryomyces hansenii* | 0 | | 35 | Phaff *et al.* (1966) |
| *Hansenula suaveolens* | 3 | 30 | 30–35 | Lund (1958) |
| *Hansenula valbyensis* | 5 | | 32–33 | Lund (1958) |
| *Kloeckera apiculata* | 3 | 30 | 35 | Lund (1958) |
| *Nadsonia elongata* | 0 | | 25 | Phaff *et al.* (1966) |
| *Pichia membranaefaciens* | 3 | | 30 | Phaff *et al.* (1966) |
| *Rhodotorula glutinis* | | | 39 | Fries (1963) |
| *Rhodotorula glutinis* | 0 | 23 | < 30 | Clinton (1968) |
| *Rhodotorula gracilis* | 5 | 27 | 37–42 | Lund (1958) |
| *Saccharomyces carlsbergensis* | 0 | 25 | 33·5 | Lund (1958) |
| *Saccharomyces cerevisiae* | 1–3 | | 40 | Guilliermond (1920) |
| *Saccharomyces cerevisiae* | 0–5 | | 42 | Sherwood and Fulmer (1926) |
| *Saccharomyces cerevisiae* | | 28 | | Richards (1934) |
| *Saccharomyces cerevisiae* | | 36 | | White and Munns (1951) |
| *Saccharomyces cerevisiae* | | | 40–41 | Battley (1964) |
| *Saccharomyces fragilis* | 5 | | 45 | Phaff *et al.* (1966) |
| *Saccharomyces intermedius* | 0·5 | | 40 | Guilliermond (1920) |
| *Saccharomyces ludwigii* | 1–3 | | 37–38 | Guilliermond (1920) |
| *Saccharomyces marxianus* | 0·5 | | 46–47 | Guilliermond (1920) |

| | | | | |
|---|---|---|---|---|
| *Saccharomyces mellis* | 23 | | 35 | Phaff *et al.* (1966) |
| *Saccharomyces octosporus* | 17 | | 33 | Phaff *et al.* (1966) |
| *Saccharomyces pastorianus* | 0·5 | | 24 | Guilliermond (1920) |
| *Saccharomyces turbidans* | 0·5 | | 40 | Guilliermond (1920) |
| *Saccharomyces validus* | 0·5 | | 39–40 | Guilliermond (1920) |
| *Saccharomycopsis guttulata* | 35 | | 40 | Phaff *et al.* (1966) |
| *Torulopsis candida* | 5 | 22 | 32 | Lund (1958) |
| *Torulopsis molischiana* | 5 | 22 | > 42 | Lund (1958) |
| Obligately Psychrophilic Yeasts | | | | |
| *Candida frigida*, strain P8 | –5– –7 | 15 | 20 | Sinclair and Stokes (1965); di Menna (1966a); Larkin and Stokes (1968) |
| *Candida gelida*, strain P16 | –5– –7 | 15 | 20 | Sinclair and Stokes (1965); di Menna (1966a); Larkin and Stokes (1968) |
| *Candida nivalis*, strain P7 | –5– –7 | 15 | 20 | Sinclair and Stokes (1965); di Menna (1966a); Larkin and Stokes (1968) |
| *Candida scottii* | | 4–15 | 15–20 | di Menna (1960) |
| *Candida scottii* | 0 | | 20 | Phaff *et al.* (1966) |
| *Candida* sp. | 0 | 11 | 17 | Lawrence *et al.* (1959) |
| *Candida* sp., strain 3E-2 | –7 | 5–20 | 20 | Straka and Stokes (1960) |
| *Candida* sp., strain 25 | –4·5 | 15 | 20 | Sinclair and Stokes (1965); Larkin and Stokes (1968) |
| *Rhodotorula infirmo-miniata* | –2 | 14–18 | 26–28 | Ahearn and Roth (1966) |

psychrophilic yeasts as those which grow reasonably well below 0° in order to differentiate them clearly from mesophilic yeasts. Further differentiation of psychrophilic yeasts into facultative and obligate types may then be possible in a manner similar to that used for psychrophilic bacteria. The data in Table I indicate also that growth temperatures vary for different strains of a species, and that some yeasts have a narrow temperature range for growth, e.g. *Saccharomycopsis guttulata*.

In a very extensive analysis, Cooke (1965) tested 2,300 strains of 70 species of yeast for their ability to grow on yeast-autolysate agar at 37°. The strains were isolated from sewage, sewage-polluted water, and related sources. Most strains of *Candida guilliermondii*, *C. krusei*, *C. tropicalis*, *Sacch. cerevisiae*, and *Torulopsis candida* grew well at 37°. In contrast, most strains of *C. curvata*, *C. humicola*, *Cryptococcus albidus*, *Hansenula californica* and *Rhodotorula glutinis* made little or no growth at 37°. Forty-two of the species included strains both capable and incapable of growth at 37°. The finding emphasizes again the fact that growth-temperature variations occur among different strains.

Temperature adaptation and special nutrients may be necessary for yeast growth at elevated temperatures. Sherman (1959a) observed that *Sacch. cerevisiae* could not grow at 40° in media containing 0·5% yeast extract unless the inoculum consisted of cells previously grown at 40°. When the concentration of yeast extract was increased to 1%, growth proceeded only after an initial death phase which could be eliminated by addition of oleic acid. The latter finding may be related to decreased sterol synthesis at elevated temperatures (Starr and Parks, 1962). Ergo sterol stimulated growth of *Sacch. cerevisiae* at 40°. Many osmophilic yeasts, such as *Sacch. rouxii*, are unusual in that they multiply at the relatively high temperature of 40° only in the presence of more than 3–4% sodium chloride or 30% sucrose or galactose (Onishi, 1959, 1963).

Growth of *Sacch. cerevisiae* at maximal temperatures, e.g. 40°, leads to a large increase in the rate of formation of respiratory deficient "petite" mutants (Sherman, 1959b). Also, temperature-sensitive mutants of *Sacch. cerevisiae* have been isolated which form colonies at 23° but not at 36°, although the wild type grew at both temperatures (Hartwell, 1967).

## B. BIOCHEMISTRY

The factors which determine the growth temperatures of microorganisms are numerous and complex. They include enzyme synthesis and activity, cellular control mechanisms, transfer of anabolic information from genes to ribosomes, substrate uptake and the composition and integrity of the semipermeable cell membrane (Ingraham and Maaløe, 1967; Stokes, 1967; Farrell and Rose, 1967). Most investigations have been made with bacteria, although several have dealt with yeasts. There is no reason to believe, at present, that there are any basic differences in this respect between these two groups of micro-organisms.

### 1. *Minimum Growth Temperature*

Mesophilic bacteria and yeasts stop growing usually above 0° and frequently above 5° or 10°. This is perhaps surprising, since one might expect that the chemical reactions in the cell would slow down but not cease entirely with decrease in temperature. Growth, therefore, might be expected to continue slowly at low temperatures, at least until the cell protoplasm froze.

Comparative biochemical studies with mesophilic and psychrophilic yeasts have been made in order to determine the mechanisms which permit psychrophiles to grow at considerably lower temperatures. Baxter and Gibbons (1962) reported that a psychrophilic species of *Candida* oxidized glucose and transported glucosamine to an appreciable extent at 0°, whereas mesophilic *C. lipolytica* was virtually inactive at this temperature, and even at 10°. Also, Cirillo *et al.* (1963) found that the transport of glucose, glucosamine, and sorbose into the same psychrophilic strain of *Candida* was less affected by low temperatures than it was in a mesophilic strain of *Sacch. cerevisiae.* More recently, Sinclair and Stokes (1965) observed relatively more rapid fermentation of glucose at low temperatures (10° and 15°) by psychrophilic *C. gelida* P16 than by mesophilic *Sacch. cerevisiae.* Rose and Evison (1965) reported that another psychrophilic strain of *Candida* (Straka and Stokes, 1960), which grew at 0°, also oxidized glucose and accumulated glucosamine at this temperature. A mesophilic strain of *C. utilis,* however, which had a minimum growth temperature of 5°–10°, ceased to oxidize glucose and accumulate glucosamine at approximately the same temperatures. Also, the uptake of uric acid and xanthine by *C. utilis* is prevented at temperatures below 4° (Quetsch and Danforth, 1964). In contrast, glutamic acid and lysine were taken up by psychrophilic *Rh. glutinis* at −4°, and this was accompanied by amino acid-pool formation and protein synthesis (Clinton, 1968).

These data suggest that enzyme activity and substrate transport are involved in the establishment of the minimum growth temperatures of yeasts, and suggest an explanation for the lower growth temperatures of psychrophilic yeasts. Other factors such as enzyme synthesis also may be involved, as in bacteria, but this has not yet been determined.

### 2. *Optimum Growth Temperature*

The optimum growth temperature is defined, usually, as the temperature at which the growth rate is highest. It must be assumed, therefore, that the overall biochemical processes and associated enzymes involved in growth proceed most rapidly at this temperature. There is very little specific biochemical information available. Sometimes the optimum temperature is defined as the temperature which yields the largest cell crop. It may be lower than the temperature for most rapid growth. Hagen and Rose (1961) obtained the largest crop of a psychrophilic *Cryptococcus* strain at 21° although growth was equally as rapid at 25°. According to Merritt (1966) an unidentified distiller's yeast grew most rapidly at 35° but gave maximal cell yields at 30°. This phenomenon had been observed frequently with bacteria and is due to the greater availability of oxygen to the bacteria at the lower temperatures (Sinclair and Stokes, 1963). This aspect merits further investigation with yeasts.

### 3. *Maximum Growth Temperature*

There is considerable evidence that the upper temperature limit for the growth of bacteria is controlled, at least in part, by the thermolability of their enzymes and enzyme-forming systems (Stokes, 1967; Farrell and Rose, 1967). Enzyme destruction occurs also in yeasts at temperatures close to the maximal growth temperature. Hagen and Rose (1962) reported that a psychrophilic *Cryptococcus* sp. could not grow at 30° unless α-ketoglutarate, citrate or isocitrate was added to the growth medium. This indicated that tricarboxylic acid-cycle enzymes were inactivated in the yeast at 30°. Likewise, the respiratory activity of a psychrophilic *Candida* decreased when the organism was incubated at temperatures 3°–5° above its maximal growth temperature, and concomitantly there was a decrease in the activities of its tricarboxylic acid-cycle enzymes (Evison and Rose, 1965).

TABLE II. *Effect of Heating of Cell Suspensions of* Candida *sp. P*16 *and* Saccharomyces cerevisiae *at* 35° *on the Fermentation and Oxidation of Glucose. From Sinclair and Stokes* (1965).

| Organism | Heated at 35° (min) | $Q_{CO_2}$* | $Q_{O_2}$* |
|---|---|---|---|
| *Candida* sp. P16 | 0 | 73 | 67 |
| | 20 | 3 | 48 |
| | 40 | 0 | 37 |
| | 60 | 0 | 11 |
| *Saccharomyces cerevisiae* | 0 | 101 | 100 |
| | 60 | 107 | |
| | 180 | 113 | 107 |

* Values for endogenous respiration have been subtracted.

The fermentative and oxidative activities of psychrophilic *C. gelida* P16 (Sinclair and Stokes, 1965), which cannot grow at 25°, were rapidly destroyed at 35°, whereas these metabolic activities in mesophilic *Sacch. cerevisiae,* which grew well at 35°, were unaffected (Table II). The principal fermentative lesion in *C. gelida* at 35° involves destruction of pyruvate decarboxylase (Grant *et al.*, 1968). In contrast, this enzyme in mesophilic *C. utilis* was not denatured at 35°. Also, psychrophilic *C. nivalis,* which cannot grow at 25°, is severely damaged at that temperature, as indicated by loss of viability, increased nutritional demands of surviving cells, and leakage of soluble components including inorganic phosphate, amino acids or short-chain polypeptides, and nucleotide monophosphate (Nash and Sinclair, 1968).

These results suggest that enzyme destruction and cell membrane damage are involved in setting the upper temperature limits for yeast growth. Other vital cell functions also may be impaired, especially at higher temperatures or longer heating periods (Farrell and Rose, 1967; Morita, 1968). The temperature-sensitive mutants of *Sacch. cerevisiae* (Hartwell, 1967), with a maximal growth temperature of 23°, when exposed to 36° exhibited a preferential loss of ability to carry out synthesis of protein, RNA and DNA, cell division and cell-wall formation. In two of the mutants, the thermolabile protein was identified as isoleucyl-t-RNA synthase (Hartwell and McLaughlin, 1968a). Defects in energy metabolism may also be involved (Hartwell and McLaughlin, 1968b).

## III. Sporulation

Temperature is one of the important environmental factors in yeast sporulation. Sporulation can take place over as wide a range of temperatures as growth itself. The minimum, optimum, and maximum sporulation temperatures for a number of species of *Saccharomyces* are given in Table III. Sporulation can occur, although slowly, at temperatures close to 0°. The optimum temperature, which is in the range of 25°–30°, is usually defined as the temperature at which definite sporulation occurs in the shortest time (Phaff and Mrak, 1949) and in most cases is close to the maximum temperature. Also, as with growth, strains may differ in their temperature range for sporulation.

## IV. Survival

### A. HEAT

The thermal death times of vegetative cells and spores of various yeasts suspended in a variety of liquids have been determined by numerous investigators. Some of their data are presented in Table IV. In general,

TABLE III. *Effect of Temperature on Sporulation of* Saccharomyces *Species*

| Species | Sporulation temperatures (°) | | | Reference |
|---|---|---|---|---|
| | Minimum | Optimum | Maximum | |
| *Saccharomyces cerevisiae* | 9–11 | 30 | 35–37 | Guilliermond (1920) |
| *Saccharomyces cerevisiae* | 3 | 24–27·5 | 33·3 | Adams and Miller (1954) |
| *Saccharomyces ellipsoideus* | 4·7–5 | 25 | 30·5–32·5 | Guilliermond (1920) |
| *Saccharomyces ellipsoideus* | 7·5 | | 29·5 | Phaff and Mrak (1949) |
| *Saccharomyces ellipsoideus* | 8 | | 33 | Phaff and Mrak (1949) |
| *Saccharomyces intermedius* | 0·5–4 | 25 | 27–29 | Guilliermond (1920) |
| *Saccharomyces pastorianus* | 0·5–4 | 27·5 | 29–31·5 | Guilliermond (1920) |
| *Saccharomyces turbidans* | 4–8 | 29 | 33–35 | Guilliermond (1920) |
| *Saccharomyces validus* | 4·8–5 | 25 | 27–29 | Guilliermond (1920) |

TABLE IV. *Thermal Death Time of Cells and Spores of Yeasts*

| Species | Suspending medium | Lethal temperature (°) | Death time (min) | | Reference |
|---|---|---|---|---|---|
| | | | Cells | Spores | |
| *Debaryomyces globosus* | Glucose broth | 55 | 25 | | Beamer and Tanner (1939) |
| | | 60 | 5 | | |
| *Monilia candida* | Glucose broth | 55 | 25 | | Beamer and Tanner (1939) |
| | | 60 | 5 | | |
| *Saccharomyces cerevisiae* | Water | 54 | 5 | | Ingram (1955) |
| | | 62 | | 5 | |
| *Saccharomyces ellipsoideus* | Beer | 54 | 20 | | Ingram (1955) |
| | | 56 | | 15 | |
| *Saccharomyces ellipsoideus* | Grape juice | 54 | 120 | | Aref and Cruess (1934) |
| | | 57·5 | 10 | | |
| *Saccharomyces ellipsoideus* | Glucose broth | 60 | 15 | | Beamer and Tanner (1939) |
| *Saccharomyces odessa* | Beer | 62–64 | 20 | | Ingram (1955) |
| | | 62–64 | | 15 | |
| *Saccharomyces turbidans* | Beer | 52 | 20 | | Ingram (1955) |
| | | 58 | | 20 | |
| *Torula monosa* | Glucose broth | 60 | 35 | | Beamer and Tanner (1939) |
| | | 50 | 20 | | |
| *Willia anomola* | Glucose broth | 55 | 10 | | Beamer and Tanner (1939) |
| | | 60 | 5 | | |

vegetative cells of yeasts are destroyed at temperatures and times similar to those which are lethal for the bacterial cell. Yeast cells are usually killed by exposure to 50°–60° for less than 30 min. The pH value of the suspending medium might be expected to influence the rate of death. However, the death temperature and death rate of *Sacch. ellipsoideus* were essentially the same in the range of pH 1·45–7·0 in grape juice (Aref and Cruess, 1934). Also, the time required to kill the yeasts investigated by Beamer and Tanner (1939) was only slightly decreased in glucose broth at pH 3·8 compared to broth at pH 6·8.

The spores of yeasts are considerably more thermolabile than those of bacteria and are only a few degrees of temperature more heat resistant than the vegetative cells which produce them.

Before the yeast cell is killed by heat, injury may appear. Fries (1963) found that cells of *Rh. glutinis* when heated at 48° for 1–4 min grew when subsequently incubated at 20° but not at 28°. Obligately psychrophilic yeasts are much more thermolabile than other yeasts, as might perhaps be expected because of their low maximal growth temperatures. The effect of temperature on survival of several such yeasts (*C. nivalis, C. frigida, C. gelida,* and *Candida* sp. strain P25) has been investigated (Sinclair and Stokes, 1965). The strains were inoculated into tryptone-yeast extract-glucose broth to give initial populations of approximately $10^6$ cells per ml. The cultures were incubated at 25°, 30° and 35°, and periodically plated to determine the number of survivors. Virtually no cells of any of the *Candida* species could be recovered from the broth cultures after exposure to 30° for 24 h. Under identical conditions, a mesophilic strain of *Sacch. cerevisiae* multiplied rapidly at 30° and exhibited extensive death only when exposed to 45° and 50° for 24 h.

The obligately psychrophilic *Candida* sp. of Lawrence *et al.* (1959) also was rapidly killed at 30° (Baxter and Gibbons, 1962). Likewise, cells of *Rh. infirmo-miniata* were destroyed by exposure to 30° for 24 h (Ahearn and Roth, 1966). Injury as well as death may occur in psychrophilic yeasts at these low temperatures, as indicated by increased nutritional requirements of the surviving cells (Nash and Sinclair, 1968). Thermal death in yeasts is undoubtedly due to the same types of cellular lesions which set the upper temperature limits for growth, namely, denaturation of enzymes and proteins, destruction of the cell membrane, and related processes.

### B. Effect of Subzero Temperatures

Yeasts are resistant to low temperatures, although some injury and death usually occurs. Resistance to cold is indicated by the survival of yeasts in polar regions and also in temperate regions during severe winters (Lund, 1958). Campbell (1943) found that a single freezing of

*Sacch. cerevisiae* (yeast cake) suspended in water at −30° or −50° destroyed 48–94% of the cells. Death was increased by repeated freezing and thawing. These results were confirmed by Baum (1943) with a pure culture of *Sacch. cerevisiae*. Storage of baker's yeast at −23·5° had little effect on the ability of the cells to produce a satisfactory bread; also yeast ascospores seemed more resistant to freezing injury than vegetative cells (Godkin and Cathcart, 1949). Stille (1950) reported that 29·2% of the cells of *Sacch. cerevisiae* were killed after freezing at −24° but only 18·4% at −193°. Young cultures of *Saccharomyces*, *Hansenula* and other yeasts, on agar medium, survived −15° for 48 h apparently without being injured (Lund, 1958). Also, Adams (1966) has shown that wine yeasts can be frozen and stored at −29° or lower for several years without total loss of viability or essential characteristics. The manifold effects of freezing on yeast cells are described by Mazur (1966).

## V. Other Aspects

The effects of temperature on the activities of yeasts are so diverse that it is necessary to place some limits on the extent of coverage of this subject in the present review. The following areas, however, merit some discussion.

### A. FERMENTATION

According to Slator (1960), brewer's yeast ferments sugar in the range of 0°–43°, and fermentation is most rapid at 40°–42°. The average $Q_{10}$ value in the range 10°–35° was 2·97 which is in good agreement with $Q_{10}$ values obtained by other early investigators (*cf.* Sherwood and Fulmer, 1926). With baker's yeast, White and Munns (1951) found that the rate of sucrose fermentation increased almost linearly between 20° and 40° and then decreased rapidly at higher temperatures. Ingram (1955) obtained maximal initial glucose fermentation rates with several yeasts at 37°–45°. At 47°, the fermentation rate decreased or was nil. Also, as previously indicated, psychrophilic yeasts ferment glucose more rapidly at low temperatures than do mesophilic yeasts, and the process is inactivated at more moderate temperatures in the psychrophiles.

It has been observed with bacteria that the optimum temperature for fermentation is different from that for growth (Dorn and Rahn, 1939). This is true also for yeasts. The growth rate for a strain of *Sacch. cerevisiae* was maximal at 30° but fermentation proceeded most rapidly at 40° (White and Munns, 1951). Psychrophilic *C. gelida* P16 grew most rapidly at 15° but fermented glucose best at 25° (Sinclair and Stokes, 1965). Merritt (1966) found that a distiller's yeast grew maximally at 35° but exhibited maximal maltase activity and ethanol production at 25° and optimum rates of glycerol and higher alcohol formation at 30°.

### B. CELL COMPOSITION

Sterol synthesis is maximal in *Sacch. cerevisiae* at 30° and progressively decreases as the growth temperature is increased (Starr and Parks, 1962). At 40° sterol synthesis is limited and, above 40°, it is severely restricted. Kates and Baxter (1962) compared the lipids of three psychrophilic strains of *Candida* grown at 10° with those of mesophilic *C. lipolytica* grown at 25° or 10°. The total fatty acids of the psychrophiles and of the mesophile grown at 10° had much higher proportions of linoleic acid and lower proportions of oleic acid than those of the mesophilic yeast grown at 25°. This difference appears to be related to the observation of Meyer and Bloch (1963) that cell-free extracts of *Torulopsis* (= *Candida*) *utilis*, which oxidatively desaturates oleic acid to linoleic acid, are especially active when the yeast is grown at suboptimum temperatures. Such cells also contain a much higher proportion of the more highly unsaturated fatty acids.

Recently, Brown and Rose (1969) determined the effect of temperature on the chemical composition and cell volume of *C. utilis*. The yeast was grown in steady-state culture in a chemostat under glucose or $NH_4^+$ limitation at temperatures of 15°–30°. The deoxyribonucleic acid content of the cells remained fairly constant. However, under $NH_4^+$ limitation, ribonucleic acid, protein and cell volume increased and carbohydrate decreased with decrease in temperature. It has been conjectured that the fatty-acid composition of the cell, especially the cell membrane, may be a controlling factor in determining the temperature limits for growth of bacteria and yeasts. There is very little evidence, however, to support this hypothesis and considerable evidence against it (Marr and Ingraham, 1962; Kates and Baxter, 1962; Shaw and Ingraham, 1965).

## VI. Acknowledgements

This review was written by the author while he held a Special Research Fellowship of the U.S. Public Health Service from the National Institutes of General Medical Sciences. I am indebted also to Dr. H. A. Barker, Department of Biochemistry, University of California, Berkeley, California for the hospitality of his laboratory during this period.

## References

Adams, A. M. (1966). Horticultural Research Institute of Ontario, Ontario, Canada, pp. 114–117.

Adams, A. M. and Miller, J. J. (1954). *Can. J. Bot.* **32**, 320–334.

Ahearn, D. G. and Roth, F. J., Jr. (1966). *In* "Developments in Industrial Microbiology", pp. 301–309. American Institute of Biological Sciences, Washington, D.C.

Aref, H. and Cruess, W. V. (1934). *J. Bact.* **27**, 443–452.

Battley, E. H. (1964). *Antonie van Leeuwenhoek* **30**, 81–96.
Baum, H. M. (1943). *Biodynamica* **4**, 71–74.
Baxter, R. M. and Gibbons, N. E. (1962). *Can. J. Microbiol.* **8**, 511–517.
Beamer, P. R. and Tanner, F. W. (1939). *Zentbl. Bakt. ParasitKde* (*Abt II*) **100**, 202–211.
Brown, C. M. and Rose, A. H. (1969). *J. Bact.* **97**, 261–272.
Campbell, J. D. (1943). *Biodynamica* **4**, 65–70.
Cirillo, V. P., Williams, P. O. and Anton, J. (1963). *J. Bact.* **86**, 1259–1264.
Clinton, R. H. (1968). *Antonie van Leeuwenhoek*. **34**, 99–105.
Cooke, W. B. (1965). *Mycopath. Mycol. appl.* **25**, 195–200.
Cooney, D. G. and Emerson, R. (1964). "Thermophilic Fungi". W. H. Freeman and Co., San Francisco.
di Menna, M. E. (1960). *J. gen. Microbiol.* **23**, 295–300.
di Menna, M. E. (1966a). *Antonie van Leeuwenhoek* **32**, 25–28.
di Menna, M. E. (1966b). *Antonie van Leeuwenhoek* **32**, 29–38.
Dorn, F. L. and Rahn, O. (1939). *Arch. Mikrobiol.* **10**, 6–12.
Evison, L. M. and Rose, A. H. (1965). *J. gen. Microbiol.* **39**, 555–570.
Farrell, J. and Rose, A. H. (1967). *A. Rev. Microbiol.* **21**, 101–120.
Fell, J. W. and Phaff, H. J. (1967). *Antonie van Leeuwenhoek* **33**, 464–472.
Fries, N. (1963). *Physiologia Pl.* **16**, 415–422.
Godkin, W. J. and Cathcart, W. H. (1949). *Food Technol., Champaign* **3**, 139–146.
Grant, D. W., Sinclair, N. A. and Nash, C. H. (1968). *Can. J. Microbiol.* **14**, 1105–1110.
Guilliermond, A. (1920). "The Yeasts", (trans. F. W. Tanner). John Wiley, London.
Hagen, P. O. and Rose, A. H. (1961). *Can. J. Microbiol.* **7**, 287–294.
Hagen, P. O. and Rose, A. H. (1962). *J. gen. Microbiol.* **27**, 89–99.
Hartwell, L. H. (1967). *J. Bact.* **93**, 1662–1670.
Hartwell, L. H. and McLaughlin, C. S. (1968a). *Proc. natn. Acad. Sci. U.S.A.* **59**, 422–428.
Hartwell, L. H. and McLaughlin, C. S. (1968b). *J. Bact.* **96**, 1664–1671.
Ingraham, J. L. and Maaløe, O. (1967). *In* "Molecular Mechanisms of Temperature Adaptation", (C. L. Prosser, ed.), pp. 297–309. American Association for the Advancement of Science, Washington, D.C.
Ingraham, J. L. and Stokes, J. L. (1959). *Bact. Rev.* **23**, 97–108.
Ingram, M. (1955). "An Introduction to the Biology of Yeasts". Isaac Pitman and Sons, London.
Kates, M. and Baxter, R. M. (1962). *Can. J. Biochem. Physiol.* **40**, 1213–1227.
Larkin, J. M. and Stokes, J. L. (1968). *Can. J. Microbiol.* **14**, 97–101.
Lawrence, N. L., Wilson, D. C. and Peterson, C. S. (1959). *Appl. Microbiol.* **7**, 7–11.
Lund, A. (1958). *In* "The Chemistry and Biology of Yeasts", (A. H. Cook, ed.), pp. 84–87. Academic Press, New York.
Marr, A. G. and Ingraham, J. L. (1962). *J. Bact.* **84**, 1260–1267.
Mazur, P. (1966). *In* "Cryobiology", (H. T. Meryman, ed.), pp. 214–315. Academic Press, New York.
Merritt, N. R. (1966). *J. Inst. Brew.* **72**, 374–383.
Meyer, F. and Bloch, K. (1963). *Biochim. biophys. Acta* **77**, 671–672.
Morita, R. Y. (1968). *Bulletin Misaki Marine Biological Institute, Kyoto University* **12**, 163–177.
Nash, C. H. and Sinclair, N. A. (1968). *Can. J. Microbiol.* **14**, 691–697.
Norkans, B. (1966). *Arch. Mikrobiol.* **54**, 374–392.
Onishi, H. (1959). *Bull. agric. Chem. Soc., Japan* **23**, 351–359.
Onishi, H. (1963). *Adv. Fd Res.* **12**, 53–94.

Phaff, H. J., Miller, M. W. and Mrak, E. M. (1966). "The Life of Yeasts". Harvard University Press, Cambridge.
Phaff, H. J. and Mrak, E. M. (1949). *Wallerstein Labs Commun.* **12**, 29–44.
Quetsch, M. F. and Danforth, W. F. (1964). *J. cell. comp. Physiol.* **64**, 123–129.
Richards, O. W. (1934). *Cold Spring Harb. Symp. quant. Biol.* **2**, 157–166.
Rose, A. H. and Evison, L. M. (1965). *J. gen. Microbiol.* **38**, 131–141.
Shaw, M. K. and Ingraham, J. L. (1965). *J. Bact.* **90**, 141–146.
Sherman, F. (1959a). *J. cell. comp. Physiol.* **54**, 29–35.
Sherman, F. (1959b). *J. cell. comp. Physiol.* **54**, 37–52.
Sherwood, F. F. and Fulmer, E. L. (1926). *J. phys. Chem.* **30**, 729–756.
Sinclair, N. A. and Stokes, J. L. (1963). *J. Bact.* **85**, 164–167.
Sinclair, N. A. and Stokes, J. L. (1965). *Can. J. Microbiol.* **11**, 259–269.
Slator, A. (1960). *J. chem. Soc.* **89**, 128–142.
Stanley, S. O. and Rose, A. H. (1967). *Phil. Trans. R. Soc. Ser. B* **252**, 199–207.
Starr, P. R. and Parks, L. W. (1962). *J. cell. comp. Physiol.* **59**, 107–110.
Stille, B. (1950). *Arch. Mikrobiol.* **14**, 554–587.
Stokes, J. L. (1963). *In* "Recent Progress in Microbiology", Symposia VIII. International Congress for Microbiology, Montreal (N. E. Gibbons, ed.), pp. 187–192. University of Toronto Press, Toronto.
Stokes, J. L. (1967). *In* "Molecular Mechanisms of Temperature Adaptation", (C. L. Prosser, ed.), pp. 311–323. American Association for the Advancement of Science, Washington, D.C.
Straka, R. P. and Stokes, J. L. (1960). *J. Bact.* **80**, 622–625.
White, J. and Munns, D. J. (1951). *J. Inst. Brew.* **57**, 280–284.

## Chapter 5

# Structure and Biosynthesis of the Yeast Cell Envelope

H. J. Phaff

*Department of Food Science and Technology and Department of Bacteriology, University of California, Davis, U.S.A.*

I. INTRODUCTION . . . . . 135
II. CHEMISTRY AND BIOCHEMISTRY OF YEAST CELL WALLS . . . 137
A. Glucans . . . . . 137
B. Mannans . . . . . 145
C. Chitin . . . . . 162
D. Protein and Carbohydrate-Protein Complexes . . . . 164
E. Lipids . . . . . 173
III. CHEMISTRY AND BIOCHEMISTRY OF CAPSULAR POLYSACCHARIDES . 174
A. α-D-Linked Glucans . . . . . 175
B. β-D-Linked Glucans . . . . . 177
C. Mannans and Phosphomannans . . . . . 177
D. Phosphogalactans . . . . . 180
E. Galactomannan . . . . . 180
F. Pentosylmannans . . . . . 181
G. Acidic Heteropolysaccharides . . . . . 182
H. Miscellaneous Heteropolysaccharides . . . . . 185
IV. ENZYMIC LYSIS OF YEAST CELL WALLS . . . . . 186
A. Gastric Juice of the Garden Snail *Helix pomatia* . . . . 186
B. Lysis of Walls by Microbial Enzymes . . . . . 189
V. REGENERATION OF CELL WALLS FROM SPHAEROPLASTS . . . . 193
A. On the Question of Protoplasts versus Sphaeroplasts . . . . 193
B. Cell Wall Regeneration . . . . . 194
VI. MORPHOLOGY . . . . . 198
A. Localization of Wall Components . . . . . 198
B. Expansion Growth . . . . . 200
VII. CELL WALL COMPOSITION AS A BASIS FOR DIFFERENTIATING YEASTS . 200
REFERENCES . . . . . 202

## I. Introduction

Current knowledge of the nature of the cell envelope of yeasts is based on the integration of data obtained by chemists, biophysicists, enzymologists and cytologists. Early investigators generally subjected whole cells (nearly always baker's or brewer's yeast) to more or less drastic chemical treatments in order to obtain cell-wall residues or to extract specific wall components. However, it is now well appreciated that some cell-wall components were not recognized at all and discarded in alkaline extracts; furthermore, extensive chemical degradation of certain molecules and rupture of linkages occurred.

With the advent of procedures to prepare purified cell walls with various instruments that effect cell breakage, much progress has been made in the study of actual walls. These procedures, which will not be discussed in detail, include the use of sonic energy, mechanical energy (usually coupled with the inclusion of glass beads in the cell suspension), or combinations thereof. Many of these procedures have been described in detail in "Methods in Enzymology" (Wood, 1966) to which the reader is referred. In our own laboratory we have found the following three methods the most effective for yeasts: (i) the Bronwill homogenizer (manufactured by Bronwill Scientific Inc., Rochester, N.Y., U.S.A.), which employs high speed shaking with 0·45 mm diam. glass beads and cooling with liquid carbon dioxide (Merkenschlager *et al.*, 1957), for the majority of species 10 g of yeast paste being broken quantitatively in a matter of minutes; (ii) the Eppenbach colloid mill (Model QV-6: manufactured by Gifford-Wood, Co., Hudson, N.Y., U.S.A.) for large quantities of yeast paste (500–1,000 g), requiring at least 1 h to obtain a high percentage of cell breakage, and employing 120 μm diam. glass beads, cooling being done with ice water (Garver and Epstein, 1959); (iii) a modified French press as described by Simpson *et al.* (1963). In all cases, after homogenization is complete, cell walls should be quickly purified by washing because of possible degradation of wall components by yeast enzymes (Abd-El-Al and Phaff, 1968, 1969).

From purified cell walls a number of wall components have been isolated and studied in detail. Considerable progress has also been made on the elucidation of the types of linkages which hold the major components together in the intact cell, although much remains to be done in this area. Similarly, on the subject of biosynthesis of wall components, only sketchy information is as yet available. Studies on yeast species other than those belonging to the genus *Saccharomyces* have shown great diversity in structure and composition, not only in the cell wall, but in the capsular polysaccharides of capsulated yeasts as well. The present chapter will deal with the actual wall and with capsular materials where applicable, but will not consider the cytoplasmic membrane *per se* as this is taken up in Chapter 6 (p. 211).

Other reviews on the cell wall of yeasts and related aspects have been published by Northcote (1963), Phaff (1963), Clarke and Stone (1963),

Bull and Chesters (1966), Gorin and Spencer (1968a) and Bartnicki-Garcia (1968).

## II. Chemistry and Biochemistry of Yeast Cell Walls

In this section chemical, biochemical and physical analyses which have been made of the principal components of the walls of different yeasts will be reviewed. These components induce glucans, mannans, chitin, protein and lipids. Capsular polysaccharides will be covered in a separate section, even though in some instances there is evidence that the wall itself may contain the same polysaccharides as those found in the capsule.

### A. GLUCANS

#### 1. *Chemical Studies*

Salkowski (1894a) prepared from yeast an insoluble polysaccharide which was termed "yeast cellulose". Zechmeister and Toth (1934, 1936) re-investigated this substance and showed by methylation analysis that it contained a preponderance of (1→3) linkages. All other polysaccharides, including cellulose, which had been investigated up to that time, contained (1→4) bonds between the glucose units. The product also differed from cellulose by not being soluble in ammoniacal copper oxide solution (Schweitzer's reagent) and by not giving a characteristic blue colouration with a solution of iodine in potassium iodide and treatment with strong acid. In order to avoid confusion with cellulose, Zechmeister and Toth referred to their material as "yeast polyose". Hassid *et al.* (1941) confirmed the presence of (1→3) linkages in baker's yeast glucan by methylation analysis. Since 2,4,6-trimethylglucose was obtained as the sole product of hydrolysis and no end group(s) could be detected, they suggested that the molecule was probably of the closed chain type. Low specific rotations of acetylated and methylated derivatives and an upward rotation during hydrolysis suggested a predominance of $\beta$-linkages in the glucan. Its molecular weight was estimated by the Staudinger viscosity method to be approximately 6,500 (degree of polymerization, 40). Very similar results were obtained by Barry and Dillon (1943) who also worked with baker's yeast. The presence of (1→3) linkages was shown by the lack of oxidation by periodic acid, except for one end group per 28 glucose units. In addition, they characterized the osazone of laminaribiose which occurred among the hydrolysis products of the glucan, and showed that these products were degraded by the $\beta$-glucosidase emulsin but not by an $\alpha$-glucosidase. Although the presence of $\beta$-(1→3) linkages as well as end groups was therefore well established, Barry and Dillon (1943) only vaguely implied the possibility of a branched structure in the glucan. Instead, they compared yeast glucan with laminarin from brown algae and stated that the former had a longer average chain length (degree of polymerization, 28) than the latter (degree of polymerization, 16).

In nearly all of these early experiments, yeast glucan was isolated after drastic treatment of whole yeast first with hot dilute alkali, followed by heating for several hours with dilute hydrochloric acid. This left an insoluble residue which was washed with water, ethanol and ether, and dried.

Bell and Northcote (1950) used a similar approach for the preparation of glucan from baker's yeast, except that they replaced mineral acids by 0·5 N-acetic acid at 75–80° to avoid degradation of glucan during extraction of cell glycogen. In addition, oxygen was excluded during methylation of the polysaccharide as long as the system was alkaline. They were the first, on the basis of methylation analysis, to obtain evidence for a highly branched polysaccharide of large molecular weight. Their evidence indicated unit chains with an average of nine glucose units linked by $\beta$-(1→3) bonds and with interchain links of the (1→2) type.

Further work was done by Peat *et al.* (1958a) who purified baker's yeast glucan as recommended by Bell and Northcote (1950), except that they found it necessary to autoclave the residue, after acetic acid extraction, with 0·02 *M*-sodium acetate at pH 7·0 and several times with water to remove glycogen completely. They partly hydrolysed the glucan by heating in 90% formic acid (in which the glucan dissolved) followed by hot dilute sulphuric acid. The following products of hydrolysis were identified: glucose, gentiobiose, laminaribiose, gentiotriose, 6-*O*-$\beta$-laminaribiosyl-glucose, laminaritriose, 3-*O*-$\beta$-gentiobiosylglucose, and gentiotetraose. No 2-*O*-$\beta$-D-glucopyranosyl-D-glucose (sophorose) was found, as might have been expected from the work of Bell and Northcote (1950), and they suggested that the results of the last mentioned authors could have been due to analysis of incompletely methylated glucan. Instead, $\beta$-(1→6) linkages were demonstrated for the first time based on the occurrence of oligosaccharides of the gentiobiose series. Since no trisaccharide represented by 3, 6-di-*O*-$\beta$-glucosylglucose was detected among the hydrolysis products, Peat *et al.* (1958a) felt that the glucan molecule was probably unbranched. Thus yeast glucan was pictured as a linear molecule in which (1→3) and (1→6) linkages occur at random or in sequences such that a group of at least three (1→6) linkages is flanked on either side by (1→3) bonds. Periodate oxidation to which only end groups and glucose units substituted in the 6-position are susceptible, showed that about 10% of the glucose residues were in this category. In a subsequent publication (Peat *et al.*, 1958b) glucan was subjected to the technique of toluene-*p*-sulphonylation by which the free primary hydroxyl groups are esterified much faster than the secondary ones. The primary tosyl groups can be replaced quantitatively by iodine under conditions which leave the secondary tosyl groups unaffected. An analysis of the iodine content of the product showed that 10–20% of the primary hydroxyl groups of the glucose units were involved in (1→6) linkages, thus confirming the results with periodate oxidation.

Work by Misaki and Smith (1963), initially described in a brief abstract and much later in a full length paper by Misaki *et al.* (1968), resulted in a return to the notion of a highly branched glucan, but with branching residues of D-glucose having linkages at positions 0–1, 0–3 and 0–6. Their glucan was made from baker's yeast in the same manner as was done by Peat and his coworkers but, in addition, the dried glucan was dissolved in methyl sulphoxide, from which it was precipitated with water. Analysis of the purified product by periodate treatment (Smith degradation) and a study of the fully methylated D-glucan revealed that the repeating unit of the glucan consists of one terminal non-reducing residue, seven β-(1→3)-linked non-terminal residues and one branching residue joined through positions 0–1, 0–3 and 0–6. The absence of a significant number of unsubstituted glucose residues in the backbone linked by β-(1→6) bonds was deduced: (i) by the formation of only trace amounts of 2,3,4-tri-*O*-methyl-D-glucose among the hydrolysis products of the essentially fully methylated glucan; and (ii) by the fact that treatment of the original glucan with periodate yielded a breakdown product which still had a degree of polymerization of 150 instead of being an oligosaccharide. This treatment will cleave all unsubstituted β-(1→6)-linked glucose residues in the backbone. Misaki *et al.* (1968) proposed, on the basis of the above evidence and on the lack of solubility of the glucan in water or alkali, the structures shown in Fig. 1 as working models for their parental yeast glucan and that subjected to degradation by periodate.

There is, however, some difficulty in reconciling this structure entirely with the finding of gentiobiose and up to gentiotetraose by Peat and his coworkers (1958a) in partial acid hydrolysates of their yeast glucan preparation. Confirmation for the presence of D-glucose triply-linked at positions 0–1, 0–3 and 0–6, and for the absence of 4,6-di-*O*-methyl-D-glucose (which would indicate (1→2) bonds) was supplied by Manners and Patterson (1966), who investigated three independently prepared samples of yeast glucan, including one from Peat and his coworkers. Their methylation analyses indicated a greater proportion of unsubstituted β-(1→6)-linked glucose residues in the backbone than fits the model proposed by Misaki *et al.* (1968). Manners and Patterson (1966) therefore suggested that different samples of glucan differ in the degree of substitution of the main chain.

A possible explanation to reconcile the observed differences was recently offered by Bacon and Farmer (1968) and Bacon *et al.* (1969). During the extraction of alkali-treated yeast with hot dilute acetic acid, followed by autoclaving in water, a water-soluble $\beta$-(1→6) glucan appeared with the glycogen. They did not study this glucan in detail, but its degree of removal very likely depends on the thoroughness with which the glucan is treated with acetic acid and water. Manners and Masson (1969) extracted baker's yeast first with hot alkali to remove mannan and, next, the residue was extracted 27 times with 0·5 M-acetic acid for three-hour periods at 90°. After removal of glycogen from the

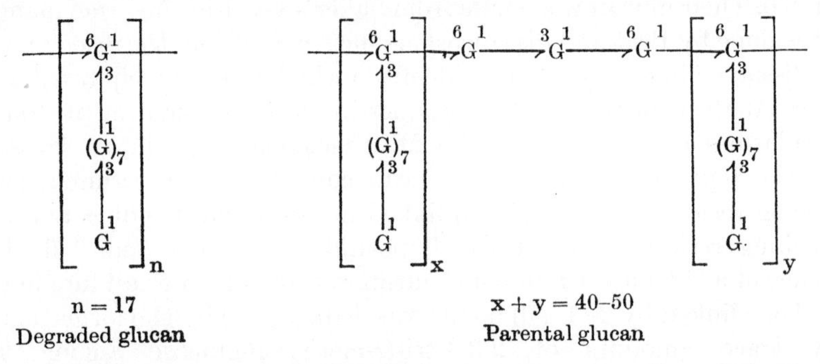


FIG. 1. Schematic representation of baker's yeast glucan according to Misaki *et al.* (1968). The formula on the left represents native glucan after treatment and resulting cleavage by periodate. G indicates a $\beta$-D-glucopyranosyl residue.

extract, they characterized the remaining polysaccharide as a linear $\beta$-(1→6) glucan with a degree of polymerization of $140 \pm 10$ and $[\alpha]_D$ −32°. A very small proportion of $\beta$-(1→3)-linked glucose residues was detected in a partial acid hydrolysate, but the polysaccharide was not hydrolysed by endo-$\beta$-(1→3) glucanase. The latter finding is indicative of the fact that no consecutive $\beta$-(1→3)-linked glucose residues were present.

The residual insoluble cell-wall glucan was also investigated by Manners and Masson (1969). Since this residue still appeared to contain remnants of $\beta$-(1→6) glucan, it was first treated with endo-$\beta$-(1→6) glucanase to remove this impurity. It should be noted that such enzymes probably do not cleave two unsubstituted glucose units connected by $\beta$-(1→6) bonds because gentiobiose is not a substrate of this enzyme. The treated glucan remained insoluble and had a degree of polymerization of approximately 1,450. Periodate oxidation studies and methylation analysis gave results which suggested that this glucan molecule is less highly branched than proposed by Misaki *et al.* (1968) and that the chains of β-(1→3)-linked glucose residues may be much longer, perhaps 15 to 30 units long. Manners and Masson (1969) point out that the observed degree of branching is a statistical value, and that individual chains or side branches may vary greatly in length. Further details of its structure remain to be clarified.

A possibly similar type of glucan with a high proportion (63–73%) of unsubstituted β-(1→6) linkages was described earlier by Bishop *et al.* (1960) in *Candida albicans* serotype A, and by Yu *et al.* (1967a) in *C. albicans* serotype B and in *C. parapsilosis*. In contrast to the extraction procedure of Bacon and Farmer, the glucan of *C. albicans* and *C. parapsilosis* was obtained in solution by boiling with 3% sodium hydroxide. However, besides representing a different species, *C. albicans* cells were first autoclaved, dried, ground in a ball mill, extracted with petroleum ether and digested with trypsin. These treatments could cause different solubility properties of the respective glucans, as is suggested also by the work of Bacon *et al.* (1969) for walls of baker's yeast. Methylation analysis and periodate oxidation of the alkali-extracted glucan of *C. albicans* (the insoluble residue was not studied) showed it to be a highly branched molecule (through glucose residues substituted in positions 0–1, 0–6 and 0–3) with approximately 73% β-(1→6) linkages and the remainder constituting β-(1→3) bonds. The glucan of *C. parapsilosis* was quite similar except that, in addition to branching at C-3, a smaller proportion of branches at C-4 was detected also.

A number of different glucans have been described from species other than those discussed above. In 1953, Houwink and Kreger briefly stated that the X-ray diffraction patterns of cell walls of *Sacch. cerevisiae* and *Schizosacch. octosporus*, after boiling for 2 h with 2% hydrochloric acid, were clearly different. These wall preparations were prepared by mechanical rupture of cells and the boiling process caused the residual glucan to become soluble in dilute alkali, to appear as fibrils in electon micrographs, and to yield a sharp X-ray diagram. Kreger (1954) found that walls of *Schizosacch. octosporus*, *Schizosacch. pombe* and *Schizosacch. versatilis* all contained 30–35% of an alkali-soluble micro-crystalline glucan, which had an X-ray diagram quite different from "hydroglucan" of baker's yeast (glucan made alkali-soluble by boiling with hydrochloric acid). In addition, the *Schizosaccharomyces* species contained a decreased amount (about 10%) of glucan resembling that of *Saccharomyces* spp. Kreger (1954) also identified this alkali-soluble glucan in several higher fungi. More recently Kreger (1965) found that the alkali-extracted atypical glucan from *Schizosacch. octosporus* precipitated upon neutralization of the alkaline extract. Based on an alkali-soluble extract from the fungus *Schizophyllum* (Wessels, 1965) with similar properties (including its X-ray diffraction pattern) and which was shown to have

$\beta$-(1→6)-linked glucose residues, Kreger assumed that the glucan from *Schizosaccharomyces* was also $\beta$-linked. Recently this material was restudied by Bacon *et al.* (1968) who also found it as the major component in the cell wall of *Cryptococcus terreus* and *Cr. albidus*. By the techniques of infrared spectroscopy, enzymic analysis and a comparison of numerical values for the spacings of X-ray diffraction maxima, the conclusion was reached that the alkali-soluble glucan contained $\alpha$-(1→3)-linked glucose residues. The proportion of $\alpha$-glucan in *Cryptococcus* walls appears to be strongly affected by the composition of the growth medium (Jones *et al.*, 1969). The $\alpha$-(1→3) glucan appears to occur fairly widespread among higher fungi belonging to either the Basidiomycetes or Ascomycetes as well as some yeasts (Kreger, 1954; Bacon *et al.*, 1968). Hasegawa *et al.* (1969) studied this polysaccharide as it was extracted from mycelium of *Aspergillus niger*. If this polymer has the same structure as that occurring in *Schizosaccharomyces* it may be assumed, until more precise structural analyses have been performed, that the new glucan is an unbranched polymer of high molecular weight composed entirely or nearly entirely of $\alpha$-(1→3)-linked glucopyranosyl units (Hasegawa *et al.*, 1969). As far as yeasts are concerned, the polysaccharide has been detected in *Endomyces decipiens*, three species of *Schizosaccharomyces*, and two of *Cryptococcus*, but it appeared absent from *Lipomyces starkeyi*. Kanetsuna *et al.* (1969) found that the yeast phase of the dimorphic fungus, *Paracoccidioides brasiliensis*, contained almost exclusively $\alpha$-linked alkali-soluble glucan, whereas the mycelial form of this pathogenic fungus contained only 60–65% of this type of glucan. It appears possible that the extracellular neutral polysaccharide isolated from the culture fluid of *Cryptococcus laurentii* by Abercrombie *et al.* (1960a) may be the same or related to that studied by Bacon *et al.* (1968). The *Cr. laurentii* polysaccharide contained D-glucose only, had an $[\alpha]_D$ value of +180° (suggesting a preponderance of $\alpha$-linkages), but contained (1→3), (1→4), (1→2) and (or) (1→6) linkages.

A presumably different type of alkali-soluble glucan was described by Eddy and Woodhead (1968), who extracted cell walls of *Sacch. carlsbergensis* with 3% sodium hydroxide under nitrogen gas at 4° for a number of days. The glucan precipitated when the unboiled extract was brought to pH 9 with acetic acid. When the purified polysaccharide was subjected to acid hydrolysis, glucose was the main sugar identified. It constituted approximately 20% of the weight of the cell walls, was not hydrolysed by a mixture of $\alpha$- and $\beta$-amylases, had a molecular weight of about $5{\cdot}2\ (\pm 1{\cdot}2) \times 10^5$, and formed a single boundary during ultracentrifugation. No conclusive evidence was given as to the nature of the bonds between the glucose residues, but the authors considered the possibility that the polysaccharide might be a metabolic precursor of alkali-insoluble yeast glucan. It is conceivable that the β-(1→6)-linked glucan described by Bacon and Farmer (1968) is similar or related.

2. *Enzymic Studies*

Enzymic analysis of whole cell walls or of purified wall fractions with various glucanases has contributed significantly to chemical studies. The use of enzymes for the preparation of sphaeroplasts will be covered later on. Tanaka and Phaff (1965) found that a strain of *Bacillus circulans* grown on baker's yeast cell walls produced extracellularly a mixture of endo-β-(1→3) and endo-β-(1→6) glucanases, which could be separated by column chromatography on DEAE-cellulose. Treatment of baker's yeast cell walls with either of the enzymes caused lysis, although the β-(1→3) glucanase was by far the more effective. Tanaka (1963) treated baker's yeast cell walls with the two separate endoglucanases, and observed that the ratio of linkages in the glucan susceptible to hydrolysis by β-(1→6) glucanase and by β-(1→3) glucanase was 1 : 2, whereas in walls of *Hansenula anomala* the respective ratio was approximately 1·3 : 1, suggesting a majority of β-(1→6) bonds in cell walls of this species. Buecher (1968) found that the activity of the above two endoglucanases on cell walls was greatly accelerated and much more complete in the presence of 2-mercaptoethanol. He repeated Tanaka's analyses of baker's yeast walls and found a ratio of linkages susceptible to β-(1→6) and β-(1→3) glucanases of 1 : 1 for baker's yeast and of 1·5 : 1 for cell walls of *Saccharomycopsis guttulata*. Since the endo-β-(1→6) glucanase of *Bacillus circulans* cannot hydrolyse single β-(1→6) bonds, as in gentiobiose, and hydrolyses gentiotriose very slowly, it would appear that there exist long blocks of β-(1→6)-linked glucose chains in these yeast walls. This is supported by the demonstration of gentiobiose, gentiotriose and gentiotetraose as hydrolysis products of cell walls of *H. anomala* hydrolysed by β-(1→6) glucanase (Tanaka, 1963). The possibility should be considered that *H. anomala* may contain a much higher proportion of β-(1→6)-linked glucan than does *Sacch. cerevisiae*. These findings are still meaningful even though Parrish *et al.* (1960) postulated, on the basis of experiments with oat glucan, that, for polysaccharidases which hydrolyse polysaccharides with mixed linkages, the *position of substitution* in the glycosyl unit to be cleaved determines the enzyme specificity for the hydrolysis of the glycosidic bond connecting C-1 to the next glucose unit. The endo-β-(1→3) glucanase of *B. circulans* follows this rule since it was shown to split β-(1→4) bonds in oat glucan, provided the glucosyl unit to be split was itself substituted in the 3-position.

The findings of Tanaka (1963) and Buecher (1968) do not fit the model for baker's yeast glucan proposed by Misaki *et al.* (1968), because this glucan does not possess a significant number of β-(1→6)-linked glucose residues without side chains. However, the isolation of glucans with a high proportion of β-(1→6) bonds in certain *Candida* species (Bishop *et al.*, 1960; Yu *et al.*, 1967a) and the discovery by Bacon and Farmer (1968) of a predominantly β-(1→6)-linked glucan component in the cell wall of baker's yeast, reconciles the experimental data of the various investigators. It would appear therefore that, during the hydrolysis of cell walls by a mixture of the two endoglucanases, oligosaccharides of the laminaribiose series arise from the side chains of the classical yeast glucan molecule, and those of the gentiobiose series from the β-(1→6)-linked glucan. The different ratios of susceptible β-(1→3) and β-(1→6) bonds in various species (see above) can then be explained by the presence of different proportions of predominantly β-(1→6)-linked glucan in the walls of these species.

Manners and Patterson (1966) treated a baker's yeast glucan preparation obtained by the procedure of Bell and Northcote (1950) with a commercial endo-β-(1→3) glucanase which was free of (1→6) glucanase activity. They found oligosaccharides of the laminaribiose series together with about 10% of a water-soluble glucan with mainly β-(1→6) bonds. These linkages were identified by the oligosaccharides formed in a partial acid hydrolysate, by methylation analysis and by enzymic hydrolysis with a β-(1→6) glucanase. It seems very possible that this polymer represents the predominantly β-(1→6)-linked glucan described by Bacon and Farmer (1968) and later characterized by Manners and Masson (1969), because it appears unlikely that the laminarinase used by Manners and Patterson would have removed all of the glucose residues attached by β-(1→3) bonds to the β-(1→6)-linked backbone of their glucan preparation, unless their enzyme preparation was contaminated with exo-β-(1→3) glucanase.

Tanaka *et al.* (1965) compared a number of yeast species and found that the cell walls of the yeasts studied showed profound differences in their susceptibilities to the separate or combined action of the endo-β-glucanases from *B. circulans*. The different types of reaction can be grouped as follows: (i) complete lysis by β-(1→3) glucanase but weak and incomplete lysis by β-(1→6) glucanase (e.g. *Saccharomyces cerevisiae*); (ii) complete lysis by either of the glucanases (e.g. *Hansenula anomala*, *H. ciferrii*, *Nadsonia elongata*, *Ashbya gossypii*); (iii) weak or incomplete lysis by either of the glucanases separately or combined (e.g. species of *Schizosaccharomyces*, *Lodderomyces* (*Saccharomyces*) *elongisporus*); (iv) no action by the combined glucanases (e.g. *Rhodotorula rubra*). *Debaryomyces hansenii* showed a unique behaviour in that its walls were only partially and weakly digested by the individual enzymes, but quite extensively by the combined action of both β-glucanases. Tanaka *et al.* (1965) also developed a new technique of cross induction, whereby the lytic organism, *B. circulans*, is inoculated in a straight line on a rectangular agar block containing the walls of species A (the inducer). Pushed against the agar strip are several small square test blocks of agar containing the walls of species B, C, D and so on. By observing the clearing of the central strip and the action of the induced enzymes as they diffuse into the satellite blocks, similarities and differences in cell-wall composition can be determined. For example, it gave evidence of significant differences in the cell-wall composition of *Lodderomyces elongisporus*, *Sacch. cerevisiae* and *Schizosaccharomyces* species. Major proportions of an $\alpha$-(1→3)-linked glucan in *Schizosaccharomyces* (Bacon *et al.*, 1968) is undoubtedly responsible for the very weak lysis of the walls of the fission yeasts by the $\beta$-glucanases of *B. circulans*. Hasegawa *et al.* (1966, 1969) observed enzyme activity specific toward $\alpha$-(1→3) glucosidic bonds in the culture fluid of *Trichoderma viride* grown on $\alpha$-(1→3) glucan (pseudonigeran). The enzyme was obtained in a state of high purity, and it hydrolysed the linear water-insoluble polymer by a random mechanism at an optimum pH value of 4·5 to dimer (nigerose) and glucose. Its $K_m$ value was found to be $4{\cdot}6 \times 10^{-2}$ M glucose equivalents, and the optimum temperature was 50°.

The structure of this $\alpha$-(1→3)-linked glucan was also subjected to enzymic analysis by Bacon *et al.* (1968). It was first freed from small amounts of $\beta$-glucan by treatment with $\beta$-glucanase from *Cytophaga johnsonii*, yielding a pure $\alpha$-glucan. This was then treated with an $\alpha$-(1→3) glucanase from a *Streptomyces* species, which catalysed hydrolysis by a random pattern to oligosaccharides. These fungal and bacterial endo-$\alpha$-D-(1→3) glucanases would appear to be useful in comparative structural studies of yeast cell walls, or for the preparation of sphaeroplasts of yeasts refractory to the action of snail enzymes.

## B. MANNANS

### 1. *Chemical Studies on Baker's Yeast Mannan*

Yeast mannan, which at the time of its discovery by Salkowski (1894b) was termed "yeast gum", was first studied in detail by Haworth *et al.* (1937, 1941). It could be extracted from whole yeast by boiling with 6% sodium hydroxide, and freed from other soluble polysaccharides by precipitation as its insoluble copper complex (addition of Fehling's solution to the alkaline extract). The precipitate was then washed with water, dissolved by adding hydrochloric acid and finally precipitated with alcohol. This procedure, with some variations, has been used by most subsequent investigators up to the present. The purified powder was composed exclusively of mannose residues and was thought to be essentially homogeneous. The homogeneity of such preparations (at least from a single species of yeast, *Sacch. cerevisiae* in this case) was later confirmed by Northcote (1954), who showed that it behaved as a single molecular species when subjected to electrophoresis in a Tiselius apparatus.

Haworth and coworkers examined the mannan by methylation analysis and found it to be a highly branched molecule with a degree of polymerization of 200–400. Its high dextrorotation indicated the presence of $\alpha$-linkages, although the presence of some $\beta$-linkages could not be excluded. Based on an analysis of the methylated sugars present after hydrolysis, the three repeating structures shown in Fig. 2 fitted the experimental data. Thus, the side chains could be single mannose

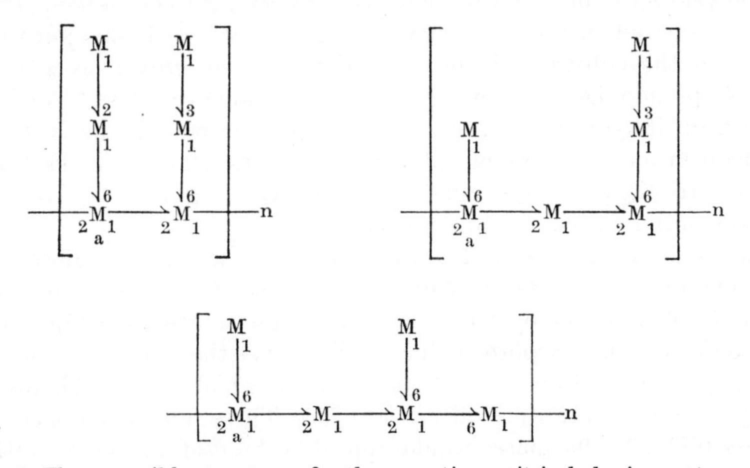

FIG. 2. Three possible structures for the repeating unit in baker's yeast mannan according to Haworth *et al.* (1937, 1941). See text for details. M indicates an $\alpha$-D-mannopyranosyl residue.

residues, or two, or a combination of one and two as shown in the upper right structure in Fig. 2. The finding of a small amount of 2,3,4-trimethyl mannose (indicating the presence of 1→6 linkages in unsubstituted mannose residues) was explained by assuming that the terminal non-reducing end of the chain occurs at position (a) in the figure, and this mannose residue would yield that particular methyl derivative.

Lindstedt (1945) analysed yeast mannan by the technique of periodate oxidation and confirmed the presence of (1→6), (1→2) and (1→3) linkages in the proportion of 2 : 3 : 1. Cifonelli and Smith (1955) repeated the methylation analysis of alkali-extracted mannan and confirmed in every particular the earlier results of Haworth *et al.* (1941).

Peat *et al.* (1961a, b), in order to avoid degradation of the mannan by hot alkali, extracted it from baker's yeast by autoclaving in citrate buffer at pH 7·0. The purified product was then subjected to linkage analysis, i.e. partial acid hydrolysis, fractionation, and examination of the fragments. Evidence was obtained showing that bonds other than α-(1→6) were preferentially hydrolysed. Taking into account the possible complication of acid reversion during hydrolysis, a homologous series of α-(1→6)-linked mannosyl oligosaccharides with a degree of polymerization of between 2 and 5 was identified in a hydrolysate with an apparent conversion to mannose of 67%. From these results, Peat and his coworkers concluded that the backbone of the highly branched mannan must contain α-(1→6) and not α-(1→2) linkages as had been postulated by Haworth and his coworkers. Enzymic proof for the existence of an α-(1→6)-linked backbone (Jones and Ballou, 1968) will be discussed in Section B.5 (p. 157). In an accompanying article (Peat *et al.*, 1961b), further results were described based on fragmentation of mannan by

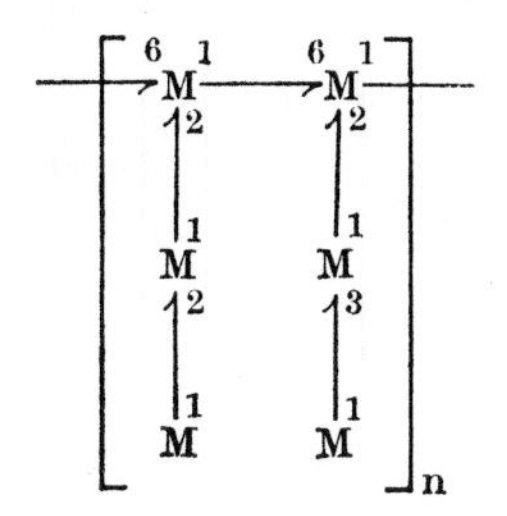

Fig. 3. Repeating unit in baker's yeast mannan according to Peat *et al.* (1961b). M indicates an α-D-mannopyranosyl residue.

acetolysis. This milder treatment, earlier applied to yeast mannan by Gorin and Perlin (1956) and by Jones and Nicholson (1958), yielded two different disaccharides in the acetolysate and provided direct evidence for the presence of α-(1→2) as well as α-(1→6) bonds in these disaccharides, but no trace of an α-(1→3) linkage was found in any of the fractions including higher oligosaccharides. Since methylation analysis was also done, showing that (1→2) linkages (other than those involved in the branching) were roughly equal in number to (1→3) bonds, they explained the low percentage of (1→2) bonds and the absence of (1→3) bonds in the acetolysate on the basis of relative stabilities of the linkages towards acid, in the order (1→6) >> (1→2) > (1→3). Based upon the combined evidence the authors arrived at the structure shown in Fig. 3 to represent the simplest repeating unit in baker's yeast mannan. Since Peat *et al.* (1961b) confirmed the earlier findings of 2,3,4-tri-*O*-methyl-D-mannose in the hydrolysate of methylated mannan (one mole in 45), they postulated that there may be an occasional mannose residue in the backbone not carrying a side chain, or that it constitutes the reducing unsubstituted end group of the backbone chain. As shown below, this conclusion proved to be correct.

Lee and Ballou (1965) obtained significantly different results when baker's yeast was subjected to controlled acetolysis. They found, in agreement with Gorin and Perlin (1956), that acetolysis selectively cleaved the α-(1→6) linkages of the backbone, giving high yields of oligosaccharides, none of which contained (1→6) bonds. Thus, in these oligosaccharides, the mannose unit at the reducing end originally was a residue in the backbone. After deacetylation, the oligosaccharides were separated on a Sephadex column and identified as mannobiose, mannotriose (both with α-(1→2) linkages) and a novel tetrasaccharide, *O*-α-D-mannopyranosyl-(1→3)-*O*-α-D-mannopyranosyl-(1→2)-*O*-α-D-

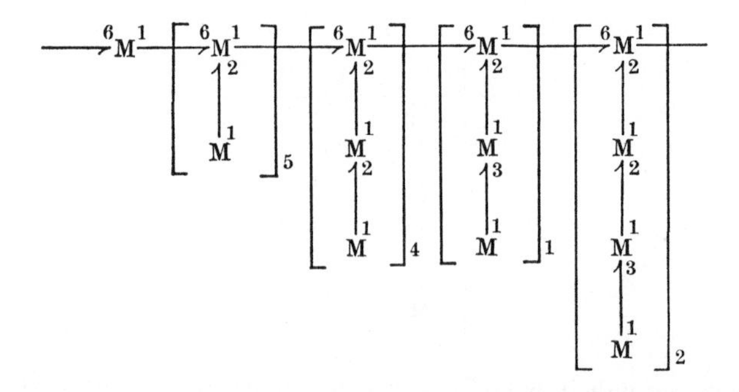

FIG. 4. Schematic structure of baker's yeast mannan according to Stewart and Ballou (1968). The order in which the side-chains occur is unknown. The subscripts outside the brackets indicate the average molecular proportions of the various types of side-chains. M indicates an α-D-mannopyranosyl residue.

mannopyranosyl-(1→2)-D-mannopyranose. Later Stewart *et al.* (1968) identified another trisaccharide, *O*-α-D-mannopyranosyl-(1→3)-*O*-α-D-mannopyranosyl-(1→2)-D-mannopyranose, as an integral part of the mannan molecule. It constituted 20% of the trisaccharide fraction. They refined the conditions of acetolysis so that further degradation of the oligosaccharides was minimized, and thus the ratio of small oligosaccharides obtained from Sephadex-column analysis very likely represents that present in the intact mannan. Not implying any particular ratio or order of the units in the chain, Stewart and Ballou (1968) proposed the basic structure for baker's yeast mannan shown in Fig. 4. The number of (1→6)-linked mannose residues in the backbone not substituted in the secondary hydroxyl groups was very small in the polymer from *Sacch. cerevisiae*, which agrees with previous reports. They also compared the mannans of several different strains of *Sacch. cerevisiae*, and one strain was grown under different conditions. It appears that there are differences in the ratios of oligosaccharides between different strains of *Sacch. cerevisiae*, but that growing conditions had no significant effect, even though the glucan : mannan ratio may be affected by nutritional factors such as biotin concentration in the medium (Dunwell *et al.*, 1961).

2. *Mannans from Yeasts other than* Saccharomyces cerevisiae

Because of its ready availability, nearly all of the earlier structural studies of yeast mannan were done with baker's yeast. However, Garzuly-Janke (1940) had shown by the Fehling's precipitation test of alkaline whole-cell extracts that mannan is widely distributed among the yeasts and only a few species seem to be devoid of it. In recent years, a great deal of interesting information has come to light showing numerous variations in structure as additional species were investigated.

Gorin and Perlin (1956, 1957) studied a mannan which was released into the culture medium when *Sacch. rouxii* (an osmophilic non-capsulated yeast) was grown under conditions favouring polyhydric alcohol formation. Presumably the mannan represented cell-wall matter which was either released during active growth or liberated during autolysis during later stages in the growth cycle. The mannan, which was studied by acetolysis (yielding α-(1→2)-linked mannobiose and mannotriose) and by methylation analysis, differed from baker's yeast mannan by either lacking, or having at most a very small number of, α-(1→3) linkages. Furthermore the side chains were limited to only one or two mannose residues linked by α-(1→2) bonds. The mannan of *Sacch. carlsbergensis*, a close relative of *Sacch. cerevisiae*, differs from that of the latter species (Stewart and Ballou, 1968) only in a quantitative way in that it has a considerably smaller number of trisaccharide side chains, a somewhat greater proportion of unsubstituted α-(1→6)-linked mannose units in the backbone, and a higher phosphate content (see Section II.B.3, p. 154).

The mannan of *Candida albicans*, a potentially pathogenic yeast, has been studied extensively in a number of laboratories. Bishop *et al.* (1960) extracted mannan with alkali from dried cells which had been previously ground in a ball-mill, extracted with solvent and digested with trypsin. The mannan was finally purified by precipitation with Fehling's solution. A highly branched structure was found in which relatively short chains of α-(1→2)-linked mannose units were linked to a presumed α-(1→6)-linked backbone in the 2-position of the backbone. Assumption of an α-(1→6)-linked backbone may lead to erroneous conclusions regarding structures, since Gorin and Spencer (1970) isolated a mannan from a *Candida* species which contained, in the main chain, short blocks of α-(1→3)-linked mannose residues alternating with α-(1→6)-linked blocks. No evidence was obtained by Bishop *et al.* (1960) for the presence of α-(1→3) bonds as are found in side chains of baker's yeast mannan. Yu *et al.* (1967b), who studied the mannans of *C. albicans*, *C. stellatoidea*, *C. parapsilosis* and *C. tropicalis*, found the polysaccharides rather similar in structure, although there was some variation from species to species. Surprisingly they found that some of the mannose residues appeared to be in the furanose configuration, since they found 3,5-di-*O*-methyl-D-mannose among the hydrolysis products from the methylated polysaccharide. However, Bhattacharjee and Gorin (1969) gave proof that the presumed 3,5-di-*O*-methyl-D-mannose was misidentified, and was actually 2,4-di-*O*-methylmannose. Kocourek and Ballou (1969) also questioned the occurrence of mannofuranose linkages in yeast mannans, since mannans are not unusually sensitive to acid hydrolysis.

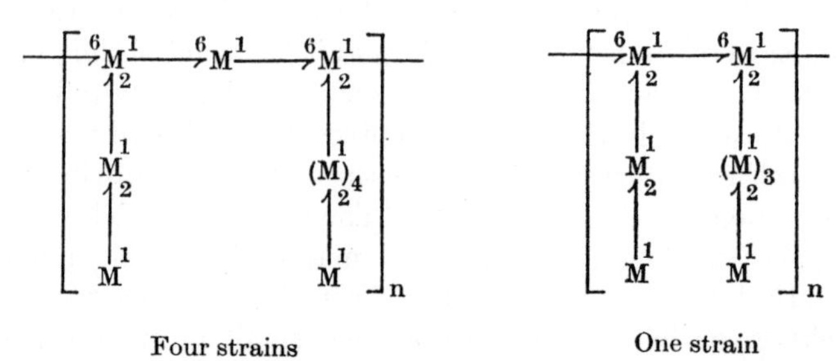


FIG. 5. Schematic structures of two types of mannan from strains of *Candida albicans* according to Šikl *et al.* (1967). M indicates an α-D-mannopyranosyl residue

Šikl *et al.* (1964, 1967) prepared mannan by alcohol precipitation from filtrates from cultures in which *C. albicans* had been grown, and from the washings of the harvested cells (cf. *Sacch. rouxii* mannan, above). Mannans from a number of strains of *C. albicans* were compared, and two types were characterized by methylation analysis, partial hydrolysis, periodate oxidation and Smith degradation. The repeating units were deduced as shown in Fig. 5. Although Šikl *et al.* confirmed the absence of α-(1→3) bonds, they found much longer side chains (up to a pentasaccharide) and, in the majority of strains, there were unsubstituted mannose residues alternating with those carrying branches. It is possible that the rather drastic treatment to which the *C. albicans* mannan of Bishop *et al.* (1960) was subjected had caused some degradation.

Stewart and Ballou (1968) also studied the mannan from *C. albicans*. By the technique of acetolysis and methylation analysis they demonstrated the presence of a high proportion of unsubstituted mannose units in the backbone and, in addition, side branches ranging from a single mannose unit to a pentasaccharide, thus confirming the long side chains discovered by Šikl *et al.* (1967). Later, Kocourek and Ballou (1969) mentioned the presence of hexaose-containing side chains in the mannans from *C. albicans*. In contrast to earlier studies on this type of mannan, Stewart and Ballou (1968) found a very small proportion of α-(1→3) linkages in the tetra- and pentasaccharide branches. In a closely related species, *C. stellatoidea*, the structure of the mannan was very similar, except that the proportion of unsubstituted mannose residues in the backbone was much smaller, but the proportion of α-(1→3) bonds in the tetra- and pentasaccharide side chains was higher. Interestingly, in this species, some of the disaccharide side chains contained α-(1→3) bonds but none was detected in the trisaccharide branches.

Stewart and Ballou (1968) also studied in detail the mannan of an apiculate yeast, *Kloeckera brevis* (syn. *Kl. apiculata*). It had a very high phosphate content (see Section II.B.3, p. 154) and side chains containing mannose, mannobiose, and mannotriose exclusively linked by α-(1→2) bonds or containing at most a minute proportion of (1→3) bonds. The backbone carried a significant proportion of α-(1→6)-linked mannose residues not involved in branching. Spencer and Gorin (1968), who studied the mannans of apiculate yeasts by proton magnetic resonance spectra, showed that *Hanseniaspora valbyensis* and *Ha. uvarum*, which are considered to be the ascosporogenous forms of *Kl. apiculata*, had identical H-1 regions in their mannan spectra as the last mentioned species, and thus presumably the same chemical structure. Similarly *Ha. osmophila* and its asporogenous equivalent, *Kl. magna*, had nearly identical spectra, but they were very different from those of the first group. It is very interesting that *Kl. africana*, which is considered a species different from *Kl. magna* on the basis of physiological properties, exhibited the same spectrum in its mannan. The spectrum of *Kl. javanica* had features some of which resembled the first group and some of which resembled the second group. The chemical structure of the last two types of mannan has not been elucidated as yet.

Quite recently, Gorin *et al.* (1969a) discovered a new class of mannans containing both α- and β-linkages in the side-chains. An earlier discovery by Gorin and his coworkers of a capsular mannan with exclusively β-bonds will be discussed in Section III (p. 177) dealing with capsular polysaccharides. Gorin *et al.* (1969a), while screening (finger-printing) a very large number of yeast species for various types of mannan by their proton magnetic resonance spectra, noted that a limited number of species exhibited a substantial proportion of H-1 signals at higher field than the usual range $\tau 4{\cdot}0$–$4{\cdot}55$. Since the high-field signals suggested the presence of β-D-pyranose units, specific rotations, ultracentrifugations, and chemical analyses of several of the mannans were performed. Each of the mannans examined gave only a single boundary upon ultracentrifugation, and had specific rotations intermediate between those with exclusively α- or β-linkages, suggesting the occurrence of both α- and β-D-linkages in the same molecule. A detailed chemical investigation revealed that the main chain or backbone consisted of the usual α-(1→6)-linked mannose residues. The β-linkages (together with some α-linkages), apparently all in the form of (1→2) bonds, occurred in the side-chains. Their findings are illustrated in Fig. 6 by the repeating units of the mannans of two species most thoroughly studied.

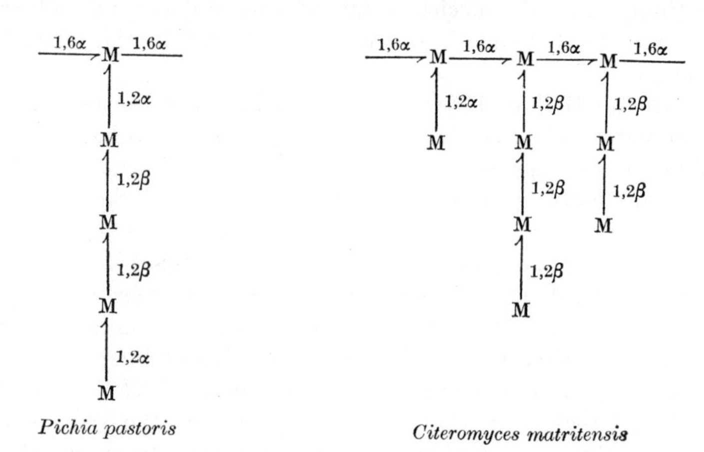


FIG. 6. Probable structure of the repeating units of the mannans of *Pichia pastoris* and *Citeromyces matritensis* according to Gorin *et al.* (1969a). M indicates an α-D-mannopyranosyl residue.

Comparable side-chains were detected by a preliminary structural investigation in mannans from *Brettanomyces anomalus* and *Sacch. rosei*, and by indirect evidence based on the proton magnetic resonance spectra of the mannans in 24 additional species.

### 3. *Presence of Phosphate in Mannan*

Northcote and Horne (1952) reported approximately 0·3% phosphorus in a mannan which they had isolated from baker's yeast. Lindquist (1953), who prepared a mannan with a molecular weight of 74,000, found a phosphorus content of about 0·2%. Eddy (1958) separated mannan from walls of *Sacch. cerevisiae* by exposure to papain. After subsequent treatment with alkali he found the phosphorus content to be 1% or less. Since the phosphate could not be removed by various treatments, Eddy postulated that the phosphate forms an integral part of the mannan molecule. Two cell-wall fractions obtained by Korn and Northcote (1960) contained mannose, phosphorus and nitrogen, which could mean that the mannan itself was phosphorylated or was complexed with a phosphoprotein. Mill (1966) was able to isolate and identify mannose 6-phosphate as a product of acid hydrolysis of mannan from *Sacch. cerevisiae*, indicating that at least in baker's yeast the phosphate appears to be linked to the 6-position of the terminal non-reducing mannose residue of the main chain, or to the (1→2)- or (1→3)-linked residues of the side-chains.

Mill (1966) found from titration curves that the phosphate in the native cell wall is di-esterified, and that the second (unknown) linkage was very labile, possibly representing a hemi-acetal bond to C-1 of another mannose residue. The somewhat higher phosphorus content in walls of flocculent yeast (0·4%) as compared to a non-flocculent yeast (0·3%) was considered a possible, although perhaps only a partial, explanation for the difference in flocculation during fermentation with brewing strains. The phosphate content of mannans appears to vary greatly between species, strains of the same species, and even in separate experiments with the same strain of yeast. Growing conditions, isolation procedure, and/or phosphatase activity could be factors controlling the mannose/phosphorus ratio. Some examples taken from the literature are given in Table I.

Kocourek and Ballou (1969) reported the phosphorus content of the mannans of ten other yeast species to range from 0·04% in a strain of *Saccharomyces lactis* (syn. *Fabospora lactis*) to 4·4% in *Candida atmosphaerica*.

Stewart and Ballou (1968) obtained both neutral oligosaccharides and acidic products after acetolysis of various mannans. Controlled acetolysis splits the α-(1→6) bonds of the mannan backbone, freeing the intact side-chains with a reducing mannose residue of the backbone attached. The acidic oligosaccharides owe their acidity to phosphate esters, and they could be separated on DEAE Sephadex A-25. Stewart and Ballou (1968) were able to identify from the acetolysis products of mannan from *Kl. brevis* (having a high phosphorus content) the monophosphates of mannobiose and mannotriose. Digestion with alkaline phosphatase from *Escherichia coli* released all of the phosphorus. Several analytical approaches indicated that the phosphate was attached to the reducing mannose unit, the one originally present in the backbone. Total acid hydrolysis with N-hydrochloric acid of mannan from *Kl. brevis* yielded a number of mannose phosphates but mannose 6-phosphate was not among them, in contrast to the report by Mill (1966). When mannan from *Kl. brevis* was digested with an exo-α-mannosidase from an *Arthrobacter* species (Jones and Ballou, 1968, 1969b), single mannose units were removed sequentially from the side-chains, giving products of greatly increased phosphate content. This provides further support for the finding that the phosphate (at least in *Kl. brevis*) must be located in the backbone on the hydroxyls of C-3 and/or C-4, the only ones not involved in other linkages.

TABLE I. *Variations in the Mannose/Phosphorus Ratio in Walls of Different Yeasts*

| Yeast strain | Moles mannose/ moles phosphorus | Reference |
|---|---|---|
| *Saccharomyces cerevisiae* (baker's yeast; strain 1) | 120 | Stewart and Ballou (1968) |
| *Saccharomyces cerevisiae* (baker's yeast; strain 2) | 78 | Stewart and Ballou (1968) |
| *Saccharomyces cerevisiae* (238 C); 7-hour culture | 120 | Stewart and Ballou (1968) |
| *Saccharomyces cerevisiae* (238 C); 24-hour culture | 144 | Stewart and Ballou (1968) |
| *Saccharomyces cerevisiae* | 15 | Mill (1966) |
| *Saccharomyces carlsbergensis* (ATCC 9080) | 21 | Stewart and Ballou (1968) |
| *Kloeckera brevis* (Phaff 55-45) | 9 | Stewart and Ballou (1968) |
| *Candida albicans* (B-792) | 18 | Stewart and Ballou (1968) |
| *Candida albicans* (AB-311) | 35 | Jones and Ballou (1969b) |
| *Candida stellatoidea* | 46 | Stewart and Ballou (1968) |
| *Candida stellatoidea* | 70 | Jones and Ballou (1969b) |

When mannan from *Kl. brevis* was isolated by extraction with citrate buffer (pH 7·0) at 120°, the mannan contained phosphate in the diester form as shown by titration curves with a single inflexion (Stewart and Ballou, 1968). Treatment at 100° for 15 min with N-hydrochloric acid or for 60 min with N-sodium hydroxide converted the phosphate to the monoester form. Alkaline degradation did not release any small molecular-weight carbohydrate material. It is noteworthy that the molecular weight of one of the mannan fractions with di-esterified phosphate ($4{\cdot}5 \times 10^4$) was not greatly different from that of alkali-extracted mannan containing mono-esterified phosphate ($3{\cdot}6 \times 10^4$). At this time the nature of the second linkage is not known.

Jones and Ballou (1969b) found that the extents to which various mannans were digested by their exo-$\alpha$-mannosidase preparation varied inversely with their phosphorus contents, and presumably the phosphate groups inhibit the action of the enzyme by steric or other interference.

Other factors, such as the distribution of the phosphate groups, probably are involved also in the susceptibility of a mannan to the action of α-mannosidase, as are the presence of β-linkages in some species of yeast (Gorin *et al.*, 1969a).

Recently Cawley and Letters (1968) treated *Sacch. cerevisiae* walls, after delipidization, with the proteolytic enzyme Pronase, and fractionated the solubilized portion (about half of the weight of the walls) on a column of DEAE-cellulose. They obtained a phosphoglycopeptide in which the phosphate was still present as a diester. Periodate oxidation of this material, followed by hydrolysis with weak and more concentrated alkali, gave results which suggested that the phosphodiester groups were linked between C-6 of one mannose unit (cf. Mill, 1966) and C-1 of another, which in turn was thought to be linked glycosidically at C-2 to another mannose residue. This arrangement would be somewhat similar to that found in the capsular phosphomannan of *Hansenula* (Jeanes and Watson, 1962) which will be discussed in Section III.C (p. 177). Phosphate esterified at C-6 of mannose in a purified mannan fraction from baker's yeast was also identified by Sentandreu and Northcote (1968). It seems most likely therefore that, in species of *Saccharomyces*, the phosphate is linked to C-6 of mannose, whereas in *Kloeckera* C-3 or C-4 of the mannose in the backbone appears to be involved.

#### 4. *Galactomannans*

Not all of the mannans which have been extracted with alkali from yeast cells or cell walls, and which were subsequently purified *via* their insoluble copper complexes, contain exclusively mannose as the component sugar.

Early reports in the literature indicated that species of *Nadsonia* lacked mannan (Garzuly-Janke, 1940; Kreger, 1954; Miller and Phaff, 1958). Later, Gorin and Spencer (1968b) and Spencer and Gorin (1968) were able to isolate, *via* their copper complexes, polysaccharides in relatively low yields, which turned out to be galactomannans. The proton magnetic resonance spectra in deuterium oxide (cf. Section VII, p. 200) of the galactomannans of *Nadsonia elongata* and *Nadsonia fulvescens* were not the same, that of the latter being more complex. Structural details are as yet unknown.

Species of the genus *Trichosporon* appear to be heterogeneous as far as their cell-wall mannan is concerned, since mannan, pentosylmannan and galactomannan have been demonstrated in different species of that genus. Galactomannan has been found in *Trich. penicillatum*, *Trich. hellenicum* (a doubtful member of this genus) and *Trich. fermentans* (Gorin and Spencer, 1968b). Since the galactomannan of *Trich. fermentans* appeared to have a simpler structure than those from the other two species, it was subjected to structural analysis. Its principal features are illustrated in Fig. 7. Its mannose : galactose ratio was 65 : 35. Side-chains terminating in a galactosyl residue were much more abundant than those ending in a mannosyl unit. The galactomannan occurring in *Candida lipolytica* was found to be very similar in structure to that of *Trich. fermentans* (Gorin *et al.*, 1969c), except that a higher proportion of the α-(1→6)-linked main chain is not substituted with side-chains, and *O*-α-D-mannopyranosyl side units occur more frequently.

Four species of *Schizosaccharomyces* all contained an apparently identical galactomannan with a major signal at τ4·26. It differed significantly in its proton magnetic resonance spectrum from that of

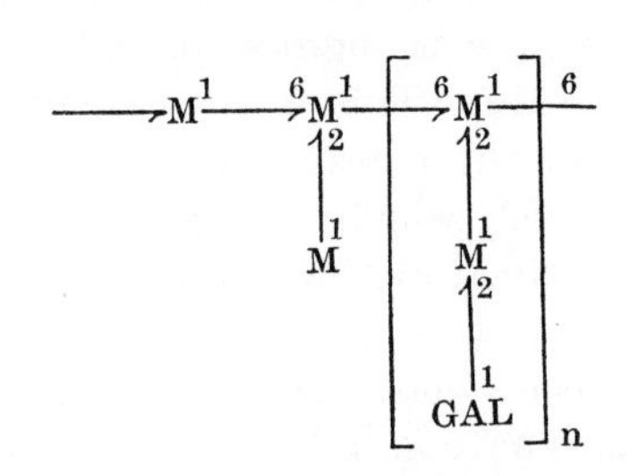


Fig. 7. Principal features of the galactomannan from *Trichosporon fermentans* according to Gorin and Spencer (1968b). Its mannose : galactose ratio was approximately 65 : 35. M indicates an α-D-mannopyranosyl residue, and GAL an α-D-galactopyranosyl residue.

*Trich. fermentans* and *Nadsonia*. Gorin *et al.* (1969c) reported later that mannan from *Schizosacch. octosporus* contains a (1→6)-linked α-D-mannopyranosyl main-chain substituted at C-2 by α-D-galactopyranosyl and some galactobiosyl units, presumably connected by (1→3) bonds.

Other yeasts in which Gorin *et al.* (1969b, c) demonstrated galactomannans were:

*Torulopsis gropengiesseri*; galactose : mannose ratio, 21 : 69. It has α-(1→2)-linked mannosaccharide side-chains of one or two units, and β-linked galactopyranosyl non-reducing end-units.

*Torulopsis magnoliae*; galactose : mannose ratio, 24 : 76. It has α-(1→2)-linked mannosaccharide side-chains of 1–9 units, and β-linked galactopyranosyl non-reducing end-units.

*Torulopsis lactis-condensi*; galactose : mannose ratio, 34 : 66. The mannan has side units of single mannosyl residues linked to the main chain by α-(1→2) bonds, and side-chains of *O*-α-D-galactosyl-(1→6)-*O*-α-D-mannosyl-(1→2).

*Torulopsis nodaensis* is another species containing galactomannan. It appears to resemble that of *T. magnoliae* but its structure has not been determined as yet by chemical analysis.

### 5. *Enzymic Studies*

Direct proof for the existence in mannan of a backbone of α-(1→6)-linked mannose residues was furnished by Jones and Ballou (1968). They isolated a soil bacterium, later identified as a species of *Arthrobacter*, which, when grown on yeast mannan, produced an extracellular exoglycosidase which was able to remove sequentially the mannose residues of the side-chains, leaving an essentially unsubstituted linear α-(1→6)-linked mannan with a molecular weight of about 8,000. Later, the enzyme was studied in more detail (Jones and Ballou, 1969a) and partially purified. Synthesis of the enzyme was induced by yeast mannan and by oligosaccharides of mannan with α-linkages, but not by mannose, glucose or glycerol. It required calcium ions for activity and hydrolysed α-(1→2), α-(1→3) and α(1→6) bonds in oligosaccharides starting from the non-reducing end. However, it was unable to hydrolyse α-(1→6) bonds of the polymer forming the backbone of the mannan molecule, even after the side-chains were hydrolysed completely by the enzyme (Jones and Ballou, 1969b).

This enzyme was used also by Gorin *et al.* (1969b) to study the structure of mannans from less well-known species. The α-D-mannans extracted from *Candida parapsilosis*, *Torulopsis apicola*, *T. bombi* (now named *T. bombicola*), *Endomycopsis fibuligera* and *Sacch. rouxii*, which have side-chains containing exclusively α-(1→2) and α-(1→3) bonds, were hydrolysed readily to linear polymers with α-(1→6)-linked mannose residues as discussed above. However, species containing galactomannans (cf. Section II.B.4, p. 155) were not hydrolysed unless the terminal galactosyl residues were first removed by selective hydrolysis with 0·33 N-sulphuric acid at 100°. This applied both to species with galactose residues attached by α- or β-linkages.

Similarly, mannans which contained mannose residues connected by β-linkages in the side-chains were refractory to the action of exo-α-D-mannanase (*Pichia pastoris*, *Sacch. lodderi*, and *Citeromyces matritensis*). Upon treatment with hot sulphuric acid, the β-linked D-mannopyranosyl side-chain residues were largely removed, upon which the remainder of the side-chain units could be removed enzymically. As is characteristic for exo-enzymes, the *Arthrobacter* enzyme caused inversion of the anomeric configuration upon hydrolysis. Gorin *et al.* (1969b) suggested that this enzyme should be designated as α-1→(2, 3, or 6)-D-mannanase. Although it does not readily hydrolyse mannan polymers with α-(1→6) bonds, it can hydrolyse oligosaccharides with such linkages. This low rate with the polymer may be related to inadequate substrate concentration.

Nagasaki *et al.* (1966) described an enzyme which they detected in the culture fluid of a strain of *Bacillus circulans* and which aided in the degradation of walls of living cells of baker's yeast, speeding up the formation of sphaeroplasts (see Section IV.B, p. 192). The enzyme was later partially purified (McLellan and Lampen, 1968) and shown to cleave a limited number of mannosidic bonds in phosphomannans in which the phosphate was still present as a diester. As a result, yeast cells harvested during the logarithmic phase of growth released, upon exposure to the enzyme, mannan and mannan-proteins into the medium, still containing the phosphodiester bonds. Upon mild acid hydrolysis,

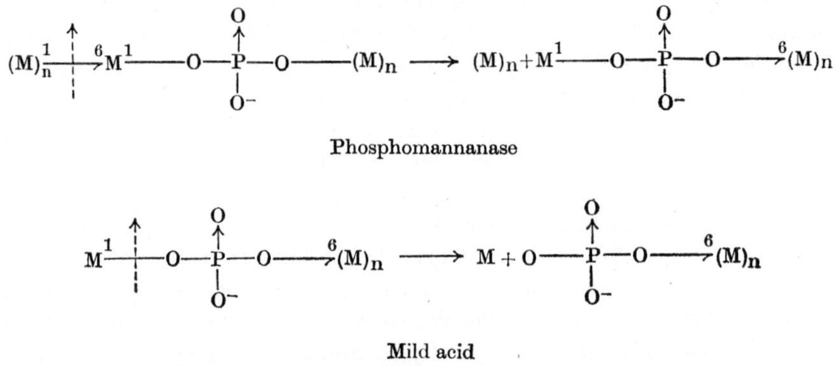

$$(M)_n \xrightarrow{1 \to 6} M \overset{1}{—} O — \underset{O^-}{\overset{O}{\overset{\uparrow}{P}}} — O — (M)_n \longrightarrow (M)_n + M \overset{1}{—} O — \underset{O^-}{\overset{O}{\overset{\uparrow}{P}}} — O \xrightarrow{6} (M)_n$$

Phosphomannanase

$$M \overset{1}{—} O — \underset{O^-}{\overset{O}{\overset{\uparrow}{P}}} — O \xrightarrow{6} (M)_n \longrightarrow M + O — \underset{O^-}{\overset{O}{\overset{\uparrow}{P}}} — O \xrightarrow{6} (M)_n$$

Mild acid

FIG. 8. Linkages of phosphomannan split by phosphomannanase followed by mild acid, according to McLellan and Lampen (1968). $(M)_n$ represents a branched mannan such as illustrated in Fig. 4. M indicates an α-D-mannopyranosyl residue.

the diester bond was split, releasing mannose and polysaccharide terminating in mannose 6-phosphate. From these products the authors concluded that this enzyme cleaves the very few mannosidic bonds adjacent to a mannose which is also phosphodiester-linked through the acid-labile carbon, and named it phosphomannanase. The linkages split are schematically shown in Fig. 8. The enzyme also splits comparable bonds in the capsular phosphomannans produced by certain *Hansenula* species (cf. Section III.C, p. 179).

### 6. *Immunochemistry of Yeast Mannans*

Hasenclever and Mitchell (1961), Summers *et al.* (1964) and various other investigators have established that, in species of the genus *Candida*, mannan is the major antigen. The above mentioned investigators have shown that in the important pathogen *Candida albicans* there are at least two serologic groups: group A mannan contains all of the determinants present in group B, plus additional determinants. Tsuchiya *et al.* (1959) and Biguet *et al.* (1962), however, regarded *C. albicans* as antigenically homogeneous. In addition, much work has been done on cross reactions between antisera to one species and polysaccharides extracted from other species (cf. Hasenclever and Mitchell, 1964). Suzuki *et al.* (1968) and Suzuki and Sunayama (1968) established, on the basis of inhibition of the precipitin reaction by oligosaccharides obtained from acetolysis of mannans from *Sacch. cerevisiae* and *C. albicans*, that the main antigenic determinants of the mannans are the α-(1→2)- and α-(1→3)-linked side-chains. Kocourek and Ballou (1969), from their studies of the structure of the side-chains of mannans from *C. albicans*, predicted that the serotypes A and B of *C. albicans* might well correlate with the presence or absence of (1→3) linkages in certain of the oligosaccharide side-chains of the cell-wall mannans of these serotypes. This prediction was based on the conclusions of Suzuki *et al.* (1968) that the terminal α-(1→3)-linked mannose in mannan from *Sacch. cerevisiae* was the immunodominant group. For example, α-(1→3)-linked mannobiose side-chains showed greater inhibitory power in the precipitin reaction than α-(1→2)-linked mannobiose.

Suzuki and Sunayama (1968) isolated by acetolysis the oligosaccharide side-chains (including the residue originally in the backbone) from mannan of *C. albicans* and partially characterized them (for the structure of this mannan, see Section II.B.2, p. 149). Their inhibition studies indicated that the hexaose was somewhat more effective as an inhibitor of the precipitin reaction than the heptaose, which is presumably due to the presence of one (1→3) linkage in the hexaose and none in the heptaose.

Ballou (1970) extended these studies with mannans and oligosaccharides from *Kloeckera brevis*, *Sacch. cerevisiae* and *Sacch.* (syn. *Fabospora*) *lactis*. The lengths of the side-chains in the mannans of these three species can be illustrated schematically as follows:

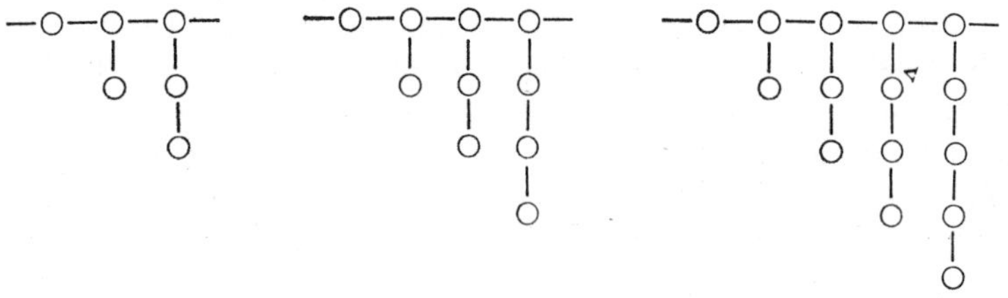

*Kloeckera brevis* *Saccharomyces cerevisiae* *Fabospora lactis*

Each antiserum (prepared with heat-killed whole yeast) cross-reacted with the mannans of the other yeasts, indicating some common determinants, namely similar short side-chains. The $\alpha$-(1→6)-linked backbone or unsubstituted portions thereof did not produce antibodies when whole cells were used as antigens. Ballou's findings confirmed those of Suzuki and his coworkers in that the side-chains containing an $\alpha$-(1→3)-linked terminal mannose unit are the immunodominant groups in the mannan antigens. In addition, he showed that the amount of antibody protein precipitated in homologous precipitin reactions increased with the length of the side-chains in the mannan used, *Fabospora* (*Saccharomyces*) *lactis* being the most effective. It was also demonstrated by inhibition experiments that mannan from *Sacch. lactis* induced the formation of at least two specific antibodies, which were directed against the tri-and tetrasaccharide side-chains respectively. The trisaccharide side-chains in mannans from *Sacch. cerevisiae* and *Sacch. lactis* appeared to be equivalent in this respect, as was also the case with the disaccharide components of these two mannans. Possibly because of the high phosphate content in mannan from *K. brevis*, somewhat anomalous results were obtained during inhibition studies with this type of mannan. Limited experiments with acetolysis products of the mannans of *Candida parapsilosis*, *Hansenula angusta* and *Pichia bispora* showed that the last two species, which are known to contain a proportion of $\beta$-linkages in the side-chains (cf. Gorin *et al.*, 1969a), gave very little or no cross reactions with anti-*Saccharomyces cerevisiae* serum.

The methodology sketched above promises to yield important results concerning the structures and immunochemistry of yeast cell-wall mannans as additional species are investigated. It is clear already that the unique pattern of the side-chains of the mannan of different species and the different linkages which have been shown to exist in them are responsible for the immunogenicity of these mannans, and that there exist both qualitative and quantitative differences between the various side-chains as antigenic determinants.

### 7. *Biosynthesis of Mannan*

Recently some information has become available on the manner in which yeasts synthesize the mannan molecule. After a preliminary announcement by Algranati *et al.* (1963), Behrens and Cabib (1968) made a more detailed study of a particulate enzyme preparation from *Sacch. carlsbergensis* which incorporated the mannose portion of guanosine diphosphate mannose (GDP-Man) into a fully branched mannan without the apparent need of an added primer. Since the enzyme was made by centrifugation of lysed protoplasts of the yeast, some endogenous primer could be expected and, in fact, the authors reported that traces of mannan were detected in the particulate sediment. The optimum pH value for the activity was between 5·5 and 7·2, and $Mn^{2+}$ was found to be a nearly absolute requirement. Acetolysis of the synthesized mannan (see Section II.B.1, p. 148) and study of the radioactivity distribution in the released oligosaccharides showed that a branched mannan, characteristic of that established by Stewart *et al.* (1968), was synthesized by the enzyme preparation. The initial product of synthesis remained attached to the particles of the enzyme preparation.

Tanner (1969) recently showed that in *Sacch. cerevisiae,* as shown earlier for *Micrococcus lysodeikticus* (Scher *et al.*, 1967, 1968), a lipophilic mannosyl intermediate is the immediate precursor for mannan biosynthesis. Mannosyl residues were transferred to the lipid intermediate, and some evidence was provided that the linkage to the lipid might be through a phosphodiester. Incorporation of $^{14}C$-mannose from GDP-Man into the lipid requires metal ions, but in contrast to the $Mn^{2+}$ requirement for mannan biosynthesis, $Mg^{2+}$ appears able to replace $Mn^{2+}$ fully. Tanner (1969) showed that a particulate enzyme system, optimal in $Mg^{2+}$ but deficient in $Mn^{2+}$ concentration, created a higher than normal steady-state level of $^{14}C$-mannosyl-lipid and a decreased rate of mannan synthesis, thus giving evidence for the intermediate role of mannosyl-lipid for its biosynthesis. Possibly the lipid intermediate provides a mechanism for passage of the complex through the cyto-plasmic membrane, after which the next step in mannan biosynthesis could occur.

The interesting and as yet unanswered question is how branching is effected by the enzyme system.

Sentandreu and Northcote (1968), on the other hand, showed that mannose and small oligosaccharides were linked by *O*-mannosyl bonds to the hydroxyls of serine and/or threonine in a peptide. *O*-α-D-Manno-pyranosyl-(1→6)-D-mannose, *O*-α-D-mannopyranosyl-(1→2)-D-mannose and *O*-α-D-mannopyranosyl-(1→3)-D-mannose have been identified so far (Sentandreu and Northcote, 1969a). A large mannan molecule was linked to the *same* peptide, presumably by an aspartamide-N-acetyl-glucosamine linkage (cf. Section II.D.3, p. 169). According to these authors, the first stage in the synthesis could be the transfer of manno-syl units from GDP-Man to serine or threonine in the mannan peptide, to give low molecular-weight oligosaccharides or even single mannosyl residues attached to the protein. This would then be followed by a transfer of these units by a transglycosylation reaction to C-2 of the mannose units in the backbone. Alternately or in addition, these units (in particular the α-(1→6)-linked mannobiose) might be attached to C-6 of the non-reducing terminal of the mannan, elongating the chain and in certain instances adding a side-chain. In the latter case addition of a trisaccharide would result in a mannobiose side-chain, the terminal mannosyl residue becoming part of the backbone. In subsequent experiments, however, Sentandreu and Northcote (1969b) obtained results contradicting the theory of transglycosylation of mannose and mannose oligosaccharides as described above. In pulse-chase experiments with radioactive glucose and whole cells for 60 sec followed by incubation in non-radioactive media, it was found that the radioactivity of the mannose and mannose oligosaccharides attached to the glycopeptide by *O*-glycosylhydroxy-amino acid bonds remained constant during the chase. If transglycosylation were to operate, the radioactivity would be expected to move into the large mannan molecule also attached to the peptide. Since the period of the unlabelled chase was only 6 min, it is possible that the time was not sufficient to achieve replacement of the labelled oligosaccharides by non-labelled molecules.

### C. CHITIN

Schmidt (1936) was the first investigator who detected chitin in certain filamentous yeasts (*Endomycopsis capsularis*, *Endomycopsis fibuligera* and *Eremascus fertilis*). This polymer, at least in the exo-skelton of crustaceans where it exists in a highly insoluble form, consists of long unbranched molecules made up of N-acetyl-D-glucosamine residues linked together by $\beta$-(1→4) bonds. Its early demonstration in these species probably stemmed from the fact that, as Kreger (1954) later confirmed, their chitin content is much higher than in non-filamentous species. The early demonstration of chitin (Schmidt, 1936; Nabel, 1939) was based on the isolation of glucosamine and on the chitosan (deacetylated chitin) iodine test. Houwink *et al.* (1951) and Roelofsen and Hoette (1951) detected chitin in many more species of yeast by a modification of the chitosan sulphate crystallization test. They also demonstrated it chromatographically as glucosamine in the acid hydrolysate of yeast residue previously extracted with hot dilute alkali and acid. The best procedure, however, was found to be extraction with cold 30% hydrochloric acid in which the chitin dissolves; then, following neutralization, it precipitates. Chitin was present, usually in low concentration, in 28 species of (mainly) non-filamentous yeasts. The only organism in which chitin could not be detected was the fission yeast *Schizosaccharomyces octosporus*. X-Ray analysis of cell walls treated in various ways further confirmed the presence of chitin in yeast (Houwink and Kreger, 1953; Kreger, 1954). When cell walls of *Sacch. cerevisiae* or *C. tropicalis* are boiled with dilute hydrochloric acid, the glucan becomes soluble in sodium hydroxide. After extraction with hot dilute alkali, an insoluble substance remains which appears granular in electron micrographs.

This residue gave X-ray diffraction patterns similar to that of crustacean chitin. In the filamenous yeasts and in *Nadsonia fulvescens*, *Rhodotorula glutinis* and *Sporobolomyces roseus* the chitin content was much greater than in most of the typical budding, ascomycetous species, whose walls have usually been reported to contain about 1–2% chitin. Confirming earlier results with *Schizosaccharomyces octosporus* Kreger (1954) demonstrated the absence of chitin in *Schizosacch. pombe* and *Schizosacch. versatilis* as well.

Since early investigators generally inferred the presence of chitin by testing for glucosamine or chitosan, in which the acetyl group is removed by drastic treatments with hot acid and alkali, Eddy's (1958) observation that snail digestive fluid and a malt enzyme released acetylglucosamine from untreated cell walls of baker's yeast is important. Glucosamine was the only hexosamine which Eddy could detect chromatographically among the products of acid hydrolysis. Based on the amounts of enzyme-liberated and residual N-acetylglucosamine, Eddy estimated the apparent glucosamine content of the wall to be about 0·8–0·9%. This approach suggests fuller use of purified chitinase preparations (Jeuniaux, 1966) for the analysis of yeast walls.

The physical nature of chitin in the wall is only partially understood. Korn and Northcote (1960) criticized the severe chemical treatments used by many investigators to fractionate wall components. They used a milder procedure and fractionated yeast cell walls into three fractions by extraction with ethylenediamine. Since crustacean chitin is completely insoluble in this solvent, they concluded that only 9% (the insoluble portion) of the apparent glucosamine content of the wall had solubility properties of typical chitin. Bacon *et al.* (1966) also developed a relatively mild fractionation procedure. They first extracted compressed baker's yeast repeatedly with dilute (3–6%) sodium hydroxide and this was followed by extraction with sodium acetate buffer at pH 5·0 for 3 h at 75–80°. This treatment extracted most of the glucan and left a residue consisting almost exclusively of the bud-scars of the original walls. Its infrared spectrum showed a high chitin content, and chemical analysis revealed about 50% chitin and 30–35% glucan. Subsequent treatment with an enzyme preparation from *Cytophaga johnsonii* (Bacon *et al.*, 1965) further enriched the bud-scar preparation in chitin content by digestion of the glucan component. Although the bud-scar region is very high in chitin content, Bacon *et al.* (1966) could account for only 20% of the original chitin in the bud-scar preparation, the rest being lost during preliminary extractions. It may be assumed that the non-extractable chitin reported by Korn and Northcote (9%) and that found by Bacon *et al.* (20%) represents the same material. This would mean that the bulk of the chitin in baker's yeast walls represents a structural element not associated with scar tissue unless part of the bud-scar chitin were to have different solubility characteristics. Houwink and Kreger (1953) interpreted the granular material in electron micrographs of baker's yeast, boiled successively with acid and alkali, and occurring both in the bud-scar areas and outside, as chitin. If this is correct, glucosamine-yielding material not occurring in the bud-scar must be more soluble in various solvents, possibly due to lower degrees of polymerization and crystallinities. Evidence for this was recently supplied by Sentandreu and Northcote (1968). Following earlier suggestions by Eddy (1958) and Korn and Northcote (1960) they obtained direct proof that N-acetylglucosamine forms one type of linkage between mannan and protein (see Section II.D.3, p. 169).

In *Cryptococcus albidus* and *Cr. terreus*, Jones *et al.* (1969) found the chitinous residue after extraction of the glucan by acid and alkali to be composed of microfibrils (cf. also Streiblová, 1968) rather than of a granular residue as described above. Species of *Cryptococcus* appear to have a high chitin content (Bacon *et al.*, 1968), but apparently it is not found in bud-scar regions since Jones *et al.* (1969) reported that *Cryptococcus* lacks well defined bud-scars in chemically extracted and enzymically degraded preparations.

Chattaway *et al.* (1968) worked with a strain of *Candida albicans* which could be grown as true hyphae on some media and as a budding yeast on others. In contrast to the work just discussed, they found that 96% of the glucosamine of either the mycelial or the budding forms, after extraction according to the procedure of Korn and Northcote (1960), had solubility properties of chitin, However, the level in the mycelial form was about three times higher than in the budding form.

Buecher (1968) determined the glucosamine content in cell walls of a budding and a hyphal form of *Saccharomycopsis guttulata*, and found 1·6–1·9% for the budding strain depending on the growing conditions and 2·3% for the hyphal strain. It is interesting that a budding variant, derived from the hyphal strain, had only 1·4% glucosamine in its walls.

## D. PROTEIN AND CARBOHYDRATE-PROTEIN COMPLEXES

### 1. *Protein Content of Cell Walls*

Before techniques were developed to prepare cell walls in relatively pure condition, it was not possible to study the protein components of the wall with any degree of accuracy. Early data by Northcote and Horne (1952) on the protein content (13%) of washed cell walls of mechanically ruptured baker's yeast may be considered too high due to the presence of small particles entrapped in the cell envelope. A number of later studies showed that the protein content of baker's yeast walls is approximately 5–7% (Roelofsen, 1953; Falcone and Nickerson, 1956; Miller and Phaff, 1958; Buecher, 1968). Data for other yeasts are not numerous and, when based on values for total nitrogen converted to protein by multiplying by 6·25, may be subject to some error due to variable amounts of non-protein nitrogen, such as N-acetyl-glucosamine. The following are some uncorrected values for the protein content (N × 6·25) in dry walls of several yeasts: *Hanseniaspora uvarum*, 7% (Miller and Phaff, 1958); a flocculating brewer's bottom yeast, 10·2% (Masschelein, 1959); *Trigonopsis veriabilis*, 7·1% for the ellipsoidal form and 14·6% for the triangular form (SentheShanmuganathan and Nickerson, 1962). Shifrine and Phaff, (1958) prepared cell walls of *Saccharomycopsis guttulata* by sonic oscillation, and measured a total protein content of 39·6% as compared to 6·5% in *Sacch. cerevisiae* used as a control. However, Buecher (1968) found that this high value for *Saccharomycopsis guttulata* was due to the failure at that time to grow large masses of healthy living cells. The very short life span of cells of this species causes the cells to become granular at a very early stage of the growth, at which point walls are very difficult to free from precipitated protein. Buecher, by sparging the growth medium with carbon dioxide (high concentrations of this gas are required for the growth of this yeast) obtained cells which were over 99% viable, and these yielded very pure cell-wall material with a total protein content of 16·3–17·6%. An as yet undescribed filamentious variant of *Saccharomycopsis guttulata* contained 14·8% protein in its walls, and a budding form, which can be isolated from the filamentous form under certain growth conditions, had walls with 21·8% protein.

Hence, it seems clear that the protein content varies in different species, because several investigators used baker's yeast as a control and, at least visually, the wall preparations were of comparable purity.

## 2. *Principal Amino Acids in Cell-wall Proteins*

A number of studies have been made of the kinds of amino acids which occur in proteins of the cell wall. Kessler and Nickerson (1959) obtained several fractions of cell walls of *Sacch. cerevisiae* by cold-alkali extraction followed by precipitation with ammonium sulphate. Their glucomannan-protein fraction I contained 17 amino acids, with glutamic acid (17·8% of the recovered amino acids) and aspartic acid (13·1%) in major amounts. Leucine, lysine and alanine were next (9·1%, 8·1% and 6·9%, respectively), while serine and threonine accounted for 4·1% and 5%, respectively. The glucomannan-protein fraction II yielded only 13 amino acids. Lacking were proline, methionine, histidine and phenylalanine. In this protein, aspartic acid (31·1%), glutamic acid (9·2%), cysteic acid (8·4%), leucine (7·0%) and alanine (6·5%) were the most abundant amino acids, whereas serine and threonine accounted for 4·6% and 5·9%, respectively.

Masschelein (1959) identified 12 amino acids in the hydrolysate of walls of a flocculent brewer's bottom yeast, but found quite different proportions depending on whether the yeast was grown anaerobically or aerobically. However, serine and threonine were among the most abundant amino acids.

A few amino-acid analyses have been reported for the cell-wall protein of yeasts other than *Saccharomyces*. A semi-quantitative study of the walls of *Trigonopsis variabilis* by SentheShanmuganathan and Nickerson (1962) showed ten amino acids with glycine, threonine and alanine in major amounts. Buecher (1968) quantitatively analysed the amino acids in highly purified walls of a budding and a filamentous strain of *Saccharomycopsis guttulata*. In both strains threonine was most abundant, followed by glutamic acid, serine, alanine, valine and aspartic acid, in that order of prevalence. Eleven other amino acids were present in smaller proportions. For this species different growing conditions did not affect the composition. A semi-quantitative assay of amino acids in a cell-wall hydrolysate of *Sacch.* (*Fabospora*) *fragilis* by Reuvers *et al.* (1969) showed 16 (possibly 17) amino acids, with serine, threonine, aspartic and glutamic acids being the most abundant. The analytical data of the various species generally agree by showing high proportions of hydroxy amino acids and dicarboxylic amino acids.

### 3. *Mannan-Protein Complexes*

This section is concerned primarily with the chemical aspects of mannan-protein complexes whereas, in the next section, enzymic properties of such complexes are covered. Although the protein of yeast cell walls cannot be removed by washing with water or neutral buffers at room temperature, some protein may be lost either during the cultivation of the cells or during their breakage. For example, Masler *et al.* (1966) obtained water-soluble extracellular polysaccharide-protein complexes when *Candida albicans* was grown in a semi-synthetic medium on a shaker (see p. 150). The earlier studies, however, were mainly concerned with the manner in which protein and mannan formed water-insoluble structural elements of the wall.

Northcote and Horne (1952) prepared purified cell walls from baker's yeast by mechanical disintegration. It was possible to extract mannan to a limited extent using boiling water, and this product after ethanol precipitation had relatively high nitrogen (2·5%) and phosphorus (0·23%) contents. Much more rapid and complete extraction was obtained by heating the walls for 6 h at 100° in 3% sodium hydroxide. The ethanol-precipitated product contained 1·3% nitrogen and 0·26% phosphorus; even after further purification *via* its copper complex, the recovered mannan still contained 1% nitrogen and 0·2% phosphorus. Although no conclusive proof was offered that the mannan preparation was homogeneous, these observations were important in showing the probable association of protein and phosphorus with yeast mannan. Many subsequent investigators have confirmed and expanded these results.

Cifonelli and Smith (1955) showed that the polysaccharide associated with purified invertase, an enzyme located in the cell wall (see Section II.D.4, p. 171), was identical with the mannan obtained by autolysing yeast.

Falcone and Nickerson (1956) prepared clean cell walls of baker's yeast and determined 6·7% protein, 0·14% sulphur and 0·34% phosphorus in the wall material. The high sulphur: protein ratio suggested that the protein might be of a pseudokeratin type. Nickerson and Falcone (1956a,b) isolated a particulate enzyme preparation from baker's yeast which was able to reduce disulphide linkages in cell-wall protein to sulphydryl groups. This disulphide reductase system was believed to play a role in softening of the cell wall during bud formation. Falcone and Nickerson (1956) treated their wall preparation with N-potassium hydroxide under slight warming (rather than by boiling as had been done by others), which solubilized 75% of the walls. After dialysis against water and lyophilization, a white powder was obtained of which the major part was water soluble. The material had a nitrogen content of 1·1% and gave upon hydrolysis only mannose and amino acids. Since the material was monodisperse upon ultracentrifugation and gave no precipitate with cold Fehling's solution, Falcone and Nickerson concluded that the mannan and protein were tightly bound. The two components were present in a weight ratio of approximately 12 : 1, respectively.

Further work on cell-wall fractionation by cold alkali was reported by Kessler and Nickerson (1959). They used walls of *Candida albicans* and baker's yeast which were first thoroughly freed of lipids, including treatment with ethanol: ether (1 : 1) containing 1% hydrochloric acid at 50° for 5 h. Extraction with alkali under nitrogen gas at room temperature yielded two types of glucomannan-protein complexes, one precipitable by saturated ammonium sulphate and the other not. The main difference between the two types was a much higher protein content in the former. Kessler and Nickerson felt that the highly acidic proteins of the walls (cf. Section II.D.2, p. 165) were linked by carboxyl groups of the dicarboxylic amino acids to hydroxyl groups of the polysaccharide.

Indirect evidence for such ester linkages was based according to them on the non-precipitability of the glucomannan-protein complexes with Fehling's solution. Since mannan extracted with boiling sodium hydroxide solution readily precipitates with Fehling's solution, they suggested that certain hydroxyl groups had been freed by hot alkali. However, since Northcote and Horne (1952) found still an appreciable protein content in mannan extracted by hot alkali (even after precipitation by Fehling's solution), the protein-polysaccharide linkages appear to be more complex than indicated by Kessler and Nickerson. A significant observation was their finding of a glucan fraction soluble in cold alkali. The reason why Falcone and Nickerson (1956) did not find glucan in their alkali extract is not clear, but it is possible that the delipidization steps used in the later work increased the solubility of part of the glucan. Although glucan was clearly demonstrated in the extract, I agree with Bishop *et al.* (1960) that Kessler and Nickerson used insufficient criteria to prove the homogeneity of the "glucomannan–protein complexes", which could have been mixtures of an alkali-soluble glucan (cf. Eddy, 1958) and a mannan–protein complex. The reader is referred to a comprehensive review by Nickerson *et al.* (1961) for further details.

Further chemical fractionation of baker's yeast cell walls was reported by Korn and Northcote (1960) who used a mild reagent, anhydrous ethylenediamine, at 37°. Fraction A was soluble in ethylenediamine and water, and consisted of a mannan-protein complex with 1·7% glucosamine and 0·3% phosphorus. Fraction B was soluble in ethylenediamine but insoluble in water, and was considered possibly a glucan–mannan–protein complex with 0·8% glucosamine and 0·12% phosphorus. Fraction C was the residue not soluble in ethylenediamine, and consisted of mannose, glucose, protein and glucosamine. Korn and Northcote (1960), who measured a number of physical properties to insure homogeneity, suggested that non-chitinous glucosamine might well provide the link between protein and carbohydrate. Later Sentandreu and Northcote (1968) re-investigated the chemical nature of the linkages which unite protein and carbohydrate in the cell wall. They used the same extraction procedure as was used by Korn and Northcote (1960), leading to Fractions A, B, and C (see above). These fractions were further purified by Pronase digestion followed by fractionation on Sephadex columns, yielding the corresponding fractions $A_2$, $B_2$, and $C_2$, respectively. The water-soluble $A_2$ (mol. wt. 76,000) contained no sulphur-containing amino acids, but was rich in serine and threonine and also contained glucosamine (the total amino acid content was 4%). The hydroxy amino acids were found to be linked through *O*-mannosyl bonds to mannose and small mannose oligosaccharides. Evidence for this was that treatment of fraction $A_2$ with 0·1 N-sodium hydroxide for one week at 4°, or for 18 h at 21°, resulted in a $\beta$-elimination reaction yielding double-bonded dehydroamino acids and mannose plus oligosaccharides. This reaction is illustrated for mannose linked to serine:

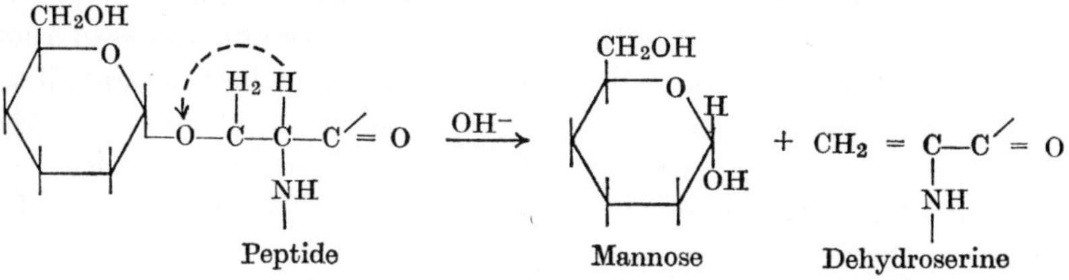

$CH_2OH$ … O—C($H_2$)—C(H)(NH)—C$\diagup$=O $\xrightarrow{OH^-}$ $CH_2OH$ … H, OH + $CH_2$ = C(NH)—C$\diagup$=O

Peptide    Mannose    Dehydroserine

Oligosaccharides released included mannobiose sugars linked by $\alpha$-(1→2), $\alpha$-(1→3) and $\alpha$-(1→6) bonds (Sentandreu and Northcote, 1969a). However, after this treatment, the large mannan molecule was still covalently linked to the peptide. It was then found that aspartic acid was the only amino acid in the alkali-treated mannan that occurred in equimolar concentration with glucosamine. Periodate oxidation studies on the mannan revealed that the amino sugar was not destroyed during oxidation of the glycopeptide. Thus, the other type of linkage between polysaccharide and peptide best fitting the experimental data is probably a nitrogen glycosyl bond between N-acetylglucosamine and aspartamide, *viz.* N-($\beta$-aspartyl)-$\beta$-D-(N-acetyl)-glucosaminide. Since the N-acetylglucosamine was not cleaved by periodate, it is thought that the linkage to mannan is through C-3 or C-4 or both. The chemical structure is as follows:

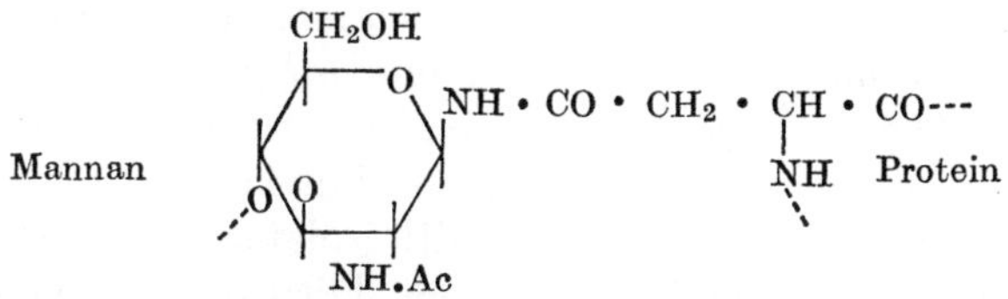

$CH_2OH$

Mannan … O, O … NH · CO · $CH_2$ · CH(NH) · CO--- Protein

NH.Ac

This structure could account for the high proportion of glucosamine not present as true chitin in yeast cell walls (see Section II.C. p. 164). Cawley and Letters (1969) found that extraction of cell walls with ethylenediamine was unnecessary, since they confirmed the results of Sentandreu and Northcote (1968) in every particular by treating cell walls directly with Pronase and thereby releasing, by a simple procedure, a similar phosphoglycopeptide as described above. The possible role of mannosyl and manno-oligosaccharides attached to the peptide has been discussed under mannan biosynthesis (Section II.B.7, p. 161).

Eddy (1958) treated walls of *Sacch. cerevisiae* with various hydrolytic enzymes and found that papain released in the initial stages (after about ten minutes) mannan residues almost exclusively. This mannan was precipitable by Fehling's solution and, after recovery in the usual way, it was found to contain almost exclusively mannose and about 2% nitrogen. It was then fractionated by paper electrophoresis in borate buffer and analysed. The results suggested that a mannan–protein complex had been isolated with a ratio of the two components of about 12 : 1, giving a molecule of slightly larger dimension than mannan itself. In the hydrolysate, small amounts of glucosamine were detected, and the phosphorus content, depending on the yeast strain used, varied from 0·06% to 1%. The phosphorus remained attached to the mannan after removal of the protein by alkaline hydrolysis.

Eddy and Longton (1969) isolated mannan–protein complexes from yeast cell walls by treatment with snail-gut enzymes. The preparations obtained from two different yeast strains were homogeneous upon ultracentrifugation, and had molecular weights ($33{\cdot}0 \times 10^4$ and $18{\cdot}9 \times 10^4$, respectively) which were about two to three times as high as similar preparations obtained by traditional alkali extraction ($12{\cdot}1 \times 10^4$ and $8{\cdot}5 \times 10^4$, respectively). These findings suggested subunits linked by alkali-labile bonds.

### 4. *Enzymes which are Located in the Cell Wall*

A number of hydrolytic enzymes in yeast are found external to the cytoplasmic membrane, and bound tightly or loosely to certain wall components. Such enzymes are frequently excreted into the medium during growth. The finding of extracellular enzymes in the culture medium is no proof in itself that the enzymes are located in the wall. For example, the exo- and endo-β-glucanases of *Fabospora fragilis*, *Hansenula anomala* and *Hanseniaspora valbyensis* (Abd-El-Al and Phaff, 1968, 1969) have not been proven to occur specifically in the cell wall prior to release into the medium. Furthermore, the finding of certain enzyme activities associated with isolated cell walls constitutes no proof that the enzyme is wall bound. For example, Nurminen and Suomalainen (1969) showed that a $Mg^{2+}$-dependent ATP-ase in cell walls of baker's yeast was actually present in the plasma membrane, which is difficult to remove from the wall proper.

Nevertheless, a number of enzymes, mainly hydrolases, occur undoubtedly as wall constituents. Four lines of evidence may be mentioned to prove their location in the wall: (i) their removal in soluble form upon conversion of yeast cells into sphaeroplasts; (ii) in the case of oligosaccharidases the products of hydrolysis may be trapped outside the plasma membrane as non-transportable compounds (e.g. oxidation of hexoses by specific oxidases or their phosphorylation by hexokinase + ATP); (iii) the *in vivo* pH-activity curve is similar to that of the isolated enzyme *in vitro*; (iv) sensitivity of a hydrolase in thin films of yeast to low-voltage electron beams of increasing penetrating power. By the last technique, Preiss (1958) showed that invertase was located just below the outer layer of the wall.

The enzymes which are definitely or likely to be found in the cell wall of yeast and yeast-like organisms include those listed in Table II.

TABLE II. *Enzymes which are Definitely or Likely to be Found in the Cell Wall of Yeasts and Yeast-like Organisms*

| Enzyme | Organism | Reference |
|---|---|---|
| Invertase | *Saccharomyces* (hybrid) | Friis and Ottolenghi (1959a) |
| Melibiase | *Saccharomyces carlsbergensis* | Friis and Ottolenghi (1959b) |
| Glucamylase | *Saccharomyces diastaticus* | Hopkins (1958) |
| *Alpha*-Amylase | *Endomycopsis fibuligera* | Wickerham *et al.* (1944) |
| Inulinase | *Fabospora fragilis* | Snyder and Phaff (1960) |
| Trehalase | *Prototheca* sp. | H. J. Phaff (unpublished data) |
| Acid phosphatase | *Saccharomyces cerevisiae* (Strain LK2G12) | McLellan and Lampen (1963) |
| Acid phosphatase | *Saccharomyces cerevisiae* (commercial baker's yeast) | Tonino and Steyn-Parvé (1963) |
| Catalase | *Saccharomyces cerevisiae* (commercial baker's yeast) | Kaplan (1965a) |
| Aryl-$\beta$-glucosidase | Yeast strain c isolated from commercial baker's yeast | Kaplan (1965b) |

Of these enzymes, invertase ($\beta$-fructofuranosidase) has been studied most widely. Most yeasts do not excrete more than traces of invertase into the medium, even though the invertase level in fully-induced *Saccharomyces* cells may be more than 100-fold in excess of that needed to supply the cells with hexoses from sucrose for their maximum rate of uptake (DelaFuente and Sols, 1962). A few species and strains of *Saccharomyces*, however, have the ability to release appreciable amounts of invertase into the culture medium (Wickerham and Dworschack, 1960; Dworschack and Wickerham, 1961). *Fabospora* (*Sacch.*) *fragilis* excretes large amounts of an inducible inulinase, an enzyme different from, but closely related to, invertase (Snyder and Phaff, 1960). Since inulinase also hydrolyses sucrose, it is possible that the hydrolase termed invertase in *Fabospora fragilis* by many authors is in fact inulinase. The rather loose anchorage of *Fabospora fragilis* sucrose hydrolase reported by Burger *et al.* (1961) may be related to its ready excretion.

Friiss and Ottolenghi (1959a) prepared sphaeroplasts of a *Saccharomyces* hybrid with the aid of snail digestive fluid. In sucrose-induced cells of this strain, which contained a single $R_2$ gene for invertase, nearly three-quarters of the invertase activity was lost by the remaining sphaeroplasts (about 90% of the cells had lost their walls). On the other hand, glucose-grown cells contained no invertase in those portions of the wall which were removed by the lytic enzymes. Burger *et al.* (1961) concluded from experiments with living and with ethyl acetate-washed cells that invertase occurs in a soluble form exterior to the plasma membrane, but inside the outer wall regions and not in combination with insoluble cell structures. Preiss (1958) also obtained evidence for such a location (see p. 171). Sutton and Lampen (1962) provided further information on the location of invertase in strains of *Sacch. cerevisiae.* First they demonstrated that isolated walls had high levels of invertase, and second the sphaeroplasts in isotonic medium failed to ferment sucrose since this disaccharide cannot be transported into the cells and the hydrolase had been removed. Glucose fermentation, on the other hand, was but little affected by removal of the cell walls. Evidence for the lack of sucrose transport was the demonstration of β-fructofuranosidase activity inside the sphaeroplasts, as shown in a lysate. Such sphaeroplasts however, were able to adapt to sucrose in the presence of small amounts of glucose as energy source (Islam and Lampen, 1962). Under such conditions the wall-less cells synthesized invertase, which was released into the medium rather than being trapped in the wall structure. The newly synthesized enzyme hydrolysed sucrose in the medium, causing fermentation of the hexoses formed.

Gascón and Ottolenghi (1967) and a number of earlier investigators (*loc. cit.*) established that, in derepressed cells of *Saccharomyces*, the cell walls contained at least two molecular forms of invertase with large molecular weights, whereas the cytoplasm contained another form of the enzyme with a much lower molecular weight. The latter form appeared to be constitutive, since its concentration was about the same in glucose-repressed cells when the amount of the large molecular-weight forms in the wall was decreased by as much as a thousand-fold. Neumann and Lampen (1967) prepared a highly purified invertase from a hybrid *Saccharomyces* strain which was mutated so that glucose repression was largely abolished. The enzyme showed a high degree of homogeneity, and had a molecular weight of about 270,000. It was a glycoprotein, containing about 50% mannan and 3% glucosamine. Since they found that the protein had a high amide content and high proportions of aspartic acid, serine and threonine, the carbohydrate–protein linkage is probably similar to that discussed in detail on p. 169. The enzyme also contained three sulphydryl groups per mole. These groups are not involved in enzymic activity but probably in additional linkages or disulphide bridges (Kidby and Davies, 1968). Lampen (1968) reported that the internal "small" invertase had a molecular weight of 130,000, and that it was nearly free of mannan and contained no glucosamine. Although the catalytic parameters of the large and small invertases were surprisingly similar, fundamental differences between the two proteins were found, such as amino-acid composition. Hence, there is no clear proof that the small internal invertase is converted into the large form after excretion through the plasma membrane, followed by complexing with mannan. The picture is further complicated by reports in the literature of enzymically active invertases with molecular weights considerably lower than 130,000 (Neumann and Lampen, 1967; *loc. cit.*). Since Metzenberg (1964) has shown the presence of active subunits in the invertase of *Neurospora*, it seems conceivable that these may also occur in yeast invertase, and this could explain the different molecular weights of the protein moiety which have been reported. If the enzyme were to combine in addition with inactive subunits involved in anchoring of the enzyme to mannan in the wall, this could explain the difference in amino acid composition between the internal and external invertase reported by Lampen (1968).

At least two other external enzymes have been shown to be mannan proteins, acid phosphatase (Boer and Steyn-Parvé, 1966) and glucamylase (Lampen, 1968; *loc. cit.*) and it would seem probable that others may be anchored in a similar manner.

Limitation of space does not permit me to cover other enzymes which have been shown to occur or are thought to occur in the cell wall in fuller detail, and the reader is referred to the references cited. In the case of *Prototheca wickerhamii* the evidence from our laboratory was based on the fact that cells of this species hydrolyse trehalose much more rapidly than they are able to utilize the glucose formed as the product of the reaction.

Aspects on the excretion of enzymes and invertase biosynthesis are covered in comprehensive reviews by Lampen (1965) and Lampen *et al.* (1967).

### E. LIPIDS

Various investigators have reported the lipid content of isolated yeast cell walls, usually in terms of bound and free lipids, the latter being extractable by organic solvents without previous hydrolysis. Masschelein (1959) determined as much as 13·5% lipid in walls of beer yeast after acid hydrolysis. Eddy (1958), on the other hand, found less than 2% total lipids in wall preparations from various *Saccharomyces* species. There is also great divergence in the values for nitrogen and phosphorus in the isolated lipids by various authors, which would be an index for their phospholipid content. Since in the washing and purification of yeast cell walls there is no assurance that the lipid-rich plasma membrane is removed, much of the reported variation in lipid contents is probably due to the extent of removal of these membranes which, according to Longley *et al.* (1968), may account for 13–20% of the dry weight of the yeast cell. Properties of the cytoplasmic membrane are covered elsewhere in this volume (Chapter 6, p. 211).

Aside from small amounts of intracellular lipids which may be trapped in the washed hulls, some species of *Torulopsis*, *Candida*, *Cryptococcus* and *Rhodotorula* are endowed with the capacity to synthesize and excrete into the medium various types of glycolipids (for a detailed review see Stodola *et al.*, 1967). Staining with lipophilic dyes has shown that these compounds may be trapped to some extent in walls or capsules while *en route* from their site of synthesis in the cell interior or membrane to the medium.

Lastly, certain species of yeast, and in particular *Hansenula ciferrii*, are known to form on their cell surfaces, and occasionally in the medium due to oversynthesis, acetylated phytosphingosines. Phytosphingosines, which are hydrophobic compounds, have the basic structure $R\cdot(CHOH)_2\cdot CHNH_2\cdot CH_2\cdot OH$ and may be responsible, at least in part, for the tendency of such yeasts to form pellicles in liquid media and for the mat appearance of the colonies.

Two extracellular sphingosines have been identified from the yeast *H. ciferrii*, although others have been found intracellularly in several species (cf. Stodola *et al.*, 1967). Their chemical structures are as follows:

$$CH_3(CH_2)_{13}\cdot\overset{\displaystyle H}{\overset{|}{C}}(OCOCH_3)\cdot\overset{\displaystyle H}{\overset{|}{C}}(OCOCH_3)\cdot\overset{\displaystyle H}{\overset{|}{C}}(NHCOCH_3)\cdot CH_2OCOCH_3$$

tetra-acetyl $C_{18}$-phytosphingosine

$$CH_3(CH_2)_{14}\cdot\overset{\displaystyle H}{\overset{|}{C}}(OCOCH_3)\cdot\overset{\displaystyle H}{\overset{|}{C}}(NHCOCH_3)\cdot CH_2OCOCH_3$$

tri-acetyl $C_{18}$-dihydrosphingosine

Other film-forming species with mat colonies remain to be investigated, but may also contain phytosphingosines or related compounds.

## III. Chemistry and Biochemistry of Capsular Polysaccharides

Many species of micro-organisms, including yeasts, synthesize exocellular polysaccharides which give rise to capsules of varying thickness. When such organisms are grown in liquid media, especially under shaking conditions, the culture media often become extremely viscous, due to the release of excess capsular material into the medium. Whether these exo- and extracellular polysaccharides constitute components of the normal cell wall is not clear. It seems quite evident, however, that in many species their synthesis is not subject to metabolic control, judging by conversions of glucose to extracellular polysaccharides of 42% in the case of phosphomannans of *Hansenula holstii* (Slodki *et al.*, 1961) and 30–35% for a heteropolysaccharide of *Cryptococcus laurentii* (Cadmus *et al.*, 1962).

It has been noted that even non-capsulated yeasts may excrete into the medium wall-related polysaccharides in shake cultures, but this happens primarily upon extended incubation, for example after 1–2 weeks (Gorin and Perlin, 1956, 1957; Šikl *et al.*, 1967; do Carmo-Sousa and Barroso-Lopes, 1970). The last two authors compared the hydrolysis products of the cell walls of three non-capsulated and two capsulated *Candida* species with the hydrolysis products of the extracellular slime of the two capsulated *Candida* species. Since the same qualitative composition was found, consisting of five sugars, they concluded that the cell-wall composition in these species is basically the same as that of the capsules and that both are synthesized by the same enzyme system. Formation of a capsule would then represent surplus production of the external layer of the cell wall, and ultimately the capsular material would be released into the medium.

Similar conclusions were reached by Šikl *et al.* (1969). On the other hand, Spencer and Gorin (1965) concluded that in certain species, such as *Rhodotorula* and *Sporobolomyces*, the capsular and cell-wall polysaccharides were distinct and different. It would seem to me that possibly a particular component of the cell wall might be deposited in a capsule or in the culture medium due to oversynthesis, but that other components would be limited to the wall. It should be mentioned in this connection that do Carmo-Sousa and Barroso-Lopes (1970), upon hydrolysis of cell walls, obtained an acid-resistant residue which was not investigated further.

It seems best to discuss the capsular polysaccharides of yeast on the basis of their structural chemistry.

### A. α-D-LINKED-GLUCANS

Formation of starch-like polysaccharides by yeasts was first reported by Aschner *et al.* (1945) and Mager (1947). They found that some capsulated species, *Torulopsis rotundata* and *T. neoformans* (syn. *Cryptococcus neoformans*), produced an amylose-like polysaccharide when grown in a simple glucose-inorganic salts medium with ammonium sulphate as the nitrogen source. Both the medium and cells turned blue on addition of Lugol iodine solution. Mager and Aschner (1947) suggested that acidification of the medium to a pH value of less than 5·0, due to growth at the expense of ammonium sulphate, was necessary for starch formation. The actual pH value at which the starch begins to be excreted was later shown to be closer to 3·0 (Foda and Phaff, 1969). Hehre *et al.* (1949) crystallized pure amylose from the culture fluid of *Cryptococcus neoformans*. The weight-average molecular weight of the amylose from a strain of *Cryptococcus laurentii* (originally named *Rhodotorula peneaus*) was determined to be approximately $2 \times 10^6$ by Gorin *et al.* (1966). A much lower value for the molecular weight (approx. 7,000), reported earlier by Kooiman (1963), was probably the result of degradation of the polysaccharide by alkali in the presence of oxygen. Chemical and enzymic ($\beta$-amylase) analysis of the polymer by Gorin *et al.* (1966) showed it to be a linear glucan with the glucose residues linked by $\alpha$-(1→4) bonds. Other yeast species listed by Lodder and Kreger-van Rij (1952) which form starch-like compounds in acidic media included *Bullera alba*, *Candida curvata*, *C. humicola*, *Trichosporon cutaneum*, *Lipomyces* (some strains) and all species of *Cryptococcus*. Species of *Tremella* (a genus belonging to the heterobasidiomycetes and apparently closely related to *Cryptococcus*) are also positive (Slodki *et al.*, 1966). Some red yeasts, including *Rhodotorula macerans* and *Rh. infirmo-miniata*, produce starch independent of the pH value of the medium (Ahearn, 1964; Ahearn and Roth, 1966). These two species are now considered as members of the genus *Cryptococcus*.

Abercrombie *et al.* (1960a) grew *Cr. laurentii* in well-buffered media at pH values of 5, 7 and 9, and reported the presence in the medium of a glucan with an $[\alpha]_D + 180°$. It did not give a blue colour with iodine solution, was resistant to the action of $\beta$-amylase, and contained (1→3), (1→4), (1→2) and/or (1→6) bonds. The proportion of this glucan increased as the pH value of the medium was lowered from 9 to 5. Although the authors considered this polymer to be extracellular, the long growing periods on a shaker (12–14 days) could have resulted in some cell breakdown, and this glucan may have originated from the wall itself. It seems possible that this type of glucan is similar in composition, and possibly structure, to the $\alpha$-linked glucan recently discovered in other species of *Cryptococcus* by Bacon *et al.* (1968) and Jones *et al.* (1969).

An extracellular $\alpha$-D-linked glucan (pullulan) has been isolated from the "black yeast" *Pullularia pullulans* (syn. *Dematium pullulans*, *Aureobasidium pullulans*). This linear glucan, which produces viscous aqueous solutions, appears to contain mainly maltotriose residues connected by $\alpha$-(1→6) linkages. Because this organism is not considered a yeast by most investigators, only brief mention of this polysaccharide is made. Further information can be found in a detailed review by Gorin and Spencer (1968a).

### B. β-D-LINKED GLUCANS

Recently the first β-linked glucan was described as a component of the extracellular slime of *Torulopsis ingeniosa* (Masler *et al.*, 1968). After removal of a mannan component (see below) by precipitation with Fehling's solution the deionized residue yielded a water-insoluble glucan in addition to a water-soluble fraction, which was not studied further. Based on the results of methylation analysis and periodate oxidation, the glucan was shown to be branched, with a backbone of β-(1→4)-linked glucose residues. On the average, every fourth glucose unit in the backbone was substituted at C-2 or C-3 by a single glucosyl unit.

### C. MANNANS AND PHOSPHOMANNANS

Gorin *et al.* (1965) studied the composition of the mucilaginous polysaccharide elaborated by *Rhodotorula glutinis* in the capsule and culture filtrate. The linear polysaccharide had a specific rotation $[\alpha]_D$ −78° and a degree of polymerization of at least 90. Periodate oxidation, Smith degradation and partial hydrolysis showed that the mannan probably contained D-mannose residues alternately linked by β-(1→4) and β-(1→3) bonds. A similar polysaccharide was excreted by *Rh. mucilaginosa*, *Rh. minuta* and *Rh. pallida*, the latter represented by a strain designated as 62-506 by Gorin *et al.* (1965).

Masler *et al.* (1968) discovered another novel type of β-linked mannan in the extracellular slime produced in high yields by *Torulopsis ingeniosa*. The polymer could be separated from other slime components by precipitation with Fehling's solution *via* its insoluble copper complex. It had an optical rotation $[\alpha]_D$ −57°, suggesting that β-linkages predominate. Methylation analysis and Smith degradation revealed that the water-soluble mannan had a backbone consisting of β-(1→4)-linked mannose residues. Every sixth residue of the backbone was substituted at C-3 by side-chains with an average chain length of three mannose residues linked by β-(1→3) bonds. Different results on the slime of *Torulopsis ingeniosa* were reported briefly by Gorin and Spencer (1970). Their data indicated a linear structure in which mannose residues were connected by β-(1→3) and β-(1→4) bonds exclusively. An $[\alpha]_D$ −72° indicated β linkages. Such a polysaccharide appears related or similar to the capsular mannan of *Rhodotorula* species (see p. 177). It is possible that the discrepancy in results can be ascribed to different strains or misidentification of a culture.

A number of yeasts, representing the so-called primitive species of the genera *Hansenula*, *Pichia* and *Pachysolen*, produce very viscous media due to the formation of capsular and extracellular phosphomannans of widely varying mannosyl : phosphoryl ratios. Conditions for the formation of phosphomannans by *H. holstii* have been analysed, and yields of 43% polysaccharides on the basis of glucose utilized have been obtained (Anderson *et al.*, 1960; Rogovin *et al.*, 1961; Wickerham and Burton, 1961). The mannose:phosphate ratio (Slodki *et al.*, 1961) in the different species varied from 2·5 : 1 in *Hansenula capsulata* to 27·5 : 1 in *Hansenula minuta* (a relatively non-slimy species of the genus). The biological properties of these phosphomannans and their use in taxonomy have been discussed by Wickerham and Burton (1962).

Limited structural studies on the phosphomannans of *H. holstii* and *H. capsulata* have been made. Jeanes *et al.* (1961) studied the phosphomannan of *Hansenula holstii*, which was isolated as the potassium salt by ethanol precipitation of the cell-free culture filtrate. It had a D-mannose:phosphorus:potassium ratio of 5 : 1 : 1 and an optical rotation of $[\alpha]_D$ +103°, indicating a majority of α-linkages. Potentiometric titration of the decationized phosphomannan showed a single inflexion point at pH 7·2, indicating that the phosphorus was present as a phosphodiester. The secondary phosphoryl was stable to alkali at 100° but was readily hydrolysed by weak acid, such as autohydrolysis (Slodki, 1962). Slodki identified D-mannose 6-phosphate as the sole phosphorylated compound in acid hydrolysates of the mannan from *H. holstii*, and hydrolysis rates and other evidence showed that the polymer contained phosphoryl cross linkages in which one secondary phosphoryl is linked as mannose 6-phosphate and the other as α-hemiacetal phosphate, thus representing a di-D-mannose 1,6′-phosphodiester. Periodate oxidation studies have shown that 33% of the D-mannose units are linked through C-1 and C-2, and 47% through C-1 and C-3 (Jeanes and Watson, 1962). The remaining 20% were either non-reducing end-units or linked through (1→6) linkages, or both. Jeanes and Watson (1962) presented additional evidence that the phosphate of the polymer does not occur as pyrophosphate, but only in the orthophosphate form. There are on average five mannosyl units distributed between phosphodiester bonds. In subsequent work (Jeanes *et al.*, 1962) partial acid hydrolysates of both native and periodate-oxidized phosphomannan yielded 3-*O*-α-D-mannopyranosyl-D-mannose and *O*-α-D-mannopyranosyl-(1→3)-*O*-α-D-mannopyranosyl-(1→3)-D-mannose. This constitutes the first manno-oligosaccharide with α-(1→3) linkages isolated from natural products. Although the absence of branching and some details in sequence have not been worked out, a possible structure based on available evidence is given schematically below:

$$\left[ M\overset{1\;2}{—}M\overset{1\;3}{—}M\overset{1\;3}{—}M\overset{1\;2}{—}M\overset{1}{—}O—\underset{\displaystyle O^-}{\overset{\displaystyle O}{\overset{\|}{\underset{|}{P}}}}—O—{}^{6}M\overset{1\;3}{—}M\overset{1\;3}{—}M\overset{1\;3}{—}M\overset{1\;2}{—}M\overset{1}{—}O—\underset{\displaystyle O^-}{\overset{\displaystyle O}{\overset{\|}{\underset{|}{P}}}}— \right]_n$$

M = an α-D-mannopyranosyl residue

The phosphomannan of *Hansenula capsulata* is very different. It has a mannose to phosphate ratio of 2·5 : 1 and an optical rotation $[\alpha]_D$ −2°, indicating mainly $\beta$-linkages (Slodki *et al.*, 1961). The phosphate appears to be present as a phosphodiester, as in the case of the phosphomannan from *H. holstii*. Isolation by Slodki (1963) of di- and trisaccharide phosphomonoesters following mild acid hydrolysis and periodate oxidation studies, revealed that the disaccharide ester was 2-O-(6-O-phosphoryl-$\beta$-D-mannopyranosyl)-D-mannose, and the trisaccharide appeared to contain an additional mannosyl residue as shown in Fig. 9. The postulated

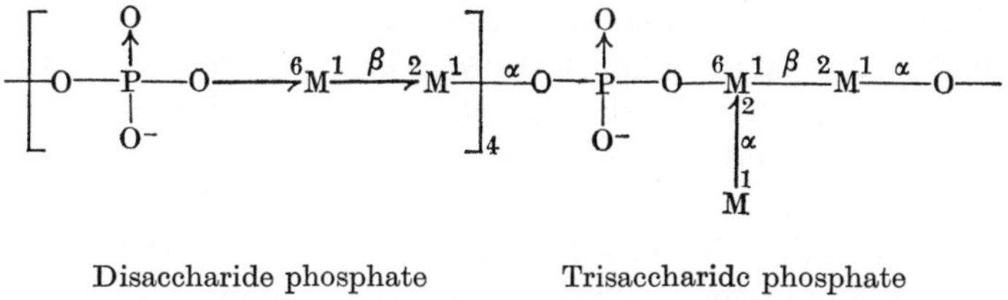

$$\left[-O-\underset{O^-}{\overset{O\uparrow}{P}}-O\longrightarrow{}^{6}M^{1}\xrightarrow{\ \beta\ }{}^{2}M^{1}\right]_4\overset{\alpha}{\longrightarrow}O-\underset{O^-}{\overset{O\uparrow}{P}}-O-{}^{6}\underset{\substack{\uparrow 2\\ \alpha\\ 1\\ M}}{M}{}^{1}\overset{\beta}{\longrightarrow}{}^{2}M^{1}\overset{\alpha}{\longrightarrow}O-$$

Disaccharide phosphate Trisaccharidc phosphate

FIG. 9. Proposed structure for the repeating units of the phosphomannan from *Hansenula capsulata* according to Slodki (1963). M indicates an $\alpha$-D-mannopyranosyl residue.

structure provides that every fifth disaccharide unit carries an appended $\alpha$-mannosyl unit. Slodki (1963) suggested that the biosynthesis of the repeating disaccharide phosphate backbone occurs by alternate mannosyl and mannose l-phosphate transfers from a donor such as GPD-mannose.

McLellan and Lampen (1968) investigated the structures of the two phosphomannans discussed above by enzymic analysis, using the phosphomannanase produced by *Bacillus circulans* (see Section II.B.5, p. 158). Both phosphomannans were depolymerized by the enzyme, as shown by a marked decrease in the viscosity of the polymer solutions and by the formation of various low molecular-weight fragments. In agreement with their results on baker's yeast mannan, the enzyme split mannosidic bonds and left the phosphodiester bonds intact. However, the apparent degree of polymerization of the fragments was greater than might have been expected from the stuctures depicted above. The high proportion of $\beta$-linkages in the mannan from *H. capsulata* could have been responsible for the finding that the fragments of this mannan were greater rather than smaller than those of the mannan from *H. holstii*, and there is as yet no precise information on the specificity of the phosphomannanase as to the anomeric linkage it can split. Incomplete enzymic hydrolysis may therefore have been responsible for the failure of McLellan and Lampen (1968) to correlate the breakdown products of the two polymers by phosphomannanase, autohydrolysis (boiling the decationized products at pH 2·5 for 20 min) or a combination of these treatments, with the structures proposed for these phosphomannans.

### D. PHOSPHOGALACTANS

Extracellular phosphogalactans, which appear to be analogous in many ways to the phosphomannans of certain *Hansenula* species, have been discovered in two unidentified strains of *Sporobolomyces* (Slodki, 1966b). The polysaccharides, which were also lightly acetylated, showed an $[\alpha]_D$ +118 to +127°, indicating a preponderance of α-linkages. The ratio of galactose : phosphate was approximately 9. Total acid hydrolysis yielded D-galactose 6-phosphate and D-galactose. Potentiometric titration of the decationized forms of the polysaccharides with alkali gave single inflexion points, characteristic of phosphodiester linkages. Mild acid hydrolysis cleaved the secondary phosphoryl linkages and released free galactose, indicating that galactosyl phosphate occurs exclusively as an end group. Galactose 1-phosphate was isolated as a product of alkaline hydrolysis at 100° in the complete absence of air. Periodate oxidation indicated that the phosphogalactans contained about equal proportions of galactose residues joined by α-(1→3) and α-(1→6) linkages. Although the overall structure as well as the site of acetylation are still unknown, part of the polymer molecule can be represented schematically as follows:

$$\mathrm{Ga}\overset{1}{—}\mathrm{O}—\underset{\underset{\mathrm{O}^-}{|}}{\overset{\overset{\mathrm{O}}{\|}}{\mathrm{P}}}—\mathrm{O}—{}^{6}\mathrm{Ga}\overset{1}{—}{}^{6}\mathrm{Ga}\overset{1}{—}{}^{3}\mathrm{Ga}$$

Ga = an α-D-galactopyranosyl residue

### E. GALACTOMANNAN

Although galactomannans have been shown to exist as cell-wall components of certain *Trichosporon* species (see Section II.B.4, p. 155), their occurrence in the extracellular slime of *Lipomyces starkeyi* has now been demonstrated by Šikl *et al.* (1968). Slodki and Wickerham (1966) had previously shown that galactose was a component of the extracellular slime of *L. starkeyi* but not of *L. lipofer* (see also Slooff *et al.*, 1969). However, Slodki and Wickerham (1966) did not fractionate their crude mixtures of polysaccharides, which contained in addition mannose and glucuronic acid. Thus, the component sugars and glucuronic acid could not be assigned to specific polysaccharides. Šikl *et al.* (1968) fractionated the slime of *L. starkeyi* by precipitating the cell-free culture filtrate with Fehling's solution. The homogeneous product purified *via* its insoluble copper complex turned out to be a galactomannan with α-(1→6)-linked mannose residues in a backbone, partially substituted at C-2 by galactose and mannose as shown in Fig. 10.

The low specific rotation ($[\alpha]_D$ +41°) follows from the occurrence of both $\alpha$- and $\beta$-linkages in the molecule.

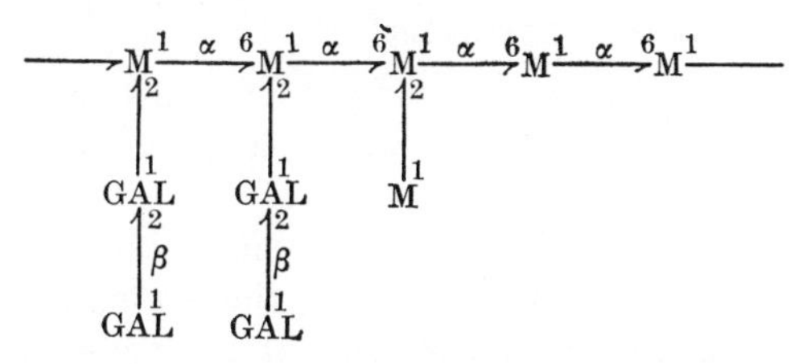


FIG. 10. Proposed structure for the galactomannan component of the extracellular slime of *Lipomyces starkeyi*, according to Šikl *et al.* (1968). Configurations of linkages not designated by $\alpha$ or $\beta$ have not been established. M indicates an $\alpha$-D-mannopyranosyl residue, and GAL an $\alpha$-D-galactopyranosyl residue.

## F. PENTOSYLMANNANS

Although many strains of *Trichosporon cutaneum* are capsulated and produce extracellular starch in media of low pH value (see Section III.A, p. 175), the capsular material released into the medium is different from that formed by cryptococci. Gorin and Spencer (1967) grew a strain of *Trich. cutaneum* in a medium containing glucose and the dialysed portion of yeast extract. The extracellular polysaccharides consisted of a glucan, which was not studied further, and a heteropolysaccharide which could be partially purified by precipitation with Fehling's solution. The branched polysaccharide had a backbone of $\alpha$-(1→3)-linked mannose units, some of which were substituted by D-mannosyl (linked by either (1→2) or (1→6) bonds), D-xylopyranosyl and 4-*O*-$\alpha$-L-arabinopyranosyl-D-xylopyranosyl units. Periodate oxidation of the heteropolymer indicated that, on a molar basis, about 45% of the anhydroaldose units were present as mannopyranosyl, xylopyranosyl and arabinopyranosyl non-reducing end units. However, the pattern of distribution of the side-chains along the main chain, the presence or absence of single arabinose units as side-chains, the exact position of attachment of the various side-chain types to the main chain, and the chemical homogeneity of the polysaccharide preparation remain to be worked out. Radioactive tracer experiments showed that D-xylose and L-arabinose units arose from glucose *via* a C-6 decarboxylation mechanism. The occurrence of a microbial polysaccharide containing L-arabinosyl units appears to be unique for *Trich. cutaneum*.

Gorin and Spencer (1968a) reported xylomannans in *Trichosporon inkin*, *Trich. sericeum* and *Trich. undulatum*. *Trichosporon sericeum* is now considered identical to *Endomycopsis ovetensis*, and *Trich. undulatum* represents a strain of *Trich. cutaneum*. Although *Trich. pullulans* formed a polysaccharide mainly consisting of mannose and xylose, this polymer contained small proportions of fucose and galactose as well.

### G. ACIDIC HETEROPOLYSACCHARIDES

These compounds form an important constituent of the capsular polysaccharides of a number of yeasts, in particular some of the species of *Cryptococcus*.

*Cryptococcus neoformans* has received much attention because the capsular polysaccharides are responsible for the serological characteristics of particular strains. Evans and his coworkers (Evans and Kessel, 1951) recognized three serological types, A, B, and C, among the strains they studied, and found that the capsules were responsible for the type specificity on the basis of precipitin, agglutination and "quellung" reactions. Most of the studies on the capsular materials of *Cryptococcus* were done with the polysaccharides which were recovered from the cell-free medium of growth, but some involved polysaccharides extracted directly from the cells (see Gorin and Spencer, 1968a, for details). In many of the studies, no strict criteria were used to ensure homogeneity of the polysaccharides, and this lack unquestionably accounted for some of the controversies regarding the nature of the component sugars in the isolated polysaccharides. Although Drouhet *et al.* (1950) reported xylose, mannose and glucuronic acid in an acid hydrolysate of polysaccharide from *Cr. neoformans*, Evans and Mehl (1951) found that galactose was present in addition as a minor component of the capsule. However, both Evans and Theriault (1953) working with serotype B, and Rebers *et al.* (1958), who analysed serotype A, came to the conclusion that there appeared to be a mixture of two polysaccharides, one containing galactose and the other not. Both groups of coworkers were able to achieve quantitative or semi-quantitative separation by appropriate precipitation procedures into polysaccharides with galactose contents ranging from less than normal to none at all. However, both types of polysaccharide were found to contain glucuronic acid. Blandamer and Danishefsky (1966) again reported the presence of galactose in the extracellular polysaccharide of serotype B, but their criteria for polysaccharide purification appear not to have been as critical as those of the above-mentioned investigators.

Gadebusch and Johnson (1961) and Gadebusch (1960a, b) obtained a partially purified enzyme mixture from a species of *Alcaligenes* which could degrade the capsular polysaccharide of a particular strain of *Cr. neoformans* both *in vitro* and *in vivo*, and they identified galactose, mannose, xylose and glucuronic acid as hydrolysis products. Again there was no assurance that the substrate was a pure compound. Capsules from other strains of this yeast, surprisingly, were not hydrolysed by the enzyme. The partially decapsulated sensitive strain (*in vitro* by the enzyme) was found to be strongly antigenic, but completely uncapsulated cells were no longer antigenic, thus confirming the antigenic role of the capsule.

The chemical structure of the polysaccharides from *Cr. neoformans* was investigated by Drouhet *et al.* (1950). Upon acid hydrolysis, xylose was released first, indicating its terminal location. Later an aldobiuronic disaccharide appeared among the hydrolysis products, which required drastic treatment with acid before it hydrolysed to mannose and glucuronic acid. In a detailed chemical study, Miyazaki (1961a, b, c) determined the molar ratio of glucuronic acid : xylose : mannose to be 1 : 1 : 3 in the acidic heteropolymer of an untyped *Cr. neoformans* strain. Its molecular weight was estimated at 6,600. By means of methylation analysis and periodate oxidation, he proposed that the polymer had a main chain with mannose residues connected by (1→2) linkages. Mannose also occurred as side units attached by (1→4) linkages, and xylose and glucuronic acid occurred as terminal groups as shown in Fig. 11.

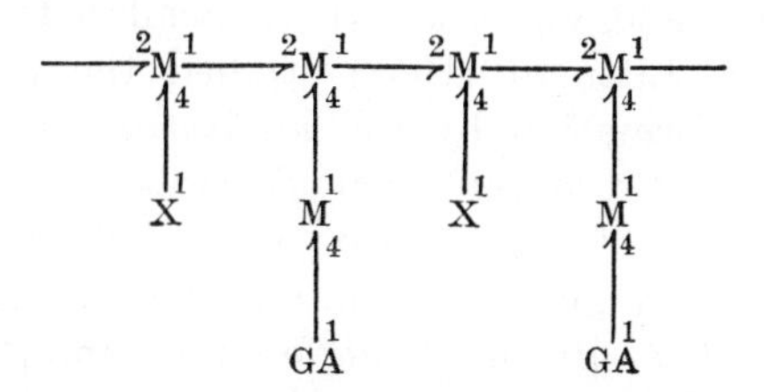


FIG. 11. Proposed structure for the capsular polysaccharide of *Cryptococcus neoformans* according to Miyazaki (1961a, b, c). M indicates an α-D-mannopyranosyl residue, X an α-D-xylopyranosyl residue, and GA an α-D-glucuronopyranosyl residue.

Miyazaki's results were largely confirmed by the chemical studies of Blandamer and Danishefsky (1966).

Detailed studies have also been made on the capsular polysaccharide of *Cr. laurentii*, a non-pathogenic highly capsulated species. Jeanes *et al.* (1964), working with *Cr. laurentii* var. *flavescens*, noted the extremely high viscosity of the extracellular slime. The polymer, which may have industrial applications, was shown to be thixotropic and rapidly regained viscosity after application of shear forces.

Abercrombie *et al.* (1960a) fractionated the extracellular polysaccharides of *Cr. laurentii* by the use of cetyltrimethylammonium bromide into an acidic heteropolymer (which precipitated with this reagent) and a neutral glucan (see Section III.A, p. 176). The heteropolymer had $[\alpha]_D$ +21°, and consisted of a mannose-containing main chain with xylose and glucuronic acid present in side branches. The ratio of mannose :

xylose was 5 : 2, with about 12–14% glucuronic acid. Jeanes *et al.* (1964), on the other hand, using the same strain, obtained an approximate molar ratio of 4 : 1 : 1 for mannose : xylose : glucuronic acid, and in addition about 7% alkali-labile *O*-acetyl substituents. Slodki *et al.* (1966) observed that, in different strains of *Cr. laurentii*, not only was there a range in $[\alpha]_D$ from $-8°$ to $+52°$, but also the molar ratios of mannose : xylose varied from 4 : 1 to 4·6 : 3.

They confirmed the presence of acetyl groups in the heteropolysaccharides of most strains, but the concentration was variable as well. Species of the heterobasidiomycetous genus *Tremella*, which appears closely related to *Cryptococcus*, contain very similar acidic heteropolysaccharides (Slodki *et al.*, 1966). One strain of *Tremella mesenterica* with an *O*-acetyl : xylose : mannose : glucuronic acid ratio of 0·5 : 4·4 : 3·8 : 1 in its heteropolysaccharide (Slodki, 1966a) formed, upon partial hydrolysis, oligosaccharides containing mannosyl units attached to 2-*O*-(β-D-glucuronopyranosyl)-D-mannose. It is evident that there is much variation in the composition of the acidic heteropolysaccharides, and that much remains to be learned of their chemical structures.

Studies on the physiology of production of acidic heteropolysaccharides are rather limited. Littman (1958) studied the nutritional factors governing capsule formation by *Cr. neoformans*, and Cadmus *et al.* (1962) those affecting *Cr. laurentii* var. *flavescens*.

Information on the biosynthesis of acidic heteropolysaccharides is beginning to accumulate, especially in recent years. Abercrombie *et al.* (1960b) grew *Cr. laurentii* in a buffered casamino acid-containing medium, using labelled D-glucose, D-galactose, D-xylose or L-arabinose as carbon source, and showed that mannose and glucuronic acid were formed from hexoses without breakdown of the carbon skeleton, whereas xylose was formed from hexose mainly by a process involving the loss of C-6. They predicted the presence of two enzymes, one to oxidize UDP-glucose to UDP-glucuronic acid and a second one which would decarboxylate the last compound to UDP-D-xylose. Ankel and Feingold (1964) subsequently demonstrated the latter enzyme in crude soluble extracts of *Cr. laurentii*.

Later Ankel and Feingold (1966) partially purified the enzyme UDP-D-glucuronate carboxylase, and reported it to have an absolute requirement for NAD and a pH optimum between 7·0 and 7·5. The enzyme responsible for formation of the glucuronic acid component of the capsular polysaccharide, uridine diphosphate D-glucose dehydrogenase, also was found in the soluble fraction (105,000 *g* supernatant) of homogenized *Cr. laurentii* (Ankel *et al.*, 1966). The enzyme is specific for UDP-glucose, whereas diphosphoglucose derivatives of adenosine, cytidine, guanosine and thymidine were inactive; NAD served as hydrogen acceptor. The pH optimum was between 7·3 and 7·8, and $K_m = 0{\cdot}6$ m*M* for both UDP-glucose and NAD. The enzyme was reported to be allosteric, since it was inhibited specifically by UDP-D-xylose, and this inhibition could be reversed by increasing the concentration of either NAD or UDP-glucose.

The macromolecular synthesis of the acidic heteropolysaccharide is essentially still unknown. Cohen and Feingold (1967) reported that a particulate enzyme in the cell homogenate of *Cr. laurentii* transferred D-xylosyl units from UDP-D-xylose to an acceptor prepared from the capsular polysaccharide of the same organism, but only after hydrolytic removal of part of the original xylose. Unidentified oligosaccharides with a degree of polymerization of 2–8, prepared by partial acetolysis from *Cr. laurentii* native polysaccharide and various cell-wall fractions from other yeasts, were not active as acceptors in this system, but dexylosylated polysaccharides from types A, B and C of *Cr. neoformans* and from *Tremella mesenterica* served as acceptors.

Later, Schutzbach and Ankel (1969) reported that α-1,3-mannobiose functioned as xylosyl acceptor from UDP-xylose, but α-1,2-mannobiose, mannose, glucose, xylose, galactose and L-arabinose were inactive in this capacity. Only a single xylosyl unit was transferred to α-1,3-mannobiose.

## H. MISCELLANEOUS HETEROPOLYSACCHARIDES

Some species of yeast produce extracellular heteropolymers with many component sugars. Gorin and Spencer (1968c) studied the capsular polysaccharide of *Candida bogoriensis* and identified glucuronic acid, fucose, rhamnose, mannose, galactose and a trace of glucose in hydrolysates. Since an aqueous-alkali cell extract gave the same sugars it was implied that the cell walls contained the same sugars. Attempts to fractionate the extracellular polysaccharide into components of different compositions were unsuccessful, which does not, however, prove that the material studied constituted a single molecular species. Gorin and Spencer deduced from various analytical approaches that fucopyranosyl, rhamnopyranosyl and galactofuranosyl residues were present as end units. Degradation analysis showed that α-(1→3)-linked D-mannopyranosyl units formed the main chain of the heteropolymer, similar to that occurring in *Cryptococcus laurentii* and *Trichosporon cutaneum*. Fragments arising from partially hydrolysed side-chains included 3-*O*-α-L-rhamnopyranosyl-L-rhamnose, 2-*O*-α-L-rhamnopyranosyl-L-fucose, 3-*O*-α-L-fucopyranosyl-L-fucose and 4-*O*-β-D-glucuronopyranosyl-L-fucose.

Other yeasts in which Gorin and Spencer (1968c) identified heteropolymers with unusual sugars were:

*Candida foliarum*: glucose, mannose, rhamnose, galactose (trace), fucose.
*Candida diffluens*: glucose (trace), mannose, rhamnose, galactose, fucose.
*Candida scottii*: glucose (trace), mannose, galactose, fucose.
*Candida humicola*: glucose, mannose, xylose, uronic acid.

The component sugars of the four species listed may represent extracellular or cell-wall polysaccharides, or both. Do Carmo-Sousa and Barroso-Lopes (1970) investigated the following *Candida* species: *C. bogoriensis*, *C. buffonii*, *C. diffluens*, *C. foliarum* and *C. javanica*. They reported that the composition of the hydrolysate of the extracellular polysaccharide was the same as that of isolated cell-wall material, and that it consisted in all cases of galactose, glucose, mannose, fucose and rhamnose. Their results differ somewhat from those of Gorin and Spencer (1968c) in that the latter found glucuronic acid and only traces of glucose in the heteropolysaccharide of *C. bogoriensis*, while *C. javanica* was reported to contain glucan with a trace of galactose (Spencer and Gorin, 1969a). Whether these and other quantitative differences in results are due to the use of different strains remains to be clarified.

Finally, mention is made of the brief statement by Schutzbach and Ankel (1969) that they detected a neutral heteropolysaccharide in the cell wall of *Cr. laurentii* var. *flavescens*, composed of D-mannose, D-galactose, D-xylose, and L-arabinose. Presumably this polysaccharide occurs in addition to the neutral glucan (see Section III.A, p. 176) reported by Abercrombie *et al.* (1960a).

## IV. Enzymic Lysis of Yeast Cell Walls

Interest in the enzymic digestion of yeast cell walls has been stimulated for three main reasons: (a) the preparation of sphaeroplasts, i.e. cells devoid of their characteristic rigid cell walls, for physiological studies; (b) studies on the biosynthesis and regeneration of cell walls from sphaeroplasts; and (c) analysis of native and isolated cell walls. Wall analysis with selected and specific enzymes avoids the often used treatments with strong acids and alkali, and can release certain compounds or groups of components intact or with minimal modification. Information can also be obtained on the nature of the bonding between different components, and differences may be revealed in the chemical composition of cell walls of different species or genera. First we shall discuss the use of snail digestive fluid and next the application of microbial and other enzymes.

### A. GASTRIC JUICE OF THE GARDEN SNAIL, *Helix pomatia*

Giaja (1914, 1919, 1922) was probably the first to observe that the digestive fluid of garden snails, which is found in a small vesicle of the alimentary canal, had the ability to digest yeast cells. He realized the great possibilities of cell-wall removal for cytologic studies, but did not use isotonic medium for sphaeroplast stabilization. Moreover, biochemists had not yet become interested in subcellular fractions at that time. Eddy and Williamson (1957, 1959) re-investigated the use of gastric juice of snails, but in conjunction with osmotic stabilizers. When young cultures of certain strains of *Sacch. carlsbergensis* and *Sacch. cerevisiae* were treated with snail enzyme in the presence of 0·55 M-rhamnose or mannitol, over 90% of the cells were rapidly converted into protoplasts which remained stable for at least several hours. Holter and Ottolenghi (1960) prepared protoplasts of *Saccharomyces* strains and of *Schizosaccharomyces pombe* and obtained better stability of the protoplasts by increasing the concentration of the stabilizing sugars or polyols to 0·8–1·2 M. They introduced sorbitol for this purpose since it has a higher solubility than mannitol. Sodium and potassium chlorides at 0·6 *M* were also suitable. Gascón and Villanueva (1965) found that the osmotic pressure was not the only requirement of an osmotic stabilizer, and they concluded that 0·8–1·0 *M* solutions of magnesium sulphate were by far the best of all stabilizers tested by them. Rost and Venner (1965a), who did not examine magnesium sulphate, found ammonium sulphate best for *Sacch. cerevisiae* and *Sacch.* (= *Fabospora*) *fragilis*, but for other species of yeast potassium chloride was equally satisfactory.

Holter and Ottolenghi (1960) confirmed the observation by Eddy and Williamson that log-phase cells were much more susceptible to protoplasting than were stationary-phase cells but, in addition, they noted considerable variation in susceptibility between different strains of *Saccharomyces*, some refusing to transform into sphaeroplasts altogether. Millbank and Macrae (1964) and Rost and Venner (1965a) also noted that there existed great differences between strains of *Saccharomyces* in their susceptibility to attack by snail enzyme. Millbank and Macrae (1964) presented evidence that such differences were caused by the ease with which the outer mannan-protein layer of the cell could be opened up or degraded by specific enzyme components.

Svihla *et al.* (1961), who were interested in obtaining cells with a high level of S-adenosylmethionine by growing them in media supplemented with L-methionine and related sulphur-containing compounds, observed that *Candida utilis* grown in the presence of sulphur-containing amino acids was much more susceptible to digestion by snail enzyme than control cells grown in their absence. However, even the latter cells were more readily transformed into sphaeroplasts than those of baker's yeast, thus confirming the differences in susceptibility between various species. The sphaeroplasts so obtained were stable for many weeks. Burger *et al.* (1961) reported that cysteine accelerated the formation of sphaeroplasts from baker's yeast by snail enzyme. Duell *et al.* (1964) obtained excellent conversion into sphaeroplasts when *Sacch. cerevisiae* cells, from either the log phase or the stationary phase, were pretreated for 30 min with a solution of 2-mercaptoethylamine and ethylenediaminetetraacetate and then, after washing, exposed in high concentrations to snail enzyme. Nurminen *et al.* (1965) observed a comparable effect by pretreatment of baker's yeast suspensions with 2-mercaptoethanol. *Saccharomyces fragilis* formed no sphaeroplasts in the absence of 2-mercaptoethanol (Davies and Elvin, 1964). Here, too, pretreatment of the cells with the sulphydryl reagent was sufficient to render the cells susceptible, and the effect therefore is on the cell wall rather than on the enzyme. On the other hand, Rost and Venner (1965a), who employed a different strain of *Sacch. fragilis*, found it to be by far the most susceptible species, even without the use of sulphydryl compounds. The stimulation by sulphydryl compounds may consist in the rupture of disulphide bonds of the wall protein (Falcone and Nickerson, 1956).

Many investigators have noted that snail enzyme does not digest the cell wall evenly. Initial breaks are usually first seen in the polar and equatorial regions of the cell. In *Candida utilis*, the sphaeroplast normally leaves the residual hull through an equatorial rupture (Svihla *et al.*, 1961). Rost and Venner (1965a), working with *Schizosaccharomyces pombe*, and Zvjagilskaja and Afanasjewa (1965) using *Endomyces magnusii*, observed that initial attack (and as a result sphaeroplast emergence) by snail enzyme on the cylindrical cells of these two species occurred nearly always near one of the poles. In *Saccharomycodes ludwigii* (a yeast which buds at the poles only), sphaeroplasts emerge at one of the poles and, in *Sporobolomyces*, it always takes place at the location where a ballistospore has been discharged (Rost and Venner, 1965a). Darling *et al.* (1969) studied the kinetics of sphaeroplast formation by a strain of *Sacch. cerevisiae* with elongated cells. By digestion of the walls with snail enzyme, they observed that there were two phases during sphaeroplast formation. First the cells lost their walls but retained their elongated form if osmotically protected. Cells in this phase were termed prosphaeroplasts. During the second phase the cells rapidly attained a spherical form and thus became typical sphaeroplasts. There was no distinct difference in ultrastructure between prosphaeroplasts and sphaeroplasts, and the rigidity of the former is probably due to a very thin remnant of cell wall which is shed during the final transformation.

When yeast cells transform into asci, the wall becomes much more susceptible to enzymic digestion. The original cell wall then functions as an envelope for the ascospores, and in some species of yeast it is lysed by glucanases elaborated by the yeast itself (Abd-El-Al and Phaff, 1968,

1969). Johnston and Mortimer (1959) introduced treatment of asci with snail digestive fluid to facilitate isolation of single ascospores for genetic studies. The thick ascospore walls do not appear to be susceptible to snail fluid.

Gastric juice of snails has been reported to contain some 30 enzymes (Holden and Tracey, 1950), including glucanase, mannanase, cellulase, chitinase, lipase and polygalacturonase, but it appears to be very low in proteolytic activity. Anderson and Millbank (1966) have fractionated the snail enzyme complex by gel filtration on Sephadex G-100. Four major fractions were separated (A to D), of which three exhibited very limited or no attack on cell walls of log-phase *Sacch. carlsbergensis*. Fraction C, however, had the ability to produce sphaeroplasts and to bring about extensive degradation of cell walls derived from such cells. As this fraction was devoid of the lipase, the phosphatases and exo-$\beta$-glucanases found in Fractions A and B (none of the fractions contained proteinase), the authors concluded that these enzymes are not essential in cell-wall degradation. Fraction C, on the other hand, contained endo-$\beta$-(1→3)- and endo-$\beta$-(1→6)-glucanases, yielding glucose and disaccharides. Some evidence for a debranching enzyme was also obtained. Mannan and protein were solubilized from log-phase cell walls, but mannan was not extensively hydrolysed. From resistant stationary-phase cell walls, Fraction C released very little protein and mannan but, as was shown for whole cells by a number of investigators, pretreatment of such walls with 2-mercaptoethanol rendered them highly susceptible to lysis. In general, therefore, sulphydryl reagents appear to facilitate the access of polysaccharidases to cell-wall polysaccharides by rupturing protein disulphide bonds, but in some species or in log-phase cells this stimulatory activity is minimal or absent.

### B. LYSIS OF WALLS BY MICROBIAL ENZYMES

Salton (1955) was one of the first investigators to isolate a number of actinomycetes and myxobacteria which showed lytic activity toward cell walls of a yeast. The organisms were strains of *Streptomyces*, *Myxococcus fulvus* and *Cytophaga johnsonii*. Lytic activities were tested by streaking the organisms on agar plates containing autoclaved cells of *Candida pulcherrima*. Lysed zones were formed around the growth of strains which had lytic activity. Later, Webley *et al.* (1967) with another strain of *Cytophaga johnsonii* showed that this myxobacterium also lyses the walls of autoclaved baker's yeast and produces an active glucanase. However, no activity could be demonstrated on the walls of live yeast cells or even yeast heated for 15 min at 75°. This lack of lytic activity was ascribed to the presence of an outer barrier of the wall, which presumably is destroyed upon autoclaving.

Horikoshi and Sakaguchi (1958) isolated from soil a strain of *Bacillus circulans* which showed lytic activity toward cell walls of *Aspergillus oryzae* and *Sacch. sake*. Synthesis of the enzyme was induced by growing the organism on *Aspergillus* mycelium, and at pH 6·5, which is the optimum. The enzyme produced only hexose-containing degradation products from *Aspergillus oryzae* walls even though the walls contained almost 40% of a hexosamine polymer. The enzyme also lowered the turbidity of a heated suspension of *Sacch. sake*, but a suspension of living cells was attacked only poorly. Apparently no proteolytic activity was associated with the lytic enzymes.

Tanaka and Phaff (1965) isolated another strain of *Bacillus circulans* from soil in addition to a number of other micro-organisms, all of which showed lytic activity towards cell walls of baker's yeast. The new strain of *Bacillus circulans* (WL-12) produced under conditions of induction a mixture of endo-β-(1→3)- and endo-β-(1→6)-glucanases when grown aerobically in a mineral medium with baker's yeast cell walls as the inducer. Maximal lytic activity occurred at pH 6·5 as for the strain reported by Horikoshi and Sakaguchi (1958). Although the two endo-glucanases mentioned above were produced under conditions of induction, the culture fluid showed no hydrolytic activity on mannan, chitin or protein of the wall. Tanaka and Phaff (1965) were able to effect a quantitative separation of the two endoglucanases on DEAE-cellulose. Treatment of baker's yeast walls with either of the two glucanases caused lysis, and the products from baker's yeast glucan included laminaribiose and higher homologues on treatment with β-(1→3)-endoglucanase, and gentiobiose and higher homologues with β-(1→6)-endoglucanase. In both cases, small amounts of unknown oligosaccharides were also formed. For further details on the action of these enzymes see Section II.A.2 (p. 143).

The ability of either the separate or combined enzymes of *Bacillus circulans* to produce sphaeroplasts of baker's yeast was extremely limited in the absence of sulphydryl compounds (Tanaka *et al.*, 1965). In their presence, the production of sphaeroplasts from *Sacch. cerevisiae* was somewhat improved, although certain other yeast-like organisms (e.g. *Eremothecium ashbyi* and *Ashbya gossypii*) yielded sphaeroplasts readily with a mixture of the two β-glucanases. Tanaka *et al.* (1965) examined many species of yeast with respect to their susceptibility to the lytic activity by the separate or the combined action of β-(1→3)- and β-(1→6)-glucanases. They observed profound differences, indicating significant variation in the composition of the glucan portions of their walls. For example, *Bacillus circulans* WL-12 appears to lack the ability to produce enzymes able to hydrolyse the α-(1→3)-linked glucan which has been demonstrated in species of *Schizosaccharomyces* by Bacon *et al.*

(1968). Later Schwencke *et al.* (1969) reported that they obtained 100% true "protoplasts" from several species of *Saccharomyces* with an enzyme preparation from *Streptomyces* WL-6, one of the lytic organisms isolated by Tanaka and Phaff. Pre-incubation in tris buffer (pH 8) including EDTA, and addition of sulphydryl compounds during digestion was necessary for satisfactory results. From *C. albicans* only "sphaeroplasts" were obtained, bodies which were spherical but resistant to osmotic shock in hypotonic medium. In all instances the enzyme was induced by the glucan of autolysed washed baker's yeast cells.

Satomura *et al.* (1960) obtained an active glucanase preparation from *Sclerotinia libertiana* grown on bran. Effective digestion of the glucan in intact yeast was not achieved with glucanases alone or with a combination of glucanase and papain. However, the digestibility of the glucan in intact cells became significantly greater by addition of lipase and phospholipase to the glucanase. This finding is contrary to the results of Anderson and Millbank (1966), who concluded that lipase played no role in the protoplast-forming ability of a purified fraction of snail enzyme. Few other investigators have used fungi for obtaining enzymes able to digest yeast cell walls. Jones and Webley (1967) isolated two strains of *Fusarium* and a sterile (unidentified) fungus all of which had lytic activity toward baker's yeast cell walls, the last organism being the most active. It would appear, however, from the long digestion times needed (up to 24 days) that the enzyme solutions were not very active. Kuroda *et al.* (1968) obtained lytic enzymes, which hydrolysed baker's yeast cell walls, from a species of *Rhizopus*. Only log-phase cells of *Sacch. cerevisiae* pretreated with 2-mercaptoethanol were transformed into sphaeroplasts by the enzyme complex, of which the identified components were $\beta$-(1→3)- and $\beta$-(1→6)-glucanase, and protease. The endo-$\alpha$-(1→3)-glucanase from *Trichoderma viride* (Hasegawa *et al.*, 1969) has not been tested as yet on yeasts containing $\alpha$-(1→3) glucan for the purpose of sphaeroplast production. All of the fungal enzymes studied have significantly lower optimum pH values (about 4·5–5·0) than bacterial enzymes.

Furuya and Ikeda (1960a, b) isolated from soil numerous microorganisms which produced clear zones on agar plates containing heated baker's yeast. A strain of *Streptomyces*, which showed very high activity, produced extracellular lytic enzymes when it was grown in liquid medium with aeration on autoclaved baker's yeast as the carbon source. The enzyme had a pH optimum of 7·6 and caused lysis of isolated cell walls of baker's yeast. The concentrated enzyme solution was also tested for its lytic activity toward suspensions of various strains and species. As had been found for snail enzyme by others (see p. 187) there were striking differences in susceptibility between different strains of *Sacch. cerevisiae* and between different species. For example, *Hansenula anomala* and *Endomyces magnusii* were extremely susceptible, while *Candida utilis*, *Saccharomycodes ludwigii* and *Rhodotorula glutinis* were fairly resistant. Present knowledge on the glucan and mannan composition of these species can explain, at least in part, why an enzyme induced by baker's yeast cell walls may have limited capacity to attack cell walls of other species. The lytic activity of the enzyme toward *Sacch. cerevisiae* was greatly decreased in the presence of a protease inhibitor from potatoes, providing indirect evidence that proteolytic activity of the enzyme complex played a role in the lytic phenomenon. Furuya and Ikeda also succeeded with their Streptomyces enzyme to transform (for the first time with a microbial enzyme) young cells of a susceptible strain of *Sacch. cerevisiae* into sphaeroplasts.

Villanueva and his coworkers compared numerous species of actinomycetes for lytic activity on yeast walls, and concluded that species of *Streptomyces* and *Micromonospora* were most suitable (Gascón and Villanueva, 1963). An enzyme preparation from a *Streptomyces* species digested cell walls from *Sacch. fragilis*, *Sacch. rosei*, *Sacch. cerevisiae*, *Candida utilis* and *Kloeckera apiculata*, and converted young cells of *C. utilis* into sphaeroplasts (Garcia Mendoza and Villanueva, 1962). Later, an enzyme preparation obtained from a species of *Micromonospora* was found to be useful in the preparation of sphaeroplasts from *C. utilis* (Ochoa *et al.*, 1963). Gascón *et al.* (1965) also used the Micromonospora enzyme to prepare sphaeroplasts from the yeast-like fungi *Oospora suaveolens* and *Geotrichum lactis*. Log-phase cells were much more susceptible to lysis than older cells.

Domanski and Miller (1968) obtained excellent formation of osmotically-sensitive sphaeroplasts when *C. albicans* cells, pretreated with mercaptoethanol and EDTA, were treated with an enzyme solution containing both chitinase and $\beta$-(1$\rightarrow$3)-glucanase and in the absence of apparent proteolytic activity. Since the purity of the enzyme solutions was not scrutinized, it is difficult to assess their results in terms of enzyme requirements for sphaeroplast formation (cf. Schwencke *et al.*, 1968).

Lastly, McLellan and Lampen (1968) identified in culture filtrates of the strain of *Bacillus circulans* of Horikoshi and Sakaguchi a factor which they found to be essential for sphaeroplast formation of log-phase *Saccharomyces* cells, and which was termed PR-factor for its role in protoplast formation. As has been explained in fuller detail in Section II.B.5 (p. 158) this enzymic factor was later renamed phosphomannanase, and its activity has been ascribed to its ability to split a limited number of mannosidic bonds so that the outer mannan-protein layer becomes solubilized or opened up, thus allowing the glucanases to break down the previously protected glucan components. McLellan and Lampen (1968) offer the hypothesis that their phosphomannanase was probably an unrecognized component in those enzyme preparations which were successful in producing sphaeroplasts from susceptible yeasts.

In summary, it would appear that, for sphaeroplast formation of *Sacch. cerevisiae*, phosphomannanse together with sulphydryl reagents open up the glucan layers for attack by endo-$\beta$-(1→3)-glucanase (possibly in collaboration with endo-$\beta$-(1→6)-glucanase). Chitinase may be beneficial to obtain true protoplasts. The behaviour of other species is largely confusing, some being more and others less susceptible to lytic enzymes. It is possible that the many variations in mannan structure now known to exist in yeast are primarily responsible for this variation in susceptibility.

Secondarily, the difference in glucan structure and the possibility of mixtures of several types of glucan in yeast walls further complicate the susceptibility of yeast cells to lytic enzymes. Proteolytic enzymes, phosphatases and phospholipases appear to play a lesser if any role in the process.

## V. Regeneration of Cell Walls from Sphaeroplasts

### A. ON THE QUESTION OF PROTOPLASTS VERSUS SPHAEROPLASTS

Before considering cell-wall regeneration, it is essential to define the osmotically fragile starting material. The term "protoplast" is normally applied to cells whose walls are completely removed, and which are therefore held intact solely by the cytoplasmic membrane or plasmalemma. If parts of the cell wall remain attached to the membrane, even though the cells are osmotically fragile, they should be referred to as sphaeroplasts. It is clear from the literature that these osmotically sensitive bodies have been poorly defined, and it seems likely that in most instances the material represented sphaeroplasts rather than protoplasts. Ottolenghi (1966), who raised this question, reported that annular rings of bud scars remain attached to a membranous matter after cells have been treated with snail enzyme. This membranous matter may represent remnants of the innermost wall layer (Streiblová, 1968). Garcia Mendoza *et al.* (1968) obtained immunological evidence that protoplast membranes of *Candida utilis* contained traces of cell-wall material not detectable by electron microscopy of thin sections.

In ultrathin sections no evidence of wall remnants has been found (see, for example, Havelkova, 1966) but, by the freeze-etching technique of Moor *et al.* (1961), Streiblová (1968) showed that the entire wall substance was not removed by snail enzyme in all cells, and that the osmotically sensitive population represented a mixture of sphaeroplasts and protoplasts. Based on the electron micrographs of baker's yeast, bud scars and associated wall remnants prepared by Bacon *et al.* (1966), Streiblová (1968) suggested that the innermost enzyme-resistant wall layer consisting in part of very fine fibrils might represent chitin, possibly in association with protein. Such a thin fibrillar layer attached to the cytoplasmic membrane was also observed in sphaeroplasts of *Sacch. cerevisiae* by Darling *et al.* (1969), who obtained in addition electron micrographs of wall remnants plus bud scars very similar to those reported by Bacon *et al.* (1966).

### B. Cell wall regeneration

Early experiments on yeast cell-wall regeneration by Nečas (1955, 1956) involved sphaeroplasts obtained by mechanical pressure on cells or by autolytic processes. When these plasma droplets were placed in liquid medium, they died but on a nutrient-agar medium a very small fraction of them increased in size (sometimes a thousand times as big as the original yeast cell), became highly vacuolated, and some of these became multinucleate while the vacuoles were replaced by dense plasma formations. When these bodies were transferred to a fresh nutrient agar, a wall developed around the giant multinucleate protoplasts and after about three days budding began, gradually resulting in normally appearing yeast cells.

Eddy and Williamson (1959) cultured sphaeroplasts of *Sacch. carlsbergensis* (obtained by the action of snail enzyme) in osmotically balanced nutrient media. They also observed irregularly shaped outgrowths with greatly increased volume. Regeneration was observed, but the walls so formed were aberrant in that they were almost devoid of amino-acid residues, although a high content of N-acetylglucosamine was found; glucose and mannose were detected qualitatively. The high level of N-acetylglucosamine is in agreement with the composition of the innermost wall layer discussed by Streiblová (1968) as mentioned above. However, Eddy and Williamson (1959) were unable to obtain further growth leading to normal cells. On the other hand, the cells with aberrant walls were able, on suitable media, to produce ascospores, which in turn germinated into normal cells.

Nečas (1961) discovered that, if protoplasts prepared by autolysis are embedded in a high concentration of gelatin (e.g. 30%), most of the sphaeroplasts regenerate a wall and become normal budding cells. These results have been confirmed by Rost and Venner (1965b) for sphaeroplasts of *Sacch. fragilis*. Apparently it is necessary that the protoplasts are completely embedded in the gelatin for regeneration to occur; cultivation on the surface of such a gelatin medium is ineffective (Nečas and Svoboda, 1965). An important condition for regeneration appears to be the minimal module of elasticity of the gelatin medium; for example, mass regeneration at 25° required at least 30% gelatin, but at 12–15° only 15% gelatin was needed. The overall mechanism of the budding of regenerating protoplasts corresponds in principle to normal budding (Nečas and Svoboda, 1967). However, sections of regenerating sphaeroplasts clearly showed an early break in the continuity of the wall of the mother cell and bud, indicating that the synthesis of the bud wall appears to be independent from that of the mother cell.

Svoboda (1965) compared eight different species of yeast for their ability to regenerate cell walls and found that six required high concentrations of gelatin. These were *Sacch. cerevisiae*, *Sacch. chevalieri*, *Sacch. lactis* (=*Fabospora lactis*), *C. utilis*, *Rh. glutinis* and *S. ludwigii*. Two species, *Nadsonia elongata* and *Schizosacch. pombe*, regenerated even on the surface of agar or in liquid nutrient media. Svoboda (1965) ascribed the regeneration of the sphaeroplasts of the last two species to the absence of typical mannan in their walls. However, the finding of galactomannans in their walls (cf. Section II.B.4, p. 155) makes the difference in the mannan composition of their walls and that of *Saccharomyces* less profound than originally thought. Of greater importance may be the very different $\alpha$-(1→3)-linked glucan in *Schizosaccharomyces* (cf. Section II.A.1, p. 141). There is, however, a different course in cell-wall regeneration between *Schizosaccharomyces* in gelatin and in liquid medium (Nečas *et al.*, 1968a). In gelatin, the process is more or less similar to that in *Sacch. cerevisiae*, but in liquid medium a very dense network composed of bundles of elementary fibrils is formed, and it takes more than 20 h before matrix is deposited in the vicinity of the plasma membrane. According to Havelková (1969) the reason why *Sacch. cerevisiae* does not, and *Schizosacch. pombe* does, regenerate in liquid medium may be the appearance of large numbers of dictyosomes in regenerating sphaeroplasts of the latter species. These dictyosomes appear to be involved in the deposition of matrix material in the fibrillar ground layer. Havelková (1969) also noted that, in contrast to *Sacch. cerevisiae* which becomes initially multinuclear during regeneration, *Schizosacch. pombe* remains uninucleate throughout (cf. also Nečas and Svoboda, 1967).

Kopecká *et al.* (1965) studied regenerating walls of *Sacch. cerevisiae* in liquid media by electron microscopy and found, after about 1–6 h, an irregular network of bundles of fibrils, occasionally interspersed by single fibrils and areas of high fibrillar density from which fibrils radiated in all directions. Since they found some amorphous matter associated with these dense areas, these could be starting areas of complete wall regeneration. However, in liquid medium, regeneration only reaches the stage of a dense network of flat bundles of fibrils. Mannan-protein complexes are synthesized by regenerating sphaeroplasts in liquid media, but these are excreted into the surrounding fluid as was shown for invertase by

Lampen *et al.* (1967) and by Sentandreu and Northcote (1969b) for a glycopeptide. In gelatin, on the other hand, the fibrils soon become masked by an amorphous cementing substance (Kopecká *et al.*, 1965). They showed that the fibrils formed during 10 h of regeneration were soluble in warm N-sodium hydroxide but not in boiling N-hydrochloric acid or upon treatment with snail enzyme. This behaviour in alkali and acid is reminiscent of the hydroglucan described by Houwink and Kreger (1953). Nevertheless, its resistance to digestion by the snail enzyme complex is puzzling, since Svoboda and Nečas (1968) showed that protoplasts cultivated in liquid medium or in 30% gelatin in the presence of snail enzyme formed neither a fibrillar network nor a wall. However, Svoboda and Nečas confirmed the observation of Kopecká and her colleagues that protoplasts cultivated for more than 10 h in gelatin acquired walls which were partially resistant to snail enzyme; such cells lost the matrix material but retained the fibrillar network. Only in the case of very young protoplasts, in the process of recovery for 3–6 h, was the wall completely destroyed by the enzyme. The resistance of older fibrils to snail enzyme may be caused by a modified physical condition, such as hydrogen bonding.

Nečas and Svoboda (1965) determined that only four to five hours of precultivation in 30% gelatin was necessary before sufficient regeneration had occurred so that the remainder, leading to regeneration and growth, could occur even in liquid medium or on the surface of agar. During this five-hour period not only a fibrillar network, but in addition the beginning of outer matrix deposition, had taken place. Conversely, if protoplasts were transferred from *liquid* nutrient medium (in which only a fibrillar network had formed on their surface) to gelatin within 4–6 h after inoculation, regeneration took place, while it did not occur if the time of cultivation in liquid medium was longer. Interestingly, a similar critical time period was observed (Svoboda and Nečas, 1968) for the recovery of normal morphogenesis after blocking the process by snail enzyme. Thus, if protoplasts are blocked from cell-wall regeneration by exposure to snail enzyme for 4–6 h, complete regeneration and budding occur after removal of the enzyme and embedding in 30% gelatin. However, if the blocking is extended beyond this time, a wall forms around a greatly enlarged protoplast, but regeneration (i.e. budding) does not take place. At present there is no precise information as to the difference in physical or chemical composition of regenerating and non-regenerating walls.

Svoboda (1966) found that regeneration could also occur when *Sacch. cerevisiae* sphaeroplasts were imbedded in agar gels, provided the agar concentration was sufficiently high, 2% (w/v) being optimal. Sphaeroplasts of the yeast-like fungus *Geotrichum candidum* also regenerated in stabilizing thin-layer agar (Fukui *et al.*, 1969). For this organism the regeneration process went through a lag phase followed by a logarithmic recovery phase. However, these authors were unable to observe wall development in ultrathin sections by electron microscopy, as a regenerated wall was difficult to recognize.

Based on the observation by Šosková *et al.* (1968) that cycloheximide prevents regeneration of *Sacch. cerevisiae* protoplasts into normal budding cells, Nečas *et al.* (1968b) determined which aspects of the overall regeneration process were affected. By following the regeneration process in liquid medium and in 30% gelatin in the presence of the antibiotic, they found that under either of the two conditions only a fine fibrillar network was formed and that only the formation of the amorphous matrix was blocked. This indicates the constitutive nature of the enzymes involved in fibril synthesis, but the *de novo* synthesis of wall proteins appears inhibited. This was also shown by Sentandreu and Northcote (1969b) in a recent study on cell-wall synthesis. Whether synthesis of the water-soluble mannan component occurs in the presence of cycloheximide has not been determined as yet. If the answer is yes, the mannan might be released into the medium in the absence of protein synthesis.

Following the findings of Megnet (1965) and Johnson (1968a) of selected killing of *Schizosaccharomyces* and *Saccharomyces* cells during growth in the presence of 2-deoxyglucose and subsequent demonstration that the action of this glucose analogue causes lysis of yeast cells at the sites of extensive cell-wall synthesis (Johnson, 1968a, b), Farkaš *et al.* (1969) studied the effect of 2-deoxyglucose on cell-wall regeneration in *Sacch. cerevisiae* sphaeroplasts. Synthesis of the fibrillar glucan in liquid medium was strongly decreased at a glucose : 2-deoxyglucose ratio of 4 : 1 and virtually completely suppressed at a ratio of 2 : 1. However, the formation of the amorphous mannan-protein matrix in gelatin was even more sensitive to the presence of 2-deoxyglucose, and was greatly decreased at a glucose : 2-deoxyglucose ratio of 20 : 1. Sphaeroplasts did not complete their regeneration under such conditions. In both of the synthetic processes, 2-deoxyglucose appears to act as a competitive inhibitor in the various enzymic steps, but the compound was not incorporated in the glucan fibrils.

Recently, synthesis of the fibrillar network of *Sacch. cerevisiae* sphaeroplasts in liquid medium was followed by autoradiography (Nečas and Kopecká, 1969). They used tritium-labelled glucose and the freeze-etching technique to elucidate the role of the cytoplasmic membrane in fibril formation. It was found that the fibrils were synthesized randomly but in specific locations over the whole sphaeroplast surface, and that the fibrils were always separated from the plasmalemma by a thin layer of non-etchable material. Autoradiographs showed that the fibrillar network grows by interposition of new elementary fibrils, the only labelled elements of the developing outer surface. It was postulated that sphaeroplasts excrete high molecular-weight glucans of low solubility and that the polymers organize themselves first into elementary fibrils by crystallization, followed by aggregation of the fibrils into more highly organized bundles. The matrix material of higher solubility cannot be incorporated into the cell envelope unless it is temporarily supported by gel media of appropriate elasticity.

Observations on the cytoplasmic membrane of freeze-etched baker's yeast cells by Moor and Mühlethaler (1963) revealed hexagonally arranged clusters of particles which are thought to be involved in the production of glucan fibrils, which subsequently form bundles of fibrils. These patterns of particles could well correspond to the locations on the sphaeroplast surface where Nečas and Kopecká (1969) demonstrated initiation of fibril formation in autoradiographs. *Candida utilis* also forms bundles of elementary fibrils (Uruburu *et al.*, 1968), but *Pichia polymorpha* produces single fibrils during regeneration, suggesting that, in the latter species, the particles on the plasmalemma are randomly distributed.

## VI. Morphology

### A. LOCALIZATION OF WALL COMPONENTS

The cell wall of baker's yeast is approximately $70 \pm 10$ nm thick (Moor and Mühlethaler, 1963), very young cells being thinnest. Upon cell breakage, yields of 15–16% (dry weight basis) have been obtained by various workers (cf. Bacon *et al.*, 1969). Walls of other species may be thicker and capsular material may exceed in thickness the diameter of a cell producing it.

Many attempts have been made to demonstrate the different polysaccharides, discussed earlier, in particular strata by the use of ultrathin sections in electron micrographs. The results are not highly definitive, although with proper fixatives (such as permanganate) three layers could be recognized in cells with thick walls. Fracturing deep-frozen baker's yeast cells on a freezing ultramicrotome has never resulted in splitting of the wall into subunits (Moor and Mühlethaler, 1963), but a perpendicular cross-fracture may show in some instances the presence of layers (Matile *et al.*, 1969; Vol. I, p. 226, Fig. 4b, showing *Saccharomycodes ludwigii*). The apparent continuous nature of the wall components is probably caused by the presence of protein throughout the wall and the partial overlapping of wall layers known to exist on the basis of other evidence. On the other hand, in some species (e.g. *Candida utilis*) osmic acid-fixed cells showed the presence of a laminar structure in thin sections where the layers are bordered by three electron-dense strata (Sentandreu and Villanueva, 1965).

It is now generally accepted that mannan-protein complexes form the outer regions of the cell wall of *Saccharomyces* species. Because of an abundance of (1→2) and (1→6) bonds in yeast mannan, the hydroxyls on C-3 and C-4 are unsubstituted. The latter are cleaved by periodate, and the aldehyde formed stains intensely with leucofuchsin. Mundkur (1960) used this cytochemical technique to demonstrate that mannan forms the outer layer of the wall, although he incorrectly thought that (1→4) bonds formed the main linkage in mannan. The very few unsubstituted neighbouring secondary hydroxyl groups in yeast glucan causes this polysaccharide to be unreactive with the reagent, and Mundkur showed it to form the inner layer. The experiments of Preiss (1958), which were mentioned in Section II.D.4 (p. 171), also showed that the mannan-protein complex invertase was located directly below the outer boundary of the wall. Bacon *et al.* (1969) pointed out that the $\beta$-(1→6)-linked glucan discovered by them in cell walls of baker's yeast would also react with periodate-leucofuchsin, and could have contributed to the intensely stained outer layer. However, this would depend in part on the proportion of $\beta$-(1→6) glucan in the wall. Bacon *et al.* (1969) have postulated, on the basis of glucan solubility of whole cells and cell walls after various treatments, that insolubility of glucan may be due to retention by a semi-permeable membrane formed by some elements of the original wall. This membrane is thought to enclose the glucan layer and to consist of chitin and one or more glucan components. It is interesting that Mundkur (1960) also suggested a chitinous membrane which became visible in thin sections after removal of the mannan layer by acid and treatment with periodate-leucofuchsin. Still another layer, closely adhering to the plasmalemma, has been demonstrated by Streiblová (1968). Presumably this layer is made up of protein in association with chitin, and this complex also fills the invaginations of the cytoplasmic membrane. As far as is known, this thin layer is not removed during the transformation of whole vegative cells into sphaeroplasts.

The only conspicuous features of the cell surface are the various kinds of bud scars and fission scars resulting from vegetative reproduction. Since this subject has been reviewed in detail by Beran (1968) and Matile *et al.* (1969), the reader is referred to these publications for further details. The high concentration of chitin in the bud scar plugs has been discussed in Section II.C (p. 163).

In capsulated species, such as *Cryptococcus laurentii* and *Rhodotorula glutinis*, the wall is initially very thin but becomes much thicker when the cell ages (Ruinen *et al.*, 1968). In these two species, the wall substance merges gradually with the layer of capsular polysaccharides.

Recent studies have made it clear that much remains to be learned about the structural organization of the wall components even in such an extensively studied species as *Saccharomyces cerevisiae*. A particularly puzzling aspect concerns the solubility properties of its glucan. Even less is known about the localization and bonding of the numerous novel polysaccharides which have been discovered in recent years in the walls of other species.

### B. EXPANSION GROWTH

Lastly, brief mention is made of the factors which play a role in the expansion growth of yeast cells in the absence of cell duplication. Yanagishima (1963) and Shimoda *et al.* (1967) discovered that auxin (indole 3-acetic acid) induces cell elongation in auxin-responsive mutants of *Saccharomyces ellipsoideus* when added to the culture medium. Since sphaeroplasts did not respond to this hormone, it was concluded that auxin plays an essential role in expansion of the cell wall itself. The phenomenon was inhibited by *trans*-cinnamic acid, an antagonist of auxin (Yanagishima and Shimoda, 1968). Similarly chloramphenicol, an inhibitor of protein synthesis, inhibited cell-expansion growth. Next, Shimoda and Yanagishima (1968) investigated the possibility that enzymic degradation of the glucan portion of the cell wall might be involved in wall loosening, and thus cause cell expansion. They used an exo-$\beta$-(1→3) glucanase of fungal origin and showed that $\beta$-(1→3) glucanase induced cell expansion more rapidly than auxin, but only in auxin-responsive strains of yeast. In addition, they found that cell walls of auxin-responsive strains were more susceptible to this hydrolytic enzyme than were walls not responsive to auxin. Shimoda and Yanagishima (1968) suggest, on the basis of preliminary additional evidence, that auxin stimulates formation and/or activity of $\beta$-(1→3) glucanase and, in this way, promotes cell elongation. The exo-$\beta$-(1→3) glucanases which have been discovered in cells, as well as extracellularly in *Sacch. cerevisiae*, *Fabospora fragilis* and *H. anomala* (Abd-El-Al and Phaff, 1968), may then be the native enzymes of these yeasts involved in cell expansion, but no information is as yet available on their control.

## VII. Cell Wall Composition as a Basis for Differentiating Yeasts

Whereas traditionally yeasts have been separated into species and genera primarily on the basis of differences in physiological and morphological properties, one of the new approaches to taxonomy and phylogeny has made use of comparisons in the cell-wall composition, in particular that of the glucans, mannans and capsular polysaccharides. It seems logical that synthesis of such complex polysaccharides is controlled by a greater portion of the genome than simple fermentation or assimilation reactions. Since the systematics of yeast is taken up in detail in Volume I of this series (p. 5), only a brief mention of this subject will be made.

Tanaka *et al.* (1965) used the susceptibility of the glucans of various species to digestion by the lytic enzymes of *Bacillus circulans* induced by the different glucans as a basis of showing differences and similarities between the various species. Some of their data have been presented in more detail in Section II.A.2 (p. 144).

Kocourek and Ballou (1969), on the basis of earlier chemical studies of yeast mannan by Ballou and his coworkers, adapted the method of acetolysis of yeast mannans for the purpose of "fingerprinting" these polysaccharides. With controlled acetolysis, the $\alpha$-(1$\rightarrow$6) linkages of the mannan backbone are selectively cleaved, and the oligosaccharides thus formed (representing the side-chains including one mannose residue originally part of the backbone) can be separated according to size by gel filtration. The elution pattern thus obtained appears to be reproducibly characteristic of a particular species. The molar ratios of the fragments can be determined from the areas under the elution peaks. A comparison of the elution patterns of a significant number of species indicated that this approach will be useful in yeast systematics. Although this method provides information on the proportion and size of side-chains in various mannans, it does not reveal the types and distribution of the linkages between the mannose units in the side-chains.

Another approach to the fingerprinting of mannans was developed by Gorin and Spencer (1968b), Gorin *et al.* (1968, 1969a, c) and Spencer and Gorin (1968, 1969a, b). They determined and made a comparison of the proton magnetic resonance spectra of the isolated mannans of a large number of yeast species. These spectra furnished very sensitive criteria for distinguishing between the mannans of different species or for showing their similarity. They obtained optimum resolution with a 100-megacycle spectrometer using a 20% mannan solution in deuterium oxide and a tetramethylsilane external standard. The H-1 protons give complex signals in the $\tau$4·0–5·0 region. As the H-1 and DOH signals are often very close or superimposed at room temperature, the measurements are made at 70° since the temperature-dependent DOH signal is then shifted upfield. With this technique, a large number of water-soluble mannose-containing polysaccharides can be distinguished. Gorin and Spencer (1968b) pointed out that the chemical shifts of the H-1 proton signals depend largely on: (i) the structure of the parent anhydrohexose unit; (ii) the position(s) substituted in the unit; (iii) the structure of the substituent unit; (iv) the structure of the aglycone unit if one is present; and (v) the position(s) substituted in such an aglycone unit. Gorin *et al.*

(1969b) concluded that polysaccharides with similar proton magnetic resonance spectra have related chemical structures, and thus mannans from different species with similar spectra may indicate that these species have a close phylogenetic relationship.

This technique has been very useful in confirming or rejecting postulated imperfect and perfect forms of the same species of yeast. The former are usually placed in such genera as *Candida*, *Torulopsis*, *Trichosporon* and *Kloeckera*, and the latter in one of the ascosporogenous genera (cf. Spencer and Gorin, 1968). The technique has also been helpful in forming tentative groups of seemingly related species in heterogeneous asporogenous genera such as *Candida* (Spencer and Gorin, 1969a). Proton magnetic resonance spectra have also been used to support or reject various kinds of evidence for proposed phylogenetic lines of *Hansenula* and *Pichia* species (Spencer and Gorin, 1969b); to obtain presumptive evidence for the occurrence of $\beta$-linkages in the side-chains of the mannans, as their presence produces proton magnetic resonance spectra with distinctive H-1 signals at higher field than $\tau 4{\cdot}55$ (Gorin *et al.*, 1969a); and to detect sugars other than mannose in the mannans (Gorin and Spencer, 1968b). Gorin and Spencer (1970) have prepared a comprehensive review of their numerous studies to which the reader is referred for further details.

The composition of capsular polysaccharides has also been found useful as a systematic aid. For example, the slimy phosphomannans found in certain species of *Hansenula* and *Pichia* point to their common ancestry (Wickerham and Burton, 1962). The presence of linear mannans with alternating $\beta$-(1$\rightarrow$3) and $\beta$-(1$\rightarrow$4) bonds in *Rhodotorula* species, and an acidic heteropolysaccharide containing xylose, mannose and glucuronic acid in *Cryptococcus*, has supported the transfer of several species from *Rhodotorula* to *Cryptococcus* (Phaff and Spencer, 1966) and has established a relationship between *Cryptococcus* and the heterobasidiomycetous genus *Tremella* (Slodki *et al.*, 1966). Similarly certain species of *Lipomyces*, which were difficult to separate by classical criteria, could be readily separated by the sugars formed on hydrolysis of their extracellular polysaccharides (Slodki and Wickerham, 1966).

## References

Abd-El-Al, A. T. and Phaff, H. J. (1968). *Biochem. J.* **109**, 347–360.

Abd-El-Al, A. T. and Phaff, H. J. (1969). *Can. J. Microbiol.* **15**, 697–701.

Abercrombie, M. J., Jones, J. K. N., Lock, M. V., Perry, M. B. and Stoodley, R. J. (1960a). *Can. J. Chem.* **38**, 1617–1624.

Abercrombie, M. J., Jones, J. K. N. and Perry, M. B. (1960b). *Can. J. Chem.* **38**, 2007–2014.

Ahearn, D. G. (1964). Ph.D. Thesis: University of Miami.

Ahearn, D. G. and Roth, J. F. (1966). *Devel. ind. Microbiol.* **7**, 301–309.
Algranati, I. D., Carminatti, H. and Cabib, E. (1963). *Biochem. Biophys. res. Commun.* **12**, 504–509.
Anderson, F. B. and Millbank, J. W. (1966). *Biochem. J.* **99**, 682–687.
Anderson, R. F., Cadmus, M. C., Benedict, R. G. and Slodki, M. E. (1960). *Archs Biochem. Biophys.* **89**, 289–292.
Ankel, H. and Feingold, D. S. (1964). *Abstr. VIth. Int. Congr. Biochem.* **vi**, 19.
Ankel, H. and Feingold, D. S. (1966). *Biochemistry, N.Y.* **5**, 182–189.
Ankel, H., Ankel, E. and Feingold, D. S. (1966). *Biochemistry, N.Y.* **5**, 1864–1869.
Aschner, M., Mager, J. and Leibowitz, J. (1945). *Nature, Lond.* **156**, 295.
Bacon, J. S. D. and Farmer, V. C. (1968). *Biochem. J.* **110**, 34P–35P.
Bacon, J. S. D., Milne, B. D., Taylor, I. F. and Webley, D. M. (1965). *Biochem. J.* **95**, 28C–30C.
Bacon, J. S. D., Davidson, E. D., Jones, D. and Taylor, I. F. (1966). *Biochem. J.* **101**, 36C–38C.
Bacon, J. S. D., Jones, D., Farmer, V. C. and Webley, D. M. (1968). *Biochim. biophys. Acta.* **158**, 313–315.
Bacon, J. S. D., Farmer, V. C., Jones, D. and Taylor, I. F. (1969). *Biochem. J.* **114**, 557–567.
Ballou, C. E. (1970). *J. biol. Chem.* **245**, 1197–1203.
Barry, V. C. and Dillon, T. (1943). *Proc. R. Ir. Acad., Sect. B.* **49**, 177–185.
Bartnicki-Garcia, S. (1968). *A. Rev. Microbiol.* **22**, 87–108.
Behrens, N. H. and Cabib, E. (1968). *J. biol. Chem.* **243**, 502–509.
Bell, D. J. and Northcote, D. H. (1950). *J. chem. Soc.* 1944–1947.
Beran, K. (1968). *Adv. microbial. Physiol.* **2**, 143–171.
Bhattacharjee, S. S. and Gorin, P. A. J. (1969). *Can. J. Chem.* **47**, 1207–1215.
Biguet, J., Tran van Ky, P. and Andrieu, S. (1962). *Mycopath. Mycol. appl.* **17**, 239–254.
Bishop, C. T., Blank, F. and Gardner, P. E. (1960). *Can. J. Chem.* **38**, 869–881.
Blandamer, A. and Danishefsky, I. (1966). *Biochim. biophys. Acta* **117**, 305–313.
Boer, P. and Steyn-Parvé, E. P. (1966). *Biochim. biophys. Acta* **128**, 400–402.
Buecher, Jr., E. J. (1968). Ph.D. Thesis: University of California, Davis.
Bull, A. T. and Chesters, C. G. C. (1966). *Adv. Enzymol.* **28**, 325–364.
Burger, M., Bacon, E. E. and Bacon, J. S. D. (1961). *Biochem. J.* **78**, 504–511.
Cadmus, M. C., Lagoda, A. A. and Anderson, R. F. (1962). *Appl. Microbiol.* **10**, 153–156.
Carmo-Sousa, L. do and Barroso-Lopes, C. (1970). *Antonie van Leeuwenhoek*, **36**, 209–216.
Cawley, T. N. and Letters, R. (1968). *Biochem. J.* **110**, 9P–10P.
Cawley, T. N. and Letters, R. (1969). *Biochem J.* **115**, 9P.
Chattaway, F. W., Holmes, M. R. and Barlow, A. J. E. (1968). *J. gen. Microbiol.* **51**, 367–376.
Cifonelli, J. A. and Smith, F. (1955). *J. Am. chem. Soc.* **77**, 5682–5684.
Clarke, A. E. and Stone, B. A. (1963). *Rev. pure appl. Chem.* **13**, 134–156.
Cohen, A. and Feingold, D. S. (1967). *Biochemistry, N.Y.* **6**, 2933–2939.
Darling, S., Theilade, J. and Birch-Andersen, A. (1969). *J. Bact.* **98**, 797–810.
Davies, R. and Elvin, P. A. (1964). *Biochem. J.* **93**, 8P.
DeLaFuente, G. and Sols, A. (1962). *Biochim. biophys. Acta* **56**, 49–62.
Domanski, R. E. and Miller, R. E. (1968). *J. Bact.* **96**, 270–271.
Drouhet, E., Segretain, G. and Aubert, J. P. (1950). *Annls Inst. Pasteur, Paris* **79**, 891–900.
Duell, E. A., Inoue, S. and Utter, M. F. (1964). *J. Bact.* **88**, 1762–1773.

Dunwell, J. L., Ahmad, F. and Rose, A. H. (1961). *Biochim. biophys. Acta* **51**, 604–607.
Dworschack, R. G. and Wickerham, L. J. (1961). *Appl. Microbiol.* **9**, 291–294.
Eddy, A. A. (1958). *Proc. R. Soc. B.* **149**, 425–440.
Eddy, A. A. and Longton, J. (1969). *J. Inst. Brew.* **75**, 7–9.
Eddy, A. A. and Williamson, D. H. (1957). *Nature, Lond.* **179**, 1252–1253.
Eddy, A. A. and Williamson, D. H. (1959). *Nature, Lond.* **183**, 1101–1104.
Eddy, A. A. and Woodhead, J. S. (1968). *F.E.B.S. Letters* **1**, 67–68.
Evans, E. E. and Kessel, J. F. (1951). *J. Immun.* **67**, 109–114.
Evans, E. E. and Mehl, J. W. (1951). *Science, N.Y.* **114**, 10–11.
Evans, E. E. and Theriault, R. J. (1953). *J. Bact.* **65**, 571–577.
Falcone, G. and Nickerson, W. J. (1956). *Science, N.Y.* **124**, 272–273.
Farkaš, V., Svoboda, A. and Bauer, Š. (1969). *J. Bact.* **98**, 744–748.
Foda, M. S. and Phaff, H. J. (1969). *Antonie van Leeuwenhoek* **35**, H9–H10.
Friis, J. and Ottolenghi, P. (1959a). *C. r. Trav. Lab. Carlsberg* **31**, 259–271.
Friis, J. and Ottolenghi, P. (1959b). *C. r. Trav. Lab. Carlsberg* **31**, 272–281.
Fukui, K., Sagara, Y., Yoshida, N. and Matsuoka, T. (1969). *J. Bact.* **98**, 256–263.
Furuya, A. and Ikeda, Y. (1960a). *J. gen. appl. Microbiol., Tokyo* **6**, 40–48.
Furuya, A. and Ikeda, Y. (1960b). *J. gen. appl. Microbiol., Tokyo* **6**, 49–60.
Gadebusch, H. H. (1960a). *J. Infect. Dis.* **107**, 402–405.
Gadebusch, H. H. (1960b). *J. Infect. Dis.* **107**, 406–409.
Gadebusch, H. H. and Johnson, J. D. (1961). *Can. J. Microbiol.* **7**, 53–60.
Garcia Mendoza, C. and Villanueva, J. R. (1962). *Nature, Lond.* **195**, 1326–1327.
Garcia Mendoza, C., Garcia Lopez, M. D., Uruburu, F. and Villanueva, J. R. (1968). *J. Bact.* **95**, 2393–2398.
Garver, J. C. and Epstein, R. L. (1959). *Appl. Microbiol.* **7**, 318–319.
Garzuly-Janke, R. (1940). *Zentbl. Bakt. ParasitKde* (*Abt. II*) **102**, 361–365.
Gascón, S. and Ottolenghi, P. (1967). *C. r. Trav. Lab. Carlsberg* **36**, 85–93.
Gascón, S. and Villanueva, J. R. (1963). *Can. J. Microbiol.* **9**, 651–652.
Gascón, S. and Villanueva, J. R. (1965). *Nature, Lond.* **205**, 822–823.
Gascón, S., Ochoa, A. G. and Villanueva, J. R. (1965). *Can. J. Microbiol.* **11**, 573–580.
Giaja, J. (1914). *C. r. Séanc. Soc. Biol.* **77**, 2–4.
Giaja, J. (1919). *C. r. Séanc. Soc. Biol.* **82**, 719–720.
Giaja, J. (1922). *C. r. Séanc. Soc. Biol.* **86**, 708–709.
Gorin, P. A. J. and Perlin, A. S. (1956). *Can. J. Chem.* **34**, 1796–1803.
Gorin, P. A. J. and Perlin, A. S. (1957). *Can. J. Chem.* **35**, 262–267.
Gorin, P. A. J. and Spencer, J. F. T. (1967). *Can. J. Chem.* **45**, 1543–1549.
Gorin, P. A. J. and Spencer, J. F. T. (1968a). *Adv. carb. Chem.* **23**, 367–414.
Gorin, P. A. J. and Spencer, J. F. T. (1968b). *Can. J. Chem.* **46**, 2299–2304.
Gorin, P. A. J. and Spencer, J. F. T. (1968c). *Can. J. Chem.* **46**, 3407–3411.
Gorin, P. A. J. and Spencer, J. F. T. (1970). *Adv. appl. Microbiol.* **12**, 25–89.
Gorin, P. A. J., Horitsu, K. and Spencer, J. F. T. (1965). *Can. J. Chem.* **43**, 950–954.
Gorin, P. A. J., Spencer, J. F. T. and MacKenzie, S. L. (1966). *Can. J. Chem.* **44**, 2087–2090.
Gorin, P. A. J., Mazurek, M. and Spencer, J. F. T. (1968). *Can. J. Chem.* **46**, 2305–2310.
Gorin, P. A. J., Spencer, J. F. T. and Bhattacharjee, S. S. (1969a). *Can. J. Chem.* **47**, 1499–1505.
Gorin, P. A. J., Spencer, J. F. T. and Eveleigh, D. E. (1969b). *Carb. Res.* **11**, 387–398.
Gorin, P. A. J., Spencer, J. F. T. and Magus, R. J. (1969c). *Can. J. Chem.* **47**, 3569–3576.

Hasegawa, S., Kirkwood, S. and Nordin, J. H. (1966). *Chemy Ind.* 1033.
Hasegawa, S., Nordin, J. H. and Kirkwood, S. (1969). *J. biol. Chem.* **244**, 5460–5470.
Hasenclever, H. F. and Mitchell, W. O. (1961). *J. Bact.* **82**, 570–573.
Hasenclever, H. F. and Mitchell, W. O. (1964). *J. Immun.* **93**, 763–771.
Hassid, W. Z., Joslyn, M. A. and McCready, R. M. (1941). *J. Am. chem. Soc.* **63**, 295–298.
Havelková, M. (1966). *Folio Microbiol., Praha* **11**, 453–458.
Havelková, M. (1969). *Folio Microbiol., Praha* **14**, 155–164.
Haworth, W. N., Hirst, E. L. and Isherwood, F. A. (1937). *J. chem. Soc.* 784–791.
Haworth, W. N., Heath, R. L. and Peat, S. (1941). *J. chem. Soc.* 833–842.
Hehre, E. J., Carlson, A. S. and Hamilton, D. M. (1949). *J. biol. Chem.* **177**, 289–293.
Holden, M. and Tracey, M. V. (1950). *Biochem. J.* **47**, 407–414.
Holter, H. and Ottolenghi, P. (1960). *C. r. Trav. Lab. Carlsberg* **31**, 409–422.
Hopkins, R. H. (1958). *Wallerstein Labs Commun.* **21**, 309–322.
Horikoshi, K. and Sakaguchi, K. (1958). *J. gen. appl. Microbiol., Tokyo* **1**, 1–11.
Houwink, A. L. and Kreger, D. R. (1953). *Antonie van Leeuwenhoek* **19**, 1–24.
Houwink, A. L., Kreger, D. R. and Roelofsen, P. A. (1951). *Nature, Lond.* **168**, 693–694.
Islam, M. F. and Lampen, J. O. (1962). *Biochim. biophys. Acta* **58**, 294–302.
Jeanes, A. and Watson, P. R. (1962). *Can. J. Chem.* **40**, 1318–1325.
Jeanes, A., Pittsley, J. E., Watson, P. R. and Kimler, R. J. (1961). *Archs Biochem. Biophys.* **92**, 343–350.
Jeanes, A., Pittsley, J. E., Watson, P. R. and Sloneker, J. H. (1962). *Can. J. Chem.* **40**, 2256–2259.
Jeanes, A., Pittsley, J. E. and Watson, P. R. (1964). *J. appl. polymer Sci.* **8**, 2775–2787.
Jeuniaux, C. (1966). In "Methods in Enzymology", (E. F. Neufeld and V. Ginsberg, eds.), vol. VIII, pp. 644–650. Academic Press, New York.
Johnson, B. F. (1968a). *J. Bact.* **95**, 1169–1172.
Johnson, B. F. (1968b). *Expl Cell Res.* **50**, 692–694.
Johnston, J. R. and Mortimer, R. K. (1959). *J. Bact.* **78**, 292.
Jones, D. and Webley, D. M. (1967). *Trans. Br. mycol. Soc.* **50**, 149–154.
Jones, D., Bacon, J. S. D., Farmer, V. C. and Webley, D. M. (1969). *Soil Biol. Biochem.* **1**, 145–151.
Jones, G. H. and Ballou, C. E. (1968). *J. biol. Chem.* **243**, 2443–2446.
Jones, G. H. and Ballou, C. E. (1969a). *J. biol. Chem.* **244**, 1043–1051.
Jones, G. H. and Ballou, C. E. (1969b). *J. biol. Chem.* **244**, 1052–1059.
Jones, J. K. N. and Nicholson, W. H. (1958). *J. chem. Soc.* 27–33.
Kanetsuna, F., Carbonell, L. M., Moreno, R. E. and Rodriquez, J. (1969). *J. Bact.* **97**, 1036–1041.
Kaplan, J. G. (1965a). *Nature, Lond.* **205**, 76–77.
Kaplan, J. G. (1965b). *J. gen. Physiol.* **48**, 873–886.
Kessler, G. and Nickerson, W. J. (1959). *J. biol. Chem.* **234**, 2281–2285.
Kidby, D. K. and Davies, R. (1968). *J. gen. Microbiol.* **53**, v.
Kocourek, J. and Ballou, C. E. (1969). *J. Bact.* **100**, 1175–1181.
Kooiman, P. (1963). *Antonie van Leeuwenhoek* **29**, 169–176.
Kopecká, M., Čtvrtnícek, O. and Nečas, O. (1965). *Abh. dt. Akad. Wiss. Berl., Klasse f. Medizin 1966* 73–75.
Korn, E. D. and Northcote, D. H. (1960). *Biochem. J.* **75**, 12–17.
Kreger, D. R. (1954). *Biochim. biophys. Acta* **13**, 1–9.
Kreger, D. R. (1965). *Abh. dt. Akad. Wiss. Berl. Klasse f. Medizin 1966* 113–129.
Kuroda, A., Tokumaru, Y. and Tawada, N. (1968). *J. ferm. Technol.* **46**, 926–937.
Lampen, J. O. (1965). *Symp. Soc. gen. Microbiol.* **16**, 115–133.

Lampen, J. O. (1968). *Antonie van Leeuwenhoek* **34**, 1–18.
Lampen, J. O., Neumann, N. P., Gascón, S. and Montenecourt, B. S. (1967). *In* "Organizational Biosynthesis", (H. J. Vogel, J. O. Lampen and V. Bryson eds.), pp. 363–372. Academic Press, New York.
Lee, Y. C. and Ballou, C. E. (1965). *Biochemistry, N.Y.* **4**, 257–264.
Lindquist, W. (1953). *J. Inst. Brew.* **59**, 56–61.
Lindstedt, G. (1945). *Arkiv. Kemi Mineral. Geol. 20A*, 1–22.
Littman, M. L. (1958). *Trans. N.Y. Acad. Sci. Ser. I*, **20**, 623–648.
Lodder, J. and Kreger-van Rij, N. J. W. (1952). "The yeasts, a taxonomic study". North Holland Publ. Co., Amsterdam.
Longley, R. P., Rose, A. H. and Knights, B. A. (1968). *Biochem. J.* **108**, 401–412.
Mager, J. (1947). *Biochem. J.* **41**, 603–609.
Mager, J. and Aschner, M. (1947). *J. Bact.* **53**, 283–295.
Manners, D. J. and Masson, A. J. (1969). *F.E.B.S. Letters* **4**, 122–124.
Manners, D. J. and Patterson, J. C. (1966). *Biochem. J.* **98**, 19C–20C.
Masler, L., Šikl, D., Bauer, Š. and Šandula, J. (1966). *Folio Microbiol., Praha* **11**, 373–378.
Masler, L., Šikl, D. and Bauer, Š. (1968). *Colls. Czech. chem. Commun.* **33**, 942–950.
Masschelein, C. A. (1959). *Rev. ferment. ind. Aliment.* **14**, 59–69, 87–112.
Matile, Ph., Moor, H. and Robinow, C. F. (1969). *In* "The Yeasts", (A. H. Rose and J. S. Harrison, eds.), Vol. I, 219–302. Academic Press, London.
McLellan, Jr., W. L. and Lampen, J. O. (1963). *Biochim. biophys. Acta* **67**, 324–326.
McLellan, Jr., W. L. and Lampen, J. O. (1968). *J. Bact.* **95**, 967–974.
Megnet, R. (1965). *J. Bact.* **90**, 1032–1035.
Merkenschlager, M., Schlossmann, K. and Kurz, W. (1957). *Biochem. Z.* **329**, 332–340.
Metzenberg, R. L. (1964). *Biochim. biophys. Acta* **89**, 291–302.
Mill, P. J. (1966). *J. gen. Microbiol.* **44**, 329–341.
Millbank, J. W. and Macrae, R. M. (1964). *Nature, Lond.* **201**, 1347.
Miller, M. W. and Phaff, H. J. (1958). *Antonie van Leeuwenhoek* **24**, 225–238.
Misaki, A. and Smith, F. (1963). *Abstr. Am. chem. Soc.*, 144th Meet., p. 14c.
Misaki, A., Johnson, Jr., J., Kirkwood, S., Scaletti, J. V. and Smith, F. (1968). *Carb. Res.* **6**, 150–164.
Miyazaki, T. (1961a). *Chem. Pharm. Bull., Tokyo* **9**, 715–718.
Miyazaki, T. (1961b). *Chem. Pharm. Bull., Tokyo* **9**, 826–829.
Miyazaki, T. (1961c). *Chem. Pharm. Bull., Tokyo* **9**, 829–833.
Moor, H. and Mühlethaler, K. (1963). *J. Cell Biol.* **17**, 609–628.
Moor, H., Mühlethaler, K., Waldner, H. and Frey-Wyssling, A. (1961). *J. biophys. biochem. Cytol.* **10**, 1–13.
Mundkur, B. (1960). *Expl Cell. Res.* **20**, 28–42.
Nabel, K. (1939). *Arch. Mikrobiol.* **10**, 515–541.
Nagasaki, S., Neumann, N. P., Arnow, P., Schnable, L. D. and Lampen, J. O. (1966). *Biochem. Biophys. res. Commun.* **25**, 158–164.
Nečas, O. (1955). *Folio Biol., Praha* **1**, 19–28.
Nečas, O. (1956). *Nature, Lond.* **177**, 898–899.
Nečas, O. (1961). *Nature, Lond.* **192**, 580–581.
Nečas, O. and Kopecká, M. (1969). *Antonie van Leeuwenhoek* **35** (suppl.), B7–B8.
Nečas, O. and Svoboda, A. (1965). *Abh. dt. Akad. Wiss. Berl., Klasse f. Medizin, 1966*, 67–71.
Nečas, O. and Svoboda, A. (1967). *Folio Biol., Praha* **13**, 379–385.
Nečas, O., Svoboda, A. and Havelková, M. (1968a). *Folio Biol., Praha* **14**, 80–85.
Nečas, O., Svoboda, A. and Kopecká, M. (1968b). *Expl Cell Res.* **53**, 291–293.
Neumann, N. P. and Lampen, J. O. (1967). *Biochemistry, N.Y.* **6**, 468–475.

Nickerson, W. J. and Falcone, G. (1956a). *Science, N.Y.* **124**, 318–319.

Nickerson, W. J. and Falcone, G. (1956b). *Science, N.Y.* **124**, 722–723.

Nickerson, W. J., Falcone, G. and Kessler, G. (1961). *Symp. Soc. gen. Physiol.* 205–228.

Northcote, D. H. (1954). *Biochem. J.* **58**, 353–358.

Northcote, D. H. (1963). *Pure appl. Chem.* **7**, 669–675.

Northcote, D. H. and Horne, R. W. (1952). *Biochem. J.* **51**, 232–236.

Nurminen, T. and Suomalainen, H. (1969). *Antonie van Leeuwenhoek* **35** (suppl.) I, 13–14.

Nurminen, T., Oura, E. and Suomalainen, H. (1965). *Suomen Kemistilehti B 38*, 282–285.

Ochoa, A. G., Garcia Acha, I., Gascón, S. and Villanueva, J. R. (1963). *Experientia* **19**, 581–582.

Ottolenghi, P. (1966). *C. r. Trav. Lab. Carlsberg* **35**, 363–368.

Parrish, F. W., Perlin, A. S. and Reese, E. T. (1960). *Can. J. Chem.* **38**, 2094–2104.

Peat, S., Whelan, W. J. and Edwards, T. E. (1958a). *J. chem. Soc.* 3862–3868.

Peat, S., Turvey, J. R. and Evans, J. M. (1958b). *J. chem. Soc.* 3868–3870.

Peat, S., Whelan, W. J. and Edwards, T. E. (1961a). *J. chem. Soc.* 29–34.

Peat, S., Turvey, J. R. and Doyle, D. (1961b). *J. chem. Soc.* 3918–3923.

Phaff, H. J. (1963). *A. Rev. Microbiol.* **17**, 15–30.

Phaff, H. J. and Spencer, J. F. T. (1966). *Proc. 2nd. Intern. Symp. on Yeasts* (A. Kockova-Kratochvilova, ed.), pp. 59–67. *Vydavatelstvo* Slovenskej Akademie Vied, Bratislava.

Preiss, J. W. (1958). *Archs Biochem. Biophys.* **75**, 186–195.

Rebers, P. A., Barker, S. A., Heidelberger, M., Dische, Z. and Evans, E. E. (1958). *J. Am. chem. Soc.* **80**, 1135–1137.

Reuvers, T., Tacoronte, E., Garcia-Mendoza, C. and Novaes-Ledieu, M. (1969). *Can. J. Microbiol.* **15**, 989–993.

Roelofsen, P. A. (1953). *Biochim. biophys. Acta* **10**, 477–478.

Roelofsen, P. A. and Hoette, I. (1951). *Antonie van Leeuwenhoek* **17**, 27–43.

Rogovin, S. P., Anderson, R. F. and Cadmus, M. C. (1961). *J. biochem. microbiol. Technol. Engng* **3**, 51–63.

Rost, K. and Venner, H. (1965a). *Arch. Mikrobiol.* **51**, 122–129.

Rost, K. and Venner, H. (1965b). *Arch. Mikrobiol.* **51**, 130–139.

Ruinen, J., Deinema, M. H. and van der Scheer, C. (1968). *Can. J. Microbiol.* **14**, 1133–1138.

Salkowski, E. (1894a). *Ber. dt. chem. Ges.* **27**, 3325–3329.

Salkowski, E. (1894b). *Ber. dt. chem. Ges.* **27**, 497–502; 925–926.

Salton, M. R. J. (1955). *Can. J. Microbiol.* **12**, 25–30.

Satomura, Y., Ono, M. and Fukumoto, J. (1960). *Bull. agr. Chem. Soc., Japan* **24**, 317–321.

Scher, M., Kramer, K. and Lennarz, W. J. (1967). *Abstr.* 154*th. National Meet. Am. Chem. Soc.*, Chicago, Ill. Sept. 11–14, D43.

Scher, M., Lennarz, W. J. and Sweeley, C. C. (1968). *Proc. natn. Acad. Sci. U.S.A.* **59**, 1313–1320.

Schutzbach, J. S. and Ankel, H. (1969). *F.E.B.S. Letters* **5**, 145–148.

Schmidt, M. (1936). *Arch. Mikrobiol.* **7**, 241–260.

Schwencke, J., Gonzales, G. and Farias, G. (1969). *J. Inst. Brew.* **75**, 15–19.

Sentandreu, R. and Northcote, D. H. (1968). *Biochem. J.* **109**, 419–432.

Sentandreu, R. and Northcote, D. H. (1969a). *Carb. Res.* **10**, 584–585.

Sentandreu, R. and Northcote, D. H. (1969b). *Biochem. J.* **115**, 231–240.

Sentandreu, R. and Villanueva, J. R. (1965). *Arch. Mikrobiol.* **50**, 103–110.

SentheShanmuganathan, S. and Nickerson, W. J. (1962). *J. gen. Microbiol.* **27**. 451–464.
Shifrine, M. and Phaff, H. J. (1958). *Antonie van Leeuwenhoek* **24**, 274–280.
Shimoda, C. and Yanagishima, N. (1968). *Physiologia Pl.* **21**, 1163–1169.
Shimoda, C., Masuda, Y. and Yanagishima, N. (1967). *Physiologia Pl.* **20**, 299–305.
Šikl, D., Masler, L. and Bauer, Š. (1964). *Experientia* **20**, 456.
Šikl, D., Masler, L. and Bauer, Š. (1967). *Chemické Zvesti* **21**, 3–12.
Šikl, D., Masler, L. and Bauer, Š. (1968). *Colls. Czech. chem. Commun.* **33**, 1157–1164.
Šikl, D., Masler, L. and Bauer, Š. (1969). *Antonie van Leeuwenhoek* **35** (suppl.) A9–A10.
Simpson, K. L., Wilson, A. W., Burton, E., Nakayama, T. O. M. and Chichester, C. O. (1963). *J. Bact.* **86**, 1126–1127.
Slodki, M. E. (1962). *Biochim. biophys. Acta* **57**, 525–533.
Slodki, M. E. (1963). *Biochim. biophys. Acta* **69**, 96–102.
Slodki, M. E. (1966a). *Can. J. Microbiol.* **12**, 495–499.
Slodki, M. E. (1966b). *J. biol. Chem.* **241**, 2700–2706.
Slodki, M. E. and Wickerham, L. J. (1966). *J. gen. Microbiol.* **42**, 381–385.
Slodki, M. E., Wickerham, L. J. and Cadmus, M. C. (1961). *J. Bact.* **82**, 269–274.
Slodki, M. E., Wickerham, L. J. and Bandoni, R. J. (1966). *Can. J. Microbiol.* **12**, 489–494.
Slooff, W. Ch., Nieuwdorp, P. J. and Bos, P. (1969). *Antonie van Leeuwenhoek* **35** (suppl.), A23–A24.
Snyder, H. E. and Phaff, H. J. (1960). *Antonie van Leeuwenhoek* **26**, 433–452.
Sošková, L., Svoboda, A. and Soška, J. (1968). *Folio Microbiol., Praha* **13**, 240–244.
Spencer, J. F. T. and Gorin, P. A. J. (1965). *Abh. dt. Akad. Wiss. Berl., Klasse f. Medizin*, 1966, 105–110.
Spencer, J. F. T. and Gorin, P. A. J. (1968). *J. Bact.* **96**, 180–183.
Spencer, J. F. T. and Gorin, P. A. J. (1969a). *Antonie van Leeuwenhoek* **35**, 361–378.
Spencer, J. F. T. and Gorin, P. A. J. (1969b). *Can. J. Microbiol.* **15**, 375–382.
Stewart, T. S. and Ballou, C. E. (1968). *Biochemistry, N.Y.* **7**, 1855–1863.
Stewart, T. S., Mendershausen, P. B. and Ballou, C. E. (1968). *Biochemistry, N.Y.* **7**, 1843–1854.
Stodola, F. H., Deinema, M. H. and Spencer, J. F. T. (1967). *Bact. Rev.* **31**, 194–213.
Streiblová, E. (1968). *J. Bact.* **95**, 700–707.
Summers, D. F., Grollman, A. P. and Hasenclever, H. F. (1964). *J. Immun.* **92**, 491–499.
Sutton, D. D. and Lampen, J. O. (1962). *Biochim. biophys. Acta* **56**, 303–312.
Suzuki, S. and Sunayama, H. (1968). *Jap. J. Microbiol.* **12**, 413–422.
Suzuki, S., Sunayama, H. and Saito, T. (1968). *Jap. J. Microbiol.* **12**, 19–24.
Svihla, G., Schlenk, F. and Dainko, J. L. (1961). *J. Bact.* **82**, 808–814.
Svoboda, A. (1965). *Abh. dt. Akad. Wiss. Berl., Klasse. f. Medizin, 1966*, 31–35.
Svoboda, A. (1966). *Expl Cell Res.* **44**, 640–642.
Svoboda, A. and Nečas, O. (1968). *Folio Biol., Praha* **14**, 390–397.
Tanaka, H. (1963). Ph.D. Thesis: University of California, Davis.
Tanaka, H. and Phaff, H. J. (1965). *J. Bact.* **89**, 1570–1580.
Tanaka, H., Phaff, H. J. and Higgins, L. W. (1965). *Abh. dt. Akad. Wiss. Berl. Klasse f. Medizin, 1966, Nr. 6*, 113–129.
Tanner, W. (1969). *Biochem. Biophys. res. Commun.* **35**, 144–150.
Tonino, G. J. M. and Steyn-Parvé, E. P. (1963). *Biochim. biophys. Acta* **67**, 453–469.
Tsuchiya, T., Fukazawa, Y. and Kawakita, S. (1959). *Mycopath. Mycol. appl.* **10**, 191–206.
Uruburu, F., Elorza, V. and Villanueva, J. R. (1968). *J. gen. Microbiol.* **51**, 195–198.

Webley, D. M., Follett, E. A. C. and Taylor, I. F. (1967). *Antonie van Leeuwenhoek* **33**, 159–165.
Wessels, J. G. H. (1965). *Wentia* **13**, 1–113.
Wickerham, L. J. and Burton, K. A. (1961). *J. Bact.* **82**, 265–268.
Wickerham, L. J. and Burton, K. A. (1962). *Bact. Rev.* **26**, 382–297.
Wickerham, L. J. and Dworschack, R. G. (1960). *Science, N.Y.* **131**, 985–986.
Wickerham, L. J., Lockwood, L. B., Pettijohn, O. G. and Ward, G. E. (1944). *J. Bact.* **48**, 413–427.
Wood, W. A. (ed.) (1966). *In* "Methods in Enzymology", (S. P. Colowick and H. O. Kaplan, eds.), Vol. IX, pp. 743–744. Academic Press, New York.
Yanagishima, N. (1963). *Plant and Cell Physiol.* **4**, 257–264.
Yanagishima, N. and Shimoda, C. (1968). *Physiologia Pl.* **21**, 1122–1128.
Yu, R. J., Bishop, C. T., Cooper, F. P., Blank, F. and Hasenclever, H. F. (1967a). *Can. J. Chem.* **45**, 2264–2267.
Yu, R. J., Bishop, C. T., Cooper, F. P., Hasenclever, H. F. and Blank, F. (1967b). *Can. J. Chem.* **45**, 2205–2211.
Zechmeister, L. and Toth, G. (1934). *Biochem. Z.* **270**, 309–316.
Zechmeister, L. and Toth, G. (1936). *Biochem. Z.* **284**, 133–138.
Zvjagilskaja, R. A. and Afanasjewa, T. P. (1965). *Abh. dt. Akad. Wiss. Berl. Klasse f. Medizin, 1966*, 27–29.

*Note added in proof*

Recent studies by T. R. Thieme and C. E. Ballou (personal communication) suggest a structure for the phosphomannan of *Kloeckera brevis* 55–45 which is different from that proposed by Stewart and Ballou (1968). Rather than being esterified to the mannose units of the backbone, Thieme and Ballou now find that the phosphate is attached to C-6 of mannose units in the side chains of the mannan. This conclusion was confirmed by the observation that essentially all of the mannose in the phosphomannan is destroyed on oxidation with sodium periodate, a result that is inconsistent with the structure previously proposed (see Section II.B.3, p. 152).

Slodki *et al.* (1970) have now found that the composition of extracellular phosphomannans produced by certain *Hansenula* species (see Section III.C, p. 177) is strongly influenced by the phosphate concentration of the medium in which the yeasts are grown. For *H. capsulata* the mannose:phosphate ratio was increased from 2.5 to 27.5 when the phosphate concentration in the medium was decreased from 0.5 to 0.05%. Without phosphate in the medium, this species formed a phosphate-free $\alpha$-mannan, whereas in a phosphate-rich medium most of the linkages were of the $\beta$-type. *Hansenula holstii* also produced a neutral extracellular mannan in a phosphate-free medium. Similarly, phosphogalactan production by *Sporobolomyces* (see Section III.D, p. 180) was strongly affected by phosphate. Without $KH_2PO_4$ in the medium the polymer had a greatly decreased phosphate content and an increased *O*-acetyl content. In addition, the percentage of glucose in the polymer increased from 21 to 54%. These findings suggest that the mannose:phosphate ratio of cell-wall mannans (see Table I, p. 154) and the anomeric linkages connecting the mannose units might also be influenced by the phosphate concentration in the medium. Kidby and Davies (1970) have recently proposed that invertase is not chemically bound to the yeast cell wall, but that this mannan-protein complex is retained in the mannan layer by an outer network of phosphomannan-protein containing numerous disulphide bridges (see Section II.D.4, p. 170). Treatment of *Fabospora* (*Saccharomyces*) *fragilis* with thiol reagents releases the hydrolase (in this case inulinase) by rupture of the S-S linkages in the outer wall.

With regard to the question of cell-wall residues in yeast protoplast preparations (see Section V.A., p. 193), Bacon *et al.* (1969) have presented further evidence that protoplasts prepared from yeast treated with snail enzyme for moderate times and with moderate enzyme concentrations may retain groups of bud scars attached by thin membranes and contain traces of glucan and mannan. Their conclusions were based on infrared spectra of frozen-dried lysed and washed protoplast preparations. Exhaustive treatment of cells with snail enzyme apparently leads to pure protoplasts. Nečas *et al.* (1969) have pointed out the danger of mistaking artificial eutectic structures at the protoplast surface for wall residues. Such structures constitute non-etchable areas during freeze-etching. These authors presented convincing evidence that the thin non-etchable layer adhering to the plasma membrane reported by Streiblová (1968; see p. 193) was an artifact created by a thin eutectic shell of frozen osmotic stabilizer plus background eutecticum which was thought to have diffused out of the protoplast between the last washing and final freezing.

The question as to excretion into the medium of soluble mannan during protoplast regeneration when protein synthesis is blocked by cycloheximide (see Section V.B, p. 194) was recently resolved by Farkaš *et al.* (1970). Protoplasts of *Sacch. cerevisiae* suspended in osmotically stabilized liquid medium synthesize glucan fibrils in the presence of cycloheximide, but they do not excrete either mannan or mannan-protein complexes under such conditions. Similarly, in the presence of 2-deoxy-D-glucose (see Section V.B, p. 194) in appropriate concentrations, a parallel inhibition of both mannan and protein excretion into the medium was observed. These authors therefore suggest that selective blocking of the synthesis of one component of the glycoprotein prevents extracellular appearance of the other component. There thus appears to exist a firm control mechanism over the synthesis of mannan-protein complexes of the cell wall.

Bacon, J. S. D., Jones, D. and Ottolenghi, P. (1969). *J. Bact.* **99**, 885–887.
Farkaš, V., Svoboda, A. and Bauer, Š. (1970). *Biochem. J.* **118**, 755–758.
Kidby, D. K. and Davies, R. (1970). *J. gen. Microbiol.* **61**, 327–333.
Nečas, O., Kopecká, M. and Brichta, J. (1969). *Expl Cell Res.* **58**, 411–419.
Slodki, M. E., Safranski, M. J., Hensley, D. E. and Babcock, G. E. (1970). *Appl. Microbiol.* **19**, 1019–1020.

## Chapter 6

# Yeast Lipids and Membranes

K. Hunter and A. H. Rose

*Microbiological Laboratories, School of Biological Sciences, Bath University, Bath, England*

I. INTRODUCTION . . . . . . . . . . . 211
II. YEAST LIPIDS . . . . . . . . . . . 212
A. Total Lipid Contents of Yeasts . . . . . . . 212
B. Composition of Yeast Lipids . . . . . . . . 216
III. LIPID METABOLISM . . . . . . . . . . 234
A. Biosyntheses . . . . . . . . . . 234
B. Catabolism . . . . . . . . . . . 252
IV. COMPOSITION, STRUCTURE AND FUNCTION OF YEAST MEMBRANES . . 254
A. Plasma Membranes . . . . . . . . . 254
B. Mitochondrial Membranes . . . . . . . . 262
C. Other Membranes . . . . . . . . . . 263
V. ACKNOWLEDGEMENTS . . . . . . . . . . 264
REFERENCES . . . . . . . . . . . . 264

## I. Introduction

Although yeasts, especially strains of *Saccharomyces cerevisiae*, are among the most intensively studied micro-organisms from a biochemical standpoint, information on the lipid composition of yeasts and on the structure and composition of yeast membranes is meagre. Lipids were among the first natural products to be examined in detail (Chevreul, 1823), and the fact that their study has lagged far behind that of other types of biological molecule is probably due largely to the unattractive physical properties of lipids. In this respect, yeasts are not unique. Indeed it has only been since accurate and sensitive methods for analysis of lipids, particularly thin layer and gas-liquid chromatography have been available, that reliable data on the lipid composition of cells have been reported. Fortunately, these analytical methods became available at an opportune time, for during the past two decades biochemists and cell physiologists have come to appreciate the importance of lipids in cell membranes. In many ways, this appreciation has bridged the gap between biochemistry and cell physiology, for membranes give integrity to cells as well as a vectoral as opposed to a scalar component to metabolism.

It is clear, nevertheless, that both biochemists and cell physiologists are still largely ignorant of the precise roles of various types of lipid and protein in cell membrances. In striving for an understanding of relationships between structure and function in membranes, it is conceivable that yeasts, probably strains of *Sacch. cerevisiae*, could provide useful model systems for studying these relationships in eukaryotic as opposed to prokaryotic organisms.

## II. Yeast Lipids

### A. TOTAL LIPID CONTENTS OF YEASTS

#### 1. *Species Variation*

There are numerous data in the literature on the total lipid contents of different species of yeast, some of which are presented in Table 1. These data show that the total lipid contents of yeasts vary widely with the species. In general, it is possible to recognize two groups. The majority of yeasts contain between about 7 and 15% dry weight of lipid, but there is a smaller class, sometimes referred to as the "fat yeasts", which contain much more lipid, ranging from around 30% to over 60% of the dry weight. We shall have occasion to refer again to fat yeasts, and to the nature of the lipid in these yeasts. The vast majority of analyses reported on the lipid contents of yeast cells were obtained gravimetrically, with methods that require relatively large amounts of cells and which also place a premium on the accuracy of weighing milligram quantities of extract. In our laboratory we have obtained satisfactory results with a microgravimetric method which requires as little as 1–2 mg lipid (Rouser *et al.*, 1967). Another method which merits examination is based on the intensity of the colour obtained when lipids are charred with sulphuric acid (Marsh and Weinstein, 1966).

#### 2. *Methods Used for Extracting Lipids*

Lipids are generally taken to be compounds that are more or less soluble in organic solvents. Workers who have determined the total lipid contents of yeasts have used a wide range of organic solvents or mixtures of solvents. Unfortunately, little attention has been given to the relative efficiency of different solvents and solvent mixtures as extractants.

A useful report on the conditions needed to extract lipids from yeasts was by Nyns *et al.* (1968), who examined the efficiency of lipid extraction

TABLE 1. *Lipid Contents of Selected Yeasts*

| Strain | Total lipid content (% dry wt) | Reference |
| --- | --- | --- |
| *Blastomyces dermatitidis* | 6–14 | Di Salvo and Denton (1963) |
| *Candida scottii* AL 25 | 8·2 | Kates and Baxter (1962) |
| *Candida scottii* $5AAP_2$ | 10·7 | Kates and Baxter (1962) |
| *Candida lipolytica* (grown at 25°) | 6·6 | Kates and Baxter (1962) |
| *Candida lipolytica* (grown at 10°) | 8·5 | Kates and Baxter (1962) |
| *Lipomyces starkeyi* | 55–65 | Starkey (1946) |
| *Pullularia pullulans* | 10·9 | Merdinger and Cwiakala (1968) |
| *Rhodotorula glutinis* L 10 | 12 | Kates and Baxter (1962) |
| *Rhodotorula gracilis* | 60 | Enebo *et al.* (1946) |
| *Rhodotorula graminis* l 10 | 31·7 | Hartman *et al.* (1959) |
| *Rhodotorula graminis* L 23 | 37·2 | Hartman *et al.* (1959) |
| *Saccharomyces cerevisiae* NCYC 366 | 11·7 | Longley *et al.* (1968) |
| *Saccharomyces cerevisiae* | 10 | Jollow *et al.* (1968) |
| *Saccharomyces cerevisiae* | 8 | Letters (1966) |
| *Saccharomyces cerevisiae* (grown at 30°) | 12·5 | Hunter and Rose (1971) |
| *Saccharomyces cerevisiae* (grown at 15°) | 14·4 | Hunter and Rose (1971) |

from *Candida lipolytica* using several solvents and solvent mixtures and either fresh or freeze-dried cells. The most efficient extraction was obtained using freeze-dried cells and a cold 1:1 mixture of chloroform and methanol. This method was described by Pedersen (1962) in his work on the lipids of *Cryptococcus terricolus*. Other methods which were compared by Nyns *et al.* (1968) were extraction of freeze-dried material with a hot mixture of ethanol and benzene (1:4; Kahane, 1963), and extracting a wet paste of cells with a hot mixture of ethanol and petroleum ether (1:1; Kumagawa and Suto, 1908), or ethanol and diethylether (1:1; Kahane and Regord, 1964), or a cold mixture of chloroform and methanol (2:1; Letters, 1962), or ethanol and diethylether (3:1; Hanahan and Jayko, 1952). These solvent systems are listed approximately in the order of decreasing efficiency for extracting lipids from *C. lipolytica*. Harrison and Trevelyan (1963) showed that neutral-solvent mixtures would extract all of the lipid from cells only if the cells were disrupted, although Letters (1966) contended that neutral solvents will suffice if acidified with hydrochloric acid. According to the latter worker the lipid not extracted with unsupplemented neutral solvents contained a high proportion of phospholipid, although he could not detect a preferential retention of any particular type of phospholipid.

An especially worrying problem which has arisen from work on comparative methods of extracting lipids from yeasts has been the report that, when cells are treated with certain organic solvents, there is an activation of phospholipases (Letters and Snell, 1963). Harrison and Trevelyan (1963) suggested that the major lipolytic enzyme in *Sacch. cerevisiae* is phospholipase C which catalyses the conversion of phospholipids into diglycerides. Because of the possible presence of phospholipases in yeasts, Letters (1968b) has advocated an initial treatment of cells with 80% (v/v) ethanol at 80° for 15 min to inactivate the enzymes; this treatment is followed by extraction of the lipids with a mixture of chloroform and methanol. This method extracts more phospholipid (5–10%) from brewer's strains of *Sacch. cerevisiae* than extraction of disrupted cells.

It is clear that in the past far too little attention has been given to the efficiency of different methods for extracting lipids from yeasts. Any newcomer to the field is well advised to compare the relative merits of different extraction methods for the strain under investigation before he attempts to report on the lipid contents of cells.

### 3. *Effect of Growth Conditions on Lipid Content*

Although there are many reports on the effects of growth conditions on the lipid composition of yeasts, with the exception of work on lipid production by 'fat' yeasts, relatively few workers have examined the effects of these conditions on the total lipid contents of cells, a reflection no doubt of the problems encountered in making accurate determinations of the total lipid contents of yeasts.

The total lipid content of *C. utilis* grown in batch culture in a chemically defined medium increases as the culture ages until the stationary phase of growth is reached (Dawson and Craig, 1966). Thereafter, following the exhaustion of glucose in the medium, there is a rapid decrease in the lipid contents of the cells.

Growth temperature is known to affect the lipid content of yeasts. A decrease in the growth temperature from 25° to 10° increased the lipid content of a strain of *C. lipolytica* from 6·6% to 8·5% (Kates and Baxter, 1962). Hunter and Rose (1971) confirmed this effect using batch-grown cultures of *Sacch. cerevisiae*. A decrease in growth temperature from 30° to 15° was accompanied by an increase from 12·5 to 14·4% in the total lipid content (percent dry weight) of cells harvested from early logarithmic-phase cultures. These workers went on to discover whether this increased synthesis of lipid was a direct result of the decrease in growth temperature or whether it was a result of the decreased growth rate which accompanies a decrease in growth temperature in batch cultures. They found that the lipid content of chemostat-grown cells tended to be greater the more slowly the cells were grown at 30°.

Castelli and his colleagues (Castelli *et al.*, 1969) studied the effect of pH value and carbon dioxide concentration on the lipid content of a strain of *Sacch. cerevisiae*. When the pH value of the medium was maintained constant at 5·5, a three-fold increase in the concentration of bicarbonate and $pCO_2$ caused a 27% increase in the total lipid content of the yeast. On the other hand, if the pH value was increased to 6·0, and the $pCO_2$ maintained constant by addition of bicarbonate, there was hardly any increase in the total lipid content of the cells.

The lipid contents of yeast also vary with the concentration of sodium chloride in the medium. As the concentration of sodium chloride in the medium was increased from zero to 10% (w/v), the lipid contents of *C. albicans* increased from 0·32 to 6·29%. Growth of the yeast was inhibited in the presence of the high concentrations of salt. It is known that $Na^+$ ions increase the rate of fermentation of glucose by yeast (Lasnitzki and Szorenyi, 1935) and Combs *et al.* (1968) suggested that, in *C. albicans*, this leads to an accumulation of lipid at the expense of other cellular compounds. They acknowledged however that, although all cells were harvested from cultures after 48-h incubation, the cells when grown in the presence of high concentrations of salt may have been from different stages in batch growth. A similar increase in the lipid content following growth in the presence of sodium chloride has been reported with *Sacch. cerevisiae* (White and Werkman, 1948).

There are a few reports of the effect of vitamin deficiency on the lipid contents of yeasts. Haskell and Snell (1965) found that the lipid content of pyridoxine-deficient *Hanseniaspora valbyensis* was about 40% less than that of cells grown in a medium containing a sufficient amount of the vitamin. The lipid content of pantothenate-deficient cells was also lower than in control cells, an observation which has also been made with *Sacch. cerevisiae* (Klein and Lipmann, 1953). Growth of yeasts under conditions of inositol deficiency also affects the total lipid content, and in a somewhat surprising fashion. It might be expected that starvation of inositol would lead simply to a decreased synthesis of phosphatidylinositol. In fact, the effect of inositol deprivation is much more dramatic in that it causes an approximately three-fold increase in the lipid content of cells (Challinor and Daniels, 1955). This effect will be referred to again in Section III.A.1.d. (p. 243).

Fears of a world-wide shortage of fat prompted several groups of microbiologists to examine factors that affect the lipid content of "fat" yeasts. This work received some justification since microbial fat was manufactured on a commercial scale in Germany during the 1914–18 and 1939–45 wars. The background and literature on this subject have been extensively reviewed by Woodbine (1959). The most important environmental factors which affect lipid synthesis by "fat" yeasts were reported by the pioneer workers in the field. Lindner (1911) found that high concentrations of carbohydrate in the medium stimulate fat production in *Endomyces* spp., and this effect was later shown by Heide (1939) to be enhanced in media containing suboptimum concentrations of a nitrogen source. Phosphate deficiency in *Sacch. cerevisiae* also gives rise to cells with a lipid content higher than cells grown in phosphate-sufficient media (Nielsen and Nilsson, 1953; Maas-Förster, 1955). Variations in medium composition, and especially in the ratio of the concentrations of the carbon and nitrogen sources and phosphate, have permitted workers to induce a very extensive synthesis of lipid in fat yeasts. Strains of *Rhodotorula gracilis* have been grown under conditions that yield cells with as much as 60% of their dry weight as lipid (Enebo *et al.*, 1944, 1946; Tornqvist and Lundin, 1951).

## B. COMPOSITION OF YEAST LIPIDS

### 1. *Intracellular Lipids*

*a. Chemistry of yeast lipids.* The lipids extracted from yeast cells contain a wide range of components the chemistry of which is discussed in the following paragraphs.

i. *Triacylglycerols* (otherwise known as triglycerides), together with phospholipids and sterols, account for the bulk of the cell lipids in yeasts. These three classes of lipid can be separated by thin-layer chromatography (Mangold, 1965) or by chromatography on paper or columns of silicic acid (Marinetti, 1967). Triacylglycerols are triesters of glycerol and long-chain fatty acids, and representatives differ in the nature and positional distribution of the fatty-acid residues. By convention (Hirschmann, 1960; IUPAC-IUB Commission on Biochemical Nomenclature, 1967), triacylglycerols are designated triacyl-*sn*-glycerols. An example is:

$$
\begin{array}{rl}
 & ①CH_2.O.CO.(CH_2)_{14}.CH_3 \\
 & | \\
CH_3.(CH_2)_7.CH{=}CH(CH_2)_7.CO.O & ②CH \\
 & | \\
 & ③CH_2.O.CO.(CH_2)_{14}.CH_3
\end{array}
$$

which is the structural formula for 1-palmityl-2-oleyl-3-palmityl-*sn*-glycerol. In yeast lipids, the fatty acids range in length from $C_8$ or occasionally lower to $C_{24}$, with the major representation coming from $C_{16}$ and $C_{18}$ acids. The acids include unsaturated as well as saturated representatives, in particular $C_{16:1}$ and $C_{18:1}$ acids. Mono-enoic acids in yeast lipids are usually $\Delta^{9,10}$. Strains of *C. utilis* contain $C_{18:3}$ acids, but these acids are not usually detected in appreciable amounts in extracts from *Sacch. cerevisiae* (but see Castelli *et al.*, 1969). In mammalian lipids, unsaturated fatty-acid residues preferentially occupy the 2-position in the glycerol residues, and this holds true also for lipids from yeasts (Meyer and Block, 1963a).

Triacylglycerols have been isolated from lipid extracts of *Sacch. cerevisiae* by chromatography on silicic acid (Barron and Hanahan, 1961), and identified by the product of acid hydrolysis (Letters, 1962). They have also been identified in extracts of a wide variety of other yeasts.

Bergel'son *et al.* (1966) reported the presence of two new types of neutral lipid in *Lipomyces starkeyi*; these are long-chain diesters of dihydric alcohols and 1-alkenyl ethers of their mono-esters. The latter class of compounds are analogues of neutral plasmalogens. These lipids are not usually separated from triacylglycerols by column chromatography or thin-layer chromatography, although they can be separated by gas-liquid chromatography. The diols detected in these lipids included ethylene glycol, propane-1,2-diol, propane-1,3-diol, butane-1,3-diol, butane-1,4-diol and probably a $C_5$ diol; they have since been further characterized by Bergel'son and his colleagues (Vaver *et al.*, 1967a). The $C_5$ diol was the only one to occur to any great extent in the diesters, the ratio of the amounts of $C_5$ diol to glycerol being 1:5. In the diols from the 1-alkenyl ethers of the mono-esters, there was five times as much butane-1,3-diol as glycerol, and three times as much ethylene glycol. Bergel'son and his colleagues have also reported the presence of various dihydric alcohols in the phospholipids of mammalian tissues (Vaver *et al.*, 1967b) but as yet these compounds have not been reported in micro-organisms.

Diacylglycerols and mono-acylglycerols, which are diesters and monoesters respectively of glycerol with long-chain fatty acids, have also been detected in extracts from yeasts (Kates and Baxter, 1962). The diacylglycerols are mostly 1,2-diacyl-*sn*-glycerols, although they include small amounts of 1,3-diacyl-*sn*-glycerols. The mono-acylglycerols which are found in lipid extracts of yeasts are probably 1-acyl-*sn*-glycerols. Diacylglycerols and mono-acylglycerols in lipid extracts of yeasts probably represent products of degradation of triacylglycerols, a point which will be discussed further in Section III.B.1 (p. 253).

ii. *Phospholipids* are substituted diesters of *sn*-glycero-3-phosphoric acid with long-chain fatty acids. The major phospholipids in extracts of yeasts are phosphatidylcholine, phosphatidylethanolamine, phosphatidylinositol, and phosphatidylserine. These classes of phospholipids can be separated from one another by thin-layer or paper chromatography. It should be emphasized that each of these names denotes a class of compound individual members of which are distinguished by the nature of the fatty-acid residues and their positional distribution. Derivatives of phosphatidic acid have a centre of symmetry on C-2; naturally occurring derivatives have been shown to be stereochemically related to *sn*-glycero-3-phosphoric acid.

Phosphatidylcholine is usually the major phospholipid in all yeast-lipid extracts so far examined (Fig. 1). Normally it represents between 35 and 50% of the total phospholipids in the extract. Hanahan and Jayko (1952) isolated 1,2-dipalmitoleyl-*sn*-glycero-3-phosphorylcholine from an extract of *Sacch. cerevisiae* by chromatography on an alumina column. This phospholipid has also been detected in extracts of *Candida* sp. (Kates and Baxter, 1962), *Hanseniaspora valbyensis* (Haskell and Snell, 1965) and *Sacch. carlsbergensis* (Letters, 1968a; Shafai and Lewin, 1968).

```
              CH2.O.CO.R1
              |
R2.CO.O.CH    O
              |         ||
              CH2.O.P.O.CH2.CH2.N.(CH3)3        Phosphatidylcholine
                    |
                    OH
```

Another major phospholipid in all of the yeasts so far examined (Fig. 1) is phosphatidylethanolamine which often accounts for 25–32%

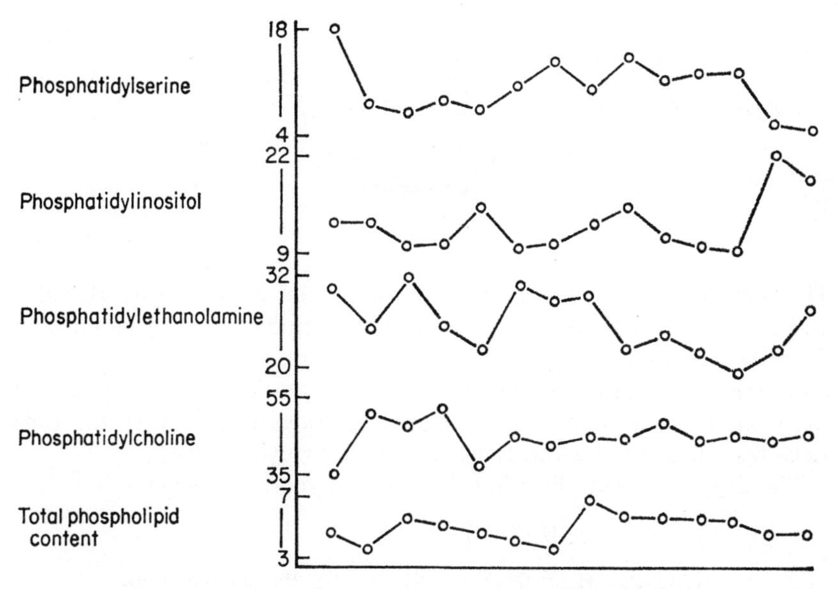


FIG. 1 Distribution of the major phospholipids in some species of yeasts. The total phospholipid contents (at the bottom of the figure) are expressed as percentages of the dry weight of the yeast. The contents of individual phospholipids are expressed as a percentage of the total lipid phosphorus. From Letters (1968b).

of the total phospholipids. Letters (1968b) characterized this phospholipid from extracts of *Sacch. cerevisiae*, after isolating it by column chromatography on silicic acid, DEAE cellulose and alumina, by chromatography of the deacylation product and of products from

```
         CH2.O.CO.R1
         |
R2CO.O.CH       O
         |      ||
         CH2.O.P.O.CH2.CH2.NH2
                |
                OH
```

Phosphatidylethanolamine

the periodate-dimethylhydrazine degradation of the deacylated compound. Phosphatidylethanolamine has also been detected in *Candida* sp. (Kates and Baxter, 1962), *Hanseniaspora valbyensis* (Haskell and Snell, 1965) and *Sacch. carlsbergensis* (Letters, 1968a; Shafai and Lewin, 1968). Letters (1966) isolated from extracts of *Sacch. cerevisiae* small quantities of N,N-dimethylphosphatidylethanolamine and N-methylphosphatidylethanolamine. It is not yet clear whether these lipids represent intermediates in the synthesis of phosphatidylcholine from phosphatidylethanolamine (see Section III.A, p. 236) or products of degradation of phosphatidylcholine.

Another characteristic yeast phospholipid is phosphatidylinositol.

$$\begin{array}{l} CH_2.O.CO.R_1 \\ R_2.CO.O.CH \quad O \\ CH_2.O.\overset{\|}{P}.O\text{—} \\ \qquad\quad OH \end{array} \quad \begin{array}{cccc} OH & & H & \\ H & & HO & OH \\ H & & HO & H \\ OH & & HO & \end{array} \qquad \text{Phosphatidylinositol}$$

This phospholipid was isolated from yeast and identified by Hanahan and Olley (1958). It was characterized by periodate-dimethylhydrazine degradation of its deacylation product (Letters, 1966). It is usual to find quite large proportions (about 20%) of phosphatidylinositol in extracts from *Sacch. cerevisiae* and *Sacch. carlsbergensis* (Longley *et al.*, 1968; Letters, 1968a; Shafai and Lewin, 1968; Trevelyan, 1968).

The proportions of phosphatidylserine in yeast lipids vary.

$$\begin{array}{l} \qquad\qquad\quad CH_2.O.CO.R_1 \\ R_2.CO.O.CH \qquad O \qquad\quad NH_2 \\ \qquad\qquad\quad CH_2.O.\overset{\|}{P}.O.CH_2.CH.COOH \\ \qquad\qquad\qquad\qquad OH \end{array} \qquad \text{Phosphatidylserine}$$

Some yeasts, such as *Lipomyces starkeyi*, contain relatively large proportions (about 18%) whereas *Sacch. cerevisiae* and *Sacch. carlsbergensis* contains as little as 4% (Letters, 1968a).

Cardiolipin, or diphosphatidylglycerol, was identified as a constituent

$$\begin{array}{l} \qquad\qquad\quad CH_2.O.CO.R_1 \qquad\qquad\qquad\qquad\qquad\qquad CH_2.O.CO.R_4 \\ R_2.CO.O.CH \qquad O \qquad\qquad\qquad\qquad\qquad\qquad O \quad HC.O.CO.R_3 \\ \qquad\qquad\quad CH_2.O.\overset{\|}{P}.O.CH_2.CHOH.CH_2.O.\overset{\|}{P}.O.CH_2 \\ \qquad\qquad\qquad\quad OH \qquad\qquad\qquad\qquad\qquad\qquad\quad OH \end{array} \qquad \text{Cardiolipin}$$

of *Sacch. cerevisiae* by Letters (1966) using chromatography on columns of silicic acid and DEAE cellulose and two-dimensional paper chromatography. Deacylation of the lipid yielded triglycerol phosphate which was degraded using periodate-dimethylhydrazine to glycerol-1,3 diphosphate. The proportion of cardiolipin in yeast lipids is usually small. *Candida scottii* and *C. lipolytica* were reported by Kates and Baxter (1962) to contain cardiolipin, and the lipid has also been detected in extracts from cells (Letters, 1968a) and mitochondria (Lukins *et al.*, 1968) of *Sacch. cerevisiae.*

Another phospholipid found in extracts from yeast and in about the same proportions as cardiolipin is phosphatidylglycerol. Naturally occurring phosphatidylglycerols are usually 3-*sn*-phosphatidyl-1-*sn*-glycerols. Phosphatidylglycerol was isolated from a lipid extract of *Sacch. cerevisiae* by Letters (1966) in a similar manner to cardiolipin. The lipid was characterized by its deacylation product, glycerylphosphorylglycerol, and the product of degradation by periodate-dimethylhydrazine. Deierkauf and Booij (1968) used similar methods to characterize the lipid isolated from a thin-layer chromatogram of an extract from another strain of *Sacch. cerevisiae.*

Amino-acid esters of phosphatidylglycerol, the so-called lipo-amino acids, have been found along with the corresponding phosphatidylglycerols in extracts of several bacteria (Macfarlane, 1964). However, we are not aware of any report, published or otherwise, of the existence of these lipids in yeasts.

Phosphatidylglycerol phosphate has recently been detected in lipid extracts of a mutant strain of *Sacch. cerevisiae* by Deierkauf and Booij (1968). This lipid had the same mobility as phosphatidylinositol on the paper chromatograms run by these workers, but they were able to show the presence of both lipids on the spot. Mild alkaline hydrolysis of the material yielded small amounts of glycero-3-phosphoric acid and inositol phosphate together with a product identified as glycerylphosphorylglyceryl phosphate which is the deacylation product of phosphatidylglycerol phosphate. Further evidence for the presence of phosphatidylglycerol phosphate on the spot was provided by digestion with phospholipase C, which yielded unchanged phosphatidylglycerol phosphate and 1,2-diacylglycerol.

Lysophosphoglycerides are related to phospholipids; they lack the 2-acyl group and have a free hydroxyl group on the glycerol residue at C-2. Lysophospholipids are more polar than the corresponding fully acylated phospholipid. Letters and Snell (1963) isolated lysophosphatidylcholine from *Sacch. cerevisiae*, and they characterized the lipid by converting it to a phosphatidylcholine by acylation with myristoyl chloride and treated the product with phospholipase A which gave a quantitative yield of myristic acid. Small amounts of lysophosphatidylglycerides can be detected on thin-layer chromatograms of lipids from most yeasts. It is generally thought that at least some lysophosphoglycerides are artefacts which are formed during extraction of the lipid. Presumably, they arise as a result of the action of phospholipases on the phospholipids. There is evidence that phospholipases may be more active in cells grown anaerobically rather than aerobically. Letters (1968b) reported that lysophosphoglycerides could not be detected in lipids from aerobically grown *Sacch. carlsbergensis* but that cells grown anaerobically contained large proportions of lysophosphatidylcholine (27·2% of the total phospholipid) and lysophosphatidylethanolamine (8·5%).

Lipid extracts of some yeasts also contain small amounts of phosphatidic acid and free fatty acids. Both classes of compound probably arise to some extent as a result of the action of phospholipases on phosphoglycerides. Phosphatidic acid has been detected in lipid extracts of *Sacch. cerevisiae* (Longley *et al.*, 1968) and *Sacch. carlsbergensis* (Shafai and Lewin, 1968) but not in extracts from *C. lipolytica* or *C. scottii* (Kates and Baxter, 1962). Free fatty acids rarely account for more than a few percent of the total lipid. This fraction contains the same range of fatty acids as are found esterified with glycerol. Free fatty acids may be degradation products, but it is possible that they are also true membrane constituents.

Plasmalogens would not appear to be common constituents of yeast lipids. Letters and Snell (1963) examined the phosphatidylcholine fraction from *Sacch. cerevisiae* for plasmalogen forms of this phospholipid, but were unable to detect any. Reference has already been made to detection in *Lipomyces starkeyi* of plasmalogen analogues of neutral lipids (see p. 217).

iii. *Hydrocarbons.* Probably the least understood of the various classes of yeast lipids are the hydrocarbons. Quantitatively, these compounds may account for 2–20% of the total lipid in yeast cells (Barron and Hanahan, 1961; Kováč *et al.*, 1967; Baraud *et al.*, 1967) but accurate determinations of the total hydrocarbon contents of yeast lipids are not easily obtained because of the lack of reliable assay methods. Some of the hydrocarbon contents quoted are obtained simply by subtracting the total sterol content from the value for the total non-saponifiable lipid (Kovac *et al.*, 1967).

Very little indeed has been reported on the composition of the hydrocarbon fraction of yeast lipids. Squalene, the $C_{30}$ hydrocarbon precursor of sterols, has been detected in lipid extracts of *Sacch. cerevisiae* (Jollow *et al.*, 1968) and *Sacch. carlsbergensis* (Shafai and Lewin, 1968). But the most extensive analysis so far reported on yeast hydrocarbons is by

Baraud and his colleagues (1967). These workers separated the hydrocarbons in lipids from *Sacch. oviformis*, grown aerobically or anaerobically, and from *Sacch. ludwigii* grown anaerobically, using gas–liquid chromatography. They detected more than 40 different hydrocarbons, ranging in chain length from $C_{10}$ to $C_{31}$. Both straight-chain and branched-chain hydrocarbons were detected, although the majority had chain lengths between $C_{10}$ and $C_{19}$.

iv. *Polyprenols* are compounds with the general structure:

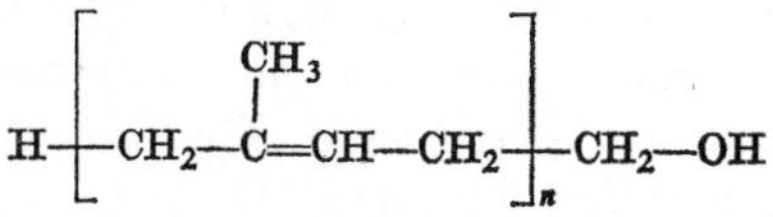

$$H\left[-CH_2-\underset{}{\overset{CH_3}{\overset{|}{C}}}=CH-CH_2-\right]_n-CH_2-OH$$

in which *n* has a value greater than 5 and may reach 24. Natural polyprenols are of three types: all *trans* alcohols such as farnesol, *cis-trans* allylic alcohols which are known as prenols, and *cis-trans* alcohols with a saturated unit at the hydroxyl end of the molecule (these are referred to as dolichols; Hemming, 1970). The shape of the molecule varies according to the number of *cis*, *trans* and saturated isoprene residues. Polyprenols have been isolated from the non-saponifiable lipids of a variety of plants, animals and micro-organisms. General biochemical interest in polyprenols has increased during the past few years because they have been found to be involved in the biosynthesis of certain bacterial cell-wall polymers (Wright *et al.*, 1967; Higashi *et al.*, 1967; Douglas and Baddiley, 1968; Scher *et al.*, 1968).

The only report of yeast polyprenols has come from Dunphy and his colleagues (1967) who extracted from baker's yeast a mixture of dolichols, ranging in chain length from $C_{70}$ to $C_{105}$. The amounts of dolichols present in the yeast lipids were small; Dunphy *et al.* (1967) reported the isolation of only 5 mg dolichol from 1 kg of pressed baker's yeast. The general formula of yeast dolichols is:

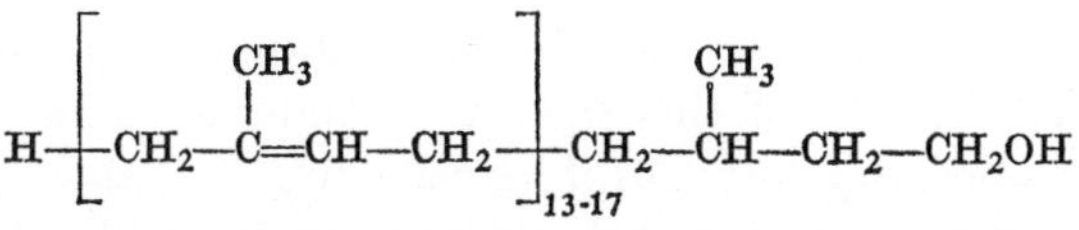

$$H\left[-CH_2-\overset{CH_3}{\overset{|}{C}}=CH-CH_2-\right]_{13\text{-}17}-CH_2-\overset{CH_3}{\overset{|}{CH}}-CH_2-CH_2OH$$

It is believed (Dunphy *et al.*, 1967) that yeast dolichols contain three of the internal isoprene residues in the *trans* configuration.

v. *Sphingolipids* are hydroxy fatty-acid esters of long-chain amino alcohols or sphingosines. Sphingolipids are of three main types, known as sphingomyelins, cerebrosides and gangliosides. Representatives of all

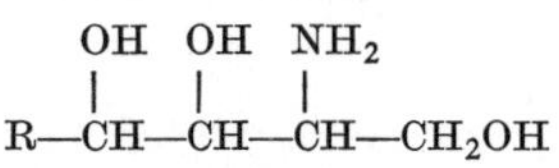

$$R-\overset{OH}{\overset{|}{CH}}-\overset{OH}{\overset{|}{CH}}-\overset{NH_2}{\overset{|}{CH}}-CH_2OH \quad \text{Phytosphingosine}$$

three types of sphingolipid have been detected in animal tissue, but the lipids from yeasts so far examined have revealed only cerebrosides and then only in very small amounts. Yeast sphingolipids are derivatives of phytosphingosines.

The first report of the presence of a cerebroside in yeast was by Reindel and his colleagues (1940). The structure of yeast cerebrosides was elucidated only after the chemistry of the mammalian counterparts had been studied. The cerebroside from *Torulopsis* (= *Candida*) *utilis* was examined by Stanacev and Kates (1963). On hydrolysis, this lipid gave $C_{18}$ and $C_{20}$ phytosphingosines in the ratio of 5:1; small amounts of a $C_{18}$ dihydrosphingosine were also detected. The fatty acids were identified as $C_{18}$, $C_{24}$ and $C_{26}$ α-hydroxy acids in the ratio 1:1:10 together with small amounts of n-$C_{18}$, n-$C_{24}$ and n-$C_{26}$ saturated acids and oleic acid. Similarly, Oda and Kamiya (1958) detected $C_{18}$ and $C_{20}$ phytosphingosines in the ratio of 2:1, and $C_{24}$ and $C_{26}$ α-hydroxy acids in the ratio of 1:0·3, in the cerebroside from a strain of baker's yeast.

In 1966, Wagner and Zofcsik (1966a) reported the presence of identical cerebrosides in the sphingolipid fractions from *Candida utilis* and *Sacch. cerevisiae*. On hydrolysis, the compound from *C. utilis* yielded three sphingosine bases, α-hydroxystearic acid and galactose, and the authors proposed the following structure for the compound:

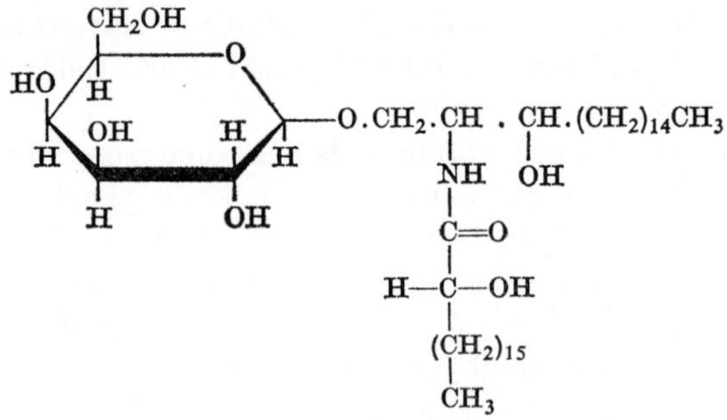

```
     CH2OH
       |-----------O
 HO  / H            \
   |   OH       H    \-----O.CH2.CH . CH.(CH2)14CH3
   H  |          |  H          |       |
       ----------                NH     OH
       H        OH               |
                                 C=O
                                 |
                              H--C--OH
                                 |
                               (CH2)15
                                 |
                                 CH3
```

Wagner and Zofcsik (1966b) isolated a second sphingolipid from these two yeasts. Hydrolysis of this second compound yielded $C_{18}$ and $C_{20}$ phytosphingosines, $C_{18}$ dihydrosphingosine, $C_{20}$, $C_{22}$ and $C_{24}$ fatty acids, and inorganic phosphate; inositol and mannose were also detected in the hydrolysate. Wagner and Zofcsik (1966b) termed this compound a "mycoglycolipid", and proposed for it the following tentative structure:

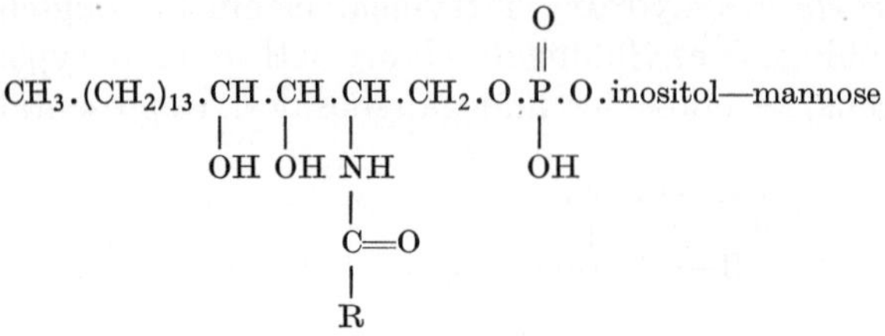

```
                                           O
                                           ||
CH3.(CH2)13.CH.CH.CH.CH2.O.P.O.inositol—mannose
              |   |   |          |
              OH  OH  NH         OH
                      |
                      C=O
                      |
                      R
```

Trevelyan (1968) has reported the presence in baker's yeast of a sphingolipid containing inositol and mannose. This compound accumulated in toluene-autolysed yeast, but Trevelyan was unable to propose a definite structure for the compound. Evidence for a yeast lipid containing an inositol mannoside residue has come from Tanner (1968). As this lipid is resistant to mild ammonolysis, it would appear to preclude the possibility that it is an inositol mannoside-containing glycolipid of the type found in mycobacteria (Lee and Ballou, 1965). Tanner (1968) suggested therefore that the inositol mannoside residue forms part of a mycoglycolipid as suggested by Wagner and Zofcsik (1966b).

vi. *Glycolipids.* Evidence for the presence of bacterial-type glycolipids (Shaw and Baddiley, 1968) in yeast has come from Baraud and his colleagues (1970), who detected monogalactosyldiacylglycerol in lipids from stationary-phase *Sacch. cerevisiae*, together with sulphatides and sterol glycosides. It would be interesting to learn more about the chemical structures of these hitherto unrecognized types of yeast lipid.

vii. *Sterols*, along with other lipids, are extracted, free or esterified, by organic solvents. They can easily be isolated from the saponified lipid extract in the form of digitonides. Separation of sterols in a mixture has in the past proved troublesome, and so little in the way of firm data on yeast sterols was available until the introduction of gas–liquid chromatography.

Sterols are based structurally on the cyclopentanoperhydrophenanthrene nucleus (Fieser and Fieser, 1959; Klyne, 1965; Shoppee, 1964). The two commonest yeast sterols are ergosterol and zymosterol (Fig. 2). Most yeasts that have been examined contain either or both of these sterols (Wieland and Benend, 1942; Usden and Burrell, 1952; Dulaney *et al.*, 1954; Longley *et al.*, 1968). There are also reports of minor amounts of other sterols in yeasts. Lanosterol (Fig. 2) is frequently detected in *Sacch. cerevisiae* (Wieland and Stanley, 1931) and in one strain of this yeast lanosterol has been found to occur along with dimethyl-4,4′-zymosterol and methyl-4α-zymosterol (Ponsinet and Ourisson, 1965). These methylated sterols may be intermediates in the synthesis of other yeast sterols (see Section III.A.2, p. 246). Another minor sterol component in some strains of *Sacch. cerevisiae* is a tetraethenoid sterol ($\Delta^{5,7,22,24(28)}$ ergostatetraen-3β-ol) which Breivik and his colleagues (1954) detected in baker's yeast. In the strain of *Sacch. cerevisiae* (NCYC 366) examined by Longley *et al.* (1968), this tetraethenoid sterol was the major component of the sterol fraction, and was accompanied by zymosterol and minor amounts of ergosterol, episterol (or fecosterol) and an unidentified $C_{29}$ di-unsaturated sterol. Other sterols which have been detected in *Sacch. cerevisiae* include ascosterol and fecosterol (Wieland *et al.*, 1941), cerevisterol and 14-dehydroergosterol (Fieser and Fieser,

1959), 5-dihydroergosterol (Callow, 1931), episterol (Wieland and Gough, 1930), and 4-α-methylene-24, 25-dihydrozymosterol (Barton *et al.*, 1968).

Sterols occur in yeasts both free and esterified with long-chain fatty acids. The presence of sterol esters in yeasts was reported as early as 1940

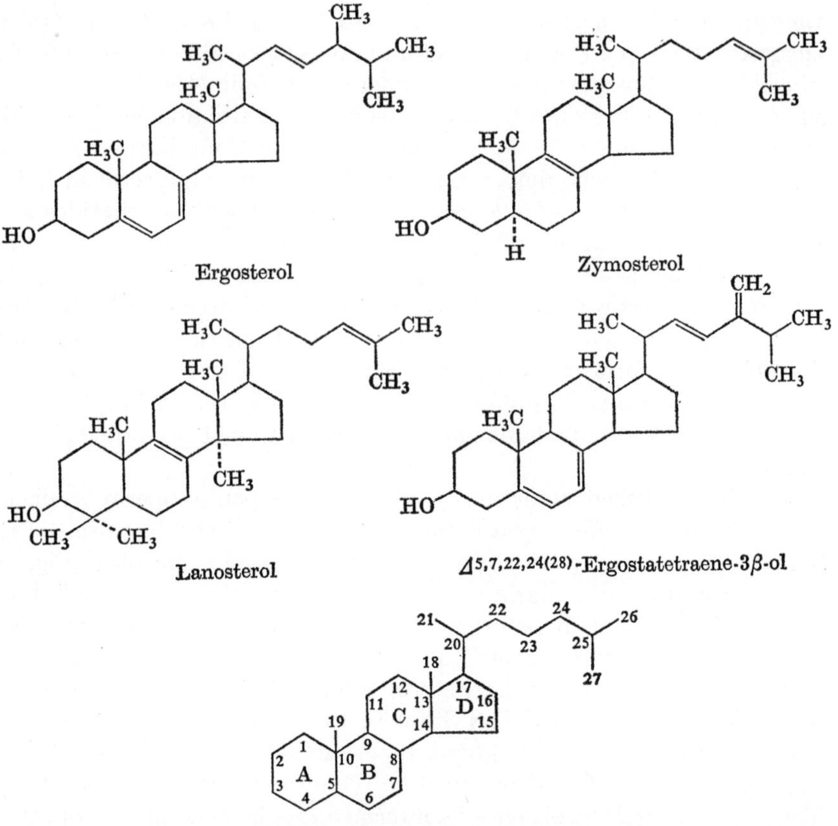

Ergosterol

Zymosterol

Lanosterol

$\Delta^{5,7,22,24(28)}$-Ergostatetraene-3β-ol

Fig. 2. Structural formulae of certain yeast sterols.

by Maguigan and Walker (1940). More recently, Adams and Parks (1967) showed that a strain of *Sacch. cerevisiae* contains ergosterol palmitate, a report which has since been extended by Madyasthia and Parks (1969). A report by Adams and Parks (1968) of the isolation of an ergosterol-polysaccharide complex from baker's yeast indicates that not all sterol derivatives are fatty-acid esters.

Estimates of the amounts of total sterol (free and combined) in yeast cells are in the range 0·1–1·0% (Shaw and Jefferies, 1953). Dulaney *et al.*

(1954) have reported a strain of *Sacch. cerevisiae* which, by suitably adjusting the cultural conditions, may contain as much as 7–10% of the cell dry weight as ergosterol.

*b. Distribution of lipids in the yeast cell.* Except with the fat yeasts, it is generally assumed that the bulk of the lipids in yeasts are located in membranes of one sort or another. Lipid droplets have been observed in cells of fat yeasts, and it has been assumed that they are made up mainly of triacylglycerols. There are many reports in the older literature (e.g. Nageli and Loew, 1878) and even in more recent publications (Dawson and Craig, 1966) of lipid droplets in cells of non-fat yeasts including *Sacch. cerevisiae* and *C. utilis*. Whether these droplets are indeed composed of lipid is not known. Indirect evidence is that triacylglycerols are not accumulated in appreciable quantities by non-fat yeasts. In a comparison of the lipid composition of whole cells and of sphaeroplast membranes from *Sacch. cerevisiae* NCYC 366, Longley *et al.* (1968) could not detect any qualitative differences, while the small quantitative differences were mainly in certain phospholipid fractions. These data argue against the accumulation of droplets of triacylglycerol in logarithmic-phase *Sacch. cerevisiae*. However, the point is far from proven, and it could be that non-fat yeasts accumulate lipid droplets only in the stationary phase of growth as suggested by the data of Dawson and Craig (1966) and Baraud *et al.* (1970).

Reports on the lipid contents of yeast cell walls range from 1 to 12% of the dry weight of the wall, or about 0·1–1% of the cell dry weight. The differences reported in the lipid contents of walls from different yeasts may be physiologically significant, although they may in part be explained by the inaccuracies of the gravimetric method when used with small quantities of cell wall. There is some disagreement among the data reported on the lipid content of walls of *Saccharomyces* spp. Kessler and Nickerson (1959) reported that walls of *Sacch. cerevisiae* contain 8–10% lipid, although Eddy (1958) found as little as 2% lipid in walls from strains of *Sacch. cerevisiae*, *Sacch. carlsbergensis* and *Sacch. oviformis*. Other data from *Sacch. cerevisiae* reported by Northcote and Horne (1952) and McMurrough and Rose (1967) lend support to the higher value. The lipid contents of walls examined by McMurrough and Rose (1967) varied with the nature of the substrate limitation imposed on the chemostat cultures of the yeast. The contents were greatest in walls from $NH_4^+$-limited cells (about 8% of the wall dry weight) compared with cells grown under glucose limitation.

Although the available evidence is scanty, it would seem that the different types of cell lipid are not evenly distributed among the various types of cell membrane (plasma, vacuolar, mitochondrial, and nuclear membranes). The only firm indication which has emerged from published data is that the intracellular membranes in yeast are somewhat richer in cardiolipin and phosphatidylcholine compared with the plasma membrane. Letters (1968b) examined the distribution of phospholipids in each of three fractions prepared from disrupted *Sacch. cerevisiae*; these fractions were composed of cells walls, mitochondria, and particles which sedimented at 100,000 *g*. The mitochondrial fraction was much richer in cardiolipin than the 100,000 *g* fraction, which in turn was richer in this phospholipid than the cell-wall fraction. There is also evidence that membranes in intracellular organelles in *Sacch. cerevisiae* are richer in phosphatidylcholine compared with the plasma membrane (Longley *et al.*, 1968). Also, it appears that the plasma membrane in *Sacch. cerevisiae* contains proportionately more unsaturated lipids than intracellular membranes (Hunter and Rose, 1971) when the cells are grown at 30° but not when grown at 15°.

c. *Effect of growth conditions on lipid composition.* The lipid composition of yeasts is very susceptible to variations in growth conditions. Indeed, in terms of chemical composition, membranes appear to be one of the most variable organelles in the yeast cell. The physiological significance of these environmentally induced changes in lipid composition is largely unknown, and this promises to be a profitable area for future research.

i. *Growth rate.* Several workers have observed quantitative changes in the lipid composition of yeasts with the age of cells in batch culture (Dawson and Craig, 1966). The main changes which have been detected affect the degree of unsaturation in the fatty-acid residues of the lipids and the relative proportions of different phospholipids. More rigorous studies using chemostat cultures, in which the effects of medium composition and growth rate can be separated, have shown that with *Sacch. cerevisiae* growing at 30° there is a tendency towards synthesis of a higher proportion of $C_{16}$ as compared with $C_{18}$ acids, and also towards production of proportionately more unsaturated acids, the slower the rate at which the cells are grown (Hunter and Rose, 1971). Also, as this yeast is grown more slowly in chemostat cultures at 30°, it synthesizes proportionately more phosphatidylcholine at the expense of phosphatidylethanolamine and phosphatidylserine (Hunter and Rose, 1971).

ii. *Growth temperature.* The lipid composition of micro-organisms is also known to be influenced by the temperature at which the organisms are grown (Farrell and Rose, 1967a, b). The main effect, one which has been known for over half a century and is found in almost all types of living organism, is a proportionate increase in the synthesis of lipids containing unsaturated fatty-acid residues as the growth temperature is lowered below the optimum. The effect has been reported with batch-grown cultures of *Sacch. cerevisiae* (Hunter and Rose, 1971) and *C. utilis* (Farrell and Rose, 1971) and *C. lipolytica* (Kates and Baxter, 1962).

Experiments using a chemostat with a device for control of dissolved oxygen tension showed that, in *C. utilis*, the proportionately increased synthesis of unsaturated fatty acids is indeed caused by a decrease in growth temperature (Brown and Rose, 1969) although the results obtained with batch cultures may be explained in part by the lowering of the growth rate which accompanies a decrease in incubation temperature of batch cultures.

A decrease in growth temperature below the optimum also leads, at least in the yeasts so far examined, to changes in the proportions of phospholipids synthesized. Kates and Baxter (1962) reported this effect when the growth temperature of *C. lipolytica* was lowered from 25° to 10°, although they considered the effect not to be sufficiently significant to merit detailed comment. Hunter and Rose (1971) discovered a similar effect in batch-grown *Sacch. cerevisiae* NCYC 366 when the growth temperature was decreased from 30° to 15°. The increase in phospholipid synthesis was mainly accounted for by phosphatidylcholine and to a lesser extent phosphatidic acid. As already mentioned, analyses of cells of this yeast grown at different temperatures, but at constant rate, in a chemostat suggested that some of the effects caused by lowering the growth temperature in batch culture may well be growth-rate effects rather than effects caused by incubation temperature (Hunter and Rose, 1971).

iii. *Oxygen tension*. The technical problems encountered in growing micro-organisms under conditions in which the dissolved oxygen tension in the culture is controlled probably explain the paucity of data on the effect of dissolved oxygen tension on the lipid composition of micro-organisms. Jollow *et al.* (1968) reported that *Sacch. cerevisiae* grown under anaerobic conditions in batch culture synthesizes lipids containing a greater proportion of saturated fatty-acid residues, particularly short-chain ($C_{10}$–$C_{14}$) saturated acids, compared with aerobically grown cells. A study of *C. utilis* grown in chemostat cultures showed that an increase in the degree of saturation and a shortening of the chain length occurred as the oxygen tension in cultures grown at a constant rate was lowered below 75 mmHg (Brown and Rose, 1969). It would seem therefore that the changes in fatty-acid composition observed when batch cultures of yeast are grown at sub-optimum temperatures are probably due to a decrease in growth temperature and growth rate, and also to an increase in dissolved oxygen tension.

When *Sacch. cerevisiae* is grown under completely anaerobic conditions, it becomes auxotrophic for sterols (Andreasen and Stier, 1954) and for fatty acids (Jollow *et al.*, 1968; Light *et al.*, 1962). Andreasen and Stier (1954) indicated that the fatty acid supplied to anaerobically growing yeast should be unsaturated, and most workers provide this in the form of Tween 80 (polyethylene sorbitan mono-oleate). The only study reported on the specificity of the fatty-acid requirement for anaerobically grown yeast is by Light and his colleagues (1962). They surveyed the growth-promoting activity of a wide range of $C_{16}$ and $C_{18}$ acids for *Sacch. cerevisiae* in ergosterol-containing medium. Acids which were active contain either a hydroxyl group, a *cis* double bond, a triple bond, or a cyclopropane ring in the middle of the molecule. Amino, epoxy, dihydroxy and *trans*-olefinic derivatives did not support growth. The anaerobically growing cells converted 9- (and 10-) hydroxystearic acids into acetoxy derivatives.

Mutant strains of *Sacch. cerevisiae* which have a nutritional requirement for unsaturated fatty acids when growing aerobically have been isolated by Resnick and Mortimer (1966). Keith and his colleagues (1969) made a more detailed study of these mutants, and showed that certain of them were deficient in a $\Delta^9$ desaturase activity, and were therefore incapable of converting palmitate to palmitoleate and stearate to oleate. Mutants of this type promise to be useful organisms with which to study the physiological effects of variations in membrane composition.

The yeast *Pityrosporum ovale* is exceptional in that it requires lipid for growth. This requirement was first studied in detail by Benham (1939, 1941, 1947) who found that it could be met by a variety of fatty materials and even by oleic acid. This suggested that *Pit. ovale* is unable to synthesize long-chain fatty acids. In a later study on the specificity of the fatty-acid requirement, Shifrine and Marr (1963) agreed with the conclusion of Benham, and they found that the requirement could be met by fatty acids with a chain length longer than $C_{10}$. Curiously, however, Shifrine and Marr (1963) found that oleic acid, which previously had been found to be very effective, was unable to meet the requirement when pure. In an attempt to resolve the problem, Wilde and Stewart (1968) only added to the confusion by finding that their strain of *Pit. ovale* converted palmitate into 9-hydroxypalmitate and stearate into 9-hydroxystearate; they could not detect the formation of oleate from saturated acids.

A similar low order of specificity to that encountered with the fatty-acid requirement for anaerobic yeast growth has been reported for the sterol requirement. Proudlock and his colleagues (1968) established the following requirements for a sterol to be active in promoting growth of anaerobically growing *Sacch. cerevisiae*; the molecule must: (a) be planar with an $\alpha$-configuration at C-5; (b) have a long alkyl side chain at C-17; and (c) have a hydroxyl group at C-3. It appears that the sterol that is included in the medium may be incorporated intact into the cellular membranes which suggests a relatively non-specific role for

these compounds in membrane function, possibly as a type of "filler".
iv. *Vitamin deficiencies.* Yeasts which are auxotrophic for vitamins can undergo marked changes in lipid composition when grown under conditions of vitamin deficiency. Reference has already been made to the effect of inositol deprivation on the total lipid content of *Sacch. cerevisiae* (Section II.A.3, p. 216). In *Sacch. carlsbergensis* inositol deficiency leads to the synthesis of excessive amounts of triacylglycerols and to a decreased synthesis of phosphatidylinositol (Shafai and Lewin, 1968). The effect of biotin deficiency on aerobically grown *Sacch. cerevisiae* is to restrict synthesis of $C_{18}$ acids, and in particular oleic acid, while synthesis of $C_{16}$ acids is correspondingly increased (Suomalainen and Keränen, 1968). This effect of biotin deficiency can be explained in part at least by the need for biotin in the synthesis of cocarboxylase, an enzyme involved in the chain-lengthening process in fatty-acid synthesis. Growing *Hanseniaspora valbyensis* under conditions of nicotinic acid, pantothenic acid or pyridoxin deficiency also alters the composition of the cell lipids (Haskell and Snell, 1965). Further experiments on the effect of vitamin deficiencies on yeast lipid composition could further elucidate the roles of vitamin-containing coenzymes in lipid synthesis.

### 2. *Extracellular Lipids*

Although polysaccharides and proteins are excreted or secreted by many micro-organisms, this is not true on the whole of lipids which are elaborated extracellularly by relatively few micro-organisms. It is somewhat surprising to discover, therefore, that over the past decade production of extracellular lipids has been reported with over 200 strains of yeast (Ruinen, 1963a, b). Priority in the discovery of extracellular lipid production by yeasts should go to Spencer who, in 1954, observed this phenomenon with a strain of *Torulopsis*. Work on lipids excreted by yeasts has been confined very largely to three laboratories, namely those of Ruinen in Wageningen in Holland, Spencer at the Prairie Regional Laboratory in Saskatoon, Saskatchewan, Canada, and Stodola at the Northern Regional Research Laboratory of the U.S. Department of Agriculture in Peoria, Illinois, U.S.A. The chemistry of the lipids produced extracellularly by yeasts has been comprehensively reviewed by Stodola *et al.* (1967). Structurally, they are rather different from the lipids located inside yeast cells, as will be apparent from the account that follows. With the exception of a strain which was isolated from the stomach of a trout, all of the yeasts which have been found to produce lipids extracellularly are from plant sources. Many of them were isolated from the phyllosphere, particularly by Ruinen and her colleagues (Ruinen, 1956, 1961, 1963a, b, 1966). The physiological significance of the production of extracellular lipids by yeasts is an utter mystery.

The extracellular lipids produced by yeasts fall into four quite distinct groups from the standpoint of chemical structure.

*a. Sphingolipids.* Small amounts of sphingolipids, particularly cerebrosides, have been detected in the intracellular lipids of several yeasts (see Section II.B.1a, p. 223). Wickerham and his colleagues (Wickerham and Stodola, 1960) found that the mating types of *Hansenula ciferrii* (NRRL Y-10131) produced extracellularly two sphingolipids. The major component is tetra-acetyl $C_{18}$-phytosphingosine (Stodola and Wickerham, 1960):

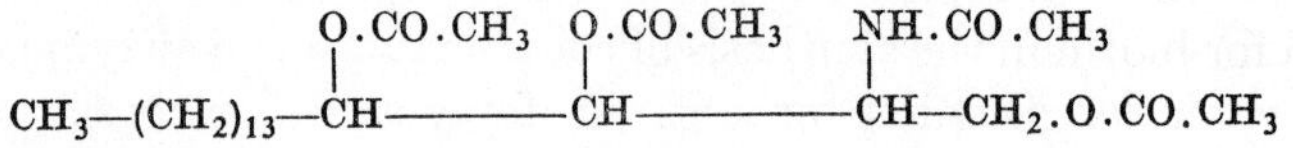

$$\begin{array}{cccc} & O.CO.CH_3 & O.CO.CH_3 & NH.CO.CH_3 \\ & | & | & | \\ CH_3—(CH_2)_{13}— & CH———— & CH———— & CH—CH_2.O.CO.CH_3 \end{array}$$

A minor component which accompanies the tetra-acetyl $C_{18}$-phytosphingosine was found to be triacetyl $C_{18}$-dihydrosphingosine (Stodola *et al.*, 1962):

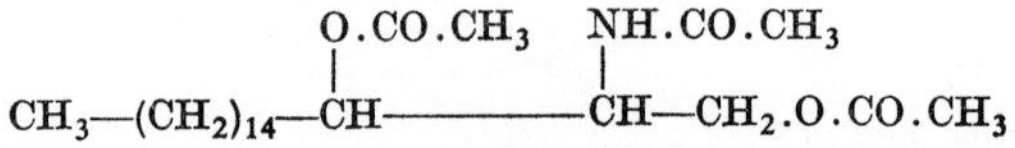

$$\begin{array}{ccc} & O.CO.CH_3 & NH.CO.CH_3 \\ & | & | \\ CH_3—(CH_2)_{14}— & CH———— & CH—CH_2.O.CO.CH_3 \end{array}$$

*b. Polyol fatty-acid esters.* Glycerides, which are found intracellularly in yeasts, are fatty-acid esters of the trihydric alcohol glycerol. Glycerides are not produced extracellularly by yeasts, as far as is known, but fatty-acid esters of other polyols are. The extracellular lipid synthesized by a strain of *Rhodotorula graminis*, which was isolated by Ruinen from the leaf surfaces of citrus fruit trees in Indonesia and Surinam (Ruinen, 1956), gives on hydrolysis a mixture of polyhydric alcohols and fatty acids, and a closely similar strain (6 CB; Tulloch and Spencer, 1964) produces extracellularly lipids made up of D-mannitol, D-arabitol and xylitol esterified with 3-D-hydroxyhexadecanoic and 3-D-hydroxyoctadecanoic acids. Similar results have been reported for analysis of the extracellular lipids elaborated by *Rh. glutinis* (Deinema, 1961) except that palmitic and oleic acids make up a significant proportion of the fatty acids. The extracellular polyol esters produced by these yeasts are acetylated, and from the analyses obtained it is thought that each polyol molecule is esterfied with one long-chain acid and that most of the remaining hydroxyl groups, including that on the fatty acid, are acetylated (Tulloch and Spencer, 1964). The yeast-like organism, *Aureobasidium pullulans*, also produces an extracellular fatty-acid ester of hexitol (Ruinen and Deinema, 1964). This class of yeast lipids appears to be somewhat similar to the acetylated sugar derivatives that occur in the cell lipids of some bacteria, such as *Streptococcus faecalis* (Welsh *et al.*, 1968).

*c. Sophorosides of hydroxy fatty acids.* Two yeasts, *Candida bogoriensis* and *Torulopsis apicola*, produce extracellularly lipids in which a hydroxyl group of a fatty acid is linked glycosidically to a carbohydrate. Compounds of this type are also produced by *Pseudomonas pyocyanea* and *Ustilago zeae* (Stodola *et al.*, 1967). *Candida bogoriensis*, which was isolated from the leaf surface of the shrub *Randia malleifera* grown in Indonesia (Deinema, 1961), produces two extracellular lipids, one as a liquid and the other a crystalline solid at ambient temperatures. The crystalline solid was shown by A. P. Tulloch, M. H. Deinema and J. F. T. Spencer (unpublished observations) to be made up of the disaccharide sophorose (2-0-β-glucopyranosyl-D-glucopyranose) linked glycosidically to 13-hydroxydocosanoic acid. The 6 and 6′ positions in the sophorose residue are thought to be acetylated:

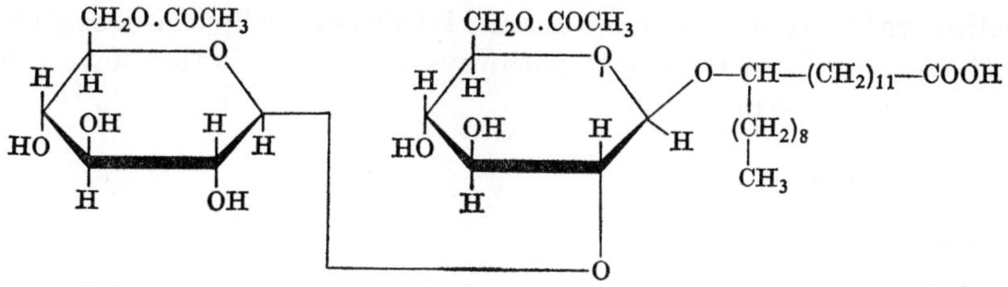

$CH_2O.COCH_3$    $CH_2O.COCH_3$

H H O OH H HO H H OH — H H O OH H HO H H — O—CH—$(CH_2)_{11}$—COOH, $(CH_2)_8$, $CH_3$ — O

As part of a research programme on osmophilic yeasts, Spencer and his colleagues examined the extracellular lipids produced by a strain of *T. apicola*. One of the lipids formed is a hydroxy fatty-acid ester of sophorose, very similar to that produced by *C. bogoriensis* except that the fatty acids are 17-L-hydroxyoctadecanoic acid and 17-L-hydroxyoctadenoic acids. A neutral glycoside, which is produced along with this acidic sophoroside, was shown to have a very unusual structure, namely a macrocyclic lactone in which the carboxyl group of the hydroxy acid is esterified with a hydroxyl group of sophorose. The formation of sophorosides by *T. apicola* is unusual in that the yield of lipid, and the nature of the fatty-acid portion, can be varied to a considerable extent by altering the nature of the substrate.

*d. Substituted acids.* Stodola and his colleagues (1965) found that an unidentified yeast (NRRL YB-250) isolated from frass produced extracellularly a lipid which, by analysis of the saponification products, was shown to be erythro-8,9,13-triacetoxydocosanoic acid:

$$CH_3{-}(CH_2)_8{-}\underset{}{\overset{O.CO.CH_3}{\overset{|}{CH}}}{-}CH_2{-}CH_2{-}CH_2{-}\overset{O.CO.CH_3}{\overset{|}{CH}}{-}\overset{O.CO.CH_3}{\overset{|}{CH}}{-}(CH_2)_6COOH$$

The same organism produces smaller amounts of erythro-8,9,-dihydroxy-13-oxodocosanoic acid.

A species of *Rhodotorula* (strain 62-506), which was isolated from the stomach of a trout, was found by A. P. Tulloch and J. F. T. Spencer (unpublished observations) to produce 8,9,13-trihydroxydocosanoic acid, which was partially acetylated and also partially esterified with long-chain acids. The free trihydroxy acid was shown to be identical with that described by Stodola *et al.* (1965).

## III. Lipid Metabolism

The principal pathways involved in lipid metabolism by living organisms are now relatively well understood. The contributions to the text edited by Greenberg (1968) should be perused for general information on these pathways. However, data specifically pertaining to yeasts are limited. Therefore, in the following account, a general outline of the pathways involved in the synthesis and breakdown of lipids will be given, and special reference made to pathways and reactions that have been shown to operate in yeasts.

### A. BIOSYNTHESES

#### 1. *Glycerides*

*a. Triacylglycerols* are synthesized by living organisms by the sequential acylation of *sn*-glycero-3-phosphoric acid. Figure 3 shows the main reactions involved in the synthesis of glycerides from $C_2$ and $C_3$ precursors.

At the hub of these metabolic pathways lies *sn*-glycero-3-phosphoric acid. In yeasts growing in media containing mono- or oligosaccharides, *sn*-glycero-3-phosphoric acid is synthesized as an intermediate on the Embden-Meyerhof-Parnas glycolytic pathway (see Chapter 7, p. 277), and can be tapped off and used as a precursor in lipid synthesis. When yeasts are grown in media containing $C_2$ compounds, such as acetate, gluconeogenesis occurs and the reactions involved are essentially a reversal of the glycolytic pathway. Again, *sn*-glycero-3-phosphoric acid can be tapped off and used as a lipid precursor.

Some yeasts, such as *Candida* spp., can use glycerol as a carbon source. The first reaction involved in glycerol utilization is catalysed by glycerol kinase, leading to synthesis of *sn*-glycero-3-phosphoric acid. This enzyme is not normally detectable in extracts of *Sacch. cerevisiae* that has been grown in the absence of glycerol, but it is readily synthesized by yeasts grown in the presence of glycerol (Wieland and Suyter, 1957). Similar findings have been reported by Gancedo *et al.* (1968) with *C. utilis* although this yeast synthesizes some glycerol kinase even when grown in glycerol-free media. These workers concluded that synthesis of glycerol kinase is induced by glycerol rather than repressed by glucose.

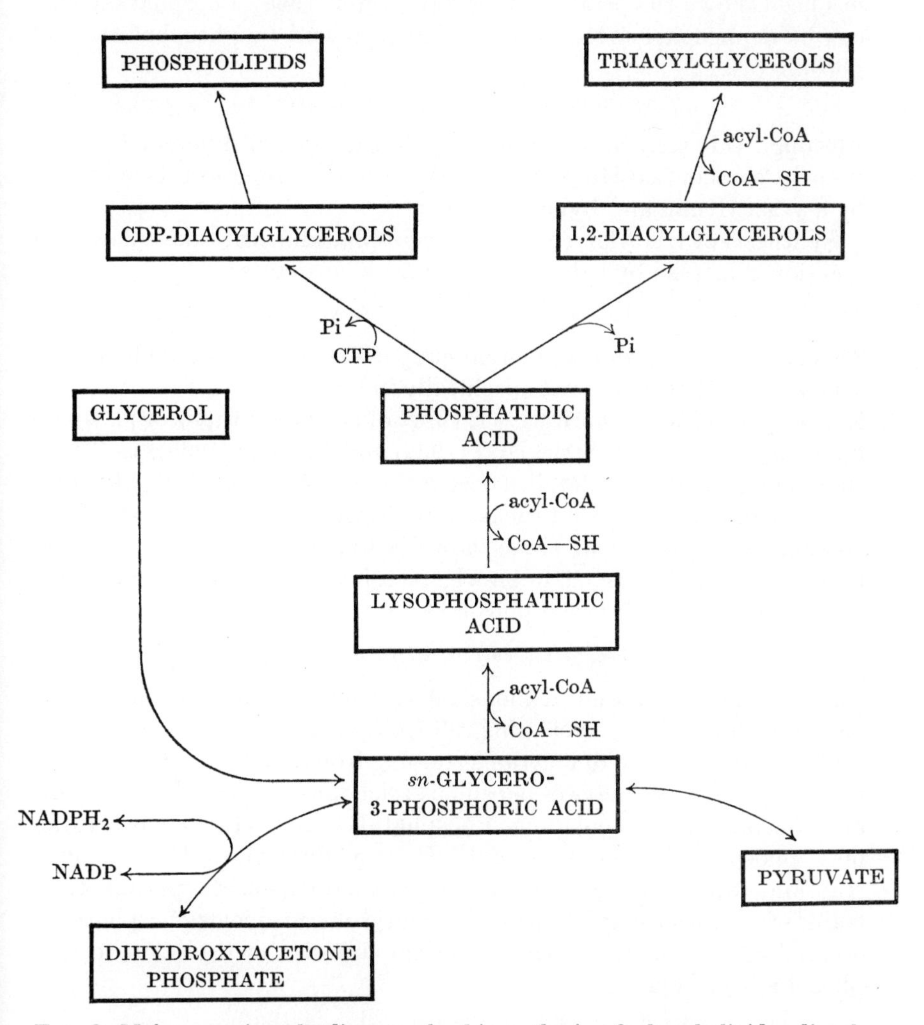


FIG. 3. Major reactions leading to the biosynthesis of phospholipids, diacylglycerols and triacylglycerols.

Acylation of *sn*-glycero-3-phosphoric acid takes place in two separate reactions, catalysed by acyl transferases, the product of the first reaction being lysophosphatidic acid:

$$\textit{sn}\text{-Glycero-3-phosphoric acid} + \text{Acyl-CoA} \rightarrow \text{Lysophosphatidic acid} + \text{CoA-SH}$$

$$\text{Lysophosphatidic acid} + \text{Acyl-CoA} \rightarrow \text{Phosphatidic acid} + \text{CoA-SH}$$

In *Sacch. cerevisiae,* these activities are associated with particles which sediment at 100,000 × *g* (Kuhn and Lynen, 1965). Alternative pathways leading to synthesis of phosphatidic acid have been reported, but none has been shown to operate in yeasts (Rossiter, 1968). Phosphatidic acid can be dephosphorylated in a reaction catalysed by phosphatidic acid phosphatase:

$$\text{Phosphatidic acid} \rightarrow \textit{sn}\text{-1,2-Diacylglycerol} + \text{Pi}$$

Although this enzyme has been detected in several different types of tissue (Coleman and Hübscher, 1962), there is no report of its synthesis by a yeast (Kuhn and Lynen, 1965).

Synthesis of triacylglycerols from phosphatidic acid takes place in a reaction catalysed by a diacylglycerol acyltransferase:

$$\textit{sn}\text{-1,2-Diacylglycerol} + \text{Acyl-CoA} \rightarrow \text{Triacylglycerol} + \text{CoA-SH}$$

This enzyme has been detected in many different tissues (Goldman and Vagelos, 1961) although not specifically in a yeast.

*b. Phospholipids.* Authoritative reviews of the biosynthesis of phospholipids have come from Strickland (1967) and Rossiter (1968), and these should be consulted for detailed accounts of these biochemical pathways. It is generally believed that the main pathways involved in phospholipid biosynthesis in yeasts are those shown in Fig. 4. The central compound in these pathways is CDP-diacylglycerol which is synthesized in a reaction involving CTP and phosphatidic acid:

$$\text{CTP} + \text{Phosphatidic acid} \rightarrow \text{CDP-Diacylglycerol} + \text{PPi}$$

The reaction has been demonstrated using extracts from several bacteria, though not as yet using cell-free extracts of yeasts.

The main sequence of reactions starting from CDP-diacylglycerol are those leading to synthesis of phosphatidylcholine *via* phosphatidylserine and phosphatidylethanolamine. Phosphatidylserine is formed by transfer of a phosphatidyl group from CDP-diacylglycerol to L-serine in a reaction catalysed by a CDP-diacylglycerol:L-serine phosphatidyl transferase. Decarboxylation of phosphatidylserine yields phosphatidylethanolamine, which can then undergo stepwise methylation to give phosphatidylcholine:

$$\begin{array}{ccc} \text{Phosphatidylethanolamine} & \rightarrow & \text{N-Methylphosphatidylethanolamine} \\ & & \downarrow \\ \text{Phosphatidylcholine} & \leftarrow & \text{N,N-Dimethylphosphatidylethanolamine} \end{array}$$

Direct evidence for the operation of these reactions in the biosynthesis of phosphatidylcholine in yeasts is slender. However, Letters (1966) detected N,N-dimethylphosphatidylethanolamine (2·4% of the total lipid phosphorus) and traces of N-methylphosphatidylethanolamine in addition to phosphatidylcholine in the lipid extract of a strain of *Sacch.*

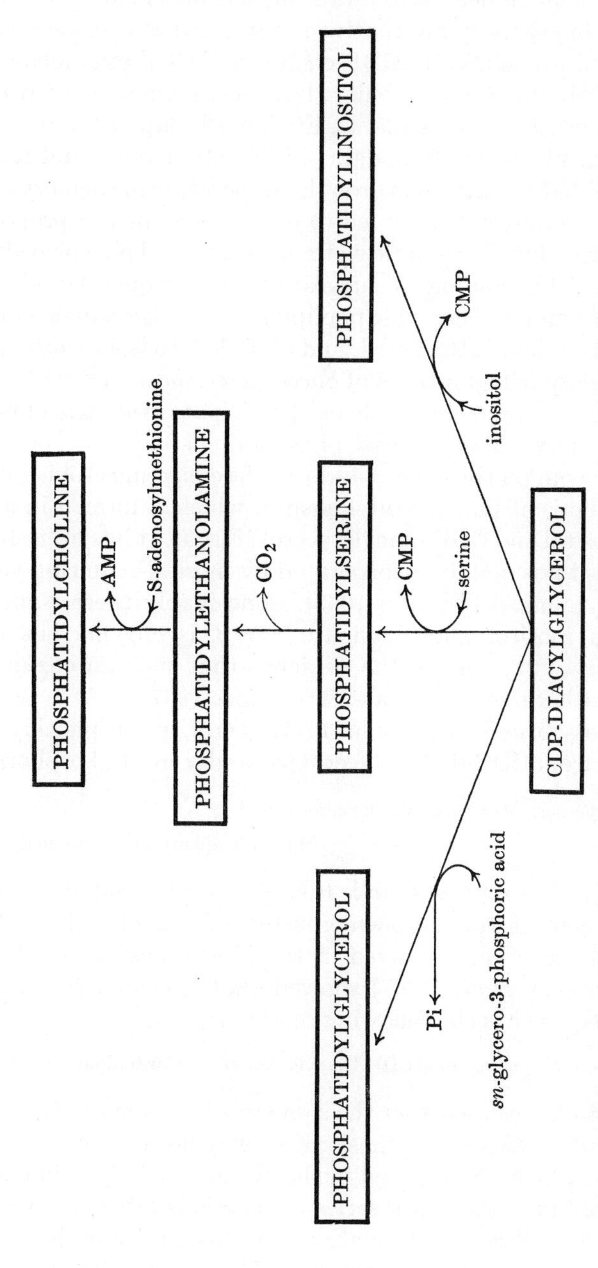


FIG. 4. Major reactions leading to the biosynthesis of the major yeast phospholipids from CDP-diacylglycerol.

*cerevisiae*. The main need is to establish that phosphatidylcholine is not synthesized in yeasts by alternative pathways that operate in mammalian tissue and which involve CDP-choline and CDP-ethanolamine. Firm evidence precluding the possibility that these mammalian pathways are operative in yeasts is not available. Evidence in support of the pathways shown in Fig. 4 has come from data obtained by Hunter and Rose (1971) who showed that variations in growth temperature in chemostat cultures of a strain of *Sacch. cerevisiae* led to fluctuations in the proportions of phosphatidylserine, phosphatidylethanolamine and phosphatidylcholine synthesized. This finding is suggestive of a sequential biosynthetic relationship among these phospholipids. It is also worth noting that several workers have attempted, and all failed, to isolate ethanolamine- and choline-requiring mutants of *Sacch. cerevisiae* a fact which suggests that ethanolamine and choline do not participate in the free or nucleotide form in the biosynthesis of yeast phospholipids.

It would seem on the other hand that free *myo*-inositol is involved in the biosynthesis of phosphatidylinositol, which is formed in a reaction between inositol and CDP-diacylglycerol (Fig. 4). It has been shown that labelled inositol is mainly incorporated by inositol-requiring yeasts into phosphatidylinositol (Tanner, 1968). Conceivably there is an involvement of an inositol phosphokinase, but this enzyme has not been detected in yeasts; indeed, the evidence that such an enzyme is synthesized in other tissues is weak (Strickland, 1967).

In the biosynthesis of phosphatidylglycerol, a phosphatidyl group is transferred from CDP-diacylglycerol to *sn*-glycero-3-phosphoric acid:

$$\text{CDP-Diacylglycerol} + \textit{sn}\text{-Glycero-3-phosphoric acid} \rightarrow \text{Phosphatidylglycerophosphate} + \text{CMP}$$

Phosphatidylglycerophosphate is then dephosphorylated to yield phosphatidylglycerol. Little has been reported, even in organisms other than yeast, of the reactions involved in the biosynthesis of cardiolipin. In *E. coli*, Stanacev *et al.* (1967) showed that the final step in the biosynthetic sequence is the following reaction:

$$\text{Phosphatidylglycerol} + \text{CDP-Diacylglycerol} \rightarrow \text{Cardiolipin} + \text{CMP}$$

It remains to be seen whether the same reaction is operative in yeasts.

*c. Fatty acid synthesis.* Synthesis of coenzyme-A esters of long-chain fatty acids, which are required in the synthesis of glycerides and other yeast lipids, takes place in a series of reactions often referred to as the *malonyl-CoA pathway*. Information on this pathway has been well reviewed by Green and Allmann (1968). The pathway involves repetitive cycles in which $C_2$ units, provided by malonyl-CoA, are added on to a primer acetyl-CoA molecule. The pathway has been intensively studied in several mammalian and plant tissues and in a few micro-organisms. Fortunately, the micro-organisms include *Sacch. cerevisiae* due largely to the efforts of Lynen and his colleagues. Fatty-acid synthesis in yeast has been lucidly reviewed by Lynen (1967) in his Third Jubilee Lecture to the Biochemical Society.

The reactions leading to fatty-acid synthesis in yeast, which include a priming reaction, chain-lengthening reactions, and a terminal reaction (Fig. 5), can be described by the overall equation ($n = 7$–$8$):

$$\text{Acetyl-CoA} + n\text{Malonyl-CoA} + 2n\text{NADPH}_2 \rightarrow$$
$$CH_3.[CH_2.CH_2.]_n.CO.CoA + nCO_2 + n\text{CoA-SH} + 2n\text{NADP} + nH_2O$$

Although the fatty-acid synthesizing system in *E. coli* can be separated into several subunits, repeated attempts to effect a similar fractionation with the yeast system have been unsuccessful. In this respect, the yeast system resembles avian and mammalian systems (Lynen, 1967). Some very elegant experimentation on the yeast system led Lynen (1967) to postulate that it consists of a multi-enzyme system made up of seven distinct enzyme units. Figure 6 shows Lynen's conception of the multi-enzyme complex. The complex has an estimated molecular weight of 2·3 million, and evidence for its existence has come from negatively stained electron micrographs (Lynen, 1967).

The first, priming, reaction (Fig. 5) is catalysed by the biotin enzyme acetyl-CoA carboxylase. The involvement of a biotin coenzyme in this reaction explains earlier reports of the biotin-sparing action of certain fatty acids for *Sacch. cerevisiae* (Ahmad and Rose, 1962; Rose, 1963). In aerobically grown *Sacch. cerevisiae*, acetyl-CoA synthetase, which provides the acetyl-CoA for the priming reaction, is associated with the microsomal fraction, but in aerobically grown cells the enzyme is detectable on the mitochondria (Klein and Jahnke, 1968). The priming reaction involves the transfer of an acetyl-CoA to a peripheral thiol group on the complex (Fig. 6). The priming reaction is followed by the transfer of a malonyl residue from malonyl-CoA to the central thiol group on the complex, which in turn is followed by a condensation between the enzyme-bound acetyl and malonyl groups to give aceto-acetyl-enzyme and liberation of carbon dioxide (Fig. 5). The stepwise conversion of the $\alpha$-oxo acid into the saturated acid is effected by a $NADPH_2$-dependent reduction to give D(-)$\alpha$-hydroxybutyryl-enzyme, followed by dehydration to crotonyl-enzyme and another $NADPH_2$-linked reduction to give the saturated butyryl-enzyme. In this second reduction, FMN serves as a hydrogen carrier. All of the acyl residues involved in these reactions remain bound to the central thiol group on the complex. But when the saturated acid has been formed, the butyryl group is transferred to the peripheral thiol group, so freeing the central

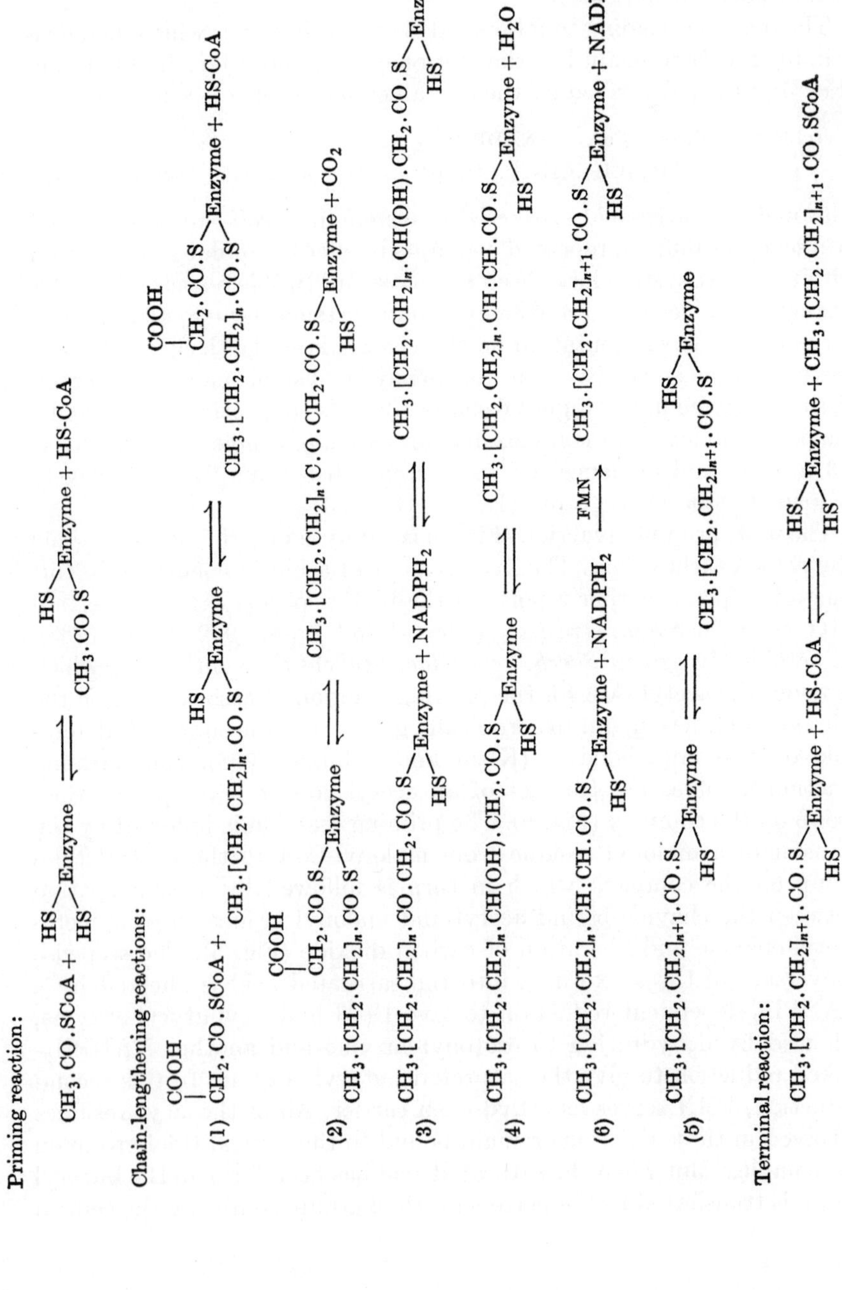

Priming reaction:

$$CH_3.CO.SCoA + \begin{matrix}HS\\HS\end{matrix}\!>\!Enzyme \rightleftharpoons \begin{matrix}HS\\CH_3.CO.S\end{matrix}\!>\!Enzyme + HS\text{-}CoA$$

Chain-lengthening reactions:

$$(1)\quad \begin{matrix}COOH\\|\\CH_2.CO.SCoA\end{matrix} + \begin{matrix}HS\\CH_3.[CH_2.CH_2]_n.CO.S\end{matrix}\!>\!Enzyme \rightleftharpoons \begin{matrix}COOH\\|\\CH_2.CO.S\\CH_3.[CH_2.CH_2]_n.CO.S\end{matrix}\!>\!Enzyme + HS\text{-}CoA$$

$$(2)\quad \begin{matrix}COOH\\|\\CH_2.CO.S\\CH_3.[CH_2.CH_2]_n.CO.S\end{matrix}\!>\!Enzyme \rightleftharpoons \begin{matrix}CH_3.[CH_2.CH_2]_n.C.O.CH_2.CO.S\\HS\end{matrix}\!>\!Enzyme + CO_2$$

$$(3)\quad \begin{matrix}CH_3.[CH_2.CH_2]_n.CO.CH_2.CO.S\\HS\end{matrix}\!>\!Enzyme + NADPH_2 \rightleftharpoons \begin{matrix}CH_3.[CH_2.CH_2]_n.CH(OH).CH_2.CO.S\\HS\end{matrix}\!>\!Enzyme + NADP$$

$$(4)\quad \begin{matrix}CH_3.[CH_2.CH_2]_n.CH(OH).CH_2.CO.S\\HS\end{matrix}\!>\!Enzyme \rightleftharpoons \begin{matrix}CH_3.[CH_2.CH_2]_n.CH{:}CH.CO.S\\HS\end{matrix}\!>\!Enzyme + H_2O$$

$$(6)\quad \begin{matrix}CH_3.[CH_2.CH_2]_n.CH{:}CH.CO.S\\HS\end{matrix}\!>\!Enzyme + NADPH_2 \xrightarrow{FMN} \begin{matrix}CH_3.[CH_2.CH_2]_{n+1}.CO.S\\HS\end{matrix}\!>\!Enzyme + NADP$$

$$(5)\quad \begin{matrix}CH_3.[CH_2.CH_2]_{n+1}.CO.S\\HS\end{matrix}\!>\!Enzyme \rightleftharpoons \begin{matrix}HS\\CH_3.[CH_2.CH_2]_{n+1}.CO.S\end{matrix}\!>\!Enzyme$$

Terminal reaction:

$$\begin{matrix}CH_3.[CH_2.CH_2]_{n+1}.CO.S\\HS\end{matrix}\!>\!Enzyme + HS\text{-}CoA \rightleftharpoons \begin{matrix}HS\\HS\end{matrix}\!>\!Enzyme + CH_3.[CH_2.CH_2]_{n+1}.CO.SCoA$$

FIG. 5. Reactions involved in the biosynthesis of coenzyme-A esters of saturated fatty acids.

thiol group for introduction of the next malonyl residue. Then, the reaction cycle starts again with butyryl-malonyl-enzyme, and is repeated until long-chain saturated fatty acids are formed.

The carrier of the central thiol group on the yeast fatty-acid synthetase complex is 4′-phosphopantetheine which is linked through a phosphodiester linkage with the hydroxyl group in a serine residue in the protein (Wells *et al.*, 1966, 1967). This type of linkage is also found in the bacterial system in which the carrier has been termed *acyl carrier protein* (Majerus *et al.*, 1965; Pugh and Wakil, 1965).

The bulk of the fatty acids in yeast lipids contain either 16 or 18 carbon atoms (see p. 217). An intriguing problem in the biosynthesis of

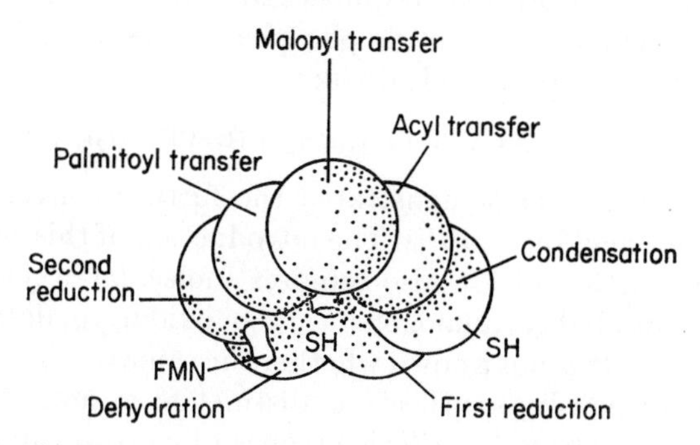


FIG. 6. A hypothetical structure for the multi-enzyme fatty acid synthetase in *Saccharomyces cerevisiae*. The seven enzyme units shown refer to the seven reactions listed in FIG. 5. From Lynen (1967).

yeast fatty acids is the mechanism which regulates chain length and which allows synthesis of only very small proportions of acids with fewer than 16 and more than 18 carbon atoms. One possible explanation is that this limitation on chain length arises from the specificity of the enzyme which catalyses termination of the process by transferring the fatty-acid residue from the enzyme to coenzyme-A (Fig. 5). This theory was discounted when Schweizer (1963) found that saturated acids ranging in chain length from $C_6$ to $C_{20}$ were transferred at about the same rate by the yeast complex. Lynen (1967) has suggested that the environment of the peripheral thiol group in the yeast complex is hydrophilic whereas that of the central thiol group is more hydrophobic. With increasing chain length, the tendency for moving from the central to the peripheral group would then gradually decrease. Conceivably, in the yeast complex, the tendency to move ceases when fatty acids with 16 or 18 carbon atoms have been synthesized.

The pathways described so far lead to synthesis of coenzyme-A esters of saturated fatty acids. Synthesis of unsaturated acids in yeasts, which are principally $C_{16:1}$, $C_{18:1}$ and $C_{18:2}$, involves the introduction of one or two double bonds into stearate or palmitate. This process differs from that which operates in many bacteria in which a double bond is introduced at the $C_{10}$ stage during the chain-lengthening process (Green and Allmann, 1968).

The enzymology of the reactions involved in the desaturation of yeast fatty acids was unravelled by Bloch and his colleagues. Using a cell-free system from *Sacch. cerevisiae*, Bloomfield and Bloch (1958, 1960) described the formation of palmitoleic and oleic acids from their saturated counterparts. For introduction of a double bond into the 9,10 position in the chain, the enzyme acts on the coenzyme-A ester of the saturated fatty acid, and requires molecular oxygen and reduced nicotinamide adenine dinucleotide phosphate. The overall equation for the conversion of stearate to oleate is:

$$CH_3, (CH_2)_{16}.CO.S.CoA \rightarrow CH_3.(CH_2)_5CH{=}CH(CH_2)_9.CO.S.CoA$$

Yuan and Bloch (1961) demonstrated the further conversion of oleic acid into linoleic acid by *C. utilis*. The introduction of this second double bond into the 12,13 position also requires molecular oxygen, and the mechanism is probably analogous to the oxidative desaturation of stearate to oleate. It is not known whether the same or different enzymes are involved. Curiously, synthesis as distinct from activity of the desaturating enzymes from *C. utilis* is favoured by growth under partially anaerobic conditions (Meyer and Bloch, 1963b).

*d. Regulation of glyceride synthesis.* The regulatory mechanisms which operate during the synthesis of glycerides have not yet been studied to the same extent as those that are involved in the synthesis of other types of cell component, a reflection no doubt of the lack of detailed information on the enzymes involved in glyceride synthesis and of the problems encountered in dealing with the enzymology of water-insoluble substrates. However, several workers have reported on compounds that regulate fatty acid synthesis in animal cells, and these studies have been extended to *Sacch. cerevisiae* principally by Klein and his colleagues. Using the cell-free lipid-synthesizing system first described by Klein (1957), these workers have shown that *sn*-glycero-3-phosphoric acid and citrate stimulate fatty acid synthesis (White and Klein, 1965, 1966). *sn*-Glycero-3-phosphoric acid is most effective at a concentration of 2 mM, and stimulates synthesis of both saturated and unsaturated fatty acids. Palmityl-CoA, on the other hand, inhibits synthesis of fatty acids in this system, the effect being greater on synthesis of unsaturated than on saturated acids (White and Klein, 1966). The data suggest that both *sn*-glycero-3-phosphoric acid and palmityl-CoA act at the site of carboxylation of acetyl-CoA (Rasmussen and Klein, 1968).

Since the bulk of yeast lipids are usually thought to be in cellular membranes, it is obvious that there must exist mechanisms for controlling the relative syntheses of triacylglycerols, phospholipids, and other membrane components in order that the cell can produce biologically active membranes. Only the faintest possible clues are as yet available as to the mechanisms involved in membrane synthesis. A tight coupling between the synthesis of phospholipids and desaturation of fatty acids has been observed in *Chlorella vulgaris* (Gurr *et al.*, 1968), but similar studies have not so far been made on yeasts. The effect of inositol deficiency in causing increased triacylglycerol synthesis in *Sacch. cerevisiae* (Challinor and Daniels, 1955) and *Sacch. carlsbergensis* (Shafai and Lewin, 1968) suggest that inositol, or phosphatidylinositol, may act as an effector in the biosynthesis of triacylglycerols in these yeasts. This possibility merits investigation.

2. *Sterols*

Several reviews have been published on sterol biosynthesis (e.g. Staple, 1967; Danielsson and Tchen, 1968) and these should be perused by the reader who requires more detailed information than the space limitations of this review allow. Fortunately, some of the pioneer work on sterol biosynthesis was done with strains of *Sacch. cerevisiae*, so that the data available are reasonably complete. The biosynthetic pathways leading to sterol biosynthesis can conveniently be divided into three parts, namely synthesis of mevalonate, conversion of mevalonate into squalene, and cyclization of squalene to give yeast sterols.

*a. Synthesis of mevalonate.* One of the main achievements in the early studies on sterol biosynthesis was the finding that the $C_6$ compound, mevalonic acid, is a key intermediate on the pathway. The reactions involved in mevalonate biosynthesis are shown in Fig. 7. Most of the enzymes which catalyse the reactions have been shown to be present in extracts of *Sacch. cerevisiae.* Sonderhoff and Thomas (1937), in one of the first studies to employ radioactive-labelled compounds, found that deuterium-labelled acetate was incorporated extensively into the sterol-rich unsaponifiable fraction of the lipids from the yeast. Later, brewer's solubles were shown to contain a compound which could replace acetate in sterol biosynthesis, and this compound was shown to be mevalonate (Tavormina *et al.*, 1956). Synthesis of hydroxymethyl-glutaryl-CoA from acetoacetyl-CoA and acetyl-CoA by extracts of *Sacch. cerevisiae* was described by Rudney (1956). Ferguson *et al.* (1959) then established that extracts of *Sacch. cerevisiae* also contain an enzyme that catalyses the conversion of hydroxymethylglutaryl-CoA to mevalonate. Enzymes which catalyse reactions leading to mevalonate biosynthesis in animal cells are microsomally located; it is not known whether they are similarly located in yeasts.

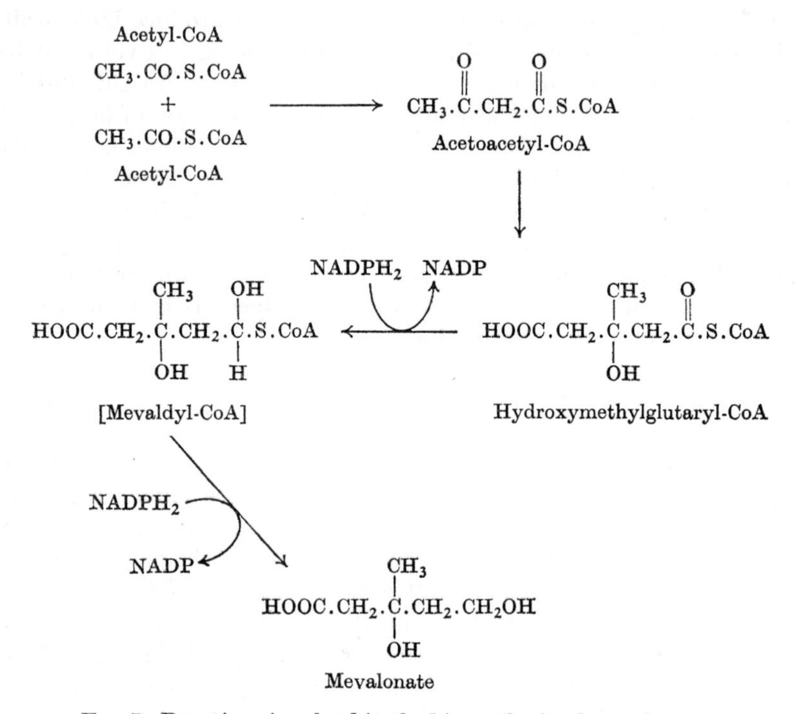


FIG. 7. Reactions involved in the biosynthesis of mevalonate.

*b. Synthesis of squalene from mevalonate.* The first steps in the incorporation of mevalonate into squalene involve the formation of the isoprenoid unit in the form of isopentenyl pyrophosphate. Mevalonic acid is converted to mevalonic acid 5-phosphate in a reaction catalysed by a kinase, and subsequently to mevalonic acid 5-pyrophosphate. This last compound then undergoes a concerted dehydration and decarboxylation to give isopentenyl pyrophosphate, which in turn is reversibly converted into dimethylallyl pyrophosphate. Formation of squalene occurs by successive condensation of $C_5$ units to give the $C_{30}$ hydrocarbon. Isopentenyl pyrophosphate condenses with dimethylallyl pyrophosphate to give the $C_{10}$ monoterpene *trans*-geranyl pyrophosphate, which then condenses with an additional molecule of isopentenyl pyrophosphate to yield the $C_{15}$ sesquiterpene *trans,trans*farnesyl pyrophosphate. Some farnesyl pyrophosphate is isomerized to nerolidol pyrophosphate. Then farnesyl pyrophosphate and probably nerolidol pyrophosphate condense to give dehydrosqualene, which is reduced to squalene in a reaction catalysed by a dehydrogenase. Further details on these reactions can be found in the general reviews of Staple (1967) and Danielsson and Tchen (1968).

Several of the enzymes which catalyse reactions involved in the formation of squalene from mevalonate have been detected in yeasts, mainly in extracts of *Sacch. cerevisiae.* It is likely that some of the intermediates remain bound to enzymes in yeasts since they cannot be detected in the free form in reaction mixtures. Sofer and Rilling (1969) have reported evidence which suggests that nerolidol pyrophosphate may remain bound to an enzyme in yeast. Also, Krishna *et al.* (1964) showed that a $C_{30}$ enzyme-bound intermediate, presumably dehydro-squalene, is formed from farnesyl pyrophosphate in extracts of *Sacch. cerevisiae,* and is converted to squalene on incubation with reduced NADP.

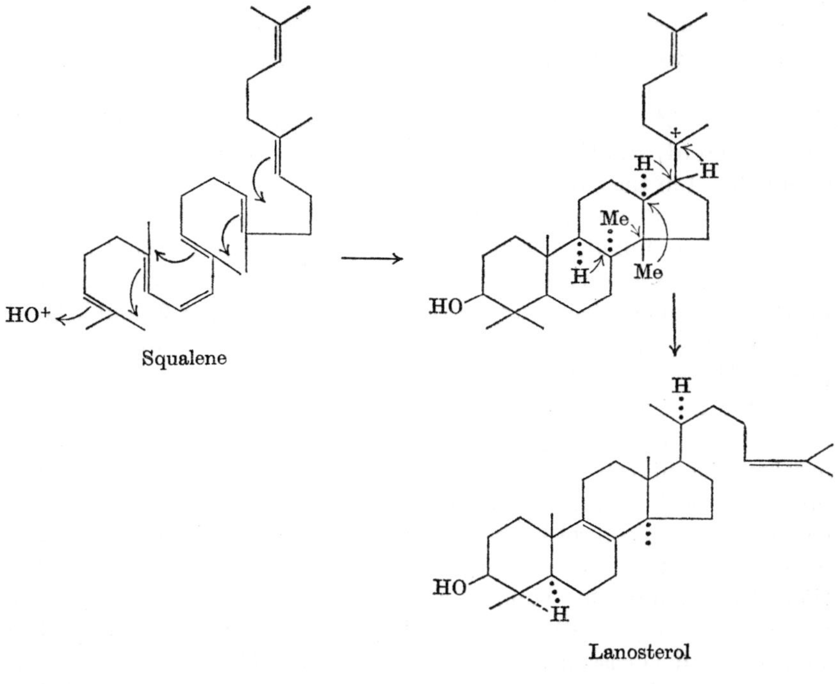

HO+

Squalene

HO

H

H

Me

H

Me

H

HO

H

Lanosterol

*c. Synthesis of yeast sterols from squalene.* Lanosterol is the first sterol formed by cyclization of the long-chain hydrocarbon squalene. This reaction was predicted on purely chemical grounds, and the first indication that squalene is cyclized to lanosterol in *Sacch. cerevisiae* came from the data of Kodicek and Ashby (1954). These workers grew their yeast under anaerobic conditions to produce sterol-deficient cells (see p. 229), and then incubated these cells aerobically with $^{14}C$-acetate. The first sterol to incorporate the label was lanosterol, whence the label was transferred to ergosterol and zymosterol which were labelled to about the same extent after 24-h incubation. Lanosterol is detectable in small amounts in lipid extracts of many yeasts, but whether it is a true membrane component or solely a precursor of ergosterol and zymosterol is unknown.

The mechanism of the conversion of squalene into lanosterol has been studied in a number of laboratories. Early experiments indicated that the conversion requires molecular oxygen, $NADPH_2$ and iron, and that a number of enzymes are involved. Recent studies on the cyclization mechanism have been done mainly with microsomal preparations from animal cells (Sih and Whitlock, 1968). These studies have furnished evidence for the role of 2,3-oxidosqualene as an intermediate, and the presumed hydroxylation at C-3 by a mixed-function oxidase.

The principal molecular modifications needed in the conversion of lanosterol into ergosterol (Fig. 2, p. 226) are: (a) loss of two methyl groups at C-4, and one at C-14; (b) isomerization of the double bond in ring B from $\Delta^8$ to $\Delta^7$; (c) introduction of a second (5,6) double bond in ring B; and (d) addition of a methyl group at C-24 and migration of the double bond in the side chain.

The carbon atoms in the methyl groups at C-4 and C-14 are probably lost as carbon dioxide. Olson *et al.* (1957) have suggested that the 14$\alpha$-methyl group is first oxidized to a carboxyl group and that the activation for the subsequent decarboxylation is provided by the 8,9-double bond. The sequence suggested by Olson and his colleagues (1957) was (I) → (II) → (III) → (IV) as shown in Scheme 1.

Akhtar and his coworkers (1969a) have proposed an alternative mechanism for removal of the C-14 methyl group in lanosterol, which involves the intermediate formation of a 8,14 diene. In support of their proposed mechanism, they showed that 5$\alpha$-ergosta-8,14-dien-3$\beta$-ol is converted to ergosterol by intact cells of *Sacch. cerevisiae*. Thus, yeast would appear to possess an enzymic system that can catalyse the reduction of a 14,15-double bond in a 8,14-diene, which suggests that the route may be (I) → (V) → (VI) → (IV) in Scheme I. Data reported by Ponsinet and Ourisson (1965) suggest that the C-14 methyl group of lanosterol may be lost before the two methyl groups on C-4. This evidence came from the isolation from *Sacch. cerevisiae* of 4,4-dimethylzymosterol and 4-methylzymosterol. Isomerisation of the $\Delta^8$-double bond of lanosterol to the $\Delta^7$-position is suggested from other data reported by the Southampton group. They (Akhtar *et al.*, 1970) synthesized [7$\alpha$-$^3H$, 26 27-$^{14}C_2$] lanosterol, and showed its conversion by intact yeast

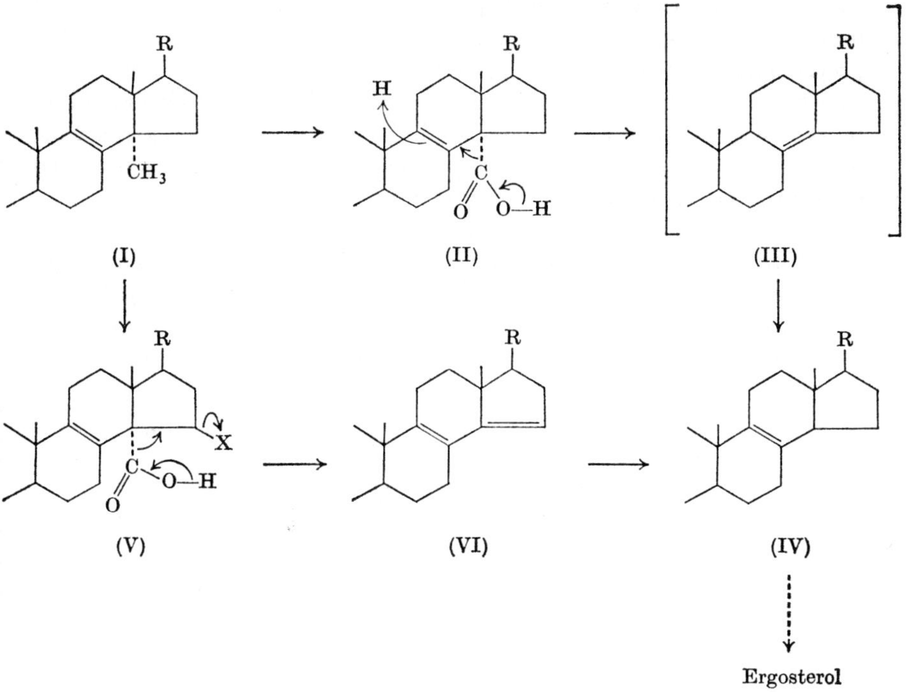

(I) (II) (III)

(V) (VI) (IV)

Ergosterol

SCHEME 1.

cells to ergosterol with the loss of the 7$\alpha$-hydrogen atom. Interestingly, when the doubly labelled lanosterol was incubated with a rat-liver homogenate, cholesterol was formed with removal predominantly of the 7$\beta$-hydrogen atom. It would seem that C-7 hydrogen atoms with opposite stereochemistry are labilized by the rat-liver and yeast $\Delta^8$-$\Delta^7$ sterol isomerases.

Introduction of the second (5,6)-double bond into ring B of lanosterol is thought to proceed by a hydroxylation-dehydration mechanism. Topham and Gaylor (1967) showed that the postulated hydroxylated intermediate, ergosta-7, 22-diene-3$\beta$, 5$\alpha$-diol, was converted to ergosterol under anaerobic conditions by a cell-free extract of *Sacch. cerevisiae.* These workers (Topham and Gaylor, 1970) have now isolated an enzyme from *Sacch. cerevisiae* which catalyses the conversion of the hydroxylated intermediate into ergosterol. They have reiterated their belief that hydroxysterols may be intermediates in the formation of the $\Delta^5$-bond in ergosterol, and they point out that this would account for the requirements for molecular oxygen and reduced nicotinamide nucleotide. An alternative mechanism for the introduction of a 5,6 double bond has come from Akhtar and Parvez (1968). They were unable to demonstrate conversion of the diol used by Topham and Gaylor (1967) into ergosterol under anaerobic conditions, although they discovered that the conversion proceeded in the presence of oxygen. Using dihydro [$5\alpha,6\alpha^{3}H_2$]ergosterol, they suggested that the introduction of the 5,6 double bond in ergosterol biosynthesis involves an overall *cis*-elimination of two hydrogen atoms, which indicates that an oxygen-dependent dehydrogenation is involved rather than a hydroxylation-dehydration mechanism.

The most extensive data pertaining to the conversion of lanosterol into ergosterol by yeasts are concerned with modifications to the side chain. Parks (1958) first suggested that the additional methyl group at C-24 is furnished by S-adenosylmethionine. Later, Lederer (1964), using mass spectrometry, found that the step was not a simple methylation for,

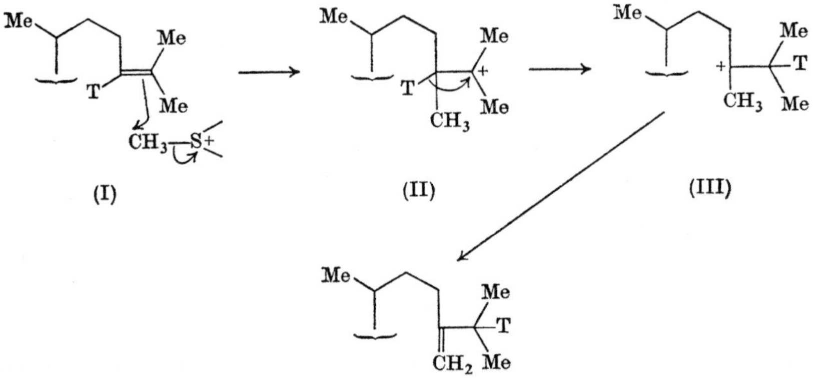

(I) (II) (III) (IV)

SCHEME 2. Me indicates a methyl group, and T a tritium atom.

when ergosterol was biosynthesized in the presence of [$CH_3$-$^{2}H_3$]-methionine, only two of the deuterium atoms from the methyl group were incorporated into ergosterol. A clue to the mechanism of the alkylation at C-24 came from the discovery by Akhtar *et al.* (1967) that, in the biosynthesis of ergosterol from labelled lanosterol, the hydrogen atom at C-24 was retained, while degradation of the labelled product indicated that the alkylation step was accompanied by migration of a hydrogen atom from C-24 to C-25. This led to the formulation of Scheme 2 as a proposed mechanism for the first stages of the alkylation. Scheme 2 predicts that a 24-methylene-sterol is an intermediate. Support for this contention has come from two sources. Firstly, Barton *et al.* (1968) have isolated from *Sacch. cerevisiae* a sterol with the side-chain structure of (IV) in Scheme 2. Secondly, Akhtar and his coworkers (1969b) synthesized 24-methylene-[23,25-$^{3}H_3$]dihydrolanosterol and showed that this compound was converted into ergosterol in good yield by intact *Sacch. cerevisiae.*

Akhtar's group have also addressed themselves to the mechanism whereby the double bond between C-22 and C-23 arises in the ergosterol side chain. They (Akhtar *et al.*, 1968, 1969b) suggested three possible mechanisms for the conversion which are designated A, B and C in Scheme

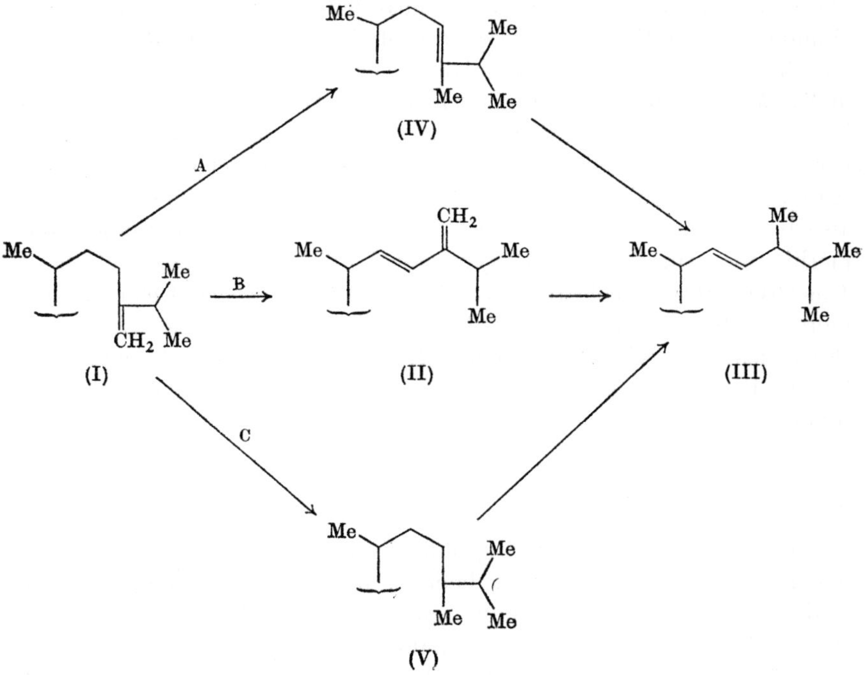

SCHEME 3. Me indicates a methyl group.

3. Mechanism A involves a series of 1,2 shifts of ethylenic linkages. Mechanism B is a two-step process with the formation of a double bond between C-22 and C-23 followed by reduction of a double bond between C-24 and C-28. Mechanism C is essentially the reverse of the oxidation-reduction sequence. Evidence in favour of mechanism B has come from a report by Katsuki and Bloch (1967) that $\Delta^{5,7,22,24(28)}$-ergostatetraene-3$\beta$-ol was converted into ergosterol by cell-free extracts of *Sacch. cerevisiae.* The enzymes which catalyse reactions leading to formation of the C-22-C-23 double bond apparently do not require the activation of a neighbouring double bond (Akhtar *et al.*, 1969b).

Very little indeed is known about the biosynthesis of zymosterol by yeasts. Katsuki and Bloch (1967) demonstrated the formation of zymosterol from mevalonate and L-methionine by a cell-free extract of *Sacch. cerevisiae*. However, convincing evidence that zymosterol is indeed a percursor of ergosterol, and the individual reactions involved in zymosterol biosynthesis, have yet to be described.

*d. Regulation of sterol biosynthesis.* As yet, there are only preliminary reports on the mechanisms that operate during the control of sterol biosynthesis in yeasts. An early report that reduction of hydroxymethylglutaryl-CoA to mevalonate was controlled in animals by bile acids (Fimognari and Rodwell, 1965) prompted Kawaguchi *et al.* (1968) to search for a similar type of control on the pathway to ergosterol synthesis in *Sacch. cerevisiae*. They found that the activity of the enzyme was inhibited by certain acidic derivatives of ergosterol which can be isolated from the yeast; these derivatives have not yet been identified. A regulatory role for transmethylase in sterol biosynthesis in yeast has also been proposed. Moore and Gaylor (1969) found that S-adenosylmethionine stimulated, and zymosterol inhibited, demethylation of 4-methylsterols by cell-free extracts of *Sacch. cerevisiae*. Clearly, until more has been discovered about the individual reactions involved in sterol biosynthesis in yeast, particularly in the later stages of the process, few workers will be tempted to research on the regulatory mechanisms involved.

### 3. *Polyprenols*

Nothing has been reported specifically on the biosynthesis of yeast dolichols. It is reasonable to believe, however, that the pathways leading to synthesis of yeast polyprenols are similar to those involved in the synthesis of squalene (p. 244), namely the condensation of isoprene units. Many polyprenols, including the yeast dolichols, contain different numbers of *cis* and *trans* residues, and one obvious question is whether these residues are formed from isopentenyl pyrophosphate or whether they isomerize later (Hemming, 1970). The availability of [4R, $4^3H_1$]— and [4S, $4^3H_1$] mevalonic acids has enabled a study to be made of the problem, not specifically with yeast dolichols but with rat-liver dolichols which the yeast polyprenols resemble. The incorporation data are consistent with: (a) chemically *trans* residues being biogenetically *trans*; (b) chemically *cis* residues being biogenetically *cis*; (c) the $\omega$-residue being biogenetically *trans*; (d) the saturated residue being biogenetically *cis*. Thus, with rat-liver dolichols, the stereochemistry of the prenols is decided when they are synthesized and no isomerization occurs. Dolichols are probably formed by *cis* addition to *trans-trans* farnesyl pyrophosphate followed by saturation of the $\alpha$-residue. Yeast dolichols may function as phosphates in the incorporation of sugar residues into yeast cell-wall polysaccharides (see p. 161). Polyprenol kinases have been detected in *Micrococcus lysodeikticus* (Allen *et al.*, 1967) and in *Staphylococcus aureus* (Higashi *et al.*, 1968) but not as yet in yeast.

### 4. *Hydrocarbons*

In view of the commercial importance of aliphatic hydrocarbons in Man's economy, it is surprising how ignorant we are of the biochemical pathways which lead to the biosynthesis of these compounds. The first significant contribution came in the report from Sandeman and Schwers (1960) who demonstrated incorporation of $^{14}$C-acetate into *n*-heptane produced by *Pinus jeffreyi*, apparently through a condensation of four acetate units followed by decarboxylation. Later studies, also with plant tissues (*Brassica oleracea*), indicated that long-chain hydrocarbons are formed from fatty acids after decarboxylation (Kolattukudy, 1966). Data reported by Tornabene and Oro (1967) using *Sarcina lutea* support the notion that bacterial aliphatic hydrocarbons are also produced from long-chain fatty acids, as do those reported by Albo and Dittmer (1969a–d). Nothing has been reported on the origin of the aliphatic hydrocarbons synthesized by yeasts, but the finding that the chain lengths of yeast hydrocarbons are similar to the principal yeast fatty acids (Baraud *et al.*, 1967) lends support to the idea that the biosynthetic origin of hydrocarbons is the same in these organisms.

### 5. *Other Lipids*

Very little has been reported on the biosynthesis of the minor components of yeast cell lipids or of the extracellular lipids synthesized by yeasts (see Section II.B.2, p. 231) although rather more has been reported on the biosynthesis of similar compounds in mammalian and plant tissues.

Experiments on animal tissues have shown that sphingosines arise by condensation of serine or a serine derivative with a derivative of palmitate or palmitaldehyde (Zabin and Mead, 1953, 1954; Sprinson and Coulon, 1954). Greene and his colleagues (1965) examined extracts of *H. ciferrii* for the ability to effect this synthesis, but obtained equivocal results. In a further study on the biosynthesis of sphingolipids in *H. ciferrii*, Thorpe and Sweeley (1967) showed that the hydroxyl group on C-4 (see p. 232) does not arise from either water or molecular oxygen; the origin of this atom still remains unknown.

Yeast sphingolipids, both intracellular and extracellular, contain fatty acids of much longer chain length ($C_{24}$,$C_{26}$) than those found in the bulk of yeast cell lipids. A system capable of synthesizing these long-chain acids has been reported in *C. utilis* (Fulco, 1967).

## B. CATABOLISM

### 1. *Lipolytic Enzymes*

Although lipolytic activity has, for some years, been recognized as a characteristic property of a few yeasts (e.g. *C. lipolytica*), very little indeed has been reported on the lipolytic enzymes of yeasts.

Triacylglycerols are hydrolysed by lipases (glycerol ester hydrolases) to give glycerol and fatty acids, the three fatty-acid residues being removed in stepwise fashion. Production of extracellular lipases is confined largely to anascosporogenous yeasts, particularly to species of *Candida* (Werner, 1960). Among the active producers of extracellular lipases that have been studied are *C. lipolytica* (Vickery, 1936a, b; Peters and Nelson, 1948a, b), *C. cylindricaceae* (Yamada and Machida, 1962), *C. paralipolytica* (Ota and Yamada, 1966), *C. humicola* (Bours and Mossel, 1969) and a *Torulopsis* sp. (Motai *et al.*, 1966).

It is likely that all yeasts synthesize an intracellular lipase which is involved in the turnover of membrane triacylglycerols. Gerbach and Güntner (1932) showed that autolysed baker's yeast and brewer's yeast hydrolysed olive oil, but the only detailed report on the lipase of *Sacch. cerevisiae* has come from Nurminen and Suomalainen (1970). These workers showed that lipase activity in *Sacch. cerevisiae* is associated with the plasma membrane obtained by enzymic digestion of isolated cell envelopes. This lipase activity can be completely released from the envelopes by repeated washing with water. It would be interesting to know more about the factors which control synthesis and activity of this membrane-bound lipase.

Five main types of enzyme are known to catalyse hydrolysis of phospholipids. These are designated, as recommended by Contardi and Ercoli (1933), as phospholipases A, B, C and D; the fifth enzyme is lysophospholipase. Phospholipase A cleaves the ester linkage in the 2-position to yield a lysophospholipid, which in turn can be subject to the action of a lysophospholipase to yield a further fatty acid and the appropriate glyceryl phosphoryl derivative. Phospholipase B combines the functions of phospholipase A and lysophospholipase. Phospholipase C catalyses the hydrolysis of a phosphate ester linkage in, for example, phosphatidylcholine to yield phosphatidic acid and choline. Action of phospholipase D on phosphatidylcholine gives rise to *sn*-1,2-diacylglycerol and phosphorylcholine.

There is no report of the production of extracellular phospholipases by yeasts, although to be fair no systematic search for this activity appears to have been made.

Evidence for the involvement of a phospholipase A in the catabolism of phosphatidylcholine by *Sacch. cerevisiae* was reported by Kokke and his colleagues (1963), and was followed by a report from the same laboratory (Van den Bosch *et al.*, 1967) of a lysophospholipase which catalyses the removal of a second fatty-acid residue from lysophosphatidylcholine by extracts oi *Sacch. cerevisiae*. An especially interesting property of phospholipases is that their activity is stimulated in the presence of diethylether and certain other organic solvents (Hanahan, 1952). Dawson (1963) suggested that this stimulatory effect is the result of penetration of the molecules of organic solvent into the phospholipid micelles so causing a wider spacing of the molecules at the lipid–water interface and therefore an easier access of the enzyme to the acyl ester bond. An apparent stimulation of phospholipase A activity in *Sacch. cerevisiae* by organic solvents was reported by Letters (1968b), and this led him to recommend the use of hot aqueous ethanol as the first solvent when extracting lipids from yeast in order to inactivate the phospholipase activity.

Extracts of *Sacch. cerevisiae* were reported by Hoffmann-Ostenhof *et al.* (1961) to contain phospholipase C activity. This was later confirmed by Harrison and Trevelyan (1963) and was claimed by these workers to be responsible for the appearance of diacylglycerols in lipid extracts of yeasts.

Lipases capable of catalysing hydrolysis of the quantitatively minor lipids in yeasts (e.g. sphingolipids) are found in extracts of many different types of animal tissue. However, there has been no report of these activities in yeasts. It is conceivable that lipase activity, and in particular phospholipase activity, is important in certain differentiation processes in yeasts, such as ascospore formation, which involve alterations in membrane content and location. But as yet nothing has been reported which implicates lipase activity in any differentiation process in yeasts.

### 2. *Further Metabolism of Hydrolysis Products*

Many yeasts can oxidize glycerol, which would be formed by the action of lipases on triacylglycerols, and from the glycerol phosphoryl derivatives that result from the action of phospholipases (see p. 295). Of the other products formed by catabolism of phospholipids, it is known that some yeasts, and particularly strains of *Cryptococcus*, can use inositol as a sole source of carbon and energy; this property is used in yeast systematics. Serine from phosphatidylserine could, presumably, be incorporated into the amino-acid metabolism of a yeast. Nothing has been reported on the utilization of ethanolamine by yeasts. A recent report by Kortstee (1970) showed that, of 10 strains of *Candida*, 15 of *Cryptococcus*, 6 of *Hansenula*, 27 of *Rhodotorula* and 6 of *Saccharomyces*, none had the ability to utilize extracellular choline. This might suggest that the ability to degrade choline is not widespread among yeasts, although it may be that the ability of these strains to transport choline into the cell is the limiting factor. Nevertheless, it is known that a few yeasts not only incorporate choline but indeed are auxotrophic for this compound. *Torulopsis pintolopesii* and *T.* (=*Candida*) *bovina* were found by Cury and his colleagues (1960) to require choline; the requirement could also be met by methionine in higher concentrations. This report suggested that these yeasts may have an alternative pathway for synthesis of phosphatidylcholine which involves CDP-choline. However, Cruz and Travassos (1970) have recently shown that this is unlikely since they could not find any evidence for incorporation of labelled choline into phosphatidylcholine.

Long-chain fatty acids can be oxidized by many and probably all yeasts. However, little has been reported on the enzymes that catalyse this oxidation in yeasts, although it is presumed that the oxidation proceeds by the β-oxidation pathway (Chapter 7, p. 295). Lebeault and his colleagues (1970) have reported that *C. tropicalis* growing on *n*-tetradecane synthesizes four different types of acyl-CoA synthase, the enzyme catalysing the formation of coenzyme-A esters of long-chain fatty acids being a mitochondrial enzyme whereas the others, which use shorter chain acids as substrates, appear in the soluble fraction of the cell.

## IV. Composition, Structure and Function of Yeast Membranes

Electron micrographs of thin sections of yeasts have furnished ample evidence for the existence of an extensive system of membranes in the yeast cell. The evidence has been well reviewed by Marchant and Smith (1968) and Matile *et al.* (1969). Many workers have experienced difficulties in preparing yeasts for electron microscopy. Direct fixation with osmium tetroxide results in poor preservation (Agar and Douglas, 1957), and potassium permanganate seems on the whole to be the best preservative of membrane architecture. More recently, electron micrographs of freeze-etched specimens of yeast cells have provided even further insight into the intracellular membrane systems in yeast cells, particularly in the hands of Matile and Moor and their colleagues at the Swiss Federal Institute in Zurich (see Matile *et al.*, 1969). A valuable comment on the interpretation of electron micrographs of freeze-etched membranes of yeast has come from Bauer (1968).

### A. PLASMA MEMBRANES

The membrane which lies immediately adjacent to the cell wall in the yeast cell, and which bounds the cytoplasm, is variously referred to as the cytoplasmic membrane, cell membrane, plasmalemma, protoplast membrane, and plasma membrane. We prefer the name "plasma membrane" (Gr. πλάσμα, to form, mould) since this surface membrane is essentially the one which gives form to the cytoplasm.

### 1. *Structure*

The reviews by Marchant and Smith (1968), Matile *et al.* (1969) and Matile (1970) have dealt comprehensively with the structure of the yeast plasma membrane, and no attempt will be made in this chapter to embellish these accounts. It should be stressed, too, that as yet nothing has been reported on mixed lipid or black membranes which might apply specifically to the yeast plasma membrane or indeed in any way illuminate the subject.

### 2. *Composition*

Two main types of method have been used to prepare intact plasma membranes from yeast cells prior to chemical analysis. Boulton (1965) pioneered the use of osmotic shock of yeast sphaeroplasts, prepared by removing the cell wall with the digestive juice of *Helix pomatia* (Eddy and Williamson, 1957; Phaff, 1971). He subjected these osmotically sensitive sphaeroplasts to osmotic lysis by mixing with ice-cold 0·025 *M*-tris buffer (pH 7·2) containing m*M*-magnesium chloride. After standing for 30 min at 0°, a pellet of intact plasma membranes and debris was obtained by centrifuging at 20,000 *g* for 30 min. The material was further separated into two fractions (designated 1·5 p 5 and 20 p 30) and each fraction was subjected to chemical and enzymic analysis (Table 2). Both fractions contained a small amount of DNA and some RNA, but about 90% of the dry weight was accounted for by lipid and protein in about equal amounts. Later, Garcia Mendoza and Villanueva (1967) also used osmotic lysis of sphaeroplasts to prepare plasma membranes from *C. utilis* using substantially the same procedure as Boulton (1965). The analytical data reported by Garcia Mendoza and Villanueva (1967) were however rather few and consisted mainly of determinations of total protein and lipid contents of the membranes (Table 2).

More detailed analyses of plasma membranes obtained by osmotic lysis of yeast sphaeroplasts were obtained for a strain of *Sacch. cerevisiae* by Longley *et al.* (1968; Table 2) and more recently from our laboratory by Hunter and Rose (1971). About 50% of the dry weight of the membrane was accounted for by protein, and approximately 40% by lipid. Other components in the membrane included RNA, carbohydrate and sterols; DNA was not assayed. The carbohydrate was shown to contain glucose and mannose, and suggested that the membranes were from sphaeroplasts rather than from true protoplasts. Longley *et al.* (1968)

TABLE 2. *Composition of Plasma Membranes from Various Yeasts*

| Component | Boulton (1965) *Saccharomyces cerevisiae* | Garcia Mendoza and Villanueva (1967) *Candida utilis* | Longley *et al.* (1968) *Saccharomyces cerevisiae* | Suomalainen *et al.* (1967) *Saccharomyces cerevisiae* |
|---|---|---|---|---|
| Protein | 46–47·5 | 38·5 | 49·3 | approx. 30–40 |
| Lipid | 37·8–45·6 | 40·4 | 39·1 | approx. 30–40 |
| DNA | 0·97 | 0 | — | — |
| RNA | 6·7 | 1·1 | 7·0 | — |
| Carbohydrate | 3·2 | 5·2 | 4·0–6·0 | approx. 25 |
| Sterols (as ergosterol) | approx. 5·6 | — | 6·0 | — |

reported that the membranes which they analysed accounted for 13–20% of the dry weight of the yeast cell, but this figure is almost certainly too high; a more correct figure would probably be nearer 10%. The membrane lipid consisted of neutral lipid (mainly triacylglycerols with small amounts of di- and monoacylglycerols), phospholipids and sterols. There were about equal amounts of neutral lipids and phospholipids, and the phospholipids that were detected were phosphatidylcholine, phosphatidylethanolamine, phosphatidylinositol + phosphatidylserine, and phosphatidic acid. Data were also reported on the fatty-acid composition of the membranes, and on the distribution of fatty acids among the various types of lipid. Sterols (assayed spectrophotometrically as ergosterol equivalent) accounted for about 6% of the dry weight of the membrane. One of the most interesting features of the data reported by Longley *et al.* (1968) was the discovery that the major sterol in the plasma membrane of their yeast is the tetraethenoid sterol, $\Delta^{5,7,22,24(28)}$-ergostatetraene-3$\beta$-ol, a sterol which had only once previously been detected in yeast (Breivik *et al.*, 1954) and one which has been shown to be a biosynthetic precursor of ergosterol (Katsuki and Bloch, 1967; see p. 249). The other main sterol in the membrane was identified as zymosterol. The minor sterol components were identified only in lipid extracts from whole cells, and included ergosterol, episterol (or fecosterol) and an unidentified $C_{29}$ di-unsaturated sterol. An extension of this study (Hunter and Rose, 1971) has shown that ergosterol may be a major sterol component when the composition of the medium is slightly changed, and that a large proportion of the cellular sterols is esterified. Longley *et al.* (1968) also showed that the fatty acids in membrane lipids were principally $C_{16:1}$ and $C_{18:1}$. Hunter and Rose (1971) have made a more detailed phospholipid analysis of *Sacch. cerevisiae* NCYC 366, and have identified cardiolipin in addition to the phospholipids identified by Longley *et al.* (1968). Another finding made by Hunter and Rose (1971) is that the lipids from this yeast probably contain a small proportion of hydrocarbon, which includes squalene.

Isolated plasma membranes from *Sacch. cerevisiae* NCYC 366 have been analysed in our laboratory for characteristic mitochondrial components, which conceivably might remain attached to the plasma membrane during osmotic lysis. The analyses have concentrated on those mitochondrial components which are known to remain tightly bound to membranes, principally cytochromes and succinate dehydrogenase. The membranes have been shown to be free from detectable amounts of cytochromes and to contain barely detectable amounts of succinate dehydrogenase, and this has led us to conclude that the plasma membranes are substantially free from mitochondrial contamination (R. J. Diamond and A. H. Rose, unpublished observations).

A recent paper by Baraud and his colleagues (1970) reports a very extensive chemical analysis of plasma membranes obtained from stationary-phase *Sacch. cerevisiae* by the sphaeroplast technique. One of the most interesting features of their paper is the detection of glycolipids including galactosyldiglycerides in the plasma membranes.

Other groups of workers are of the opinion that yeast plasma membranes prepared by osmotic lysis are heterogeneous and may contain molecular species that are not components of the *in vivo* plasma membrane. This belief has led them to search for alternative methods for preparing yeast plasma membranes for analysis. Matile *et al.* (1967) homogenized cells of *Sacch. cerevisiae* ETH 1022 in a solution of 0·25 *M*-sucrose, 0·05 *M*-tris-HCl buffer, and m*M*-EDTA (pH 7.2). After removal of the Ballotini beads, the cell-free extract was submitted to differential centrifugation. After removing the mitochondrial fraction, the microsomal fraction (150,000 *g* for 30 min) was obtained and layered on a density gradient of Urografin. On centrifugation, the plasma membrane formed a distinct band at 1·165–1·170 $g/cm^2$. The band was collected, and the material washed and analysed. The reported analyses (Matile *et al.*, 1967) were not extensive, and were confined mainly to reporting on the contents of lipid, protein, polysaccharide and phospholipid phosphorus. However, some of Matile's data have since been briefly tabulated by him (Matile, 1970). Of particular interest was the finding of a large amount of carbohydrate (0·75 mg per mg protein), which was shown to contain only mannose. Treatment of the plasma membrane with detergents liberated globular particles composed of mannan-protein. An electron micrograph of a freeze-etched membrane prepared in this way, and showing the globular particles, appears on p. 238 in Volume 1.

Workers in Suomalainen's laboratory share the reservations regarding the use of osmotic lysis to prepare isolated plasma membranes from yeast. They have developed a method for preparing these membranes from *Sacch. cerevisiae* which involves digesting isolated walls with snail-gut juice (Suomalainen *et al.*, 1967a, b). To date, the analytical data reported on plasma membranes prepared by this technique have been few (Table 2, p. 256), although more extensive data have been reported on the enzymic activities of these membranes (Nurminen *et al.*, 1970).

There are reasons to believe that using snail-gut juice to obtain yeast plasma membranes, either in preparing sphaeroplasts or in digesting isolated walls, should be viewed with caution. Sphaeroplasts prepared using snail juice are stable and retain their osmotic sensitivity on storage at least for periods of a few days. It is possible, however, that during digestion of cells with the juice, some of the plasma-membrane components may be digested by juice enzymes. Nurminen and his colleagues (1970) have appreciated this possibility and have employed snail juice from which the lipase and phosphatase were removed by gel filtration on a column of Sephadex G 100 in 0·85% (w/v) sodium chloride.

### 3. *Function*

Although the plasma membrane probably functions in a great variety of ways in yeast cell metabolism, basically it can be considered to have three main functions: (a) to form an expandable cover and protective barrier to the protoplast; (b) to act as an organelle which controls the exit and entry of solutes; and (c) to serve as an organelle on which components of the cell wall, and extramural layers where appropriate, are synthesized. There are numerous descriptive reports in the literature on these various functions, especially on the role of the plasma membrane in solute transport; here the reader is referred to Chapter 2 (p. 39) of this volume.

In any discussion of these various functions of the yeast plasma membrane, questions inevitably arise as to the precise role (if any) of the various membrane components in these individual functions. A few data on this point are available, but not nearly sufficient to enable the yeast physiologist to arrive at a coherent account of the involvement of membrane components in individual functions. It is worth stressing that yeasts, and especially strains of *Sacch. cerevisiae* many of which readily form sphaeroplasts after digestion with enzymes, provide excellent organisms with which to study composition-function relationships in cells. Varying the nature and composition of the environment (e.g. by altering the growth temperature or growth rate, or growing the yeast anaerobically in media containing different sterols) can be used to alter the composition of the yeast plasma membrane in a predictable manner. By studying the ways in which various membrane functions are changed when the membrane composition is altered, it is possible to gain an insight into the physiological roles of individual membrane components. An extensive programme of study along these lines has been initiated in our Laboratory in the University of Bath, although, as yet, only preliminary data are available.

*a. Extensibility.* When the yeast cell wall is removed by enzymic digestion (see Phaff, 1971, for a review of the methods used), the protoplasts or sphaeroplasts that are formed behave as osmometers in that they are able to change volume in response to changes in the osmotic pressure of the suspending liquid. Yeast sphaeroplasts therefore furnish an ideal experimental system with which to investigate the ways in which the composition of the plasma membrane affects its ability to stretch.

When yeast sphaeroplasts are suspended in solutions of considerably lower osmotic pressure than those (around 0·8 *M*) used to stabilize them, they burst as a result of osmotic swelling. It is usually assumed that osmotic bursting, or osmotic lysis, occurs when the plasma membrane is unable to stretch further as water continues to flow into the sphaeroplast. This may be so, but Corner and Marquis (1969) have suggested that this does not take place when protoplasts from *Bacillus megaterium* are subjected to osmotic lysis. Their data indicate that, when protoplasts swell in hypotonic solutions, their surface membranes become so stretched that pores in the membrane are extended and allow entrance into the protoplast of molecules of the stabilizing solute. Thus, osmotic bursting of these bacterial protoplasts can occur without the membrane becoming fully extended.

Data reported on yeast sphaeroplasts relate only to osmotic lysis. Using sphaeroplasts from *Sacch. cerevisiae*, Indge (1968a) showed that resistance to lysis was lowered by EDTA and citrate, both apparently acting as chelating agents, but only when the sphaeroplasts were undergoing osmotic stress. Indge's (1968a) data also indicated that binding of cations by the membrane confers resistance to osmotic stress, possibly with divalent ions acting as bridges between adjacent negatively charged groups on phospholipids or proteins. These findings were confirmed by Diamond and Rose (1970) who used sphaeroplasts from another strain of *Sacch. cerevisiae* grown in batch culture at 30° or 15°. Growth at the lower temperature led to an increase in the phosphatidylcholine content of the cellular lipid (Hunter and Rose, 1971) and was accompanied by an increase in the resistance to osmotic lysis, as well as an increase in the ability of the membrane to retain ions when extracted with EDTA. It would now be interesting to discover whether other alterations in membrane composition, such as changes in the sterol composition, affect resistance to osmotic lysis. These data should contribute significantly to our understanding of the function of sterols in membranes.

*b. Solute transport.* The process of solute uptake in yeasts has been reviewed by Suomalainen and Oura (1971) in this volume. Their account indicates the lack of data on the role of individual plasma-membrane components in transport processes. The inhibitory effects of nickelous and uranyl ions on monosaccharide transport in *Sacch. cerevisiae* implicate phosphoryl groups, possibly polyphosphate, in the uptake process (Van Steveninck and Rothstein, 1965), but do not provide evidence for the involvement of any particular membrane component.

Evidence for a role for phosphatidylglycerol in monosaccharide transport by *Sacch. cerevisiae* has come from Deierkauf and Booij (1968), a role similar to that played by this phospholipid in sugar transport by *Escherichia coli* (Milner and Kaback, 1970). Actively transported glucose could only be detected in the phosphorylated form in *Sacch. cerevisiae* by Van Steveninck (1969), while, Farrell and Rose (1971) found that glucosamine is accumulated by *C. utilis* as a mixture of glucosamine 1- and 6-phosphates. These findings are consistent with the scheme put forward by Deierkauf and Booij (1968) which involves phosphorylation of the transported solute by phosphatidylglycerol phosphate.

Another yeast membrane component which has been implicated in a transport process is phosphatidylinositol. Cerbón (1969, 1970) reported experiments on arsenate transport in *Sacch. carlsbergensis* which suggest a role for this phospholipid in the transport process.

Active transport of solutes across membranes is thought to involve the activity of adenosine triphosphatases (ATPases), particularly ATPases activated by $Na^+ + K^+$ (Stein, 1967). Analysis of isolated yeast plasma membranes (Matile *et al.*, 1967; Nurminen *et al.*, 1970) has revealed the presence of only a $Mg^{2+}$-activated ATPase, which leaves unresolved the question of the possible role of $[Na^+ + K^+]$-activated ATPases in solute uptake by yeasts.

*c. Cell-wall synthesis.* The ability of sphaeroplasts to regenerate cell-wall macromolecules when incubated under appropriate conditions was reviewed by Phaff (1971). That yeast sphaeroplasts are rarely if ever able to regenerate a wall similar in composition to that in the intact cell emphasizes the subtle and delicate nature of the process of cell-wall growth. One would imagine that the enzymes which catalyse synthesis of yeast cell-wall macromolecules are arranged in a rather critical pattern in the plasma membrane, a pattern that allows the newly synthesized macromolecules to be located in the correct positions in the growing wall. However, nothing is known of the role of individual plasma membrane components in regulating the activity of cell-wall synthesizing enzymes. Production of excessive amounts of chitin by regenerating protoplasts of *C. utilis* (Garcia Mendoza and Novaes Ledieu, 1968) might be taken to indicate that, during protoplast formation, certain of these regulatory processes are deranged. It would be interesting to known how specific alterations in the composition of the plasma membrane (e.g. in sterol or fatty-acid composition) affect the pattern of cell-wall regeneration, for these data could begin to illuminate the nature of membrane components that regulate cell-wall synthesis. The involvement of polyprenols in the synthesis of bacterial cell-wall polysaccharides (see p. 223) leads one to speculate on the possible role of the dolichols in the yeast membrane in synthesis of wall macromolecules. Tanner (1969) reported evidence for a lipid intermediate in the *in vitro* synthesis of mannan by a preparation from *Sacch. cerevisiae.* As yet, the lipid has not been identified, although it is worth noting that rat-liver dolichols function as glycosyl donors in glucan synthesis (Behrens and Leloir, 1970).

## B. MITOCHONDRIAL MEMBRANES

Studies on the composition and metabolic activity of isolated mitochondria have been done very largely with organelles obtained from mammalian cells (see Racker, 1970, for a review of this work). However, workers in several laboratories have isolated yeast mitochondria, either by lysis of sphaeroplasts (Chappell and Hansford, 1969) or disruption of cells in solutions of appropriate osmolarity (Vitols and Linnane, 1961) and have examined, albeit less extensively than with mammalian mitochondria, the composition and metabolic activity of these isolated organelles. Mitochondria from *Sacch. cerevisiae* resemble mammalian mitochondria in their oxidative activity (Lukins *et al.*, 1968) but differ in lipid composition in that they contain a lower proportion of phospholipids and a higher proportion of triacylglycerols as well as a higher proportion of mono-unsaturated fatty acids in place of the polyunsaturated fatty acids found in mammalian mitochondria.

The main advantage to researching on yeast mitochondria is that they permit studies to be made on the effect of environmental conditions on mitochondrial composition and metabolic activity, studies which are not possible with mammalian mitochondria. So far, experiments have concentrated on the effects of catabolite repression on the composition and structure of yeast mitochondria, and on the biogenesis of mitochondria that follows aeration of anaerobically grown cells.

Depressed respiratory activity in *Sacch. cerevisiae* grown under conditions which lead to catabolite repression was first observed by Ephrussi and his colleagues (1956); later it was shown that glucose-repressed yeast cells contain fewer mitochondria than non-repressed cells (Yotsuyanagi, 1962; Polakis *et al.*, 1965; Jayaraman *et al.*, 1966). More recently, Lukins *et al.* (1968) found that glucose repression of *Sacch. cerevisiae* lowered the phospholipid content of mitochondria and also led to synthesis of a higher proportion of saturated fatty acids in the mitochondrial lipids.

However, the majority of experiments on yeast mitochondria have been prompted by the facility offered by these organisms, and in particular *Sacch. cerevisiae*, in studying the biogenesis of mitochondria (Roodyn and Wilkie, 1968). Whereas yeast cells grown aerobically produce mitochondria which can be seen in electron micrographs of thin sections of cells, comparable structures cannot usually be seen in sections of cells grown anaerobically in media lacking ergosterol and Tween 80.

However, when anaerobically-grown cells are aerated, they acquire the ability to respire and at the same time mitochondria can be seen to develop in the cells. One of the main problems has been to establish the nature of the mitochondrial precursors in anaerobically grown cells. There is now general agreement that cells grown anaerobically in supplemented media contain mitochondrial profiles that are visible in electron micrographs of thin sections of cells that have been suitably prepared. But the nature of the mitochondrial precursors in cells that have been grown in the absence of sterol and Tween supplements is very uncertain. Watson and his colleagues (1970) isolated structures from yeast cells grown in the absence of supplements. These structures were very fragile and could be isolated only from sphaeroplasts that had been treated with glutaraldehyde. The mitochondrial precursors contained a much lower proportion of unsaturated fatty acids than mitochondria from cells grown anaerobically in supplemented medium (80%) or aerobically (82%). These findings lend support to the view that unsaturated fatty acids in yeast mitochondrial membranes have a critical function in oxidative phosphorylation (Proudlock *et al.*, 1969; Haslam *et al.*, 1970).

### C. OTHER MEMBRANES

A number of other types of membrane are recognized in yeast cells, in addition to the plasma and mitochondrial membranes. The cytology of these membranes has been discussed at some length by Marchant and Smith (1968) and by Matile *et al.* (1969). So far, isolation methods are available for only two of these additonal types of yeast membrane namely the vacuolar and nuclear membranes, although there is in principle no reason why other fractions, for example one rich in endoplasmic reticulum, should not be obtained by subjecting lysed yeast sphaeroplasts to careful centrifugation regimes.

Stable vacuoles have been obtained from sphaeroplasts of *Sacch. cerevisiae* by Matile and Wiemken (1967) by lowering the concentration of the suspending mannitol solution from 0·7 *M* to 0·233 *M*. The vacuoles were isolated by flotation in Ficoll (about 8%) followed by low-speed centrifugation. Indge (1968c) obtained vacuoles from a strain of *Sacch. carlsbergensis* by metabolic lysis (Indge, 1968b) of sphaeroplasts; he isolated the vacuoles by differential centrifugation. Matile and Wiemken (1967) showed that the vacuoles contain a wide range of hydrolytic enzymes including proteases, ribonuclease, and esterase, as well as some aminopeptidase, and they made the provocative suggestion that yeast vacuoles may act as lysosomes in the cell. Recent work by the group working in University College, Cardiff (Cartledge *et al.*, 1969; D. Lloyd, unpublished observations) indicates that the vacuoles released from yeast sphaeroplasts by osmotic lysis may not contain such an extensive range of enzymes as Matile and Wiemken (1967) suggested. Isolated yeast vacuoles act as osmometers but nothing has yet been reported on the properties of the isolated vacuolar membrane. However, if the vacuole does act as a type of lysosome in the yeast cell, it is possible that the membrane may have unusual composition and structure in order that it can resist the action of the hydrolytic enzymes contained in the vacuole.

Rozijn and his colleagues (Rozijn and Tonino, 1964; Smitt *et al.*, 1970) have reported methods for isolating nuclei from lysates of yeast sphaeroplasts As yet, however, data have not been published on the composition and structure of the yeast nuclear membrane.

## V. Acknowledgements

We are greatly indebted to many colleagues for valuable discussion and for providing unpublished information on yeast lipids and membranes, and in particular to Dr. R. Letters, Arthur Guinness Son and Co. Ltd., St. James's Gate Brewery, Dublin, Eire, and Dr. F. W. Hemming, Department of Biochemistry, University of Liverpool, England. Our thanks are also due to the Science Research Council (U.K.) for a research grant (B/SR/5724) which has supported work from this laboratory on yeast lipids and membranes, and to Arthur Guinness Son & Co. Ltd., for a research studentship awarded to K.H.

### References

Adams, B. G. and Parks, L. W. (1967). *J. Cell. Physiol.* **70**, 161–168.
Adams, B. G. and Parks, L. W. (1968). *J. Lipid Res.* **9**, 8–11.
Agar, H. D. and Douglas, H. C. (1957). *J. Bact.* **73**, 365–375.
Ahmad, F. and Rose, A. H. (1962). *J. gen. Microbiol.* **28**, 147–160.
Akhtar, M., and Parvez, M. A. (1968). *Biochem. J.* **108**, 527–531.
Akhtar, M., Hunt, P. F. and Parvez, M. A. (1967). *Biochem. J.* **103**, 616–622.
Akhtar, M., Parvez, M. A. and Hunt, P. F. (1968). *Biochem. J.* **106**, 623–626.
Akhtar, M., Brooks, W. A. and Watkinson, I. A. (1969a). *Biochem. J.* **115**, 135–137.
Akhtar, M., Parvez, M. A. and Hunt, P. F. (1969b). *Biochem. J.* **113**, 727–732.
Akhtar, M., Rahimtula, A. D. and Wilton, D. C. (1970). *Biochem. J.* **117**, 539–542.
Albo, P. W. and Dittmer, J. C. (1969a). *Biochemistry, N.Y.* **8**, 394–405.
Albo, P. W. and Dittmer, J. C. (1969b). *Biochemistry, N.Y.* **8**, 953–959.
Albo, P. W. and Dittmer, J. C. (1969c). *Biochemistry, N.Y.* **8**, 1913–1918.
Albo, P. W. and Dittmer, J. C. (1969d). *Biochemistry, N.Y.* **8**, 3317–3324.
Allen, C. M., Alworth, W., Macrae, A. and Bloch, K. (1967). *J. biol. Chem.* **242**, 1895–1902.
Andreasen, A. A. and Stier, T. J. B. (1954). *J. cell. comp. Physiol.* **43**, 271–284.
Baraud, J., Cassagne, C., Genevois, L. and Joneau, M. (1967). *C. r. hebd. Séanc. Acad. Sci., Paris* **265**, 83–85.

Baraud, J., Maurice, A. and Napias, C. (1970). *Bull. Soc. chim. Biol.* **52**, 421–432.
Barron, E. J. and Hanahan, D. J. (1961). *J. biol. Chem.* **231**, 493–503.
Barton, D. H. R., Harrison, D. M. and Widdowson, D. A. (1968). *Chem. Commun.* 17–19.
Bauer, H. (1968). *J. Bact.* **96**, 853–854.
Behrens, N. H. and Leloir, L. F. (1970). *Proc. natn. Acad. Sci. U.S.A.* **66**, 153–159.
Benham, R. W. (1939). *J. invest. Derm.* **2**, 187–193.
Benham, R. W. (1941). *Proc. Soc. exp. Biol. Med.* **46**, 176–178.
Benham, R. W. (1947). *In* "Biology of Pathogenic Fungi", (W. J. Nickerson, ed.). pp. 63–70. Chronica Botanica Co., Waltham, Mass.
Bergel'son, L. D., Vaver, V. A., Prokozova, N. V., Ushakov, A. N. and Popkova, G. A. (1966). *Biochim. biophys. Acta* **116**, 511–520.
Bloomfield, D. and Bloch, K. (1958). *Biochim. biophys. Acta* **30**, 220–221.
Bloomfield, D. and Bloch, K. (1960). *J. biol. Chem.* **235**, 337–345.
Boulton, A. A. (1965). *Expl Cell Res.* **37**, 343–359.
Bours, J. and Mossel, D. A. A. (1969). *Antonie van Leeuwenhoek* **35**, Suppl. 129–130.
Breivik, O. N., Owades, J. L. and Light, R. F. (1954). *J. org. Chem.* **19**, 1734–1740.
Brown, C. M. and Rose, A. H. (1969). *J. Bact.* **99**, 371–378.
Callow, R. K. (1931). *Biochem. J.* **25**, 87–94.
Cartledge, T. G., Burnett, J. K., Brightwell, R. and Lloyd, D. (1969). *Biochem. J.* **115**, 56P–57P.
Castelli, A., Littaru, G. P. and Barbaresi, G. (1969). *Arch. Mikrobiol.* **66**, 34–39.
Cerbón, J. (1969). *J. Bact.* **97**, 658–662.
Cerbón, J. (1970). *J. Bact.* **102**, 97–105.
Challinor, S. W. and Daniels, N. W. R. (1955). *Nature, Lond.* **176**, 1267–1268.
Chappell, J. B. and Hansford, R. G. (1969). *In* "Subcellular Components; Preparation and Fractionation", (G. D. Birnie and S. M. Fox, eds.). pp. 43–56. Butterworths, London.
Chevreul, M. E. (1823). Recherches chimiques sur les corps gras d'origine animale. Levrault, Paris.
Coleman, R. and Hübscher, G. (1962). *Biochim. biophys. Acta* **56**, 479–490.
Combs, T. J., Guarneri, J. J. and Pisano, M. A. (1968). *Mycologia* **60**, 1232–1239.
Contardi, A. and Ercoli, A. (1933). *Biochem. Z.* **261**, 275–302.
Corner, T. R. and Marquis, R. E. (1969). *Biochim. biophys. Acta* **183**, 544–558.
Cruz, F. S. and Travassos, L. R. (1970). *Arch. Mikrobiol.* **73**, 111–120.
Cury, A., Suassuna, E. N. and Travassos, L. R. (1960). *An. Microbiol.* **8**, 13–64.
Danielsson, H. and Tchen, T. T. (1968). *In* "Metabolic Pathways", (D. M. Greenberg, ed.), Vol. II, pp. 117–168. Academic Press, New York.
Dawson, P. S. S. and Craig, B. M. (1966). *Can. J. Microbiol.* **12**, 775–785.
Dawson, R. M. C. (1963). *Biochem. J.* **88**, 414–423.
Deierkauf, F. A. and Booij, H. L. (1968). *Biochim. biophys. Acta* **150**, 214–225.
Deinema, M. H. (1961). Thesis: University of Wageningen.
Diamond, R. J. and Rose, A. H. (1970). *J. Bact.* **102**, 311–319.
Di Salvo, A. F. and Denton, J. F. (1963). *J. Bact.* **85**, 927–931.
Douglas, L. J. and Baddiley, J. (1968). *FEBS Letters* **1**, 114–116.
Dulaney, E. L., Stapley, E. O. and Simpf, K. (1954). *Appl. Microbiol.* **2**, 371–379.
Dunphy, P. J., Kerr, J. D., Pennock, J. F., Whittle, K. J. and Feeney, J. (1967). *Biochim. biophys. Acta* **136**, 136–147.
Eddy, A. A. (1958). *Proc. Roy. Soc. B.* **149**, 425–440.
Eddy, A. A. and Williamson, D. H. (1957). *Nature, Lond.* **179**, 1252–1253.
Enebo, L., Anderson, L. G. and Lundin, H. (1946). *Archs Biochem.* **11**, 383–395.
Enebo, L., Elander, M., Berg, F., Lundin, H., Nilsson, R. and Myrbäck, K. (1944). *Iva* **6**, 252–267.

Ephrussi, B., Slonimski, P. P., Yotsuyanagi, Y. and Tavlitzki, J. (1956). *C. r. Trav. Lab. Carlsberg, Sér. Physiol.* **26**, 86–102.
Farrell, J. and Rose, A. H. (1967a). *In* "Thermobiology", (A. H. Rose, ed.), pp. 147–219. Academic Press, London.
Farrell, J. and Rose, A. H. (1967b). *A. Rev. Microbiol.* **21**, 101–120.
Farrell, J. and Rose, A. H. (1971). *Arch. Mikrobiol.* Submitted for publication.
Ferguson, J. J., Durr, I. F. and Rudney, H. (1959). *Proc. natn. Acad. Sci. U.S.A.* **45**, 499–504.
Fieser, L. F. and Fieser, M. (1959). "Steroids". Reinhold Publishing Corp., New York.
Fimognari, G. M. and Rodwell, V. W. (1965). *Biochemistry, N.Y.* **4**, 2086–2090.
Fulco, A. J. (1967). *J. biol. Chem.* **242**, 3608–3613.
Gancedo, C., Gancedo, J. M. and Sols, A. (1968). *Eur. J. Biochem.* **5**, 166–172.
Garcia Mendoza, C. and Villanueva, J. R. (1967). *Biochim. biophys. Acta* **135**, 189–195.
Garcia Mendoza, C. and Novaes Lidieu, M. (1968) *Nature, Lond.* **220**, 1035.
Gerbach, G. and Güntner, H. (1932). *Sber. Akad. Wiss. Wien Abt. IIb* 415–428.
Goldman, P. and Vagelos, P. R. (1961). *J. biol. Chem.* **236**, 2620–2623.
Green, D. E. and Allmann, D. W. (1968). *In* "Metabolic Pathways", (D. M. Greenberg, ed.), 3rd edition, Vol. II, pp. 37–67. Academic Press, New York.
Greenberg, D. M. ed. (1968). "Metabolic Pathways", vol. II. Academic Press, New York.
Greene, M. L., Kaneshiro, T. and Law, J. H. (1965). *Biochim. biophys. Acta* **98**, 582–588.
Gurr, M. I., Robinson, M. P., Sword, R. W. and James, A. T. (1968). *Biochem. J.* **110**, 49P–50P.
Hanahan, D. J. (1952). *J. biol. Chem.* **195**, 199–206.
Hanahan, D. J. and Jayko, M. E. (1952). *J. Am. chem. Soc.* **74**, 5070–5073.
Hanahan, D. J. and Olley, J. N. (1958). *J. biol. Chem.* **231**, 813–828.
Harrison, J. S. and Trevelyan, W. E. (1963). *Nature, Lond.* **200**, 1189–1190.
Hartman, L., Hawke, J. C., Shorland, F. B. and di Menna, M. E. (1959). *Archs Biochem. Biophys.* **81**, 346–352.
Haskell, B. E. and Snell, E. E. (1965). *Archs Biochem. Biophys.* **112**, 494–505.
Haslam, J. M., Proudlock, J. W. and Linnane, A. W. (1970). *J. Bioenergetics* **1**, in press.
Heide, S. (1939). *Arch. Mikrobiol.* **12**, 131–141.
Hemming, F. W. (1970). *Symp. Biochem. Soc. No.* 29. 105–117.
Higashi, Y., Strominger, J. L. and Sweeley, C. C. (1967). *Proc. natn. Acad. Sci. U.S.A.* **57**, 1878–1884.
Higashi, Y., Siewert, G. and Strominger, J. L. (1968). *Fedn Proc. Fedn. Am. Socs exp. Biol.* **27**, 294.
Hirschmann, H. (1960). *J. biol. Chem.* **235**, 2762–2767.
Hoffmann-Ostenhof, O., Geyer-Fenzl, M. and Wagner, E. (1961). *In* "The Enzymes of Lipid Metabolism", (P. Desnuelle, ed.). pp. 39–42. Pergamon Press, Oxford.
Hunter, K. and Rose, A. H. (1971). *Biochim. biophys. Acta* in preparation.
Indge, K. J. (1968a). *J. gen. Microbiol.* **51**, 425–432.
Indge, K. J. (1968b). *J. gen. Microbiol.* **51**, 433–440.
Indge, K. J. (1968c). *J. gen. Microbiol.* **51**, 441–446.
IUPAC-IUB Commission on Biochemical Nomenclature (1967). *Eur. J. Biochem.* **2**, 127–131.
Jayaraman, J., Cotman, C., Mahler, H. R. and Sharp, C. W. (1966). *Archs Biochem. Biophys.* **116**, 224–251.
Jollow, D., Kellerman, G. M. and Linnane, A. W. (1968). *J. Cell Biol.* **37**, 221–230.

Kahane, E. (1963). *C. r. hebd. Séanc. Acad. Sci., Paris* **257**, 1966–1968.
Kahane, E. and Regord, M. Th. (1964). *Ann. Nutr. aliment.* **18**, 1–21.
Kates, M. and Baxter, R. M. (1962). *Can. J. Biochem. Physiol.* **40**, 1213–1227.
Katsuki, H. and Bloch, K. (1967). *J. biol. Chem.* **242**, 222–227.
Kawaguchi, A., Hatanaka, H. and Katsuki, H. (1968). *Biochem. Biophys. res. Commun.* **33**, 463–468.
Keith, A. D., Resnick, M. R. and Haley, A. B. (1969). *J. Bact.* **98**, 415–420.
Kessler, G. and Nickerson, W. J. (1959). *J. biol. Chem.* **234**, 2281–2285.
Klein, H. P. (1957). *J. Bact.* **73**, 530–537.
Klein, H. P. and Jahnke, L. (1968). *J. Bact.* **96**, 1632–1639.
Klein, H. P. and Lipmann, F. (1953). *J. biol. Chem.* **203**, 95–99.
Klyne, W. (1965). *The Chemistry of the Steroids.* Methuen & Co. Ltd., London.
Kodicek, E. and Ashby, D. R. (1954). *Biochem. J.* **57**, xii.
Kokke, R., Hooghwinkel, G. J. M., Booij, H. L., Van den Bosch, H., Zelles, L., Mulder, E. and Van Deenen, L. L. M. (1963). *Biochim. biophys. Acta* **70**, 351–353.
Kolattukudy, P. E. (1966). *Biochemistry, N.Y.* **5**, 2265–2275.
Kortstee, G. J. J. (1970). *Arch. Mikrobiol.* **71**, 235–244.
Kováč, L., Šubík, J., Russ, G. and Kollár, K. (1967). *Biochim. biophys. Acta* **144**, 94–101.
Krishna, G., Feldbruegge, D. H. and Porter, J. W. (1964). *Biochem. Biophys. res. Commun.* **14**, 363–369.
Kuhn, N. J. and Lynen, F. (1965). *Biochem. J.* **94**, 240–246.
Kumagawa, M. and Suto, K. (1908). *Biochem. Z.* **8**, 212–247.
Lasnitzki, A. and Szorenyi, E. (1935). *Biochem. J.* **29**, 580–587.
Lebeault, J. M., Roche, B., Duvnjak, Z. and Azoulay, E. (1970). *Arch. Mikrobiol.* **72**, 140–153.
Lederer, E. (1964). *Biochem. J.* **93**, 449–468.
Lee, Y. C. and Ballou, C. E. (1965). *Biochemistry, N.Y.* **4**, 1395–1404.
Letters, R. (1962). *J. Inst. Brew.* **68**, 318–321.
Letters, R. (1966). *Biochim. biophys. Acta* **116**, 489–499.
Letters, R. (1968a). In "Aspects of Yeast Metabolism", (A. K. Mills, ed.), pp. 303–319. Blackwell, Oxford.
Letters, R. (1968b). *Bull. Soc. chim. Biol.* **50**, 1385–1393.
Letters, R. and Snell, B. K. (1963). *J. chem. Soc.* 5127–5130.
Light, R. J., Lennarz, W. J. and Bloch, K. (1962). *J. biol. Chem.* **237**, 1793–1800.
Lindner, P. (1911). *Jb. Vers Anst. Brau., Berl.* **14**, 551–570.
Longley, R. P., Rose, A. H. and Knights, B. A. (1968). *Biochem. J.* **108**, 410–412.
Lukins, H. B., Jollow, D., Wallace, P. G. and Linnane, A. W. (1968). *Aust. J. exp. Biol. Med. Sci.* **46**, 651–665.
Lynen, F. (1967). *Biochem. J.* **102**, 381–400.
Maas-Förster, M. (1955). *Arch. Mikrobiol.* **22**, 115–144.
Macfarlane, M. G. (1964). *Adv. Lipid Res.* **2**, 91–125.
Madyasthia, P. B. and Parks, L. W. (1969). *Biochim. biophys. Acta* **176**, 858–862.
Maguigan, W. H. and Walker, E. (1940). *Biochem. J.* **34**, 804–813.
Majerus, P. W., Alberts, A. W. and Vagelos, P. R. (1965). *J. biol. Chem.* **240**, 4723–4726.
Mangold, H. K. (1965). *In* "Thin Layer Chromatography", (E. Stahl, ed.), pp. 137–186. Academic Press, New York.
Marchant, R. and Smith, D. G. (1968). *Biol. Rev.* **43**, 459–480.
Marinetti, G. V., ed. (1967). *Lipid Chromatographic Analysis*, Vol. 1. Edward Arnold, Ltd., London.
Marsh, J. B. and Weinstein, D. B. (1966). *J. Lipid Res.* **7**, 574–576.

Matile, Ph. (1970). *FEBS Symposium* **20**, 39–49.
Matile, Ph. and Wiemken, A. (1967). *Arch. Mikrobiol.* **56**, 148–155.
Matile, Ph., Moor, H. and Muhlethaler, K. (1967). *Arch. Mikrobiol.* **58**, 201–211.
Matile, Ph., Moor, H. and Robinow, C. F. (1969). *In* "The Yeasts", (A. H. Rose and J. S. Harrison, eds.), Vol. 1, pp. 219–302. Academic Press, London.
McMurrough, I. and Rose, A. H. (1967). *Biochem. J.* **105**, 189–203.
Merdinger, E. and Cwiakala, C. E. (1968). *Lipids* **2**, 276–277.
Meyer, F. and Bloch, K. (1963a). *J. biol. Chem.* **238**, 2654–2659.
Meyer, F. and Bloch, K. (1963b). *Biochim. biophys. Acta* **77**, 671–673.
Milner, L. S. and Kaback, H. R. (1970). *Proc. natn. Acad. Sci. U.S.A.* **65**, 683–690.
Moore, J. T. and Gaylor, J. L. (1969). *J. biol. Chem.* **244**, 6334–6340.
Motai, H., Ichishima, E. and Yoshida, F. (1966). *Nature, Lond.* **210**, 308–309.
Nageli, C. and Loew, O. (1878). *Justus Leibigs Annln Chem.* **193**, 322–348.
Nielsen, H. and Nilsson, N. G. (1950). *Archs Biochem. Biophys.* **25**, 316–322.
Northcote, D. H. and Horne, R. W. (1952). *Biochem. J.* **51**, 232–236.
Nurminen, T. and Suomalainen, H. (1970). *Biochem. J.* **118**, 759–763.
Nurminen, T., Oura, E. and Suomalainen, H. (1970). *Biochem. J.* **116**, 61–69.
Nyns, E. J., Chiang, N. and Wiaux, A. L. (1968). *Antonie van Leeuwenhoek* **34**, 197–204.
Oda, T. and Kamiya, H. (1958). *Chem. Pharm. Bull., Tokyo* **6**, 682–687.
Olson, J. A., Lindberg, M. and Bloch, K. (1957). *J. biol. Chem.* **226**, 941–956.
Ota, Y. and Yamada, K. (1966). *Agric. Biol. Chem., Tokyo* **30**, 351–358.
Parks, L. W. (1958). *J. Am. chem. Soc.* **80**, 2023–2024.
Pedersen, T. A. (1962). *Acta Chem. Scand.* **16**, 374–382.
Peters, I. I. and Nelson, F. E. (1948a). *J. Bact.* **55**, 581–591.
Peters, I. I. and Nelson, F. E. (1948b). *J. Bact.* **55**, 593–600.
Phaff, H. J. (1971). *In* "The Yeasts", (A. H. Rose and J. S. Harrison, eds.), Vol. 2. pp. 135–209. Academic Press, London.
Polakis, E. S., Bartley, W. and Meek, G. A. (1965). *Biochem. J.* **97**, 298–302.
Ponsinet, G. and Ourisson, G. (1965). *Memoirs Soc. Chim.* **5**, 3682–3684.
Proudlock, J. W., Wheeldon, L. W., Jollow, D. J. and Linnane, A. W. (1968). *Biochim. biophys. Acta* **152**, 434–437.
Proudlock, J. W., Haslam, J. M. and Linnane, A. W. (1969). *Biochem. biophys. Res. Commun.* **37**, 847–852.
Pugh, E. L. and Wakil, S. J. (1965). *J. biol. Chem.* **240**, 4727–4733.
Racker, E. ed. (1970). "Membranes of Mitochondria and Chloroplasts". Van Nostrand Reinhold Co., New York.
Rasmussen, R. K. and Klein, H. P. (1968). *J. Bact.* **95**, 157–161.
Reindel, F., Weickmann, A., Picard, S., Luber, K. and Turula, P. (1940). *Justus Leibigs Annln Chem.* **544**, 116–137.
Resnick, M. A. and Mortimer, R. K. (1966). *J. Bact.* **92**, 597–600.
Roodyn, D. B. and Wilkie, D. (1968). "The Biogenesis of Mitochondria". Methuen, London.
Rose, A. H. (1963). *J. gen. Microbiol.* **31**, 151–160.
Rossiter, R. J. (1968). *In* "Metabolic Pathways", (D. M. Greenberg, ed.), 3rd edition, Vol. II, pp. 69–115. Academic Press, New York.
Rouser, G., Kritchevsky, G. and Yamamoto, A. (1967). *In* "Lipid Chromatographic Analysis", (G. V. Marinetti, ed.), Vol. 1, pp. 99–162. Edward Arnold, London.
Rozijn, Th. H. and Tonino, G. J. M. (1964). *Biochim. biophys. Acta* **91**, 105–112.
Rudney, H. (1956). *Fedn Proc. Fedn. Am. Socs exp. Biol.* **15**, 342–343.
Ruinen, J. (1956). *Nature, Lond.* **177**, 220–221.

Ruinen, J. (1961). *Plant Soil* **15**, 81–109.
Ruinen, J. (1963a). *Antonie van Leeuwenhoek* **29**, 425–438.
Ruinen, J. (1963b). *J. gen. Microbiol.* **32**, iv.
Ruinen, J. (1966). *Ann. Inst. Pasteur, Lille Suppl.* **111**, 342–346.
Ruinen, J. and Deinema, M. H. (1964). *Antonie van Leeuwenhoek* **30**, 377–384.
Sandeman, W. and Schwers, W. (1960). *Chem. Ber.* **93**, 2266–2271.
Scher, M., Lennarz, W. J. and Sweeley, C. C. (1968). *Proc. natn. Acad. Sci. U.S.A.* **59**, 1313–1320.
Schweizer, E. (1963). Thesis: University of Munich.
Shafai, T. and Lewin, L. M. (1968). *Biochim. biophys. Acta* **152**, 787–790.
Shaw, N. and Baddiley, J. (1968). *Nature, Lond.* **217**, 142–144.
Shaw, W. H. C. and Jefferies, J. P. (1953). *Analyst* **78**, 509–528.
Shifrine, M. and Marr, A. G. (1963). *J. gen. Microbiol.* **32**, 263–270.
Shoppee, C. W. (1964). "Chemistry of the Steroids". Butterworths, London.
Sih, C. J. and Whitlock, H. W. (1968). *A. Rev. Biochem.* **37**, 661–694.
Smitt, W. W. S., Nanni, G., Rozijn, Th. H. and Tonino, G. J. M. (1970). *Expl Cell Res.* **59**, 440–446.
Sofer, S. S. and Rilling, H. C. (1969). *J. Lipid Res.* **10**, 183–187.
Sonderhoff, R. and Thomas, H. (1937). *Justus Liebigs Annln Chem.* **530**, 195–213.
Sprinson, D. B. and Coulon, A. (1954). *J. biol. Chem.* **207**, 585–592.
Stanacev, N. Z. and Kates, M. (1963). *Can. J. Biochem. Physiol.* **41**, 1330–1334.
Stanacev, N. Z., Chang, Y. Y. and Kennedy, E. P. (1967). *J. biol. Chem.* **242**, 3018–3019.
Staple, E. (1967). *In* "Biogenesis of Natural Compounds", (P. Bernfield, ed.), 2nd edition, pp. 207–245. Pergamon Press, Oxford.
Starkey, R. L. (1946). *J. Bact.* **51**, 33–50.
Stein, W. D. (1967). "The Movement of Molecules across Cell Membranes". Academic Press, New York.
Stodola, F. H. and Wickerham, L. J. (1960). *J. biol. Chem.* **235**, 2584–2585.
Stodola, F. H., Wickerham, L. J., Scholfield, C. R. and Dutton, H. J. (1962). *Archs Biochem. Biophys.* **98**, 176.
Stodola, F. H., Deinema, M. H. and Spencer, J. F. T. (1967). *Bact. Rev.* **31**, 194–213.
Stodola, F. H., Vesonder, R. F. and Wickerham, L. J. (1965). *Biochemistry, N.Y.* **4**, 1390–1394.
Strickland, K. P. (1967). *In* "Biogenesis of Natural Compounds", (P. Bernfield, ed.), 2nd edition, pp. 103–205. Pergamon Press, Oxford.
Suomalainen, H. and Kevänen, A. J. A. (1968). *Chem. Phys. Lipids* **2**, 296–315.
Suomalainen, H. and Oura, E. (1971). *In* "The Yeasts", (A. H. Rose and J. S. Harrison, eds.), Vol. 2, pp. 3–74. Academic Press, London.
Suomalainen, H., Nurminen, T. and Oura, E. (1967a). *Abstr. FEBS 4th Meeting, Oslo.* p. 111.
Suomalainen, H., Nuriminen, T. and Oura, E. (1967b). *Acta chem. fenn. B.* **40**, 323–326.
Tanner, W. (1968). *Arch. Mikrobiol.* **64**, 158–172.
Tanner, W. (1969). *Biochem. biophys. Res. Commun.* **35**, 144–150.
Tavormina, P. A., Gibbs, M. H. and Huff, J. W. (1956). *J. Am. chem. Soc.* **78**, 4498–4499.
Thorpe, S. R. and Sweeley, C. C. (1967). *Biochemistry, N.Y.* **6**, 887–897.
Topham, R. W. and Gaylor, J. L. (1967). *Biochem. biophys. Res. Commun.* **27**, 644–649.
Topham, R. W. and Gaylor, J. L. (1970). *J. biol. Chem.* **245**, 2319–2327.
Tornabene, T. G. and Oro, J. (1967). *J. Bact.* **94**, 349–358.
Tornqvist, E. and Lundin, H. (1951). *Int. sugar J.* **53**, 123–127.

Trevelyan, W. E. (1968). *J. Inst. Brew.* **74**, 365–369.
Tulloch, A. P. and Spencer, J. F. T. (1964). *Can. J. Chem.* **42**, 830–835.
Usden, V. R. and Burrell, R. C. (1952). *Archs Biochem. Biophys.* **36**, 172–177.
Van den Bosch, H., Van Den Elzen, H. M. and Van Deenen, L. L. M. (1967). *Lipids* **2**, 279–280.
Van Steveninck, J. (1969). *Archs Biochem. Biophys.* **130**, 244–252.
Van Steveninck, J. and Rothstein, A. (1965). *J. gen. Physiol.* **49**, 235–246.
Vaver, V. A., Prokazova, N. V., Ushakov, A. N., Golovkina, L. S. and Bergel'son, L. D. (1967a). *Biokhimya* **32**, 310–317.
Vaver, V. A., Shchennikov, V. A. and Bergel'son, L. D. (1967b). *Biokhimya* **32**, 1027–1031.
Vickery, J. R. (1936a). *J. Counc. scient. ind. Res. Aust.* **9**, 107–112.
Vickery, J. R. (1936b). *J. Counc. scient. ind. Res. Aust.* **9**, 196–202.
Vitols, E. and Linnane, A. W. (1961). *J. biophys. biochem. Cytol.* **9**, 701–710.
Wagner, H. and Zofcsik, W. (1966a). *Biochem. Z.* **346**, 333–342.
Wagner, H. and Zofcsik, W. (1966b). *Biochem. Z.* **346**, 343–350.
Watson, K., Haslam, J. M. and Linnane, A. W. (1970). *J. Cell Biol.* **46**, 88–96.
Wells, W. W., Schultz, J. and Lynen, F. (1966). *Biochem. Z.* **346**, 474–490.
Wells, W. W., Schultz, J. and Lynen, F. (1967). *Proc. natn. Acad. Sci. U.S.A.* **56**, 633–639.
Welsh, K., Shaw, N. and Baddiley, J. (1968). *Biochem. J.* **107**, 313–314.
Werner, H. (1960). *Zentbl. Bakt. ParasitKde. Abt. I* **200**, 113–124.
White, A. G. C. and Werkman, C. H. (1948). *Archs Biochem. Biophys.* **17**, 475–482.
White, D. and Klein, H. P. (1965). *Biochem. Biophys. res. Commun.* **20**, 78–84.
White, D. and Klein, H. P. (1966). *J. Bact.* **91**, 1218–1223.
Wickerham, L. J. and Stodola, F. H. (1960). *J. Bact.* **80**, 484–491.
Wieland, H. and Benend, W. (1942). *Z. phys. Chem.* **274**, 215–225.
Wieland, H. and Gough, G. A. C. (1930). *Justus Liebigs Annln Chem.* **482**, 36–49.
Wieland, H. and Stanley, W. M. (1931). *Justus Liebigs Annln Chem.* **489**, 31–42.
Wieland, O. and Suyter, M. (1957). *Biochem. Z.* **329**, 320–331.
Wieland, H., Rath, F. and Hess, H. (1941). *Justus Liebigs Annln Chem.* **548**, 34–49
Wilde, P. F. and Stewart, P. S. (1968). *Biochem. J.* **108**, 225–231.
Woodbine, M. (1959). *Progr. ind. Microbiol.* **1**, 179–245.
Wright, A., Dankert, M., Fennessey, P. and Robbins, P. W. (1967). *Proc. natn. Acad. Sci. U.S.A.* **57**, 1798–1803.
Yamada, K. and Machida, H. (1962). *J. agric. Chem. Soc., Japan* **36**, 858–861.
Yotsuyanagi, Y. (1962). *J. Ultrastruct. Res.* **7**, 121–140.
Yuan, C. and Bloch, K. (1961). *J. biol. Chem.* **236**, 1277–1279.
Zabin, I. and Mead, J. F. (1953). *J. biol. Chem.* **205**, 271–277.
Zabin, I. and Mead, J. F. (1954). *J. biol. Chem.* **211**, 87–93.

*Chapter 7*

# Energy-Yielding Metabolism in Yeasts

A. SOLS, C. GANCEDO AND G. DELAFUENTE

*Instituto de Enzimología, Centro de Investigaciones Biológicas, Consejo Superior de Investigaciones Científicas, Madrid, Spain*

I. INTRODUCTION . . . 271
II. ANAEROBIC METABOLISM . . . 273
A. Alcoholic Fermentation . . . 274
B. Other Types of Fermentation . . . 281
C. Hexose Monophosphate Oxidative Pathway . . . 282
D. Contribution of Different Pathways of Glucose Metabolism to Anaerobic Growth . . . 283
III. AEROBIC METABOLISM . . . 286
A. Citric Acid Cycle . . . 286
B. Ancillary Pathways . . . 289
IV. ENERGY SOURCES OTHER THAN GLUCOSE . . . 290
A. Sugars and Related Compounds . . . 290
B. Nitrogen Compounds . . . 292
C. Hydrocarbons . . . 292
V. ENERGY RESERVES AND THEIR MOBILIZATION . . . 292
A. Glycogen . . . 293
B. Trehalose . . . 294
C. Lipids . . . 295
D. Endogenous Metabolism . . . 296
VI. REGULATION OF ENERGY METABOLISM . . . 297
A. The Pasteur Effect . . . 297
B. Inhibition of Respiration by Fermentation . . . 301
C. Metabolic Integration . . . 302
VII. ACKNOWLEDGEMENTS . . . 303
REFERENCES . . . 303

## I. Introduction

Yeasts are heterotrophic organisms that can use sugars and a variety of other organic compounds as nutrient sources. From these compounds they derive the carbon skeletons needed to synthesize cellular constituents and the energy necessary to allow biosynthetic reactions to proceed. It is the aim of this chapter to consider the ways of utilization of different compounds by yeasts as sources of energy. Energy storage and its mobilization will also be described.

It will be useful to summarize first some general concepts concerning chemical energy and its metabolic harnessing. The extent to which a reaction proceeds is determined by the sign and magnitude of the change in free energy ($\Delta G$) of the reaction. A large negative value of $\Delta G$ indicates reactions which may proceed nearly to completion; a large positive value characterizes reactions which, in the absence of an external driving force, occur only to a limited extent. However, it is important to emphasize that the magnitude of $\Delta G$ does not give any information about the rate of the reaction. For example, the complete oxidation of glucose is thermodynamically possible because the value of $\Delta G$ in the reaction:

$$C_6H_{12}O_6 + 6\,O_2 \rightarrow 6\,CO_2 + 6\,H_2O$$

is −688 kcal/mole; but it is well known that glucose can be left indefinitely in contact with oxygen at room temperature without appreciable reaction. Nevertheless, this reaction is carried out readily by a very large variety of organisms that are able to utilize the energy contained in the glucose molecule. However, if the energy liberated in the oxidation of glucose were set free all at the same time, an organism could do almost nothing with it (except possibly burn itself!). If the energy is to be utilized, it needs to be delivered gradually in discrete quantities, and the organism must possess some mechanism for harnessing it. It is well established that organisms harness the energy liberated from their nutrients by synthesizing ATP, whose pyrophosphate bonds are "energy rich" bonds. Because of this energy, the terminal phosphate group of ATP may be transferred to another molecule, resulting in the production of the corresponding phosphorylated molecule and ADP. The ADP thus produced may be rephosphorylated to ATP, utilizing the energy set free in other reactions. In this way the adenylic system acts as common intermediate between energy-yielding (exergonic) and energy-requiring (endergonic) reactions. The free energy of hydrolysis of ATP places it in a favourable position to play the role of energy carrier. The standard free energy of the reaction:

$$ATP^{4-} + H_2O \rightarrow ADP^{3-} + HPO_4^{2-} + H^+$$

is about −7 kcal/mole. This value is intermediate between that of the free energy of hydrolysis of the very energy-rich compounds like phosphoenolpyruvate or 1,3-diphosphoglycerate (−12·8 and −11·8 kcal/mole, respectively), and that of compounds with low-energy phosphate bonds, for example the hexose phosphates. ATP can elevate the energy content of a molecule such as a hexose to that of a hexose phosphate, making it more reactive and able to participate in a reaction where energy is needed and where ATP is not directly implicated.

In addition to ATP, cells possess other nucleotides (GTP, ITP, UTP, CTP, etc.) which can also act as carriers of the harnessed energy for certain biosynthetic processes. ATP can phosphorylate the corresponding diphosphates (GDP, IDP, UDP, CDP, etc.) in reactions catalysed by the enzyme nucleoside diphosphokinase. The resulting triphosphate can then be used as energy carrier in certain specific reactions.

It could be thought that, owing to the diversity of nutrients that cells can utilize, they should have a great variety of energy-trapping mechanisms; but, in fact, the different types of mechanisms are astonishingly few. Either the compounds enter a metabolic pathway in which their energy is "concentrated" in such a way as to allow direct synthesis of ATP in single enzyme reactions, as in some steps of glycolysis, or they generate reduced coenzymes which, upon reoxidation *via* the respiratory chain, also generate ATP. Reactions of both types will be described.

## II. Anaerobic Metabolism

Good ability to thrive in, at least partly, anaerobic conditions is an important property of many yeasts, particularly those generally involved in fermentation processes. The classic concept of Pasteur of fermentation as "life without air" stemmed largely from his fundamental studies on the biological basis of the making of wine and beer.

Glucose and certain other sugars are widely utilized as a major source of energy by heterotrophic organisms. With few exceptions, whenever glucose is utilized as the source of energy—and eventually carbon—the glycolytic pathway is the backbone of its degradation system, as illustrated in Fig. 1, most of its steps being common to anaerobic fermentation and to aerobic utilization through the citric acid cycle.

There are oxidation processes which occur in the absence of air, where the role of molecular oxygen as terminal electron acceptor is accomplished by some other inorganic substance, such as nitrate or sulphate. Fermentations can be defined as oxido-reduction processes where the terminal electron acceptors are organic compounds. Normally, the electron donor and the electron acceptor arise from the same molecule. For this reason, in order to be fermentable, a compound should be in an intermediate state of oxidation. Carbohydrates meet these requirements almost ideally and, in fact, are at the origin of most fermentations produced by micro-organisms.

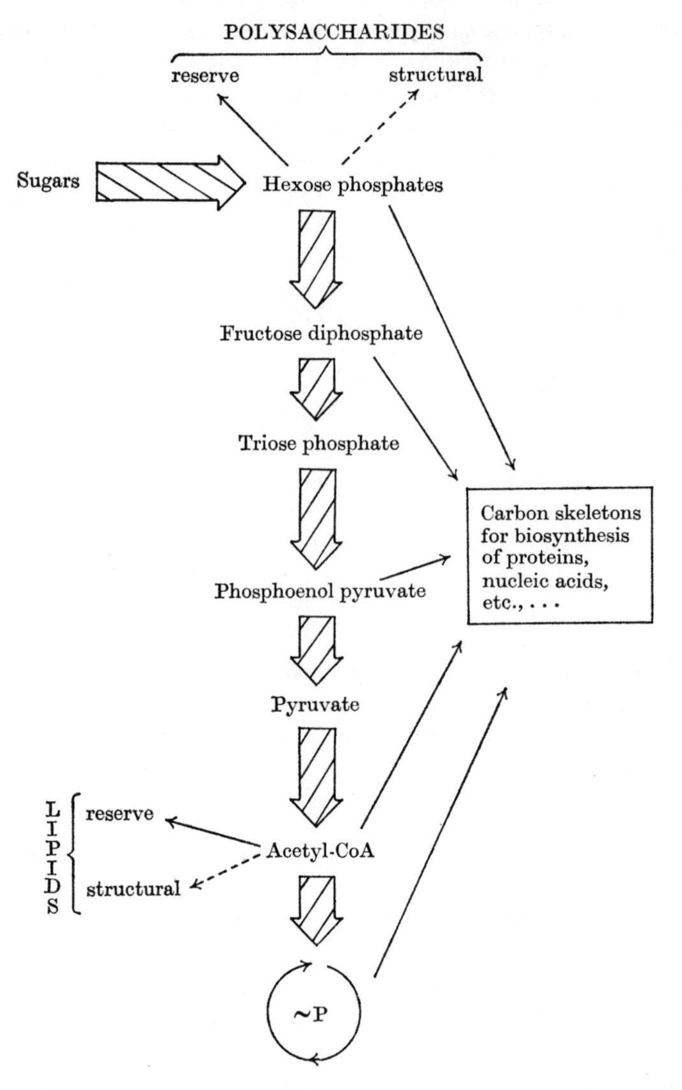


FIG. 1. Schematic outline of metabolism in yeast of sugar as energy and carbon source.

## A. ALCOHOLIC FERMENTATION

Anaerobic glycolysis yielding ethanol and carbon dioxide as final by-products (ATP being a physiological end-product of all types of glycolysis) is very typical among yeasts, although many lower plants and even some metazoa (Ward and Crompton, 1969) are also able to carry out alcoholic fermentation. The study of alcoholic fermentation by yeasts, which began in the second half of the nineteenth century, was of fundamental importance in the development of biochemistry in general and of enzymology in particular. The names of Pasteur, Buchner, Harden and Young are historical landmarks of the pioneering work from the 1850s to the early 1900s. In the following decades the realization of the very close similarities between anaerobic glycolysis in yeast and muscle, despite the difference in final by-products (lactate in muscle), opened the way to the basic and fruitful concept of unity in the living world at the biochemical (i.e. molecular) level.

The pathway of fermentation of glucose as far as the pyruvate stage is identical with that utilized in respiration. What distinguishes fermentation from respiration is the ultimate fate of pyruvate. The overall equation for alcoholic fermentation was established as early as 1815 by Gay-Lussac as:

$$C_6H_{12}O_6 \rightarrow 2\,C_2H_5OH + 2\,CO_2$$

Although this equation is fundamentally correct, it should be pointed out that in the fermentation of glucose by yeast, certain other products also appear, although in minor proportions (see Vol. 3, Chapter 6). Pasteur observed that some glycerol and succinic acid consistently accompany the ethanol formed during fermentation. We will see below the significance of these additional products in relation to anaerobic growth.

The change in free energy of the Gay–Lussac equation is about –56 kcal/mole. If we compare this figure with that for the complete combustion of glucose (–688 kcal/mole), it is obvious that fermentation is a far less economical process than respiration, since with the former the cell can obtain much less energy (less ATP) per mole of glucose utilized. For this reason, the respiratory process will, in general, be employed whenever possible. As Pasteur (1876) pointed out "if we supply yeast with sufficient free oxygen ... it ceases to be a ferment, i.e. the ratio between the weight of the plant developed and that of the sugar decomposed is similar to that in the case of fungi. If we deprive the yeast of air entirely, it will multiply as if air were present, though less actively, and in these circumstances its fermentative character will be most marked; moreover, under these circumstances we shall find the greatest disproportion ... between the weight of yeast and the weight of sugar decomposed." Aerobically the proportion of substrate assimilated is very much higher than anaerobically. The yield of cells growing aerobically may reach ten times that attained in anaerobic conditions for the same quantity of glucose used.

Although different yeasts may vary in detail, their fermentative patterns have general characters in common. Almost all can ferment glucose. Since the most thoroughly studied process has been the fermentation of glucose by *Saccharomyces cerevisiae,* we will take this case as a model.

### 1. *Uptake of Glucose*

The utilization of glucose by intact yeasts requires as the first step a catalysed transport across the membrane (Cirillo, 1961). This requirement is due to the fact that the plasma membrane of the yeast cells is highly impermeable to compounds even of the size of the commonest sugars. The constitutive hexose transport in *Sacch. cerevisiae* can take glucose across the membrane about a million times faster than it would otherwise enter. Transport takes place under equilibrium conditions (i.e. "facilitated diffusion"), and is not accumulative; that is, it is not active in the thermodynamical sense, although it is very active indeed in a physiological sense (Heredia *et al.*, 1968). Nevertheless, accumulative transport of monosaccharides (thermodynamically "active") has been reported to occur in *Rhodotorula* (Kotyk and Höfer, 1965).

### 2. *Glycolytic Enzymes*

Once glucose has been transported into the cell, it undergoes phosphorylation to glucose 6-P by a constitutive *hexokinase* (Sols *et al.*, 1958) according to Step 1 (Table I). The mechanism of this physiologically irreversible reaction has been extensively studied; in contrast to the hexokinases of higher organisms, the yeast enzyme is not sensitive to allosteric inhibition by glucose 6-P (DelaFuente, 1970).

A very active *glucosephosphate isomerase* (Step 2) maintains glucose 6-P and fructose 6-P near equilibrium (about 3/1). Of the several anomeric forms of glucose 6-P in aqueous solution, the open chain form seems to be the preferred substrate, although the enzyme is also able to catalyse the opening of $\alpha$-glucopyranose 6-P. Several metabolites of the pentose phosphate pathway can act as strong competitive inhibitors (Salas, M. *et al.*, 1965). The existence of three isoenzymes in baker's yeast, as well as in brewer's yeast, has recently been reported (Nakagawa and Noltmann, 1967).

The phosphorylation of fructose 6-P to fructose 1,6-diphosphate catalysed by *phosphofructokinase* (Step 3) is virtually irreversible *in vivo,* a fact that contributes to make this enzyme particularly suitable for the regulation of glycolysis. Phosphofructokinases are very complex enzymes, whose activities are affected by a considerable number of metabolic agents. Yeast phosphofructokinase is strongly inhibited by ATP, raising the $K_m$ for fructose 6-P (Viñuela *et al.*, 1963). This inhibition by ATP can be counteracted by AMP with great efficiency (Ramaiah *et al.*, 1964). Ions of $NH_4^+$ can also activate phosphofructokinase by counteracting the inhibitory effect of ATP, and as the $NH_4^+$ ion is a key

TABLE I. *Data on Glycolytic Enzymes from Yeast* (Saccharomyces cerevisiae)

| Enzyme | Reaction catalysed | $\Delta G_0'$[a] (kcal/mole) | $K_m$ (mM) | | Cofactors | Molecular weight | Molecular activity (mol./min) | Sub-units |
|---|---|---|---|---|---|---|---|---|
| Hexokinase | (1) glucose + ATP → glucose 6-P + ADP | −5·1 | Glucose<br>ATP | 0·1[b]<br>0·2 | $Mg^{2+}$ | 96,000[c] | 13,000 | 4 |
| Glucosephosphate isomerase | (2) glucose 6-P ⇌ fructose 6-P | +0·5 | | | | 145,000[d] | 97,000[e] | |
| Phosphofructo-kinase | (3) fructose 6-P + ATP → fructose 1,6-diP + ADP | −4·2 | Fructose 6-P<br>ATP | 0·15[f]<br>0·02 | $Mg^{2+}$ | 580,000[g] | 67,000 | 13 |
| Aldolase | (4) fructose 1,6-diP ⇌ glyceraldehyde 3-P + dihydroxyacetone-P | +5·51 | Fructose 1,6-diP<br>Glyceraldehyde 3-P<br>Dihydroxyacetone P | 0·3[h]<br>2<br>2 | $(Zn^{2+})$* | 70,000[h] | 12,000 | 2 |
| Triosephosphate isomerase | (5) dihydroxyacetone-P ⇌ glyceraldehyde 3-P | +1·83 | | | | | | |
| Glyceraldehyde-3-P dehydrogenase | (6) glyceraldehyde 3-P + $P_i$ + NAD ⇌ 1,3-diP-glycerate + $NADH_2$ | +1·5 | | | | 120,000[i] | 3,000 | 2 |
| Phosphoglycerate kinase | (7) 1,3-diP-glycerate + ADP ⇌ 3-P-glycerate + ATP | −4·75 | 1,3-DiP-glycerate<br>ADP<br>3-P-Glycerate<br>ATP | 0·002[j]<br>0·2<br>0·2<br>0·1 | $Mg^{2+}$ | 34,000[k] | 110,000[j] | |
| Phosphoglycerate mutase | (8) 3-P-glycerate ⇌ 2-P-glycerate | +1·06 | 2-P-Glycerate | 0·1[l] | 2,3-DiP-glycerate[l] | 112,000[l] | 80,000 | |
| Enolase | (9) 2-P-glycerate ⇌ P-enolpyruvate + $H_2O$ | −0·64 | 2-P-Glycerate | 0·2[m] | $Mg^{2+}$ | 67,000[m] | 5,700 | 1 |
| Pyruvate kinase | (10) P-enolpyruvate + ADP → pyruvate + ATP | −6·1 | P-Enolpyruvate<br>ADP | 2[n]<br>0·5 | | 150,000[n] | 30,000 | 2 |
| Pyruvate decarboxylase | (11) pyruvate → acetaldehyde + $CO_2$ | −5·1 | Pyruvate | 1·0[o] | Thiamin pyrophosphate<br>$Mg^{2+}$ | 175,000[p] | 9,000 | 2 |
| Alcohol dehydrogenase | (12) acetaldehyde + $NADH_2$ ⇌ ethanol + $NAD_2$ | −5·4 | Acetaldehyde<br>$NADH_2$ | 0·01[q]<br>0·78 | $(Zn^{2+})$* | 150,000[r] | 37,000 | 4[s] |

*References:* [a]Burton (1957); [b]Darrow and Colowick (1962); [c]Schulze *et al.* (1966); [d]Noltmann and Bruns (1959); [e]Nakagawa and Noltmann (1965); [f]Sols and Salas (1966); [g]Lindell and Stellwagen (1968); [h]Rutter and Hunsley (1966); [i]Krebs (1955); [j]Bücher (1955); [k]Larsson-Raznikiewicz and Malström (1961); [l]Grisolía (1962); [m]Westhead (1966); [n]Haeckel *et al.* (1968); [o]Holzer and Goedde (1957); [p]Ullrich and Kempfle (1969); [q]Wratten and Cleland (1963); [r]Racker (1955); [s]Ohta and Ogura (1965).

* Protein-bound, involved in the active structure.

substrate for yeast growth, this fact is likely to have a physiological significance (Sols and Salas, 1966). Finally citrate, in the range of physiological concentrations, enhances the inhibition by ATP, and this effect seems to play an important role in the mechanism of the Pasteur effect (Salas, M. L. *et al.*, 1965; see Section VI.A, p. 297).

Another interesting feature of phosphofructokinase is the fact that, in contrast to its high specificity for the allosteric inhibition by ATP, it has a broad specificity for the nucleotide substrate: GTP, CTP, ITP and UTP can replace ATP as phosphate donor. It has been shown that in yeast the concentration of phosphofructokinase can vary widely, depending on the carbon and energy source (Gancedo *et al.*, 1965); the decrease in phosphofructokinase observed when yeast is grown on non-sugar precursors would allow gluconeogenesis to proceed without a wasteful cycle at the level of the pair of antagonistic enzymes phosphofructokinase and fructose-1,6-diphosphatase. Conversion of yeast phosphofructokinase into a form insensitive to inhibition by ATP and citrate was observed by Viñuela *et al.* (1964), and has been further studied (Gancedo *et al.*, 1969), but its mechanism and physiological significance remain as yet unclear.

An *aldolase* catalyses the cleavage of fructose diphosphate into dihydroxyacetone-P and glyceraldehyde 3-P (Step 4). Yeast aldolase is specifically stimulated by $K^+$ and $NH_4^+$ ions, and inhibited by EDTA and other chelating agents; it belongs to Class II aldolases, which contain $Zn^{2+}$ (Rutter, 1964).

The interconversion between the two products of fructose diphosphate cleavage, dihydroxyacetone-P and glyceraldehyde 3-P, is catalysed by *triosephosphate isomerase* (Step 5). The equilibrium is displaced in favour of dihydroxyacetone-P (equilibrium constant $K = 22$), but as glyceraldehyde 3-P is removed by the next glycolytic enzyme, the reaction proceeds, in glycolysis, in the direction of glyceraldehyde 3-P formation.

The step catalysed by *glyceraldehyde 3-phosphate dehydrogenase* (Step 6) is the only oxidative step in the sequence of glycolysis leading to pyruvate. The accompanying change in free energy is channelled to the incorporation of $P_i$ into an energy-rich bond in the product 1,3-diphosphoglycerate. In the reaction, NAD is converted to $NADH_2$, and for glycolysis to proceed NAD must be regenerated in some other reaction. We shall see later that several reactions can fulfil this role of regenerating NAD. Poisoning of glycolysis by iodoacetate, a tool widely used in studies of fermentation, is due to the irreversible inhibition of glyceraldehyde 3-phosphate dehydrogenase because iodoacetate readily reacts with key –SH groups of the enzyme (Holzer *et al.*, 1955).

Until this stage the successive transformations suffered by the glucose molecule have not yielded any usable energy; in the step catalysed by *phosphoglycerate kinase* (Step 7), the energy-rich acyl phosphate bond of 1,3-diphosphoglycerate is transferred to ADP, giving 3-phosphoglycerate and ATP.

3-Phosphoglycerate is converted to 2-phosphoglycerate by a *phosphoglycerate mutase* (Step 8). The yeast enzyme needs for activity catalytic amounts of 2,3-diphosphoglycerate as specific cofactor (Rodwell *et al.*, 1957).

The dehydration of 2-phosphoglycerate catalysed by *enolase* (Step 9) yields phosphoenolpyruvate. In this step a low-energy phosphate ester is converted into an energy-rich enolphosphate. The reaction requires $Mg^{2+}$, and it seems that enolase binds one metal ion per molecule, giving a complex which in turn reacts with the substrate to form a ternary complex (Cohn, 1963). The inhibition of fermentation by fluoride ions is primarily due to inhibition of the enolase, as in the presence of phosphate a magnesium fluorophosphate complex is formed that can bind to the active site of the enzyme (Warburg and Christian, 1941).

*Pyruvate kinase* catalyses the cleavage of phosphoenolpyruvate to pyruvate, coupled with the transfer of its energy-rich phosphate to ADP to yield ATP (Step 10). The equilibrium is displaced far towards the formation of pyruvate and ATP, and the reverse reaction in gluconeogensis is by-passed in yeasts by the system pyruvate carboxylase and phosphoenolpyruvate carboxykinase, which involves oxaloacetate as intermediate. As already stated in the case of phosphofructokinase, another enzyme which acts irreversibly in physiological conditions, the glycolytic enzyme should not be active when gluconeogenesis is operative in order to avoid a cycle where ATP would be wasted. In the case of pyruvate kinase two different regulation mechanisms have been described (Gancedo *et al.*, 1967). Either the concentration of pyruvate kinase changes markedly with the carbon source available to the yeast (*C. utilis*, *Rh. glutinis*), or the pyruvate kinase depends for its activity on concentrations of fructose diphosphate that are reached in glycolysis but not in gluconeogenesis (*Sacch. cerevisiae*). In certain yeasts the two mechanisms can co-exist (Torróntegui *et al.*, 1968). The pyruvate kinase from *Sacch. cerevisiae* can be affected by some other metabolites (Haeckel *et al.*, 1968).

In anaerobiosis the pyruvate formed in the preceding reaction is mainly decarboxylated to acetaldehyde by *pyruvate decarboxylase* (Step 11), an enzyme which requires thiamin pyrophosphate and $Mg^{2+}$ as cofactors (Green *et al.*, 1941).

The acetaldehyde formed is reduced by $NADH_2$ to ethanol in a reaction catalysed by *alcohol dehydrogenase* (Step 12). It has been reported recently (Schimpfessel, 1968; Roche and Azoulay, 1969) that two isoenzymes of alcohol dehydrogenase are present in yeast. The first, which is constitutive, would be related with glucose fermentation, while the second, induced by oxygen and repressed by glucose, would have the role of oxidizing ethanol when yeast is growing on this compound. Both enzymes require NAD as coenzyme, but they differ in thermostability and substrate specificity.

### 3. *Energy Gain in Glycolysis*

As indicated above, the energy gain in glycolysis in the form of ATP is localized at two steps: the phosphoglycerate kinase and the pyruvate kinase reactions. Let us examine in some detail how this ATP synthesis is achieved.

Formation of 1,3-diphosphoglycerate involves the oxidation of an aldehyde to a carboxylic acid, coupled with the phosphorylation of ADP. The oxidation of an aldehyde to an acid is an exergonic process which liberates about 7 kcal/mole; that is, an energy of the same order as the free energy of hydrolysis of ATP. It is then possible to link both processes with a negligible net decrease in free energy. The first step in the process is the formation of a low energy thiohemiacetal between the aldehyde and a sulphydryl group of the enzyme. In the case of glyceraldehyde 3-phosphate dehydrogenase this sulphydryl group forms part of a cysteine residue. The hemiacetal is next dehydrogenated by bound nicotinamide nucleotide, resulting in a high-energy thioester bond. The thioester has about the same energy level as an acid anhydride and can undergo phosphorolysis, a process in which the acyl group is transferred from the sulphur atom to the phosphate group. The resulting compound, 1,3-diphosphoglycerate, therefore possesses an energy-rich carboxyl-phosphate bond, and the phosphate group may be transferred to ADP by a specific enzyme.

$$\mathrm{CH_2O{-}\text{Ⓟ}{-}CHOH{-}C(H){=}O} + \mathrm{E{-}SH} \longrightarrow \mathrm{CH_2O{-}\text{Ⓟ}{-}CHOH{-}C(H)(OH){-}S{-}E} \xrightarrow{\mathrm{NAD}\ \to\ \mathrm{NADH_2}}$$

$$\mathrm{CH_2O{-}\text{Ⓟ}{-}CHOH{-}C({=}O){-}S{-}E} \xrightarrow{\mathrm{H_2O}} \mathrm{CH_2O{-}\text{Ⓟ}{-}CHOH{-}C({=}O){-}O{\sim}\text{Ⓟ}} \xrightarrow{\mathrm{ADP}\ \to\ \mathrm{ATP}} \mathrm{CH_2O{-}\text{Ⓟ}{-}CHOH{-}C({=}O){-}OH}$$

$\Delta G = +7$

$\Delta G = -11{\cdot}8$ $\qquad\qquad \Delta G = -3{\cdot}1$

ATP is also generated in the conversion of phosphoenolpyruvate to pyruvate. The energy-rich phosphoenolpyruvate is generated when

2-phosphoglycerate (a low-energy phosphate ester) is dehydrated by enolase. Cleavage of phosphoenolpyruvate may then be linked with the transfer of phosphate to the acceptor ADP.

$$\begin{array}{c} CH_2OH \\ | \\ C{-}H \\ | \\ C{-}O{-}Ⓟ \\ | \\ COOH \end{array} \xrightarrow{H_2O} \begin{array}{c} CH_2 \\ \| \\ C{-}O{\sim}Ⓟ \\ | \\ COOH \end{array} \xrightarrow{ADP \quad ATP} \begin{array}{c} CH_3 \\ | \\ C{=}O \\ | \\ COOH \end{array}$$

$\Delta G = -3$ $\quad$ $\Delta G = -12{\cdot}8$ $\quad$ $\Delta G = +7$

Now we can calculate the energy balance of alcoholic fermentation. Two molecules of ATP are consumed in the earlier steps of glycolysis (hexokinase and phosphofructokinase), and two are generated from each triosephosphate. Since each glucose molecule gives two molecules of triosephosphate, four molecules of ATP are synthesized. The net gain is therefore 2 moles of ATP per mole of glucose utilized. The value of $\Delta G$ for the reaction:

$$C_6H_{12}O_6 \rightarrow 2\,C_2H_5OH + 2\,CO_2$$

is −56 kcal/mole. Of this energy only about 14 kcal are retained in the 2 moles of ATP gained. The energy yield of alcoholic fermentation is therefore about $14/56 = 0{\cdot}25$ (25%).

## B. OTHER TYPES OF FERMENTATION

As already stated, in one of the glycolytic steps (the oxidation of glyceraldehyde 3-P to 1,3-diphosphoglycerate) an oxidized coenzyme (NAD) is a participant in the reaction as acceptor of the hydrogen atoms removed from glyceraldehyde 3-P, becoming itself reduced to $NADH_2$. Since the pool of oxidized nucleotides is limited, there has to be a way of regenerating NAD for glycolysis to continue. The last step in alcoholic fermentation, reduction of acetaldehyde to ethanol, accomplishes this function. $NADH_2$ is a coenzyme in the reaction catalysed by alcohol dehydrogenase, and NAD arises as a product. Ethanol here is merely a by-product.

However, if for any reason the alcohol dehydrogenase reaction is not operative, regeneration of NAD can be carried out by an alternative reaction. Reduction of dihydroxyacetone-P to glycerol 3-P is the main alternative. This glycerophosphate is then hydrolysed by a specific phosphatase to glycerol, which is innocuous and can leak out of the cell (Gancedo *et al.*, 1968). Regeneration of NAD operates in this way at the beginning of glucose fermentation, when the concentration of acetaldehyde is not yet sufficient to support efficient activity of the alcohol dehydrogenase (Holzer *et al.*, 1963). This situation can be artificially maintained by addition of sulphite to the medium. Sulphite binds acetaldehyde and thus blocks the normal major means of NAD regeneration. The main fermentation product in this case is glycerol (Neuberg and Reinfurth, 1918). This kind of fermentation has been termed the "second form of fermentation". It should be emphasized that this process is unphysiological and completely useless for the cell, which does not obtain from it either energy or building blocks. The overall reaction is:

$$\underset{\text{glucose}}{C_6H_{12}O_6} \xrightarrow{\text{ATP, NAD}} \underset{\text{acetaldehyde}}{CH_3CHO} + CO_2 + \underset{\text{glycerol}}{C_3H_8O_3}$$

When the medium has an alkaline pH, another type of fermentation occurs, the so-called "third form of fermentation", in which ethanol, acetic acid, glycerol and carbon dioxide appear as products. The biochemical interpretation of this phenomenon is that, when the yeast is in an alkaline medium, an aldehyde dehydrogenase with an alkaline pH optimum (8·7) diverts acetaldehyde from its usual route to ethanol, to give acetic acid. This enzyme also requires NAD (Black, 1951), so that the cell must again utilize the glycerol 3-phosphate dehydrogenase reaction to re-oxidize $NADH_2$. It seems likely that the physiological role of this process is to bring the pH of the medium within a range favourable for the yeast cell. *Zygosaccharomyces acidifaciens* appears to undergo this type of fermentation as its normal glycolytic process (Nickerson and Carroll, 1945).

In certain conditions in the absence of a nitrogen source, pyruvate and glycerol can accumulate as products (Neuberg and Kobel, 1930). A possible explanation for this "fourth form of fermentation" may be the inducible character of pyruvate decarboxylase (Ruiz-Amil *et al.*, 1966). If glucose is offered to a yeast whose pyruvate decarboxylase is repressed, and in which the possibility of protein synthesis is hindered, fermentation will proceed only to the level of pyruvate; the regeneration of NAD by the alcohol dehydrogenase reaction will no longer be possible, and the glycerol 3-phosphate dehydrogenase will then fulfil a vicarious role as in the other cases mentioned above.

### C. HEXOSE MONOPHOSPHATE OXIDATIVE PATHWAY

Yeasts do not metabolize all the glucose they consume through the Embden–Meyerhof glycolytic pathway, but channel part through the pentose phosphate cycle (Blumenthal *et al.*, 1954). In this pathway energy is channelled, not to the formation of ATP, but to the reduction of NADP. A majority of the reactions require reducing power that in many cases has to be supplied as $NADPH_2$. In aerobic conditions

$NADPH_2$ can be partly oxidized through transdehydrogenation and the respiratory chain, thus contributing to the generation of ATP.

The $NADPH_2$-yielding reactions are located in the early steps of the pathway. The first, oxidation of glucose 6-P to 6-phosphogluconolactone, is catalysed by glucose 6-phosphate dehydrogenase, an enzyme widely distributed among yeasts, although differences in concentration between different species have been reported (Horecker *et al.*, 1968). The 6-phosphogluconolactone formed in this reaction is hydrolysed, giving 6-phosphogluconate. Although the hydrolysis of the lactone can be a spontaneous process, a specific lactonase has been described in yeast (Brodie and Lipmann, 1955). The second step yielding $NADPH_2$ is the oxidative decarboxylation of 6-phosphogluconate to ribulose 5-P, catalysed by phosphogluconate dehydrogenase. The other reactions of the pathway are isomerizations or rearrangements of the carbon chain and will not be considered here, except to mention that these rearrangements between phosphorylated sugars allow the possibility for the hexose monophosphate pathway to operate as a cycle through whose operation a molecule of glucose could be fully oxidized to 6 $CO_2$, producing 12 $NADPH_2$ as the physiological product (see Horecker, 1968).

### D. CONTRIBUTION OF DIFFERENT PATHWAYS OF GLUCOSE METABOLISM TO ANAEROBIC GROWTH

Most yeasts are facultative anaerobes. Anaerobic growth on glucose as the only source of energy and carbon (with the exception of certain minor factors), poses complex problems of metabolic integration of potentially conflicting needs. In order to grow and multiply in these conditions, yeasts have to use glucose to meet three major requirements: (i) the metabolic intermediates needed as starting materials for the major biosynthetic pathways; (ii) metabolically-harnessed energy (i.e. ATP); and (iii) reducing power. The last requirement arises from the fact that yeast constituents are on the average in a more reduced state than the nutrient sugar.

There are in yeast metabolic pathways that, starting from glucose, diverge at various points and can supply, proceeding to different stages, each of the three major requirements mentioned above. When facing these fundamental problems, the traditional descriptions of various forms of yeast fermentation (see Section II.B, p. 281) lead practically nowhere. Indeed, some of the special "forms" of fermentation, as usually defined, are physiologically meaningless. Clarification of the complex problem of anaerobic growth of yeast from glucose can be greatly aided by an evaluation of each of the possible pathways diverging from glucose 6-P and reaching different levels. Such individual evaluations must include, insofar as possible within the framework of the available knowledge, specification of the physiological product(s) and of the net expenditure required to obtain them. This systematic evaluation is summarized in Table II. If these various pathways, in the double sense of direction and extent, are considered as different "metabolic wheels", each turning over as many times as appropriate and possible in relation to the others, it becomes feasible to account for the three major requirements of energy, carbon skeletons and reducing power. These three factors, plus a few inorganic salts, are in principle all that is essential for the anaerobic growth of yeast on glucose. Adjustment of the turnover of the different "wheels" would depend on competition for, and availability of, common intermediates, aided in some cases by finer regulatory signals involving allosteric enzymes.

This model of "metabolic wheels" capable of being co-ordinated can be a useful framework for working hypotheses to clarify important questions on which the available information is scanty. For instance, there is the problem of the balance of reducing power. The fact that most dehydrogenases are specific for one of the two dinucleotides creates a kind of functional division that could be referred to as "chemical compartmentation". With this expression we wish to focus attention on the fact that two specific dehydrogenases, even if present in the same cellular compartment where the two dinucleotides are available, would each use a different one, despite the absence of physical separation. Oxidative reactions in catabolic pathways are typically linked to NAD, while most reductive biosynthetic pathways are linked to NADP, specifically or preferentially. There are certain key reactions that strongly tend to maintain a high $NAD/NADH_2$ ratio, needed for glycolysis, and a high $NADPH_2/NADP$ ratio, favourable for biosynthetic processes. On first approximation, yeast growing anaerobically on glucose would have considerable $NADH_2$ formation in the glycolytic pathway, while the overall requirement for the building up of new cell materials requires large amounts of $NADPH_2$ This contrast suggests the possibility of the potential involvement in anaerobically-growing yeast of certain enzymes or isoenzymes able to catalyse efficiently major biosynthetic reducing reactions with $NADH_2$ as electron donor. An additional possibility to be considered is transhydrogenation between the two dinucleotides. The $NADH_2$ disposal pathway leading to glycerol is probably more a safety device than a systematic process, that would as such be too wasteful.

Another major area where much additional information is desirable in relation to the problem of the facultative anaerobic growth of yeast is the exploration, in whole extracts of anaerobically-grown yeast, of enzymes that could be involved in the reductive (Rossi *et al.*, 1964) and oxidative pathways that can lead from pyruvate to the various dicarboxylic acids required for certain major biosynthetic pathways.

TABLE II. *Main Pathways Followed by Glucose as Source of Energy and Carbon in Anaerobically-Growing Yeast*[a]

| System | End product | Physiological products | Net expenditure (per mole of glucose) | By-products |
|---|---|---|---|---|
| A. GLYCOLYTIC PATHWAYS leading to | Ethanol | 2 ~P | | 2 Ethanol + 2 $CO_2$ |
| | Glycerol | 2 NAD | 2 ~P | 2 Glycerol |
| | Glycerol 3-P | 2 Glycerol residues + 2 NAD | 2 ~P | — |
| | 3-Phosphoglycerate | 2 3-Phosphoglycerate | 2 NAD | — |
| | Phosphoenolpyruvate | 2 Phosphoenolpyruvate | 2 NAD | — |
| | Pyruvate | 2 Pyruvate + 2 ~ P | 2 NAD | — |
| | Oxaloacetate | 2 Oxaloacetate | 2 NAD (+2 $CO_2$) | — |
| | Succinate | 2 Succinate + 2 FMN[b] | (2 $CO_2$) | — |
| | "Active acetate" | 2 Acetyl-CoA + 2 ~ P | 4 NAD | 2 $CO_2$ |
| | α-Ketoglutarate | α-Ketoglutarate (+ ~P) | 4 NAD | $CO_2$ |
| | Acetic acid | 2 $H^+$ + 2 ~ P | 4 NAD | 2 Acetate |
| B. HEXOSE MONOPHOSPHATE PATHWAYS leading to | Pentose-P | Pentose-P + 2 $NADPH_2$ | ~P | $CO_2$ |
| | Erythrose-P | Erythrose-P + 2 $NADPH_2$ | ~P | $CO_2$ |
| | (Pentose-P cycle) | 12 $NADPH_2$ | ~P | 6 $CO_2$ |
| C. POLYSACCHARIDE SYNTHESES | | Hexose residue | 2 ~ P | — |
| REDUCTIVE BIOSYNTHESES | | Cell constituents | $x$ ~ P<br>$y$ $NADPH_2$<br>(or $NADH_2$?) | |

[a] For the sake of clarity, $P_i$ changes have been omitted and some simplifications have been made, particularly in the hexose monophosphate pathways. ~P stands for the terminal phosphoryl group of ATP.

[b] Or perhaps some other oxidized coenzyme.

## III. Aerobic Metabolism

### A. CITRIC ACID CYCLE

The final pathway for the oxidation of nutrients in aerobic conditions is the citric acid, tricarboxylic acid or Krebs cycle. Carbohydrate, amino acids and lipid derivatives are oxidized to carbon dioxide and water in the cycle which, coupled to the respiratory chain, acts as the main energy producer of the cell. The citric acid cycle fulfils another important mission: several of its intermediates serve as precursors for the synthesis of many cellular constituents. In yeasts and other eucaryotic organisms, the different enzymes that catalyse the various steps of the cycle are localized in the mitochondria.

The principal fuels for the cycle are two-carbon units in the form of acetyl-CoA, although some other compounds may enter the cycle at different levels. The cycle is depicted in Fig. 2. Since this chapter is concerned with energy production, only those processes relevant to this aspect will be described, namely, the generation of acetyl-CoA and the four dehydrogenation steps of the cycle.

Acetyl-CoA arises mainly from pyruvate or from fatty acid degradation (see Section V.C, p. 295). Pyruvate in turn arises from the degradation of glucose or from certain amino acids. The system which transforms pyruvate into acetyl-CoA is a complex of enzymes arranged in an ordered way. In the first step of the reaction, pyruvate is activated by protein-bound thiamin pyrophosphate, resulting in the formation of hydroxyethylcarboxylthiamin pyrophosphate. This compound loses carbon dioxide, producing "active acetaldehyde" which undergoes oxidation by lipoate with the formation of acetyl-dihydrolipoate, which then reacts with CoA to form acetyl-CoA. Lipoate is re-oxidized by an NAD-dependent flavoprotein. Coenzyme A acts then as a carrier of acetyl units to the citric acid cycle; with its sulphydryl groups it may form thioester bonds (as in acetyl-CoA) which are high-energy bonds (acid anhydrides). The free energy of hydrolysis of acetyl-CoA is $-8{\cdot}8$ kcal/mole (compared with the value of $-7$ kcal/mole for the hydrolysis of ATP). For this reason the acetyl group may be transferred to other acetyl acceptors. Fig. 2 shows the fate of the acetyl residue in the citric acid cycle. Four dehydrogenations and two decarboxylations take place in the cycle: the four dehydrogenations are the key reactions from the energetic point of view, while the two decarboxylations liberate the acetate residue as carbon dioxide.

The first dehydrogenation (isocitrate to $\alpha$-ketoglutarate) is catalysed by isocitrate dehydrogenase. Two enzymes have been described in yeast (Kornberg and Pricer, 1951), one NAD-dependent and the other linked to NADP. The NAD-isocitrate dehydrogenase is always associated with

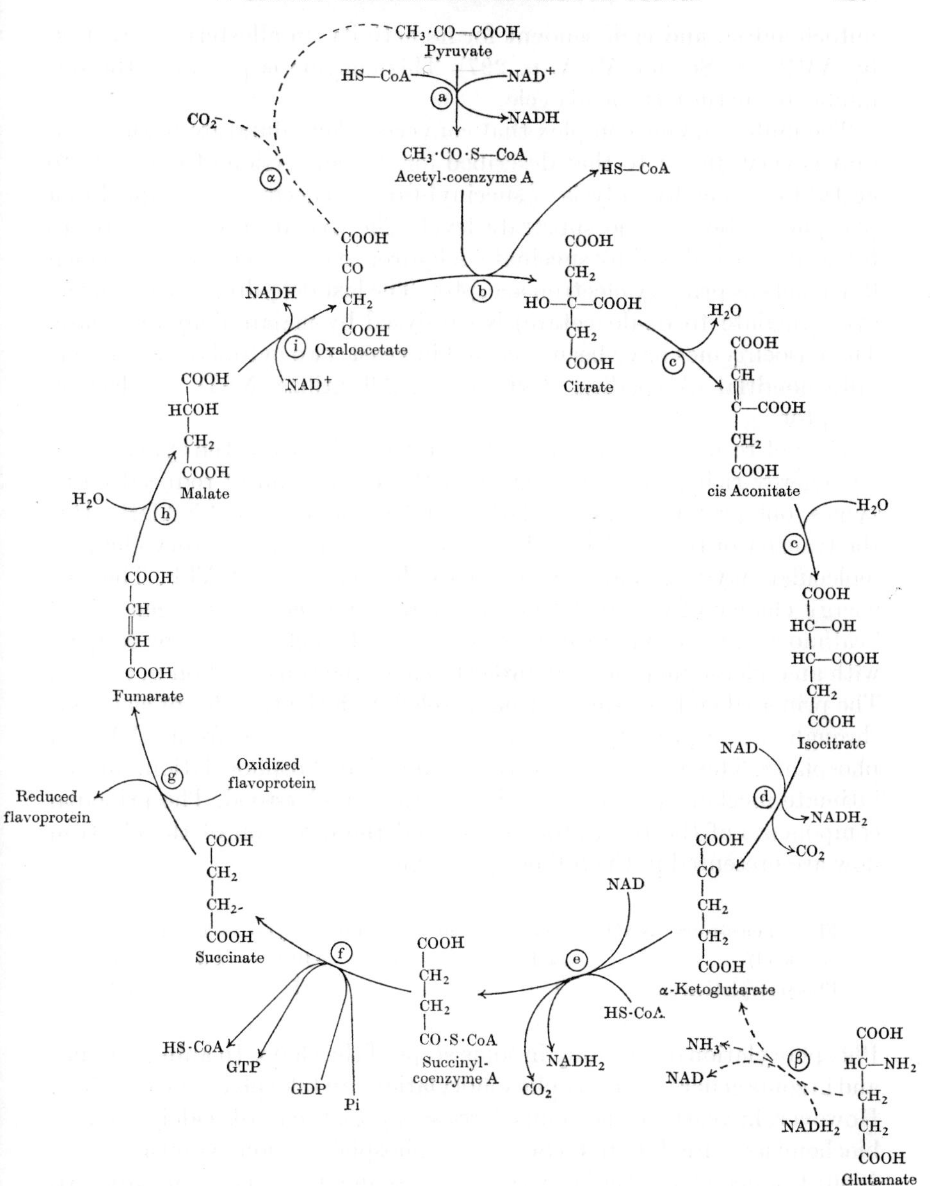


FIG. 2. Citric acid cycle. The figure includes a pathway linking glycolysis and the citric acid cycle: (a) pyruvate dehydrogenase complex; (b) citrate synthase; (c) aconitase; (d) isocitrate dehydrogenase; (e) α-ketoglutarate dehydrogenase; (f) succinate thiokinase; (g) succinate dehydrogenase; (h) fumarase; (i) malate dehydrogenase; (α) pyruvate carboxylase; (β) glutamate dehydrogenase.

mitochondria, and is dependent for its activity on allosteric activation by AMP (see Section VI.A, p. 297). This enzyme is probably the one implicated in the citric acid cycle.

The multi-enzyme complex that converts α-ketoglutarate to succinyl-CoA is very similar to that described for the conversion of pyruvate to acetyl-CoA. The hydrolysis of succinyl-CoA to succinate is coupled to a phosphorylation at the substrate level. The oxidation of succinate to fumarate is catalysed by succinate dehydrogenase, an enzyme for which flavin acts as primary electron acceptor. The last dehydrogenation of the cycle (malate to oxaloacetate) is catalysed by malate dehydrogenase. Three isoenzymes have been reported in yeast, two cytoplasmic and one mitochondrial (Atzpodien *et al.*, 1968). All utilize NAD as electron acceptor.

The net result of a full turn of the citric acid cycle is the synthesis of one energy-rich phosphate bond and the production of reduced coenzymes, one protein-bound $FADH_2$ and three molecules of $NADH_2$. It is the transfer of the displaced electrons, through the respiratory chain to molecular oxygen, that is coupled to the synthesis of ATP. The free energy change of the oxidation of $NADH_2$ by molecular oxygen is −52 kcal/mole. Liberation of this energy in partial reactions may be coupled with an endergonic process in order to effect the conservation of energy. The principal endergonic reaction coupled with the transfer of electrons through the respiratory chain is the synthesis of ATP from ADP and phosphate. This process is known as oxidative phosphorylation, and its intimate mechanism is only beginning to be understood. The principal components of the respiratory chain and the direction of the electron flow are presented in the following scheme:

| Flow of electrons | NAD | → | flavoproteins | → | cyt *b* | → | cyt *c* | → | cyt *a* | → | $O_2$ |
|---|---|---|---|---|---|---|---|---|---|---|---|
| $\Delta G$ (kcal) | | −12·4 | | −4·1 | | −10·1 | | −1·3 | | −24·4 | |
| Phosphorylation site | | I | | | | II | | | | III | |

Phosphorylation occurs only in some steps of the chain. In animal tissues and in some genera of yeasts phosphorylation can take place at three sites. However, in yeasts of the genus *Saccharomyces*, the most widely used by biochemists, Site I is not coupled to phosphorylation (Ohnishi *et al.*, 1966; Chance *et al.*, 1967), with the consequent synthesis of only two instead of three ATP molecules per $NADH_2$ re-oxidized.

Since mitochondria are not permeable to the dinucleotides (Boxer and Devlin, 1961; Ohnishi *et al.*, 1966) some device must exist to allow for the oxidation by mitochondria of cytoplasmically-produced $NADH_2$. The systems which are most likely to play a role in this process are the α-glycerophosphate shuttle and the oxaloacetate cycle, thus:

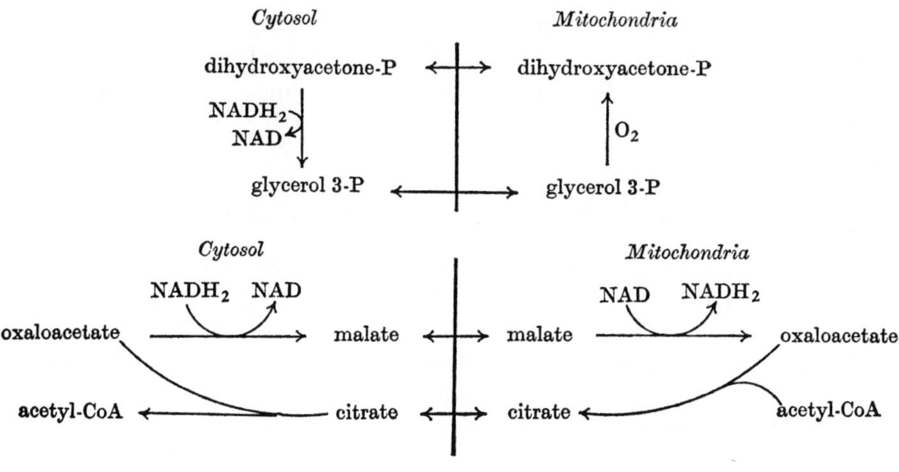


The existence in yeast of a glycerol 3-phosphate dehydrogenase and a glycerophosphate oxidase has been documented (Gancedo *et al.*, 1968), as well as the presence of a cytoplasmic and a mitochondrial malate dehydrogenase (Atzpodien *et al.*, 1968). However, the relative importance of each process in the removal of $NADH_2$ cannot be assessed from the available information.

While the oxidation of $NADH_2$ in mitochondria results in the formation of high-energy bonds (synthesis of ATP), there is little evidence that $NADPH_2$ could be oxidized in the same way. In order to serve for the generation of ATP, $NADPH_2$ must be transhydrogenated with NAD, which then can follow the pathway described above.

## B. ANCILLARY PATHWAYS

The draining of intermediates of the citric acid cycle as starting materials for a variety of biosynthetic pathways requires the existence of some pathway able to replenish the cycle intermediates. For the pathways fulfilling this replenishing role, Kornberg (1965) has introduced the term "anaplerotic pathways".

In yeasts growing on pyruvate precursors as main source of energy and carbon, the anaplerotic requirement is accomplished by an enzyme that catalyses the formation of oxaloacetate from pyruvate and carbon dioxide, and requires ATP (Ruiz-Amil *et al.*, 1965). This enzyme is known as pyruvate carboxylase, a term that should not be confused with the glycolytic pyruvate decarboxylase described above (Section II.A.2, p. 276), for which the name pyruvate carboxylase had been unwisely used for a long time until the identification of the truly carboxylating enzyme.

Metabolism of acetate and acetate precursors in the absence of pyruvate precursors uses an anaplerotic pathway, the glyoxylate cycle (Kornberg, 1965), which involves two inducible enzymes: isocitratase and malate synthase.

## IV. Energy Sources Other Than Glucose

Although in the preceding pages attention has been centred on glucose metabolism, glucose is by no means the only major carbon source commonly available to yeasts. We shall now review briefly the pathways of utilization of some of the compounds more commonly metabolized by yeasts. Most of them, after a few steps, enter the glycolytic chain, while some others are transformed into intermediates of the citric acid cycle.

### A. SUGARS AND RELATED COMPOUNDS

*a. Monosaccharides.* Fructose and mannose are readily utilized by virtually all yeasts. Both are taken up by the constitutive hexose transport system and phosphorylated by the hexokinase. Their phosphoric esters are, either directly or after isomerization by mannose-phosphate isomerase, able to follow the glycolytic pathway. Galactose is taken up by an inducible transport system and phosphorylated to galactose 1-P by an inducible galactokinase. Uridyl transferase catalyses the reaction:

$$\text{Gal 1-P} + \text{UDPG} \rightarrow \text{G 1-P} + \text{UDPGal}$$

UDPG is regenerated from UDPGal by the action of UDPG-epimerase. Finally, glucose 1-P is transformed into glucose 6-P by phosphoglucomutase.

A number of yeasts do not utilize pentoses owing to the lack of specific kinases, and eventually dehydrogenases or other enzymes that contribute to make possible their access to the common pentose phosphate pathway. The very efficient utilization of xylose by *Candida utilis* involves two specific hydrogenases and a kinase (Chakravorty *et al.*, 1962).

*b. Glycosides.* Yeasts can utilize a number of oligosaccharides, generally after adaptation. The long controversy about the "direct" or "indirect' fermentation of disaccharides has been closed by the work of DelaFuente and Sols (1962). Maltose and related $\alpha$-glucosides, as well as lactose and related $\beta$-galactosides, are transported by inducible specific systems and hydrolysed inside the cell, also by inducible glycosidases. On the other hand, sucrose and related $\beta$-fructosides are hydrolysed outside the cell by a $\beta$-fructosidase located in the outer surface of the cell membrane. The same is true for $\alpha$-galactosides such as melibiose. The liberated hexoses are then transported and metabolized by the ordinary constitutive pathway. Extracellular trehalose is not utilized by most yeasts, apparently because of lack of a suitable transport system, since trehalase, which is very common in yeasts, is an intracellular enzyme.

Trisaccharides and tetrasaccharides can be broken down as they are analogues of disaccharides which can be utilized. For instance, maltotriose is assimilated by the same system as that by which maltose is utilized, and raffinose is assimilated by the combined action of those involved in the utilization of melibiose and sucrose. Polysaccharides with more than three hexose units are not easily transported. Because of this, they are utilized only insofar as they can be attacked by surface glycosidases. For instance, inulin is slowly hydrolysed by invertase ($\beta$-fructosidase) outside the cell and the liberated fructose is fermented by the ordinary pathway. Heteroglycosides that can be hydrolysed by yeast glycosidases, after transport when necessary, can contribute at least their sugar moiety for yeast metabolism.

*c. Polyols.* Yeasts show great differences in their ability to utilize polyols, a fact that makes these compounds a routine tool in taxonomic studies. The initial step of the intracellular metabolism of most polyols is probably an oxidation to the corresponding aldose or ketose, which can then be phosphorylated and enter the glycolytic cycle or the pentose phosphate cycle. The subject has recently been reviewed by Barnett (1968).

Glycerol is outstanding as a nutrient for many yeasts; its utilization involves a specific pathway involving phosphorylation to glycerol 3-P and oxidation of the latter to dihydroxyacetone-P (Gancedo *et al.*, 1968).

*d. Ethanol and organic acids.* Most yeasts can grow on simple substances such as ethanol, lactate or pyruvate. Ethanol is assimilated *via* acetaldehyde (see Section II.A.2, p. 276), acetate and acetyl-CoA. Lactate can be oxidized to pyruvate, which is itself an intermediate in glycolysis. Pyruvate utilization possibly depends on some inducible transport system. Without a controllable transport system, leakage of pyruvate would pose a serious interference to fermentation. Medrano *et al.* (1969) have found that addition of glucose stops pyruvate uptake by *Rhodotorula glutinis*. Nevertheless, at very low pH values and high pyruvic acid concentration in the medium, there would be a reasonable possibility of significant entrance by diffusion of the undissociated molecule, as is the case with other uncharged compounds of similar size (Gancedo *et al.*, 1968). Indeed, a vigorous decarboxylation of pyruvic acid by intact yeast at pH 2 under anaerobic conditions has been observed by Suomalainen *et al.* (1969).

### B. NITROGEN COMPOUNDS

A number of amino acids can serve as sources of carbon and energy as well as of nitrogen. Glutamate undergoes an oxidative deamination, yielding α-ketoglutarate and $NH_4^+$; the α-ketoglutarate formed enters the Krebs cycle and provides energy as well as carbon skeletons. It has been shown that yeast possesses two glutamate dehydrogenases (Hierholzer and Holzer, 1963). One, NAD-dependent, has high activity in cells grown on media with glutamate or other amino acids as nitrogen source, having therefore the characteristics of a catabolic enzyme. On the other hand, the NADP-dependent glutamate dehydrogenase is believed to be implicated in biosynthesis, although it shows no repression by glutamate.

Some other amino acids can also be deaminated directly: threonine and serine produce α-ketobutyrate and pyruvate respectively (Boll and Holzer, 1965), but in order to be assimilated most amino acids are first transaminated with α-ketoglutarate yielding glutamate and a keto acid. Alanine, for instance, gives pyruvate, and aspartate oxaloacetate which can be metabolized further by the glycolytic and tricarboxylic acid pathways. α-Ketoglutarate is regenerated from glutamate by the glutamate dehydrogenase described previously.

### C. HYDROCARBONS

The ability of some yeasts to grow on hydrocarbons as sole carbon and energy source is mentioned here, on account of their potential interest as food supplements. Vigorous investigations are being carried out in several laboratories in order to develop this possibility, *Rhodotorula* and *Candida* being the genera most used (Johnson, 1967; Roche and Azoulay, 1969; see Vol. 3).

## V. Energy Reserves and Their Mobilization

In most yeasts the energy reserves consist largely of carbohydrates, which account for about one-fourth of the dry matter. Almost the whole of the reserve carbohydrate comprises glycogen and trehalose. Baker's yeast usually contains 16–20% of glycogen and 6–10% of trehalose. Brewer's yeast has been reported to have usually less trehalose, about 1%. In any case, the storage of reserve carbohydrate is markedly dependent on the metabolic state of the cell, as shown by the following arguments. Polyphosphates, which can accumulate to over 10% of the dry weight in yeast, seem to be mainly a phosphate store rather than an energy store of the "phosphagen" type (Harold, 1966).

A. GLYCOGEN

The molecular structure of yeast glycogen, as well as its biosynthesis, is considered elsewhere (Chapter 10, p. 419). The existence has been reported of yeast mutants unable to store glycogen in conditions in which normal yeasts accumulate large amounts (Chester, 1968). These mutants, when grown in a medium with 8% glucose instead of the 1% normally used, accumulate glycogen and become indistinguishable from normal cells. Therefore, the failure must be at the glucose-transport step. In the mutants, a loss of affinity of the transport system (Heredia *et al.*, 1968) seems to be a plausible explanation of these observations.

Glycogen occurs as a heterogenous material. Samples of glycogen isolated from *Sacch. cerevisiae*, after disruption of the cells, have been separated by centrifugation into two components (Eaton, 1961). Over 80% of the fermentable glycogen is derived from the heavier component. Its greater susceptibility to fermentation has been related to the length of the glucose chains external to the branching points.

Parnas (1937) demonstrated that the initial step in glycogen breakdown was a phosphorolysis. After the Coris' discovery of the phosphorylase reaction in muscle, the similarities between the phosphorolytic activities of muscle and yeast extracts were well established (Cori, 1956; Cori *et al.*, 1938). Early work on the mechanisms of glycogen phosphorolysis was obscured by the fact that preparations of yeast glycogen phosphorylase were contaminated with variable amounts of amylase and other enzymes involved in the breakage of $(1 \rightarrow 4)$ and $(1 \rightarrow 6)$ glucosidic linkages.

In contrast with the well studied phosphorylases from animal tissues, yeast phosphorylases have received very little attention. The action of phosphorylase on glycogen is incomplete because of the inability of the enzyme to split or by-pass the $(1 \rightarrow 6)$ branch linkages. Then, after phosphorolysis of a number of external $(1 \rightarrow 4)$ linkages, a limit dextrin results.

The existence of isoamylase activity has been reported, and it has been suggested that this is responsible for the debranching of glycogen in yeast (Maruo and Kobayashi, 1951; Gunja *et al.*, 1961; Kjølberg and Manners, 1963). The isoamylase described by Manners and coworkers hydrolyses $(1 \rightarrow 6)$ linkages of the outer branches of glycogen and amylopectin. It is thus less specific than animal amylo-$(1 \rightarrow 6)$ glycosidase or the R enzyme from plants. On the other hand, Lee *et al.* (1967) have described a debranching enzyme in *Sacch. cerevisiae* similar to the amylo 1–6/oligo 1–4 $\rightarrow$ 1–4 glucan transferase of animal tissues (Brown and Illingworth, 1964). This transferase activity removes maltotriosyl residues from the outer branches of the limit dextrins, exposing the branch points as (1–6) glucosyl stubs which are then split off by the amylo (1–6) glycosidase found also in yeasts. The available evidence suggests that both the debranching activities mentioned exist in yeasts, although studies are lacking with genetically well-defined material.

Through these phosphorolytic and hydrolytic reactions yeast glycogen can be converted to various proportions of glucose-P and free glucose, thus joining the common glycolytic pathway.

## B. TREHALOSE

As in the case of glycogen, the molecular structure and biosynthesis of trehalose is reviewed by Manners (see Chapter 10, p. 419). With regard to the role of trehalose as carbohydate reserve in yeast, a number of observations have accumulated which give a puzzling picture. Only after an extensive investigation of the regulatory properties of both the synthesis and degradation pathways of trehalose (DelaPeña and DelaFuente, 1969) has a reasonable explanation of the apparently anomalous behaviour of trehalose as reserve carbohydate become possible.

Homogenates of baker's or brewer's yeast contain a trehalase able to split trehalose, to give two molecules of glucose. While the levels of trehalose can vary from virtually nothing to about 10% of the dry weight, no appreciable variations in the trehalase activity of homogenates can be detected. In actively-dividing populations no trehalose has been found. In resting cells with high external glucose, trehalose accumulates rapidly to a stationary level of about 0·1 *M*, in which conditions Avigad (1960) has shown that a rapid turnover of trehalose of about 5 μmoles glucose/min.g (wet weight) takes place. This figure is slightly higher than the trehalase activity found in homogenates in optimal conditions (DelaPeña and DelaFuente, 1969). If an assimilable nitrogen source is added, trehalose virtually disappears in about one hour. On the contrary, when cells are suspended in water or in a buffer, the trehalose concentration decreases by only about 20% in the first two hours and then remains practically unchanged for hours, or even days.

In order to explain the above observations, a "compartmentation" which keeps trehalose and trehalase out of contact has been invoked (Souza and Panek, 1968). Nevertheless, the observations on the turnover of trehalose in the presence of glucose (Avigad, 1960), and the observations on the effect of dinitrophenol and other cell poisons (see p. 296) renders this hypothesis very improbable.

The results of DelaPeña and DelaFuente (1969) indicate that trehalose-P synthetase is an inducible enzyme, apparently induced (or activated) by its substrates. Moreover, when the level of the substrates is low, the activity of the enzyme decreases abruptly within a few minutes. Trehalase is inhibited by AMP, and this inhibition is sharply dependent on the pH value. Since the level of trehalose depends on two pathways, those of synthesis and degradation, it is conceivable that considerable variations can occur, according to the metabolic state of the cell. In the absence of any external sugar, when the pool of triphosphates and that of sugar phosphates becomes low, and AMP accumulates, trehalose synthesis stops and so does its degradation; in these conditions the level remains almost steady. In the presence of external assimilable sugar both pathways operate, synthesis holding some initial advantage, so that trehalose accumulates. In the presence of an assimilable nitrogen source, when the synthesis of precursors for division starts, again the triphosphate pool is reduced and degradation pathways become predominant, so that the stored trehalose disappears.

### C. Lipids

In certain conditions yeasts accumulate lipids as potential reserves of energy (and carbon). Excess of air and abundant sugar supply are the principal requirements for such a process. Species of yeasts exist which can accumulate lipids to an extent that may exceed 50% of the dry weight (e.g. *Rhodotorula*, *Lipomyces*). Brewer's and baker's yeasts normally contain much less lipid (about 5%). The fatty acid synthetase complex from yeast has been isolated and thoroughly studied by Lynen (1968). The energy value of the accumulated lipid lies in the possibility of its mobilization and oxidation.

The reactions involved in the oxidation of fatty acids are:

$$(1)\quad R{-}CH_2{-}CH_2{-}COOH + HS{-}CoA + ATP \xrightarrow[\text{thiokinase}]{} R{-}CH_2{-}CH_2{-}CO{-}S{-}CoA + AMP + PP_i$$

$$(2)\quad R{-}CH_2{-}CH_2{-}CO{-}S{-}CoA + FP \xrightarrow[\text{acyl-CoA dehydrogenase}]{} R{-}CH{=}CH{-}CO{-}S{-}CoA + FPH_2$$

$$(3)\quad R{-}CH{=}CH{-}CO{-}S{-}CoA + H_2O \xrightarrow[\text{enoyl hydrase}]{} R{-}CHOH{-}CH_2{-}CO{-}S{-}CoA$$

$$(4)\quad R{-}CHOH{-}CH_2{-}CO{-}S{-}CoA + NAD \xrightarrow[\text{3-hydroxyacyl-CoA dehydrogenase}]{} R{-}CO{-}CH_2{-}CO{-}S{-}CoA + NADH_2$$

$$(5)\quad R{-}CO{-}CH_2{-}CO{-}S{-}CoA + HS{-}CoA \xrightarrow[\beta\text{-ketoacyl-CoA thiolase}]{} R{-}CO{-}S{-}CoA + CH_3{-}CO{-}S{-}CoA$$

↘ to Step 2

It must be pointed out that the reactions outlined here have been studied mainly in animal tissues. It is assumed that they hold in a similar fashion for yeast, but work in this field is needed in order to clarify the situation. As shown in the above outline, the complete oxidation of a fatty acid involves the repetition of a cycle at the end of which an acetyl-CoA unit and an acyl-CoA, shorter by two carbon atoms than the initial fatty acid, are formed. This oxidation also involves the operation of the citric acid cycle. Energy is gained by the oxidation of acetyl-CoA produced *via* the citric acid cycle and by the coupling of the oxidative reactions, Steps 2 and 4, with the respiratory chain.

Step 2 is catalysed by a flavoprotein (acyl-CoA dehydrogenase) which transforms the fatty acyl-CoA ester into an unsaturated ester. Transfer of each pair of electrons through the respiratory chain to oxygen results in the synthesis of two molecules of ATP. In Step 4 a hydroxyacyl-CoA ester is dehydrogenated to the corresponding keto-ester, a process in which NAD acts as electron acceptor. Oxidation *via* the respiratory chain of the $NADH_2$ produced allows the synthesis of two molecules of ATP for each pair of electrons transferred, as explained elsewhere (Section III.A, p. 286).

### D. ENDOGENOUS METABOLISM

Intact yeast, in the absence of external nutrient, uses its reserves only sparingly. Earlier studies (Stier and Stannard, 1936; Spiegelman and Nozawa, 1945) established that yeast cells are able to respire endogenously, but that endogenous anaerobic fermentation could not be observed unless certain "inducer" compounds (dinitrophenol or sodium azide) were present. More recent reports (Chester, 1959) showed that yeast cells can both respire and ferment their endogenous reserves even in the absence of inducers.

Although the occurrence of utilization of endogenous reserves appears to be well established, the nature of the materials utilized has been the subject of controversy. Lipids, amino acids and carbohydrates have been suggested as possible substrates. Spiegelman and Nozawa (1945) carried out careful respiratory quotient determinations on several strains of yeast, and came to the conclusion that it corresponds to the oxidation of carbohydrates. Chester (1959) established that decreases in the glycogen and trehalose fractions accounted for 85–100% of the carbon dioxide evolved. Respiratory quotients much greater than 1·0 were concomitant with accumulation of ethanol. A curious observation was that aerobic conditions seem to stimulate the rate of endogenous fermentation. Eaton (1959, 1960) showed that all the $^{14}CO_2$ released by uniformly labelled cells could be accounted for by a depletion of glycogen. From these results, the existence of two glycogen pools was postulated, one with long external chains, the other with short external chains, the former being degraded at a higher rate than the latter. By contrast, the reserve of trehelose is depleted to the extent of about 20% in the first two hours, and very slowly after this time.

The effect of dinitrophenol or sodium azide on endogenous fermentation deserves some comment. The former was extensively studied by Berke and Rothstein (1957), and recently Stoppani *et al.* (1968) have resolved the mechanism by which it uncouples oxidative phosphorylation. In the presence of $3 \times 10^{-4}$ $M$-dinitrophenol a slow but constant decrease in glycogen takes place, not very different from that occurring in yeast cells suspended in non-metabolizable buffer. On the contrary, trehalose decreases rapidly, so that in less than an hour the reserve is reduced to one half of that initially present, and after four hours it is practically exhausted. Similar results have been reported using sodium azide (Brady *et al.*, 1961), arsenate, arsenite, iodoacetate and other inhibitors.

Since the biosyntheses of glycogen and trehalose have essentially the same precursors, during active fermentation by resting cells both are synthesized. In the absence of nutrients, the ATP pool becomes depleted, while the level of AMP rises. The increase in AMP has two effects: (i) activation of the phosphorolytic degradation of glycogen; and (ii) inhibition of the hydrolysis of trehalose. Therefore, the substrate for endogenous fermentation must be preferentially glucose P derived from glycogen. Trehalose is less degraded, and even synthesized to some extent in so far as the levels of glucose 6-P and UDPG allow.

When dinitrophenol is added to starved yeast, the ATP pool becomes even more depleted. While the increase in the inhibition of trehalose hydrolysis by AMP does not change significantly, the synthesis of trehalose P could stop and, as a consequence, the intracellular store of trehalose would largely disappear within the first hour.

## VI. Regulation of Energy Metabolism

### A. THE PASTEUR EFFECT

The fundamental work of Pasteur on alcoholic fermentation by yeast opened the way that led, during the early decades of the present century, to the identification and detailed unravelling of the glycolytic pathway, the backbone of glucose metabolism in the large majority of organisms. Moreover, by studying the effect of the presence of oxygen on fermentation, Pasteur discovered the inhibition of fermentation by respiration, so that the "Pasteur effect" became the earliest and longest standing puzzle in metabolic regulation.

In alcoholic fermentation, and in anaerobic glycolysis in general, much of the chemical energy of the sugar is wasted by the cell in the elimination of ethanol (or other by-products). Obviously, if a cell could oxidize glucose completely to carbon dioxide and water, it could get much more profit per unit consumed. As a consequence, whenever oxidation is possible, *the cell uses glucose at a lower rate* than in anaerobic conditions. This decrease of the rate of glucose utilization is the most basic and complex problem of regulation, and is generally referred to as the "Pasteur effect". In marked contrast to the reasonableness of the purpose of the Pasteur effect, understanding of the mechanism has been most elusive indeed. Since the main lines of the enzymic basis of fermentation were completed in the 1920s, a large variety of claims have been made, including some very peculiar ones. Dickens (1951) ended a review by saying that, as first suggested by D. Burk, "the certainty with which various authors claim to have 'explained' the Pasteur effect is almost as general as the effect itself". Outstanding among these theories were those involving substrate competition between fermentation and respiration for ADP, inorganic phosphate, or both, formulated independently in the early 1940s by Lynen (1941) and Johnson (1941), and vigorously championed in the 1950s by opposing schools; one centred on the competition for ADP, and the other favoured competition for phosphate. The pursuit of this line, although too unsophisticated, gave considerable "side benefits" in the unravelling of several basic aspects of the working of some key reactions in the pathways involved.

Of fundamental importance for the understanding of the puzzling mechanism of the Pasteur effect was the advent of what may be described as "the third dimension in enzymology"; that is, the recognition of specific regulatory mechanisms for the control of enzyme activities, the heteropic allosteric effects, involving specific binding of certain metabolites by key enzymes at a site (or sites) distinct from the active one(s). In addition, in the particular case of yeasts, a complete understanding would not have been possible without the full realization, gained during the last decade, that for the utilization of most macronutrients a catalysed transport across the cell membrane is the first step, and may well be a key feature.

The Pasteur effect can now be explained as a consequence of basic regulatory mechanisms involving different key enzymes of glucose metabolism. These mechanisms make possible a precise and immediate adjustment of the rate of utilization of glucose to the metabolic requirements of the cell for energy and carbon skeletons. The recognition of the fact that the long pathway of glucose degradation generally involves two major pathways in anaerobiosis, and three in aerobiosis, is crucial for the understanding of the metabolic regulation of the whole system. These pathways (see Fig. 3) are: (i) from extracellular sugar to glucose 6-P, a major metabolic crossroads; (ii) from glucose 6-P to ATP and a carbon by-product in anaerobiosis, or to ATP and citrate in aerobiosis, followed in the latter case by; (iii) from citrate to more ATP and carbon dioxide plus water. Each pathway can be controlled by allosteric feed-back mechanisms affecting either the first step, or the first irreversible step, of each pathway. The Pasteur effect involves the integration of a series of feedback mechanisms acting in sequence (Sols, 1967).

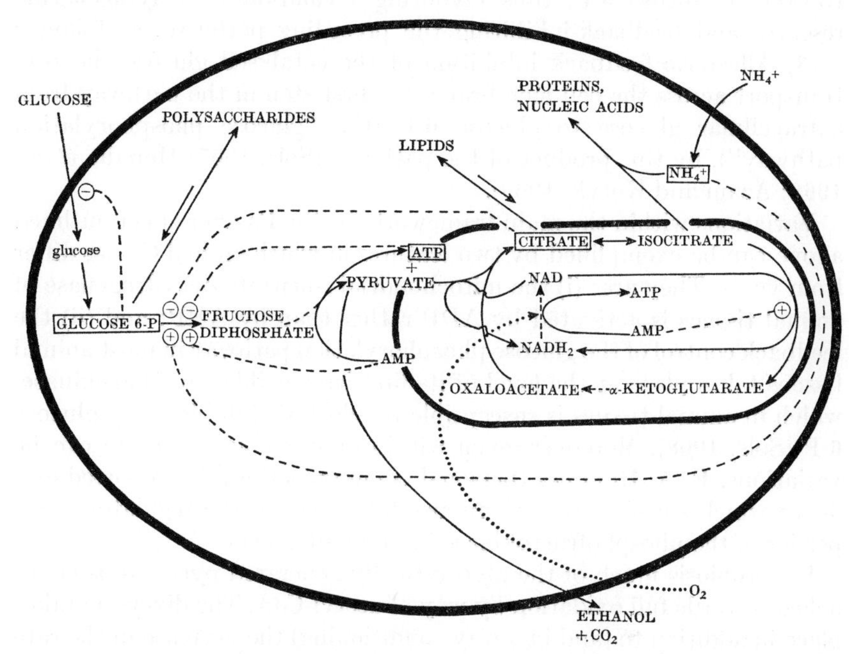


FIG. 3. Allosteric feedback regulation of glucose metabolism in yeast, and Pasteur effect. Mitochondrial "compartmentation" of the citric acid cycle and certain metabolites is indicated in the figure. ⊖ indicates inhibition, and ⊕ activation.

Within the general principle thus outlined, the specific regulatory mechanisms implicated in the Pasteur effect in baker's yeast include the following regulated steps and regulatory metabolites:

(1) Feedback control of the NAD-dependent isocitrate dehydrogenase, the first irreversible step in the pathway of oxidation of citrate through the citric acid cycle, by its dependence for activity in physiological conditions on allosteric activation by AMP (Hathaway and Atkinson, 1963). When the energy level in the cell is high, with a very low AMP/ATP ratio, the slowing down of isocitrate dehydrogenase activity allows the citrate concentration to increase and be used for the building up of fat reserves and/or to increase the feedback inhibition of the preceding pathway, as follows:

(2) Feedback inhibition of phosphofructokinase, the first irreversible step of the common glycolytic pathway from glucose 6-P, by the physiological products of this pathway: ATP in anaerobiosis, and ATP *and* citrate in aerobiosis (Salas *et al.*, 1965). The allosteric control by ATP is reinforced by a strong allosteric activating effect of AMP (Salas *et al.*, 1968). The inhibition of phosphofructokinase raises the concentration of glucose 6-P, thus favouring formation of polysaccharide reserves and feedback-inhibiting the preceding pathway, as follows:

(3) Allosteric feedback inhibition of the catalysed glucose (hexose) transport across the cell membrane, the first step in the pathway from extracellular glucose to glucose 6-P (the "glucose phosphorylation pathway"), by this product of the pathway (Sols, 1967; Heredia *et al.*, 1968; Azam and Kotyk, 1969).

Variations within the main framework for the Pasteur effect outlined above can be exemplified by two features in which animal tissues differ from yeast. They are: (i) the mitochondrial isocitrate dehydrogenase of animal tissues is activated by ADP rather than by AMP; and (ii) the feedback control of the glucose phosphorylation pathway in most animal tissues takes place at the level of its first irreversible step, hexokinase, which in animal tissues is susceptible to allosteric inhibition by glucose 6-P (Sols, 1968). Moreover, even within different yeasts, there can be variations. F. A. Hommes (personal communication) has reported evidence which suggests that there are differences in the regulatory properties of the phosphofructokinase in different yeasts.

In aerobiosis much of the glycolytically generated pyruvate is channelled towards full oxidation by way of acetyl-CoA. The diversion takes place in addition to (and in a way conditioning) the decrease in the rate of glucose expenditure. The ready *visibility* of the change in carbon dioxide evolution has often interfered with the awareness of the fundamental slowing down of the rate of glucose degradation. The mechanism of the diversion of pyruvate from by-products to oxidation can be accounted for by simple enzyme competition (Holzer, 1961). Feedback controls involving long pathways tend to produce oscillations in the rate of flow of metabolites. This basic tendency is reinforced in transition states involving major readjustment in the levels of certain coenzymes and intermediary metabolites. Very marked oscillations in yeast glycolysis, both *in vitro* and *in vivo*, have been observed and studied by Chance, Hess and their colleagues (Chance *et al.*, 1964; Hess and Boiteux, 1968; Hommes, 1965; Pye, 1969). These observations simulate a "multisite" control of the common glycolytic pathway, and may blur awareness of the fundamental role of some of the basic allosteric mechanisms specific for the regulation of glycolysis. The additional steps that can affect the flux of metabolites within certain sectors of the glycolytic pathway have been recently reviewed by Rose and Rose (1969).

The regulatory mechanisms described above seem to account sufficiently for the Pasteur effect in yeast and animal tissues. Additional regulatory mechanisms affecting key glycolytic enzymes, including pyruvate kinase (Gancedo *et al.*, 1967), are not likely to play a significant role in the Pasteur effect. Warburg's hypothetical "Pasteur reaction" ("Atmung und Gärung sind durch eine chemische Reaktion verbunden die ich Pasteursche Reaktion nenne"; Warburg, 1926) has thus been replaced by the integration of a series of allosteric feedback signals by intermediates that are used as metabolic messengers.

### B. INHIBITION OF RESPIRATION BY FERMENTATION

Many yeasts show marked impairment of respiratory capability when abundant glucose is available in the medium. This phenomenon, first studied by Ephrussi and coworkers (Ephrussi *et al.*, 1956), bears some resemblance to the inhibition in tumours of (endogenous) respiration produced by the addition of glucose, which was first observed by Crabtree (1929) and referred to thereafter as the "Crabtree effect", or "reverse Pasteur effect". Later work on the phenomenon of "aerobic fermentation of glucose" (i.e. alcohol formation despite availability of air) and on glucose-induced respiratory impairment, has led to the identification of a "glucose effect" (or "catabolite effect") on the (functional?) levels of a variety of mitochondrial enzymes and members of the respiratory chain. These decreased levels are reached within two to four hours after addition of glucose; return to normal high aerobic levels takes place within a similar time after the exhaustion of glucose from the medium (Utter *et al.*, 1967; Chapman and Bartley, 1968). The changes in enzymic activities related to respiration accompany marked morphological changes in mitochondrial structure and integrity (Linnane *et al.*, 1967).

There is considerable uncertainty as to whether the impairment of respiration caused by glucose is: (i) a case of the "catabolite repression" that affects the synthesis of many catabolic enzymes (Polakis *et al.*, 1965; DeDeken, 1966; Görts, 1967); (ii) related to the disassembly of normal mitochondrial structures; or (iii) involves a combination of factors. The mechanism(s) of catabolite repression in general is far from clear, and is currently under study in several laboratories. In any case, there is little reason to attempt to correlate the impairment of yeast respiratory ability by certain fermentable sugars with the inhibition of respiration provoked acutely by glucose in certain animal tissues (ably reviewed by Ibsen, 1961). What is well established is that abundance of rapidly-fermentable sugars tends to impair the respiratory machinery even under aerobic growth conditions, leading to a situation more or less similar to that of anaerobically-grown yeast. The formation of the dicarboxylic acids needed for anabolic pathways can be achieved by a branched non-cyclic pathway diverging from oxaloacetate. This branched pathway includes a reductive pathway involving an adaptive fumarate reductase (Rossi *et al.*, 1964) that leads to succinate, and an oxidative pathway that leads to $\alpha$-ketoglutarate. The efficiency of the reductive pathway of succinate formation in anaerobic yeast can be inferred from the fact that substantial amounts of succinate accumulate in the medium during yeast fermentation.

### C. METABOLIC INTEGRATION

The growth of yeasts, like that of all heterotrophic organisms, requires large amounts of chemical energy. The number of biosynthetic reactions that are energy-dependent probably runs into hundreds. The almost universal immediate metabolic energy donor is ATP. Energy requirements for growth have been estimated as about 1 $\mu$mole of ATP for each 10 $\mu$g of newly synthesized dry-cell material (Gunsalus and Schuster, 1961), or about 20,000 $\mu$moles ATP per gram of wet yeast. To obtain this energy from anaerobic glycolysis, yeast needs to ferment, to ethanol and carbon dioxide, some ten units of sugar for each unit transformed into new cellular material. This large figure is 10,000 times greater than the average content of ATP in yeast. The magnitude of this contrast leads to the metabolic requirement for a very fast turnover of the energy-rich bonds of the cell ATP. In fact, the half-life of ATP in actively-growing yeast is about one second. ATP can fulfil its role as universal cellular current insofar as the adenyl nucleotides behave as a "battery" that is as readily coupled to a variety of energy-requiring reactions as to reactions that can "recharge the battery". If ATP represents the fully charged battery ADP is half-charged, and AMP is in the completely discharged state. While many energy-requiring reactions use only the terminal bond of ATP, there are also many others that involve the $\beta$-P-bond, ending thus in fully discharged AMP. These three adenyl nucleotides are very readily interconverted by an adenylate kinase (or "myokinase"), which catalyses a reaction whose equilibrium is very close to one. Atkinson and Walton (1967) have introduced the concept of "energy charge", defined as the ratio (ATP + 1/2 ADP)/(ATP + ADP + AMP). This ratio varies between 0 and 1, and is numerically equal to half the average number of anhydride-bound phosphate groups per adenosine moiety.

The relatively low *capacity* of the adenylate "battery" to store energy makes fine adjustment in the metabolic integration of energy-spending and -yielding pathways of paramount importance. This crucial integration is largely due to the fact that the generality of major energy-yielding pathways have a key step geared to the "energy charge" through sensitivity to allosteric activation by AMP (or ADP), inhibition by ATP, or both. Moreover, some major biosynthetic pathways involving the formation of non-essential products (storage materials) have also a key step geared to the "energy charge" through allosteric inhibition by AMP. Outstanding illustrations of these generalizations are found in the antagonistic pair of enzymes phosphofructokinase-fructosediphosphatase, key steps respectively in energy-yielding glycolysis and in a potential pathway for accumulation of polysaccharide stores from non-sugar nutrients. Yeast phosphofructokinase is allosterically inhibited by ATP and allosterically activated by AMP (Salas *et al.*, 1968) while yeast fructosediphosphatase is allosterically inhibited by AMP (Gancedo *et al.*, 1965). While coarse controls of induction and repression of many enzymes of potential catabolic importance are of very great significance for the long-term economic efficiency of cells facing different nutritional conditions, it is the network of signals involving AMP and ATP as allosteric regulators of the activity of key enzymes of major pathways for energy production or storage that makes possible a highly efficient, rapid and smooth adjustment between expenditure and production of metabolic energy.

## VII. Acknowledgements

The authors are indebted to Dr. J. M. Gancedo for her very valuable advise and criticism, to Miss Clotilde Estévez for very able assistance in the preparation of the manuscript, and to Mr. Lorenzo Seguido for the preparation of the figures.

### References

Atkinson, D. E. and Walton, G. M. (1967). *J. biol. Chem.* **242**, 3239–3241.
Atzpodien, W., Gancedo, J. M., Duntze, W. and Holzer, H. (1968). *Eur. J. Biochem.* **7**, 58–62.
Avigad, G. (1960). *Biochim. biophys. Acta* **40**, 124–134.
Azam, F. and Kotyk, A. (1969). *FEBS Letters* **2**, 333–335.
Barnett, J. A. (1968). *In* "The Fungi", (G. C. Ainsworth and K. S. Sussman, eds.), Vol. III, pp. 557–595. Academic Press, New York.
Berke, H. L. and Rothstein, A. (1957). *Arch. Biochem. Biophys.* **72**, 380–395.
Black, S. (1951). *Archs Biochem. Biophys.* **34**, 86–97.
Blumenthal, H. J., Lewis, K. F. and Weinhouse, S. (1954). *J. Am. chem. Soc.* **76**, 6093–6097.
Boll, M. and Holzer, H. (1965). *Biochem. Z.* **343**, 504–518.
Boxer, G. E. and Devlin, T. M. (1961). *Science, N.Y.* **134**, 1495–1501.
Brady, T. G., Duggan, P. F., McGann, C. and Tully, E. (1961). *Archs Biochem. Biophys.* **93**, 220–230.
Brodie, A. F. and Lipmann, F. (1955). *J. biol. Chem.* **212**, 677–685.
Brown, D. H. and Illingworth, B. (1964). *In* "Control of Glycogen Metabolism", (W. J. Whelan and M. P. Cameron, eds.), pp. 139–150. J. & A. Churchill, London.

Bücher, T. (1955). *In* "Methods in Enzymology", (S. P. Colowick and N. O. Kaplan, eds.), Vol. I, pp. 415–422. Academic Press, New York.
Burton, K. (1957). *In* "Energy Transformations in Living Matter", (H. A. Krebs and H. L. Kornberg), pp. 275–285. Springer Verlag, Berlin.
Chakravorty, M., Veiga, L. A., Bacila, M. and Horecker, B. L. (1962). *J. biol. Chem.* **237**, 1014–1020.
Chance, B., Hess, B. and Betz, A. (1964). *Biochem. biophys. Res. Commun.* **16**, 182–187.
Chance, B., Lee, C.-P. and Mela, L. (1967). *Fedn Proc. Fedn. Am. Socs exp. Biol.* **26**, 1341–1354.
Chapman, C. and Bartley, W. (1968). *Biochem. J.* **107**, 455–465.
Chester, V. E. (1959). *Nature, Lond.* **183**, 902–903, and **184**, 1956–1957.
Chester, V. E. (1968). *J. gen. Microbiol.* **51**, 49–56.
Cirillo, V. P. (1961). *A. Rev. Microbiol.* **15**, 197–218.
Cohn, M. (1963). *Biochemistry, N.Y.* **2**, 623–629.
Cori, C. F. (1956). *In* "Enzymes: Units of Biological Structure and Function", (O. H. Gabler, ed.), pp. 573–583. Academic Press, New York.
Cori, G. T., Colowick, S. P. and Cori, C. F. (1938). *J. biol. Chem.* **123**, 375–380.
Crabtree, H. G. (1929). *Biochem. J.* **23**, 536–545.
Darrow, R. A. and Colowick, S. P. (1962). *In* "Methods in Enzymology", (S. P. Colowick and N. O. Kaplan, eds.), Vol. V, pp. 226–235. Academic Press, New York.
De Deken, R. H. (1966). *J. gen. Microbiol.* **44**, 149–156.
DelaFuente, G. (1970). *In* "Metabolic Regulation and Enzyme Action", (A. Sols and S. Grisolía, eds.), pp. 249–262. Academic Press, London.
DelaPeña, R. and DelaFuente, G. (1969). To be published.
DelaFuente, G. and Sols, A. (1962). *Biochim. biophys. Acta* **56**, 49–62.
Dickens, F. (1951). *In* "The Enzymes", (J. B. Sumner and K. Myrbäck, eds.), Vol. II, Part 1, pp. 624–683. Academic Press, New York.
Eaton, N. R. (1959). *Biochim. biophys. Acta* **36**, 259–262.
Eaton, N. R. (1960). *Archs Biochem. Biophys.* **88**, 17–25.
Eaton, N. R. (1961). *Archs Biochem. Biophys.* **95**, 464–469.
Ephrussi, B., Slonimski, P. P., Yotsuyanagi, Y. and Tavlitzki, J. (1956). *C. r. Trav. Lab. Carlsberg, Sér. physiol.* **26**, 87.
Gancedo, C., Salas, M. L., Giner, A. and Sols, A. (1965). *Biochem. biophys. Res. Commun.* **20**, 15–20.
Gancedo, C., Gancedo, J. M. and Sols, A. (1968). *Eur. J. Biochem.* **5**, 165–172.
Gancedo, J. M., Gancedo, C. and Sols, A. (1967). *Biochem. J.* **102**, 23C–25C.
Gancedo, J. M., Atzpodien, W. and Holzer, H. (1969). *6th FEBS Meet. Abstr.* Madrid, p. 668.
Görts, C. P. (1967). *Antonie van Leeuwenhoek* **33**, 451–463.
Green, D. E., Herbert, D. and Subrahmanyan, V. (1941). *J. biol. Chem.* **138**, 327–339.
Grisolía, S. (1962). *In* "Methods in Enzymology", (S. P. Colowick and N. O. Kaplan, eds.), Vol. V, pp. 236–242. Academic Press, New York.
Gunja, Z. H., Manners, D. J. and Khin Maung (1961). *Biochem. J.* **81**, 392–398.
Gunsalus, I. C. and Schuster, C. W. (1961). *In* "The Bacteria", (I. C. Gunsalus and R. Y. Stanier, eds.), Vol. II, pp. 1–58. Academic Press, New York.
Haeckel, R., Hess, B., Lauterborn, W. and Wüster, K. H. (1968). *Hoppe-Seyler's Z. physiol. Chem.* **349**, 699–714.
Harold, F. M. (1966). *Bact. Rev.* **30**, 772–794.

Hathaway, J. A. and Atkinson, D. E. (1963). *J. biol. Chem.* **238**, 2875–2881.
Heredia, C. F., Sols, A. and DelaFuente, G. (1968). *Eur. J. Biochem.* **5**, 321–329.
Hess, B. and Boiteux, A. (1968). *Hoppe-Seyler's Z. physiol. Chem.* **349**, 1567–1574.
Hierholzer, G. and Holzer, H. (1963). *Biochem. Z.* **339**, 175–185.
Holzer, H. (1961). *Cold Spring Harb. Symp. quant. Biol.* **26**, 277–288.
Holzer, H. and Goedde, H. W. (1957). *Biochem. Z.* **329**, 192–208.
Holzer, H., Holzer, E. and Schultz, G. (1955). *Biochem. Z.* **326**, 385–404.
Holzer, H., Bernhardt, W. and Schneider, S. (1963). *Biochem. Z.* **336**, 495–509.
Hommes, F. A. (1965). *Comp. Biochem. Physiol.* **14**, 231–238.
Horecker, B. L. (1968). *In* "Carbohydrate Metabolism and its Disorders", (F. Dickens, P. J. Randle and W. J. Whelan, eds.), Vol. 1, pp. 139–167. Academic Press, London.
Horecker, B. L., Rosen, O. M., Kowal, J., Rosen, S., Scher, B., Lai, C. Y., Hoffee, P. and Cremona, T. (1968). *In* "Aspects of Yeast Metabolism", (A. K. Mills, and H. A. Krebs, eds.), pp. 71–103. Blackwell Scientific Publications, Oxford.
Ibsen, K. (1961). *Cancer Res.* **21**, 829–841.
Johnson, M. J. (1941). *Science, N.Y.* **94**, 200–202.
Johnson, M. J. (1967). *Science, N.Y.* **155**, 1515–1519.
Kjølberg, O. and Manners, D. J. (1963). *Biochem. J.* **86**, 258–262.
Kornberg, A. and Pricer, W. (1951). *J. biol. Chem.* **189**, 123–136.
Kornberg, H. L. (1965). *In* "Essays in Biochemistry", (P. N. Campbell and G. D. Greville, eds.), Vol. 2, pp. 1–31. Academic Press, London.
Kotyk, A. and Höfer, M. (1965). *Biochim. biophys. Acta* **102**, 410–422.
Krebs, E. G. (1955). *In* "Methods in Enzymology", (S. P. Colowick and N. O. Kaplan, eds.), Vol. I, pp. 407–411. Academic Press, New York.
Larsson-Raznikiewicz, M. and Malström, B. G. (1961). *Archs Biochem. Biophys.* **92**, 94–99.
Lee, E. Y. C., Nielson, L. D. and Fisher, E. H. (1967). *Archs Biochem. Biophys.* **121**, 245–247.
Lindell, T. J. and Stellwagen, E. (1968). *J. biol. Chem.* **243**, 907–912.
Linnane, A. W., Biggs, D. R., Huang, M. and Clark-Walker, G. D. (1967). *In* "Aspects of Yeast Metabolism", (A. K. Mills and H. A. Krebs, eds.), pp. 217–242. Blackwell Scientific Publications, Oxford.
Lynen, F. (1941). *Justus Liebigs Annln Chem.* **546**, 120–141.
Lynen, F. (1968). *In* "Aspects of Yeast Metabolism", (A. K. Mills and H. A. Krebs, eds.), pp. 271–302. Blackwell Scientific Publications, Oxford.
Maruo, B. and Kobayashi, T. (1951). *Nature, Lond.* **167**, 606–607.
Medrano, L., Ruiz-Amil, M. and Losada, M. (1969). *Arch. Mikrobiol.* **66**, 239–249.
Nakagawa, Y. and Noltmann, E. A. (1965). *J. biol. Chem.* **240**, 1877–1881.
Nakagawa, Y. and Noltmann, E. A. (1967). *J. biol. Chem.* **242**, 4782–4788.
Neuberg, C. and Kobel, M. (1930). *Biochem. Z.* **219**, 490–494.
Neuberg, C. and Reinfurth, E. (1918). *Biochem. Z.* **92**, 234–266.
Nickerson, W. J. and Carroll, W. R. (1945). *Archs Biochem.* **7**, 257–271.
Noltmann, E. A. and Bruns, F. H. (1959). *Biochem. Z.* **331**, 436–445.
Ohnishi, T., Sottocasa, G. and Ernster, L. (1966). *Bull. Soc. Chim. biol.* **48**, 1189 1203.
Ohta, T. and Ogura, Y. (1965). *J. Biochem., Tokyo* **58**, 73–89.
Parnas, J. K. (1937). *Ergebn. Enzymforsch.* **6**, 57–110.
Pasteur, L. (1876). "Études sur la Bière". Gauthieur-Villars éd., Paris.
Polakis, E. S., Bartley, W. and Meek, G. A. (1965). *Biochem. J.* **97**, 298–302.

Pye, E. K. (1969). *Can. J. Bot.* **47**, 271–285.
Racker, E. (1955). *In* "Methods in Enzymology", (S. P. Colowick and N. O. Kaplan, eds.), Vol. I, pp. 500–503. Academic Press, New York.
Ramaiah, A., Hathaway, J. A. and Atkinson, D. E. (1964). *J. biol. Chem.* **239**, 3619–3622.
Roche, B. and Azoulay, E. (1969). *Eur. J. Biochem.* **8**, 426–434.
Rodwell, V. W., Towne, J. C. and Grisolía, S. (1957). *J. biol. Chem.* **228**, 875–890.
Rose, I. A. and Rose, Z. B. (1969). *In* "Carbohydrate Metabolism", (M. Florkin and E. H. Stotz, eds.), Vol. 17, pp. 93–161. Elsevier, Amsterdam.
Rossi, C., Hauber, J. and Singer, T. P. (1964). *Nature, Lond.* **204**, 167–170.
Ruiz-Amil, M., Torróntegui, G. de, Palacián, E., Catalina, L. and Losada, M. (1965). *J. biol. Chem.* **240**, 3485–3492.
Ruiz-Amil, M., Fernández, M. J., Medrano, L. and Losada, M. (1966). *Arch. Mikrobiol.* **55**, 46–53.
Rutter, W. J. (1964). *Fedn Proc. Fedn. Am. Socs exp. Biol.* **23**, 1248.
Rutter, W. J. and Hunsley, J. R. (1966). *In* "Methods in Enzymology", (W. A. Wood, ed.), Vol. IX, pp. 480–486. Academic Press, New York.
Salas, M. L., Salas, J. and Sols, A. (1968). *Biochem. biophys. Res. Commun.* **31**, 461–466.
Salas, M., Viñuela, E. and Sols, A. (1965). *J. biol. Chem.* **240**, 561–568.
Salas, M. L., Viñuela, E., Salas, M. and Sols, A. (1965). *Biochem. biophys. Res. Commun.* **19**, 371–376.
Schimpfessel, L. (1968). *Biochim. biophys. Acta* **151**, 317–329.
Schulze, I. T., Gazith, J. and Gooding, R. H. (1966). *In* "Methods in Enzymology", (W. A. Wood, ed.), Vol. IX, pp. 376–381. Academic Press, New York.
Sols, A. (1967). *In* "Aspects of Yeast Metabolism", (A. K. Mills and H. A. Krebs, eds.), pp. 47–66. Blackwell Scientific Publications, Oxford.
Sols, A. (1968). *In* "Carbohydrate Metabolism and its Disorders", (F. Dickens, P. J. Randle and W. J. Whelan, eds.), Vol. 1, pp. 53–87. Academic Press, London.
Sols, A. and Salas, M. L. (1966). *In* "Methods in Enzymology", (W. A. Wood, ed.), Vol. IX, pp. 436–442. Academic Press, New York.
Sols, A., DelaFuente, G., Villar-Palasí, C. and Asensio, C. (1958). *Biochim. biophys. Acta* **30**, 92–101.
Souza, N. O. and Panek, A. D. (1968). *Archs Biochem. Biophys.* **125**, 22–28.
Spiegelman, S. and Nozawa, M. (1945). *Archs Biochem. Biophys.* **6**, 303–321.
Stier, T. J. B. and Stannard, J. N. (1936). *J. gen. Physiol.* **19**, 461–477.
Stoppani, A. O. M., Claisse, L. M. and De Pahn, E. M. (1968). *Revta Soc. argent. Biol.* **44**, 45–53.
Suomalainen, H., Kouttinen, K. and Oura, E. (1969). *Arch. Mikrobiol.* **64**, 251–261.
Torróntegui, G. de, Palacián, E., Tresguerres, E. F. and Losada, M. (1968). *Arch. Mikrobiol.* **62**, 192–197.
Ullrich, J. and Kempfle, M. (1969). *FEBS Letters* **4**, 273–274.
Utter, M. F., Duell, E. A. and Bernofsky, C. (1967). *In* "Aspects of Yeast Metabolism", (A. K. Mills and H. A. Krebs, eds.), pp. 197–215. Blackwell Scientific Publications, Oxford.
Viñuela, E., Salas, M. L. and Sols, A. (1963). *Biochem. biophys. Res. Commun.* **12**, 140–145.
Viñuela, E., Salas, M. L., Salas, M. and Sols, A. (1964). *Biochem. biophys. Res. Commun.* **15**, 243–249.
Warburg, O. (1926). "Stoffwechsel der Tumoren". Springer, Berlin.

Warburg, O. and Christian, W. (1941). *Biochem. Z.* **310**, 384–421.
Ward, P. F. V. and Crompton, D. W. T. (1969). *Proc. R. Soc., Series B.* **172**, 65–88.
Westhead, E. W. (1966). *In* "Methods in Enzymology", (W. A. Wood, ed.), Vol. IX, pp. 670–679. Academic Press, New York.
Wratten, C. C. and Cleland, W. W. (1963). *Biochemistry, N.Y.* **2**, 935–941.

*Chapter 8*

# The Properties and Composition of Yeast Nucleic Acids

JEAN-CLAUDE MOUNOLOU

*Institut de Biologie Moleculaire, Faculté des Sciences, Paris, France*

I. INTRODUCTION . . . . . . . . . . . . . . 309
II. DEOXYRIBONUCLEIC ACIDS . . . . . . . . . . . . 310
A. Isolation of Yeast DNA . . . . . . . . . . 310
B. Structure and Composition of DNA . . . . . . . 311
C. Heterogeneity of Yeast DNA . . . . . . . . . 311
D. Composition and Structure of Nuclear DNA . . . . . 312
E. Yeast DNA and Evolution . . . . . . . . . 313
F. Response to Mutagens and Radiations . . . . . . 313
G. DNA and Chromosomes . . . . . . . . . . 314
H. DNA Replication . . . . . . . . . . . 315
III. RIBONUCLEIC ACIDS . . . . . . . . . . . . 316
A. Bulk RNA . . . . . . . . . . . . 316
B. Messenger RNA . . . . . . . . . . . 317
C. Ribosomal RNA . . . . . . . . . . . 318
D. Transfer RNA . . . . . . . . . . . 319
E. Control of RNA Synthesis . . . . . . . . . 322
IV. MITOCHONDRIAL NUCLEIC ACIDS . . . . . . . . . 324
A. Mitochondrial DNA . . . . . . . . . . 324
B. Mitochondrial RNAs . . . . . . . . . . 326
V. CONCLUDING REMARKS . . . . . . . . . . . 326
REFERENCES . . . . . . . . . . . . . . 327

## I. Introduction

The subject of nucleic acids has been one of the main topics in biology for the last thirty years. Knowledge in this field developed through various stages and yeasts have brought their contribution to each of them. In the 1940s the main problem concerned the identification of the genetic role of DNA and the correlated analysis of its structure (Zamenhof and Chargaff, 1948). The next step consisted of the study of the mechanisms of transcription and traduction of the genetic information and the deciphering of the genetic code. Yeasts provided an important contribution; yeast alaninyl-t-RNA was the first nucleic acid for which the complete sequence was worked out (Holley *et al.*, 1965). Biologists did not, however, restrict themselves to handling yeasts in the same way as bacteria; in fact, yeasts are eucaryotic organisms with organelles present in the cytoplasm. Through the discovery of cytoplasmic inheritance of mitochondrial characters (Ephrussi and Slonimski, 1955) and the identification of specific mitochondrial nucleic acids yeast has recently become the most accurate tool for the study of nucleocytoplasmic interactions at the molecular level.

## II. Deoxyribonucleic Acids

Yeasts were used to a great extent in early work on molecular biology to determine the amount of DNA per cell and the structure of the molecule. Ogur and Rosen (1950) developed a technique for the separation and characterization of yeast DNA and RNA, and they published the first estimations of the DNA content per cell. Although it was possible to carry out a quantitative determination, the DNA obtained was partially hydrolysed, and not suitable for the study of molecular structure. This problem was solved by Zamenhof and Chargaff (1948), who succeeded in preparing highly polymerized and purified DNA from yeast, and were able to determine its composition. Yeast DNA is composed exclusively of deoxyribonucleotides. The four bases are adenine (A), guanine (G), thymine (T) and cytosine (C); no significant amount of methylcytosine can be detected. This was one of the first examples used by Chargaff to challenge the tetranucleotide hypothesis. He observed the molar equivalence of adenine with thymine, and guanine with cytosine. The overall base composition of yeast DNA has been established by Vischer *et al.* (1949) as: $A = 31{\cdot}4\%$, $T = 32{\cdot}7\%$, $G = 18{\cdot}6\%$, $C = 17{\cdot}3\%$.

### A. Isolation of Yeast DNA

Most of the procedures in use today for the preparation of yeast DNA are derived from that of Marmur (1961). This technique starts with the disruption of yeast cells by mechanical means, or of protoplasts by osmotic shock. The homogenate thus obtained is heated in the presence of a detergent, then thoroughly deproteinized, and finally DNA is isolated by alcohol precipitation. In the course of the purification, treatment with ribonuclease eliminates most of the contaminating

RNA. This procedure yields a purified preparation of DNA, but unfortunately the many manipulations cause some shearing of the DNA molecule, although it is still highly polymerized. In order to obtain a greater degree of integrity, phenol extraction has been suggested by several authors (Lamonthezie and Guerineau, 1965; Sinclair *et al.*, 1967). Technical simplifications have been introduced by Smith and Halvorson (1966). The yield obtained by all these techniques is about 80% of the total DNA of the cell.

More recently, in an attempt to isolate intact DNA, Moustacchi and Williamson (1966) and Sinclair *et al.* (1967) succeeded in preparing DNA from yeast protoplasts by disrupting them in a detergent and centrifuging the whole lysate in a CsCl density gradient. This technique provides highly polymerized DNA molecules which are so long that they can only be photographed in their entirety with difficulty by electron microscopy. However, the yield of such a preparation is somewhat less than when more exhaustive procedures are used.

### B. STRUCTURE AND COMPOSITION OF DNA

By all the biochemical criteria available, yeast DNA behaves as a classical double-stranded DNA. This fact has been confirmed by the study of the thermal denaturation and of the buoyant density in CsCl. When heating yeast DNA Marmur and Doty (1962) observed the transition from the native double-stranded structure to the denatured state. This was monitored by a rise in absorbance of 40%, and they determined its midpoint ($T_m$) as 85° in a solution containing 0·15 *M*-NaCl and 0·015 *M*-Na citrate. From this $T_m$ determination they predicted that yeast DNA contained 37% GC. The combined guanine and cytosine content of yeast DNA has also been estimated from its buoyant density in CsCl. Schildkraut *et al.* (1962) determined the buoyant density as 1·698 g/cm$^3$, and estimated the GC value for yeast DNA as 39%. Strand separation produced an increase in density of 0·015 g/cm$^3$.

The base composition of yeast DNA has been determined chemically by several authors (Vischer *et al.*, 1949; Belozerski and Spirin, 1960). Their estimations of the GC percentage agree completely at a value of 35·7%. There is, however, a slight discrepancy between this value and the composition predicted from the buoyant density determination. This ambiguity has been removed by a more accurate re-evaluation of the homogeneity of yeast DNA.

### C. HETEROGENEITY OF YEAST DNA

Re-examination of yeast DNA has been stimulated in the 1960s by the discovery of DNA in yeast mitochondria by Schatz *et al.* (1964). Both thermal denaturation profiles and CsCl ultra-centrifugation tracings showed yeast DNA to be heterogeneous (Tewari *et al.*, 1965, 1966; Corneo *et al.*, 1966; Moustacchi and Williamson, 1966; Sinclair, 1966; Mounolou *et al.*, 1966). The bulk of yeast DNA contains two classes of DNA; a major component of nuclear origin that represents about 90% of the total, and a minor component which is mitochondrial DNA (Fig. 1).

Mitochondrial DNA has a buoyant density of 1·683 g/cm$^3$ in CsCl, and its $T_m$ is 75°; it therefore has a very low GC content (17–20%). The discovery of mitochondrial DNA shed a new light on the function and genetics of yeast mitochondria. These problems will be discussed later; but attention is first focused on yeast nuclear DNA.

### D. COMPOSITION AND STRUCTURE OF NUCLEAR DNA

Nuclear DNA is usually separated from mitochondrial DNA by CsCl centrifugation or by hydroxyapatite chromotography (Bernardi *et al.*, 1968). The characterization of nuclear DNA has been achieved by classical techniques. Its $T_m$ is 85°, with a hyperchromic shift of 44%; its buoyant density in CsCl is 1·699 g/cm$^3$, with an increase of 0·015 g/cm$^3$ when denatured. From these values the predicted GC contents are respectively 38·5 and 40·0%. Chemical analysis is now in good agreement with physical determinations: 39·7% (A = 29·5%, T = 30·8%, G = 20·1%, C = 19·6%; Bernardi *et al.*, 1969).

Even after separation from mitochondrial DNA, nuclear DNA still appears heterogeneous. The peak of nuclear DNA run in CsCl gradient centrifugation often appears skew on its heavy side; sometimes one can even distinguish a shoulder, as shown in Fig. 1 (Moustacchi and Williamson, 1966; Corneo *et al.*, 1966). This observation suggests the presence in nuclear DNA of a heavy satellite. However, this class of DNA is very close to nuclear DNA as judged by its density (1·704 g/cm$^3$), but it has not yet been possible to separate nuclear DNA into two molecular species. The proportion of this heavy satellite seems to vary greatly according to the experimental conditions used for cell culture and DNA isolation. Nothing is yet known about the precise localization or function of this DNA.

Observation by electron microscopy of yeast nuclear DNA by the Kleinschmidt and Zahn (1959) technique has revealed some circular molecules among the DNA population. The proportion of the circular filaments is smaller than 1%. Measurements of these circles showed a variation in length from 0·3 to 7 μm, with a main grouping at 1·95 μm (Sinclair *et al.*, 1967). It is not possible to assume a specific localization and function for these molecules. Whether this class of circular DNA is related or not to the heavy satellite is still an unsolved problem.

Our knowledge of the sequence of yeast nuclear DNA is still very limited, although as early as 1950 Zamenhof and Chargaff (1950) observed some differences in the base composition of total DNA, and some fragments resistant to a controlled DNase treatment. The subject of yeast nuclear DNA is therefore an open field for biologists and biochemists.

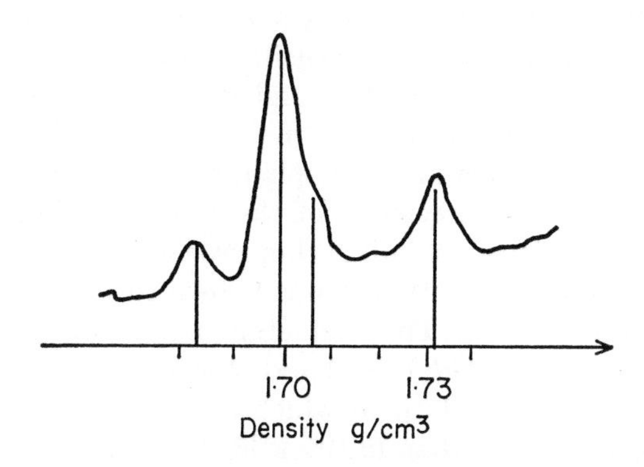


FIG. 1. Density profile in a CsCl gradient ultracentrifugation of total DNA of *Saccharomyces cerevisiae*. [*Micrococcus lysodeikticus* DNA (1·731 g/cm$^3$) used as reference standard.]

## E. YEAST DNA AND EVOLUTION

The DNA base composition has been suggested as a useful tool for the investigation of the taxonomy of yeasts. Extensive studies have been carried out on a great number of different yeast species (Stenderup and Bak, 1968; Meyer and Phaff, 1969). The base composition of yeast DNA ranges from 35 to 57% GC. These observations support the division of yeasts into ascosporogenous and anascosporogenous species. Moreover, some re-organization of the *Candida* species has been undertaken, and the possible relationship between *Cryptococcus* and *Rhodotorula* with basidiomycetes has been strengthened (Villa and Storck, 1968).

## F. RESPONSE TO MUTAGENS AND RADIATION

Many mutagens have been applied to yeast for genetic purposes. Two particular aspects of their action are interesting in relation to the structure and composition of DNA. One is the molecular mechanism of mutation, the other the process by which mutation becomes effective. Because of its genetic complexity yeast does not provide an easy system for the study of the former problem. In fact, the chemical action of various mutagens (Freese, 1963) is assumed to be the same as in other organisms. However, repair mechanisms were discovered in yeast long ago, and they appeared especially effective in correcting the damage caused by ultraviolet radiation. Many attempts have been made to study this molecular process. A photoreactivating enzyme which has been isolated (Muhammed, 1966) causes *in vitro* the disappearance of thymine dimers in ultraviolet-irradiated DNA (Wulff and Rupert, 1962). The same mechanism is thought to be effective *in vivo* (Patrick and Haynes, 1968). In order to study such an enzyme, many mutants have been obtained that are sensitive to ultraviolet radiation (Moustacchi, 1964; Snow, 1967; Cox and Parry, 1968). At least 22 loci seem to confer ultraviolet sensitivity on yeast cells, and many interpretations have been put forward to explain such a situation. The locus (or loci) controlling the enzyme(s) of the repair mechanism, and the relation with the genetic recombination process have not been determined, although ultraviolet radiation is known to affect recombination (Wilkie and Lewis, 1963). The first step in this direction has been achieved by Moustacchi *et al.* (1967) and Moustacchi (1969), who found an ultraviolet-sensitive mutant in which the frequency of the mitotic recombination was higher than in the wild-type cells.

The different mutagens that are effective in yeast have been divided into two classes, according to the form of the mutation in the first cell generation after mutagenesis. Replicating instabilities, which continue to give rise to mutational events and replicate as an unstable condition in a particular locus, have been observed with ethylmethane sulphonate, nitrosomethylguanidine and ultraviolet radiation (Nasim, 1967; Nasim and Auerbach, 1967). Hydroxylamine and nitrous acid are not effective in this respect. Two hypotheses are consistent with the observations: either (i) replicating instabilities are transformed into complete mutations by a repair mechanism; or (ii) they arise by different molecular mechanisms involving the chemical structure of DNA and its folding in the chromosome. This again provides a wide field for investigation by biochemists and geneticists.

### G. DNA AND CHROMOSOMES

The yeast nucleus has presented a puzzling question to the biologists for years, and there has been some controversy about the presence and number of the chromosomes, which are usually very difficult to observe (Tamaki, 1965) because the chromatin appears coagulated, and the nuclear membrane persists at cell division (Rozijn and Tonino, 1964). In order to compare yeast chromosomes with those of other eucaryotic cells, biochemical investigations of proteins associated with DNA have been made; the presence of histones has been demonstrated (Tonino and Rozijn, 1966). Among the basic proteins of this group, Tonino and Rozijn noticed the absence of a lysine-rich fraction, which is supposed to play an important role in chromosome condensation, and they correlated this observation with the special behaviour of yeast chromosomes. Fogel and Mortimer (1968) were able to postulate that fragments of a DNA strand is transferred to its homologue in the process of meiotic gene conversion. In spite of its complexity, genetic analysis is still the best tool available for the study of chromosome structure.

The amount of DNA per yeast cell has been carefully studied for a considerable time in order to determine the quantity of genetic information present in each nucleus, as it is most important to correlate these facts with the state of ploidy determined by genetic methods. Ogur *et al.* (1952) found that the DNA content per cell was proportional to the degree of ploidy. The amount of DNA per cell has been re-examined by Moustacchi *et al.* (1967) and Schweizer *et al.* (1969); it is of the order $1{\cdot}2 \times 10^{10}$ daltons per haploid cell. The quantity of DNA is independent of growth conditions in *Saccharomyces cerevisiae* (Williamson and Scopes, 1961; Polakis and Bartley, 1966). The research was extended by Dastidar (1966) to *Candida albicans*, which can be grown in single-cell or mycelial forms.

### H. DNA REPLICATION

DNA replication has been extensively investigated in bacteria; such studies are far more difficult in yeast, mainly owing to the size of the pool of precursors and the lack of efficient thymine or thymidine incorporation. In spite of these problems, Corneo *et al.* (1966) stated that they observed semiconservative replication of yeast DNA. Two main experimental approaches have been followed for the study of DNA synthesis in yeast: one concerns the turnover of DNA in resting cells, the other involves synchronous cultures.

Resting cells are obtained by transferring yeast from a growing culture into buffered medium where lack of a nitrogen source prevents growth. Under these conditions nuclear DNA synthesis very soon ceases, a 0·2–0·3% increase in DNA is attributable to cessation of the replication process. Two hours later no appreciable turnover can be detected (Mounolou *et al.*, 1968; Rabinowitz *et al.*, 1968, 1969).

Applying the second method, Williamson (1965) showed that DNA is synthesized quite soon after bud emergence in synchronized cells, and the S period does not last more than a quarter of the generation time. These results, obtained with *Saccharomyces cerevisiae*, have been confirmed by similar observations on *Schizosaccharomyces pombe* (Bostock *et al.*, 1966) and on *Saccharomyces lactis* (Smith *et al.*, 1968). In a study of the molecular mechanism of mitotic recombination, Esposito (1968) showed that the ability to recombine in response to ultraviolet radiation is maximal at the beginning of DNA duplication, and declines during this process. Therefore, recombination may involve initiation of replication. Moreover, yeast cells become synchronized when placed in conditions where they undergo meiosis and DNA synthesis has occurred during ascopore formation. In fact, replication proceeds just before meiosis, and ends with prophase I (Croes, 1966).

The molecular control of DNA replication is not fully understood, it does not appear to proceed through DNA polymerase (Eckstein *et al.*, 1967). However, the inhibition of DNA synthesis by cycloheximide, which blocks protein synthesis, suggests the need for permanent protein elaboration (de Kloet, 1966). In this respect DNA replication in yeast seems to be controlled by a mechanism different from that observed in bacteria.

## III. Ribonucleic Acids

When compared with other cells, yeasts appear very rich in ribonucleic acids, and have provided an industrial source of the various molecular species. The amount of RNA may reach one-third of the total protein, and is some fifty to a hundred times greater than the DNA content (Fukuhara, 1967a). Early studies aimed at the determination of the base composition and the stability of bulk RNA, but interest was very rapidly focused on several problems. One is the heterogeneity of bulk RNA and the localization and characterization of different species (especially messenger RNA); another concerns the RNAs considered as elements of the protein synthesizing machinery – ribosomal RNA and transfer RNA. Thus yeast presented the possibility of examining RNA synthesis and its control.

### A. BULK RNA

Many procedures have been employed to isolate pure RNA from yeast for biochemical purposes. With the help of a detergent, Crestfield *et al.* (1955) separated sufficient RNA to determine the overall base composition: A = 25%; U = 27%; G = 28%; C = 20%. However, this technique causes some degradation of the molecules, and RNA fragments of very low molecular weight are formed. Osawa (1960) introduced a procedure that provided RNA in an undegraded form, and obtained a very similar composition: A = 25·6%; U = 26·2%; G = 28·6%; C = 19·6%. The possibility of isolating intact RNA leads immediately to a study of the heterogeneity of this product. The identification of several molecular species can be achieved by methods involving differences in base composition (Jones *et al.*, 1967), but usually the separation is made according to the various molecular weights of the RNA species (Osawa, 1960; Otaka *et al.*, 1961; Shortman and Fukuhara, 1963). Sedimentation analysis revealed the occurrence of three main classes of RNA: 24 S,

17 S, 4 S. Ribosomal RNA consists of the 24 S and 17 S forms; 4 S RNA contains all the transfer RNA.

The majority of bulk RNA is metabolically stable, and subject only to a small turnover under growth conditions (Halvorson, 1958).

### B. MESSENGER RNA

The preceding techniques dealing with bulk RNA were unable to reveal directly the existence of messenger RNA in yeast. Special procedures had to be applied to study the product of DNA transcription. Using $^{32}P$ pulses during the exponential growth of yeast cells, Ycas and Vincent (1960) and Kitazume and Ycas (1963) detected the occurrence of a messenger-type RNA the composition of which was intermediate between that of DNA and bulk DNA. This observation was strengthened by the study of the $^{32}P$ uptake by yeast cells, which demonstrated the very rapid labelling of nucleotides (Borst-Pauwels, 1962; Borst-Pauwels *et al.*, 1962), and by careful examination of the RNA synthesized under such conditions (Ycas and Vincent, 1960). It was of special importance to show that the base sequence in $^{32}P$-labelled RNA was approximately random, and that the base composition obtained was independent of the label position in the nucleotides (2′–3′ or 5′).

Once the occurrence of some messenger RNA has been demonstrated, several problems arose concerning this product; these are concerned with the heterogeneity and specificity of this type of RNA. Two different approaches have been followed to answer these questions: (i) physico-chemical methods to fractionate pulse-labelled RNA; (ii) transcription when cells are placed under conditions in which they express their functions sequentially, or develop a very specific function.

Using the first approach, Westphal *et al.* (1966) showed that messenger RNA was heterogenous both on the basis of molecular weight and base composition. Sedimentation analysis revealed the existence of RNA species ranging from 4 S to 33–36 S. Yeast messenger RNA appears more stable than bacterial messenger RNA, but less so than yeast ribosomal and transfer RNA.

Several attempts have been made to study the transcription and the specificity and heterogeneity of its product. An interesting approach to this problem has been tried with synchronized yeast cells. Many experiments have shown that certain enzymes are synthesized during a brief and definite period of the cell cycle (Sylven *et al.*, 1959; Gorman *et al.*, 1964). A further investigation of this experimental system led Tauro *et al.* (1968) to the conclusion that the transcription of the chromosome is not related to the distance of the genes from their centromere. It seems rather to be unidirectional for all the chromosomes, or at least for some regions of them, for the expression of linked genes occurs at close intervals and reflects the order of their position on the chromosome. However, in such conditions all the genes involved in growth are expressed at each cell generation, and the molecular pattern of transcription is very complex.

To avoid the complexity involved during growth, some authors tried to study messenger RNA in resting cells. In the experiments of Shortman and Fukuhara (1963) the metabolism of RNA was investigated during the respiratory adaption of resting anaerobically propagated cells of *Saccharomyces cerevisiae*. These cells were transferred from an anaerobic culture to a buffered glucose medium; the induction of the respiratory enzymes was triggered by oxygen in the absence of growth. Under these conditions, which combine a step-down effect and respiratory adaptation, a special class of RNA with a high turnover rate was detected. This messenger RNA is different from DNA and bulk RNA in its base composition, and shows a marked heterogeneity in molecular size. This RNA is associated with membrane fractions which bear active 80 S ribosomes, and seems to be correlated with the nucleus (Fukuhara, 1967b). A step-down effect has also been examined in *Schizosaccharomyces pombe* by Mitchison and Gross (1965). The overall elaboration of RNA was stopped for about one hour, during which lag the synthesis of a special messenger RNA proceeded. Again this RNA is found to be heterogenous in size, and different from DNA and bulk RNA in its base composition.

With regard to the complexity of yeast metabolism, none of these attempts to study messenger RNA could determine the specificity of the information carried by these species at the molecular level.

### C. RIBOSOMAL RNA

Interest in yeast ribosomal RNA centres on two problems: (i) the specificity of its composition and structure in relation to its function; and (ii) the control of its synthesis.

Unlike messenger RNA, ribosomal RNA represents the majority of the total RNA of yeast cells, and is composed of definite molecular species. The two RNAs are easy to characterize by sucrose-gradient centrigufation (Shortman and Fukuhara, 1963). Sedimentation coefficients vary only slightly from one publication to another; they are close to 24 S and 17 S (Otaka *et al.*, 1961; Fukuhara, 1967; Beck *et al.*, 1966; Rogers *et al.*, 1967). Thus they appear to differ from the values obtained with bacterial ribosomal RNA. Gel electrophoresis has recently been found to be a very sensitive method of fractionation and characterization for RNA. Using this technique, Loening (1968) showed that yeast ribosomal RNA behaves like all plant ribosomal RNA. The molecular weights of the two components are respectively $1{\cdot}3 \times 10^6$ and $7{\cdot}2 \times 10^5$. The base composition of ribosomal RNA has been established as follows: $A = 26{\cdot}6\%$,

U = 26·6%, G = 26·9%, C = 20·0% (Beck *et al.*, 1966; Mitchison and Gross, 1965). In addition to these two RNAs, a third species has been found in yeast ribosomes. This has a small molecular size (5 S), and its composition differs from those of transfer RNA, 24 S RNA and 17 S RNA (Marcot-Queiroz *et al.*, 1965). The 5 S RNA is bound to the 60 S particle of the ribosome.

These three species represent the RNA components of the ribosomes existing in the cytoplasm. Indeed, Rogers *et al.* (1967) indicated that yeast mitochondria contain specific ribosomes different from the cytoplasmic type. This observation will be discussed later; however, in bulk preparations from entire cells, cytoplasmic ribosomes represent an enormous majority.

Ribosomal RNA is intimately packed with ribosomal proteins *in vivo*; it represents about 40% of the ribosome (Chao, 1957) and contributes to its architecture (Otaka and Uchida, 1963; McPhie and Gratzer, 1966). In order to study the role of ribosomal RNA its structure has been investigated. Physical techniques have provided information of a very general nature; this lack of precision is probably relevant to the high molecular weight and the complexity of the two molecular species: ribosomal RNA exhibits some degree of double-stranded conformation, though it is less important than in transfer RNA (Gratzer, 1966), and protonation is as effective on this RNA as on any other RNA (Sarkar and Yang, 1965). Mild and controlled action of ribonuclease has been used as an analytical tool; it reveals a definite number of labile points which separate 12 characteristic fragments. In fact, very little is known about the precise association of ribosomal RNA and proteins, and the structure of yeast ribosomes requires further investigation.

## D. TRANSFER RNA

Transfer RNA (t-RNA) is a key participant in the process of elaboration of polypeptide chains, and its double specificity with respect to the amino acids and the genetic code has focused interest on its composition and structure. Yeast has proved to be a convenient material for the study of such problems because of the abundance of RNA in the cells. A great deal of effort has been devoted to this subject, and has led to the deciphering of the first complete sequence of a specified nucleic acid species (Holley *et al.*, 1965). This section will be restricted to general statements about yeast t-RNA.

### 1. *General Features of t-RNA*

The isolation procedure for t-RNA is fairly simple (Holley, 1963; Cantoni and Richards, 1966). It consists mainly of phenol extraction followed by alcohol precipitations and DEAE-cellulose chromatography. The RNA obtained has a molecular weight close to 26,000 (Lindahl *et al.*, 1965) and its sedimentation velocity is 4 S. As a result of measurements of both optical rotatory dispersion (Gratzer, 1966) and thermal denaturation (Fresco *et al.*, 1966; Goldstein, 1966), t-RNA is known to assume conformations that contain organized helical regions. The extent of single-chain areas does not exceed about 10 nucleotides out of 80 (Mommaerts *et al.*, 1964; Seidel and Cramer, 1965). This organization has been confirmed by the cross-linking effect of formaldehyde (Axelrod *et al.*, 1969). However, cytosine is still available for protonation (Sarkar and Yang, 1965).

Chemical studies of t-RNA have revealed the existence in these molecules of unusual nucleotides besides the four main ones; most contain methylated bases (Borek, 1963) and it appears that some inosine and pseudouridine are present (Holley *et al.*, 1963). New methods which combine electrophoresis and chromatography have been developed to determine the composition of t-RNA and characterize the rare bases (Gebicki and Freed, 1965; Hiby and Kroger, 1967; Robins *et al.*, 1967). Pseudouridine has been the most intensively studied (Chambers, 1966; Goldwasser and Heinrikson, 1966). Other derivatives of uracil also occur, such as 5-hydroxyuridine or dihydro-uracil (Lis and Passarge, 1966; Magrath and Shaw, 1967). The methylation process of the bases has been studied by the use of yeast mutants (Peetrissant and Gaye, 1966; Kjellin-Straby and Boman, 1965; Phillips and Kjellin-Straby, 1967). Methylation also affects the ribose moieties (Gray and Lane, 1967). The presence of the rare nucleotides was correlated with the ability to bind amino acids some years ago (Osawa and Otaka, 1959).

### 2. *The Structure of a Specific t-RNA*

The first steps in protein synthesis involve the activation of the amino acids, followed by their transfer to specific t-RNA acceptors (Berg, 1961). Such specificity must reside in the sequence of each t-RNA molecule, but also in its overall conformation. The study of this property of t-RNA requires the isolation of each amino acid acceptor species. Countercurrent distribution with various solvent systems is the most widely used technique (Apgar *et al.*, 1962). Recently, chromatographic methods have appeared (Miyazaki *et al.*, 1966, 1967; Wimmer *et al.*, 1968). In practice repeated purification steps have to be carried out in order to yield a reasonably pure preparation of a specific t-RNA.

Once a particular species has been isolated, it is possible to study its specificity. The first aim is to determine the base composition and sequence. The overall composition appears to differ from one species to another (Holly *et al.*, 1963). Interest then turns to the sequence of each t-RNA molecule. This deciphering process is very laborious! Mild RNase treatments are used to degrade the t-RNA into characteristic oligonucleotides (Asano, 1961; Holley *et al.*, 1965; Madison *et al.*, 1966, 1967; Madison and Kung, 1967; Mirzabekov *et al.*, 1965; Li *et al.*, 1966; Apgar *et al.*, 1966; Baev *et al.*, 1967a; Venkstern *et al.*, 1967). The composition of the oligonucleotides is then determined, and the sequence of the t-RNA is reconstructed for successive nucleotides (Holley, 1968). These procedures led Holley *et al.* (1965) to establish the first sequence of a nucleic acid – alanine-t-RNA. Very soon this was followed by the elucidation of the sequences of tyrosine-t-RNA (Madison *et al.*, 1966), phenylalanine-t-RNA (Rajbhandary *et al.*, 1966), valine-t-RNA (Baev *et al.*, 1967b) and serine-t-RNA (Zachau *et al.*, 1966).

The sequence having been determined, the question of the structure of the molecule immediately arises, and models have been built assuming the occurrence of helical regions. Cantoni *et al.* (1963) suggested a hairpin-like form for serine-t-RNA. Holley *et al.* (1965) favoured a clover-leaf structure for alanine-t-RNA; this configuration appears to agree with the existence of at least two loops revealed by RNase treatment (Armstrong *et al.*, 1966) and by optical rotation dispersion measurements (Vournakis and Scheraga, 1966; Cantor *et al.*, 1966). By comparing aspartic acid, lysine-, and glycine-t-RNAs, Sarin *et al.* (1966) showed that each species exhibits a specific conformation. Chemical modifications have been produced on the t-RNAs to study how the structure contributes to the specificity of function. The occurrence of the terminal A unit is necessary to amino-acid acceptance (Makman and Cantoni, 1966), although *in vivo* this nucleotide is subjected to some turnover (Rosset and Monier, 1965). Cyanoethylation of pseudouridine (Rake and Tener, 1966) and chlorination of G and U (Duval and Ebel, 1966) decrease the ability to bind amino acids. The presence of high concentrations of $Mg^{2+}$ produces a reversible change from an active to an inactive form of t-RNA (Fresco *et al.*, 1966; Adams *et al.*, 1967). On the other hand, by comparing all the known sequences, the anticodon has been located in a loop of the molecule. Chemical modification of the nucleotides adjacent to the anticodon decreases the binding to the messenger RNA-ribosome complex (Fittler and Hall, 1966); this function is also affected when polyamines are associated with the t-RNA molecule (Tanner, 1967).

In connection with the determinaton of the universality of the genetic code, many studies have been concerned with the species specificity of t-RNAs. Some similarities have been observed in the composition of homologous t-RNA from yeast and other sources (Doctor *et al.*, 1966; Yoshida and Ukita, 1966; Wagner and Ingram, 1966). This was extended to codon recognition by Soll *et al.* (1966, 1967). The synthetase of one organism can bind an amino acid to the specific t-RNA of another organism (Berg *et al.*, 1961). Yet this universality is somewhat restricted, and Yamane and Sueoka (1963) found a partial conservation of specificity between yeast t-RNA and yeast aminoacyl-t-RNA synthetases. The restriction is complete for tyrosine-t-RNA (Hayashi, 1966). The observations have been supported by accurate comparisons of various t-RNAs. Even though t-RNAs from *Saccharomyces* and *Torula* species may be similar (Fujimura and Miura, 1966), they are clearly different from t-RNAs from other sources (Cerna *et al.*, 1966; Tomlinson and Campbell, 1966).

As soon as the isolation of a specific t-RNA was possible, a heterogeneity was apparent with respect to the binding of a particular amino acid, and more accurate methods of isolating iso-accepting t-RNAs have been introduced (Artamonova *et al.*, 1967; Winstein, 1969; Sekiya *et al.*, 1969). Two species of serine-t-RNA have been characterized and, although they exhibit the same anticodon, their sequences are different (Zachau, 1967; Zachau *et al.*, 1967; Lebovitz *et al.*, 1966). Four glycine-t-RNAs and four lysine-t-RNAs have been isolated (Bergquist, 1966), three valine-t-RNAs, two phenylalanine-t-RNAs (Kawata *et al.*, 1967), two histidine-t-RNAs (Takeishi *et al.*, 1967) and a second alanine-t-RNA (Reeves *et al.*, 1968). Using the ribosomal binding technique, this puzzling situation was examined by Soll *et al.* (1966, 1967), who observed that a change in the first letter of a codon requires a new t-RNA, and that a single t-RNA molecule is capable of multiple codon recognition. In fact, each cell contains more t-RNA species than would be necessary for the recognition of all the codons. This redundancy is believed to account for the numerous genetic suppressors known in yeast (Bergquist *et al.*, 1968).

## E. CONTROL OF RNA SYNTHESIS

Recent investigations on the effect of growth conditions on RNA synthesis led to the discovery of a different process of control in different classes of RNA. A change from starvation to growth of cells or protoplasts induces a preferential burst of ribosomal RNA synthesis (Kudo and Imahori, 1965; Hutchinson and Hartwell, 1967; Retel and Planta, 1967). Synthesis of t-RNA seems to be less directly dependent on possible changes of growth conditions. This is stressed too by the study of the stability of different RNA species in the absence of growth. Halvorson (1958) pointed out that the turnover of bulk RNA was higher in the absence of cell division than in growing cells. Under starving conditions t-RNA is relatively stable, and some kind of messenger

RNA is synthesized, but ribosomal RNA is progressively degraded (Fukuhara, 1967b). This can be correlated with the rapid degradation of ribosomal RNA in Mg-deficient or Sr-treated cells (Beck *et al.*, 1967; Yamamoto, 1967). Inhibitors of protein and RNA synthesis have been used to investigate the different control mechanisms of the three RNA classes. Cycloheximide blocks protein synthesis, but does not affect RNA formation to the same extent (Kerridge, 1958). In *Saccharomyces cerevisiae* and *Sacch. carlsbergensis* some messenger- and transfer-RNA formation still proceeds in the presence of the drug, while ribosomal RNA synthesis is greatly depressed (de Kloet, 1965, 1966a, b; Fukuhara, 1965). However, a stimulation of bulk RNA synthesis has been reported in *Saccharomyces pastorianus* under the same conditions (Siegel and Sisler, 1964). The permanent formation of some proteins is necessary for continuous elaboration of ribosomal RNA. This conclusion is strengthened by the specific effect of 5-fluorouracil and 6-azauracil, which stop ribosomal RNA formation (de Kloet and Strijkert, 1966) and therefore ribosomal activity (de Kloet, 1967; Kempner, 1961).

Such a tight control of ribosomal RNA synthesis suggests the existence of a complex molecular mechanism of elaboration. When transferring starving cells to a growing state, Retel and Planta (1967) showed that a ribosomal RNA precursor of high molecular weight is first elaborated in the nucleus, then this RNA is converted into 26 S and 18 S RNAs. Taber and Vincent (1969a, b) confirmed the occurrence of this process, and observed that the precursor (38 S) is accumulated in the presence of cycloheximide. The drug appears to block the synthesis of the protein which converts 38 S RNA into 26 S and 18 S RNAs. *In vivo*, Lochmann (1967) discovered that this process is susceptible to the photodynamic action of dyes.

Finally the regulation of RNA synthesis must be examined at the level of transcription. Owing to the development of hybridization, it is now possible to study the molecular relationship between DNA and specific RNAs. Separate cistrons have been found in the genome for the three types of RNA (26 S, 18 S and 4 S). Segments of nuclear DNA complementary to ribosomal RNAs represent 2% of the total DNA; and segments complementary to t-RNA, 0·08% (Fukuhara, 1967b; Retel and Planta, 1968; Schweizer *et al.*, 1969). Therefore, the yeast genome exhibits a pronounced redundancy of ribosomal- and transfer-RNA genes; this is supposed to explain the high RNA content of the cells and the continuous synthesis of ribosomal RNA throughout the cell cycle (Tauro *et al.*, 1968). Such a situation should induce yeast biochemists and geneticists to speculate, and undertake research in order to fill the obvious gaps in our knowledge in the control of RNA synthesis and the significance of genetic redundancy.

## IV. Mitochondrial Nucleic Acids

Since the concomitant discovery of the role of DNA in the transfer of genetic information, and of the cytoplasmic heredity of mitochondrial characters, the question of the existence of specific mitochondrial nucleic acids has remained latent. However, the observations of Schatz *et al.* (1964) on the occurrence of DNA associated with yeast mitochondria initiated many experimental approaches to this problem.

### A. MITOCHONDRIAL DNA

The aim of the first studies on mitochondrial DNA was to identify and characterize this species. In the 1960s Appelby and Morton (1960) described the association of a small piece of DNA with the crystalline yeast cytochrome $b_2$; this proved to be the consequence of the crystallization process (Jackson *et al.*, 1965). Caesium chloride density-gradient centrifugation has been the most useful tool for the unambiguous separation of mitochondrial from nuclear DNA, as shown in Fig. 1 (Tewari *et al.*, 1965, 1966; Sinclair, 1966; Corneo *et al.*, 1966; Guerineau *et al.*, 1966; Moustacchi and Williamson, 1966; Mounolou *et al.*, 1966; Rabinowitz, 1968). Yeast mitochondrial DNA is very different from nuclear DNA: its buoyant density in CsCl is 1·683 g/cm$^3$, its $T_m$ is 75° under standard conditions, and chemical analysis reveals a GC content close to 20%. *In situ* this DNA is associated with the inner membrane of the organelle (Yotsuyanagi, 1966). Mitochondrial DNA accounts for a small proportion of the total DNA of the cell (about 5%). Renaturation studies reveal a marked homogeneity of the molecules (Corneo *et al.*, 1966). The structure of native mitochondrial DNA has been investigated by electron microscopy (Sinclair *et al.*, 1967; Van Bruggen *et al.*, 1968). In spite of many technical difficulties, two classes of molecules have been found; a linear one which contains two populations of 4·5 μm and 10 μm length and also very long fragments (>25 μm) (Guerineau *et al.*, 1968a; Schapiro *et al.*, 1968; Borst *et al.*, 1968), and a second type which consists of circular molecules of various sizes from 1 μm to 10 μm (Avers, 1967; Avers *et al.*, 1968; Guerineau *et al.*, 1968b; Schapiro *et al.*, 1968). Hence it now appears that yeast-mitochondrial DNA is heterogeneous. The meaning of this observation will be found in the correlation of genetic and microscopic studies.

Mitochondrial mutants of yeast belong to two classes; one exhibits Mendelian inheritance, the other does not. Mutants of the second group are called cytoplasmic *petites*, they are respiratory deficient. This mutation, which affects a cytoplasmic hereditary factor, occurs at a high frequency (about 1%) and is not reversible. It causes very pronounced alterations in the structure and function of the mitochondria.

On comparing cytoplasmic petites to normal cells (*grandes*) it has been possible to show that the mutation is not due to the loss of mitochondrial DNA. Furthermore, other changes in the mitochondrial DNA sequence have been found in some cytoplasmic mutants (Mounolou *et al.*, 1966, 1967, 1968a; Carnevali *et al.*, 1966). Mitochondrial DNA is the molecular basis of cytoplasmic inheritance of mitochondrial characters.

Cytoplasmic mutation often induces an important modification in the sequence of mitochondrial DNA: many mutants studied exhibited a lower GC content (as low as 4%) (Bernardi *et al.*, 1968; Mehrotra and Mahler, 1968). Bernardi *et al.* (1969) even demonstrated the existence of AT sequences along the DNA molecule, and the relative length of these fragments increases in some cytoplasmic mutants. Carnevali *et al.* (1969) postulated a different replication rate for the different molecules of a heterogeneous mitochondrial DNA population. However, this does not explain the molecular process of mutation.

Whatever the explanation of cytoplasmic mutation may be, it must account for the different efficiencies of various mutagens. Some agents like acridines (Ephrussi *et al.*, 1949; Marcovitch, 1951; Slonimski and de Robichon-Szulmajster, 1957), ethidium bromide (Slonimski *et al.*, 1968), ultraviolet radiation (Raut and Simpson, 1955; Wilkie, 1963) induce cytoplasmic mutations with a much higher frequency than chromosomal ones. On the other hand, some mutagens effective on nuclear information do not produce cytoplasmic mutants (Nagai *et al.*, 1961; Schwaier *et al.*, 1968). Target analysis indicates the existence of a score of genetic units in each aerobic derepressed cell; this value falls when the cells are in a repressed state (Wilkie, 1963; Sugimura *et al.*, 1966; Allen and McQuillan, 1969). The sensitivity of mitochondrial DNA to ultraviolet radiation is the consequence of the lack of any repair mechanism in mitochondria (Nakai, 1968; Moustacchi, 1969). In all yeasts mitochondrial DNA is very poor in GC, 20–25% (Villa and Storck, 1968); this particular composition does not explain the sensitivity to some mutagens, because respiratory deficient mutants are known only in a small number of species (Nagai *et al.*, 1961).

The existence of specific genetic information within the mitochondria brings a new interest to the question of the autonomy and biogenesis of the organelles. In this connection the demonstration that mitochondrial DNA replication is not controlled in the same way as nuclear DNA, has been of great importance. Mitochondrial DNA is replicated in both aerobic and anaerobic cells, although the mitochondrial structures are not developed and the respiratory enzymes are not synthesized (Swift *et al.*, 1967; Fukuhara, 1968). When anaerobic cells are transferred from growing to non-growing conditions, mitochondrial DNA synthesis still proceeds at an appreciable rate in spite of the rapid decrease in nuclear

DNA replication (Rabinowitz *et al.*, 1969). In cells in this state, a burst of mitochondrial DNA synthesis can be triggered by induction with oxygen (i.e. respiratory adaptation) (Rabinowitz *et al.*, 1968, 1969; Mounolou *et al.*, 1968b). In synchronized aerobic cells mitochondrial DNA replication does not occur at the same time as that of nuclear DNA. According to the yeast species and to the technique of synchronization, it appears to be either continuous throughout the cycle (Williamson, 1969) or limited to a specific period (Smith *et al.*, 1968).

### B. MITOCHONDRIAL RNAS

Once mitochondrial DNA has been identified as the genetic material of the organelle, it is necessary to know what the content of this information is. On account of the difference between mitochondria of cytoplasmic petites and normal grandes (Yotsuyanagi, 1962), it has been suggested that a special structural protein is under the control of mitochondrial DNA. This hypothesis has not yet been clearly supported by experimental results.

As early as 1965, Wintersberger (1965) reported the existence of specific 23 S, 15 S and 4 S RNAs in mitochondria, and Fukuhara (1967b) showed that yeast cells contain some metabolically stable RNAs which hybridize with mitochondrial DNA. Mitochondrial ribosomal RNAs have been characterized and chemically analysed (Fauman *et al.*, 1969; Steinschneider, 1969). They clearly differ from the RNAs of the cytoplasmic ribosomes, and appear very similar to bacterial RNAs. Among the 4 S RNAs of the cell, some species hybridize with mitochondrial DNA and represent mitochondrial t-RNAs (Fukuhara *et al.*, 1969), and a specific *N*-formylmethionyl t-RNA has been identified (Smith and Marker, 1968). Cross-hybridization experiments indicate that the information contents of normal mitochondrial DNA and of the DNA of cytoplasmic petites can be very different (Wintersberger and Viehhauser, 1968; Fukuhara *et al.*, 1969). A new approach to the study of the genetic load of mitochondrial DNA has been initiated by the discovery of a new type of cytoplasmic mutant which exhibits mitochondrial resistance to antibiotics (Wilkie, 1968; Slonimski *et al.*, 1969). The existence of a protein-synthesizing mechanism in mitochondria, composed of specific ribosomes, specific t-RNAs and initiator, clearly depends on the mitochondrial genetic information.

## V. Concluding Remarks

Although the overall picture of the functions of yeast nucleic acids may seem coherent, some aspects are still very puzzling. Among these is the meaning of the high level of genetic redundancy in the genome.

Experimental approaches of this problem will not rely only on the technology of nucleic acids. For example, the classification of the function of the many suppressors needs close correlation of genetic studies, and accurate examination of the different t-RNA species.

A future topic of biology deals with cell differentiation and the controlled expression of genetic information. In this respect yeasts present most interesting possibilities; namely, the existence of reversible yeast and mycelial forms in *Candida albicans*, aerobic and anaerobic states of *Saccharomyces cerevisiae*, and the occurrence of a killer character (Somers and Bevan, 1969). The more intensive studies now in progress are focused on the genetic content of mitochondrial DNA and the interaction of this information with the message of the nucleus for the development of mitochondria. In years to come yeast nucleic acids will undoubtedly be the main contributors to our rapidly growing knowledge of the behaviour of organelles in the cell.

## References


Adams, A., Lindahl, T. and Fresco, J. R. (1967). *Proc. natn. Acad. Sci. U.S.A.* **57**, 1684–1691.

Allen, N. E. and McQuillan, A. M. (1969). *J. Bact.* **97**. 1142–1148.

Apgar, J., Holley, R. W. and Merrill, S. H. (1962). *J. biol. Chem.* **237**, 796–802.

Apgar, J., Everett, G. A. and Holley, R. W. (1966). *J. biol. Chem.* **241**, 1206–1211.

Appelby C. A. and Morton, R. K. (1960). *Biochem. J.* **75**, 258–269.

Armstrong, A., Hagopian, H. and Ingram, V. M. (1966). *Biochemistry, N.Y.* **5**, 3027–3036.

Artamonova, V. A., Frolova, L. Y. and Kiselev, L. L. (1967). *Molec. Biol.* **1**, 530–538.

Asano, K. (1961). *J. Biochem., Tokyo* **50**, 544–545.

Avers, C. J. (1967). *Proc. natn. Acad. Sci. U.S.A.* **58**, 620–627.

Avers, C. J., Billheimer, F. E., Hoffmann, H. P. and Paul, R. M. (1968). *Proc. natn. Acad. Sci. U.S.A.* **61**, 90–97.

Axelrod, V. D., Feldman, M. Y. and Chugueo, I. I. (1969). *Biochim. biophys. Acta* **186**, 33–45.

Baev, A. A., Mirzabekov, A. D., Aksel'rod, V. D., Venkstern, T. V., Li, L., Krutilina, A. I., Fodor, I. and Kazarinova, L. (1967a). *Dokl. Akad. Nauk SSSR* **173**, 204–207.

Baev, A. A., Venkstern, T. V., Mirzabekov, A. D., Krutilina, A. I., Li, L. and Aksel'rod, V. D. (1967b). *Molec. Biol.* **1**, 754–766.

Beck, G., Duval, J., Aubel-Sadron, G. and Ebel, J. P. (1966). *Bull. Soc. Chim. biol.* **48**, 1205–1219.

Beck, G., Aubel-Sadron, G. and Ebel, J. P. (1967). *Bull. Soc. Chim. biol.* **49**, 349–360.

Belozersky, A. N. and Spirin, A. S. (1960). *In* "Nucleic Acids" (E. Chargaff and J. N. Davidson, eds.) Vol. 3, p. 147. Academic Press, New York.

Berg, P. (1961). *A. Rev. Biochem.* **30**, 293–312.

Berg, P., Bergmann, F. H., Ofengand, E. J. and Dieckmann, M. (1961). *J. biol. Chem.* **236**, 1726–1734.

Bergquist, P. L. (1966). *Cold Spring Harb. Symp. quant. Biol.* **31**, 435–447.

Bergquist, P. L., Burns, D. J. W. and Plinston, C. A. (1968). *Biochemistry, N.Y.* **7**, 1751–1761.
Bernardi, G., Carnevali, F., Nicolaieff, A., Piperno, G. and Tecce, G. (1968). *J. molec. Biol.* **37**, 493–505.
Bernardi, G., Faures, M., Piperno, G. and Slonimski, P. P. (1970). *J. molec. Biol.* **48**, 23–42.
Borek, E. (1963). *Cold Spring. Harb. Symp. quant. Biol.* **28**, 139–148.
Bostock, C. J., Donachie, W. D., Masters, M. and Mitchison, J. M. (1966). *Nature, Lond.* **210**, 808–810.
Borst, P., Van Bruggen, E. F. J. and Ruttenberg, G. J. C. M. (1968). *In* "Biochemical Aspects of the Biogenesis of Mitochondria" (E. C. Slater, J. M. Tager, S. Papa and E. Quagliariello, eds.), pp. 51–69. Adriatica Editrice, Bari.
Borst-Pauwels, G. W. F. H. (1962). *Biochim. biophys. Acta* **65**, 403–406.
Borst-Pauwels, G. W. F. H., Loef, H. W. and Havinga, E. (1962). *Biochim. biophys. Acta* **65**, 407–411.
Cantoni, G. L. and Richards, H. H. (1966). *In* "Procedures Nucleic Acid Research" (G. L. Cantoni and D. R. Davies, eds.), pp. 624–628. Harper and Row, New York.
Cantoni, G. L., Ishikura, H., Richards, H. H. and Tanaka, K. (1963). *Cold Spring Harb. Symp. quant. Biol.* **28**, 123–132.
Cantor, C. R., Raskunas, S. R. and Tinoco, I. (1966). *J. molec. Biol.* **20**, 39–62.
Carnevali, F., Piperno, G. and Tecce, G. (1966). *Atti. Accad. naz. Lincei Rc.* **41**, 194–196.
Carnevali, F., Morpurgo, G. and Tecce, G. (1969). *Science, N.Y.* **163**, 1331–1333.
Cerna, J., Kalousek, F. and Rychlik, I. (1966). *Coll. Czech. chem. Commun.* **31**, 2794–2802.
Chambers, R. W. (1966). *In* "Progress in Nucleic Acid Research" (J. N. Davidson and W. E. Cohn, eds.) Vol. 5, pp. 350–398. Academic Press, New York.
Chao, F. H. (1957). *Archs Biochem. Biophys.* **70**, 426–431.
Corneo, G., Moore, C., Sanadi, D. R., Grossman, L. I. and Marmur, J. (1966). *Science, N.Y.* **151**, 687–689.
Cox, B. S. and Parry, J. M. (1968). *Mutation Res.* **6**, 37–55.
Crestfield, A. M., Smith, K. C. and Allen, F. W. (1955). *J. biol. Chem.* **216**, 185–193.
Croes, A. F. (1966). *Exptl Cell Res.* **41**, 452–454.
Dastidar, S. G. (1966). *Indian J. Exptl Biol.* **4**, 51–52.
Doctor, B. P., Loebel, J. E. and Kellog, D. A. (1966). *Cold Spring Harb. Symp. quant. Biol.* **31**, 543–548.
Duval, J. and Ebel, J. P. (1966). *C. r. hebd. Séanc. Acad. Sci., Paris* **263**, 1773–1776.
Eckstein, H., Paduch, V. and Hilz, H. (1967). *Eur. J. Biochem.* **3**, 224–231.
Ephrussi, B. and Slonimski, P. P. (1955). *Nature, Lond.* **176**, 1207–1208.
Ephrussi, B., Hottinguer, H. and Chimenes, A. M. (1949). *Annls Inst. Pasteur, Paris* **76**, 351–367.
Esposito, R. E. (1968). *Genetics* **59**, 191–210.
Fauman, M., Rabinowitz, M. and Getz, G. S. (1969). *Biochim. biophys. Acta* **182**, 355–360.
Fittler, F. and Hall, R. H. (1966). *Biochem. biophys. Res. Commun.* **25**, 441–446.
Fogel, S. and Mortimer, R. K. (1968). *Genetics, Princeton* **60**, 178.
Freese, E. (1963). *In* "Molecular Genetics" (J. H. Taylor, ed.) Vol. I, pp. 207–269. Academic Press, New York.
Fresco, J. R., Adamas, A., Ascione, R., Henley, D. and Lindahl, T. (1966). *Cold Spring Harb. Symp. quant. Biol.* **31**, 527–537.
Fujimura, Y. and Miura, K. (1966). *Biochim. biophys. Acta* **129**, 409–411.
Fukuhara, H. (1965). *Biochem. biophys. Res. Commun.* **18**, 297–301.

Fukuhara, H. (1967a). *Biochim. biophys. Acta* **134**, 143–164.
Fukuhara, H. (1967b). *Proc. natn. Acad. Sci. U.S.A.* **58**, 1065–1072.
Fukuhara, H. (1968). *In* "Biochemical Aspects of The Biogenesis of Mitochondria" (E. C. Slater, J. M. Tager, S. Papa and E. Quagliariello, eds.), pp. 303–325. Adriatica Editrice, Bari.
Fukuhara, H., Faures, M. and Genin, C. (1969). *Molec. gen. Genetics* **104**, 264–281.
Gebicki, J. M. and Freed, S. (1965). *Anal. Biochem.* **13**, 505–509.
Goldstein, J. (1966). *Biochim. biophys. Acta* **123**, 620–623.
Goldwasser, E. and Heinrikson, R. L. (1966). *In* "Progress in Nucleic Acid Research" (J. N. Davidson and W. E. Cohn, eds.) Vol. 5, pp. 399–416. Academic Press, New York.
Gorman, J. Tauro, P. Laberge, M. and Halvorson, H. (1964). *Biochem. biophys. Res. Commun.* **15**, 43–49.
Gratzer, W. B. (1966). *Biochim. biophys. Acta* **123**, 431–434.
Gray, M. W. and Lane, B. G. (1967). *Biochim. biophys. Acta* **134**, 243–257.
Guerineau, M., Gosse C. and Paoletti, C. (1966). *C. r. hebd. Séanc. Acad. Sci., Paris* **262**, 1901–1904.
Guerineau, M., Grandchamp, C., Yotsuyanagi, Y. and Slonimski, P. P. (1968a). *C. r. hebd. Séance Acad. Sci., Paris* **266**, 1884–1887.
Guerineau, M., Grandchamp, C., Yotsuyanagi, Y. and Slonimski, P. P. (1968b). *C. r. hebd. Séance Acad. Sci., Paris* **266**, 2000–2003.
Halvorson, H. O. (1958). *Biochim. biophys. Acta* **27**, 255–276.
Hayashi, H. (1966). *J. molec. Biol.* **19**, 161–173.
Hiby, W. and Kroger, H. (1967). *J. Chromat.* **26**, 545–548.
Holley, R. W. (1963). *Biochem. biophys. Res. Commun.* **10**, 186–188.
Holley, R. W. (1968). *In* "Progress in Nucleic Acid Research" (J. N. Davidson and W. E. Cohn, eds.) Vol. 8, pp. 37–48. Academic Press, New York.
Holley, R. W., Apgar, J., Everett, G. A., Madison, J. T., Merrill, S. H. and Zamir, A. (1963). *Cold Spring Harb. Symp. quant. Biol.* **28**, 117–121.
Holley, R. W., Apgar, J., Everett, G. A., Madison, J. T., Marquisee, M., Merrill, S. H., Penswick, J. R. and Zamir, A. (1965). *Science, N.Y.* **147**, 1462–1465.
Hutchinson, H. T. and Hartwell, L. H. (1967). *J. Bact.* **94**, 1697–1705.
Jackson, J. F., Kornberg, R. D., Berg, P. Bhandary, U. L. R., Stuart, A., Khorana, H. G. and Kornberg, A. (1965). *Biochim. biophys. Acta* **108**, 243–248.
Jones, A. S., Parsons, D. G. and Roberts, D. G. (1967). *Eur. Polym. J.* **3**, 187–198.
Kawata, M., Miyazaki, M. and Takemura, S. (1967). *J. Biochem., Tokyo* **62**, 287–292.
Kempner, E. S. (1961). *Biochim. biophys. Acta* **53**, 111–122.
Kerridge, D. (1958). *J. gen. Microbiol.* **19**, 497–506.
Kitazume, Y. and Ycas, M. (1963). *Biochim. biophys. Acta* **76**, 391–400.
Kjellin-Straby, B. and Boman, H. (1965). *Proc. natn. Acad. Sci. U.S.A.* **53**, 1346–1352.
Kleinschmidt, A. and Zahn, R. K. (1959). *Z. Naturf.* **14b**, 770.
Kloet, S. R. de (1965). *Biochem. biophys. Res. Commun.* **19**, 582–586.
Kloet, S. R. de (1966). *Biochem. J.* **99**, 566–581.
Kloet, S. R. de (1967). *Biochem. J.* **106**, 167–178.
Kloet, S. R. de and Strijkert, P. J. (1966). *Biochem. biophys. Res. Commun.* **23**, 49–55.
Kudo, Y. and Imahori, K. (1965). *J. Biochem., Tokyo* **58**, 364–372.
Lamonthezie, N. and Guerineau, M. (1965). *Bull. Soc. Chim. biol.* **47**, 807–819.
Lebovitz, P. Ipata, P. L., Makman, M. H., Richard, H. H. and Cantoni, G. L. (1966). *Biochemistry, N.Y.* **5**, 3617–3625.

Li, L., Venkstern, T. V., Mirzabekov, A. D., Krutilina, A. I. and Baev, A. A. (1966). *Biokhimiya* **31**, 117–124.

Lindahl, T., Henley, D. D. and Fresco, J. R. (1965). *J. Am. chem. Soc.* **87**, 4961–4963.

Lis, A. W. and Passarge, W. E. (1966). *Arch. Biochem. Biophys.* **114**, 593–595.

Lochmann, E. R. (1967). *Z. Naturf.* **22**, 196–200.

Loening, U. E. (1968). *J. molec. Biol.* **38**, 355–365.

Madison, J. T. and Kung, H. K. (1967). *J. biol. Chem.* **242**, 1324–1330.

Madison, J. T., Everett, G. A. and Kung, H. K. (1966). *Science, N.Y.* **153**, 531–534.

Madison, J. T., Everett, G. A. and Kung, H. K. (1967). *J. biol. Chem.* **242**, 1318–1323.

Magrath, D. I. and Shaw, D. C. (1967). *Biochem. biophys. Res. Commun.* **26**, 32–37.

Makman, M. H. and Cantoni, G. L. (1966). *Biochemistry, N.Y.* **5**, 2246–2254.

Marcot-Queiroz, J., Julien, J., Rosset, R. and Monier, R. (1965). *Bull. Soc. Chim. biol.* **47**, 183–195.

Marcovitch, H. (1951). *Annls Inst. Pasteur, Paris* **81**, 452–468.

Marmur, J. (1961). *J. molec. Biol.* **3**, 208–218.

Marmur, J. and Doty, P. (1962). *J. molec. Biol.* **5**, 109–118.

McPhie, P. and Gratzer, W. B. (1966). *Biochemistry, N.Y.* **5**, 1310–1315.

McPhie, P., Hounsell, J. and Gratzer. W. B. (1966). *Biochemistry, N.Y.* **5**, 988–993.

Mehrotra, B. D. and Mahler, H. R. (1968). *Archs Biochem. Biophys.* **128**, 685–703.

Meyer, S. A. and Phaff, H. J. (1969). *J. Bact.* **97**, 52–56.

Mirzabekov, A. D., Venkstern, T. V. and Baev, A. A. (1965). *Biokhimiya* **30**, 825–835.

Mitchison, J. M. and Gross, P. R. (1965). *Exptl Cell Res.* **37**, 259–277.

Miyazaki, M., Kawata, M. and Takemura, S. (1966). *J. Biochem., Tokyo* **60**, 519–525.

Miyazaki, M. Kawata, M., Nakazawa, K. and Takemura, S. (1967). *J. Biochem., Tokyo* **62**, 161–169.

Mommaerts, W. F. H. M., Brahms, J., Weil, J. H. and Ebel, J. P. (1964). *C. r. hebd. Séance. Acad. Sci., Paris* **258**, 2687–2689.

Mounolou, J. C., Jakob, H. and Slonimski, P. P. (1966). *Biochem. biophys. Res. Commun.* **24**, 218–224.

Mounolou, J. C., Jakob, H. and Slonimski, P. P. (1967). *In* "The Control of Nuclear Activity" (L. Goldstein, ed.), pp. 413–431. Prentice-Hall, Englewood-Cliffs, N.J.

Mounolou, J. C., Jakob, H. and Slonimski, P. P. (1968a). *In* "Biochemical Aspects of the Biogenesis of Mitochondria" (E. C. Slater, J. M. Tager, S. Papa and D. Quagliariello, eds.), pp. 473–474. Adriatica Editrice, Bari.

Mounolou, J. C., Perrodin, G. and Slonimski, P. P. (1968b). *In* "Biochemical Aspects of the Biogenesis of Mitochondria" (E. C. Slater, J. M. Tager, S. Papa and E. Quagliariello, eds.), pp. 133–148. Adriatica Editrice, Bari.

Moustacchi, E. (1964). Thèse: Faculté des Sciences, Paris.

Moustacchi, E. (1969). *Mutation Res.* **7**, 171–185.

Moustacchi, E. and Williamson, D. H. (1966). *Biochem. biophys. Res. Commun.* **23**, 56–61.

Moustacchi, E., Hottinguer-Margerie, H. and Fabre, F. (1967). *Genetics, Princeton* **57**, 909–918.

Muhammed, A. (1966). *J. biol. Chem.* **241**, 516–523.

Nakai, S. (1968). Abstracts of presentations at the yeast genetics conference, Osaka, p. 26.

Nagai, S., Yanaguishima, N. and Nagai, H. (1961). *Bact. Rev.* **25**, 404–426.

Nasim, A. (1967). *Mutation Res.* **4**, 753–763.

Nasim, A. and Auerbach, C. (1967). *Mutation Res.* **4**, 1–14.
Ogur, M. and Rosen, G. (1950). *Arch. Biochem.* **25**, 262–276.
Ogur, M., Minckler, S., Lindegren, G. and Lindegren, C. C. (1952). *Archs Biochem. Biophys.* **40**, 175–184.
Osawa, S. (1960). *Biochim. biophys. Acta* **43**, 110–122.
Osawa, S. and Otaka, E. (1959). *Biochim. biophys. Acta* **36**, 549–551.
Otaka, E., Oota, Y. and Osawa, S. (1961). *Nature, Lond.* **191**, 598–599.
Otaka, Y. and Uchida, K. (1963). *Biochim. biophys. Acta* **76**, 94–104.
Patrick, M. H. and Haynes, R. H. (1968). *J. Bact.* **95**, 1350–1354.
Peetrissant, G. and Gaye, P. (1966). *Annls Biol. anim. Biochim. Biophys.* **6**, 333–349.
Phillips, J. H. and Kjellin-Straby, B. (1967). *J. molec. Biol.* **26**, 509–518.
Polakis, E. S. and Bartley, W. (1966). *Biochem. J.* **98**, 883–887.
Rabinowitz, M. (1968). *Bull. Soc. Chim. biol.* **50**, 311–348.
Rabinowitz, M., Getz, G. S. and Swift, H. H. (1968). *In* "Biochemical Aspects of the Biogenesis of Mitochondria" (E. C. Slater, J. M. Tager, S. Papa and E. Quagliariello, eds.), pp. 155–169. Adriatica Editrice, Bari.
Rabinowitz, M., Getz, G. S., Casey, J. and Swift, H. H. (1969). *J. molec. Biol.* **41**, 381–400.
Rajbhandary, U. L., Stuart, A., Faulkner, R. D., Chang, S. H. and Khorana, H. G. (1966). *Cold Spring Harb. Symp. quant. Biol.* **31**, 425–434.
Rake, A. V. and Tener, G. M. (1966). *Biochemistry, N.Y.* **5**, 3992–4003.
Raut, C. and Simpson, W. L. (1955). *Archs Biochem. Biophys.* **57**, 218–228.
Reeves, R. H., Imura, N., Schavam, H., Weiss, G. B., Schulman, L. H. and Chambers, R. W. (1968). *Proc. natn. Acad. Sci. U.S.A.* **60**, 1450–1457.
Retel, J. and Planta, R. J. (1967). *Eur. J. Biochem.* **3**, 246–258.
Retel, J. and Planta, R. J. (1968). *Biochim. biophys. Acta* **169**, 416–429.
Robins, M. J., Hall, R. H. and Thedford, R. (1967). *Biochemistry, N.Y.* **6**, 1837–1847.
Rogers, P. J., Preston, B. W., Titchener, E. B. and Linnane, A. W. (1967). *Biochem. biophys. Res. Commun.* **27**, 405–411.
Rosset, R. and Monier, R. (1965). *Biochim. biophys. Acta* **108**, 376–384.
Rozijn, T. H. and Tonino, G. J. M. (1964). *Biochim. biophys. Acta* **91**, 105–112.
Sarin, P. S., Zamecnick, P. C., Bergquist, P. L. and Scott, J. F. (1966). *Proc. natn. Acad. Sci. U.S.A.* **55**, 579–585.
Sarkar, P. K. and Yang, J. T. (1965). *Archs Biochem. Biophys.* **112**, 512–518.
Schapiro, L., Grossman, L. I., Marmur, J. and Kleinschmidt, A. (1968). *J. molec. Biol.* **33**, 907–922.
Schatz, G., Haslbrunner, E. and Tuppy, H. (1964). *Biochem. biophys. Res. Commun.* **15**, 127–132.
Schildkraut, C. L., Marmur, J. and Doty, P. (1962). *J. molec. Biol.* **4**, 430–443.
Schwaier, R., Nashed, N. and Zimmermann, F. K. (1968). *Mol. gen. Genetics* **102**, 290–300.
Schweizer, E., MacKechnie, C. and Halvorson, H. O. (1969). *J. molec. Biol.* **40**, 261–268.
Seidel, H. and Cramer, F. (1965). *Biochim. biophys. Acta* **108**, 365–375.
Sekiya, T., Takeishi, K. and Ukita, T. (1969). *Biochim. biophys. Acta* **182**, 411–426.
Shortman, K. and Fukuhara, H. (1963). *Biochim. biophys. Acta* **76**, 501–524.
Siegel, M. R. and Sisler, H. D. (1964). *Biochim. biophys. Acta* **87**, 70–82.
Sinclair, J. H. (1966). Thesis: University of Chicago.
Sinclair, J. H,, Stevens, B. J., Sangavi, P. and Rabinowitz, M. (1967). *Science, N.Y.* **156**, 1234–1237.

Slonimski, P. P. and Robichon-Szulmajster, H. de (1957). *In* "Drug Resistance in Microorganisms", Ciba Found. Symp., pp. 210–230.
Slonimski, P. P., Perrodin, G. and Croft, J. H. (1968). *Biochem. biophys. Res. Commun.* **30**, 232–239.
Slonimski, P. P., Cohen, D., Deutsch, J., Netter, P. and Petrochilo, E. (1969). *In* "The Development and Inter-relationships of Cell Organelles", Symposium 24 of the Society for Experimental Biology, London.
Smith, A. E. and Marker, K. A. (1968). *J. molec. Biol.* **38**, 241–243.
Smith, D., Tauro, P., Schweizer, E. and Halvorson, H. O. (1968). *Proc. natn. Acad. Sci. U.S.A.* **60**, 936–942.
Smith, J. D. and Halvorson, H. O. (1966). *Bact. Proc.* 34.
Snow, R. (1967). *J. Bact.* **94**, 571–576.
Soll, D., Jones, D. S., Ohtsika, E., Faulkner, R. D., Lolvimann, R., Hayatu, H., Khorana, H. G., Cherayil, J. D., Hampel, A. and Bock, R. M. (1966). *J. molec. Biol.* **19**, 556–573.
Soll, D., Cherayil, J. D. and Bock, R. M. (1967). *J. molec. Biol.* **29**, 97–112.
Somers, J. M. and Bevan, E. A. (1969). *Genet. Res.* **13**, 71–84.
Steinschneider, A. (1969). *Biochim. biophys. Acta* **186**, 405–408.
Stenderup, A. and Bak, A. L. (1968). *J. gen. Microbiol.* **52**, 231–236.
Sugimura, T., Okabe, K. and Imamura, A. (1966). *Nature, Lond.* **212**, 304–305.
Swift, H., Rabinowitz, M. and Getz, G. (1967). *J. biophys. biochem. Cytol.* **35**, 131A.
Sylven, B., Tobias, C. A., Malmgren, H., Ottoson, R. and Thorell, B. (1959). *Exptl Cell Res.* **16**, 77–87.
Taber, R. L. and Vincent, W. S. (1969a). *Biochem. biophys. Res. Commun.* **34**, 488–494.
Taber, R. L. and Vincent, W. S. (1969b). *Biochim. biophys. Acta* **186**, 317–325.
Takeishi, K., Nishimura, S. and Ukita, T. (1967). *Biochim. biophys. Acta* **145**, 605–612.
Tamaki, H. (1965). *J. gen. Microbiol.* **41**, 93–98.
Tanner, M. J. A. (1967). *Biochemistry, N.Y.* **6**, 2686–2694.
Tauro, P., Halvorson, H. O. and Epstein, R. L. (1968). *Proc. natn. Acad. Sci. U.S.A.* **59**, 277–284.
Tewari, K. K., Jayaraman, J. and Mahler, H. R. (1965). *Biochem. biophys. Res. Commun.* **21**, 141–148.
Tewari, K. K., Votsch, W., Mahler, H. R. and Mackler, B. (1966). *J. molec. Biol.* **20**, 453–481.
Tomlinson, G. A. and Campbell, J. J. R. (1966). *Biochim. biophys. Acta* **123**, 337–344.
Tonino, G. d. M. and Rozijn, T. H. (1966). *Biochim. biophys. Acta* **124**, 427–429.
Van Bruggen, E. F. J., Runner, C. M., Borst, P., Ruttenberg, G. J. C. M., Kroon, A. M. and Shuurmans-Stekhoven, F. M. A. H. (1968). *Biochim. biophys. Acta* **161**, 402–414.
Venkstern, T. V., Li, L., Krutilina, A. I., Mirzabekov, A. D., Aksel'rod, V. D. and Baev, A. A. (1967). *Dokl. Akad. Naut SSSR* **173**, 459–462.
Villa, V. D. and Storck, R. (1968). *J. Bact.* **96**, 184–190.
Vischer, E., Zamenhof, S. and Chargaff, E. (1949). *J. biol. Chem.* **177**, 429–438.
Vournakis, J. N. and Scheraga, H. A. (1966). *Biochemistry, N.Y.* **5**, 2997–3006.
Wagner, E. K. and Ingram, V. H. (1966). *Biochemistry, N.Y.* **5**, 3019–3027.
Wehr, C. T. and Parks, L. W. (1969). *Bact. Proc.* 28.
Westphal, H., Oeser, A. and Holzer, H. (1966). *Biochem. Z.* **346**, 252–263.
Wilkie, D. (1963). *J. molec. Biol.* **7**, 527–533.

Wilkie, D. (1968). *In* "Biochemical Aspects of the Biogenesis of Mitochondria" (E. C. Slater, J. M. Tager, S. Papa and E. Quagliariello, eds.), pp. 457–468. Adriatica Editrice, Bari.
Wilkie, D. and Lewis, D. (1963). *Genetics, Princeton* **48**, 1701–1716.
Williamson, D. H. (1965). *J. biophys. biochem. Cytol.* **25**, 517–528.
Williamson, D. H. (1969). *In* "The Development and Inter-relationships of Cell Organelles", Symposium 24 of the Society for Experimental Biology, London.
Williamson, D. H. and Scopes, A. W. (1961). *Exptl Cell Res.* **24**, 338–349.
Wimmer, E., Maxwell, I. H. and Tener, G. M. (1968). *Biochemistry, N.Y.* **7**, 2629–2663.
Winstein, W. A. (1969). *Biochim. biophys. Acta* **182**, 402–410.
Wintersberger, E. (1965). *In* "Regulation of the Metabolic Process in Mitochondria" (J. M. Tager, S. Papa, E. Quagliariello and E. C. Slater, eds.), pp. 439–453. Elsevier Publishing Co., Amsterdam.
Wintersberger, E. and Viehhauser, G. (1968). *Nature, Lond.* **220**, 699–702.
Wulff, D. L. and Rupert, C. S. (1962). *Biochem. biophys. Res. Commun.* **7**, 237–240.
Yamamoto, T. (1967). *Biochem. biophys. Res. Commun.* **29**, 21–27.
Yamane, T. and Sueoka, N. (1963). *Proc. natn. Acad. Sci. U.S.A.* **50**, 1093–1100.
Ycas, M. and Vincent, W. S. (1960). *Proc. natn. Acad. Sci. U.S.A.* **46**, 804–811.
Yoshida, M. and Ukita, C. (1966). *Biochim. biophys. Acta* **123**, 214–216.
Yotsuyanagi, Y. (1962). *J. Ultrastruct. Res.* **7**, 141–150.
Yotsuyanagi, Y. (1966). *C. r. hebd. Séance. Acad. Sci., Paris* **262**, 1348–1351.
Zachau, H. G. (1967). *Umschau* **67**, 17–20.
Zachau, H. G., Duetting, D., Feldmann, H., Melchers, F. and Karau, W. (1966). *Cold Spring Harb. Symp. quant. Biol.* **31**, 417–424.
Zachau, H. G., Duetting, D. and Feldmann, H. (1967). *Hoppe-Seyler's Z. physiol. Chem.* **347**, 212–235.
Zamenhof, S. and Chargaff, E. (1948). *J. biol. Chem.* **173**, 325–327.
Zamenhof, S. and Chargaff, E. (1950). *J. biol. Chem.* **187**, 1–14.

# Chapter 9

# Nucleic Acid and Protein Synthesis in Yeasts: Regulation of Synthesis and Activity

H. DE ROBICHON-SZULMAJSTER AND Y. SURDIN-KERJAN

*Laboratoire d'Enzymologie, Centre National de la Recherche Scientifique, 91-Gif-sur-Yvette, France*

I. INTRODUCTION . . . . . . . . . . . . . 336

II. BIOSYNTHESIS OF PURINE AND PYRIMIDINE DERIVATIVES AND ITS REGULATION . . . . . . . . . . . 338
- A. Purine Derivatives . . . . . . . . . . . 338
- B. Pyrimidine Derivatives . . . . . . . . . . 343
- C. Utilization of Purines and Pyrimidines by Yeasts . . . . 347

III. BIOSYNTHESIS OF AMINO ACIDS AND ITS REGULATION . . . . . 348
- A. Introductory Remarks . . . . . . . . . . 348
- B. Histidine . . . . . . . . . . . . 348
- C. Arginine . . . . . . . . . . . . 353
- D. Proline . . . . . . . . . . . . 356
- E. Lysine . . . . . . . . . . . . 358
- F. Glutamate . . . . . . . . . . . . 361
- G. Threonine, Methionine and Cysteine . . . . . . 362
- H. Isoleucine, Valine and Leucine . . . . . . . 376
- I. Tryptophan, Tyrosine and Phenylalanine . . . . . 382

IV. PROTEIN BIOSYNTHESIS . . . . . . . . . . . 388
- A. Introductory Remarks . . . . . . . . . . 388
- B. Aminoacyl-t-RNA Synthetases . . . . . . . . 388
- C. Transfer Ribonucleic Acid . . . . . . . . 392
- D. Messenger Ribonucleic Acid . . . . . . . . 396
- E. Ribosomes . . . . . . . . . . . 399
- F. Inhibition of Protein Synthesis . . . . . . . 401

V. NUCLEIC ACID BIOSYNTHESIS . . . . . . . . . . 402
- A. Enzymic Systems for Polynucleotide Synthesis . . . . 402
- B. Ribosomal RNA Synthesis . . . . . . . . 405
- C. Methylation of Nucleic Acids . . . . . . . . 407

REFERENCES . . . . . . . . . . . . . . 409

## I. Introduction

Since the dawn of Molecular Biology, bacteria, and most particularly *Escherichia coli*, seemed to be *the* organisms which could provide the answers to all problems. Fortunately, less fashionable work continued to be done with other less primitive organisms, including yeast. The recent transition of general interest from Molecular towards Developmental Biology has rendered some nobility to studies involving eucaryotic organisms, among which yeast remains the simplest and easiest with which to work. It appears then, that accumulation of knowledge which in the last decade arose from studies in genetics, cellular organization, metabolism and regulatory mechanisms in yeast and, especially in *Saccharomyces cerevisiae*, has reached a point of sophistication that designates yeast as one of the best candidates for many studies in Developmental Biology.

In this chapter an attempt has been made to summarize a very abundant literature relating to nucleic acids and protein biosynthesis, starting with pathways which ensure fabrication of their building blocks, namely amino acids, and purine and pyrimidine derivatives, and including the mechanisms which regulate their production. Analogies and differences between yeasts and other organisms are also emphasized, in the hope that this chapter may provide a tool for scientists not familiar with yeast metabolism. Scientists already working with this organism might find regrettable omissions, or even errors! We hope to be forgiven, as the amount of relevant material is enormous. The authors are thankful to all their colleagues in France, the United States, Canada, and so on, who provided recent publications, information or help in particular parts of the manuscript.

More information on different aspects of the present chapter can be found either in the other volumes of this series, and especially in the chapter on yeast genetics (Mortimer and Hawthorne, Vol. 1, chapter 8, pp. 385–460) or in the following books and reviews: Mahler and Cordes (1967), Cohen (1968), Umbarger (1969), Gross (1969), Grunberg-Manago (1962), Lengyel and Söll (1969).

Generally, the gene nomenclature used in this chapter is that adopted by the author quoted. Besides the recognized forms such as Ala, ATP CoA, NAD, RNA (see *Biochemical Journal* 1970, **116**, 1), the following abbreviations are used throughout the chapter.

| | |
|---|---|
| AA | Aminoacyl adenylate. |
| AcHS | Acetyl homoserine. |
| αAHB | α-Acetohydroxybutyrate. |
| AICAR | 5-Aminoimidazole-4-carboxyamide ribonucleotide. |

| | |
|---|---|
| AIC-$RP_a$ | *N*-(5′-Phospho-D-1′-ribosyl formimino-5-amino-1-(5″-phosphoribosyl)-4-imidazole carboxamide; BBM-II. |
| AIC-$RP_b$ | BBM-III. |
| AIR | 5-Aminoimidazole ribonucleotide. |
| αAL | α-Acetolactate. |
| anth | Anthranilate. |
| anthDrib | 1-(*O*-Carboxylphenylamino)-1-deoxyribulose 5-phosphate. |
| anthRP | *N*-(5′-Phosphoribosyl) anthranilate. |
| APS | Adenine monophosphate-$SO_4^{2-}$. |
| ASA | Aspartate semialdehyde. |
| asp-$PO_4$ | β-Aspartyl phosphate. |
| BBM | Bound Bratton-Marshall compound. |
| CAIR | 5-Aminoimidazole-4-carboxylic acid ribonucleotide. |
| CDMC | *cis*Dimethyl citraconate. |
| $CH_3FH_4$ | 5-Methyl tetrahydrofolate. |
| $CH_2OHFH_4$ | 5-Hydroxymethyl tetrahydrofolate. |
| $CH_3$-X | Methylated derivative of unknown compound. |
| CP | Carbamyl phosphate. |
| $C(\text{-SH})_2$ | Low molecular weight protein, reduced form (see pp. 364 and 372). |
| C-SS | Low molecular weight protein, oxidized form (see pp. 364 and 372). |
| DAHP | 3-Deoxy-D-arabino heptulosonate 7-phosphate. |
| DEAE | *O*-(Diethylaminoethyl)cellulose. |
| DHMV | Dihydroxy-β-methyl valerate. |
| 5DHQ | 5-Dehydroquinate. |
| 5DHSh | 5-Dehydroshikimate. |
| DHV | Dihydroxyvalerate. |
| EDTA | Ethylenediaminetetra-acetate. |
| EIGP | D-Erythroimidazoleglycerol phosphate. |
| endo $SO_4^{2-}$ | Endogenous sulphate ion. |
| exo $SO_4^{2-}$ | Exogenous sulphate ion. |
| enz. | Enzyme. |
| E4P | Erythrose 4-phosphate. |
| 3EP-Sh5P | 3-Enolpyruvyl 5-phosphoshikimate. |
| exo $SO_4^{2-}$ | Exogenous sulphate ion. |
| FAICAR | 5-Formamidoimidazole-4-carboxamido ribonucleotide. |
| FGAM | *N*-Formylglycinamidine ribonucleotide. |
| FGAR | α-*N*-Formylglycinamide ribonucleotide. |
| $FH_4$ | Tetrahydrofolate. |
| GAR | Glycinamide ribonucleotide. |

| | |
|---|---|
| HC | Homocysteine. |
| HP | Histidinol phosphate. |
| HS | Homoserine. |
| $HS\text{-}PO_4$ | Homoserine phosphate. |
| IAP | Imidazole acetolphosphate. |
| IGP | Indoleglycerol phosphate. |
| αIPM | α-Isopropyl malate. |
| βIPM | β-Isopropyl malate. |
| αKB | α-Ketobutyrate. |
| αKG | α-Ketoglutarate. |
| KIC | α-Ketoisocaproate. |
| KMV | α-Keto-β-methyl valerate. |
| KV | α-Ketoisovalerate. |
| me-enz. | Methylated enzyme. |
| *O*-acS | *O*-Acetylserine. |
| *O*-acHS | *O*-Acetylhomoserine. |
| *O*-succHS | *O*-Succinylhomoserine. |
| PABA | p-Aminobenzoic acid. |
| PAPS | Phosphoadenosine phosphosulphate. |
| PEP | Phosphoenolpyruvate. |
| P-glyc | Phosphoglycerate. |
| phpyr | Phenyl pyruvate. |
| pOHphpyr | p-Hydroxyphenyl pyruvate. |
| PRA | 5-Phosphoribosyl-1-amine. |
| PR-AMP | *N*′-(5′-Phosphoribosyl)-adenosine monophosphate. |
| PR-ATP | *N*′-(5′-Phosphoribosyl)-adenosine triphosphate. |
| PRPP | 5-Phosphoribosyl-1-pyrophosphate. |
| pyr | pyruvate. |
| SAICAR | 5-Aminoimidazole-4-*N*-succinocarboxamide ribonucleotide. |
| SAE | *S*-Adenosylethionine. |
| SAHC | *S*-Adenosylhomocysteine. |
| SAM | *S*-Adenosylmethionine. |
| Sh | Shikimate. |
| Sh5P | 5-Phosphoshikimate. |
| X | Unknown compound. |
| Y | Unknown compound. |

## II. Biosynthesis of Purine and Pyrimidine Derivatives and its Regulation

### A. PURINE DERIVATIVES

In yeast, the purine biosynthetic pathway appears to be identical with that described for other organisms (see Magasanik, 1962). Five loci concerned with adenine biosynthesis *in Saccharomyces cerevisiae* were first described by Roman (1956). Two are tightly linked ($ad_5$ and $ad_7$); the others segregate independently, although $ad_3$ and $ad_6$ are located on the same chromosome. The steps involved are summarized in Table I.

TABLE I. *Biosynthesis of Purine Derivatives*

| Step | Reaction |
|---|---|
| 1 | *PRPP* + glutamine → *PRA* + glutamate |
| 2 | PRA + glycine + ATP $\xrightarrow{Mg^{2+}}$ *GAR* + ADP + $P_i$ |
| 3 | GAR + $N^5$,$N^{10}$-methenyl-$FH_4$ + $H_2O$ → *FGAR* + $FH_4$ + $H^+$ |
| 4 | FGAR + glutamine + $H_2O$ + ATP $\xrightarrow{Mg^{2+}}$ *FGAM* + glutamate + ADP + $P_i$ |
| 5 | FGAM + ATP $\xrightarrow{Mg^{2+}}$ *AIR* + ADP + $P_i$ |
| 6 | AIR + $CO_2$ → *CAIR* |
| 7 | CAIR + aspartic acid + ATP $\xrightarrow{Mg^{2+}}$ *SAICAR* + ADP + $P_i$ |
| 8 | SAICAR → *AICAR* + fumarate |
| 9 | AICAR + $N^{10}$-formyl-$FH_4$ $\xrightarrow{K^+}$ *FAICAR* + $FH_4$ |
| 10 | FAICAR → *IMP* + $H_2O$ |
| 11 | IMP + aspartate → *adenylosuccinate* |
| 12 | Adenylosuccinate → [AMP] + fumarate |
| 13 | IMP + $NAD^+$ → *XMP* + NADH + $H^+$ |
| 14 | XMP + glutamine + ATP + $H_2O$ → [GMP] + glutamate + AMP + $PP_i$ |

Compounds in italics are those in the main sequence of reactions; compounds enclosed in boxes are the end-products of the synthetic pathways. For abbreviations see pp. 336–338.

Only a few genetic blocks have been correlated with enzyme deficiency. Gene-enzyme relationships which are known for *Saccharomyces cerevisiae* and *Schizosaccharomyces pombe* are given in Table IIa and b, respectively.

Although enzymic activities cannot be measured, accumulation data indicate that, in the $ad_2$ mutants of *Sacch. cerevisiae*, the enzymic activity which corresponds to Step 6 is missing and that catalysis of Step 7 does not occur in $ad_1$ mutants (Silver and Eaton, 1968, 1969; Fisher, 1969). Identical defects correspond to $ad_6$ and $ad_7$ genes, respectively, in *Schizosacch. pombe* (Fisher, 1969). Adenylosuccinase, catalysing Step 12, is absent in mutants $ad_{13}$ of *Sacch. cerevisiae* and $ad_8$ of *Schizosacch. pombe*. These mutants accumulate SAICAR, the product of Step 7. It seems generally admitted (Mahler and Cordes, 1967) that the enzyme catalysing the elimination of fumaric acid from adenylosuccinate (Step 12) also catalyses the analogous reaction of Step 8. If this is the case in yeast, it would explain the accumulation of SAICAR.

In *Schizosacch. pombe* the first step has been extensively studied (Heslot *et al.*, 1966a, b). The enzyme catalysing this reaction (PRPP amido transferase) has been purified. It is missing in mutant $ad_4$ of this organism (Heslot *et al.*, 1966a, b; Nagy, 1969). IMP dehydrogenase catalysing Step 13 has also been purified (Pourquié, 1969). The $ad_{10}$ mutation in *Schizosacch. pombe* leads to accumulation of AICAR. Two enzymes are known to be absent in cell-free extracts of this mutant: those catalysing Steps 9 and 10. This mutation leads to a double requirement in adenine and histidine; this same phenotype has been observed for the $ad_3$ mutants of *Sacch. cerevisiae* (Roman, 1956; Clavilier *et al.*, 1960). These requirements have been explained in *Schizosacch. pombe* by a regulatory effect due to the accumulated intermediary compound AICAR; this will be analysed further.

The other purine-requiring mutants of *Sacch. cerevisiae* (Table II) have been related to the different steps in the pathway by studies of the accumulation of intermediary compounds (Dorfman, 1963; Silver and Eaton, 1968). More recently, Dorfman (1969) has shown that the loss of adenylosuccinate synthetase (Step 11) leads, by a single genetic event, to a derepressed accumulation of red pigment in $ad_1$ or $ad_2$ mutants. It is suggested that the enzyme might play a role in regulation of the pathway.

PRPP is the first substrate of two related pathways—purine and histidine (Fig. 1). No mutants seem to be known for the PRPP synthesis. AICAR, one of the intermediates in inosinic acid synthesis, is also generated as a by-product of histidine biosynthesis. It has been shown that in *Schizosacch. pombe*:

(i) AICAR inhibits the activity of the enzyme catalysing the first step in histidine biosynthesis (1-phosphoribosyl-ATP pyrophosphorylase). This inhibition is synergistic with histidine (Whitehead *et al.*, 1966).

(ii) Insertion of another mutation blocking a step previous to AICAR formation suppresses AICAR accumulation and simultaneously the histidine requirement of $ad_{10}$ mutants (Whitehead *et al.*, 1966). An identical regulatory effect might account for the histidine requirement of the $ad_3$ mutants of *Sacch. cerevisiae*. However, the enzymes catalysing Steps 9 and 10 are present in this mutant; its deficiency lies in an inability to synthesize *N*-10-formyltetrahydrofolic acid (Jones and Magasanik, 1967). Moreover, in this yeast, the first enzyme of the histidine pathway does not appear to be inhibited by AICAR in cell-free

TABLE II. *Gene–Enzyme Relationship for Purine Biosynthesis*

| Step | Enzyme studied | Intermediate accumulated | Phenotype | Gene | Ref. |
|---|---|---|---|---|---|
| a. *Saccharomyces cerevisiae* | | | | | |
| 1 | | | | | |
| 2 | | | | | |
| 3 | | | | $ad_4$ | a |
| 4 | | | | $ad_6$ | a |
| 5 | | | | $ad_7$ | a |
| 6 | | AIR | Pink | $ad_2$ | a, b, c |
| 7 | | CAIR | Pink | $ad_1$ | a, b, c |
| 8 | | | | $(ad_{13})$ | |
| 9 | | AICAR<br>SAICAR | Requires adenine + histidine | | |
| 10 | | | | | |
| 11 | | Inosine | | $ad_{12}$ | e |
| 12 | + | SAICAR | | $ad_{13}$ | d, e |
| 13 | | | | | |
| 14 | | | | $gu_1$ | d |
| b. *Schizosaccharomyces pombe* | | | | | |
| 1 | + | | | $aza_1$<br>$ad_4$ | f, g |
| 2 | | | | $ad_1$ | l |
| 3 | | | | $ad_5$ | m |
| 4 | | | | | |
| 5 | | | | | |
| 6 | | AIR | Pink | $ad_6$ | c |
| 7 | | CAIR | Pink | $ad_7$ | c |
| 8 | | | | | |
| 9 | + | | | | |
| | one protein? (steps 9, 10) | AICAR | Requires adenine + histidine | $ad_{10}$ | h |
| 10 | + | | | | |
| 11 | | IMP | | $ad_2$ | i |
| 12 | + | SAICAR | | $ad_8$ | j |
| 13 | | | | $gu_1$ | i |
| 14 | | | | $gu_2$ | i, k |

For abbreviations see pp. 336–338.

*References*: a, Silver and Eaton (1968); b, Silver and Eaton (1969); c, Fisher (1969); d, von Borstel (1966); e, Dorfman (1969); f, Heslot *et al.* (1966); g, Nagy (1969); h, Whitehead *et al.* (1966); i, Leupold (1961), Heslot (personal communication); j, Megnet (1959); k, Pourquié (1969); l, Heslot *et al.* (1966b); m, Heslot *et al.* (unpublished results).

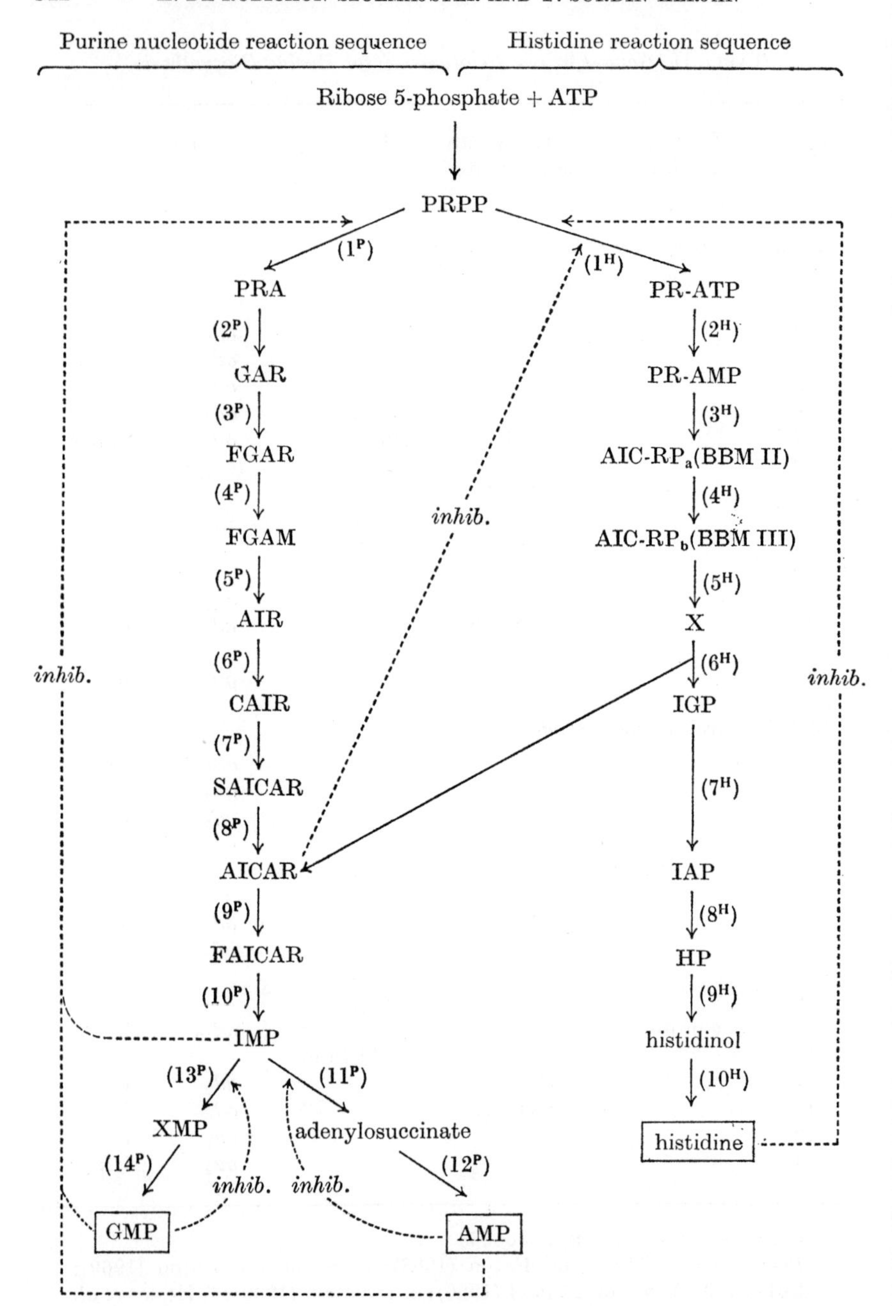


FIG. 1. Biosynthesis of purine nucleotides and histidine. For abbreviations see pp. 336–338. Superscripts P and H indicate, respectively, steps in the reaction sequences leading to purine nucleotides and to histidine.

extracts (Mazlen and Eaton, 1966)*. Inability to synthesize *N*-10-formyltetrahydrofolic acid and $N^5,N^{10}$-methenyl tetrahydrofolate provides an explanation of the mutation $ad_9$ in *Schizosacch. pombe* (Nagy *et al.*, 1969). A *Schizosacch. pombe* mutant, $aza_1$, resistant to 8-azaguanine, which maps close to the $ad_4$ locus, has been isolated. It has been shown to synthesize a modified PRPP amido transferase desensitized to the inhibitory effects normally exerted by IMP and GMP. This mutant excretes inosine and hypoxanthine, which are supposedly derived from IMP (Heslot *et al.*, 1966; Nagy, 1969).

(iii) In *Sacch. cerevisiae* (Burns, 1964) IMP, GMP and AMP seem to exert a control on the synthesis of purine nucleotides. In *Schizosacch. pombe* IMP, GMP and AMP, to a lower extent, participate in the feedback inhibition of the enzyme catalysing the first step in the purine pathway—PRPP amido transferase (Nagy, 1969). In addition, the first enzyme in the GMP synthesis from IMP (IMP dehydrogenase) is inhibited allosterically by GMP (Pourquié, 1969). The fact that the mutant $aza_1$ does not excrete AMP seems to indicate that regulation occurs also on Step 11 or Step 12. These regulations are summarized in Fig. 1.

An interesting feature of the biosynthesis of purine nucleotides is the GTP requirement for the adenylosuccinase catalyzed reaction and, in turn, the ATP requirement for GMP synthesis. This interaction of both end-products on each other's biosynthesis might provide the cell with a means of equilibrating the production of these two compounds continuously.

### B. PYRIMIDINE DERIVATIVES

The pyrimidine biosynthetic pathway in yeast is the same as that in bacteria. It involves nine steps (Fig. 2 and Table III). The enzymes catalysing Steps 1, 2, 3, 4 and 6 have been characterized. Mutants corresponding to these steps have been isolated (Lacroute *et al.*, 1965; Mortimer and Hawthorne, 1966; Lacroute, 1968). The gene-enzyme relationships can be seen in Table III. The *cpu* and $ura_2$ loci are linked, the other *ura* loci are independent.

The first reaction is common to both pyrimidine and arginine biosynthetic pathways. In *E. coli*, monogenic mutants showing a double auxotrophy for arginine and uracil have been obtained (see Umbarger and Davis, 1962). On the other hand, in *Neurospora crassa* (see Davis, 1965; Williams and Davis, 1968) and *Coprinus radiatus* (Cabet *et al.*, 1967) there are two distinct enzymes. In the latter two organisms, two independent types of mutation affecting carbamyl phosphate synthesis, and leading respectively to requirements for arginine or uracil, can be obtained, showing that some "compartmentalization" prevents carbamyl phosphate synthesized by one enzyme from being used in the other pathway. In *Sacch. cerevisiae* two carbamyl phosphate synthetases

*Note added in proof: Recent findings demonstrate that there are two distinguishable peaks concerned with methylenetetrahydrofolate dehydrogenase activity, and that only one of these is missing in $ad_3$ mutants (Lazowska and Luzzati, 1970a, b).

TABLE III. *Pyrimidine Biosynthesis and Gene–Enzyme Relationship in* Saccharomyces cerevisiae

| Step | Reaction | Gene | Enzyme studied | Ref. |
|---|---|---|---|---|
| 1 | *Glutamine* + ATP + $CO_3H^-$ $\xrightarrow{Mg^{2+}}$ *CP* + glutamate + ADP | { *cpu* / $ura_2$ | { One enzyme carrying both activities (steps 1 and 2) | a |
| 2 | CP + aspartate → *ureidosuccinate* | $ura_2$ | | b |
| 3 | Ureidosuccinate → *dihydro-orotate* + $H_2O$ | $ura_4$ | + | b |
| 4 | Dihydro-orotate + $NAD^+$ → *orotate* + NADH + $H^+$ | $ura_1$ (fragments) | + | b |
| 5 | Orotate + PRPP → *orotidine* 5-$PO_4$ + PP | | | |
| 6 | Orotidine 5-$PO_4$ → *UMP* + $CO_2$ | $ura_3$ (chrom. V) | + | b |
| 7 | UMP + ATP → *UDP* + ADP | | | |
| 8 | UDP + ATP → $\boxed{\text{UTP}}$ + ADP | | | |
| 9 | UTP + $NH_3$ + ATP $\xrightarrow[Mg^{2+}]{(GTP)}$ $\boxed{\text{CTP}}$ + ADP + $P_i$ | | — | |

Compounds in italics are those in the main sequences of reactions; compounds enclosed in boxes are end-products of the synthetic pathways. For abbreviations see pp. 336–338.

*References:* a, Lacroute *et al.* (1965); b, Lacroute (1968).

(CPases) have also been demonstrated. However, mutants lacking one of these two enzymes have no nutritional requirements, indicating that in

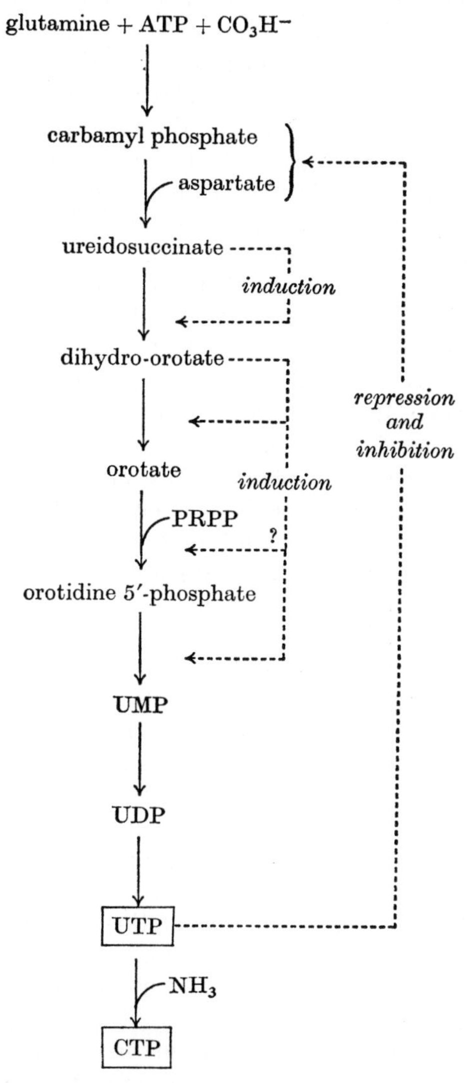


FIG. 2. Regulation of pyrimidine biosynthesis in *Saccharomyces cerevisiae* (according to Lacroute, 1968). For abbreviations see pp. 336–338.

this organism carbamyl phosphate from one pathway is freely available for the other (Lacroute *et al.*, 1965).

Regulatory properties of these enzymes help to explain the selective basis of the mutants. The synthesis of CPase (arginine) is subject to potent repression by exogenous arginine, while synthesis of CPase (uracil) is repressed by exogenous uracil. In addition, the latter enzyme is very sensitive to feedback inhibition exerted by UTP (Lacroute *et al.*, 1965). The activity of the second enzyme, aspartate transcarbamylase (ATCase), is also inhibited by UTP: this inhibition can be overcome by ATP (Davis, 1964). The two following observations have led to the conclusion that the two activities (CPase and ATCase) are borne by the same complex enzyme molecule: (i) mutants lacking one or other enzyme (CPase or ATCase) map at the same locus; (ii) concomitant loss of feedback inhibition by UTP for these two enzymes is found in mutants which also map at the same locus (Lacroute, 1968).

Purification of this multifunctional enzyme complex (mol. wt., 600,000) has been achieved. Omission of UTP during purification causes dissociation into a complex of mol. wt. 300,000 which still carries the two activities. Other steps permit the recovery of ATCase activity alone (mol. wt., 140,000; Lue and Kaplan, 1969). Desensitization of this enzyme towards an UTP inhibitory effect is accompanied by a 4- to 5-fold increase of activity (Kaplan and Messmer, 1967). The loss of feedback inhibition can be obtained *in vivo*, either by specific mutation at the ATCase locus (Kaplan *et al.*, 1967) or during derepression of Enzymes 1 and 2, due to pyrimidine starvation (Kaplan *et al.*, 1969).

Synthesis of aspartate transcarbamylase is placed under repressive control, which is mediated by exogenously-supplied uracil (Lacroute, 1968). The true co-repressor might be UTP formed directly from it (see below). A series of ingenious experiments carried out with mutants blocked at different steps in the pathway, either grown in excess uracil, under uracil starvation, or in the presence of intermediate compounds (provided specific mutations are introduced to allow their incorporation), have shown that the enzyme catalysing Step 3 is induced by the second product of the pathway (ureidosuccinate), and that enzymes catalysing Steps 4 and 6 (Step 5 has not been examined) are induced by the third product, dihydro-orotate. Regulation of this pathway shows, then, a combination of a classical repression scheme and a sequential induction scheme first found in *Pseudomonas* for mandelic acid catabolism (Stanier, 1947).

In a gene dosage study concerned successively with enzymic activities 2, 3, 4 and 6, about 50% of the enzyme level present in homozygous diploids has been found in heterozygous diploids, in repressed as well as derepressed cultures. These results have been taken as an indication that there is, even in the repressed stage, an excess of each of the enzymes studied, and that, consequently, regulation of pyrimidine production *in vivo* is mostly attained through feedback inhibition of the first enzyme (Lacroute, 1968).

### C. UTILIZATION OF PURINES AND PYRIMIDINES BY YEASTS

As a general permeability barrier seems to exist towards most phosphorylated compounds, mutants blocked in purine or pyrimidine biosynthesis are generally fed with bases or nucleosides (Kerr *et al.*, 1951; Pomper, 1952a). Adenine is used by adenine- and guanineless mutants. On the other hand, guanine can only be used by guanineless mutants (Pomper, 1952a). However, it appears from the observation of both pathways, that free bases or even nucleosides are not normal intermediates (see Figs. 1 and 2). The following types of reactions, called "salvage mechanisms" have been described in micro-organisms:

(1) base + PRPP $\rightarrow$ nucleotide + $PP_i$
(2) base + ribose 1-phosphate $\rightarrow$ nucleoside + $P_i$
(3) nucleoside + ATP $\rightarrow$ nucleotide + ADP

In the yeast *Schizosacch. pombe*, a mutant resistant to 2-6-diaminopurine ($dap_1$) has been shown to be devoid of AMP pyrophosphorylase which catalyses the reaction:

$$\text{adenine} + \text{PRPP} \rightarrow \text{AMP} + \text{PP}_i$$

In another mutant ($pur_1$) resistant to 8-aza-2-thioxanthine, IMP, XMP and AMP pyrophosphorylase activities are lost together, indicating that a single enzyme is most probably responsible for the formation of nucleotides from guanine, hypoxanthine and xanthine (de Groodt *et al.*, 1969).

These findings suggest that, at least for purines, reactions of Type 2 are not operative. It follows that free base and nucleoside utilization might not involve the same mechanisms. If this observation were also true for pyrimidines, it could throw some light on the report that a particular pyrimidineless mutant of *Sacch. cerevisiae* requires addition of *meso*inositol when uracil is the pyrimidine source, but grows without it in the presence of uridine, cytosine or cytidine (de Robichon-Szulmajster, 1956).

Thymidylic acid originates from CMP through the following transformations:

$$\text{CMP} \rightarrow \text{CDP} \rightarrow \text{dCDP} \rightarrow \text{dCMP} \rightarrow \text{dUMP} \rightarrow \text{TMP}$$

The methylation of dUMP catalysed by thymidylate synthetase requires $N^5,N^{10}$-methylenetetrahydrofolate as methyl carrier (see Friedkin, 1963). Thymine or thymidine are not used by pyrimidineless mutants of yeast, and thymineless mutants have not been found. A partial explanation of these observations lies in the absence of thymidine kinase in *Sacch. cerevisiae* as well as in *Neurospora crassa*, *Aspergillus nidulans* and *Euglena gracilis* (Grivell and Jackson, 1968).

## III. Biosynthesis of Amino Acids and its Regulation

### A. INTRODUCTORY REMARKS

As already shown, the biosynthesis of nucleic acid building blocks, and the regulatory aspects thereof, cannot be properly understood without accounting for interference with biosyntheses of other important compounds, such as histidine (in the case of purines) and arginine (in the case of pyrimidines).

Examination of the biosynthesis of amino acids and its regulation offers other examples of intricate pathways, parts of which are shared by different end-products. A simplified view of these interrelationships in yeast is presented in Fig. 3. Some pathways will then be considered as a dichotomic tree from which each part successively analysed interferes with the overall mechanism of biosynthesis and regulation. Sometimes the degradative pathways will also have to be considered, but only to the extent of their participation in a cyclic recapture of precursors. This section has been voluntarily restricted to amino acids which are commonly found in proteins, and among these, to amino acids which have received special attention from biochemists and geneticists working with yeasts.

### B. HISTIDINE

Histidine is synthesized by a linear pathway which involves ten steps (see Fig. 1, Tables IV and V). All the intermediates but one (Step 5) are known. In *Sacch. cerevisiae*, besides Steps 5 and 6 all the known mutants have been assigned to precise enzymic deficiencies, although assays are sometimes indirect (Fink, 1964, 1966). All the loci involved have been mapped (Mortimer and Hawthorne, 1966). Mutants for a few steps have also been obtained and mapped in *Saccharomyces lactis* (Tingle *et al.*, 1968). Gene–enzyme relationships are shown in Table V. As for other biosynthetic pathways in yeast, histidine genes are distributed throughout the genome. The slight linkage observed in *Sacch. lactis* does not correspond to a cluster (Tingle *et al.*, 1968). However, a cluster of genes has been characterized in *Sacch. cerevisiae*—the $his_4$ locus, which contains three regions placed in the following order, A, B, C. Each of those regions specifies for one enzymic activity, Steps 3, 2 and 10, respectively. The $his_4$ locus has some of the genetic characteristics of an operon, i.e. polarity in complementation and recombination. However, no regulation seems to affect any of these enzymes. On the other hand, the three enzymes seem to be part of a complex in which all three activities are closely associated. The same situation has been found in *Neurospora*, in which a single protein of mol. wt. 140,000 catalyses the three reactions (Ahmed, 1968). It seems, then, that close linkage between

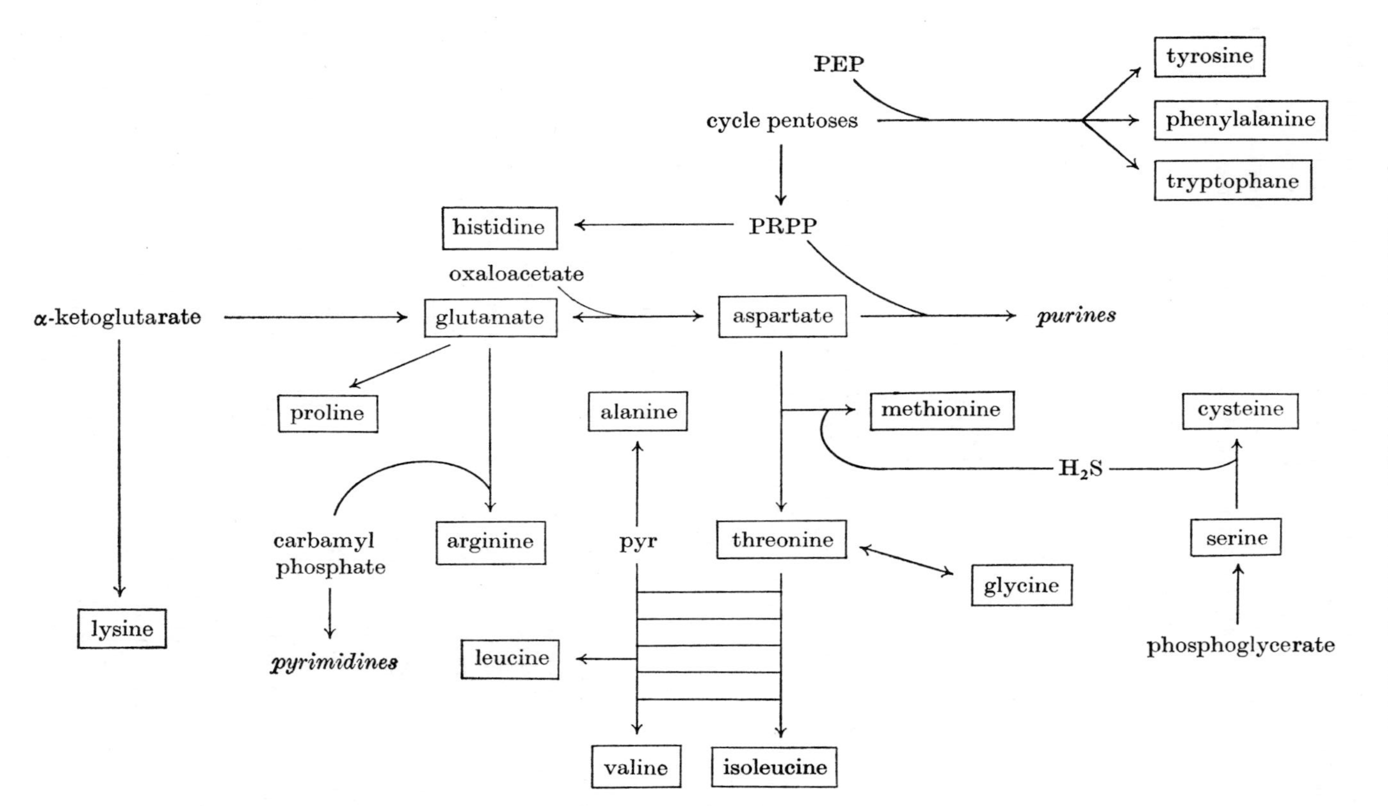


FIG. 3. Gross metabolic relationship between amino acid, purine and pyrimidine biosynthesis in yeast. For abbreviations see pp. 336–338.

structural genes specifying for enzymes involved in a common biosynthetic pathway might result from necessity in "channelling" intermediates, and have no obligatory implication in regulatory mechanisms. In this respect, genetic organization in eucaryotic organisms seems to differ greatly from that occurring in procaryotic organisms.

TABLE IV. *Biosynthesis of Histidine*

| Step | Reaction |
|---|---|
| 1 | *PRPP* + ATP → *PR-ATP* + $PP_i$ |
| 2 | PR-ATP → *PR-AMP* + $PP_i$ |
| 3 | PR-AMP → *AIC-RP*$_a$(BBM-II) + $H_2O$ |
| 4 | AIC-$RP_a$ → *AIC-RP*$_b$(BBM-III) |
| 5 | AIC-$RP_b$ + glutamine → X |
| 6 | X → AICAR + *IGP* |
| 7 | IGP → *IAP* + $H_2O$ |
| 8 | IAP + glutamate ⇌ *histidinol-PO*$_4$ + αKG |
| 9 | Histidinol-$PO_4$ → *histidinol* + $P_i$ |
| 10 | Histidinol + 2NAD$^+$ → [histidine] + 2NADH + 2H$^+$ |

Compounds in italics are those in the main sequence of reactions; compound enclosed in box is the end-product of the synthetic pathway. For abbreviations see pp. 336–338.

No repression of histidine enzyme synthesis in yeast has been described so far. Feedback inhibition of the first step, PR-ATP pyrophosphorylase, has been studied in *Schizosacch. pombe* (Heslot *et al.*, 1966; Whitehead *et al.*, 1966; Whitehead, 1967). The enzyme is inhibited by histidine. Co-operativity of histidine molecules has been observed (Whitehead, 1967). However, AICAR, the by-product of Step 6, has been shown to participate to this inhibition, in the concerted fashion (Whitehead *et al.*, 1966) summarized in Table VI.

The participation of AICAR in *in vivo* regulation of the first step of histidine biosynthesis is evidenced by the behaviour of mutants blocked in Step 9 or Step 10 of purine biosynthesis in *Schizosacch. pombe*, and which, therefore, accumulate AICAR. These mutants require histidine in addition to adenine, and the histidine requirement disappears if a mutation occurs which modifies the feedback properties of PR-ATP pyrophosphorylase (Heslot *et al.*, 1966; Whitehead *et al.*, 1966).

Two points arise from these observations concerning interference between purine derivatives and histidine biosynthesis. First, no mutant has been found for PRPP biosynthesis which is the common precursor of both pathways. It is tempting to suggest that, by analogy with what

TABLE V. *Gene–Enzyme Relationship for Histidine Biosynthesis in* Saccharomyces cerevisiae and Saccharomyces lactis

| Step | *Saccharomyces cerevisiae* | | | | | *Saccharomyces lactis* | | | |
|---|---|---|---|---|---|---|---|---|---|
| | Gene | Chromosome | Enzyme studied | Accumulated compounds | Ref. | Gene | Chromosome | Enzyme studied | Ref. |
| 1 | *his*$_1$ | V | + | None | a | | | + | d |
| 2 | *his*$_{4B}$[a] | | + | None | a, b | | | + } Indirect | d |
| 3 | *his*$_{4A}$[a] | III | + | | | | | + | |
| 4 | | | | BBM | a | *his*$_1$[b] | | | d |
| 5 | *his*$_7$ | II | | BBM | a | *his*$_2$[b] | | | d |
| 6 | *his*$_6$? | IX | | | | | | | |
| 7 | *his*$_3$ | Frag. 1 | | Imidazole, glycerol | a | *his*$_3$[b] | | + | d |
| 8 | *his*$_5$ | Frag. 4 | | Imidazole, acetol | a | | | | |
| 9 | *his*$_2$ | VI | | Histidinol $PO_4$ | a, c | | | | |
| 10 | *his*$_{4C}$[a] | III ($\alpha$) | + | Histidinol | a, b | *his*$_4$ | linked to $\alpha$, *ad*$_1$ | + | d |

For abbreviations see pp. 336–338.
[a] Linked in the order A, B, C. [b] Slightly linked.
*References:* a, Fink (1964); b, Fink (1966); c, Gorman and Hu (1969); d, Tingle *et al.* (1968).

is known for the pyrimidine and arginine pathways or for aromatic amino acid biosynthesis, there might be two PRPP synthetases. Synthesis by one of these would be subject to regulation by histidine, and the other to regulation by one of the purine derivatives. If this were the case, the absence of the mutant should mean that PRPP synthesized by

TABLE VI. *Concerted Inhibition Exerted by L-Histidine and AICAR on the Activity of PR-ATP pyrophosphorylase* Schizosaccharomyces pombe (*Whitehead* et al., *1966*)

| Compound | Concentration | Inhibition (%) | |
|---|---|---|---|
| | | Observed | Expected |
| Histidine | $7 \cdot 5 \times 10^{-5}$ *M* | 14 | |
| AICAR | $2 \times 10^{-4}$ *M* | 0 | |
| | $4 \times 10^{-4}$ *M* | 29 | |
| | $1 \times 10^{-3}$ *M* | 56 | |
| Histidine + AICAR | $7 \cdot 5 \times 10^{-5}$ *M*, $2 \times 10^{-4}$ *M* | 41 | 14 |
| Histidine + AICAR | $7 \cdot 5 \times 10^{-5}$ *M*, $4 \times 10^{-4}$ *M* | 89 | 43 |

For abbreviation see pp. 336–338.

one enzyme would remain freely available to the other pathway. Therefore a way to select such mutants might be to place the cells in conditions of complete repression of one or other of the activities.

The second point concerns the dual production of AICAR, either as an obligatory intermediate in purine biosynthesis or as a by-product of histidine biosynthesis (one mole per mole of histidine formed). The existence only of purineless mutants blocked previous to AICAR synthesis (see Table II) implies that AICAR arising from the histidine pathway is not available to enzymes involved in the synthesis of purine derivatives. It follows that some channelling must occur in the last pathway for enzymes concerned with synthesis and utilization of "purine-AICAR".

In connection with this pathway, the interesting observation has been made (Luzzati *et al.*, 1959) that histidine interferes with genic conversion occurring in heteroallelic diploids at locus $ad_3$ (adenine + histidine requirement). These results indicate that the end-product of a biosynthetic pathway might, in addition to regulation exerted at the transcriptional or translational levels, exert a specific action on the molecular events leading to gene copy. This observation remains, so far, the only one known of this type.

### C. ARGININE

In yeast, biosynthesis of arginine appears to proceed in an identical fashion to that found in most bacteria, with only one exception—

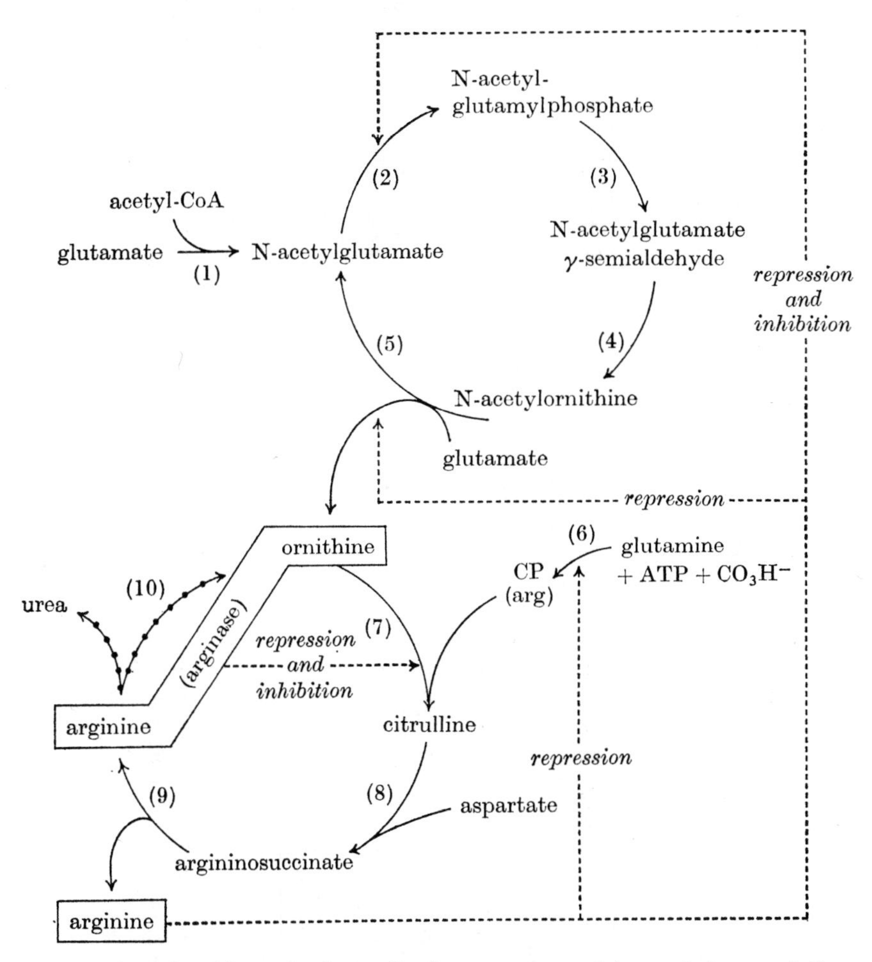


FIG. 4. Arginine biosynthesis in *Saccharomyces cerevisiae* and its regulation. —·—·—·—·—·— = induction by arginine.

Step 5 consists in a simple deacetylation in *E. coli*, whereas a transacetylation takes place in *Sacch. cerevisiae* (de Deken, 1962, 1963) as well as in *Neurospora* and *Micrococcus glutamicus* (Udaka and Kinoshita, 1968; Fig. 4 and Table VII). Two moles of glutamate are therefore necessary for the first mole of ornithine formed, but Reaction 5 regenerates acetylglutamate, providing a cyclic synthesis of ornithine.

TABLE VII. *Arginine Biosynthesis in* Saccharomyces cerevisiae: *Enzymic Steps and Gene–Enzyme Relationship*

| Step | Reaction | Gene | Enzyme studied | Ref. |
|---|---|---|---|---|
| 1 | *Glutamate* + acetyl-CoA → *N-acetylglutamate* + CoA | | | |
| 2 | *N*-Acetylglutamate + ATP → *N-acetylglutamyl phosphate* + ADP | | } + | } a |
| 3 | *N*-Acetylglutamyl phosphate + NADPH + $H^+$ → *N-acetylglutamate-γ-semialdehyde* + $NADP^+$ | $arg_5$ | | |
| 4 | *N*-Acetylglutamate-γ-semialdehyde + glutamate ↔ *N-α-acetylornithine* + α-ketoglutarate | | | |
| 5 | *N*-α-Acetylornithine + glutamate ↔ *N*-acetylglutamate + *ornithine* | $arg_7$ | + | a, b |
| 6 | Glutamine + ATP + $CO_3H^-$ → *CP* + glutamate + ADP | $cpa_1$, $cpa_2$ | + | c |
| 7 | Ornithine + CP → *citrulline* + $P_i$ | $arg_3$(*arg* F) | + | d, e, f, g,h |
| 8 | Citrulline + aspartate $\xrightarrow[Mg^{2+}]{ATP}$ *argininosuccinate* | $arg_1$, $arg_{10}$ | | i |
| 9 | Argininosuccinate → [arginine] + fumarate | $arg_4$ | | i |

Compounds in italics are those in the main sequence of reactions; compound enclosed in box is the end-product of the synthetic pathway. For abbreviations see pp. 336–338

*References:* a, de Deken (1962); b, Middlehoven (1969); c, Lacroute *et al.* (1965); d, Bechet and Wiame (1965); e, Thuriaux *et al.* (1969); f, Bechet *et al.* (1970); g, Ramos *et al.* (1970); h, Bourgeois (1969); i, von Borstel (1966)

Although a small amount of acetylornithinase activity has been detected in crude extracts, its inefficiency *in vivo* is attested by the existence of an arginine-requiring mutant devoid of ornithine transacetylase (de Deken, 1962, 1963). In *C. utilis* both enzymes have also been found, but acetylornithinase represents only 10% of the transacetylase activity (Middlehoven, 1963). *N*-Acetylation of glutamate provides complete independence between the pathways for arginine and proline biosyntheses, which both originate from glutamate and require reduction of the $\gamma$-carboxyl of glutamic acid as an intermediate step.

The existence of a cycle for ornithine biosynthesis cancels the biological importance of Reaction 1. This is evidenced by a comparison of the regulated steps: in *E. coli*, Step 1 is subject to feedback inhibition exerted by arginine (Vyas and Maas, 1963); in *Sacch. cerevisiae*, arginine represses and inhibits the synthesis of *N*-acetyl-$\gamma$-glutamyl phosphate from acetylglutamate. Additionally, arginine also represses transacetylase synthesis (de Deken, 1963). Thus, the first and last steps of the ornithine cycle are subject to regulatory effects of arginine. However, it should be pointed out that the effects are rather small (fourfold at the maximum, using an arginine auxotroph) compared to that observed at the ornithine transcarbamylase level (OTCase, see below), and that no variation of transacetylase specific activity could be detected using an arginine prototroph (Middlehoven, 1969).

From ornithine, arginine biosynthesis proceeds through three steps. The first of these, Step 7 in the pathway, is catalysed by ornithine transcarbamylase. It has been mentioned in the section on pyrimidine biosynthesis that carbamyl phosphate is synthesized by two different enzymes (Step 6), one repressed and inhibited by uracil derivatives (CPase, *ura*), the other repressed by arginine (CPase, *arg*) (Lacroute *et al.*, 1965). The CPase (*arg*) synthesis can be blocked by two independent mutations, $cpa_1$ and $cpa_2$, the respective roles of which remain unknown. The finding that both genes segregate independently from $arg_3$, which corresponds to ornithine transcarbamylase (See Table VII for gene–enzyme relationship), renders unlikely the existence of an enzyme aggregate which would parallel the situation already described for pyrimidine biosynthesis (see Section II.B, p. 343).

A potent repression of OTCase synthesis exerted by exogenous arginine was found by Bechet *et al.* (1962). More recently, as much as an 85-fold difference between completely repressed and completely derepressed levels has been observed (Middlehoven, 1969). Arginine-mediated feedback inhibition of OTCase was also observed, but very unusual conditions were required: cells had to be grown in the presence of arginine: inhibition of protein synthesis prevented inhibition of OTCase to be expressed; a mutation which abolished arginine-mediated repressibility of OTCase also suppressed the capacity of OTCase to be inhibited in crude extracts. It was concluded that a special protein, named the "epiprotein" or the "specific regulatory binding protein" was necessary, in addition to arginine, to mediate OTCase feedback inhibition (Bechet and Wiame, 1965).

Recent findings have led to a satisfactory understanding of this interesting and, so far, unique phenomenon (Messenguy and Wiame, 1969). Three unlinked loci appear to affect regulation of OTCase synthesis (Bechet *et al.*, 1957). The regulatory mutants *arg-r*1 and *arg-r*2 were found to be unable to grow on arginine or ornithine as nitrogen source, in contrast to wild type strains. On the other hand, enzymes responsible for arginine catabolism, namely arginase and ornithine transaminase, are normally induced in arginine-grown cells (Middlehoven, 1964, 1968). As might be expected, *arg-r*1 and *arg-r*2 mutants were found to carry a pleiotropic non-inducibility of arginase and ornithine transaminase (Bechet *et al.*, 1957). Arginase and "epiprotein" activities were found in the same peak after elution from a Sephadex-200 column. Arginase was able to bind OTCase only in the presence of arginine and ornithine, respectively substrate and product of arginase activity. The conditions for optimal OTCase inhibition have been found to be strictly identical to the OTCase-arginase-binding conditions.

It may be noted that induction of arginase leading to ornithine production, also leads to the closure of a second cycle in arginine biosynthesis (see Fig. 4), and thus to a coupling between anabolism and catabolism of arginine, in both metabolic and regulatory aspects.

Variation in the size of the arginine pool does not appear to be directly related to the observed OTCase level (Ramos *et al.*, 1969). Two interesting comments can be made relative to the size of the arginine pool: (i) during exponential growth the concentration of arginine in the amino acid pool is one of the highest, and represents more than 25% of the total; the arginine concentration drops suddenly at the end of the logarithmic phase; (ii) lysine prevents arginine accumulation in the pool, even during the logarithmic phase (Bourgeois, 1969). These observations might be related to the presence or absence of arginine-degrading enzymes.

It is tempting to suggest that lysine, which has been found to mimic arginine or ornithine as co-repressor of OTCase (Ramos *et al.*, 1969), might do so as a consequence of its ability to induce arginase*.

D. PROLINE

Proline seems to have been overlooked both by geneticists (no proline-requiring mutants) and by biochemists working with yeasts. Although the pathway for proline biosynthesis is quite short and simple (see

*Note added in proof: This hypothesis has recently been verified (Bourgeois, 1971).

Fig. 5) it is, nevertheless, not very well known in other organisms. In *E. coli* glutamate $\gamma$-semialdehyde biosynthesis requires ATP and NAD. If one considers that parallel transformation of acetylglutamate to acetylglutamate semialdehyde, or of aspartate to aspartate semialdehyde, requires two steps, the first catalysed by a kinase, the second by a dehydrogenase (see Section III.C and G, pp. 353 and 362), the question arises as to whether glutamate semialdehyde synthesis can be achieved by only one step. At present no answer can be given to this question, since studies have been carried out with washed cell suspensions (Baich and Pierson, 1965; Tristram and Thurston, 1966).

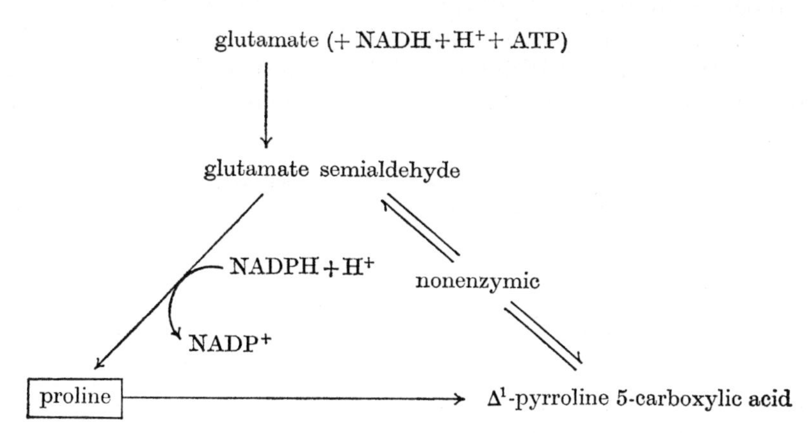


FIG. 5. Biosynthesis of proline.

*In vivo,* proline is produced by an NADP-linked oxidation of glutamate semialdehyde. *In vitro,* glutamate-$\gamma$-semialdehyde is in non-enzymatic equilibrium with a cyclic compound ($\Delta$-1-pyrroline-5-carboxylic acid) which also occurs enzymically in proline catabolism (proline oxidase), providing a cyclic resynthesis of glutamate-$\gamma$-semialdehyde.

In *E. coli,* two types of proline-requiring mutants are known, which evidently corresponds to the two steps in the pathway: one which grows on glutamate-$\gamma$-semialdehyde; the other which accumulates $\Delta$-pyrroline-5-carboxylic acid. Regulation by repression and feedback inhibition exerted by proline has been shown to occur exclusively at the first step level (Baich and Pierson, 1965; Tristram and Thurston, 1966).

In yeast, the existence of inducible arginine-degrading enzymes can provide an additional source of glutamate semialdehyde:

arginine $\rightarrow$ ornithine + urea
ornithine + $\alpha$-ketoglutarate $\rightleftharpoons$ glutamate-$\gamma$-semialdehyde + glutamate

When these enzymes are present, only the last step remains specific for proline biosynthesis. This double possibility of synthesizing glutamate semialdehyde might have prevented detection of mutants lacking the first step (or steps) in proline biosynthesis in yeast. The existence of arginaseless mutants in *Sacch. cerevisiae* (Messenguy and Wiame, 1969) might then provide better conditions to study proline biosynthesis and its regulation in this organism.

## E. LYSINE

Lysine is the only amino acid to be synthesized in yeast through a metabolic pathway entirely distinct from the pathway used in bacteria (see Fig. 6). The steps have been definitely established by the pioneer work of Strassman and Weinhouse (1953), Jones and Broquist (1965), Broquist and Trupin (1966), and seem to apply to different yeasts

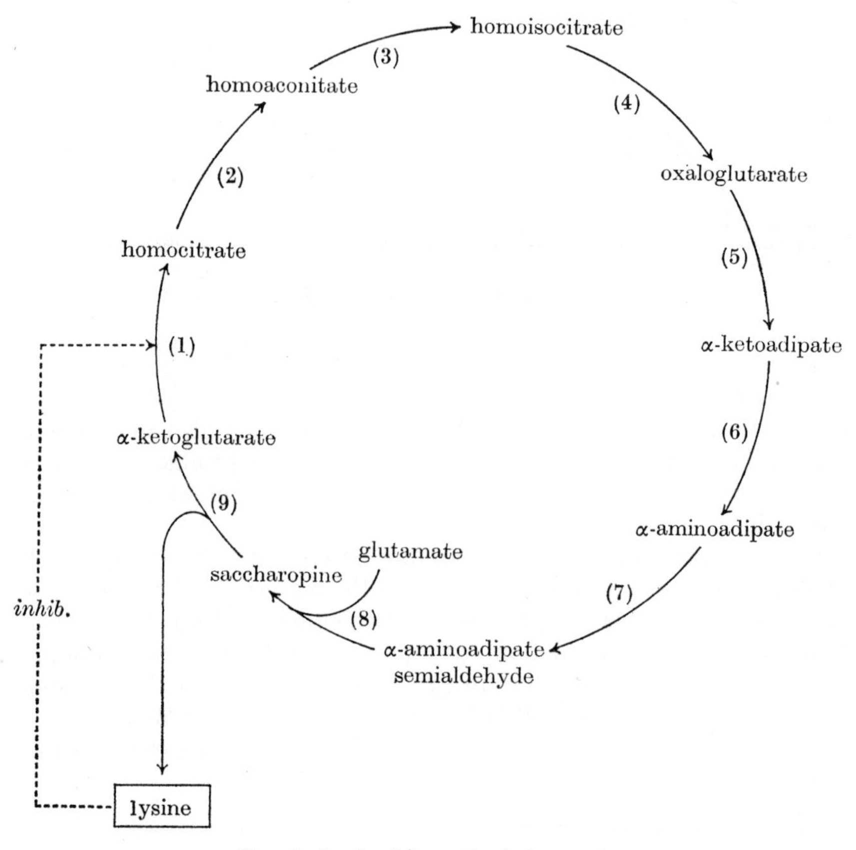


FIG. 6. Lysine biosynthesis in yeast.

TABLE VIII. *Lysine Biosynthesis and Gene–Enzyme Relationship in* Saccharomyces cerevisiae

| Step | Reaction | Gene | Enzyme studied | Ref. |
|---|---|---|---|---|
| 1 | *α-Ketoglutarate* + acetyl-CoA → *homocitrate* | | + | b |
| 2 | Homocitrate → *homoaconitate* + $H_2O$ | *lys*$_4$ } *lys*$_{6,7,11}$ | } | a, c |
| 3 | Homoaconitate + $H_2O$ → *homoisocitrate* | *lys*$_{15}$ } *lys*$_{6,7,11}$ | } | a, c |
| 4 | Homoisocitrate + $NAD^+$ → *oxaloglutarate* + NADH + $H^+$ | *lys*$_8$ } *lys*$_{6,7,11}$ | | a, d |
| 5 | Oxaloglutarate → *α-ketoadipate* + $CO_2$ | *lys*$_{12}$? } *lys*$_{6,7,11}$ | | a, d |
| 6 | α-Ketoadipate + glutamate → *α-aminoadipate* + α-ketoglutarate | | | |
| 7 | α-Aminoadipate + ATP + NADPH + $H^+$ $\xrightarrow{Mg^{2+}}$ *α-aminoadipate semialdehyde* + ADP + $P_i$ + $NADP^+$ | *lys*$_2$(chrom. 2) } *lys*$_{5,6,13,14}$ | | a, e |
| 8 | α-Aminoadipate semialdehyde + glutamate + NADH + $H^+$ → *saccharopine* + $NAD^+$ | *lys*$_9$(chrom. 14) } *lys*$_{5,6,13,14}$ | | a, e |
| 9 | Saccharopine + $NAD^+$ → [lysine] + α-ketoglutarate + NADH + $H^+$ | *lys*$_1$(chrom. 9) } *lys*$_{5,6,13,14}$ | | a, e |

Compounds in italics are those in the main sequence of reactions; compound enclosed in box is the end-product of the synthetic pathway. For abbreviations see pp. 336-338.

*References:* a, Hwang *et al.* (1966); b, Tucci (1969); c, Maragoudakis and Strassman (1966); d, Bhattacharjee and Strassman (1967); e, Jones and Broquist (1965).

(*C. utilis*, *Sacch. cerevisiae*) and fungi. Regeneration of $\alpha$-ketoglutarate at the end of the pathway makes it possible for it to be considered as a cycle. The gene-enzyme relationship presented in Table VIII is mostly based on accumulation data (Maragoudakis and Strassman, 1966; Scheifinger *et al.*, 1966; Maragoudakis *et al.*, 1966; Bhattacharjee and Strassman, 1967).

Hwang *et al.* (1966) have classified the 14 known independent lysine loci as *AAA* and *aaa* according to their ability or inability to grow on $\alpha$-aminoadipic acid. Some inconsistency has been found among $lys_4$ mutants. It appeared that an additional genetically-controlled factor is, in some cases, necessary to permit $\alpha$-aminoadipate utilization. It seems that at least seven genes are necessary to carry out Steps 1 to 5, and at least seven genes for Steps 5 to 9 of lysine biosynthesis in yeast. It would, therefore, not be surprising if more steps and intermediates than those presented in Fig. 6 and Table VIII were implicated in this pathway.

No mutants have been found, so far, for Steps 1 and 6. As far as Step 6 is concerned, it was thought that transamination of $\alpha$-ketoadipic acid could have been catalyzed by unspecific transaminases. However, glutamic-oxaloacetic transaminase has been shown to be unrelated, as $thr_5$ mutants which lack this enzyme (see Section III.G, p. 362) still possess glutamic-$\alpha$-ketoadipic transaminase, as well as $lys_6$ or $lys_8$ mutants (Piediscalzi *et al.*, 1968). Feedback inhibition exerted by lysine has been observed on homocitrate synthesis, the first step in the pathway, both *in vivo* and *in vitro* (Maragoudakis *et al.*, 1967; Tucci, 1969). Lysine-mediated repression seems also to occur at this level, although to a more limited extent (Tucci, 1969).

It is interesting to note that lysine is the only amino acid which is not used by yeast as nitrogen source (see Bourgeois, 1969). On the other hand, lysine has been found able to play a role in the regulation of threonine and methionine and of arginine biosynthesis (see Section III.G, p. 362 and Section III.C, p. 353, respectively). In both cases uptake did not appear to be involved in the observed effects. Therefore, lysine, as such or as its activated products, has to be considered as a regulator for the functioning of two unrelated pathways.

Another interesting phenomenon concerns the interference between lysine and glutamate biosynthesis in yeast (Ogur *et al.*, 1965). Both amino acids arise from $\alpha$-ketoglutarate. Mutants deficient in aconitase and, therefore, unable to synthesize $\alpha$-ketoglutarate ($glt_1$) grow on glutamate and have no lysine requirement (see following Section). However, unexpectedly, the nonallelic mutations $lys_6$ and $lys_8$ lead to a growth requirement for glutamate in addition to lysine. In turn, another mutation ($glt_2$) appears also to lead to a double lysine + glutamate requirement, and is found to be allelic to $lys_8$. From accumulation data, $lys_8$ mutants appear to be devoid of homoisocitrate dehydrogenase (Step 4 in the pathway). Nevertheless, $glt_2$, $lys_6$ or $lys_8$ mutants also lack aconitase and, additionally, cytochromes *a* and *b*. Part of their phenotype is then similar to the cytoplasmic $\rho^-$ mutants. However, their complex requirements and deficiencies segregate in a Mendelian manner, as a single gene unit. All the other lysine-requiring mutants require only lysine for growth.

Such interference between the synthesis of two amino acids and specific cytochromes synthesis has to be considered in parallel with the effect of threonine metabolism in cytochrome oxidase biosynthesis (see Section III.G, p. 362).

### F. GLUTAMATE

Glutamate biosynthesis occurring by amination of α-ketoglutarate represents the most important of the reactions which ensure ammonia fixation in yeast metabolism (Witt *et al.*, 1964). This reaction is catalysed by glutamate dehydrogenase:

$$\alpha\text{-ketoglutarate} + NH_3 + NADPH + H^+ \rightleftharpoons \text{glutamate} + NADP^+ + H_2O$$

No regulation seems to affect this enzyme in yeast. However, there is a second glutamate dehydrogenase, linked to NAD instead of NADP, which is subject to regulation. This enzyme can be found only in cultures grown in the absence of ammonium ions. Its synthesis is repressed as soon as ammonia is added back to the culture (Holzer and Hierholzer, 1963; Bernhardt *et al.*, 1966). The transfer from an ammonia-containing medium to an ammonia-free medium results in a series of oscillations in the rate of synthesis of the NAD-dependent glutamate dehydrogenase, due to changes in the intracellular level of ammonium ions (Panten *et al.*, 1967). These results seem to indicate that the NADP-dependent enzyme has a biosynthetic role, while the NAD-dependent enzyme is mostly degradative. A mutant lacking the NADP-dependent enzyme (locus $glt_6$) has been reported (see von Borstel, 1966). It would be interesting to know if such a mutant, in the absence of ammonia, could use the NAD-dependent enzyme in the biosynthetic direction.

For two independent mutations ($glt_1$ and $glt_2$) aconitase deficiency seems to be the primary block, which really concerns α-ketoglutarate synthesis and results in interruption of the tricarboxylic acid cycle (Ogur *et al.*, 1965). However, the phenotypes of both classes of mutants are much more complex, thus:

(i) In $glt_2$ mutants lysine is also required for growth in the presence of glutamate (Ogur *et al.*, 1965). α-Ketoglutarate being a common precursor of glutamate and lysine (see also previous Section), the lysine requirement might be due to an incapability of *glt*$_2$ mutants to convert glutamate to α-ketoglutarate; i.e. to the inability of NADP-linked glutamate dehydrogenase to function *in vivo* in the degradative direction, which is accompanied by an inability of the NAD-linked glutamate dehydrogenase to become derepressed under the conditions used. Growth of *glt*$_1$ mutant cells in the absence of ammonia should provide conditions for the recovery of the NAD-linked glutamate dehydrogenase, and therefore overcome the lysine requirement.

(ii) Both *glt*$_1$ and *glt*$_2$ have also an apparent lack of cytochromes *a* and *b* (Ogur *et al.*, 1964), or at least show an inability to utilize non-fermentable carbon and energy sources directly (Ogur *et al.*, 1965, 1964). In the case of *glt*$_1$ mutants, growth in a reduced concentration of glucose (0·2%) and 1% lactate promotes ability to respire ethanol, acetaldehyde, lactate and glucose. Such cells still fail to utilize acetate (Ogur *et al.*, 1964). In addition, glucose-grown cells exhibit, together with a cytochrome deficiency, an abnormal mitochondrial picture. However, in contrast with cytoplasmic respiratory-deficient mutants $\rho^-$, cristate mitochondria, typical of respiratory-sufficient cells, appear in cells grown on suboptimal glucose medium when fortified with lactate (Bowers *et al.*, 1967).

Therefore, as in the case of threonine-less mutants (see following Section), the cytochrome deficiency which accompanies *glt*$_1$ mutation lies in an inability to overcome the glucose repressive effect, in conditions which allow derepressed synthesis of such cytochromes on normal cells. Nevertheless, in both cases appropriate conditions can be found (see following Section) which permit derepressed synthesis of cytochromes to occur.

### G. THREONINE, METHIONINE AND CYSTEINE

In yeast, threonine biosynthesis occurs through the same steps as in bacteria (see Fig. 7). However, there are three main differences in this branched biosynthetic pathway when bacteria and yeast are compared.

(i) Lysine which, in bacteria, originates from the second intermediate in threonine biosynthesis (aspartate semialdehyde) is, in yeast, synthesized through an entirely distinct pathway (see Section III.E, p. 358); but the three first enzymes are still shared by threonine and methionine biosynthesis and are, consequently, subject to regulation by both end-products.

(ii) Although the carbon skeleton of methionine originates from homoserine, as in bacteria, the intermediate steps between homoserine and homocysteine are very different. First, an acetylation of homoserine occurs, instead of a succinylation. Second, a substitution of the *O*-acetyl group by a sulphydryl group (provided by $H_2S$) leads directly to homocysteine. Cystathionine, the biosynthesis and splitting of which has

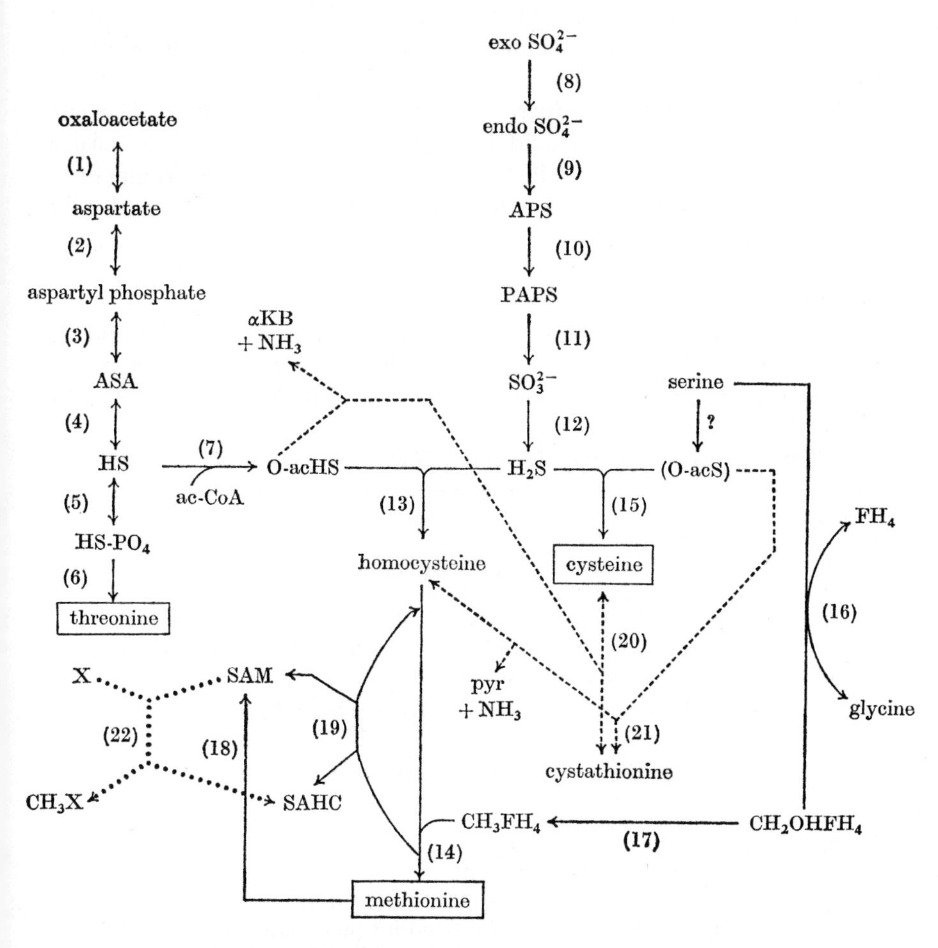


FIG. 7. Threonine, methionine and cysteine biosynthesis in *Saccharomyces cerevisiae*.

been described in yeast, does not, therefore, appear to be directly involved in methionine biosynthesis in yeast.

(iii) Cysteine biosynthesis normally occurs by sulphydrylation of serine (also activated in *O*-acetyl serine) as in bacteria. The pathway which ensures sulphate assimilation is therefore shared by methionine and cysteine biosynthesis. Cysteine can also arise from methionine.

In this transformation cystathionine might be the principal intermediate. For these two reasons cysteine biosynthesis appears, in yeast, to be metabolically less important than in bacteria. In contrast, methionine biosynthesis becomes more prominent. This conclusion is based on the participation of methionine in the regulation of sulphide production.

The complexity of this intricate pathway is such that a clear-cut separation of each amino acid biosynthetic process would be either artificial or redundant. It will therefore be examined in a fragmented fashion, with the hope that a more complete picture will eventually materialize.

### 1. *Homoserine Biosynthesis from Aspartate*

Three steps are necessary to form homoserine. Steps 2, 3 and 4, first studied by Black and Wright (1955a, b, c), using yeast, are common to threonine and methionine biosyntheses (see Fig. 7 and Table IX). As

TABLE IX. *Biosynthesis of Threonine, Methionine and Cysteine in Yeast*

| Step | Reaction |
|---|---|
| 1 | Glutamate + *oxaloacetate* $\rightleftharpoons$ *aspartate* + α-ketoglutarate |
| 2 | Aspartate + ATP $\rightleftharpoons$ *β-aspartyl phosphate* + ADP |
| 3 | β-Aspartyl phosphate + NADPH + $H^+$ $\rightleftharpoons$ *ASA* + $NADP^+$ + $P_i$ |
| 4 | ASA + NADPH (or NADH) + $H^+$ $\rightleftharpoons$ *HS* + $NADP^+$ (or $NAD^+$) |
| 5 | HS + ATP → *O-phosphohomoserine* + ADP |
| 6 | *O*-Phosphohomoserine + $H_2O$ → [threonine] + $P_i$ |
| 7 | HS + acetyl-CoA → *O-acetylhomoserine* + CoA |
| 8 | exo $SO_4^{2-}$ → endo $SO_4^{2-}$ |
| 9 | $SO_4^{2-}$ + ATP → *APS* + $PP_i$ |
| 10 | APS + ATP → *PAPS* + ADP |
| 11 | {NADPH + $H^+$ + C-SS[a] $\rightleftharpoons$ $NADP^+$ + *C-(SH)*$_2$<br>{C-$(SH)_2$ + PAPS $\rightleftharpoons$ phosphoadenosine phosphate + $SO_3^{2-}$ + C-SS[a] |
| 12 | $SO_3^{2-}$ + 3NADPH + $4H^+$ → *$H_2S$* + $3NADP^+$ + $3H_2$ |
| 13 | *O*-Acetylhomoserine + $H_2S$ → *homocysteine* |
| 14 | Homocysteine + 5-$CH_3FH_4$ → [methionine] + $FH_4$ |
| 15 | *O*-Acetylserine + $H_2S$ → [cysteine] |

Compounds in italics are those in the main sequence of reactions; compounds enclosed in boxes are the end-products of synthetic pathways For abbreviations see pp. 336–338.

[a] C-SS = low molecular weight protein (see text, p. 372).

expected, three genetically and enzymically distinct homoserineless mutants (able to grow on homoserine alone or on threonine + methionine, were found—namely *thr*$_3$, *thr*$_2$ and *thr*$_6$ blocked in Steps 2, 3 and 4, respectively (de Robichon-Szulmajster *et al.*, 1966). However, an additional block (*thr*$_5$) also leading to homoserine or threonine + methionine auxotrophy, was found to be deficient in the aspartate-producing step (Step 1), namely the aspartate aminotransferase (see Table X). *thr*$_5$ Mutants are really aspartateless mutants, and their threonine +

TABLE X. *Gene–Enzyme Relationship in Threonine, Methionine and Cysteine Biosynthesis*

| Step | Gene | Chromosome | Enzyme studied | Ref. |
|---|---|---|---|---|
| 1 | *thr*$_5$ | XII | + | a |
| 2 | *thr*$_3$ | V | + | a, b, c |
| 3 | *thr*$_2$ | Frag. 2 | + | a, d, e, f |
| 4 | *thr*$_6$ | | + | a, g, h, i |
| 5 | *thr*$_1$ | VIII | + | a, j |
| 6 | *thr*$_4$? | III | + | j, k |
| 7 | *met*$_2$ | Frag. 3 | + | l, m |
| 8 | | | | |
| 9 | *met*$_{10}$? / 3, 17 (p) | VI | + | n, o, p |
| 10 | 15 (p) | | + | n, o, p |
| 11 | 12, 22 (p) | | + | o, p, q, r, s, t |
| 12 | | | + | t, u |
| 13 | *met*$_8$ | II | + | v, w |
| 14 | | | + | x, y |
| 15 | | | + | v, z |

*References:* a, de Robichon-Szulmajster *et al.* (1966); b, Black and Wright (1955a); c, de Robichon-Szulmajster and Corrivaux (1963); d, Black and Wright (1955b); e, de Robichon-Szulmajster *et al.* (1963); f, Surdin (1967); g, Black and Wright (1955c); h, Karassevitch and de Robichon-Szulmajster (1963); i, Surdin-Kerjan (1969); j, de Robichon-Szulmajster (1967); k, Watanabe *et al.* (1957); l, de Robichon-Szulmajster and Cherest (1967); m, Nagai and Flavin (1967); n, Robbins and Lipmann (1958a, b); o, Wilson and Bandurski (1958); p, Naiki (1964); q, Wilson *et al.* (1961); r, Asahi *et al.* (1961); s, Torii and Bandurski (1964); t, Naiki (1965); u, Wainwright (1962, 1967); v, Wiebers and Garner (1967); w, Cherest *et al.* (1969); x, Pigg et al. (1962); y, Botsford and Parks (1967); z, Schlossmann *et al.* (1962).

methionine requirement becomes evident only when the other metabolic products which originate from aspartate, especially adenine, uracild an arginine, are present (de Robichon-Szulmajster *et al.*, 1966; Surdin-Kerjan, 1969). Alanine aminotransferase activity was found in extracts of $thr_5$ cells, indicating that both enzymes are coded by distinct genes (de Robichon-Szulmajster *et al.*, 1966). Another peculiarity of this part of the pathway in yeast is the capacity of yeast homoserine dehydrogenase to utilize NADPH or NADH (Black and Wright, 1955c; Karassevitch and de Robichon-Szulmajster, 1963; Surdin-Kerjan, 1969). Earlier evidence that a single enzyme was involved (Black and Wright, 1955c) has recently been confirmed. Surdin-Kerjan (1969) has shown that: (i) $thr_6$ mutants are devoid of activity towards both cofactors; and (ii) heat inactivation curves are unimodal and identical whichever coenzyme is used. Furthermore, NADH exerts a protective effect which equally applies to activities towards the two nucleotides.

As far as regulatory aspects are concerned (see Fig. 8) it has been shown that a potent repression and complete inhibition are exerted respectively by threonine on synthesis and activity of aspartokinase (Step 2 of Fig. 7, and Tables IX and X; de Robichon-Szulmajster and Corrivaux, 1963). A homotropic cooperative effect of threonine molecules ($n = 2$) has been shown (Surdin-Kerjan, 1969). On the other hand, aspartate aminotransferase was found insensitive to end-product regulation. On these bases, aspartokinase has to be considered as the first step in threonine biosynthesis.

Methionine, as well as threonine, participates in the regulation of homoserine biosynthesis. Enzymes catalysing Steps 2 and 4 are derepressed during methionine starvation. In addition, and paradoxically, aspartate semialdehyde dehydrogenase (ASA-dHase) catalysing Step 3 is "induced" when cultures are grown in the presence of $10^{-2}$ *M* methionine (de Robichon-Szulmajster *et al.*, 1963, 1966). Another interesting feature shown by this enzyme is its activation by the bicarbonate anion. The mechanism of activation appears to reside in a decreased affinity for inorganic phosphate, one of the substrates of the forward reaction. The biological interest of this effect might be to favour the reaction in the biosynthetic direction (aspartate semialdehyde production). On the contrary, the homocysteine inhibitory effect seems to have no physiological significance, as it is observed only in the reverse direction (Surdin, 1967, Surdin-Kerjan, 1969).

## 2. *Threonine Biosynthesis from Homoserine*

The conversion of homoserine to threonine in yeast has been shown to be a two-step process (Steps 5 and 6 in Fig. 7, and Table IX; Watanabe and Shimura, 1955, 1956; Watanabe *et al.*, 1955, 1957) as in bacteria.

Threonine-requiring mutants blocked at the locus $thr_1$ were found to be deficient in the first step catalyzed by homoserine kinase (de Robichon Szulmajster, 1967). Further purification (G. Talbot, unpublished results) demonstrated that threonine exerts an inhibition, competitive towards homoserine, on yeast homoserine kinase activity. Such an inhibition has previously been mentioned by Wormser and Pardee (1958).

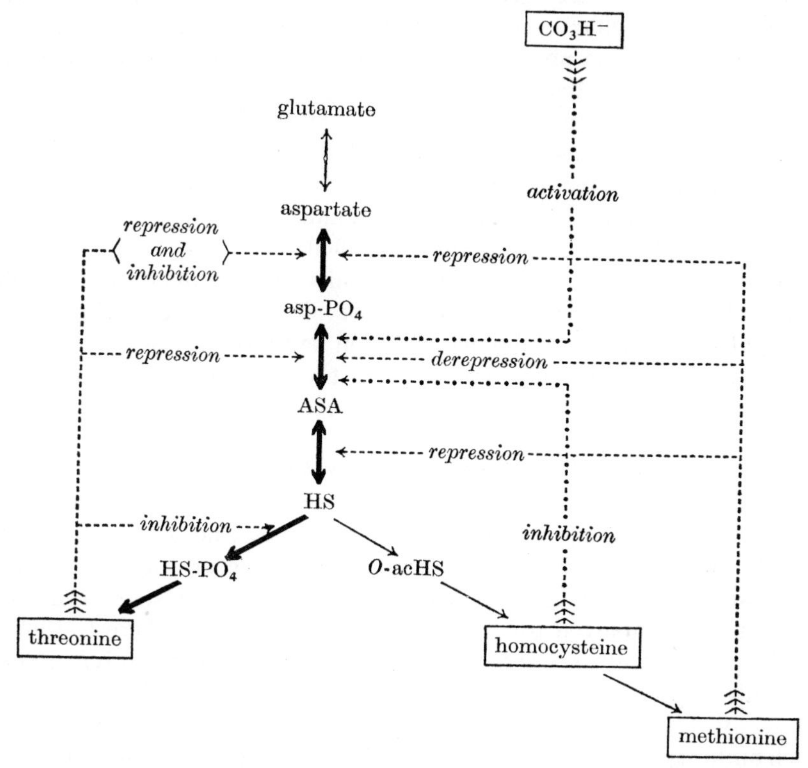


FIG. 8. Regulation of homoserine and threonine biosynthesis from aspartate in *Saccharomyces cerevisiae*. For abbreviations see pp. 336–338.

Partially purified yeast homoserine kinase (Watanabe *et al.*, 1957) was used to prepare homoserine phosphate in a study of threonine synthetase (*O*-phosphohomoserine mutaphosphatase) from *Neurospora* (Flavin and Slaughter, 1960). The corresponding enzyme has not been studied as such in yeast. The only locus involved in threonine biosynthesis which remains unassigned ($thr_4$) could, therefore, correspond to a deficiency in this last step of the pathway (de Robichon-Szulmajster, 1967).

### 3. *The Role of Threonine in Cytochrome Oxidase Biosynthesis*

The extensive study of threonine mutants in *Sacch. cerevisiae* (de Robichon-Szulmajster *et al.*, 1966; de Robichon-Szulmajster, 1967;

Surdin-Kerjan, 1969) has led to the discovery that all threonine auxotrophs show a partial or complete incapacity to synthesize cytochrome oxidase in the presence of glucose, in conditions which satisfy the auxotrophic requirements of the strains and allow cytochrome oxidase synthesis in wild type strains (Surdin *et al.*, 1969). Cytochrome oxidase synthesis requires either substitution of glucose by a carbon source used only aerobically, such as glycerol or ethanol, or the addition of $10^{-2}$ $M$ threonine to the glucose medium. No other amino acid or derivative, except the threonine precursors homoserine or aspartate semialdehyde have been found effective; and among respiratory enzymes only cytochrome oxidase has been shown to be affected. As this enzyme is a good candidate for mitochondrial DNA-directed synthesis, it seems possible that the threonine effect indicates a function as either a mitochondrial threonyl-t-RNA or a mitochondrial threonyl-t-RNA synthetase (de Robichon-Szulmajster *et al.*, 1969).

These observations, together with results already mentioned which were obtained with certain lysine auxotrophs (Section III.E, p. 358), focus attention on interference between entirely distinct metabolic pathways and, more particularly, on the existence of regulatory processes specific for the synthesis of mitochondrial enzymes.

4. *Homocysteine Biosynthesis from Homoserine*

The first step specific for methionine biosynthesis has been shown to be an activation of the hydroxyl group of homoserine. In yeast, direct acetylation from acetyl-CoA has been shown to occur only after some purification (Nagai and Flavin, 1967). Using crude extracts, an exchange reaction was employed (de Robichon-Szulmajster and Cherest, 1967). Evidence that this exchange reflects the normal enzyme activity, and that this enzyme is truly involved in methionine biosynthesis, arose from the study of a methionine auxotroph blocked at the locus $met_2$ (Table X). Such mutants grow on *O*-acetylhomoserine, homocysteine or methionine, and are devoid of homoserine-*O*-transacetylase activity (de Robichon-Szulmajster and Cherest, 1967).

Biosynthesis of homocysteine can be obtained by two processes. In the first, direct sulphydrylation of *O*-acetylhomoserine is accompanied by acetate elimination (Wiebers and Garner, 1967; see Table IX). In yeast, the activity is 20 times greater when *O*-acetylhomoserine rather than homoserine is provided as substrate. The succinyl derivative of homoserine is inefficient (Wiebers and Garner, 1967). By the second system, cystathionine synthesis by condensation of *O*-acetylhomoserine and cysteine ($\gamma$-synthetase) occurs, followed by cleavage of homocysteine and pyruvate + $NH_3$ ($\beta$-cleavage; see below). Although both types of reaction have generally been found *in vitro*, only one type seems to be operative *in vivo*, in a given organism, on the basis of mutant requirements and enzymic deficiencies.

In *Neurospora crassa*, methionine-requiring mutants have been characterized which lack cystathionine synthetase but still possess intact, and apparently inefficient, homocysteine synthetase (Kerr and Flavin, 1968). On the other hand, in *Sacch. cerevisiae*, a mutant has been studied which grows on homocysteine or methionine and is devoid of homocysteine synthetase activity (Cherest *et al.*, 1969). Reversion to prototrophy is accompanied by recovery of homocysteine synthetase activity.

A potent methionine-mediated repression is exerted on synthesis of the two enzymes homoserine-*O*-transacetylase and homocysteine synthetase, providing additional evidence that these two reactions are directly involved in methionine biosynthesis in yeast. A regulatory gene, not linked to the two independent structural genes ($met_2$ and $met_8$), has been studied. In the presence of the wild type dominant allele (*eth-2s*) and exogenous methionine, a pleiotropic repression of the synthesis of both enzymes occurs. This repression is cancelled in the presence of the mutated recessive allele, *eth-2r* (Cherest *et al.*, 1969). The implication of methionyl-t-RNA in the formation of the corresponding aporepressor-corepressor system is evidenced by the finding that, in a thermosensitive mutant of *Sacch. cerevisiae* (McLaughlin and Hartwell, 1969) in which methionyl-t-RNA synthetase is impaired, derepression of homoserine-*O*-transacetylase and homocysteine synthetase occurs, even at the permissive temperature (unpublished results from this laboratory).

Methionine also inhibits, competitively to *O*-acetylhomoserine, the activity of homocysteine synthetase (Wiebers and Garner, 1967; Cherest *et al.*, 1969). A representation of these regulatory systems is given in Fig. 9.

### 5. *The Homocysteine Methylation Process*

Although a great deal of attention has been devoted to this problem, the detailed mechanism by which final methylation of homocysteine occurs still remains controversial. In every organism the methyl group of methionine originates from serine. However, the exact nature of the folic acid derivative, and the steps necessary for its synthesis, seem to vary from one organism to the other. Also, in only one organism, *E. coli*, have a B12-dependent and a B12-independent system been characterized (Shapiro and Schlenk, 1965). The function of the B12-dependent system requires at least catalytic amounts of *S*-adenosylmethionine (SAM). It has been shown recently that SAM is necessary to ensure the methylation of the enzyme itself, the unmethylated form being inactive (Rüdiger and Jaenicke, 1969).

In yeast, there is biochemical and genetic evidence that a folic acid derivative is involved in the methylation process, although a boiled extract from yeast still remains more active than purified known derivatives (Botsford and Parks, 1967). A mutant (locus $met_5$—Parks;

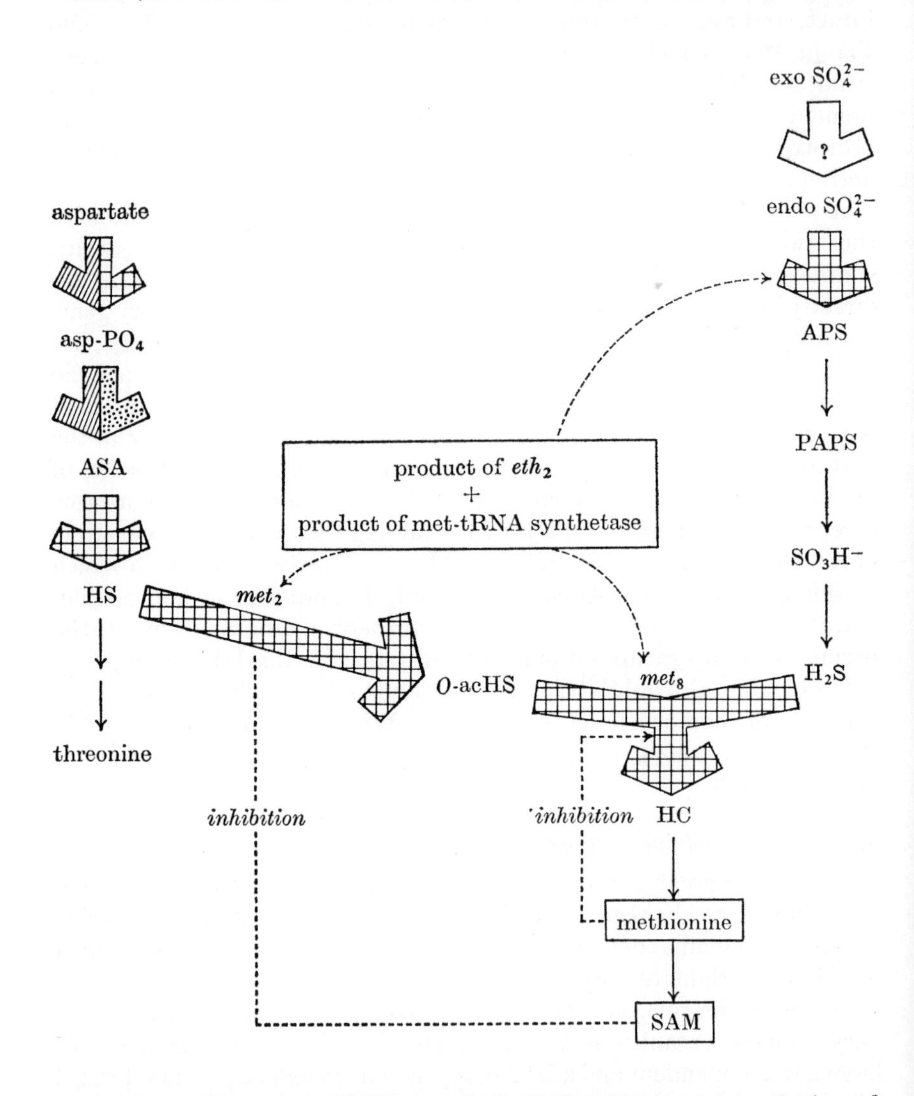


Fig. 9. The effects of methionine or *S*-adenosylmethionine on synthesis and activity of enzymes directly involved in threonine and methionine biosynthesis in *Saccharomyces cerevisiae*. ▦ derepression in methionine limitation; ▩ derepression in excess methionine; ▨ derepression in threonine limitation. For abbreviations see pp. 336–338.

Pigg *et al.*, 1962; Botsford and Parks, 1969) has been found which lacks one of the steps involved in this process ($FH_4$-serine *trans*-hydroxymethylase; see Fig. 7, Reaction 16). *S*-Adenosylmethionine (SAM) has also some indispensable role *in vivo*, as judged by the existence of another mutant (locus $met_1$—Parks; Pomper, 1952b) in which the lack of SAM-homocysteine methyl-transferase suppresses methionine biosynthesis from homocysteine (Pigg *et al.*, 1962). The hypothesis that methylation occurs rather at the level of *S*-adenosylhomocysteine (SAHC) than homocysteine itself seems unlikely, as a study using

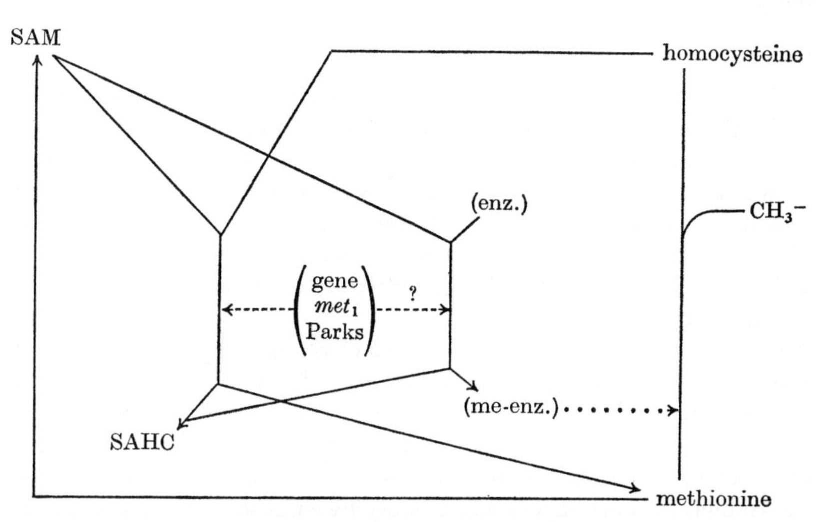


Fig. 10. Hypothetical scheme to account for the effect of gene $met_1$-Parks and *S*-adenosylmethionine requirement in the methylation of homocysteine in *Saccharomyces cerevisiae*. For abbreviations see pp. 336–338.

SAHC labelled in different positions revealed that the compound is merely used after splitting of the molecule (Duerre, 1968; Shapiro and Ehninger, 1969). In addition, a complete cell-free system was developed in which transfer of $\beta$-carbon from serine to homocysteine can occur without any $B_{12}$ or SAM requirement (Botsford and Parks, 1967). If the findings of Rüdiger and Jaenicke (1969) were to apply to the yeast system, it would be tempting to assume that the enzyme missing in "$met_1$—Parks" mutants is the one which, in yeast, also ensures methylation of the specific homocysteine transmethylase. The absence of SAM requirement *in vitro* would then indicate that an already methylated enzyme system had been obtained by Botsford and Parks (1967). These considerations are summarized in Fig. 10.

In this respect it can be recalled that, according to Maw (1966), as much as 30% of the sulphur of ethionine can be found in methionine residues after incorporation of ethionine into yeast protein. These findings have recently been confirmed (Cherest *et al.*, 1969). As such a transformation does not apparently occur in soluble fractions from ethionine-sensitive wild type strains, it can be assumed that an enzymic process exists which is able to methylate (or exchange the methyl and ethyl groups) of preformed proteins.

Every enzyme mentioned in this section shows some regulation of its synthesis either by methionine or by homocysteine; thus providing, if it is necessary, more evidence that they are all-important for methionine biosynthesis.

### 6. *The Sulphate Assimilation Pathway*

Reduction of sulphate to sulphite has been shown to require previous activation of inorganic sulphate. This activation process, first discovered in liver (Hilz and Lipmann, 1955), was also found and mainly studied in yeast. It was shown to comprise two distinct steps, the sulphate activation itself, catalyzed by ATP sulphurylase, followed by a second phosphorylation catalyzed by a kinase (Reactions 9 and 10 in Fig. 7, and Tables IX and X; Robbins and Lipman, 1958a, b; Wilson and Bandurski, 1958).

Reduction of the activated product of these two reactions, phosphoadenosine phosphosulphate (PAPS) to sulphite, requires three components. Two of these are heat-labile, high molecular weight, specific protein molecules. The third, also proteinic in nature, has a low molecular weight, is heat-resistant and apparently acts as an intermediate in the reduction process. The most likely mechanism is that presented for Reaction 11 in Table IX (Wilson and Bandurski, 1958; Wilson *et al.*, 1961; Asahi *et al.*, 1961; Torrii and Bandurski, 1964). There are indications that the sulphite anion is not free in the medium, but remains bound to the second enzyme.

Three classes of methionine auxotrophs, sulphiteless mutants of *Sacch. cerevisiae*, were isolated by Naiki (1964) and shown to lack, respectively, Steps 9, 10 and 11 (see Tables IX and X). Complementation between the three classes occurs *in vitro*. No mutants have been found for Step 12 which leads to hydrogen sulphide, the final product of this process. The enzyme sulphite reductase was thought to comprise more than one protein component (Wainwright, 1962). However, it was shown later that only one is essential (Wainwright, 1967). Sulphate assimilation is probably preceded by a sulphate permease of the kind described in bacteria (Pardee and Prestidge, 1966; Pardee *et al.*, 1966) and in *Aspergillus nidulans* (Arst, 1968).

Sulphide production in *Sacch. cerevisiae* is subject both to repression and feedback inhibition. Repression of ATP-sulphurylase has been shown to occur in cultures in which methionine is the only sulphur source. On the other hand, when the sulphur source is cysteine, enzyme synthesis is enhanced (de Vito and Dreyfuss, 1964). Later, the methionine-repressive effects were extensively examined and shown to be operative, also, in the presence of exogenous sulphate (Cherest *et al.*, 1969; Fig. 9.) Synthesis of ATP-sulphurylase was thought not to be subject to the effect of the gene $eth_2$, but more recent studies have shown that a specific response of ATP-sulphurylase to both components of the repressor system, which affects at least two other enzymes involved in methionine biosynthesis ($eth_2$ and t-$RNA^{met}$; see previous Section), can be demonstrated. Moreover, repression–derepression of ATP-sulphurylase and homocysteine synthetase seems to occur in a co-ordinated fashion (unpublished results from this laboratory).

ATP-Sulphurylase is also subject to feedback inhibition exerted by sulphide (de Vito and Dreyfuss, 1964). In addition, in cells grown on cysteine as the only sulphur source, an almost complete disappearance of sulphite reductase has been observed, which is overcome by methionine (de Vito and Dreyfuss, 1964). It is, however, difficult to be sure that these are true repressive–antirepressive effects, as the amino acids have not been tried on sulphate-grown cells and, also, possible interference at the level of amino acid permease has not been considered.

### 7. *Cysteine Biosynthesis from Serine*

Cysteine synthetase has been found to catalyse direct sulphydrylation of serine in yeast (Schlossmann *et al.*, 1962). The discovery of preferential utilization of acetylated homoserine in the parallel synthesis of homocysteine in yeast (see previous Section), and of the role of *O*-acetylserine in cysteine biosynthesis in bacteria (Kredich and Tomkins, 1966) prompted a reinvestigation of cysteine biosynthesis in yeast. A 10- to 20-fold increase in activity was found by Wiebers and Garner (1967) and by Thompson and Moore (1968) when *O*-acetylserine was supplied instead of serine as a substrate. No yeast mutants have apparently been found which lack this activation step. It has been demonstrated in spinach that homocysteine synthetase and cysteine synthetase are two different enzymes (Giovanelli and Mudd, 1968).

### 8. *Cysteine–Homocysteine Transsulphuration Pathway*

Cystathionine has been considered as an intermediate in bacterial methionine biosynthesis, as well as in *Neurospora crassa* (Delavier-Klutchko and Flavin, 1965). Biosynthesis and splitting of this compound can occur in the $\beta$- or $\gamma$-positions of the molecule, as shown in

Fig. 11. It should be noted that, as in the case of homocysteine synthetase already mentioned, succinylhomoserine in bacteria (Kaplan and Flavin, 1965), and *O*-acetylhomoserine in *Neurospora* (Nagai and Flavin, 1967; Flavin, 1967), are better substrates than free homoserine for their corresponding cystathionine $\gamma$-synthetases.

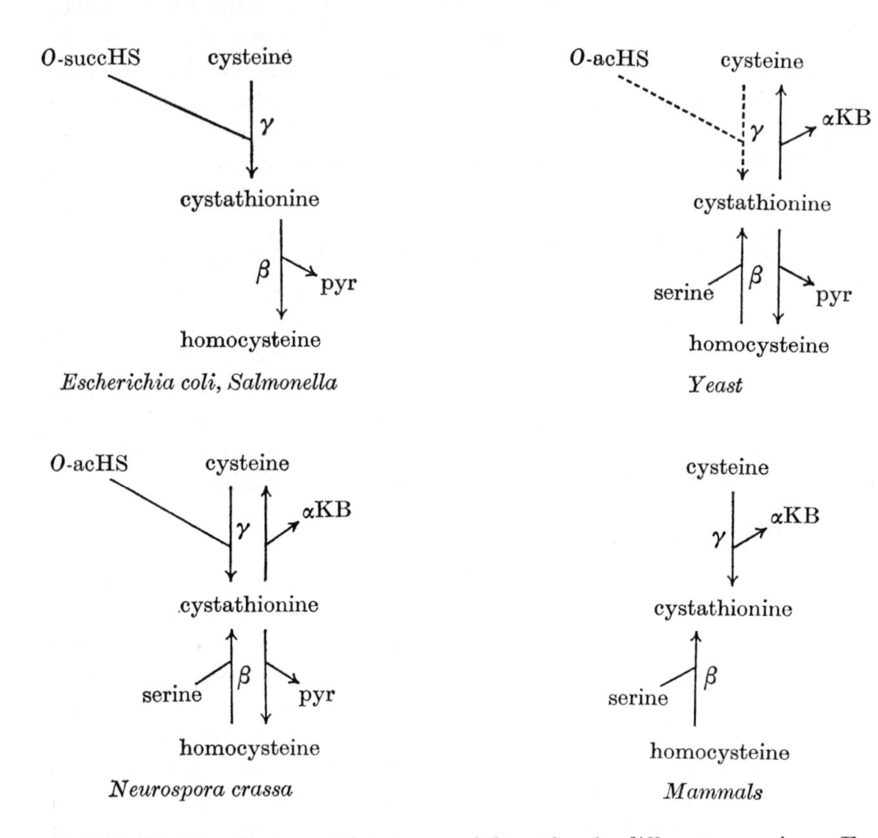


FIG. 11. Cysteine homocysteine trans-sulphuration in different organisms. For abbreviations see pp. 336–338.

There is still conflicting evidence about the "obligate" participation of cystathionine in methionine biosynthesis even in an organism like *N. crassa*, in which all four types of reaction have been shown to occur, in addition to direct sulphydrylation of *O*-acetylhomoserine (Flavin and Slaughter, 1964; Flavin, 1968; Kerr and Flavin, 1969). In the case of cystathionine participation in *Neurospora*, $me_3$ and $me_7$ mutants are both characterized by the absence of cystathionine $\gamma$-synthetase activity and complement *in vitro*. The sulphydrylase, when purified, could not utilize cysteine, and its addition to extracts of $me_3$ and $me_7$ mutants did not restore $\gamma$-cystathionine synthetase activity (Kerr and Flavin, 1968). Cystathionine $\gamma$-synthetase is repressed by exogenous methionine (Kerr and Flavin, 1968) and its activity is inhibited by *S*-adenosylmethionine (Kerr and Flavin, 1969). In addition, $\gamma$-cystathionase, which is supposed to be the same enzyme as $\gamma$-synthetase (Flavin, 1967), is derepressed in sulphur deprivation (Flavin and Slaughter, 1967b). $\beta$-Cystathionase, which does not derepress in the same conditions, is also a less specific enzyme which is able to catalyse cleavage of a variety of thio-, disulphide- and hydroxy-amino acids (Delavier-Klutchko and Flavin, 1965).

The results of the following *in vivo* experiments using differently labelled compounds do not support cystathionine participation in *Neurospora*. When exogenous cystathionine was supplied to cystathionineless *Neurospora* mutants, the 4-carbon skeleton of methionine did not originate from this compound (Wiebers and Garner, 1964). Formation of $^{35}S$ cysteine and homocysteine favoured the concept of separate routes for the synthesis of these two amino acids (Wiebers and Garner, 1966). Using spinach extracts, Giovanelli and Mudd (1966, 1967) showed that *O*-acetylhomoserine is converted at least ten times faster into *S*-adenosylmethionine than into cystathionine. On the other hand, the fact that methionine C mutants of *Salmonella* which have lost cystathionine $\gamma$-synthetase have also lost homocysteine synthetase activity (Flavin and Slaughter, 1967a), could be taken as an indication of identity between the two enzymes in this organism.

In yeast, cell-free extracts were found to be able to cleave cystathionine in the $\beta$- and $\gamma$-positions. Cystathionine $\beta$-synthetase was also found (Delavier-Klutchko and Flavin, 1965a), but $\gamma$-synthetase could never be detected. In *Sacch. cerevisiae* cystathionine, although actively transported, could not satisfy the organic sulphur requirements of any methionine auxotroph (Sorsoli *et al.*, 1968). There are, however, *in vivo* indications that the transsulphuration pathway might be operative in some conditions: (i) homocysteine synthetase-deficient mutants, but not homoserine *O*-transacetylase-deficient mutants, were found to be capable of growing in the presence of cysteine; (ii) mutants blocked in the sulphur pathway can grow on homocysteine or methionine. Utilization of the latter amino acid for cysteine synthesis must imply reformation of homocysteine, which might occur through methionine: *S*-adenosylmethionine transmethylation (see previous Section).

Altogether, these facts and the role of homocysteine synthetase described previously indicate that, at least in the case of yeast, cystathionine synthesis and degradation are not on the direct biosynthetic pathway of methionine, but can participate in "salvage" mechanisms, either for cysteineless or homocysteineless mutants.

### 9. *The Metabolic Roles of* S-*Adenosylmethionine*

Methionine is an amino acid of special importance in view of its dual role as protein component and as the direct and only source of methyl groups in cellular metabolism. *S*-Adenosylmethionine (SAM) is the activated product of methionine (sulphonium derivative) from which all kinds of methylations can occur (see Shapiro and Schlenk, 1965). A detailed study appeared recently which is concerned with uptake and utilization of SAM and *S*-adenosylhomocysteine in *Sacch. cerevisiae* (Knudsen *et al.*, 1969). So far no mutants have been found, in any organism, with a defect in SAM biosynthesis.

Each particular methylation process seems to require a specific transmethylase. Of particular interest is the case of a methionine auxotroph which has been shown specifically to lack the ability to methylate guanine in the $N^2$ position, in yeast t-RNA. All other methylated bases have been found unchanged (Björk and Svensson, 1969; Svensson *et al.*, 1969; see also Section V.C, p. 407). Submethylation or over-methylation of different types of RNA or DNA, at specific positions, is certainly of great importance for their respective metabolic functions (see Kjellin-Sträby and Phillips, 1968). In this connection it is worth mentioning that, in yeast, one of the main toxic effects of ethionine lies in its ability to be activated in *S*-adenosylethionine (SAE), resulting in transfer of ethyl groups to all compounds and positions normally methylated (Parks, 1958; see also Shapiro and Schlenk, 1965). A mutation leading to a dominant ethionine resistance in *Sacch. cerevisiae*, which manifests itself by a reduced transfer of the ethionine ethyl group in different classes of cellular components, has been presumed to be concerned with an increased ability to degrade SAE (Spence *et al.*, 1967). Such a degradation might be due to a cleavage normally effective on SAM (Mudd, 1959).

## H. ISOLEUCINE, VALINE AND LEUCINE

The pathways for isoleucine, valine and leucine have been extensively studied in *Sacch. cerevisiae*, and are identical to those found in other organisms (Fig. 12). The system comprises three parts:

(1) The first stage, catalysed by threonine deaminase, leads to the formation of $\alpha$-ketobutyrate, the latter being the first substrate (together with pyruvate) of the combined pathway leading to isoleucine and valine biosynthesis. Mutants blocked in Step 1 (Table XI) are then the only ones to require isoleucine alone.

(2) In the combined isoleucine and valine biosynthesis, each of the four steps is catalysed by only one enzyme, which latter is concerned with both the corresponding 4-carbon and 5-carbon substrates. Mutants in this part of the system require isoleucine and valine.

(3) Leucine biosynthesis, which starts with the penultimate product of valine biosynthesis, comprises the third stage. Mutants in this part of the pathway require leucine only.

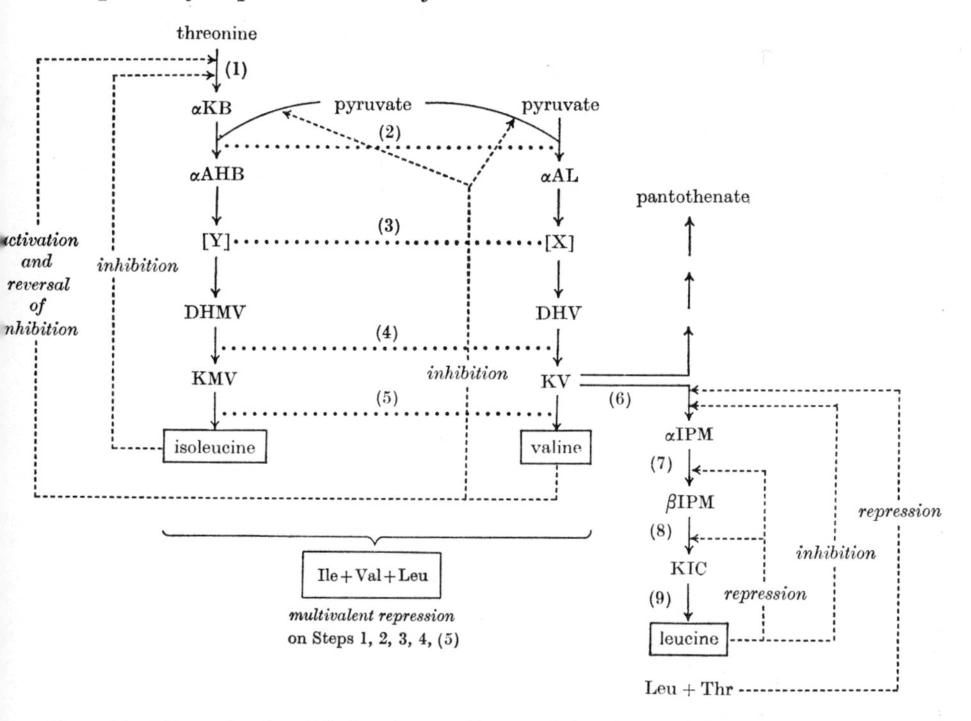


FIG. 12. Biosynthesis of isoleucine, valine and leucine and its regulation in *Saccharomyces cerevisiae*. For abbreviations see pp. 336–338.

Reactions and gene-enzyme relationships are shown in Table XI. As already mentioned for other pathways in *Sacch. cerevisiae*, genes coding for independent reactions are all unlinked.

## 1. *Threonine Deaminase*

Since the pioneer work of Changeux (1963, 1964a, b, c) and Freundlich and Umbarger (1963) on the *E. coli* enzyme, followed by the allosteric model developed by Monod *et al.* (1965) and due to the very interesting properties of threonine deaminase, numerous studies have been devoted to this enzyme as obtained from different organisms (see Umbarger, 1969 for review). In *Sacch. cerevisiae*, isoleucine-requiring mutants were

TABLE XI. *Gene–Enzyme Relationship in Isoleucine-Valine-Leucine Biosynthesis in* Saccharomyces cerevisiae

| Step | Reaction | Gene | Chromo-some | Enzyme studied | Ref. |
|---|---|---|---|---|---|
| 1 | *threonine* ⟶ $\alpha KB + NH_3$ | $is_1$ | V | + | a, b, c, d, e |
| 2a<br>2b | pyr + pyr ⟶ $\alpha AL$<br>αKB + pyr ⟶ $\alpha AHB$ | $is_2$ | | + | d, e, f |
| 3a<br>3b | αAL ⟶ X ⟶ *DHV*<br>αAHB ⟶ Y ⟶ *DHMV*<br>($NADH + H^+ \rightarrow NAD^+$) | $is_4$ + $is_5$ | linked | + | b, d, e |
| 4a<br>4b | DHV ⟶ *KV* + $H_2O$<br>DHMV ⟶ *KMV* + $H_2O$ | $is_3$ | X | + | b, d, e |
| 5a<br>5b | KV ⟶ [*valine*]<br>KMV ⟶ [*isoleucine*]<br>(glutamate ⟶ αKG) | | | + | d |
| 6 | KV + AcCoA ⟶ *αIPM* + CoA | $le_6$ | XII | + | g |
| 7 | αIPM ⟶ CDMC ⟶ *βIPM* | $le_{1,4,5,7}$<br>$le_{8,10}$ | | + | g |
| 8 | $\beta IPM + NADPH + H^+$ ⟶ *KIC* + $NADP^+ + CO_2$ | $le_{2,3,9}$ | III | + | g |
| 9 | KIC + glutamate ⟶ [*leucine*] + αKG | | | | |

Compounds in italics are those in the main sequence of reactions; compounds enclosed in boxes are the end-products of synthetic pathways.

For abbreviations see pp. 336–338.

*References:* a, Kakar and Wagner (1964); b, Cennamo *et al.* (1964); c, de Robichon-Szulmajster and Magee (1968); d, Bussey and Umbarger (1969); e, Magee and Hereford (1969); f, Magee and de Robichon-Szulmajster (1968a, b); g, Satyanarayana *et al.* (1968a, b).

characterized as lacking threonine deaminase (Kakar and Wagner, 1964). Different heteroallelic mutations have been allocated to locus $is_1$ which corresponds to this enzyme (Kakar 1963a), and different types of suppressors which act on these heteroalleles have been studied (Kakar, 1963b). Mitotic gene conversion involving locus $is_1$ and intragenic complementation have also been studied recently (Zimmerman, 1968, 1969).

Properties of the yeast enzyme had been found to be very similar to those of the corresponding enzymes obtained from other sources (Cennamo *et al.*, 1964). The following properties have been described: (i) $NH_4^+$ ions are stimulatory (Holzer *et al.*, 1964); (ii) cooperativity of threonine molecules is observed; (iii) feedback inhibition is exerted by isoleucine in a cooperative fashion; (iv) valine restores Michaelian kinetics in response to threonine concentration, and is also able to reverse isoleucine inhibition. Further studies of the yeast enzyme led to the estimation of cooperativity indices for each of the enzyme ligands, and demonstrated the effect of pH value on relative affinities and cooperativity (de Robichon-Szulmajster and Magee, 1968; Brunner and de Robichon-Szulmajster, 1969). Recently, a very interesting $is_1$ mutant, which grows on threonine, was found to possess a nearly normal amount of a threonine deaminase, the properties of which are altered in such a way that it is rendered nonfunctional *in vivo*. The molecular weights of the wild type and mutant enzymes have been found to be similar (180,000; Brunner *et al.*, 1969).

Regulation of the synthesis of threonine deaminase will next be examined, together with the second part of the pathway. However, it can be pointed out here that the complexity of regulatory controls to which threonine deaminase is subjected, and the fact that identical signals have apparently been selected, in this case, in all organisms studied, points to the metabolic importance of this enzyme, which constitutes a bridge between two pathways in amino acid biosynthesis (see Fig. 3).

Generally, in bacteria, a degradative, often inducible, threonine deaminase, which exhibits regulatory properties quite distinct from the biosynthetic enzyme, has also been found (Changeux, 1963, 1964a, b, c; Whiteley and Tahara, 1966). In *Sacch. cerevisiae,* such an enzyme has been shown to be absent, in the wild type as well as in $is_1$ mutants, even after growth in the presence of high threonine concentrations (de Robichon-Szulmajster and Magee, 1968; Brunner *et al.*, 1969).

### 2. *The Combined Isoleucine–Valine Pathway*

Kakar and Wagner (1964) characterized mutants concerned with two of the four steps involved in this part of the pathway (Steps 3 and 4).

One class of isoleucine–valine auxotrophs was found to possess all the enzymes tested. This apparent contradiction was removed by showing that the acetohydroxy acid synthetase (Step 2) was improperly assayed at pH 6·0, that the true biosynthetic enzyme had an optimum at pH 8·0 and was missing in $is_2$ mutants (Magee and de Robichon-Szulmajster, 1968a). In addition, the pH 8·0 enzyme was shown to be sensitive to feedback inhibition exerted by valine (Magee and de Robichon-Szulmajster, 1968b). No mutants have been found for the last step (transaminase B). Curiously, revertants which have kept only one of the requirements or become valine-sensitive have been found (Kakar *et al.*, 1964). Biochemical analysis of such revertants could prove to be very interesting.

Beside the end-product inhibition of acetohydroxy acid synthetase already mentioned, enzymes in this part of the pathway are subject to multivalent repression of the type first shown by Freundlich *et al.* (1962). Two recently published sets of results obtained independently by Bussey and Umbarger (1969) and Magee and Hereford (1969) have shown that a multivalent repression is caused by isoleucine, leucine and valine. However, when isoleucine is replaced by threonine, the same repression was observed (Bussey and Umbarger, 1969). As only a wild type strain was studied, it seems possible that the threonine effect was due to its endogenous transformation to isoleucine. Enzymes catalysing Steps 2, 3 and 4 show a much stronger response than those catalysing Steps 1 and 5.

Detailed comparison of the results obtained in these two different investigations are of interest, and have been presented together in Table XII. The values obtained show that the strain used by Bussey and Umbarger (1969) exhibited, on minimal medium, derepressed levels of the four enzymes studied. As a result repression, but no derepression, could be achieved experimentally. On the contrary, the strains used by Magee and Hereford (1969) showed, on minimal medium, intermediate levels of the four enzymes studied; though repression as well as derepression could be observed. It can be seen from Table XII that, despite the differences in strain behaviour, the total range of variation is very similar in both sets of data. In addition, Magee and Hereford (1969) have shown some correlation between the rates of synthesis of the enzymes studied when pairs were compared. However, annulment of synthesis was not obtained coordinately. These findings are important in view of the absence of linkage between the structural genes encoding for these enzymes. They can be interpreted if one assumes that a common repressor system is involved and that a redundant, but not exactly identical, operator-like structure is attached to each independent structural gene. Thus, different affinities towards a common repressor

| Strain | | | | | Enzyme | | | | | | | | | |
|---|---|---|---|---|---|---|---|---|---|---|---|---|---|---|
| No. | Origin | Auxotrophy | Ref. | Culture cond. | TD | | AHA Sy. | | RI | | DH | | TA | |
| | | | | | S.A. | DR/R | S.A. | DR/R | S.A. | DR/R | S.A. | DR/R | S.A. | DR/R |
| W.T. | L | — | a | M | 239 | 1·9 | 75 | 5·3 | 27 | 2·7 | 21 | 3·5 | 192 | 1·6 |
| | | | | R | 128 | | 14 | | 10 | | 6 | | 120 | |
| M13 | MH | Leu | b | DR | 150 | 2·0 | 7 | 3·5 | 22 | 3·1 | 26 | 5·2 | | |
| | | | | M | 112 | | 2 | | 18 | | 17 | | | |
| | | | | R | 75 | | 2 | | 7 | | 5 | | | |
| M6 | MH | Ile | b | DR | | | 13 | 4·3 | 23 | 3·3 | 9 | 4·5 | | |
| | | | | M | | | 7 | | 18 | | 8 | | | |
| | | | | R | | | 3 | | 7 | | 2 | | | |
| M2 | MH | Ile | b | DR | 235 | 2·5 | | | 25 | 4·2 | 30 | 7·5 | | |
| | | + | | M | 104 | | | | 8 | | 11 | | | |
| | | Val | | R | 93 | | | | 6 | | 4 | | | |
| M12 | MH | Ile | b | DR | 167 | 1·8 | 14 | 4·7 | | | 17 | 8·5 | | |
| | | + | | M | 167 | | 9 | | | | 18 | | | |
| | | Val | | R | 82 | | 3 | | | | 2 | | | |
| M7 | MH | Ile | b | DR | 214 | 1·8 | 13 | | 21 | | | | | |
| | | + | | M | 155 | | 22 | 3·1 | 32 | 2·0 | | | | |
| | | Val | | R | 119 | | 7 | | 16 | | | | | |

*Strains:* Numbers correspond to those used by the authors. L = from Dr. Lindegren's stock; MH = from Drs Mortimer and Hawthorne's stock.

*Enzymes:* TD, threonine deaminase; AHASy., acetohydroxyacid synthetase; RI, reductoisomerase; DH, dehydrase; TA, transaminase, catalysing respectively Steps 1, 2, 3, 4 and 5 in Table XI.

*Culture conditions:* DR = limitation in one of the auxotrophic requirements by the use of isoleucyl glycine instead of free isoleucine; M (ref. a) = no supplement; M (ref. b) = minimal medium + $2 \times 10^{-3}M$ L-valine and $5 \times 10^{-4}M$ L-isoleucine; R (ref. a) = minimal medium + $10^{-2}M$ L-isoleucine, $5 \times 10^{-2}M$ L-valine and $10^{-2}M$ L-leucine; R (ref. b) = minimal medium + $5 \times 10^{-}$M of each L-amino acid (isoleucine, valine, leucine).

*References:* a, Bussey and Umbarger (1969); b, Magee and Hereford (1969).

*Results:* From ref. a, M = mean value calculated from non-repressive conditions; R = mean value calculated from repressive conditions. From ref. b, for comparative presentation values have been recalculated on the basis of 1 min instead of 20 min. DR/R = ratio of maximally derepressed over maximally repressed values obtained by the authors quoted. (See text for comments.)

could result from small differences in individual operators. Such an explanation seems even more likely if one wishes to account for the complexity, either at the repressor or operator level, which underlies the multivalent type of repression.

### 3. *Leucine Biosynthesis*

An investigation of leucine biosynthesis in yeast has recently been carried out (Satyanarayana *et al.*, 1968a, b). From this study, gene–enzyme relationship appears quite difficult to establish, since no less than six independent loci, and three other independent loci, lead, respectively, to a deficiency in the second and third enzymes of this branch of the pathway (Steps 7 and 8, Fig. 12 and Table XI). As in the case of isoleucine–valine, there are no mutants for the last transamination step.

Only one locus is concerned with the first step, and the enzyme α-isopropylmalate synthetase is subject to end-product inhibition exerted by leucine. A mutant, resistant to an isoleucine analogue (5′5′5′-trifluoroleucine), has been found to possess an enzyme less sensitive to feedback inhibition. Leucine is also active as a repressor of the second and third enzyme synthesis. Repression of the first enzyme was attained in the wild type only when threonine and leucine were added together. Isoleucine and valine alone, or in combination with leucine, did not cause repression. Deviation from this pattern was observed in different leucine auxotrophs, but it was not possible to correlate it with a regulatory gene.

## I. TRYPTOPHAN, TYROSINE AND PHENYLALANINE

The pathway leading to the biosynthesis of aromatic amino acids in yeast proceeds through the same steps and branching points as in other organisms (Fig. 13 and Table XIII). Mutants have been obtained for all the steps but two (the ultimate transamination steps leading respectively to phenylalanine and tyrosine). Gene–enzyme relationships are shown in Table XIV. The metabolic complexity of this pathway has led, in every organism studied, to a very complex regulatory pattern which involves a multiplicity of enzymes catalysing the first step, a multi-enzyme aggregate catalysing at least four of the following steps, and sometimes antagonistic controls exerted by the three end-products at different points.

Owing to the existence of two excellent recent reviews dealing with comparative aspects of this pathway in different organisms (Doy, 1968b; Gibson and Pittard, 1968), the following section will be concerned strictly with *Sacch. cerevisiae*. The pattern of regulation observed in this system is presented in Fig. 13.

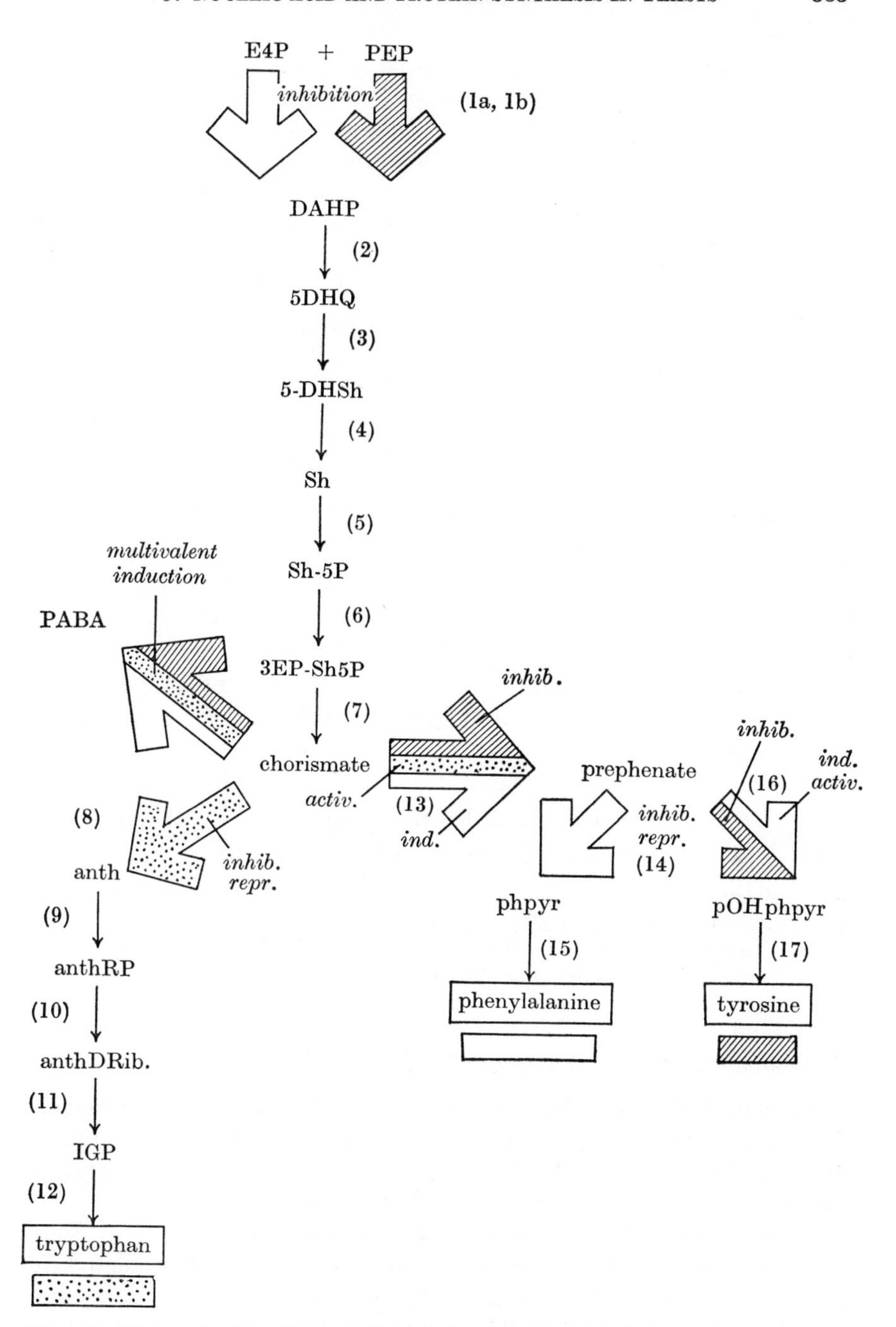


FIG. 13. Biosynthesis and regulation of aromatic amino acids in *Saccharomyces cerevisiae*. For abbreviations see pp. 336–338.

TABLE XIII. *Biosynthesis of Tryptophan, Phenylalanine and Tyrosine*

| Step | Reaction |
|---|---|
| 1a | E4P + PEP → *DAHP*(Phe) |
| 1b | E4P + PEP → *DAHP*(Tyr) |
| 2 | DAHP + NADH + $H^+$ → *5DHQ* + $NAD^+$ |
| 3 | 5DHQ → *5DHSh* + $H_2O$ |
| 4 | 5DHSh + NADPH + $H^+$ → *shikimate* + $NADP^+$ |
| 5 | Shikimate + ATP → *Sh5P* + ADP |
| 6 | Sh5P + PEP → *3EP-Sh5P* |
| 7 | 3EP-Sh5P → *chorismate* + $P_i$ |
| 8 | Chorismate + glutamine → *anthranilate* + glutamate |
| 9 | Anth + PRPP → *anthRP* + $PP_i$ |
| 10 | AnthRP → *anthDrib* |
| 11 | AnthDrib → *IGP* + $CO_2$ |
| 12 | IGP + serine → [tryptophan] + glyceraldehyde-3-P |
| 13 | Chorismate → *prephenate* |
| 14 | Prephenate → *phenylpyruvate* + $CO_2$ + $H_2O$ |
| 15 | Phpyr + glutamate → [phenylalanine] + αKG |
| 16 | Prephenate + NADH + $H^+$ → *pOHphpyr* + $NAD^+$ + $CO_2$ |
| 17 | pOHphpyr + glutamate → [tyrosine] + αKG |

Compounds in italics are those in the main sequence of reactions; compounds enclosed in boxes are the end-products of synthetic pathways. For abbreviations see pp. 336–338.

## 1. *Common Part of the Pathway: Chorismate Biosynthesis*

Lingens *et al.* (1966a, b) were able to separate two DAHP synthetases by ammonium sulphate fractionation, one being feedback-inhibited by phenylalanine, the other by tyrosine. No sensitivity towards tryptophan or key intermediates, such as chorismate or prephenate, was detected. Identical results were obtained by Doy (1968a).

Another approach, selection of mutants sensitive to amino acid inhibition (Meuris *et al.*, 1967), led to the study of phenylalanine- and tyrosine-sensitive mutants. These mutants were shown to have lost, respectively, the phenylalanine-sensitive or the tyrosine-sensitive DAHP

TABLE XIV. *Gene–Enzyme Relationship in Aromatic Amino Acid Biosynthesis in* Saccharomyces cerevisiae

| Step | Gene | Chromosome | Enzyme studied | Ref. |
|---|---|---|---|---|
| 1a | *phe*-$i_1$ | | + | a, b, c |
| 1b | *tyr*-$i_1$ | | + | a, b, c |
| 2 | $Arom_{1C}$ | | | d |
| 3 | $Arom_1$ | | Enzymes | |
| 4 | $Arom_{1D}$ | | 2, 3, 4 | d |
| 5 | $Arom_{1B}$ | Linked | (at least) | d |
| 6 | $Arom_{1A}$ $(ty_{2,4})$ | Frag. 2 | form a complex | d |
| 7 | $Arom_2$ $(ty_3)$ | VII | + | d |
| 8 | $tr_2$ | V | | |
| | $tr_3$ | Linked to $ur_1$ | + | e, f |
| 9 | $tr_4$ | Frag. 2 | + | e, f |
| 10 | $tr_1$ | IV | + | e, f |
| 11 | $tr_3$ | Linked to $ur_1$ | + | e, f |
| 12 | $tr_5$ | VII | + | e, f, g |
| 13 | KP 171, E 104 (a) | | + | a |
| 14 | HK 11(a) | Unlinked | + | a |
| 15 | $pha_1$, $pha_2$ | | + | a |
| 16 | HK 99(a) | II | + | a |
| 17 | $ty_1$ | | + | a |

*References:* a, Lingens *et al.* (1966, 1967); b, Meuris (1967), Meuris (1967a, b); c, Doy (1968a, b); d, de Leeuv (1967); e, Wegman and de Moss (1965), de Moss (1965); f, Doy and Cooper (1966); g, Manney (1964, 1968a, b), Duntze and Manney (1968), Manney *et al.* (1969).

aldolases. Genes encoding for these two isoenzymes were found to be independent. Individual mutations alone do not lead to auxotrophy. However, recombinant strains carrying the two mutations only grow in the presence of the three aromatic amino acids plus *p*-aminobenzoic acid, the principal growth factor that also originates from the same pathway (Meuris, 1967a, b). These requirements are in favour of the existence of only two DAHP aldolases in yeast. Inhibitions exerted either by phenylalanine or tyrosine on each DAHP aldolase are cooperative, and affect the $V_{max}$ only (Lingens *et al.*, 1967).

It has been shown, in all these studies, that no regulation of synthesis of these enzymes by any combination of the end-products is apparent. Mutants lacking the steps following DAHP aldolase map as a cluster, and at least the enzymes catalysing Steps 2, 3 and 4 form a multi-enzyme complex. In contrast, the last step of the common pathway is encoded by an independent gene (de Leeuw, 1967). No regulation has been reported to affect Steps 2 to 7.

### 2. *Tryptophan Biosynthesis from Chorismate*

The first step of this branch (Step 8 in Fig. 13 and Table XIII), catalysed by anthranilate synthetase, has been found to be absent in two groups of tryptophan auxotrophs corresponding to the unlinked genes $tr_2$ and $tr_3$. In addition, $tr_3$ mutants have also lost an enzyme further along the pathway, indoleglycerol phosphate synthetase (Step 11). The polypeptide chain, encoded by the $tr_3$ locus, seems then necessary for two enzyme functions. As the two activities sediment together it is believed that they form an enzyme complex. However, only anthranilate synthetase activity is subject to feedback inhibition exerted by tryptophan (Wegman and de Moss, 1965; Doy and Cooper, 1966). Synthesis of anthranilate synthetase is also repressed by tryptophan (Doy and Cooper, 1966; Lingens *et al.*, 1967).

The last step in tryptophan biosynthesis, catalyzed by tryptophan synthetase, has been extensively studied. Locus $tr_5$, encoding for this enzyme, was chosen as a suitable system for studies in suppressibility and complementation (Manney, 1964, 1968a, b; Duntze and Manney, 1968; Manney *et al.*, 1969). Some of the suppressible mutants have been found to produce fragments (mol. wt., 35,000) able to catalyse conversion of indoleglycerol phosphate to indole, while the complete enzyme (mol. wt., 160,000) catalyses the overall reaction to tryptophan, with no detectable intermediate formed. Results obtained provide evidence that super-suppressible mutations cause chain termination (Manney, 1964, 1968a). A study of the properties of the enzyme produced by *in vivo* intragenic complementation showed that one type of complementation leads to a hybrid enzyme very close to the wild type. A second type, leading to intermediate formation of indole, has been described as an example of intracellular cross-feeding. The authors favour the concept of yeast tryptophan synthetase being formed of identical subunits (Duntze and Manney, 1968). A recent fine mapping study of the $tr_5$ gene has shown that this locus can be divided into two distinct functional regions, corresponding to the two half-reactions (Manney *et al.*, 1969).

In addition, two macromolecular fractions, formed under different growth conditions, affect tryptophan synthetase activity. One fraction inactivates the enzyme; the other protects it against this inactivation (Manney, 1968b).

### 3. *Chorismate Mutase*

This enzyme constitutes the intermediate step leading to prephenate, the common precursor of phenylalanine and tyrosine. Its activity is allosterically inhibited by tyrosine (K system; Lingens *et al.*, 1967). If this retro-inhibition happened to be complete, it would create an auxotrophic requirement for phenylalanine. This inconvenience might be the evolutionary justification for the finding that phenylalanine behaves as an inducer of the enzyme. The observation that tryptophan activates chorismate mutase (Lingens *et al.*, 1967) is less easy to explain.

### 4. *Phenylalanine Biosynthesis from Prephenate*

The first step of the phenylalanine branch (Step 14 in Fig. 13 and Table XIII), catalysed by prephenate dehydrogenase, is inhibited and repressed by the end-product phenylalanine (Lingens *et al.*, 1967).

### 5. *Tyrosine Biosynthesis from Prephenate*

The first step in the tyrosine branch (Step 16 in Fig. 13 and Table XIII), catalysed by prephenate dehydrase, is subject to a more complex regulation than that exerted on the corresponding step in the phenylalanine branch. As could be expected from the position of this enzyme in the pathway, it has been found to be allosterically inhibited by tyrosine (K system). Surprisingly, prephenate dehydrase is also induced and activated by tyrosine (Lingens *et al.*, 1967).

If one tries to rationalize these findings, it seems that the antagonistic regulation mechanisms mentioned as existing in this pathway, especially those acting on chorismate mutase and prephenate dehydrase, might meet the need to keep the three end-products in a biologically acceptable balance. This conclusion is emphasized by the finding that the three amino acid end-products together are able to induce the first enzyme in *p*-aminobenzoate synthesis, which also derives from chorismate, the synthesis of which would be cancelled by the regulatory network previously described.

The need for equilibration between end-products of a common pathway might also apply to the antagonistic effects already mentioned, such as the GMP, AMP cross-effects in the purine pathway (see Section II.A, p. 343), the isoleucine–valine effects on threonine deaminase in the isoleucine–valine–leucine pathway (see Section III.H, p. 376), or the methionine mediated induction of aspartate-semialdehyde dehydrogenase in the threonine–methionine biosynthesis (see Section III.G, p. 364).

## IV. Protein Biosynthesis

### A. INTRODUCTORY REMARKS

Most of our information on protein biosynthesis has been derived from studies on *E. coli* and reticulocytes. The strong similarities between the two systems, and the apparent universality of the genetic code, leads one to think that the main features of protein synthesis in yeast are similar to those in other organisms.

The amino acid sequence of a protein is specified by the sequence of nucleotides in a particular segment of deoxyribonucleic acid (DNA). The DNA is transcribed into a messenger RNA (m-RNA), which has a sequence complementary to that of one of the strands of the DNA serving as a template (transcription). The m-RNA binds to ribosomes which are the sites of protein synthesis; there, the m-RNA determines the order of linkage of amino acids into a specific protein (translation). During translation, a group of three nucleotides in the m-RNA (codon) specifies the amino acid which is to be linked to the growing peptide chain. The amino acid is activated and transferred on a transfer ribonucleic acid (t-RNA) molecule which recognizes the corresponding codon on the messenger. The codons specifying each of the 20 amino acids have been determined (see Cold Spring Harbor Symposium on Quantitative Biology, Vol. 31, 1966). It seems that the sequence of the amino acids in the peptide chain contains all the information required for the chain folding in the three-dimensional structure of the molecule (Anfinsen, 1967). The mechanism of the biosynthesis of proteins has been extensively reviewed by Lengyel and Söll (1969). As very little work has been done in yeast on the mechanism of initiation, elongation and termination of polypeptide chains, we will deal more especially with amino acid activation, t-RNAs and m-RNA. Some aspects of protein biosynthesis in yeast have been described by Halvorson and Heredia (1968).

### B. AMINOACYL-t-RNA SYNTHETASES

The aminoacyl-t-RNA synthetases catalyse a two-step reaction:

$$\text{(I)}\quad \text{enz.} + \text{ATP} + \text{AA} \rightleftharpoons \text{enz.:AA-AMP} + \text{PP}_i$$
$$\text{(II)}\quad \text{enz.:AA-AMP} + \text{t-RNA} \rightleftharpoons \text{AA-t-RNA} + \text{enz.} + \text{AMP}$$

Reaction I is the amino acid activation reaction. In this step, the aminoacyl-t-RNA synthetase specific for one of the 20 amino acids found in protein, catalyses a reaction which yields a mixed anhydride formed between the carboxyl group of the amino acid and the 5′-phosphate of AMP. This aminoacyl adenylate does not accumulate as a free intermediate, and remains attached to the enzyme. In the second step

(Reaction II), the enzyme: aminoacyl adenylate complex reacts with a t-RNA molecule which is specific for the amino acid, and the aminoacyl moiety is transferred to the t-RNA.

The activity of an aminoacyl-t-RNA synthetase may be measured by the various methods described by Stulberg and Novelli (1962). With any method, the optimal conditions (i.e. pH value, ionic strength and $Mg^{2+}$/ATP ratio) must be worked out with each system.

### 1. *Purification and Properties of Aminoacyl-t-RNA Synthetases*

In order to obtain a better understanding of amino acid activation, it is necessary to work with pure or nearly pure enzymes. All procedures described for the purification of synthetases involve classical enzyme purification methods.

To date, a number of aminoacyl-t-RNA synthetases have been extracted from yeast and purified: seryl-, phenylalanyl-, arginyl-, and leucyl-t-RNA synthetases (Makman and Cantoni, 1965); valyl- and lysyl-t-RNA synthetases (Lagerkvist and Waldenström, 1965); methionyl-t-RNA synthetase (Berg, 1956); tyrosyl-t-RNA synthetase (Kijima *et al.*, 1968) and cysteinyl-t-RNA synthetase (James and Bucovaz, 1969). Seryl-t-RNA synthetase has been found to be homogenous, as determined by equilibrium centrifugation; so are valyl- and lysyl-t-RNA synthetases. Makman and Cantoni (1965) have obtained, along with pure seryl-t-RNA synthetase, phenylalanyl-t-RNA synthetase 90% pure, and arginyl- and leucyl-t-RNA synthetases partially purified. The molecular weights of these enzymes are given in Table XV.

TABLE XV. *Molecular Weights of the Aminoacyl-t-RNA Synthetases Purified from Yeast*

| Enzyme | Molecular weight | Method | Ref. |
|---|---|---|---|
| Seryl-t-RNA synthetase | 89,000 | A | a |
| Phenylalanyl-t-RNA synthetase | 180,000 | A | a |
| Valyl-t-RNA synthetase | 118,000[a] | | |
| | 107,000[b] | B | b |
| Lysyl-t-RNA synthetase | 113,000 | A | c |
| Cysteinyl-t-RNA synthetase | 160,000 | C | d |

*Methods:* A, equilibrium centrifugation; B, Archibald method; C, Sephadex G 200. [a] using a solution containing 6 mg/ml protein; [b] using a solution containing 3 mg/ml protein.

*References:* a, Makman and Cantoni (1965); b, Lagerkvist and Waldenström (1967); c, Lagerkvist and Waldenström (1965); d, James and Bucovaz (1969).

As in other organisms (see Lengyel and Söll, 1969), most of the yeast aminoacyl-t-RNA synthetases have a molecular weight around 100,000 daltons.

The kinetic parameters have been determined for valyl-, seryl-, tyrosyl- and cysteinyl-t-RNA synthetases; they are summarized in Tables XVI and XVII. With the purified yeast seryl-t-RNA synthetase,

TABLE XVI. *Kinetic Parameters* $K_m$ *of Some Aminoacyl-t-RNA Synthetases*

| Compound | Valyl-t-RNA synthetase[a] (ref. a) (mM) | Seryl-t-RNA synthetase[a] (ref. b) (mM) | Cysteinyl-t-RNA synthetase[b] (ref. c) (mM) |
|---|---|---|---|
| Corresponding amino-acid | 0·03 | 0·01 | 0·57 |
| ATP | 0·04 | 0·5 to 1 | 1·1 |
| $Mg^{2+}$ | 8 | — | 1 |
| t-RNA | 0·001 (crude) | 0·00027 (purified) | — |

*Assay method:* [a] aminoacyl-t-RNA formation; [b] ATP-$PP_i$ exchange reaction.
*References:* a Lagerkvist and Waldenström (1967); b, Makman and Cantoni (1965); c, James and Bucovaz (1969).

TABLE XVII. *Kinetic Parameters of Tyrosyl-t-RNA Synthetase* (*Kijima* et al., *1968*)

| Kinetic parameter | Assay method | | |
|---|---|---|---|
| | Tyrosine hydroxamate formation (m$M$) | $PP_i$-ATP exchange (m$M$) | Tyrosyl-t-RNA formation (m$M$) |
| $K_m$ ATP | 1·0 | 1·0 | 1·0 |
| $K_m$L tyrosine | 0·25 | 0.025 | 0·005 |
| $K_m$ t-$RNA^{tyr}$ | — | — | 0·0002 |
| $K_i$ tyramine | 1·1 | 0·09 | 0·12 |

Makman and Cantoni (1966) determined the $K_m$ of formation of seryl-t-RNA both with yeast and *E. coli* t-$RNA^{ser}$. They found that the affinity for yeast t-$RNA^{ser}$ is ten times higher than for the *E. coli* substrate, and that the inhibition of the reaction by 20 m$M$ phosphate ions is lower if yeast t-RNA is used.

### 2. *Aminoacyl-t-RNA Synthetase Activity*

When an aminoacyl-t-RNA synthetase is incubated in the presence of its corresponding amino acid, ATP and magnesium ions, the enzyme: aminoacyl adenylate complex is formed. This can be isolated from the reaction mixture by passing through columns of Sephadex G-50 or G-75 (Allende and Allende, 1964; Norris and Berg, 1964). The complex has been isolated in the case of the yeast valine system by Lagerkvist *et al.* (1966) and of serine by Buestein *et al.* (1968), who also studied the heat stability of the enzyme:seryl-AMP complex after absorption on nitrocellulose filters (Yarus and Berg, 1967). They found a stabilizing effect of glutathione, presumably protecting the very labile anhydride bond.

Generally the formation of the enzyme:amino acyl-AMP complex requires only the presence of the amino acid, ATP and $Mg^{2+}$. However, a few cases have been found in which the presence of t-RNA is necessary for the activation step. Glutamyl-t-RNA synthetase from yeast (Lee *et al.*, 1967) catalyses glutamyl–adenylate formation only at high concentrations of glutamate. In the presence of t-RNA, low concentrations of glutamate are sufficient. In the glutamine system, t-$RNA^{gln}$ is strictly required for the formation of glutaminyladenylate. The same phenomenon is observed in the glutamic and glutamine systems of pork liver (Lee *et al.*, 1967) and *E. coli* (Ravel *et al.*, 1964) and has been reported for a few other systems of *E. coli* (see Lengyel and Söll, 1969).

The enzyme:seryl-AMP complex can transfer its amino acid to the t-$RNA^{ser}$. This reaction requires magnesium ions which can be replaced by spermidine, or by calcium or manganese ions (Buestein *et al.*, 1968). This is in contrast to the results found with the valine system (Lagerkvist and Waldenström, 1965), where the transfer reaction proceeds in the absence of magnesium ions and even in the presence of EDTA. Svensson (1968) has shown that methionyl-t-RNA synthetase from *Sacch. cerevisiae* is activated by divalent metal ions. He has shown that the transfer step is activated by $NH_4^+$, $K^+$ or $Rb^+$, while the activation reaction is unaffected by these ions.

The binding of the aminoacyl-t-RNA synthetases to their corresponding t-RNA will be reviewed later (Section IV.C, p. 394).

### 3. *Temperature-Sensitive Mutants*

Hartwell (1967) has isolated 400 temperature-sensitive mutants of *Sacch. cerevisiae*. Among these, 21 displayed a rapid cessation of protein synthesis after a shift from the permissive (23°) to the non-permissive (36°) temperature. Hartwell and McLaughlin (1968a) have shown that two of these mutants have a lesion in the structural gene for isoleucyl-t-RNA synthetase. As this monogenic mutation leads to the loss of 99%

of the enzyme activity, they concluded that haploid yeast cells synthesize only one species of isoleucyl-t-RNA synthetase. In a further report (McLaughlin and Hartwell, 1969), evidence is presented that another of the 21 mutants is defective in methionyl-t-RNA synthetase. The authors show that the methionyl-t-RNA synthetase gene is not linked to the isoleucyl-t-RNA synthetase gene. This last is located approximately 22 units to the left of marker $p_9$, on Chromosome XI of the yeast genetic map (Mortimer and Hawthorne, 1966).

## C. TRANSFER RIBONUCLEIC ACID

A number of t-RNAs which have been purified and studied to date have been extracted from yeast. A general review on t-RNAs has appeared recently (Ebel, 1968).

### 1. *Base Composition of t-RNAs*

The chain length of the known t-RNA molecules varies from 75 to 85 nucleotides. The base composition of bulk t-RNA from different sources has been determined (Miura, 1967). Comparison between the composition of yeast and *E. coli* t-RNAs shows no essential differences between them. However, it seems that in yeast the methylated bases are more abundant than in other organisms (see Section V, p. 407).

### 2. *Purification and Structure*

*a. Primary structure.* To date, the primary structure of eight specific yeast t-RNAs is known: t-RNA$^{ala}$ (Holley *et al.*, 1965),* t-RNA$^{ser}_{I}$ and t-RNA$^{ser}_{II}$ (Zachau *et al.*, 1966), t-RNA$^{tyr}$ (Madison *et al.*, 1966), t-RNA$^{phe}$ (RajBhandary *et al.*, 1967), t-RNA$^{val}$ (Baiev *et al.*, 1967) from baker's or brewer's yeast; and from *C. utilis* t-RNA$^{val}$ (Takemura *et al.*, 1968) and t-RNA$^{ile}$ (Takemura *et al.*, 1969a, b).

Other specific t-RNAs from yeast have been partially purified. Four different kinds of t-RNA$^{gly}$ have been isolated from brewer's yeast (Bergquist *et al.*, 1968), and t-RNA$^{his}$ (Takeishi *et al.*, 1967) and t-RNA$^{lys}$ (Vold, 1969) have been purified to some extent. Using benzoylated DEAE-cellulose columns, Wimmer *et al.*, (1968a, b) purified t-RNA$^{phe}$ from brewer's yeast, and t-RNA$^{tyr}$ and t-RNA$^{trp}$ from *Sacch. cerevisiae*. By the same technique, Gillam *et al.* (1968) purified the t-RNA acceptors for aspartate, arginine, glycine, methionine and threonine. Japanese groups are currently purifying t-RNAs from *C. utilis* (Miyazaki *et al.*, 1967; Miyazaki and Takemura, 1968; Takemura and Miyazaki, 1969).

Two types of t-RNA$^{met}$ have been characterized from baker's yeast (Takeishi and Ukita, 1968; RajBhandary and Ghosh, 1969). One of

*Note added in proof: Total synthesis of the gene corresponding to this yeast t-RNA has just been accomplished by Khorana's group (Agarwal *et al.*, 1970).

them, t-RNA$_{I}^{met}$ is esterified by methionine as efficiently by the *E. coli* methionyl-t-RNA synthetase as by the yeast enzyme, while the t-RNA$_{II}^{met}$ is charged only partially by the *E. coli* enzyme. The t-RNA$_{I}^{met}$ can be converted to t-RNA$^{f-met}$ by *E. coli* methionyl-t-RNA trans-formylase in the presence of $N^{10}$-formyltetrahydrofolate, while t-RNA$_{II}^{met}$ cannot. The amino acid incorporation studies *in vitro* with a natural messenger (Takeishi and Ukita, 1968) or with poly-(UG) or poly-(AUG) (RajBhandary and Ghosh, 1969) show that yeast t-RNA$_{I}^{f-met}$ serves as an initiator of protein synthesis. Smith and Marker (1968) have reported the existence of a t-RNA$^{f-met}$ in mitochondria extracted from yeast.

*b. Secondary and tertiary structures.* A few speculative models for the secondary structure of t-RNA have been proposed. The most popular is the clover-leaf form first proposed by Holley *et al.* (1965) for t-RNA$^{ala}$. It has since been extended to other t-RNAs of known sequence. Philipps (1969) has compared 14 different t-RNAs from different origins, and has proposed a generalized clover-leaf structure which might apply to all t-RNA molecules.

Several models have been proposed for the tertiary structure: Lake and Beeman (1968) proposed one to fit their small-angle X-ray scattering data; Doctor *et al.* (1969) made theirs after analysis of fibres of t-RNA$^{tyr}$; Cramer *et al.* (1968) interpreted chemical and physical studies on t-RNA and suggested a tertiary structure. Ninio *et al.* (1969) proposed a model which meets some chemical data and their small-angle X-ray scattering data.

### 3. *Role of t-RNA in Protein Biosynthesis*

*a. Redundancy.* In DNA–RNA hybridization experiments, Schweizer *et al.* (1969) found that between 0·064% and 0·080% of nuclear yeast DNA is complementary to yeast t-RNA. Assuming a genome size of yeast nuclear DNA of $1{\cdot}25 \times 10^{10}$ daltons, these data correspond to 320–400 cistrons for t-RNA. If one assumes that there are 60 different species of t-RNA, this would amount to an average of five to seven identical cistrons encoding for each species. This value is intermediate between the numbers found in bacteria and in higher organisms.

The existence of redundant t-RNA in bacteria may explain certain mechanisms of nonsense or missense suppression (Goodman *et al.*, 1968; Gupta and Khorana, 1966). In yeast (Mortimer, 1969; Gilmore *et al.*, 1968; Hawthorne and Mortimer, 1968) several pieces of evidence favour the proposal that the super-suppressors are products of t-RNA structural genes. They would act as bacterial suppressors at the level of translation, by permitting the nonsense codon to be read by an altered t-RNA (Goodman *et al.*, 1968; see Mortimer and Hawthorne, 1969).

*b. Binding of aminoacyl-t-RNA synthetase to the corresponding t-RNA.* Fixation of the amino acid is made on the 2′- or 3′-hydroxyl of the adenosine in the CCA terminal triplet. In rat liver t-RNA, Daniel and Littauer (1963) have shown that the CCA triplet is not an obligatory feature: the amino acid is bound if CCA is replaced by CA, but at a slower rate.

(i) *Participation of the anticodon.* The participation of the anticodon in the recognition of t-RNA by the synthetase has been discussed. Some experiments have shown a parallelism between the decrease of the acceptor capacity and the susceptibility of the anticodon to certain modifications (Ebel *et al.*, 1965; Kisselev and Frolova, 1964; Harriman and Zachau, 1966).

More recent studies eliminate the participation of the anticodon in the synthetase binding site. Yoshida *et al.* (1967) have shown that the acceptor capacity remains unchanged after chemical modifications of the t-RNA molecule, resulting in complete loss of the binding to the m-RNA-ribosome complex. In *E. coli* a study of the suppressor t-RNAs leads to the same conclusion. The $\text{t-RNA}^{\text{tyr}}_{\text{suIII}}$ differs from $\text{t-RNA}^{\text{tyr}}$ only in one base of the anticodon (CUA in place of GUA). This replacement does not modify the tyrosine-binding capacity. (Goodman *et al.*, 1968).

(ii) *Other studies on the synthetase binding site.* The high specificity of the interaction t-RNA-aminoacyl-t-RNA synthetase leads one to think that the binding site should be located in parts of the t-RNA molecule which show variations from one t-RNA to another. Different approaches to the definition of the synthetase binding site were used: chemical modification of specific bases (Duval and Ebel, 1966, 1967; Yoshida *et al.*, 1966), biological substitution of uridine by 5-fluorouridine (Geige *et al.*, 1969a, b) and inhibition by polynucleotides of the acceptor activity (Hayashi and Miura, 1966). None of these experiments was conclusive, due to the concomitant modification of the structure of the t-RNA. Recently, Schulman and Chambers (1968) have suggested that the acceptor site involves the first three base pairs on the t-RNA near the CCA end, with a possible action of the other portions of the molecule.

Nevertheless, all the experiments carried out in this field suggest that the recognition between a t-RNA and its aminoacyl-t-RNA synthetase may vary from case to case (Lagerkvist *et al.*, 1966; Lagerkvist and Rymo, 1969; Ohta *et al.*, 1967; Okamoto and Kawade, 1967).

*c. Sites implicated in the transfer of the amino acid to the peptide chain.*

(i) *Codon–anticodon interaction.* Fig. 14 shows the anticodons of the different t-RNAs of known structure. As each amino acid can be specified by several codons, there are two possibilities; one t-RNA recognizes only one codon (this implies that there would be several different t-RNAs for each amino acid), or one t-RNA recognizes all the triplets coding for its specific amino acid. Several investigations have proved the second hypothesis to be correct: one t-RNA recognizes all the codons corresponding to its specific amino acid provided that they differ only in the third nucleotide of the triplet (Söll *et al.*, 1966, 1967 Söll and RajBhandary, 1967; Mirzabekov *et al.*, 1967). Crick (1966) has systematically explored the codon–anticodon pairings. He proposed his wobble hypothesis, which could explain the general nature of the genetic code degeneracy.

| t-RNA | Codon<br>→ | Anticodon<br>← |
|---|---|---|
| Alanine | U<br>GCC<br>A | CGI |
| Serine I and II | U<br>UCC<br>A | AGI |
| Valine | U<br>GUC<br>A | CAI |
| Phenylanine | UUU<br>C | AA2′OMeG |
| Tyrosine | UAC<br>U | A$\phi$G |
| Isoleucine | A<br>UAC<br>U | UA |

FIG. 14. Codon–anticodons of known yeast t-RNAs.

Changes in the translation of the code have been obtained by *in vitro* chemical modification of t-RNA (Bakes *et al.*, 1965, 1968). The suppressor mutants are also good examples, in *E. coli*, of the changes in the translation induced by the modification of the anticodon (Goodman *et al.*, 1968). Some alterations of the codon–anticodon recognition have been observed after modification of bases not contained in the anticodon. Fittler and Hall (1966) have shown that oxidation of the isopentenyladenine adjacent to the anticodon leads to an important decrease of the fixation of seryl-t-RNA to the m-RNA-ribosome complex.

(ii) *Binding of aminoacyl-t-RNA to ribosomes.* All the mechanisms involved in the protein synthesis on ribosomes have been reviewed by Lengyel and Söll (1969). In yeast, very little is known about the initiation, elongation and termination of polypeptide chains. Richter and

Klink (1967) have isolated from 105 kg of yeast a supernatant containing three transfer factors active in amino acid polymerization.

Ayuso and Heredia (1968), using a yeast hybrid (*Sacch. fragilis* × *Sacch. dobzanskii*), have studied the requirements of the enzymic binding of phenylalanyl-t-RNA to purified yeast ribosomes. At low magnesium concentrations, the binding is dependent upon one of the yeast supernatant transfer factors (Factor A). This reaction requires GTP. In addition to this enzymic type of binding, phenylalanyl-t-RNA is also bound to yeast ribosomes in the absence of GTP and Factor A if the magnesium concentration is about 20 mM. In contrast, the binding of *N*-acetylphenylalanyl-t-RNA requires GTP and other factors which are possibly bound to the ribosomes.

*d. Binding site for t-RNA-CCA-pyrophosphorylase.* The terminal triplet CCA can be bound and split by a t-RNA-CCA-pyrophosphorylase, thus:

$$\text{t-RNA-CCA} + 3\ \text{PP}_i \leftrightharpoons \text{t-RNA} + \text{ATP} + 2\ \text{CTP}$$

Little work has been done on the binding site of this enzyme on t-RNAs. It seems logical to assume that it would recognize an identical structure on all t-RNAs. It was therefore of interest to see if it was the GTψC sequence common to all t-RNAs. Geige *et al.* (1969a, b) have shown that the 5-fluorouracil incorporated in t-RNA chains replaces uridine, pseudouridine and thymidine. This incorporation has no effect on the CCA fixation. It seems, then, that the GTψC sequence is not implicated.

Yoshida *et al.* (1966) and Harriman and Zachau (1966) have shown that the acceptor site for amino acids and the fixation site of the t-RNA-CCA-pyrophosphorylase are not identical. The amino acid fixation site is more sensitive to iodination and ultraviolet irradiation than the CCA fixation.

## D. MESSENGER RIBONUCLEIC ACID

### 1. *Characterization and Base Composition*

Volkin and Astrachan (1957) observed in phage-infected *E. coli* a RNA fraction which had none of the properties of t-RNA or ribosomal RNA (r-RNA). Later, Ycas and Vincent (1960) in yeast, and Gros *et al.* (1961) in *E. coli*, observed an RNA fraction which seemed to meet all the requirements proposed by Jacob and Monod (1961) for messenger RNA (m-RNA).

In yeast, the m-RNA has recently been accurately characterized by Hartlief and Koningsberger (1968). These authors, using protoplasts of *Sacch. carlsbergensis*, showed that the rapidly-labelled RNA consisted mainly of m-RNA. Their criteria were: (i) the heterogeneous size distribution, as shown by the radioactivity pattern obtained after sucrose gradient centrifugation; (ii) the disappearance of most of this radioactivity when the protoplasts were incubated with $^{14}$C-uracil in the presence of actinomycin D; (iii) the sensitivity of this RNA to ribonuclease; and (iv) the base composition of this RNA, which is intermediate between yeast DNA and yeast bulk RNA. Table XVIII summarizes the base

TABLE XVIII. *Base Composition of Rapidly-labelled RNA Compared with Yeast DNA and Yeast Bulk RNA*

| Nucleic acid | Proportion of component (moles per cent) | | | |
|---|---|---|---|---|
| | A | U (T) | G | C |
| Yeast DNA[a] | 31·6 | 34·5 | 16·7 | 17·2 |
| Total yeast RNA[a] | 25·4 | 26·0 | 28·6 | 20·0 |
| 3-min labelled RNA[b] | 31 | 32 | 16 | 21 |
| 5-min labelled RNA[a] | 32·6 | 25·2 | 21·9 | 20·9 |
| 6-min labelled RNA[b] | 28 | 31 | 18 | 22 |

[a] from Fukuhara (1967); [b] from Hartlief and Koningsberger (1968).

composition of rapidly-labelled RNA, as given by two research groups (Fukuhara, 1967; Hartlief and Koningsberger, 1968) using different yeast sources. It can be seen that they are in good agreement and very near the DNA composition.

Hartlief and Koningsberger (1968) isolated and partially characterized the m-RNA associated with polysomes which presumably synthesize glucosidase in *Sacch. carlsbergensis*. This is of great interest, but needs further investigation. Also, using an immunological method, Warren and Goldthwait (1962) isolated a ribosomal fraction associated with triosephosphate dehydrogenase.

## 2. *Correspondence between Genes and Proteins*

The linear sequence of nucleotides in a DNA cistron specifies the linear sequence of amino acids in a polypeptide chain, the m-RNA carrying the information from DNA to the ribosomes where proteins are synthesized. Synchronously-dividing yeast cultures have been a very useful tool to answer the question whether the process of transcription and translation of genetic information is ordered. If m-RNA is unstable, and is produced by ordered transcription of cistrons, then in synchronously-dividing cells a step-wise production of a given enzyme would be expected. Gorman *et al.* (1964) observed a step-wise synthesis of $\beta$-glucosidase in a synchronously-dividing culture of a hybrid yeast

TABLE XIX. *Enzyme Synthesis in Yeast Synchronous Cultures (reproduced from Mitchison, 1969, by permission of the author and the editors of* Science, *copyright 1969 by the American Association for the Advancement of Science)*

| Enzyme | Pattern | Ref. |
|---|---|---|
| *Saccharomyces cerevisiae* | | |
| Protease | P | a |
| Peptidase | P | a |
| α-Glucosidase | S | b, c, d |
| β-Glucosidase | S(2 and 3) | d |
| Sucrase | S(2) | b |
| Alkaline phosphatase | S(2) | b |
| Histidinol dehydrogenase | S | c |
| Orotidine-5′-phosphate decarboxylase | S | c |
| Aspartokinase | S | c |
| Phosphoribosyl-ATP pyrophosphorylase | S | c |
| Threonine deaminase | S | c |
| Arginosuccinase | S | c |
| Saccharopine dehydrogenase | S | c |
| Saccharopine reductase | S | c |
| Alcohol dehydrogenase | S | e |
| Hexokinase | S | e |
| Glyceraldehyde-3-phosphate dehydrogenase | S | e |
| DNA polymerase | P | f |
| *Saccharomyces dobzhanskii* | | |
| β-Glucosidase | S | b |
| *Saccharomyces dobzhanskii* × *Saccharomyces fragilis* | | |
| α-Glucosidase | S | d |
| β-Glucosidase | S(2) | d |
| Alkaline phosphatase | S(2) | d |
| *Schizosaccharomyces pombe* | | |
| Aspartate transcarbamylase | S | g |
| Ornithine transcarbamylase | S | g |
| Tryptophan synthetase | S | h |
| Alcohol dehydrogenase | S | h |
| Homoserine dehydrogenase | S | h |
| Alkaline phosphatase | C(L) | i |
| Acid phosphatase | C(L) | i |
| Sucrase | C(L) | i |
| Maltase | (C?) | g |

S = step enzyme (numerals after S indicate more than one step per cycle); P = peak enzyme; C(L) = continuous linear enzymes.

*References:* a, Silven *et al.* (1959); b, Halvorson *et al.* (1966); c, Tauro *et al.* (1968); d, Tauro and Halvorson (1966); e, Eckstein *et al.* (1966); f, Eckstein *et al.* (1967); g, Bostock *et al.* (1966); h, A. A. Robinson, cited by Mitchison (1969): i, Mitchison (1969).

(*Sacch. fragilis* × *Sacch. dobzhanskii*) and a step-wise synthesis of α-glucosidase, invertase and alkaline phosphatase in a synchronously-dividing culture of *Sacch. cerevisiae.*

Tauro and Halvorson (1966) and Tauro *et al.* (1968) studied the effect of gene position on the timing of enzyme synthesis, and the influence of the gene position in relation to the centromere. They concluded that the periodic synthesis of enzymes observed in synchronous cultures of *Sacch. cerevisiae* was the result of an ordered process of transcription of the various structural genes. Their data are consistent with a unidirectional model in which transcription starts at the end of the chromosome, but they cannot eliminate a regional transcription of parts of chromosomes.

In synchronously-dividing cultures of *Schizosacch. pombe,* Bostock *et al.* (1966) showed that both periodic (ATCase and OTCase) and continuous (alkaline phosphatase, invertase and maltase) synthesis occurs. In this respect *Schizosacch. pombe* resembles bacteria. Recently, Mitchison (1969) reviewed the various patterns of enzyme synthesis in different yeasts dividing synchronously (see Table XIX). He pointed out that the transcription of the genes, in a sequence which corresponds to their order on the chromosomes, does not fit all the results. Nor does the theory primarily developed for procaryotes, mainly that the periods of synthesis are due to oscillations in the negative feedback system of an enzyme. These oscillations would be in phase with the cell cycle.

## E. RIBOSOMES

### 1. *Characterization*

It is now well established that protein synthesis occurs on polyribosomes in microbial and mammalian cells. Marcus *et al.* (1963, 1965, 1967) were the first to demonstrate the occurrence of ribosomes in rapidly-dividing yeast cells (*Sacch. dobzhanskii* × *Sacch. fragilis*). The ribosomes were found in clusters of varying size, called polyribosomes, apparently bound together by a strand of m-RNA. They resolved the polyribosome classes by exponential sucrose density gradients. Table XX gives the sedimentation values for ribosomes and polyribosomes. It was postulated that the different peaks consisted of dimers, trimers, and so on. This was confirmed by electron microscopy (Marcus *et al.*, 1967). A careful study showed that the polyribosomes are disaggregated to smaller units by long-term grinding, pressure-cell disruption, and incubation with dilute ribonuclease. The presence of dimers or trimers in the extract is probably due to the disruption of larger polyribosomes during preparation, as a constant amino acid incorporation per ribosome is found in cell-free systems. The larger polyribosomes contain messages sufficiently large

TABLE XX. *Sedimentation Values of Yeast Ribosomes, Polyribosomes and Ribosomal Subunits (reproduced from Marcus* et al., *1967, with permission)*

| Unit | s Value[a] |
|---|---|
| Subunit a | 40S |
| Subunit b | 60 |
| 1 | 80 |
| 2 | 110 |
| 3 | 140 |
| 4 | 164 |
| 5 | 186 |
| 6 | 204 |
| 7 | 226 |
| 8 | 244 |
| 9 | 280 |
| 10 | 297 |
| 11 and larger | 300 and greater |

[a] s values obtained from sucrose density gradients and in the Beckman Model E analytical ultracentrifuge.

to code for protein subunits. Polyribosomes containing more than ten ribosomes would then code for high molecular weight proteins.

A more interesting suggestion, made by Marcus *et al.*, is that the m-RNA binding these large polyribosomes may contain information for a number of proteins. Then it would be likely that polycistronic m-RNA exists in yeast; in fact, linkage has been shown between: (i) three genes in the histidine biosynthetic pathway; (ii) four genes in the aromatic amino acid pathway; (iii) two genes in the uracil biosynthesis, etc. (see appropriate parts of Sections II and III).

Using *Sacch. fragilis*, Cotter *et al.* (1967) studied the internal organization of the ribosome. Ribosomes have a sedimentation coefficient of 80·4s and a molecular weight of $3{\cdot}95 \times 10^6$ daltons. The study has shown that the conformation of r-RNA is similar in the free state and in the ribosome. It contains about 60% of paired bases in short helical segments; the ribosomal proteins are not associated with the double helical parts of the r-RNA, and may be packed into the nonhelical loops and associated in some form of quaternary structure. The surface of a ribosome consists chiefly of RNA and not of protein, so, by inference, the surface would be constituted of the double helical parts projecting outwards.

2. *Free Ribosomal RNA*

Free r-RNA has been isolated from ribosomes of *Sacch. lactis* and from the hybrid (*Sacch. dobzhanskii* × *Sacch. fragilis*), using detergents and phenol extractions, by Buening and Bock (1967). They dissociated the polynucleotide fragments held by non-covalent forces by heating at 80°C for 1 min. More than 60% of the 26S RNA survived this treatment. This r-RNA consisted of two components, 17S and 26S, the molecular weights of which have been determined in equilibrium sedimentation experiments (see Table XXI).

TABLE XXI. *Molecular Weights of 17S and 26S Ribosomal Potassium Ribonucleates (reproduced from Buening and Bock, 1967, with permission)*

| Parameter | 26S | 17S |
|---|---|---|
| Extrapolated experimental molecular weight | $1{\cdot}1 \times 10^6$ | $0{\cdot}74 \times 10^6$ |
| Nucleotides per strand | 3,100 | 2,100 |
| Maximal contamination by sister RNA component (%) | 10 | 15 |
| Corrected molecular weight | $1{\cdot}2 \times 10^6$ | $0{\cdot}65 \times 10^6$ |
| Nucleotides per strand | 3,300 | 1,800 |

## F. INHIBITION OF PROTEIN SYNTHESIS

### 1. *Mutants*

Hartwell and McLaughlin (1968b) investigated 16 of their temperature-sensitive mutants of *Sacch. cerevisiae* These can be grouped into four classes, on the basis of glucose incorporation and polyribosome content at the restrictive temperature as compared with the parental strain. The properties of the mutants of the first class are consistent with a defect in the initiation of polypepide chains or in RNA synthesis. Mutants of the second class probably have a defect in the elongation of the polypeptide chain. The third class corresponds to a defect in the energy metabolism, and the fourth to lesions in the membrane. From the mutants of the first class, the authors selected one which is apparently defective in the initiation of protein synthesis (Hartwell and McLaughlin, 1969). The existence of this mutant shows that in yeast, as in bacteria, there must be a process unique for the initiation of polypeptide chains. This attack on the problem seems promising.

### 2. *Inhibition of Protein Synthesis by Cycloheximide*

Cycloheximide is an antibiotic produced by *Streptomyces griseus* (Whiffen *et al.*, 1946) which inhibits growth of many different species of yeast, and it is known that *Sacch. fragilis* is resistant to it (Siegel and Sisler, 1965). Various workers have studied the effect of this inhibitor on yeast metabolism (Latuasan and Berends, 1958; Kerridge, 1958; Tsukada *et al.*, 1962; Siegel and Sisler, 1964). Siegel and Sisler (1965), in their *in vitro* experiments on *Sacch. pastorianus*, showed that cycloheximide inhibits protein synthesis in a cell-free system. Inhibition of protein synthesis probably takes place at the level of the transfer of activated amino acids on the ribosomes. However, this inhibition occurs *in vitro* only when a supernatant fraction is added during incubation. The protein-synthesizing system from the cycloheximide-resistant yeast *Sacch. fragilis* was resistant, *in vitro*, to the antibiotic. Combinations of the ribosomes and supernatant enzymes from the resistant and susceptible yeasts have shown that resistance is determined by ribosomes, and not by the enzymes in the supernatant fraction.

3. *Action of Antibiotics on Yeast*

The antibacterial antibiotics chloramphenicol, lincomycin, erythromycin, carbomycin, spiramycin and oleandomycin act by inhibiting bacterial protein synthesis. The same drugs inhibit the mitochondrial protein-synthesizing system in *Sacch. cerevisiae*, both *in vivo* and *in vitro*, without affecting the cytoplasmic system (Clark-Walker and Linnane, 1966; Lamb *et al.*, 1968). Conversely, cycloheximide is ineffective on mitochondrial systems (Siegel and Sisler, 1965). These observations lead to many experiments on yeast mitochondria.

Other antibiotics, such as puromycin, proflavin and nystatin, inhibit enzyme synthesis (Holzer *et al.*, 1964), as measured by the activity of the NAD-dependent glutamic dehydrogenase in derepression studies. An earlier study on nystatin showed that this antibiotic becomes bound to the cell membrane (Lampen *et al.*, 1962).

## V. Nucleic Acid Biosynthesis

In Section IV (p. 388) we have seen that several different RNAs are involved in the biosynthesis of protein—transfer RNA, messenger RNA and ribosomal RNA. This heterogeneity of cellular RNA contrasts with DNA, which can be regarded as a single biochemical unit. It seems logical, then, that DNA synthesis, in all organisms studied so far, should proceed by a single mechanism, whereas several enzyme systems are apparently involved in the synthesis of RNA. Several reviews on nucleic acid synthesis in bacteria have appeared (e.g. Grunberg-Manago, 1962).

### A. ENZYMIC SYSTEMS FOR POLYNUCLEOTIDE SYNTHESIS

1. *Synthesis of DNA*

In extracts of growing cells of *Sacch. cerevisiae*, Eckstein *et al.* (1967) detected an enzyme having all the characteristics of a DNA polymerase. Working with a crude extract, these authors determined several properties of the enzyme: the incorporation of ($^3$H) ATP depends on added primer DNA and on the presence of all four deoxynucleotides. The synthesized material is susceptible to DNAase and resistant to RNAase. Using synchronized cells of *Sacch. cerevisiae*, the same authors found that DNA polymerase activity oscillates during the cell cycle (peak enzyme, see Table XIX); maximal activity is found just before the onset of DNA replication. They showed that this oscillation cannot be due to the cyclic influence of effectors present in extracts, but that it is the DNA polymerase activity itself which fluctuates. They think that degradation and resynthesis, or transformation to an inactive "zymogen", are possible mechanisms of the oscillation of DNA polymerase activity.

More recently, Iwashima and Rabinowitz (1969) separated both a mitochondrial and a soluble DNA polymerase; they purified the soluble enzyme 150-fold, and the mitochondrial 50-fold. Enzymic properties of the two enzymes are similar. Study of the template showed that the activity is highest with *E. coli* DNA and then, in decreasing order, come salmon sperm DNA and calf thymus DNA. The enzyme activity is higher with a native template than with a denatured one. Density gradient centrifugation shows two components in the supernatant DNA polymerase; the mitochondrial enzyme sediments with the slower component of the supernatant DNA polymerase. It seems, then, that the two enzymes have different sedimentation coefficients, the soluble one being slightly contaminated by mitochondrial DNA polymerase.

### 2. *Synthesis of RNA*

*a. RNA polymerase.* Frederick *et al.* (1969) isolated and purified a DNA-dependent RNA polymerase from *Sacch. cerevisiae* 3,000-fold. The yeast RNA polymerase resembles the polymerases from *E. coli* and other sources (Grunberg-Manago, 1962) in its absolute requirement for a DNA primer, a divalent cation and all four nucleoside triphosphates. It differs fundamentally in its marked preference for a denatured template with a wide variety of DNA preparations, the synthesis of RNA being more than 10-fold greater with a denatured than with a native template. In contrast, also, with the *E. coli* enzyme, RNA chains formed with the yeast RNA polymerase using denatured calf thymus and T2 templates have average chain lengths of 1,500 to 2,000 nucleotides, which is much longer than the chains obtained with a denatured template and the *E. coli* enzyme (Table XXII). As it was not possible to measure accurately initiation and average lengths of RNA chains in the yeast system with native DNA because of the limited RNA synthesis under these conditions, it was impossible to determine whether initiation with yeast polymerase occurs selectively with purines. With a denatured DNA template, a preponderance of GTP over ATP initiation was observed (Table XXII). Yeast RNA polymerase catalyses the

TABLE XXII. *Comparison of Chain Initiation and Average Chain Length with RNA Polymerase from* Saccharomyces cerevisiae *and* Escherichia coli (*reproduced from Frederick* et al., *1969, with permission*)

| Organism | DNA template (denatured) | $\gamma^{32}$-P ATP incorporated ($\mu$moles) | $\gamma^{32}$-P GTP incorporated ($\mu$moles) | RNA synthesis ($\alpha^{32}$-P UTP incorporated) ($\mu$moles) | Average chain length |
|---|---|---|---|---|---|
| *Saccharomyces cerevisiae* | Calf thymus | 0·69 | 5·2 | 11,600 | 1,990 |
| | $T_2$ | 0·37 | 4·2 | 6,700 | 1,450 |
| *Escherichia coli* | Calf thymus | 1·1 | 5·1 | 1,340 | 210 |
| | $T_2$ | 1·0 | 2·0 | 720 | 240 |

formation of polyriboadenylate in a reaction analogous to that carried out by the *E. coli* polymerase (Stevens, 1964).

The authors of this study discuss the preferential use of a denatured template by yeast RNA polymerase. One of the hypotheses presented is the possible existence of a second RNA polymerase preferentially utilizing native DNA. However, attempts to detect such an enzyme have failed. Another hypothesis is that some factors might combine with native DNA to allow the enzyme to utilize a helical template. Until now, no such agents have been shown to exist.

*b. Other enzymes involved in polynucleotide synthesis.* We will only consider the t-RNA-CCA-pyrophosphorylase which binds or splits the terminal CCA triplet of a t-RNA which has been reviewed in the section dealing with t-RNA (p. 396). Some DNA-independent enzymes have been shown to exist in *E. coli*; for example, the polynucleotide phosphorylase. The physiological role of this enzyme is still unknown (Grunberg-Manago, 1962). To our knowledge it has not yet been found and studied in yeast.

## B. RIBOSOMAL RNA SYNTHESIS

### 1. *Redundancy of r-RNA Genes*

Retel and Planta (1968), using the hybridization technique of Gillespie and Spiegelman (1965) on a system isolated from *Sacch. carlsbergensis*, showed that about 2% of yeast DNA is complementary to r-RNA. There is, moreover, a great structural similarity between the sites for the 17S and 26S components of r-RNA. Using the same technique on a system extracted from *Sacch. cerevisiae*, Schweizer *et al.* (1969) obtained similar results, and concluded that there were 140 cistrons coding for r-RNA in this organism.

The high proportion of yeast DNA coding for r-RNA may explain the 10-fold higher RNA/DNA ratio in this organism compared to others (Fukuhara, 1967; Leslie, 1955). This high redundancy of r-RNA genes may be a useful means for the cell to meet the demand for ribosomes during periods of intensive protein synthesis. It might also correspond to a mechanism of resistance to mutagenic effects on basic metabolic functions.

### 2. *Existence of a Precursor in the Synthesis of r-RNA*

Shortman and Fukuhara (1963), working on respiratory adaptation in *Sacch. cerevisiae*, showed that a special class of RNA is synthesized during induction of respiratory enzymes in non-growing cells. This species of RNA was different from r-RNA and t-RNA in its molecular size. Its base composition was different from bulk yeast RNA and bulk yeast DNA. They were able to demonstrate that this high-turnover RNA was transformed into stable RNA more quickly in oxygen-induced than in non-induced cells.

Later, Retel and Planta (1967) showed that in *Sacch. carlsbergensis*, as in animal cells, the initial event in the biosynthesis of r-RNA appears to be the formation, in the nucleus, of a species of RNA sedimenting at approximately 45S. This RNA is then transformed into the 18S and 26S species by intermediates yet unknown. These authors were very careful to differentiate between ribosomal precursor RNA and messenger RNA, as the two species are rapidly-labelled nuclear RNA. As we will see later in this section, the same precursor for the 18S and 26S r-RNAs is not a commonly accepted view, and some authors have formulated other hypotheses.

Mayo *et al.* (1968), using techniques involving fractionation of RNA by phenol at different temperatures, found that, in *Sacch. carlsbergensis* in the presence of cycloheximide, small amounts of 28S RNA are synthesized. No formation of 18S RNA occurs. Their results seem to contradict the view (Retel and Planta, 1968; Perry, 1966) that 18S and 28S r-RNAs arise from a common precursor in a process where 18S is liberated together with 35S and 28S RNAs. So, the situation could be as in rat liver nuclei (Muramatsu *et al.*, 1966) or in HeLa Cells (Zimmerman and Holler, 1967). Two high molecular weight precursors are present in yeast: the 45S species reported by Retel and Planta (1967) which is converted to 35S and 28S, and another precursor for the 18S which can be formed only if protein synthesis occurs. The other alternative, proposed by Mayo *et al.* (1968), is that, indeed, the 28S and 18S arise from the same precursor, but only the 28S is split off in the absence of protein synthesis, leaving the 18S as part of the precursor molecule.

### 3. *Action of Cycloheximide*

Fukuhara (1965) studied the *in vivo* effect of cycloheximide on the synthesis of RNA and protein during induction of respiratory enzymes (occurring without growth). At low concentrations of cycloheximide protein synthesis is totally inhibited, whereas RNA synthesis is only slightly depressed. In these conditions there is an accumulation of a high-turnover RNA sedimenting at 16S, and accompanied by other cytoplasmic species.

De Kloet (1965), working with *Sacch. carlsbergensis*, reported that during the incubation of cells or protoplasts in the presence of cycloheximide, there was an accumulation of a DNA-like base composition RNA with sedimentation properties different from r-RNA, and certainly heavier than the heaviest r-RNA fraction. The differences between the results of Fukuhara and De Kloet may be due to the different yeasts used by these authors or to their different RNA extraction methods.

In a more detailed study, De Kloet (1966) reported that in cells of *Sacch. carlsbergensis* exposed to cycloheximide the synthesis of 16S and 23S r-RNAs is stopped, and that there is an accumulation of two RNA species in the cell—t-RNA and a high molecular weight RNA, the base composition of which is intermediate between DNA and r-RNA. This RNA contains less methylated bases than r-RNA, and sediments faster than the 23S r-RNA component. It is not stable, as the effect of cycloheximide on protein and RNA synthesis can be abolished by washing the cells.

It seems, as de Kloet pointed out, that the properties of this high molecular RNA species are those usually ascribed to polycistronic m-RNA. The possibility thus exists that the antibiotic inhibits the r-RNA synthesis completely while allowing m-RNA and t-RNA synthesis to continue for some time. As a matter of fact, there are strong indications that m-RNA does indeed accumulate, but it has not yet been shown that it is functional in the synthesis of proteins. In this case, the observed inequality of the base composition of RNA formed in the presence of cycloheximide and DNA might reflect an unequal base composition for the two DNA strands, one of which is copied only into RNA.

Recently Taber and Vincent (1969a, b) have reported the existence of a r-RNA (38S) precursor in *Schizosacch. pombe*. They appear to have convincing evidence that it is indeed the precursor of the r-RNA components. Especially, their hybridization experiments show that the cistrons coding for the 38S molecule are homologous to those coding for r-RNA. This accumulation of a 38S RNA, in the absence of protein synthesis, strongly suggests that a post-transcriptional control mechanism exists in the maturation of r-RNA. This is in contrast to bacteria, in which the two types of r-RNA (16S and 23S) appear to be synthesized directly (Osawa, 1968).

### C. METHYLATION OF NUCLEIC ACIDS

Nucleic acids from all organisms examined so far contain a number of methylated bases in addition to the four main bases of their primary structure. Yet, no monomeric methylated precursors were ever found within any tissue, except thymidylic acid. Now, it is well documented that the methylations catalysed by specific enzymes occur on previously synthesized nucleic acids, and that *S*-adenosylmethionine is the methyl donor (Borek and Srinavasan, 1966).

In yeast t-RNA, twelve different methylated bases have been detected (Hall, 1965; Littlefield and Dunn, 1958; Dunn, 1963). The occurrence and distribution of these components in t-RNA seem to vary from one strain to another. Björk and Svensson (1969) fractionated the t-RNA methylases from *Sacch. cerevisiae* with ammonium sulphate, Sephadex G-150 and hydroxyapatite, and obtained eight separate fractions. They could demonstrate activities for the synthesis of 1-methyladenine, 5-methylcytosine, 5-methyluracil (thymine) and for four methylated guanines. Three of these fractions possessed activity for 5-methyluracil only, and were apparently free from other methylase activities. The other fractions were heterogeneous with respect to the products formed.

Svensson *et al.* (1969) studied some of the properties of these enzyme fractions. The three fractions catalysing the formation of 5-methyluracil were shown to possess different properties. They methylate *E. coli* t-RNA to different extents; they produce 5-methyluracil in different positions on the t-RNA, and methylate the t-RNA from *Aerobacter aerogenes* and wheat germ differently. One of the three fractions turned out to contain at least two enzymes which do not select the same sites for methylation on the t-RNAs. So, yeast contains at least three or four different t-RNA methylases catalysing the formation of 5-methyluracil which differ in their specificity. Moreover, these authors have strong genetic and biochemical evidence that there is only one t-RNA methylase which catalyses the formation of $N^2$-dimethylguanine in *Sacch. cerevisiae*.

Kjellin-Sträby and Phillips (1968) have described the accumulation, in *Sacch. cerevisiae*, of methyl-deficient t-RNA during a period of high metabolic activity preceding the end of the logarithmic growth phase. It is unlikely that this phenomenon would be due to a lack of *S*-adenosylmethionine, unless this component is physically compartmentalized. So, for the time being, Kjellin-Sträby and Phillips do not ascribe any significance to their observation.

As yet, only speculations have been made about the function of methyl groups in t-RNA. Their real significance is not yet known, In *E. coli* B, Pillinger *et al.* (1969) showed that hypermethylation of the total t-RNA resulted in a decrease in the ability to accept 19 amino acids. The acceptance of cysteine is enhanced in the hypermethylated t-RNA; but this t-RNA had been chemically hypermethylated and the effects observed could be a result of the changes occurring in the secondary or tertiary structure.

At least some of these methylations are indispensible for correct cell metabolism. This is deduced from the fact that a mutant of *Sacch. cerevisiae*, lacking the $N^2$-dimethylguanine methylating enzyme (Phillips and Kjellin-Sträby, 1967), requires methionine for growth. In these conditions it contains a low but not negligible amount of $N^2$-methyl- and $N^2$-dimethylguanine in its t-RNA. The methionine requirement might then be explained if the mutation resides, not in a complete loss of the enzyme, but in a change in its affinity for *S*-adenosylmethionine; or if, in the presence of an increased *S*-adenosinemethionine concentration (as a result of the addition of exogenous methionine), the methylation of guanine residues on the t-RNAs can be catalysed by the other methylating enzymes.

## References


Agarwal, K. L., Büchi, H., Caruthers, M. H., Gupta, N., Khorana, H. G., Kleppe, K., Kumar, H., Ohtsuka, E., RajBhandary, V. L., Vandesande, J. H., Sgaramella, V., Weber, H. and Yamada, T. (1970). *Nature, Lond.* **227**, 27–34.

Ahmed, A. (1968). *Molec. gen. Genetics* **103**, 185–193.

Allende, J. E. and Allende, C. C. (1964). *Biochem. biophys. Res. Commun.* **16**, 342–346.

Anfinsen, C. B. (1967). "Harvey Lectures", Ser. 61, p. 95. Academic Press, New York.

Arst, H. N. (1968). *Nature, Lond.* **219**, 268–270.

Asahi, T., Bandurski, R. S. and Wilson, L. G. (1961). *J. biol. Chem.* **236**, 1830–1835.

Ayuso, M. S. and Heredia, C. (1968). *Eur. J. Biochem.* **7**, 111–118.

Baich, A. and Pierson, D. J. (1965). *Biochim. biophys. Acta* **104**, 397–404.

Baiev, A. A., Venkstern, T. V., Mirzabekov, A. D., Krutilina, A. I., Li, L. and Axelrod, V. D. (1967). *Molek. Biologiya* **1**, 754–766.

Bakes, J., Befort, N., Weil, J. H. and Ebel, J. P. (1965). *Biochem. biophys. Res. Commun.* **19**, 84–88.

Bakes, J., Ehresman, C., Befort, N., Weil, J. H. and Ebel, J. P. (1968). *Eur. J. Biochem.* **4**, 490–495.

Bechet, J. and Wiame, J. M. (1965). *Biochem. biophys. Res. Commun.* **21**, 226–234.

Bechet, J., Wiame, J. M. and Grenson, M. (1962). *Archs int. Physiol. Biochim.* **70**, 564–565.

Bechet, J., Grenson, M. and Wiame, J. M. (1970). *Eur. J. Biochem.* **12**, 31–39.

Berg, P. (1956). *J. biol. Chem.* **222**, 1025–1034.

Bergquist, P. L., Burns, D. J. W. and Plinston, C. A. (1968). *Biochemistry, N.Y.* **7**, 1751–1760.

Bernhardt, W., Zink, M. and Holzer, H. (1966) *Biochim. biophys. Acta* **118**, 549–555.

Bhattacharjee, J. K. and Strassman, M. (1967). *J. biol. Chem.* **242**, 2542–2546.

Björk, G. R. and Svensson, I. (1969). *Eur. J. Biochem.* **9**, 207–215.

Black, S. (1965). *A. Rev. Biochem.* **32**, 399–418.

Black, S. and Wright, N. G. (1955a). *J. biol. Chem.* **213**, 27–38.

Black, S. and Wright, N. G. (1955b). *J. biol. Chem.* **213**, 39–50.

Black, S. and Wright, N. G. (1955c). *J. biol. Chem.* **213**, 51–60.

Borek, E. and Srinivasan, P. R. (1966). *Ann. Rev. Biochem.* **35**, 275–298.

Borstel, R. C. von (Ed.) (1966). Microbial Genetics Bulletin, No. 25. Oak Ridge, U.S.A.

Bostock, C. J., Donachie, W. D., Masters, M. and Mitchison, J. M. (1966). *Nature, Lond.* **210**, 808–810.

Botsford, J. L. and Parks, L. W. (1967). *J. Bact.* **94**, 966–971.

Botsford, J. L. and Parks, L. W. (1969). *J. Bact.* **97**, 1176–1183.

Bourgeois, C. (1969). *Bull. Soc. Chim. biol.* **51**, 935–949.

Bourgeois, C. (1971). *Eur. J. Biochem.* In press.

Bowers, W. D., McClary, D. O. and Ogur, M. (1967). *J. Bact.* **94**, 482–484.

Broquist, H. P. and Trupin, J. S. (1966). *A. Rev. Biochem.* **35**, 231–274.

Brunner, A. and de Robichon-Szulmajster, H. (1969). *FEBS Letters* **5**, 141–144.

Brunner, A., Devillers-Mire, A. and de Robichon-Szulmajster, H. (1969). *Eur. J. Biochem.* **10**, 172–183.
Buening, G. and Bock, R. M. (1967). *Biochim. biophys. Acta* **149**, 377–386.
Buestein, H. G., Allende, C. C., Allende, J. E. and Cantoni, G. L. (1968). *J. biol. Chem.* **243**, 4693–4699.
Burns, V. W. (1964). *Biophys. J.* **4**, 151–166.
Bussey, H. and Umbarger, H. E. (1969). *J. Bact.* **98**, 623–628.
Cabet, D., Gans, M., Motta, R. and Prévost, G. (1967). *Bull. Soc. Chim. biol.* **49**, 1537–1543.
Cennamo, C., Boll. M. and Holzer, H. (1964). *Biochem. Z.* **340**, 125–145.
Changeux, J. P. (1963). *Cold Spring Harb. Symp. quant. Biol.* **28**, 497–504.
Changeux, J. P. (1964a). *Bull. Soc. Chim. biol.* **46**, 927–946.
Changeux, J. P. (1964b). *Bull. Soc. Chim. biol.* **46**, 947–961.
Changeux, J. P. (1964c). *Bull. Soc. Chim. biol.* **46**, 1151–1173.
Cherest, H., Eichler, F. and de Robichon-Szulmajster, H. (1969). *J. Bact.* **97**, 328–336.
Cherest, H., Talbot, G. and de Robichon-Szulmajster, H. (1969). *J. Bact.* **102**, 448–461.
Clark-Walker, G. D. and Linnane, A. W. (1966). *Biochem. biophys. Res. Commun.* **25**, 8–13.
Clavilier, L., Luzzati, M. and Slonimski, P. P. (1960). *C. r. Séanc. Soc. Biol.* **54**, 1970–1974.
Cohen, G. N. (1968). "The Regulation of Cell Metabolism", Hermann, France.
Cotter, R. I., McPhie, P. and Gratzer, W. B. (1967). *Nature, Lond.* **216**, 864–868.
Cramer, F., Doepman, H., de Haar, F., Schlimme, E. and Seidel, H. (1968). *Proc. natn. Acad. Sci. U.S.A.* **61**, 1384–1391.
Crick, F. H. C. (1966). *J. molec. Biol.* **19**, 548–555.
Daniel, V. and Littauer, U. Z. (1963). *J. biol. Chem.* **238**, 2102–2112.
Davis, R. H. (1965). *Genetics, Princeton* **107**, 54–68.
Deken, R. H. de (1962). *Biochem. biophys. Res. Commun.* **8**, 462–466.
Deken, R. H. de (1963). *Biochim. biophys. Acta* **78**, 606–616.
Delavier-Klutchko, C. and Flavin, M. (1965). *J. biol. Chem.* **240**, 2537–2549.
Doctor, B. P., Fuller, W. and Webb, N. L. (1969). *Nature, Lond.* **221**, 58–59.
Dorfman, B. (1963). *Genetics, Princeton,* **48**, 887.
Dorfman, B.-Z. (1969). *Genetics, Princeton* **61**, 377–389.
Doy, C. H. (1968a). *Biochim. biophys. Acta* **151**, 293–295.
Doy, C. H. (1968b). *Rev. pure appl. Chem.* **18**, 41–78.
Doy, C. H. and Cooper, J. M. (1966). *Biochim. biophys. Acta* **127**, 302–316.
Duerre, J. A. (1968). *Archs Biochem. Biophys.* **124**, 422–430.
Dunn, D. B. (1963). *Biochem. J.* **86**, 14P.
Duntze, W. and Manney, T. R. (1968). *J. Bact.* **96**, 2085–2093.
Duval, J. and Ebel, J. P. (1966). *C. r. hebd. Séanc. Acad. Sci., Paris* **263**, 1773–1776.
Duval, J. and Ebel, J. P. (1967). *Bull. Soc. Chim. biol.* **49**, 1665–1678.
Ebel, J. P. (1968). *Bull. Soc. Chim. biol.* **50**, 2255–2276.
Ebel, J. P., Weil, J. M., Rether, B. and Heinrich, J. (1965). *Bull. Soc. Chim. biol.* **47**, 1599–1608.
Eckstein, H., Paduch, V. and Hilz, H. (1966). *Biochem. Z.* **344**, 435–445.
Eckstein, H., Paduch, V. and Hilz, H. (1967). *Eur. J. Biochem.* **3**, 224–231.
Fink, G. R. (1964). *Science, N.Y.* **146**, 525–527.
Fink, G. R. (1966). *Genetics, Princeton* **53**, 445–459.
Fisher, C. R. (1969). *Biochem. biophys. Res. Commun.* **34**, 306–310.
Fittler, F. and Hall, R. H. (1966). *Biochem. biophys. Res. Commun.* **25**, 441–446.

Flavin, M. (1967). *J. biol. Chem.* **242**, 3884–3895.
Flavin, M. and Slaughter, C. (1960). *J. biol. Chem.* **235**, 1103–1109.
Flavin, M. and Slaughter, C. (1964). *J. biol. Chem.* **239**, 2212–2219.
Flavin, M. and Slaughter, C. (1967a). *Biochim. biophys. Acta* **132**, 400–405.
Flavin, M. and Slaughter, C. (1967b). *Biochim. biophys. Acta* **132**, 406–411.
Frederick, E. W., Maitra, V. and Hurwitz, J. (1969). *J. biol. Chem.* **244**, 413–424.
Friedkin, M. (1963). *A. Rev. Biochem.* **32**, 185–214.
Freundlich, M. and Umbarger, H. E. (1963). *Cold Spring Harb. Symp. quant. Biol.* **28**, 505–511.
Freundlich, M. ,Burns, R. O. and Umbarger, H. E. (1962). *Proc. natl. Acad. Sci. U.S.A.* **48**, 1804–1808.
Fukuhara, H. (1965). *Biochem. biophys. Res. Commun.* **18**, 297–301.
Fukuhara, H. (1967). *Biochim. biophys. Acta* **134**, 143–164.
Geige, R., Heinrich, J., Weil, J. H. and Ebel, J. P. (1969a). *Biochim. biophys. Acta* **174**, 43–52.
Geige, R., Heinrich, J., Weil, J. W. and Ebel, J. P. (1969b). *Biochim. biophys. Acta* **174**, 53–70.
Gibson, F. and Pittard, J. (1968). *Bact. Rev.* **32**, 465–492.
Gillam, I., Blew, D., Warrington, R. C., von Tigerstrom, M. and Tener, G. M. (1968). *Biochemistry, N.Y.* **7**, 3459–3468.
Gillespie, D. and Spiegelman, S. (1965). *J. molec. Biol.* **12**, 829–842.
Gilmore, R. A., Stewart, J. W. and Sherman, F. (1968). *Biochim. biophys. Acta* **161**, 270–272.
Giovanelli, J. and Mudd, S. H. (1966). *Biochem. biophys. Res. Commun.* **25**, 366–371.
Giovanelli, J. and Mudd, S. H. (1967). *Biochem. biophys. Res. Commun.* **27**, 150–156.
Giovanelli, J. and Mudd, S. H. (1968). *Biochem. biophys. Res. Commun.* **31**, 275–280.
Goodman, H. M., Abelson, J., Landy, A., Brenner, S. and Smith, J. D. (1968). *Nature, Lond.* **217**, 1019–1024.
Gorman, J., Tauro, P., La Berge, M. and Halvorson, H. O. (1964). *Biochem. biophys. Res. Commun.* **15**, 43–49.
Gorman, J. A. and Hu, A. S. L. (1969). *J. biol. Chem.* **244**, 1645–1650.
Grivell, A. R. and Jackson, J. F. (1968). *J. gen. Microbiol.* **54**, 307–317.
Groodt, A. de, Heslot, H. and Poirier, L. (1969). *C. r. hebd. Séanc. Acad. Sci., Paris* **269**, 1431–1433.
Gros, F., Hiatt, H., Gilbert, W., Kurland, C. G., Risebrough, R. W. and Watson, J. D. (1961). *Nature, Lond.* **190**, 581–585.
Gross, S. R. (1969). *A. Rev. Genetics* **3**, 395–424.
Grunberg-Manago, M. (1962). *A. Rev. Biochem.* **31**, 301–332.
Gupta, N. K. and Khorana, H. G. (1966). *Proc. natn. Acad. Sci. U.S.A.* **56**, 772–779.
Hall, R. H. (1965). *Biochemistry, N.Y.* **4**, 661–670.
Halvorson, H. O. and Heredia, C. (1968). *In* "Aspects of Yeast Metabolism" (A. K. Mills and H. Krebs, eds.), pp. 107–130. Blackwell Scientific Publications, Oxford and Edinburgh.
Halvorson, H. O., Bock, R. M., Tauro, P., Epstein, R. and La Berge, M. (1966). *In* "Cell Synchrony" (I. L. Cameron and G. M. Padilla, eds.). Academic Press, New York.
Harriman, P. and Zachau, H. (1966). *J. molec. Biol.* **16**, 387–403.
Hartlief, R. and Koningsberger, V. V. (1968). *Biochim. biophys. Acta* **166**, 512–531.
Hartwell, L. H. (1967). *J. Bact.* **93**, 1662–1670.

Hartwell, L. H. and McLaughlin, C. S. (1968a). *Proc. natn. Acad. Sci. U.S.A.* **59**, 422–428.
Hartwell, L. H. and McLaughlin, C. S. (1968b). *J. Bact.* **96**, 1664–1671.
Hartwell, L. H. and McLaughlin, C. S. (1969). *Proc. natn. Acad. Sci. U.S.A.* **62**, 468–474.
Hawthorne, D. C. and Mortimer, R. K. (1968). *Genetics, Princeton* **60**, 735–741.
Hayashi, H. and Miura, K. (1966). *Nature, Lond.* **209**, 376–378.
Heslot, H., Nagy, M. and Whitehead, E. (1966a). *C. r. hebd. Séanc. Acad. Sci., Paris* **263**, 57–58.
Heslot, H., Nagy, M. and Whitehead, E. (1966b). Proc. 2nd Symp. Yeasts, Bratislava, pp. 269–271.
Hilz, H. and Lipmann, F. (1955). *Proc. natn. Acad. Sci. U.S.A.* **41**, 880–890.
Holley, R. W. Apgar, J., Everett, G. A., Madison, J. T., Marquisee, M., Merrill, S. H., Penswick, J. R. and Zamir, A. (1965). *Science, N.Y.* **147**, 1462–1465.
Holzer, H. and Hierholzer, G. (1963). *Biochim. biophys. Acta* **77**, 329–331.
Holzer, H., Cennamo, C. and Boll, M. (1964). *Biochem. biophys. Res. Commun.* **14**, 487–492.
Hwang, Y. L., Lindegren, G. and Lindegren, C. (1966). *Can. J. Genet. Cytol.* **8**, 471–480.
Iwashima, A. and Rabinowitz, M. (1969). *Biochim. biophys. Acta* **178**, 283–293.
Jacob, F. and Monod, J. (1961). *J. molec. Biol.* **3**, 318–356.
James, H. L. and Bucovaz, E. T. (1969). *J. biol. Chem.* **244**, 3210–3216.
Jones, E. E. and Broquist, H. P. (1965). *J. biol. Chem.* **240**, 2531–2536.
Jones, E. W. and Magasanik, B. (1967). *Biochem. biophys. Res. Commun.* **29**, 600–604.
Kakar, S. N. (1963a). *Genetics, Princeton* **48**, 957–966.
Kakar, S. N. (1963b). *Genetics, Princeton* **48**, 967–979.
Kakar, S. N. and Wagner, R. P. (1964). *Genetics, Princeton* **49**, 213–222.
Kakar, S. N., Zimmermann, F. and Wagner, R. P. (1964). *Mut. Res.* **1**, 381–386.
Kaplan, M. M. and Flavin, M. (1965). *Biochim. biophys. Acta* **104**, 390–396.
Kaplan, J. G. and Messmer, I. (1967). *Can. J. Biochem. Physiol.* **47**, 477–479.
Kaplan, J. G., Duphil, M. and Lacroute, F. (1967). *Archs Biochem. Biophys.* **119**, 541–551.
Kaplan, J. G., Lacroute, F. and Messmer, I. (1969). *Archs Biochem. Biophys.* **129**, 539–544.
Karassevitch, Y. and de Robichon-Szulmajster, H. (1963). *Biochim. biophys. Acta* **73**, 414–426.
Kerr, D. S. and Flavin, M. (1968). *Biochem. biophys. Res. Commun.* **31**, 124–130.
Kerr, D. S. and Flavin, M. (1969). *Biochim. biophys. Acta* **177**, 177–179.
Kerr, S. E., Seraidarian, K. and Brown, G. B. (1951). *J. biol. Chem.* **188**, 207–216.
Kerridge, D. (1958). *J. gen. Microbiol.* **19**, 497–506.
Kijima, S., Ohta, T. and Imahori, K. (1968). *J. Biochem., Tokyo* **63**, 434–445.
Kisselev, L. L. and Frolova, L. Y. (1964). *Biokhimiya* **29**, 1117.
Kjellin-Sträby, K. and Phillips, J. H. (1968). *J. Bact.* **96**, 760–767.
Kloet, S. R. de (1965). *Biochem. biophys. Res. Commun.* **19**, 582–586.
Kloet, S. R. de (1966). *Biochem. J.* **99**, 566–581.
Knudsen, R. C., Moore, K. and Yall, I. (1969). *J. Bact.* **98**, 629–636.
Kredich, N. M. and Tomkins, G. M. (1966). *J. biol. Chem.* **241**, 4955–4965.
Lacroute, F. (1968). *J. Bact.* **95**, 824–832.
Lacroute, F., Pierard, A., Grenson, M. and Wiame, J. M. (1965). *J. gen. Microbiol.* **40**, 127–142.
Lagerkvist, U. and Rymo, L. (1969). *J. biol. Chem.* **244**, 2476–2483.
Lagerkvist, U. and Waldenström, J. (1965). *J. biol. Chem.* **240**, 2264–2265.
Lagerkvist, U. and Waldenström, J. (1967). *J. biol. Chem.* **242**, 3021–3025.

Lagerkvist, U., Rymo, L. and Waldenström, J. (1966). *J. biol. Chem.* **241**, 5391–5400.
Lake, J. A. and Beeman, W. W. (1968). *J. molec. Biol.* **31**, 115–125.
Lamb, A. J., Clark-Walker, G. D. and Linnane, A. W. (1968). *Biochim. biophys. Acta* **161**, 415–427.
Lampen, J. O., Anow, P. M., Borowska, Z. and Laskin, A. I. (1962). *J. Bact.* **84**, 1152–1160.
Latuasan, H. E. and Berends, W. (1958). *Recl. Trav. chim. Pays-Bas Belg.* **77**, 416.
Lazowska, J. and Luzzati, M. (1970a). *Biochem. biophys. Res. Commun.* **39**, 34–39.
Lazowska, J. and Luzzati, M. (1970b). *Biochem. biophys. Res. Commun.* **39**, 40–45.
Lee, L. W., Ravel, J. M. and Shive, W. (1967). *Archs Biochem. Biophys.* **121**, 614–618.
Leeuw, A. de (1967). *Genetics, Princeton* **56**, 554.
Lengyel, P. and Söll, D. (1969). *Bact. Rev.* **33**, 264–301.
Leslie, L. (1955). *In* "The Nucleic Acids" (E. Chargaff and J. M. Davidson, eds.), Vol. 2, pp. 1–50. Academic Press, New York.
Leupold, U. (1961). *In* "Microbial Genetics Bulletin", No. 25 (R. C. von Borstel, ed.), Oak Ridge, U.S.A.
Lingens, F., Goebel, W. and Uesseler, H. (1966a). *Biochem. Z.* **346**, 357–367.
Lingens, F., Sprössler, B. and Goebel, W. (1966b). *Biochim. biophys. Acta* **121**, 164–166.
Lingens, F., Goebel, W. and Uesseler, H. (1967). *Eur. J. Biochem.* **1**, 363–374.
Littlefield, J. W. and Dunn, D. B. (1958). *Biochem. J.* **70**, 642–651.
Lue, F. and Kaplan, J. G. (1969). *Biochem. biophys. Res. Commun.* **34**, 426–433.
Luzzati, M., Clavilier, L. and Slonimski, P. P. (1959). *C. r. hebd. Séanc. Acad. Sci., Paris* **249**, 1412–1414.
Madison, J. T., Everett, G. A. and Kung, H. (1966). *Science, N.Y.* **153**, 531–534.
Magasanik, B. (1962). *In* "The Bacteria", (I. C. Gunsalus and R. Y. Stanier, eds.) Vol. III, pp. 295–334. Academic Press, New York and London.
Magee, P. T. and Hereford, L. M. (1969). *J. Bact.* **98**, 857–862.
Magee, P. T. and de Robichon-Szulmajster, H. (1968a). *Eur. J. Biochem.* **3**, 502–506.
Magee, P. T. and de Robichon-Szulmajster, H. (1968b). *Eur. J. Biochem.* **3**, 507–511.
Mahler, H. R. and Cordes, E. H. (1967). *In* "Biological Chemistry". Harper International Editions, New York; Evanston, London.
Makman, M. H. and Cantoni, G. L. (1965). *Biochemistry, N.Y.* **4**, 1434–1442.
Makman, M. H. and Cantoni, G. L. (1966). *Biochemistry, N.Y.* **5**, 2246–2254.
Manney, T. R. (1964). Ph.D. Thesis. Univ. California, Berkeley, U.S.A.
Manney, T. R. (1968a). *Genetics, Princeton* **60**, 719–733.
Manney, T. R. (1968b). *J. Bact.* **96**, 403–408.
Manney, T. R., Duntze, W., Janosko, N. and Salazar, J. (1969). *J. Bact.* **99**, 590–596.
Maragoudakis, M. E. and Strassman, M. (1966). *J. biol. Chem.* **241**, 695–699.
Maragoudakis, M. E., Holmes, H., Ceci, L. N. and Strassman, M. (1966). *Fedn. Proc. Fedn. Am. Socs exp. Biol.* **25**, 710.
Maragoudakis, M. E., Holmes, H. and Strassman, M. (1967). *J. Bact.* **93**, 1677–1680.
Marcus, L., Bretthauer, R. K., Bock, R. M. and Halvorson, H. O. (1963). *Proc. natn. Acad. Sci. U.S.A.* **50**, 782–789.
Marcus, L., Bretthauer, R. K., Halvorson, H. O. and Bock, R. M. (1965). *Science, N.Y.* **147**, 615–617.
Marcus, L., Ris, H., Halvorson, H. O., Bretthauer, R. K. and Bock, R. M. (1967). *J. Cell. Biol.* **34**, 505–512.

Maw, G. A. (1966). *Biochem. J.* **98**, 28P.
Mayo, V. S., Andrean, B. A. G. and de Kloet, S. R. (1968). *Biochim. biophys. Acta* **169**, 297–305.
Mazlen, A. S. and Eaton, N. R. (1966). *Biochem. biophys. Res. Commun.* **26**, 590–595.
McLaughlin, C. S. and Hartwell, L. H. (1969). *Genetics, Princeton* **61**, 557–566.
Megnet, R. (1959). *Genetics, Princeton* **44**, 526.
Messenguy, F. and Wiame, J. M. (1969). *FEBS Letters* **3**, 47–49.
Meuris, P. (1967a). *C. r. hedb. Séanc. Acad. Sci., Paris* **264**, 1197–1199.
Meuris, P. (1967b). *Bull. Soc. chim. biol.* **49**, 1573–1578.
Meuris, P., Lacroute, F. and Slonimski, P. P. (1967). *Genetics, Princeton* **56**, 149–161.
Middlehoven, W. J. (1963). *Biochim. biophys. Acta.* **77**, 152–154.
Middlehoven, W. J. (1964). *Biochim. biophys. Acta* **93**, 650–652.
Middlehoven, W. J. (1968), *Biochim. biophys. Acta* **156**, 440–443.
Middlehoven, W. J. (1969). *Antonie van Leeuwenhoek* **35**, 215–226.
Mirzabekov, A. D., Grünberger, D., Holy, A., Baiev, A. A. and Sorms, F. (1967). *Biochim. biophys. Acta* **145**, 845–847.
Mitchison, J. M. (1969). *Science, N.Y.* **165**, 657–663.
Miura, K. (1967). *In* "Progress in Nucleic Acid Research and Molecular Biology" (J. N. Davidson and W. E. Cohn, eds.) Vol. 6, pp. 39–82. Academic Press, New York.
Miyazaki, M. and Takemura, S. (1968). *J. Biochem., Tokyo* **63**, 637–648.
Miyazaki, M., Kawata, M., Nakazawa, K. and Takemura, S. (1967). *J. Biochem., Tokyo* **62**, 161–169.
Monod, J., Wyman, J. and Changeux, J. P. (1965). *J. molec. Biol.* **12**, 88–118.
Mortimer, R. K. (1969). *Genetics, Princeton* **61**, 329–334.
Mortimer, R. K. and Hawthorne, D. C. (1966). *Proc. natn. Acad. Sci. U.S.A.* **53**, 165–173.
Mortimer, R. K. and Hawthorne, D. C. (1969). *In* "The Yeasts" (A. H. Rose and J. S. Harrison, eds.) Vol 1, pp. 385–460. Academic Press, London.
Moss, J. A. de (1965). *Biochem. biophys. Res. Commun.* **18**, 850–857.
Mudd, S. H. (1959). *J. biol. Chem.* **234**, 87–92.
Muramatsu, M., Hodnett, J. L. and Busch, H. (1966). *Biochim. biophys. Acta* **123**, 116–125.
Nagai, S. and Flavin, M. (1967). *J. biol. Chem.* **242**, 3884–3895.
Nagy, M. (1970). *Biochim. biophys. Acta* **198**, 471–481.
Nagy, M., Heslot, H. and Poirier, L. (1969). *C. r. hebd. Séanc. Acad. Sci., Paris* **269**, 1268–1271.
Naiki, N. (1964). *Pl. Cell. Physiol., Tokyo* **5**, 71–78.
Naiki, N. (1965). *Pl. Cell Physiol., Tokyo* **6**, 179–194.
Ninio, J., Favre, A. and Yaniv, M. (1969). *Nature, Lond.* **223**, 1333–1335.
Norris, A. T. and Berg, P. (1964). *Proc. natn. Acad. Sci. U.S.A.* **52**, 330–337.
Ogur, M., Coker, L. and Ogur, S. (1964). *Biochem. biophys. Res. Commun.* **14**, 193–197.
Ogur, M., Roshanmanesh, A. and Ogur, S. (1965). *Science, N.Y.* **147**, 1590.
Ohta, T., Shimada, I. and Imahori, K. (1967). *J. molec. Biol.* **26**, 519–524.
Okamoto, T. and Kawade, Y. (1967). *Biochim. biophys. Acta* **145**, 613–620.
Osawa, S. (1968). *A. Rev. Biochem.* **37**, 109–130.
Panten, K., Bernhardt, W. and Holzer, H. (1967). *Biochim. biophys. Acta* **139**, 33–39.
Pardee, A. B. and Prestidge, L. S. (1966). *Proc. natn. Acad. Sci. U.S.A.* **55**, 189–191.
Pardee, A. B., Prestidge, L. S., Whipple, M. B. and Dreyfuss, J. (1966). *J. biol. Chem.* **241**, 3962–3969.
Parks, L. W. (1958). *J. biol. Chem.* **232**, 169–176.

Perry, R. P. (1966). *In* "Progress in Nucleic Acid Research and Molecular Biology" Vol. 6, pp. 220–257.
Philipps, G. R. (1969). *Nature, Lond.* **223**, 374–377.
Phillips, J. H. and Kjellin-Sträby, K. (1967). *J. molec. Biol.* **26**, 509–518.
Piediscalzi, N., Fjellstedt, T. and Ogur, S. (1968). *Biochem. biophys. Res. Commun.* **32**, 380–384.
Pigg, J., Spence, K. D. and Parks, L. W. (1962). *Archs Biochem. Biophys.* **97**, 491–496.
Pillinger, D. J., Hay, J. and Borek, E. (1969). *Biochem. J.* **114**, 429–435.
Pomper, S. (1952a). *J. Bact.* **63**, 707–713.
Pomper, S. (1952b). *J. Bact.* **64**, 353–361.
Pourquié, J. (1969). *Biochim. biophys. Acta* **185**, 310–315.
RajBhandary, U. L., Chang, S. H., Stuart, A., Faulkner, R. D., Hoskinson, R. M. and Khorana, H. G. (1967). *Proc. natl. Acad. Sci. U.S.A.* **57**, 751–758.
RajBhandary, U. L. and Ghosh, H. P. (1969). *J. biol. Chem.* **244**, 1104–1113.
Ramos, F., Thuriaux, P., Wiame, J. M. and Bechet, J. (1970). *Eur. J. Biochem.* **12**, 40–47.
Ravel, J. M., Wang, S. F., Heinemeyer, C. and Shive, W. (1965). *J. biol. Chem.* **240**, 432–438.
Retel, J. and Planta, R. J. (1967). *Eur. J. Biochem.* **3**, 248–258.
Retel, J. and Planta, R. J. (1968). *Biochim. biophys. Acta* **169**, 416–429.
Richter, D. and Klink, F. (1967). *Biochemistry, N.Y.* **6**, 3569–3575.
Robbins, P. W. and Lipmann, F. (1958a). *J. biol. Chem.* **223**, 681–685.
Robbins, P. W. and Lipmann, F. (1958b). *J. biol. Chem.* **233**, 686–690.
Robichon-Szulmajster, H. de (1956). *Biochim. biophys. Acta* **21**, 313–320.
Robichon-Szulmajster, H. de and Corrivaux, D. (1963). *Biochim. biophys. Acta* **73**, 248–256.
Robichon-Szulmajster, H. de, Surdin, Y., Karassevitch, Y. and Corrivaux, D. (1963). Coll. Internat. C.N.R.S. No. 124, pp. 255–269.
Robichon-Szulmajster, H. de, Surdin, Y. and Mortimer, R. K. (1966). *Genetics, Princeton* **53**, 609–619.
Robichon-Szulmajster, H. de (1967). *Bull. Soc. chim. biol.* **49**, 1431–1462.
Robichon-Szulmajster, H. de and Cherest, H. (1967). *Biochem. biophys. Res. Commun.* **28**, 256–262.
Robichon-Szulmajster, H. de and Magee, P. T. (1968). *Eur. J. Biochem.* **3**, 492–501.
Robichon-Szulmajster, H. de, Surdin, Y. and Slonimski, P. P. (1969). Eur. *J. Biochem.* **7**, 531–536.
Roman, H. (1956). *C. r. Trav. Lab. Carlsberg. Sér. physiol.* **26**, 299 314.
Rüdiger, M. and Jaenicke, L. (1969). *Eur. J. Biochem.* **10**, 557–560.
Satyanarayana, T., Umbarger, H. E. and Lindegren, G. (1968a). *J. Bact.* **96**, 2012–2017.
Satyanarayana, T., Umbarger, H. E. and Lindegren, G. (1968b). *J. Bact.* **96**, 2018–2024.
Schlossmann, K., Brüggemann, J. and Lynen, F. (1962). *Biochem. Z.* **336**, 258–273.
Schulman, L. H. and Chambers, R. W. (1968). *Proc. natn. Acad. Sci. U.S.A.* **61**, 308–315.
Schweizer, E., McKechnie, C. and Halvorson, H. O. (1969). *J. molec. Biol.* **40**, 261–277.
Shapiro, S. K. and Ehninger, D. J. (1969). *Biochim. biophys. Acta* **177**, 67–77.
Shapiro, S. K. and Schlenk, F. (Eds.) (1965). "Transmethylation and Methionine Biosynthesis". University Press, Chicago, U.S.A.

Sheifinger, C., Ogur, S. and Ogur, M. (1966). *Fedn Proc. Fedn Am. Socs exp. Biol.* **25**, 710.

Shortman, K. and Fukuhara, H. (1963). *Biochim. biophys. Acta* **76**, 501–524.

Siegel, M. R. and Sisler, H. D. (1964). *Biochim. biophys. Acta* **87**, 70–82 and 83–89.

Siegel, M. R. and Sisler, H. D. (1965). *Biochim. biophys. Acta* **103**, 558–567.

Silven, B., Tobias, C. A., Malmgren, H., Ottoson, R. and Thorell, B. (1959). *Expl. Cell Res.* **16**, 75–87.

Silver, J. M. and Eaton, N. R. (1968). *Genetics, Princeton* **60**, 225–226.

Silver, J. M. and Eaton, N. R. (1969). *Biochem. biophys. Res. Commun.* **34**, 301–305.

Smith, A. E. and Marcker, K. A. (1968). *J. molec. Biol.* **38**, 241.

Söll, D. and RajBhandary, U. L. (1967). *J. molec. Biol.* **29**, 113–124.

Söll, D., Jones, D. S., Ohtsuka, E., Faulkner, R. D., Lohrman, R., Hayatsu, H., Khorana, H. G., Cherayil, J. D., Hampel, A. and Bock, R. M. (1966). *J. molec. Biol.* **19**, 556–573.

Söll, D., Cherayil, J. D. and Bock, R. M. (1967). *J. molec. Biol.* **29**, 97–112.

Sorsoli, W. A., Buettner, M. and Parks, L. W. (1968). *J. Bact.* **95**, 1024–1029.

Spence, K. D., Parks, L. W. and Shapiro, S. K. (1967). *J. Bact.* **94**, 1531–1537.

Stanier, R. Y. (1947). *J. Bact.* **54**, 339.

Stevens, A. (1964). *J. biol. Chem.* **239**, 204–209.

Strassman, M. and Weinhouse, S. (1953). *J. Amer. chem. Soc.* **75**, 1680–1684.

Stulberg, M. P. and Novelli, G. D. (1962). *In* "Methods in Enzymology" (N. O. Kaplan and S. P. Colowick, eds.) Vol. 5, pp. 703–707. Academic Press, New York.

Surdin, Y. (1967). *Eur. J. Biochem.* **2**, 341–348.

Surdin-Kerjan, Y. (1969). PhD Thesis, Paris University.

Surdin, Y., de Robichon-Szulmajster, H., Lachowicz, T. M. and Slonimski, P. P. (1969). *Eur. J. Biochem.* **7**, 526–530.

Svensson, I. (1968). *Biochim. biophys. Acta* **167**, 179–183.

Svensson, I., Björk, G. R. and Lundahl, P. (1969). *Eur. J. Biochem.* **9**, 216–221.

Taber, R. L. and Vincent, W. S. (1969a). *Biochem. biophys. Res. Commun.* **34**, 488–494.

Taber, R. L. and Vincent, W. S. (1969b). *Biochim. biophys. Acta* **186**, 317–325.

Takeishi, K. and Ukita, T. (1968). *J. biol. Chem.* **243**, 5761–5769.

Takeishi, K., Nishimura, S. and Ukita, T. (1967). *Biochim. biophys. Acta* **145**, 605–612.

Takemura, S. and Miyazaki, M. (1969). *J. Biochem., Tokyo* **65**, 159–169.

Takemura, S., Mizutani, T. and Miyazaki, M. (1968). *J. Biochem., Tokyo* **64**, 839–848.

Takemura, S., Murakami, M., Miyazaki, M. (1969a). *J. Biochem., Tokyo* **65**, 489–491.

Takemura, S., Murakami, M., Miyazaki, M. (1969b). *J. Biochem., Tokyo* **65**, 553–565.

Tauro, P. and Halvorson, H. O. (1966). *J. Bact.* **92**, 652–661.

Tauro, P., Halvorson, H. O. and Epstein, R. L. (1968). *Proc. natl. Acad. Sci. U.S.A.* **59**, 277–284.

Thomson, J. F. and Moore, D. P. (1968). *Biochem. biophys. Res. Commun.* **31**, 281–286.

Tingle, M., Herman, A. and Halvorson, H. P. (1968). *Genetics, Princeton* **58**, 361–371.

Torii, K. and Bandurski, R. S. (1964). *Biochem. biophys. Res. Commun.* **14**, 537–542.

Tristram, H. and Thurston, C. F. (1966). *Nature, Lond.* **212**, 74–75.

Tsukada, Y., Sugimori, T., Imai, K. and Katagiri, H. (1962). *J. Bact.* **83**, 70–75.

Tucci, A. F. (1969). *J. Bact.* **98**, 624.

Udaka, S. and Kinoshita, S. H. (1958). *J. gen. appl. Microbiol.* **4**, 272.
Umbarger, E. (1969). *A. Rev. Biochem.* **38**, 323–370.
Umbarger, E. and Davis, B. D. (1962). *In* "The Bacteria", (I. C. Gunsalus and R. Y. Stanier, eds.). Vol. III, pp. 168–251. Academic Press, New York.
de Vito, P. C. and Dreyfuss, J. (1964). *J. Bact.* **88**, 1341–1348.
Vold, B. S. (1969). *Biochim. biophys. Acta* **182**, 585–586.
Volkin, E. and Astrachan, L. (1957). *In* "The Chemical Basis of Heredity", p. 686. John Hopkins Press, Baltimore.
Vyas, S. and Maas, W. K. (1963). *Archs Biochem. Biophys.* **100**, 542–546.
Wainwright, T. (1962). *Biochem. J.* **83**, 39P.
Wainwright, T. (1967). *Biochem. J.* **103**, 56P.
Warren, W. A. and Goldthwait, D. A. (1962). *Proc. natn. Acad. Sci. U.S.A.* **48**, 698–709.
Watanabe, Y. and Shimura, K. (1955). *J. Biochem., Tokyo* **42**, 181–192.
Watanabe, Y. and Shimura, K. (1956). *J. Biochem., Tokyo* **43**, 283–294.
Watanabe, Y., Konishi, S. and Shimura, K. (1955). *J. Biochem., Tokyo* **42**, 837–844.
Watanabe, Y., Konishi, S. and Shimura, K. (1957). *J. Biochem., Tokyo* **44**, 299–307.
Wegman, J. and de Moss, J. A. (1965). *Bact. Proc.* 88.
Whiffen, A. J. Bohonas, J. N. and Emerson, R. L. (1946). *J. Bact.* **52**, 610.
Whitehead, E. (1967). *Bull. Soc. Chim. biol.* **49**, 1529–1535.
Whitehead, E., Nagy, M. and Heslot, H. (1966). *C. r. hebd. Séanc. Acad. Sci., Paris* **263**, 819–821.
Whiteley, H. R. and Tahara, M. (1966). *J. biol. Chem.* **241**, 4881–4884.
Wiebers, J. L. and Garner, H. R. (1964). *J. Bact.* **88**, 1798–1804.
Wiebers, J. L. and Garner, H. R. (1966). *Biochim. biophys. Acta* **117**, 403–409.
Wiebers, J. L. and Garner, H. R. (1967). *J. biol. Chem.* **242**, 5644–5649.
Williams, L. G. and Davis, R. M. (1968). *Genetics, Princeton* **60**, 238.
Wilson, L. G. and Bandurski, R. S. (1958). *J. biol. Chem.* **233**, 975–981.
Wilson, L. G., Asahi, T. and Bandurski, R. S. (1961). *J. biol. Chem.* **236**, 1822–1829.
Wimmer, E., Maxwell, I. H. and Tener, G. M. (1968a). *Biochemistry, N.Y.* **7**, 2623–2628.
Wimmer, E., Maxwell, I. H. and Tener, G. M. (1968b). *Biochemistry, N.Y.* **7**, 2629–2634.
Witt, I., Weiler, P. G. and Holzer, H. (1964). *Biochem. Z.* **339**, 331–337.
Wormser, E. H. and Pardee, A. B. (1958). *Archs Biochem. Biophys.* **78**, 416–432.
Yarus, M. and Berg. P. (1967). *J. molec. Biol.* **28**, 479–490.
Ycas, M. and Vincent, W. S. (1960). *Proc. natn. Acad. Sci. U.S.A.* **46**, 804–811.
Yoshida, H., Duval, J. and Ebel, J. P. (1966). *C.r. hebd. Séanc. Acad. Sci., Paris* **262**, 233–236.
Yoshida, M., Furuichi, Y., Ukita, T. and Kaziro, Y. (1967). *Biochim. biophys. Acta* **149**, 308–310.
Zachau, H. G., Dütting, D., Feldman, H., Melchers, F. and Karau, W. (1966). *Cold Spring Harb. Symp. quant. Biol.* **31**, 417–424.
Zimmerman, E. F. and Holler, B. W. (1967). *J. molec. Biol.* **23**, 149–161.
Zimmerman, F. K. (1968). *Molec. gen. Genetics* **101**, 171 –184.
Zimmerman, F. K. (1969). *Molec. gen. Genetics* **103**, 348–362.

## NOTE ADDED IN PROOF

Recent studies on phenotypic expression of some mutations at the locus $ade_{12}$ in *Sacch. cerevisiae* seems to indicate that, in addition to its catalytic activity, adenylosuccinate synthetase might participate in the regulation of purine nucleotide biosynthesis (Dorfman *et al.*, 1970). *In vitro* experiments provided strong evidence that one physiological function of a multi-enzyme aggregate in the pyrimidine pathway of yeast (see p. 346) resides in a metabolic compartmentation at the molecular level (Lue and Kaplan, 1970).

Data on the activity of a t-RNA synthetase, modified either genetically or by the use of an antibiotic, favour the participation of acylated t-RNAs in the negative control of enzyme synthesis in *Sacch. cerevisiae*. This is true for isoleucyl-t-RNA in the multivalent repression of enzymes involved in isoleucine–valine biosynthesis (McLaughlin *et al.*, 1969), threonyl-t-RNA in threonine-mediated repression of aspartokinase (Nass and Hasenbank, 1970) and methionyl-t-RNA in methionine-mediated repression of a group of three enzymes involved in methionine biosynthesis (Cherest and de Robichon-Szulmajster, 1970).

The mechanism of protein biosynthesis in yeast appears to be much like that in bacteria. Albrecht *et al.* (1970) further characterized the transfer factors (see pp. 395–396) FI and FII from yeast; these factors are functionally similar to, respectively, the bacterial translocase and T factor. Housman *et al.* (1970) suggest that initiation of polypeptide chains by a special t-RNA$^{met}$ may be a general phenomenon in all eukaryotic proteins. This has received some genetic support in yeast (Sherman *et al.*, 1970). On the other hand, the results of Richter and Lipmann (1970) show that yeast cannot use the same mechanism as *E. coli* for distinguishing initiator t-RNA from other amino-acyl-t-RNAs as both Met-t-RNAs$^{met}$ from yeast form a ternary complex with GTP and binding factor T from yeast or *E. coli*. A recent characterization of a messenger-like RNA from *Sacch. cerevisiae* using chromatography on methylated albumin kieselguhr has been made (Johnson, 1970) and the existence of a precursor of the 60*S* ribosomal subunit in yeast has been evidenced (Taber *et al.*, 1969).

The biosynthesis of r-RNA has recently received a great deal of work. The initial event is the formation of a high molecular-weight precursor (see p. 406). The processing of this precursor into 26*S* and 17*S* r-RNA has been shown to be non-conservative in *Sacch. carlsbergensis* (Retel and Planta, 1970) unlike the results obtained with *Schizosacch. pombe* (Taber and Vincent, 1969; cited on p. 407). Retel *et al.* (1969), with *Sacch. carlsbergensis*, have shown that methylation of r-RNA begins on the precursor and that later methylations take place at subsequent stages of the maturation and presumably after the conversion of the precursor RNA into the two species of r-RNA. In a comparison between the 26*S* and 17*S* r-RNAs, Van Den Bos and Planta (1970) have evidenced structural differences between these two species besides the striking analogies in nucleotide composition, degree of methylation and extensive cross hybridization with yeast DNA. In the same field, de Kloet (1970) used hybridization in solution to investigate the sequence homology between DNA and r-RNA of *Sacch. carlsbergensis*. In agreement with results obtained with *Sacch. cerevisiae* and *Sacch. carlsbergensis* using other techniques (see p. 405), 1·3 and 0·7% of DNA were found homologous to, respectively, the 26*S* and 17*S* r-RNA. For more information, an excellent review on the structure and biosynthesis of r-RNA has appeared recently (Attardi and Amaldi, 1970).

## Additional References


Albrecht, U., Prenzel, K. and Richter, D. (1970). *Biochemistry, N.Y.* **9**, 361–368.

Attardi, G. and Amaldi, F. (1970). *A. Rev. Biochem.* **39**, 183–226.

Cherest, H. and de Robichon-Szulmajster (1970). In "Genetics of Industrial Microorganisms", (Z. Vanek, S. Hostalek and J. Cudlin, eds.). Czech. Acad. Sc. publ. (in press).

Dorfman, B. S., Goldfinger, B. A., Berger, M. and Goldstein, S. (1970). *Science, N.Y.* **168**, 1482–1484.

Housman, D., Jacobs-Lorena, M., RajBhandary, U. L. and Lodish, H. F. (1970). *Nature, Lond.* **227**, 913–918.

Johnson, R. (1970). *Biochem. J.* **119**, 699–706.

de Kloet, S. R. (1970). *Archs Biochem. Biophys.* **136**, 402–412.

Lue, P. F. and Kaplan, J. G. (1970). *Biochim. biophys. Acta* **220**, 365–372.

McLaughlin, C. S., Magee, P. T. and Hartwell, L. H. (1969). *J. Bact.* **100**, 579–584.

Nass, G. and Hasenbank, R. (1970). *Mol. Gen. Genetics* **108**, 28–32.

Richter, D. and Lipmann, F. (1970a). *Biochem. biophys. Res. Commun.* **38**, 864–870.

Retel, J. and Planta, R. J. (1970). *Biochim. biophys. Acta* **199**, 286–288.

Retel, J., Van Den Bos, R. C. and Planta, R. J. (1969). *Biochim. biophys. Acta* **195**, 370–380.

Sherman, F., Stewart, J. W., Parker, J. H., Putterman, G. J., Agrawal, B. B. L. and Margoliash, E. (1970). *Symp. Soc. exp. Biol.* **24**, 85.

Taber, R. L., Vincent, W. S. and Coetzee, M. L. (1969). *Biochim. Biophys. Acta* **195**, 99–108.

Van Den Bos, R. C. and Planta, R. J. (1970). *Nature, Lond.* **225**, 183–184.

# Chapter 10

# The Structure and Biosynthesis of Storage Carbohydrates in Yeast

D. J. MANNERS

*Heriot-Watt University, Edinburgh, Scotland*

I. INTRODUCTION . . . 419
II. TREHALOSE . . . 420
A. Chemical Structure . . . 420
B. Biosynthesis . . . 421
III. GLYCOGEN . . . 423
A. Molecular Structure . . . 423
B. Biosynthesis . . . 430
REFERENCES . . . 437

## I. Introduction

Although a substantial amount of biochemical information on the reserve carbohydrates of yeast, trehalose and glycogen, has been reported (Trevelyan, 1958), detailed knowledge of their biosynthesis was, until recently, limited. The present review will therefore concentrate on the enzyme systems involved in the conversion of glucose into trehalose and glycogen. Since 40% of the dry weight of anaerobically grown yeast cells may be reserve carbohydrate (Chester, 1963) the importance of these metabolic pathways should not be underestimated. More general aspects of the utilization of these reserves, as a source of energy for various metabolic processes, will be considered elsewhere (Chapter 7, p. 271).

Any discussions of the structure and biosynthesis of yeast glycogen must also include some mention of the cell-wall polysaccharides composed of D-glucose and D-mannose. During the isolation of any of the three types of polysaccharide, the others will be present, to a greater or lesser extent, as impurities. It is generally accepted that the cell wall of *Saccharomyces cerevisiae* contains a β-glucan and a mannan (Trevelyan, 1958), but the molecular structures of these polymers have not been unequivocally established, although substantial progress with yeast mannan has recently been reported (Jones and Ballou, 1969).

In the case of yeast glucan, which contains a high proportion of β-(1 → 3)-D-glucosidic linkages, various structures have been proposed on the assumption that the polysaccharide was homogeneous (for details, see Manners and Patterson, 1966; Misaki *et al.*, 1968). However, recently Bacon and Farmer (1968) have suggested that yeast β-glucan is heterogeneous and is a mixture of two polysaccharides. The minor component has been identified by infrared spectroscopy as a β-(1 → 6)-glucan. This finding has been confirmed by Manners and Masson (1969), who were able to detect the β-(1 → 6)-glucan in yeast glycogen preparations, and to extract it selectively from insoluble β-glucan preparations. Many workers have reported analytical data for the gross composition of the carbohydrates of yeast cells, based on fractionation procedures devised by Trevelyan and Harrison (1956), and by Pfäffli and Suomalainen (1960). The figures for yeast β-glucan and for glycogen (and particularly where the latter is analysed as acetic acid-soluble, alkali-soluble and perchloric acid-soluble fractions, or as alkali-soluble and -insoluble fractions) will be subject to some uncertainty in view of the soluble nature of the β-(1 → 6)-glucan. Moreover, unspecific methods for the estimation of carbohydrate based on anthrone have been used.

Since more reliable data are not yet available, there is no alternative but to use the original data, with the proviso that the quantitative significance of the polysaccharide results is not as great as was originally believed.

## II. Trehalose

### A. CHEMICAL STRUCTURE

Trehalose was first isolated from pressed baker's yeast by Koch and Koch (1925). It was crystallized from ethanol extracts and was identified on the basis of melting point, optical rotation, molecular weight and hydrolysis by strong acid to two molecules of D-glucose. An improved yield of about 1–2 g of crystalline trehalose from 300 g of yeast was obtained by Steiner and Cori (1935) using *N*-sulphuric acid as an extractant. These workers also noted the variation in trehalose concentration with conditions of growth; for example, it was lowered by aeration. More recently, Trevelyan and Harrison (1952) have used trichloroacetic acid to extract trehalose from yeast cells.

The chemical structure of trehalose (α-D-glucopyranosyl-α-D-glucopyranoside; Fig. 1), originally isolated from trehala manna, was determined by the classical methods of methylation, periodate oxidation and application of the Hudson Isorotation Rules (for a detailed review, see Birch, 1963). The major piece of evidence arises from the production of two molecules of 2,3,4,6-tetra-*O*-methyl-D-glucopyranose on hydrolysis of the methylated disaccharide.

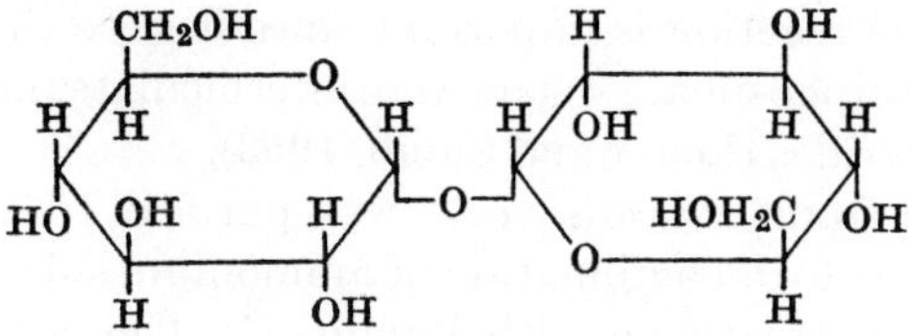

FIG. 1. Structure of trehalose.

Although it has been stated that trehalose occurs only in baker's yeast (Trevelyan, 1958) this statement appears to be incorrect. Stewart *et al.* (1950) were able to isolate 2–10 g quantities per kg from several strains of brewing yeast. Analytical data for the trehalose content of brewer's yeast are given in Table I (Chester, 1963, 1964). This also shows that the trehalose content of aerobically grown yeast is very low.

TABLE I. *The Carbohydrate Composition of Some Strains of* Saccharomyces cerevisiae (*Data adapted from Chester, 1963, 1964*)

| Carbohydrate | Carbohydrate content as proportion of total carbohydrate (%)[a] | | | | |
|---|---|---|---|---|---|
| | Baking strain | Brewing strains | | | |
| | 77 | 4236 | 4236[b] | 7000 | 7001 |
| Trehalose | 5 | 8 | 1 | 5 | 6 |
| Glycogen | 62 | 76 | 38 | 71 | 52 |
| Mannan | 16 | 7 | 36 | 12 | 24 |
| β-Glucan | 16 | 9 | 26 | 12 | 19 |

[a] Percentage expressed to nearest whole number.
[b] Yeast grown aerobically. All other yeasts were grown anaerobically at 25° and harvested at 48 h.

## B. BIOSYNTHESIS

The pathway for the biosynthesis of trehalose from glucose is shown in Fig. 2. A key enzyme is UDPG-glucose 6-phosphate transglucosylase (trehalose phosphate synthetase) which catalyses the transfer of a glucose residue from uridine diphosphate glucose (UDPG) to glucose 6-phosphate (G.6.P):

$$\text{UDPG} + \text{G.6.P} \rightarrow \text{UDP} + \text{trehalose phosphate}$$

Historically, the reaction is important since it gave the first indication that nucleoside diphosphate sugars were intermediates in the biosynthesis of higher saccharides (Leloir and Cabib, 1953).

The enzyme from brewer's yeast was purified 15–20-fold by Cabib and Leloir (1958) by a combination of ammonium sulphate and acetone fractionation and treatment with alumina $C_\gamma$. The purified enzyme had an optimum pH value of 6·6 and required $2{\cdot}5 \times 10^{-2}$ *M* magnesium ions

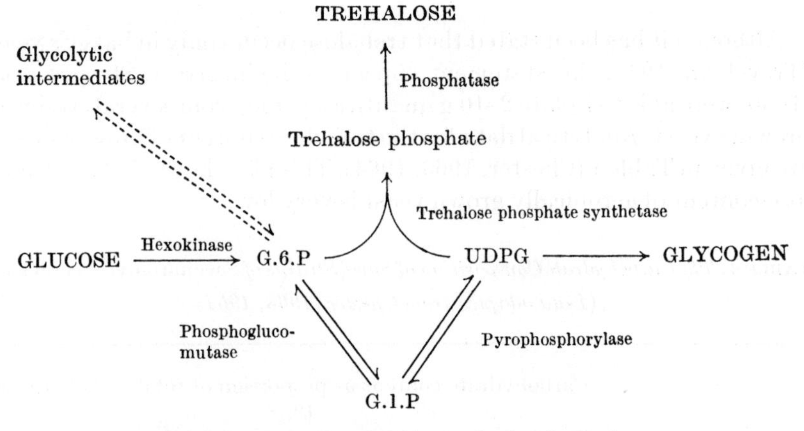


FIG. 2. Pathway for the biosynthesis of trehalose.

for maximum activity. Enzymic action could not be reversed, and glucose or glucose 1-phosphate failed to act as alternative acceptor. The reaction product was identified as trehalose phosphate on the basis of electrophoretic and chromatographic mobility, optical rotation and analysis of the products of acid hydrolysis. Incubation of trehalose phosphate with kidney or intestinal phosphatase, or with a specific phosphatase also present in brewer's yeast, gave a disaccharide which was identical with authentic trehalose. The yeast phosphatase was also activated by magnesium ions.

The pathway for trehalose synthesis in baker's yeast is generally similar to that in brewer's yeast (Panek, 1962). Yeast cells grown in the presence of $^{32}$P contained radioactive glucose 6-phosphate and UDPG, and non-radioactive trehalose. The specific trehalose phosphate phosphatase may be relatively more active in baker's yeast, since trehalose rather than the disaccharide phosphate was detected in various enzyme digests. The enzymes of the pathway are present in both resting and growing cells, but trehalose synthesis does not occur in growing cells, probably because glucose 6-phosphate is required for other metabolic purposes, e.g. as an intermediate in the formation of amino acids. In the presence of isonicotinoyl hydrazide or sodium fluoride, growth is inhibited but the ability to synthesize trehalose is not affected. It therefore appears that resting cells provide optimum conditions for the synthesis of trehalose.

Of the other enzymes involved in the biosynthesis of trehalose, the properties of hexokinase and phosphoglucomutase are well documented (Crane, 1962; Najjar, 1962). UDPG-Pyrophosphorylase was first described in 1953, and catalyses the reaction:

$$\text{UTP} + \text{G.1.P} \rightleftharpoons \text{PP} + \text{UDPG}$$

The enzyme from brewer's yeast was purified 250-fold by Munch-Petersen (1955), using a combination of ethanol and ammonium sulphate fractionations. The reaction is freely reversible, and at equilibrium the digest contains about 52% of UDPG. The enzyme has no action on other nucleoside diphosphate sugars, has an optimum pH range of 6·5–8, and also requires magnesium ions for maximum activity.

The trehalose content of yeast cells depends upon their metabolic state. During the propagation of baker's yeast and its transfer from anaerobic to aerobic conditions, the trehalose content increases stage by stage from about 1% to 8·5% in the commercial product (Suomalainen and Pfäffli, 1961). At each growth stage, the trehalose almost disappears from the cells during the first hour, but is resynthesized during the later part of the growth phase. It has been suggested the trehalose functions as an energy reserve for those metabolic steps which precede cell division (Panek, 1962).

The above changes in trehalose content are accompanied by somewhat similar changes in glycogen content, which increases from about 5% to 12% (Suomalainen and Pfäffli, 1961). At the beginning of each growth phase, some but not all of the glycogen is mobilized, and further net synthesis then takes place. These results suggest that trehalose and glycogen synthesis take place by two non-competitive and parallel pathways, and that the conversion of glucose into UDPG (see Figs. 2 and 5) will provide the initial steps.

## III. Glycogen

### A. MOLECULAR STRUCTURE

#### 1. *Introduction*

Glycogens, whether from animal or microbial cells, are now generally accepted to be multiply branched molecules of high molecular weight ($\sim 10^7$) which consist of numerous chains of $\alpha$-(1 $\rightarrow$ 4)-linked D-glucose residues. These chains normally contain an average of about 12 D-glucose residues, although individual chains vary considerably in length, and are arranged to form a tree- or bush-like structure (Fig. 3; for a review, see Manners, 1957). There are small variations in the properties and molecular structure of glycogens from different biological sources. In particular, the iodine staining power of yeast glycogen may be less than that of mammalian glycogens. For example, the iodine complex of brewer's yeast glycogen has $\lambda_{max}$ 430 nm and $E_{max}$ 0·3, whereas rabbit liver and

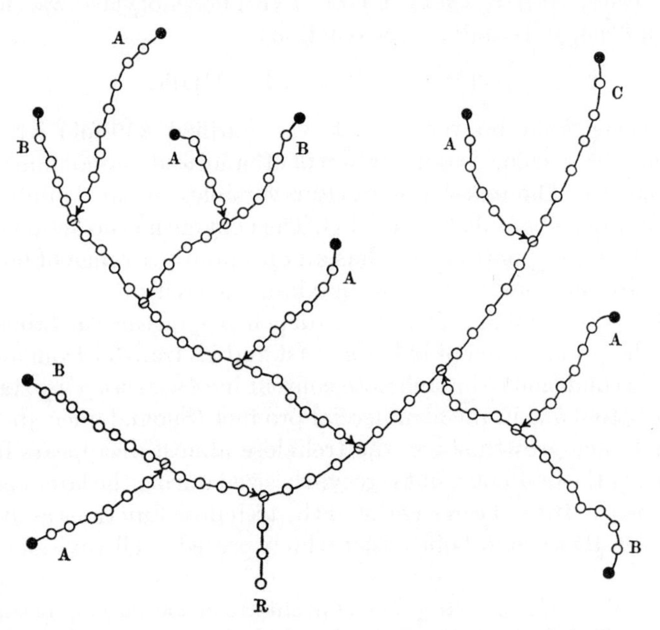


FIG. 3. Molecular structure of glycogen. ●, indicates a non-reducing end-group (which gives rise to tetra-*O*-methyl D-glucose on methylation, and to formic acid on periodate oxidation); ○—, an α-1,4-linked D-glucose residue; ⊘←, an α-1,6-linked D-glucose residue; R, free reducing group; A, B, C, types of chain (see p. 428).

muscle glycogen has $\lambda_{max}$ values of 475 and 490 nm respectively, and $E_{max}$ values of 0·3 and 0·4 (Archibald *et al.*, 1961). The structural significance of these differences is not yet known.

## 2. *Methods for the Isolation of Yeast Glycogen*

Since yeast glycogen is enclosed by the relatively stable glucan-mannan membrane, the successful extraction of glycogen involves an appropriate preliminary treatment of the yeast cells. Treatment of pressed baker's yeast with 3% sodium hydroxide solution at 100° failed to extract any glycogen, although similar treatment of dried baker's yeast extracted about 30% of the total glycogen (Northcote, 1953). By contrast, mechanical breakage of the cell walls by grinding with sand (a procedure introduced by Harden and Young, 1912) followed by exhaustive extraction with boiling water enabled all of the glycogen to be brought into solution (Northcote, 1953). It seems probable that, during the drying of yeast, the permeability of the cell wall is altered (see Vol. 3, Chapter 7); this property is, of course, eliminated altogether during mechanical breakdown.

The majority of studies have been carried out on pressed yeast using hot 3% sodium hydroxide solution to effect cytolysis, and to extract selectively the mannan. The insoluble residue is a mixture of $\beta$-glucans and glycogen, from which the latter can be extracted either with cold $N$-hydrochloric acid or 0·5 $N$-acetic acid at 75° (Northcote, 1953). It is now known that the insoluble residue remaining after acid extraction contains all the $\beta$-(1 $\rightarrow$ 3)-glucan, part of the $\beta$-(1 $\rightarrow$ 6)-glucan, and a residual trace of glycogen (Manners and Masson, 1969). The acetic-acid extract contains the major part of the glycogen and the remainder of the $\beta$-(1 $\rightarrow$ 6)-glucan. At the time when yeast glycogen was first characterized (1953–55), the presence of a $\beta$-(1 $\rightarrow$ 6)-glucan in the yeast cell wall was unknown, and it seems probable that the glycogen preparations contained a small amount of $\beta$-(1 $\rightarrow$ 6)-glucan impurity. The presence of this would not affect the general conclusions on the molecular structure of the glycogen, but some of the more quantitative aspects of the analyses may be less certain. For example, the specific rotation of some yeast glycogen preparations (+184 to +188 degrees; Northcote, 1953) is appreciably lower than that of other yeast glycogen preparations (+198 degrees) and of mammalian glycogens (+195 to +200 degrees; Manners, 1957) whilst the $\beta$-(1 $\rightarrow$ 6)-glucan has $[\alpha]_D$ −32 degrees (Manners and Masson, 1969).

Although the use of hot alkali in the preparation of yeast glycogen may be unavoidable, it should be noted that this treatment almost certainly causes some degradation of the macromolecule. There is now substantial evidence that a *limited* number of $\alpha$-(1 $\rightarrow$ 4)-D-glucosidic linkages are broken (Bryce *et al.*, 1958). Nevertheless, this behaviour should not be exaggerated; from a physico-chemical point of view, the rupture of ten linkages in a polysaccharide of molecular weight $16 \times 10^6$ may well be detectable by modern sensitive instruments, but this represents the cleavage of only one out of 10,000 linkages. The overall molecular architecture would be retained but the significance of any molecular weight estimations would be reduced.

### 3. *Methods for the Characterization of Yeast Glycogen*

The methods available for this purpose are similar to those generally employed in starch chemistry (for reviews, see Whelan, 1955, 1956; Manners, 1957).

Historically, the most important method is the methylation technique in which the polysaccharide is converted into the *O*-methyl ether and hydrolysed with acid to give a mixture of methylated D-glucoses, which are then separated, estimated and identified. Measurement of the proportion of tetra-*O*-methyl-D-glucose, which arises only from the non-reducing terminal residues, enables the average chain length to be determined, since each chain in the macromolecule, irrespective of the length, contains only one non-reducing terminal residue. The nature of the tri-*O*-methyl D-glucose characterizes the type of repeating glucosidic linkage, whilst the inter-chain linkages can be identified from the di-*O*-methyl derivative, assuming that the polysaccharide is fully methylated.

Periodate oxidation provides a simpler method for measurement of the average chain length. In a large macromolecule such as glycogen, which does not contain a significant proportion of reducing terminal residues, the only triol groups present are located in the non-reducing terminal residues. Since these groups yield one molecular proportion of formic acid, the average chain length can be calculated from the maximum production of formic acid. This method could be used on the decigram scale, in contrast to methylation, which originally required gram quantities of glycogen (Northcote, 1953). Since this work was completed, micromethods of periodate oxidation and methylation analysis (the latter involving gas–liquid chromatography) have been devised, but have not yet been applied to structural studies of yeast glycogen.

Identification of the products of partial acid hydrolysis of glycogen also enables the repeating glucosidic linkage and inter-chain linkage to be characterized. In particular, the configuration of these linkages is obtained directly, whereas in other methods (e.g. methylation analysis) this property is deduced indirectly from measurements of the specific rotation of the polysaccharide or a derivative. This method also requires gram quantities of polysaccharide, and cannot be used to determine the average chain length.

Enzymic degradation methods have been widely used in structural studies on glycogens (for a review, see Manners, 1962). The susceptibility of the polysaccharide to $\alpha$-amylase provides preliminary evidence for the presence of adjacent sequences of $\alpha$-(1 $\rightarrow$ 4)-linked D-glucose residues, whilst the extent of hydrolysis by $\beta$-amylase enables the relative length of the exterior chains to be assessed. The latter enzyme catalyses a stepwise hydrolysis of alternate linkages in a chain of $\alpha$-(1 $\rightarrow$ 4)-linked D-glucose residues, to give maltose, but is unable to hydrolyse or by-pass $\alpha$-(1 $\rightarrow$ 6)-D-glucosidic inter-chain linkages. In a branched glycogen-type polysaccharide, enzyme action is therefore confined to the exterior chains. Originally, it was believed that $\beta$-amylolysis left very short exterior chain "stubs" containing either two or three glucose residues, so that the exterior chain length is given by $n + 2{\cdot}5$, where $n$ is the number of glucose residues removed by β-amylase. However, recent evidence suggests that $(n + 2{\cdot}0)$ is a more correct factor.

Studies of the hydrolysis of the inter-chain linkages in glycogen by "debranching" enzymes are a recent development. These enzymes are α-(1 → 6)-glucosidases which may hydrolyse certain inter-chain linkages in glycogen, and in the derived α- and β-limit dextrins, but have no action on the simple disaccharide isomaltose, or on α-(1 → 4)-D-glucosidic linkages in any of the substrates. It is now clear that several distinct types of debranching enzyme exist, each with a characteristic specificity towards the various possible polysaccharide substrates. The latter include glycogen, amylopectin (the branched component of starch which resembles glycogen in many ways, except that the average chain length is 20–25 D-glucose residues) and pullulan (a linear polymer of α-(1 → 6)-linked maltotriose units, which is synthesized by *Pullularia pullulans*; Bender *et al.*, 1959).

The debranching enzymes include: (i) isoamylase (glycogen 6-glucanohydrolase) which occurs in yeast and sweet corn, and acts on both glycogen and amylopectin, but not on pullulan or branched oligosaccharide limit dextrins (Gunja *et al.*, 1961; Manners and Rowe, 1968a); (ii) R-enzyme (amylopectin 6-glucanohydrolase) which occurs in certain higher plants, but acts only on amylopectin and its β-limit dextrin (Hobson *et al.*, 1951); (iii) limit dextrinase (α-dextrin 6-glucanohydrolase) which occurs in yeast and the cereals, and hydrolyses α-(1 → 6)-glucosidic linkages in pullulan and α-limit dextrins, but has no action on glycogen or amylopectin (Manners and Rowe, 1968b and unpublished work); (iv) pullulanase, which is produced by certain bacteria (e.g. *Aerobacter aerogenes* and *Streptococcus mitis*) and acts on amylopectin, glycogen, pullulan and α-limit dextrins (Abdullah *et al.*, 1966; Walker, 1968); (v) amylo-1,6-glucosidase, which is present in mammalian and yeast cells and can hydrolyse *terminal* α-(1 → 6)-linked D-glucose residues in certain substrates (e.g. $6^3$-α-D-glucosyl-maltotetraose and α-D-glucosyl Schardinger dextrin) and may also show both transferase and hydrolytic activity towards the phosphorylase limit dextrins of glycogen and amylopectin (Lee *et al.*, 1967; Bathgate and Manners, 1968).

With regard to the *in vivo* breakdown of yeast glycogen, the presence of amylo-1,6-glucosidase in yeast implies a phosphorolytic pathway for degradation, and there is substantial evidence for an active phosphorylase in extracts of baker's yeast (see, for example, Trevelyan *et al.*, 1954). Isoamylase may form part of a non-phosphorolytic pathway of glycogen catabolism; it should be noted that there are at least two pathways for the catabolism of glycogen in mammalian liver and muscle tissues (Whelan, 1964). In addition, isoamylase has provided a useful structural tool in studies of the *in vitro* degradation of glycogen and starch-type polysaccharides (see, for example, Kjolberg and Manners, 1963a).

### 4. *Structure of Baker's Yeast Glycogen*

The glycogen isolated from baking strains of *Saccharomyces cerevisiae* has attracted rather more attention than the polysaccharide from brewing strains.

Prior to the period 1944–53, comparative studies on yeast and animal glycogens had been based on superficial properties such as optical rotation, iodine staining power, opalescence of aqueous solutions, and analytical data for carbon, hydrogen, oxygen, nitrogen and phosphorus contents. The limitations of these properties were recognized by Jeanloz (1944), who fractionated baker's yeast glycogen by electrodialysis and measured the β-amylolysis limits of the fractions. The larger fraction (73%), which was relatively insoluble in water, had a β-amylolysis limit of 46%, whereas the smaller and more soluble fraction (27%) gave 49% conversion into maltose. Various samples of animal glycogens had β-amylolysis limits of 32–54% (Meyer and Jeanloz, 1943).

The first detailed study of baker's yeast glycogen was carried out by Northcote (1953). By methylation analysis, the presence of chains of (1 → 4)-linked D-glucose residues was established, and the amount of tetra-*O*-methyl-D-glucose corresponded to an average chain length of 12 glucose residues. The latter result was confirmed by periodate oxidation; the production of formic acid corresponded to a chain length of 11–12 glucose residues. On incubation with β-amylase, 50% conversion into maltose was observed indicating exterior and interior chain lengths of eight and three glucose residues, respectively. All of these results were generally similar to those reported for rabbit liver glycogen.

The nature of the inter-chain linkages was not established from the methylation study, and this information was provided by Peat *et al.* (1955). Partial acid hydrolysis of baker's yeast glycogen (4·5 g) gave a mixture of D-glucose (2·45 g), maltose (0·98 g), isomaltose (0·17 g) and smaller amounts of maltotriose, panose and maltotetraose. The chemical characterization of these oligosaccharides gave definite evidence of chains of α-(1 → 4)-linked D-glucose residues which were inter-linked by α-(1 → 6)-D-glucosidic linkages.

The inter-chain linkages in baker's yeast glycogen are hydrolysed by yeast isoamylase. This α-(1 → 6)-glucosidase has no action on α-(1 → 3)-, α-(1 → 4)-, or β-(1 → 6)-D-glucosidic linkages (Gunja *et al.*, 1961), and this result therefore characterizes the type of inter-chain linkage. However, enzyme action on all glycogens is incomplete. In a glycogen-type polysaccharide, three types of chain may be described. A-Chains (side-chains) are attached to the macromolecule only by single α-(1 → 6)-

linkages from the reducing residue, whereas B-chains (main chains), to which one or more A-chains are attached, are also linked by the reducing group to the remainder of the macromolecule. It is probable that in the native glycogen only one free reducing group is present, and this terminates the sole C-chain. In the tree-type structure, there are approximately equal numbers of A- and B-chains (Manners, 1962). The action of isoamylase initially involves the hydrolysis of the $\alpha$-(1 $\rightarrow$ 6)-linkages attaching A-chains to the macromolecule, thereby liberating linear maltosaccharides. These are a mixture of maltopentaose (16%), maltohexaose (18%), maltoheptaose (25%), malto-octaose (31%) and maltononaose (10%), showing that rather more than half of the A-chains contain seven or eight glucose residues, and the remainder five, six or nine glucose residues (Bathgate and Manners, 1968).

### 5. *Structure of Brewer's Yeast Glycogen*

Glycogen from brewer's yeast was extensively degraded by salivary $\alpha$-amylase and partly degraded by soya bean $\beta$-amylase, giving 44% conversion into maltose. The average chain length, determined by periodate oxidation, was 13 glucose residues, hence the exterior and interior chains contained, on the average, eight and four glucose residues respectively (Manners and Khin Maung, 1955). The sedimentation constant of the glycogen was equivalent to a molecular weight of $2 \times 10^6$, and by light scattering, a weight average molecular weight of $4{\cdot}4 \times 10^6$ was obtained (Bryce *et al.*, 1958). For the reasons stated previously (p. 425), these figures represent a minimum order of magnitude. In contrast to the results of Jeanloz (1944) on baker's yeast glycogen, the brewer's yeast glycogen was not heterogeneous with respect to solubility in water. This difference may be due to differences in the methods of isolation of the glycogens, rather than to any significant difference in the state of the glycogen within the various yeast cells.

A proportion of the inter-chain linkages in brewer's yeast glycogen are hydrolysed by yeast isoamylase (Gunja *et al.*, 1961). This treatment increased the $\beta$-amylolysis limit from 44% to 68%; with rabbit liver glycogen, the corresponding results were 46% and 78%. Isoamylase action was incomplete, as in the previous experiments with baker's yeast glycogen, although in the presence of $\beta$-amylase the combined action of the two enzymes results in virtually complete degradation of the polysaccharide.

The available evidence therefore shows that glycogen isolated from both baking and brewing strains of *Saccharomyces cerevisiae* is very similar in molecular structure to the glycogens from animal tissues. Some typical analytical results are given in Table II.

As noted previously, certain fractionation procedures devised for the analysis of yeast carbohydrates give polysaccharide fractions which are described as acetic acid-soluble glycogen, alkali-soluble glycogen, etc. Detailed structural studies on these fractions have not yet been carried out. It is not known whether, in the yeast cell, there is a series of macromolecules showing some variation in degree of branching, or whether the differences in solubility of the fractions represent differences in physical state and in degree of contamination with other cell constituents (e.g. protein, nucleic acid, β-glucans, etc.). The solubility differences may also reflect variations in the *in vivo* molecular weight of glycogen which are not apparent after the isolation of the polysaccharide.

TABLE II. *Properties of Some Glycogens (Experimental data collected from the review by Manners, 1957)*

| Sample | $[\alpha]_D$ (degrees, $H_2O$) | Average chain length | β-Amylolysis limit (%) | Exterior chain length | Interior chain length |
|---|---|---|---|---|---|
| Yeast (baker's) | +187 | — | 46–49 | — | — |
| Yeast (baker's) | +184 to +188 | 12 | 50 | 8 | 3 |
| Yeast (brewer's) | +198 | 13 | 44 | 8 | 4 |
| Rabbit muscle | +196 | 13 | 45 | 8 | 4 |
| Rabbit liver | +198 | 13 | 43 | 8 | 4 |
| Horse muscle | +198 | 12 | 42 | 7 | 4 |
| Foetal sheep liver | +196 | 13 | 49 | 8–9 | 3–4 |
| Synthetic[a] | +198 | 13–14 | 47 | 8 | 4–5 |

[a] Prepared by the action of yeast branching enzyme on amylopectin (Gunja *et al.*, 1960).

## B. BIOSYNTHESIS

### 1. *Synthesis of α-(1 → 4)-D-Glucosidic Linkages*

The classical researches of Leloir and his coworkers (for a review see Leloir, 1964) have clearly established that, in mammalian tissues, the α-(1 → 4)-D-glucosidic linkages in glycogen are synthesized by the successive transfer of glucose residues from UDPG to a polysaccharide acceptor:

$$n\ \text{UDPG} + [\text{G}]_m \rightarrow n\ \text{UDP} + [\text{G}]_{n+m}$$

where $[\text{G}]_n$ and $[\text{G}]_{n+m}$ represent linear chains of $n$ and $(n + m)$ α-(1 → 4)-linked D-glucose residues. The reaction, which is not reversible, is catalysed by the enzyme glycogen-UDP glucosyl transferase (also known as UDPG α-glucan transglucosylase or glycogen synthetase), and may be assayed either from the rate of formation of UDP (using phosphoenol pyruvate and pyruvic kinase), or the rate of incorporation of $^{14}C$ from UDP-$^{14}C$-glucose into polysaccharide.

A similar enzyme is present in baker's yeast, and was purified about 500-fold by Algranati and Cabib (1960, 1962). The purification involved solubilization of the enzyme by treatment of an extract of freeze-dried yeast with 1% digitonin, followed by an ammonium sulphate fractionation, adsorption of the enzyme on retrograded amylose and then elution from amylose onto glycogen. This procedure gave a preparation devoid of trehalose phosphate synthetase, phosphorylase and branching enzyme. The enzyme was finally separated from glycogen by adsorbing the mixture on calcium phosphate gel, treatment with α-amylase (to degrade the glycogen) and then dissolution of the gel in 0·1 *M* EDTA. The purified enzyme showed a high preference for glycogen as the acceptor substrate. The activity with amylopectin and soluble starch was only 50% and 38% respectively of that towards glycogen, and with glycogen β-limit dextrin, amylose or various maltosaccharides, was only 3–10%. During the reaction, the amount of UDP formed was stoichiometrically equivalent to the amount of glucose incorporated into glycogen. Using UDP-$^{14}C$-glucose, radioactive glycogen was formed, which, on degradation with β-amylase, gave maltose as the only radioactive product. The newly transferred glucose residues were therefore attached by α-(1 → 4)-D-glucosidic linkages to the acceptor molecule.

At pH 7·5, the activity of the enzyme was increased about 1·5–2-fold by 1·0 μ*M* D-glucose 6-phosphate. At pH values above 8, an activation from 5–12-fold was observed. At pH 8·25, the concentration of glucose 6-phosphate required to give maximum activation was 5 m*M*; at this concentration, D-glucosamine 6-phosphate gave 93% of this activation and D-galactose 6-phosphate 30%, but D-glucose 1-phosphate, D-fructose 1-phosphate, D-fructose 6-phosphate and 6-phospho-D-gluconate gave little or no activation. The *in vivo* significance of the activation by glucose 6-phosphate will be considered in detail below (p. 435).

Although most glycogen synthetases show a very high degree of specificity towards the nucleoside diphosphate sugar, Biely *et al.* (1968) have shown that 2-deoxy-D-glucose may be incorporated into yeast glycogen. Incubation of UDP-2-deoxy-D-glucose with purified yeast glycogen synthetase and glycogen gave a polysaccharide which, on digestion with β-amylase, gave a mixture of maltose and a new disaccharide which was identified as 2,2$^{1}$-dideoxymaltose. This indicates that 2-deoxy-D-glucose residues had been transferred to the outer chains of glycogen. Although 2-deoxy-D-glucose is not a natural substrate for yeast metabolism, the above results show that, under certain circumstances, monosaccharides other than D-glucose might be incorporated into glycogen. In mammalian tissues there is evidence for the limited incorporation of D-galactose (Nordin and Hansen, 1963) and D-galactosamine (Keppler *et al.*, 1967) into glycogen, presumably by a special reaction involving glycogen synthetase.

### 2. *Synthesis of* α-(1 → 6)-D-*Glucosidic Linkages*

The linear chains of α-(1 → 4)-linked D-glucose residues are interlinked to form the multiply branched tree-type structure by branching enzymes. Enzyme action (see Fig. 4) involves: (a) the scission of a non-terminal α-(1 → 4)-D-glucosidic linkage; and (b) attachment of the short maltosaccharide chain (c) thus formed to an adjacent chain (A) by means of an α-(1 → 6)-D-glucosidic linkage. Branching enzyme action is therefore a special example of transglucosylation in which a short chain of

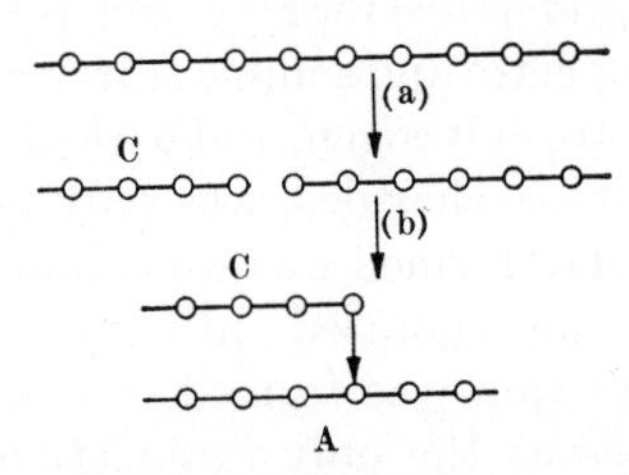


FIG. 4. Formation of inter-chain linkage by a branching enzyme. ○—, indicates an α-(1→4)-linked D-glucose residue; ↓, an α-(1→6)-inter-chain linkage.

glucose residues, rather than a single glucose residue, is transferred. For a review of branching enzymes, see Manners (1968).

Branching enzyme activity was originally detected by Cori and Cori (1943) in mammalian liver and heart tissues; more recently, the enzyme from rabbit muscle has been extensively purified (Brown and Brown, 1966). In all cases, enzyme action caused a decrease in the iodine-staining power and the extent of degradation by phosphorylase or β-amylase of the substrate, which was usually amylopectin.

The presence of branching enzyme in extracts of brewer's yeast was first described by Gunja *et al.* (1960). The enzyme was partly purified by a combination of ammonium sulphate and ethanol fractionation methods, and was free from α-amylase, isoamylase and phosphorylase. This work was carried out in 1955–59 before the discovery of glycogen synthetase, so that assays for the latter enzyme were not carried out.

Yeast branching enzyme caused a marked change in the iodine-staining properties of amylose. There was a large decrease in $E_{max}$ values at all wavelengths (e.g. 63% and 78% at 630 nm after 25 and 40 h incubation respectively) and a displacement of the $\lambda_{max}$ from 630 nm to about 530 nm.

These changes were accompanied by a 50% decrease in β-amylolysis limit, giving an amylopectin-type polysaccharide as the product. On re-incubation of the latter with additional enzyme, further branching took place, and the final end-product was a glycogen-type polysaccharide. A synthetic glycogen prepared from potato amylopectin (average chain length 22 and β-amylolysis limit 53%) had a chain length of 13·5 and a β-amylolysis limit of 47% (Table II, p. 430). The newly formed inter-chain linkages were hydrolysed by isoamylase, and were therefore characterized as α-(1→6)-D-glucosidic in type. Yeast branching enzyme had no action on glycogen itself.

Branching enzyme has also been isolated from extracts of baker's yeast (Kjolberg and Manners, 1963b). The action on amylopectin was similar to that of the enzyme from brewer's yeast. In addition, amylopectin β-limit dextrin was shown to be a substrate. This implies that branched chains of glucose residues may be transferred, under certain conditions, during the conversion of an amylopectin into a glycogen.

Although these studies have provided useful information on the specificity of the yeast branching enzymes, the *in vivo* action presumably involves a sequential action with glycogen synthetase. In this, linear chains of α-(1 → 4)-linked D-glucose residues would be formed, and the chain-lengthening action of the synthetase would continue until the chains were a suitable substrate for branching enzyme. Branching would then occur, with the formation of new A-chains, and glycogen synthetase action would then resume. The alternate action of synthetase and branching enzyme would result in the formation of a multiply-branched molecule, and the regularity of branching within the molecule would be controlled, in part, by the specificity of the branching enzymes.

In mammalian tissues the branching enzymes readily transfer malto-heptaose chains (Verhue and Hers, 1966); it is probable that the yeast branching enzymes are not greatly dissimilar in view of the demonstration of A-chains containing mainly maltoheptaose and -octaose units in baker's yeast glycogen (Bathgate and Manners, 1968).

### 3. In Vivo *Aspects of Glycogen Synthesis*

There is now substantial evidence (Smith *et al.*, 1968) that, in mammalian tissues, a major pathway for the synthesis of glycogen from glucose involves the metabolic pathway shown in Fig. 5. Key steps are: (a) the phosphorylation of D-glucose by hexokinase; (b) the conversion of glucose 6-phosphate into glucose 1-phosphate; (c) the formation of UDPG by the pyrophosphorylase; and finally (d) the combined action of glycogen synthetase and branching enzyme to produce glycogen. This statement does not rule out the possible presence of a minor alternative pathway for glycogen synthesis although, at present, only indirect evidence for this has been reported. For a discussion of this possibility, see Hue and Hers (1969).

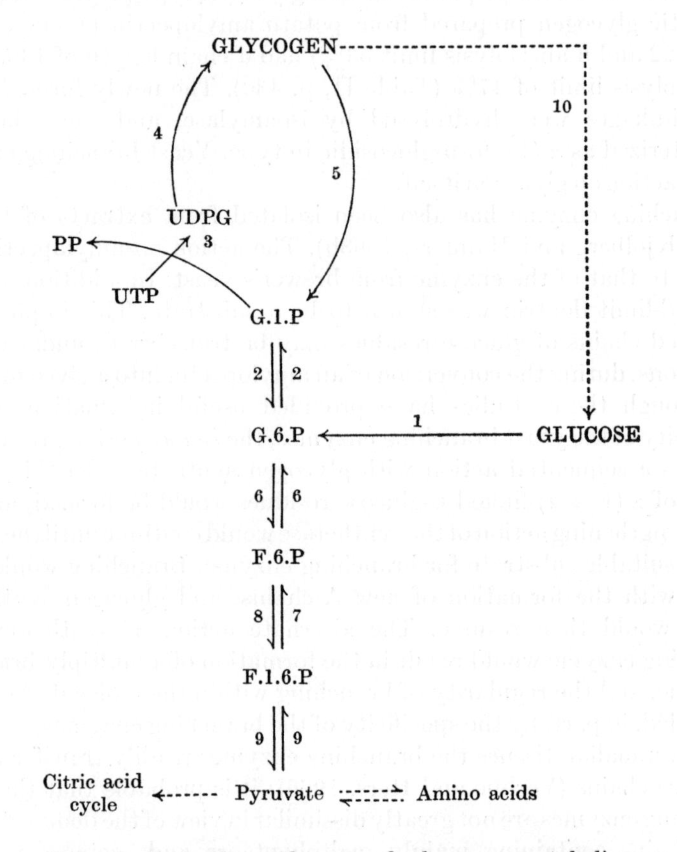


FIG. 5. Probable pathways of glycogen metabolism. Enzymes: 1. hexokinase; 2. phosphoglucomutase; 3. pyrophosphorylase; 4. glycogen synthetase and branching enzyme; 5. phosphorylase and amylo-1,6-glucosidase; 6. phosphoglucoisomerase; 7. fructose 1,6-diphosphatase; 8. phosphofructokinase; 9. glycolytic pathway; 10. non-phosphorolytic breakdown of glycogen.

In yeast cells, it is probable that the major pathway for glycogen synthesis involves the stages shown in Fig. 5, although it must be emphasized that the individual enzymes concerned are not chemically identical to their animal counterparts. For example, some of the properties of yeast and muscle glycogen synthetases are different, even though their catalytic activities are similar. Nevertheless, it is believed that the activities of the various yeast enzymes are adequate to account for the observed rates of glycogen synthesis, depending upon the metabolic state of the yeast cells, and that the whole process may be regulated, at least in part, by the activity of the glycogen synthetase.

The activation of yeast glycogen synthetase by glucose 6-phosphate has been examined in detail by Rothman and Cabib (1966, 1967a). The addition of certain anions (e.g. chloride, phosphate, pyruvate) caused marked inhibition (30–65%) which was reversed by 10 m*M* glucose 6-phosphate. Hence, the activity of the enzyme in any particular assay would be dependent on the nature and concentration of any anions which were present. Kinetic studies showed that the anions were allosteric, i.e. they interacted with the enzyme at sites which were distinct from those at which UDPG and glycogen were bound. These effects were observed with concentrations of anion inhibitors which were much greater than those existing *in vivo*. However, the reversal of this inhibition and the activation of the enzyme in the absence of any inhibitor by glucose 6-phosphate may be physiologically important in providing a regulatory system for glycogen synthesis, particularly at the *in vivo* pH value of 6.

Various nucleotides (e.g. ATP and ADP) may also serve as allosteric inhibitors, the effect of which is reversed by glucose 6-phosphate. Rothman and Cabib (1967b) have suggested that the combined concentration of the adenine nucleotides may be sufficient to ensure almost complete inhibition of the enzyme in the absence of glucose 6-phosphate. The activity of glycogen synthetase would then be determined by the concentration of glucose 6-phosphate. The latter is related to the activity of phosphofructokinase which, in turn, is determined by the ATP:AMP ratio. This kinase is inhibited by ATP and activated by AMP, so that at high ratios it would be inhibited, causing an increase in the concentration of glucose 6-phosphate, and a relief of the inhibition of glycogen synthetase by the adenine nucleotides. When the ATP:AMP ratio was low, the rate of glycogen synthesis would be reduced. Yeast phosphofructokinase is activated by ammonium ions (Muntz, 1947), so that in the presence of this ion glycogen synthetase activity would also be lowered. This regulatory control would provide an explanation for the report of Trevelyan and Harrison (1956) that yeast cells readily form glycogen when grown on a medium containing glucose and salts, but that synthesis ceases on the addition of ammonium ions.

The possible heterogeneous nature of glycogen *in vivo* is considered by Eaton (1960, 1961) to have biochemical significance. From kinetic studies of the endogenous respiration of yeast, he has concluded that there are two metabolically distinct "pools" of glycogen, one of which may be metabolized either aerobically or anaerobically (fermentable glycogen) and the other only aerobically (oxidizable glycogen). By ultracentrifugation, he was able to obtain two fractions of glycogen, one (60% of the total) having a sedimentation coefficient of 138S, and the remaining 40% had a value of 5S. A similar analysis of glycogen from cells from which one pool had been removed by fermentation (i.e. oxidizable glycogen) also showed heavy and light components with sedimentation coefficients of 121S and 5S respectively. Further analysis showed that 80% of the fermentable glycogen was derived from the heavier component. An attempt was made to examine the structure of the heavy and light components but, unfortunately, average chain lengths were determined by the periodate oxidation method of Perlin (1954). This is now known to be inaccurate (Manners and Wright, 1961). The significance of the results reported by Eaton (1961) is therefore uncertain (average chain lengths of less than five glucose residues are inconsistent with Table II), and his conclusion that the fermentable glycogen pool consists largely of the exterior chains of the heavier component requires confirmation. It is possible that the fermentable and oxidizable glycogens represent degradation by a phosphorylase-amylo-1,6-glucosidase system and a non-phosphorolytic pathway, respectively.

The isolation of two different molecular-weight forms of glycogen raises the question as to whether these are synthesized by the same or different multi-enzyme systems, and at the same or different sites in the yeast cell. Clearly, further experimental work is required on this aspect of glycogen synthesis.

It has been reported that certain cultures of brewing strains of *Sacch. cerevisiae* contained about 0·5% of mutants which were relatively deficient in glycogen and trehalose (Chester, 1967, 1968). The actual carbohydrate contents depended upon the growth conditions (Chester and Byrne, 1968; see Table III). The deficiency in glycogen was shown qualitatively by the lack of iodine staining of certain cells, although analysis showed that glycogen was not completely absent, and the properties of glycogens from the parent and mutant were identical. The effect was most marked when the yeast was grown aerobically in a medium containing 1% glucose (Table III). The mutant frequency could be increased to about 5% by ultraviolet irradiation (Chester, 1968).

The UDPG concentration of both normal and mutant yeasts was measured by Chester and Byrne (1968). The concentration of the nucleoside diphosphate sugar increased as growth progressed, but in all conditions the mutant contained significantly less UDPG than the parent.

In the presence of a high concentration of glucose (8%), glycogen synthesis in both the parent and mutant yeasts was greatly stimulated. A stimulation by glucose of liver glycogen synthesis, and the activity of

TABLE III. *Carbohydrate Composition of Normal and Mutant Strains of* Saccharomyces cerevisiae (*Data from Chester and Byrne, 1968*)

| Carbohydrate | Carbohydrate content as proportion of non-carbohydrate dry weight of yeast (%) | | | | | |
|---|---|---|---|---|---|---|
| | 8% Glucose (anaerobic) | | 1% Glucose (anaerobic) | | 1% Glucose (aerobic) | |
| | Parent | Mutant | Parent | Mutant | Parent | Mutant |
| Trehalose | 1·7 | 0·8 | 0·9 | 0·6 | 0·9 | 0·4 |
| Alkali-soluble glycogen | 7·3 | 7·4 | 3·0 | 3·6 | 1·7 | 1·5 |
| Acid-soluble glycogen | 70·0 | 54·8 | 40·0 | 25·1 | 18·6 | 7·3 |
| Mannan | 7·5 | 9·3 | 9·9 | 11·6 | 9·5 | 10·0 |
| Glucan | 9·4 | 9·5 | 8·8 | 9·1 | 6·5 | 6·5 |
| Sum of fractions | 95·9 | 81·8 | 62·6 | 50·0 | 37·2 | 25·7 |

glycogen synthetase in liver homogenates, has been observed by De Wulf and Hers (1957). This activation is distinct from the effect of glucose 6-phosphate, and is due to the conversion of the enzyme from a native or inactive form (*b*) into another form (*a*) which is active under physiological conditions (De Wulf *et al.*, 1968). It is possible that yeast glycogen synthetase also exists in more than one form, these being enzymically interconvertible; it is therefore clear that the levels of glucose 6-phosphate and glucose all play a part in controlling the rate of glycogen synthesis, in addition to their normal roles as intermediary metabolites.

## References

Abdullah, M., Catley, B. J., Lee, E. Y. C., Robyt, J., Wallenfels, K. and Whelan, W. J. (1966). *Cereal Chem.* **43**, 111–118.

Algranati, I. D. and Cabib, E. (1960). *Biochim. biophys. Acta* **43**, 141–142.

Algranati, I. D. and Cabib, E. (1962). *J. biol. Chem.* **237**, 1007–1013.

Archibald, A. R., Fleming, I. D., Liddle, A. M., Manners, D. J., Mercer, G. A. and Wright, A. (1961). *J. chem. Soc.* 1183–1190.

Bacon, J. S. D. and Farmer, V. C. (1968). *Biochem. J.* **110**, 34–35P.

Bathgate, G. N. and Manners, D. J. (1968). *Biochem. J.* **107**, 443–445.

Bender, H., Lehmann, J. and Wallenfels, K. (1959). *Biochim. biophys. Acta* **36**, 309–316.

Biely, P., Farkas, V. and Bauer, S. (1968). *Biochim. biophys. Acta* **158**, 487–488.

Birch, G. G. (1963). *Adv. Carbohyd. Chem.* **18**, 201–225.

Brown, B. I. and Brown, D. H. (1966). *Meth. Enzym.* **8**, 395–403.

Bryce, W. A. B., Greenwood, C. T., Jones, I. G. and Manners, D. J. (1958). *J. chem. Soc.* 711–715.
Cabib, E. and Leloir, L. F. (1958). *J. biol. Chem.* **231**, 259–275.
Chester, V. E. (1963). *Biochem. J.* **86**, 153–160.
Chester, V. E. (1964). *Biochem. J.* **92**, 318–323.
Chester, V. E. (1967). *Nature, Lond.* **214**, 1237–1238.
Chester, V. E. (1968). *J. gen. Microbiol.* **51**, 49–56.
Chester, V. E. and Bryne, M. J. (1968). *Archs Biochem. Biophys.* **127**, 556–562.
Cori, G. T. and Cori, C. F. (1943). *J. biol. Chem.* **151**, 57–63.
Crane, F. K. (1962). *In* "The Enzymes", (P. D. Boyer, H. Lardy and K. Myrbäck, eds.), Vol. 6, pp. 47–66. Academic Press, New York.
De Wulf, H. and Hers, H. G. (1957). *Eur. J. Biochem.* **2**, 50–56.
De Wulf, H., Stalmans, W. and Hers, H. G. (1968). *Eur. J. Biochem.* **6**, 545–551.
Eaton, N. R. (1960). *Archs Biochem. Biophys.* **88**, 17–25.
Eaton, N. R. (1961). *Archs Biochem. Biophys.* **95**, 464–469.
Gunja, Z. H., Manners, D. J. and Khin Maung (1960). *Biochem. J.* **75**, 441–450.
Gunja, Z. H., Manners, D. J. and Khin Maung (1961). *Biochem. J.* **81**, 392–398.
Harden, A. and Young, W. J. (1912). *J. chem. Soc.* 1928–1930.
Hobson, P. N., Whelan, W. J. and Peat, S. (1951). *J. chem. Soc.* 1451–1459.
Hue, L. and Hers, H. G. (1969). *FEBS Letters,* **3**, 41–43.
Jeanloz, R. (1944). *Helv. chim. Acta* **27**, 1501–1509.
Jones, G. H. and Ballou, C. E. (1969). *J. biol. Chem.* **244**, 1043–1059.
Keppler, D., Reutter, W. and Decker, K. (1967). Abstracts 4th FEBS Meeting, Oslo, p. 90.
Kjolberg, O. and Manners, D. J. (1963a). *Biochem. J.* **86**, 258–262.
Kjolberg, O. and Manners, D. J. (1963b). *Biochem. J.* **86**, 10–11P.
Koch, E. M. and Koch, F. C. (1925). *Science, N.Y.* **61**, 570–572.
Lee, E. Y. C., Nielson, L. D. and Fischer, E. H. (1967). *Archs Biochem. Biophys.* **121**, 245–246.
Leloir, L. F. and Cabib, E. (1953). *J. Am. chem. Soc.* **75**, 5445.
Leloir, L. F. (1964). *Biochem. J.* **91**, 1–8.
Manners, D. J. (1957). *Adv. Carbohyd. Chem.* **12**, 261–298.
Manners, D. J. (1962). *Adv. Carbohyd. Chem.* **17**, 371–429.
Manners, D. J. (1968). *In* "Control of Glycogen Metabolism", (W. J. Whelan, ed.), pp. 83–100. Universitetsforlaget, Oslo.
Manners, D. J. and Khin Maung (1955). *J. chem. Soc.* 867–870.
Manners, D. J. and Masson, A. J. (1969). FEBS Letters **4**, 122–124.
Manners, D. J. and Patterson, J. C. (1966). *Biochem. J.* **98**, 19–20C.
Manners, D. J. and Rowe, K. L. (1968a). *Carbohyd. Res.* **9**, 107–121.
Manners, D. J. and Rowe, K. L. (1968b). *Biochem. J.* **110**, 35P.
Manners, D. J. and Wright, A. (1961). *J. chem. Soc.* 2681–2684.
Meyer, K. H. and Jeanloz, R. (1943). *Helv. chim. Acta* **26**, 1784–1798.
Misaki, A., Johnson, J., Kirkwood, S., Scaletti, J. V. and Smith, F. (1968). *Carbohyd. Res.* **6**, 150–164.
Munch-Petersen, A. (1955). *Acta chem. scand.* **9**, 1523–1536.
Muntz, J. A. (1947). *J. biol. Chem.* **171**, 653–665.
Najjar, V. A. (1962). *In* "The Enzymes", (P. D. Boyer, H. Lardy and K. Myrbäck, eds.). Vol. 6, pp. 161–178. Academic Press, New York.
Nordin, J. H. and Hansen, R. G. (1963). *J. biol. Chem.* **238**, 489–494.
Northcote, D. H. (1953). *Biochem. J.* **53**, 348–352.
Panek, A. (1962). *Archs Biochem. Biophys.* **98**, 349–355.
Peat, S., Whelan, W. J. and Edwards, T. E. (1955). *J. chem. Soc.* 355–359.

Perlin, A. (1954). *J. Am. chem. Soc.* **76**, 4101–4103.
Pfäffli, S. and Suomalainen, H. (1960). *Suom. Kemistilehti* B **33**, 61–65.
Rothman, L. B. and Cabib, E. (1966). *Biochem. biophys. Res. Commun.* **25**, 644–650.
Rothman, L. B. and Cabib, E. (1967a). *Biochemistry, N.Y.* **6**, 2098–2106.
Rothman, L. B. and Cabib, E. (1967b). *Biochemistry, N.Y.* **6**, 2107–2112.
Smith, E. E., Taylor, P. M. and Whelan, W. J. (1968). *In* "Carbohydrate Metabolism and its Disorders", (F. Dickens, P. J. Randle and W. J. Whelan, eds.). Vol. 1, pp. 89–138. Academic Press, New York.
Steiner, A. and Cori, C. F. (1935). *Science, N.Y.* **82**, 422–423.
Stewart, L. C., Richtmyer, N. K. and Hudson, C. S. (1950). *J. Am. chem. Soc.* **72**, 2059–2061.
Suomalainen, H. and Pfäffli, S. (1961). *J. Inst. Brew.* **67**, 247–254.
Trevelyan, W. E. (1958). *In* "The Chemistry and Biology of Yeasts", (A. H. Cook, ed.), pp. 369–436. Academic Press, New York.
Trevelyan, W. E. and Harrison, J. S. (1952). *Biochem. J.* **50**, 298–303.
Trevelyan, W. E. and Harrison, J. S. (1956). *Biochem. J.* **63**, 23–33.
Trevelyan, W. E., Mann, P. F. E. and Harrison, J. S. (1954). *Archs Biochem. Biophys.* **50**, 81–91.
Verhue, W. and Hers, H. G. (1966). *Biochem. J.* **99**, 222–227.
Walker, G. J. (1968). *Biochem. J.* **108**, 33–40.
Whelan, W. J. (1955). *In* "Modern Methods of Plant Analysis", (K. Paech and M. V. Tracey, eds.), Vol. II, pp. 145–196. Springer-Verlag, Berlin.
Whelan, W. J. (1956). *In* "Encyclopaedia of Plant Physiology", (W. Ruhland, ed.), Vol. VI, pp. 154–240. Springer-Verlag, Berlin.
Whelan, W. J. (1964). *In* "Control of Glycogen Metabolism", (W. J. Whelan and M. P. Cameron, eds.), p. 402. Churchill, London.

# *Chapter 11*

# Biochemistry of Morphogenesis in Yeasts

S. BARTNICKI-GARCIA AND IAN MCMURROUGH*

*Department of Plant Pathology, University of California, Riverside, U.S.A.*

I. INTRODUCTION . . . . . . . . . . . . . 441
A. Experimental Approach . . . . . . . . 441
B. Yeast: Morphogenetic Concepts . . . . . . . 444
II. CELL WALL BIOCHEMISTRY: MORPHOGENETIC ASPECTS . . . . . 444
A. Cell Wall Composition . . . . . . . . . 444
B. Cell Wall Biogenesis . . . . . . . . . 453
III. VEGETATIVE DEVELOPMENT . . . . . . . . . 464
A. Biochemical Aspects of Cell Division . . . . . . 464
B. Environmental Control . . . . . . . . . 473
IV. SEXUAL SPORULATION . . . . . . . . . . 483
V. ACKNOWLEDGEMENTS . . . . . . . . . . 487
REFERENCES . . . . . . . . . . . . . 487

## I. Introduction

### A. EXPERIMENTAL APPROACH

Biochemistry of morphogenesis is a study of chemical and metabolic events underlying morphological differentiation. In the yeasts, cellular morphology is dictated by the shape of the cell wall. Hence, according to the experimental approach pioneered by W. J. Nickerson, biochemical studies of yeast morphogenesis may be conveniently focused on those internal and external factors which govern the architecture of the cell wall. In this school of thought, the major problems of the biochemistry of yeast morphogenesis concern the elucidation of the molecular structure of the polymers which constitute the yeast cell wall, and the

* Present address: Arthur Guinness Son & Co. Ltd., St. James Gate, Dublin, Eire.

unravelling of the cellular mechanisms which synthesize, assemble or modify these polymers. A precise understanding of the molecular structure of cell wall polymers might answer the most fundamental questions of shape determination but, at our present depth of knowledge, the composition of the cell wall alone cannot adequately serve to explain its three-dimensional geometry. For instance, the extent to which the various cell wall polymers can spontaneously associate in predetermined configurations is not known. Conceivably, the intrinsic properties of cell wall macromolecules might confer a potentiality for self-assembly. If this is shown not to be so, then the central problem of biochemical morphogenesis will be largely one of explaining the spatial orientation of cell wall polymers by their synthesizing enzymes. It seems likely that any satisfactory explanation of such mechanism will feature the participation of some subcellular structure which underlies the site of synthesis and dictates the pattern of cell wall assembly.

Much of the existing knowledge of potential significance in the biochemistry of yeast morphogenesis pertains to the influence of the environment on morphological development. Of particular interest has been the role of nutritional factors, factors which can be conveniently manipulated to effect experimental control of form development. *Bona fide* morphogenetic sequences (budding, germination, sporulation) can be regulated by the composition of the external milieu but, admittedly, some of the environmentally induced variations in yeast morphology may be regarded as aberrations caused by extreme growth conditions. These derangements could nevertheless be exploited to disclose useful information on the molecular basis of form determination.

Having defined the principal problems of yeast morphogenesis with the intention of indicating the breadth of the topic, it becomes imperative to introduce sizeable restrictions on the range of our subsequent discussions to avoid undue or lengthy repetition of material covered elsewhere in these volumes. At the risk of appearing monoideistic, most of our deliberations will be concentrated on the yeast cell wall and the exogenous control of vegetative morphogenesis. For the most part, we shall refrain from discussing the morphogenetic role of the internal biochemical machinery of the cell though we recognize that this is a promising, but essentially virgin, field harbouring some of the most fundamental secrets of biochemical differentiation.

Although fruitful inroads have been made into diverse aspects of the biochemistry of yeast morphogenesis, the overall picture is far too clouded and fragmented to allow more than the most speculative generalizations. This situation precludes a unified treatment of the topic and, instead, we must deal separately with some salient attempts at elucidating particular cases of yeast morphogenesis.

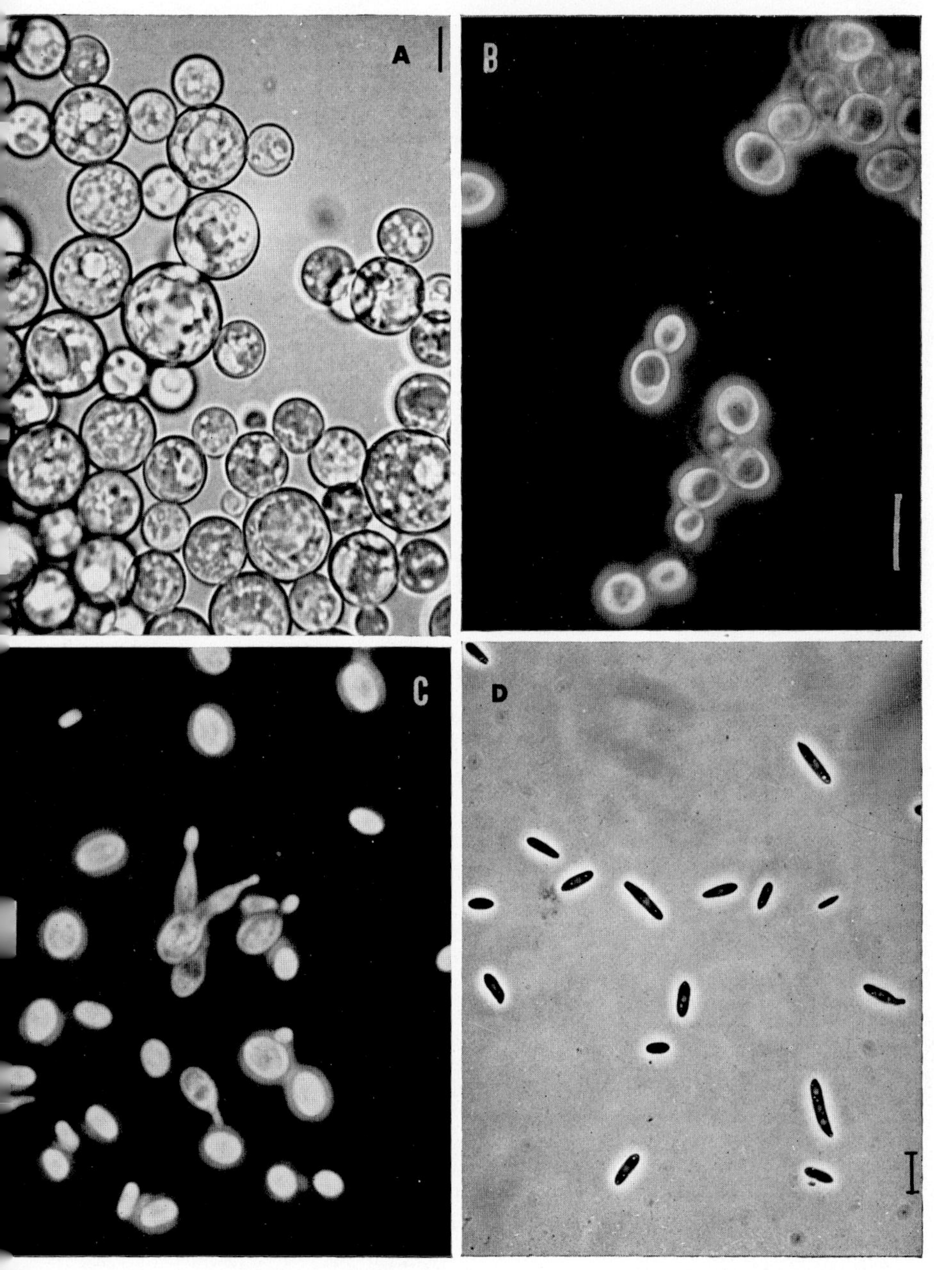


FIG. 1. Examples of yeast morphogenesis in the Fungi: A, a Zygomycete, *Mucor rouxii* artnicki-Garcia and Nickerson, 1962); B, a Hemiascomycete, *Saccharomyces cerevisiae*: an ascomycetous Deuteromycete, *Verticillium albo-atrum* (reproduced by courtesy of C. Wang); D, a Heterobasidiomycete, *Ustilago cynodontis* (Talou and Tavlitzki, 1969, reproduced by courtesy of B. Talou). Magnification marker = 10 μm.

### B. YEAST: MORPHOGENETIC CONCEPTS

It is well known that the yeasts do not constitute a natural taxonomic unit (see Kreger van Rij, 1969; Vol. I, Chapter 1) but a heterogenous, sometimes quite arbitrary, assembly of higher fungi. For our purposes the broad taxonomic definition of a yeast is unsatisfactory, and consequently we shall endeavour to adopt a simpler morphogenetic concept. The most characteristic feature of a majority of the species of the yeasts is a predominant growth habit as single cells which multiply by budding, i.e. typical yeast cells. Accordingly, throughout this chapter vegetative "yeast morphogenesis" refers specifically to the formation of single vegetative cells by a budding process. Fission yeasts have, however, also been included, as their mode of cell division by binary fission is closely related to the budding process of certain yeasts (see Section II.B.2.*c*, p. 460). Clearly this definition excludes some organisms ordinarily classified among the yeasts which show little or no tendency to form yeast cells. Conversely, yeast morphogenesis is not exclusively found in the yeasts, but is also manifested by many other fungi not usually regarded as such. Examples of yeast morphogenesis in the Zygomycetes, Ascomycetes and Basidiomycetes are shown in Fig. 1.

For obvious reasons our emphasis will be on organisms generally considered to belong to the yeasts. However, pertinent examples of yeast morphogenesis in other fungi will be included. By this treatment of yeast morphogenesis we are deliberately seeking to encourage comparisons of the biochemical processes underlying morphogenesis. It must be understood, however, that in doing so we do not imply that the various morphogenetic phenomena cited necessarily have a common biochemical foundation in all fungi. Indeed, considerations of the biochemical diversity exhibited by the major classes of fungi (*cf.* Bartnicki-Garcia, 1970) would tend to predict the contrary.

## II. Cell Wall Biochemistry: Morphogenetic Aspects

### A. CELL WALL COMPOSITION

#### 1. *General*

The yeast cell wall is an envelope consisting of intermeshed polysaccharide microfibrils embedded in a complex matrix composed of various polysaccharides, proteins and lipids. In recent years we have witnessed significant advances in our knowledge of the gross features of cell-wall chemistry of numerous yeast species, but our comprehension of the fine structure and interrelationships of the various cell-wall polymers remains quite meagre. Most studies on cell-wall structure have been conducted with cell walls derived from *Saccharomyces cerevisiae* or *Candida albicans*. Detailed information on yeast cell wall composition appears in this volume (Chapter 5, p. 135) and in previous reviews (Phaff, 1963; Nickerson, 1963; Northcote, 1963). It would not serve our present purpose to attempt a review of this literature. Instead we will select findings with potential morphogenetic implication.

Elucidation of the cell-wall components, if any, responsible for shape determination, rather than merely shape maintenance, is certainly of paramount importance, So far, however, it has not been possible to assign the role of shape determination to any one component of the yeast cell wall. Possibly, no single wall component can be solely responsible for the morphology of a cell. The isolation of much of the cell-wall substance as glycoprotein complexes (cf. Nickerson *et al.*, 1961) indicates that the cooperation of polysaccharides, proteins, and perhaps lipids as well, is essential for form development.

The importance of cell-wall composition in morphogenesis is mainly supported by indirect or circumstantial evidence. This evidence may be of an ontogenetic nature, and relates modifications in cell-wall chemistry to cellular differentiation in the life cycle of a given organism (see Sections II.A.2, p. 447; II.A.3, p. 451). In this category we could also include comparable observations made on the cell walls regenerated by yeast protoplasts or sphaeroplasts. Cells with newly regenerated cell walls often assume highly distorted shapes concomitant with pronounced aberrations in cell-wall composition (see Section II.B.3, p. 462). Protoplasts of *Sacch. cerevisiae* which are experimentally prevented from rebuilding a complete cell wall are unable to undergo morphogenesis and eventually die (Svoboda and Nečas, 1968; Farkaš *et al.*, 1969). A chemically complete cell wall is seemingly essential for normal morphogenesis. Furthermore, it seems plausible that morphological differentiation may depend to a large extent on specific variation in the proportion of, and/or the interaction among, structural components of the cell wall (*cf.* Bartnicki-Garcia, 1968a). Other types of evidence arise in a phylogenetic context, such as the observation that cell-wall composition is similar among taxonomically related organisms, and tends to differ markedly among organisms with divergent morphological features. For instance, the cell walls of budding yeasts are rich in mannans, whereas the cell walls of mycelial fungi have only small quantities of mannose polymers (Garzuly-Janke, 1940; for further discussion see Bartnicki-Garcia, 1968a; 1970).

### 2. *Polysaccharides*

Three main types of polysaccharides, glucans, mannans and chitin occur in yeasts. The generic terms "glucan" and "mannan" are used because the fine structure of these polysaccharides has not been determined in many of the examples cited. It should be remembered, however, that not only may the glucans and mannans of various yeast species differ, but there may be more than one type of glucan or of mannan present in the cell walls of a particular organism. These three types account for more than 80% of the dry weight of most yeast cell walls (Phaff, 1963), but their relative proportions may differ widely among different genera. To a great extent the taxonomic position of a particular yeast is correlated with the composition of its cell wall. This correlation seems to apply to the entire spectrum of fungi (Bartnicki-Garcia, 1968a) and provides a biochemical basis for testing phylogenetic relationships (Bartnicki-Garcia, 1970). Although only limited data are available, the cell walls of ascomycetous yeasts may be recognized as being chemically distinct from those of basidiomycetous yeasts. In ascomycetous yeasts (e.g. *Saccharomyces*, *Hansenula* spp.) and their anascosporogenous forms (e.g. *Candida* spp.), glucan and mannan are the principal cell-wall constituents (Northcote and Horne, 1952; Falcone and Nickerson, 1956; Miller and Phaff, 1958). In contrast, basidiomycetous yeasts (*Sporobolomyces*) and related forms (*Rhodotorula*) contain mainly chitin (Kreger, 1954) and mannan (Crook and Johnston, 1962) with smaller amounts of galactose and fucose. Interestingly the mannan from basidiomycetous yeasts seems to differ from that of ascomycetous yeasts in not forming insoluble copper-complexes (Nickerson, 1963). It is perhaps significant that the cell walls of yeasts with unusual morphological features also have distinctive compostion. Thus, the "triangular" yeast *Trigonopsis variabilis* contains little or no alkali-insoluble polysaccharides (e.g. "yeast glucan", chitin). The fission yeasts, *Schizosaccharomyces* spp., have "yeast glucan" but seem to lack chitin and mannan (Kreger, 1954; Crook and Johnston, 1962). Recently, however, mannan and galactan were found to be minor components of cell walls in *Schizosaccharomyces pombe* (Deshusses *et al.*, 1969). Of further interest are the seemingly parallel changes in cell-wall composition which accompany the progressive morphological variations of genera within the Hemiascomycetes. Concomitant with the transition from a typically budding yeast habit (e.g. *Saccharomyces*) to a predominantly mycelial growth form (e.g. *Endomycopsis* → *Endomyces* → *Eremascus*) is a diminution in the cell-wall mannan content (Kreger, 1954). The proportion of chitin, on the other hand, increases until the cell-wall composition approximates that of typical mycelial fungi of the Euascomycetes. These phylogenetic and biochemical observations lend credence to the belief that the morphology of the yeast cell wall may depend on certain mutual interactions of its component polysaccharides. This belief is reinforced by the observed alterations in the composition of cell-wall polysaccharides from certain organisms grown under different environmental conditions. Although no drastic departures in cell-wall composition have been recorded as a result of different incubation conditions (see, for example, Reuvers *et al.*, 1969), the relative proportion of different polysaccharides (see below) and other cell-wall components may vary within certain limits. Often these alterations in wall composition are accompanied by changes in the morphology of the cell. Hence the contention that the morphogenetic effect of the environment is mediated, at least in part, *via* alterations in the chemical structure of the cell wall.

In *Saccharomyces* and *Candida* spp., a highly insoluble glucan ("yeast glucan"), currently believed to be a highly branched β-1-3-glucan with β-1-6 linkages at the branching units (Misaki *et al.*, 1968; Manners and Patterson, 1966), has been regarded as the skeletal support of the cell wall. This is based on the observation that, after removal of essentially all other components, the residual insoluble microfibrillar network of "yeast glucan" continues to display the outline of the yeast cell (Northcote and Horne, 1952; Houwink and Kreger, 1953). The more soluble polysaccharides of these cell walls probably occur as non-fibrillar gluco-mannanprotein complexes (Kessler and Nickerson, 1959). In basidiomycetous yeasts, chitin probably bears the burden of skeletal support. It is not possible to assess the relative architectural importance of fibrillar and non-fibrillar wall components, but an analogy with reinforced concrete may be appropriate, the microfibrillar glucan being the steel framework and the alkali-soluble glycoproteins the cementing material.

As a result of disturbances in the supply of growth factors, changes in cell-wall polysaccharide composition have been noted. Concurrently with an impairment in the final steps of the cell division process (Section III.A.4, p. 472), inositol deficiency also affects the morphology and polysaccharide content of cell walls of *Schizosaccharomyces pombe*. "Abnormal" cells, characterized by marked elongation and inhibition of transverse septum formation (Schopfer *et al.*, 1962), are obtained by cultivation in the presence of anti-inositol metabolites (isomytilitol or 2-C-methylene-*myo*-inositol oxide). Cells with normal morphology contained a mannan : glucan ratio of 15 : 85 in their cell walls; in abnormal cells, mannan was partially replaced by galactan (Deshusses *et al.*, 1969).

When typical yeast forms are compared with elongated or mycelial forms of the same organism, gross differences in the relative proportions of cell-wall polysaccharides are often detected. In agreement with the correlation suggested above (p. 446) between mannan content and vegetative morphology in Hemiascomycetes, the walls from yeast cells of *Mucor rouxii* were found to be considerably richer in mannose-containing polymers than mycelial walls (Bartnicki-Garcia and Nickerson, 1962a). In *Pullularia pullulans* the content of cell-wall mannose of the yeast forms was also higher than that of filamentous cells (Brown and Nickerson, 1965). An interesting and seemingly pertinent correlation between cell shape and cell-wall polysaccharide composition, disclosed by Garcia-Mendoza and Novaes-Ledieu (1968), occurs during cell-wall regeneration by protoplasts of *Candida utilis*. Regenerating protoplasts of *C. utilis* produce initially tubular forms from which the normal ellipsoidal cells later originate (Uruburu *et al.*, 1968). The newly regenerated tubular cell walls are rich in chitin but contain only traces of mannan; in normal cell walls the reverse relative composition obtains. This relationship, artificially elicited in one organism, appears analogous to the phylogenetic correlation, previously mentioned (p. 446), between morphology and cell wall composition in different members of the Hemiascomycetes. These two separate lines of evidence both associate a predominance of mannan over chitin with an ellipsoidal morphology, while filamentous or tubular cell walls are characterized by a lower mannan and higher chitin content.

In apparent contradiction to these relationships, elongated cell walls of *Hansenula schneggii* (Sundhagul and Hedrick, 1966) were shown to contain 2·5 times as much Fehling's-precipitable mannan as ellipsoidal forms. Also, the cell walls from the yeast form of *Histoplasma capsulatum*, a dimorphic pathogenic fungus, contained less mannan and considerably more chitin than the walls of the mycelial form (Domer *et al.*, 1967). Chitin content was also higher in the yeast forms of another dimorphic pathogenic fungus, *Paracoccidioides brasiliensis* (Kanetsuna *et al.*, 1969), than in their mycelial counterparts. These last examples deny any universal correlation between overall mannan and chitin content of the cell wall and cell morphology, and question the proposed role of cell-wall mannan as a determinant of yeast morphogenesis (Bartnicki-Garcia, 1963). Such a role, however, should not be summarily dismissed. Contradictory relationships could conceivably be ascribed to the failure to discriminate between different kinds of mannans (homopolymers, heteropolysaccharides, phosphopolymers) present in yeast cell walls. Possibly, each kind of mannan may have a different architectural influence on the cell wall. Also, the extent of the interaction between mannans and other cell-wall components may play a determinant role in wall architecture, but this is yet to be ascertained.

As noted above, a certain latitude in cell wall composition is permitted without apparent changes in morphology. For instance, between extremes of cultivation conditions in continuous culture the mannan and glucan contents of walls of *Saccharomyces cerevisiae* varied from 51% to 38% and from 32% to 47%, respectively, without ostensible changes in the morphology of the ellipsoidal cells (McMurrough and Rose, 1967). Conversely, changes in morphology may not be accompanied by gross

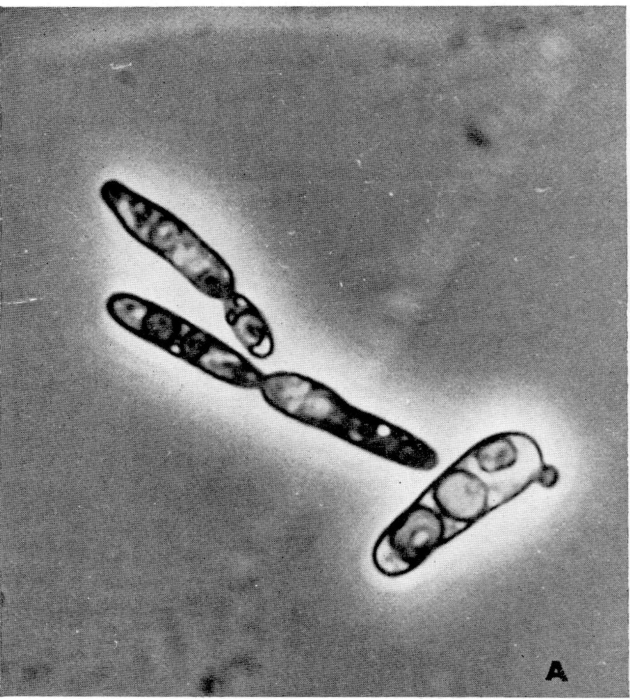


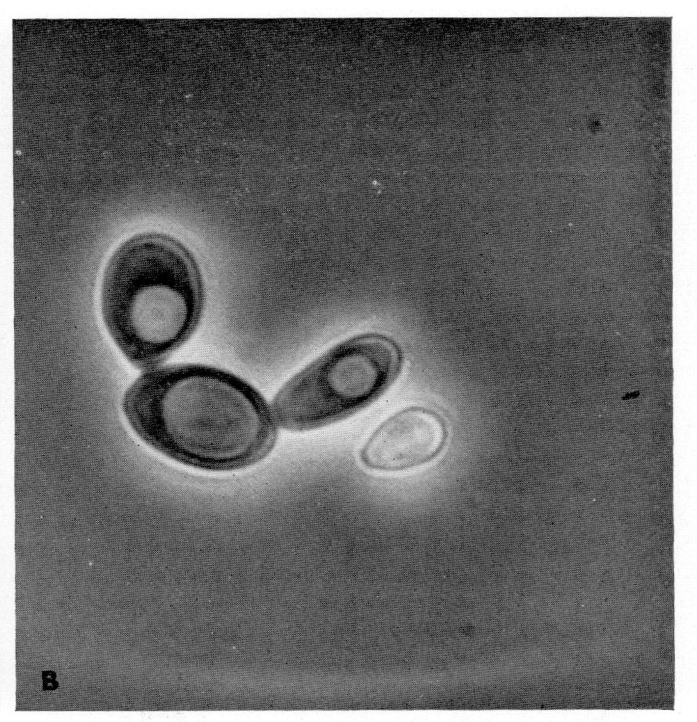


FIG. 2. Nutritional control of morphology in *Saccharomyces cerevisiae* grown in chemostat cultures (McMurrough and Rose, 1967). A, Elongated cells obtained under conditions of nitrogen ($NH_4^+$) limitation at a dilution rate of 0·33 $h^{-1}$; B, ellipsoidal cells obtained under conditions of glucose limitation at a dilution rate of 0·35 $h^{-1}$. Magnification, × 2,500.

departures in cell-wall composition. For instance, when *Sacch. cerevisiae* was grown in continuous culture under conditions of ammonium limitation, the resulting cells were distinctly elongated (Fig. 2) but the values of overall mannan and glucan content were not significantly different from those of ellipsoidal cells grown under oxygen deficiency (McMurrough and Rose, 1967). It is quite likely that gross chemical analyses may not reveal fine structural differences of importance in cell-wall morphogenesis. With this in mind, an attempt was made to fractionate the cell-wall polysaccharides of *Sacch. cerevisiae* on the basis of alkali solubility. It was found that elongated forms contained only about two thirds as much hot-alkali insoluble glucan but twice as much cold-alkali soluble mannan as the ellipsoidal cells (McMurrough and Rose, 1967). Electron microscopy also indicated differences in cell-wall structure between these two forms. On comparing mycelial and yeast cell walls of *Paracoccidioides brasiliensis*, Kanetsuna *et al.* (1969) noted a similar total glucan content but with major differences in certain important properties. The glucan from yeast walls was almost entirely soluble in alkali whereas the mycelial glucan was only 60–65% alkali-soluble, the rest being an insoluble glucan with different chemical properties and different susceptibility to enzymic digestion. In conclusion, there is reason to expect that a deeper probing of the chemical anatomy of the cell wall might reveal the chemical basis for morphological alterations.

### 3. *Proteins*

Seemingly proteins perform an important role in maintaining the integrity of the yeast cell wall, but the manner in which they function is not completely clear. There is usually about 10% protein in yeast cell walls, though in *Saccharomycopsis guttulata* 40% of the cell wall was reported to be protein (Shifrine and Phaff, 1958). Much of the yeast cell-wall protein occurs in complexes with polysaccharides (Eddy, 1958; Kessler and Nickerson, 1959; Korn and Northcote, 1960), consequently one possible role of protein would be to bind together the different cell-wall polysaccharides (Nickerson, 1963).

Lysis of the cell walls of *Saccharomyces cerevisiae* by proteolytic enzymes testifies to the prominent contribution of protein to the stability of the cell-wall structure in this yeast (Nickerson, 1963). On the other hand, the cell walls of *C. albicans*, which have a similar gross composition, are less susceptible to enzymic proteolysis. This suggests either a variable contribution of proteins to cell-wall structure in different organisms or, perhaps, different degrees of accessibility of the cell-wall proteins to enzymic attack.

One important feature of yeast-wall proteins is their high sulphur content (Falcone and Nickerson, 1956). For the most part the sulphur is in the form of disulphide bridges. These linkages are believed to play a key part in morphogenetic phenomena. Thus, it has been proposed that enzymic cleavage of the disulphide bridges in cell-wall glycoproteins is an indispensable part of the budding mechanism of *Candida albicans* (Nickerson and Falcone, 1956) and *Sacch. cerevisiae* (Brown and Hough, 1966).

Further evidence in support of the proposal that the integrity of the wall depends on cross-linked protein complexes has been gained from studies on protoplast formation. For successful protoplast formation of *Sacch. cerevisiae*, both the cell-wall mannan and glucan must be degraded (Millbank and Macrae, 1964) but, even in the presence of these two enzymes, stationary-phase yeast cells yield protoplasts with difficulty (Sutton and Lampen, 1962). This process is greatly facilitated when 2-mercaptoethanol or other thiol reagent is included prior to or during the application of the lytic enzymes (Davies and Elvin, 1964). Seemingly, reduction of disulphide links leads to a debilitating breach in the structure of the cell wall (Bacon *et al.*, 1965).

As in the case of polysaccharides, comparative studies on the protein content of the cell wall of two morphological types of the same organism often reveals marked differences which may have a bearing on shape determination. For instance, elongated cells of *Sacch. cerevisiae* have half as much protein as ellipsoidal cells (McMurrough and Rose, 1967). A similar correlation exists in *Hansenula schneggii* (Sundhagul and Hedrick, 1966). Likewise, in dimorphic fungi (e.g. *Mucor rouxii*, Bartnicki-Garcia and Nickerson, 1962a), the cell wall of yeast forms has been shown to have considerably more protein than mycelial walls, but in *Paracoccidioides brasiliensis* (Kanetsuna *et al.*, 1969) the reverse is true. As was noted above for cell-wall polysaccharides, no simple correlation can be established between overall cell-wall protein content and the cellular morphology of dimorphic fungi; yet, the existence of major variations in the proportion of cell-wall protein may be regarded as a crude manifestation of chemical changes underlying differences in shape, texture and thickness of the cell wall. Seemingly these alterations in protein content do not represent general increases or decreases in cell-wall protein, but a more selective effect on different proteins. Such may be concluded from the finding that the amino-acid composition of the cell wall also undergoes some major revisions. For instance, Power and Challinor (1969) found that the weakened and structurally different cell wall of *Sacch. cerevisiae* obtained by growing the organism under inositol deficiency, manifests an amino acid composition quite different from that of the ordinary cell walls. Likewise there were some marked differences in amino-acid composition in cell-wall proteins derived from blastospore and mycelial forms of *C. albicans* (Chattaway *et al.*, 1968).

4. *Lipids*

Lipids may play an important role in yeast morphogenesis, but convincing proof remains to be adduced. The existence of cell-wall lipids has often been doubted, and their presence in cell-wall preparations has been rationalized as cytoplasmic contamination. However, the lipids extracted from isolated cell walls of *Nadsonia elongata* differed qualitatively from the cytoplasmic lipids in having mainly saturated fatty acids and being completely devoid of palmitoleic acid (Dyke, 1964). This observation strongly supports the claim that lipids are genuine cell-wall constituents, and encourages an examination of their role, if any, in the architecture of the cell wall. Hurst (1952) noted that lipids contribute to maintain the stiffness of the cell wall. Electron diffraction studies suggested a preferred orientation of the hydrocarbon chains of surface lipids so that their long axes are normal to the plane of the cell surface. When whole cells of baker's yeast were dehydrated, relatively little flattening ensued and the cells remained as prolate spheroids. In contrast, a marked degree of flattening was observed when cells were first extracted with organic solvents to remove lipid and subsequently dehydrated. Seemingly the walls were weakened by the displacement of some of their lipid components.

Cell-wall lipids have been specifically implicated in the morphological differentiation of *Trigonopsis variabilis* (SentheShanmuganathan and Nickerson, 1962a; see Section III.B.3, p. 480). Changes in total lipid content and in the proportions of different lipid fractions have been noticed in comparative studies of the wall composition of different cellular forms of *C. albicans* (Bianchi, 1968), and *Sacch. cerevisiae* (McMurrough and Rose, 1967). The morphogenetic significance of these findings remains to be determined.

In searching for evidence on the participation of lipids in the architecture of the cell wall of yeasts, it might be profitable to entertain the possibility that they may participate as lipoprotein complexes analogous to those currently considered to take part in stabilizing the architecture of the bacterial cell wall (Braun and Schwarz, 1969). Also, some intriguing observations relating lipids to the solubility behaviour of the otherwise alkali-insoluble "yeast" glucan may be pertinent. For instance, Zeichmeister and Toth (cited by Bacon *et al.*, 1966) reported that the removal of lipid by precise solvent manipulation rendered yeast cells almost entirely soluble in alkali. Kessler and Nickerson (1959) also noted that lipid extraction was a critical stage in the isolation of gluco-mannan-protein complexes from cell walls of *Sacch. cerevisiae* and *C. albicans*.

## B. CELL-WALL BIOGENESIS

### 1. *Current Status*

The mechanism of cell-wall construction may play a crucial role in morphogenesis, yet very little is known about the manner by which a cell wall is put together. It is perhaps of value in discussing morphogenesis to separate the entire process of cell-wall biosynthesis into two stages. The first stage corresponds to the biosynthesis of low molecular-weight precursors, and the second stage to their polymerization into the insoluble fabric of the cell wall. Whereas most metabolic reactions occurring during the first stage can be adequately described by conventional *scalar* biochemistry, in the second stage the *vectorial* aspects of biosynthesis are certainly the most important for morphogenesis and, lamentably, the least understood. Much more is known about the synthesis of sugar nucleotides (polysaccharide precursors) and their incorporation into polymers *in vitro* (see Chapter 5, p. 135), but the three-dimensional characteristics of these syntheses have yet to be elucidated. For instance, one would like to know the step and manner in which the biosynthetic pathway of a cell-wall polysaccharide (e.g. glucose → → → UDP-glucose → → → glucan) acquires spatial orientation. Presumably this occurs when the biosynthesis progresses to a stage requiring the participation of membrane-bound enzymes, i.e. subsequent to the synthesis of the sugar nucleotide.

Unequivocal characterization of the subcellular sites where cell-wall material is polymerized has yet to be attained. It is widely accepted that membraneous structures, such as the endoplasmic reticulum, Golgi-bodies or the plasmalemma, are the seat of cell-wall polysaccharide synthesis in various eukaryotic organisms (Northcote, 1963; Hassid, 1969; Lamport, 1970).

There is reason to believe, however, that at least some of the cell-wall polymers, particularly those forming the cell-wall skeleton, may be polymerized *in situ* by the synthesizing enzymes which are themselves part of the cell-wall fabric. This idea has received some support from studies on the biosynthesis of two wall polysaccharides of mycelial fungi: β-1-3-, β-1-6-glucan (Wang and Bartnicki-Garcia, 1966) and chitin (McMurrough and Bartnicki-Garcia, 1969). The enzymes catalysing the polymerization of these two polysaccharides are principally located in the cell-wall fraction as compared with cytoplasmic fractions. The release into the medium of uridine diphosphate-glucose (presumably the precursor of wall glucan biosynthesis) by protoplasts of *Saccharomyces fragilis* (Rost and Venner, 1968) is consistent with the possibility that cell-wall glucan may be synthesized outside the cytoplasmic membrane barrier. Nečas and Kopecká (1969) concluded, on the basis of freeze-etching electron microscopy of regenerating walls of *Sacch. cerevisiae* protoplasts, that glucan microfibrils were always formed outside the plasma membrane. They claimed that these microfibrils were not connected to either invaginations or isodiametric granules of the plasmalemma surface, as had been previously reported by Moor and Mühlethaler (1963).

Some of the discrepant findings regarding the site of wall polymer synthesis and the pattern of wall construction (see p. 459) may reflect actual differences in the mode of deposition of each wall component. The existence of such differences may be inferred from the observation that cycloheximide selectively inhibits the synthesis of mannan-protein, the matrix material of *Sacch. cerevisiae* cell walls, without affecting the formation of glucan microfibrils (Nečas *et al.*, 1968; Elorza and Sentandreu, 1969).

Regardless of whether cell-wall polymers are synthesized *in situ*, or transported from some internal biosynthetic site to their final destination, the possibility of the pre-existing cell-wall fabric serving as a template for its own orderly assembly constitutes an attractive hypothesis. Experiments on cell-wall regeneration by protoplasts (see Section II.B.3, p. 462) might eventually afford a verdict on this hypothesis.

Some of the most fundamental morphogenetic processes of yeasts (budding, cell fusion, germination) apparently involve the differentiation of the cell wall at some localized site. An important task for the near future should be the elucidation of the factors which guide and restrict cell-wall metabolism to such areas.

### 2. *Patterns of Cell-Wall Construction in Growth and Multiplication*

Elucidation of the pattern of cell-wall growth in a developing yeast cell is a prerequisite for understanding the molecular events underlying its morphogenesis. Knowledge on the biochemical mechanism of cell-wall growth at the macromolecular level is now beginning to accumulate, but it is too early to draw any morphogenetic conclusions. In recent years, however, the foundations for understanding the molecular aspects of cell-wall growth of yeasts have been laid down by cytological studies. With the aid of fluorescent stains, fluorescent antibodies, autoradiography and electron microscopy, it has been possible to disclose some of the gross features of vegetative cell-wall differentiation in yeasts. Particularly rewarding have been the investigations of Streiblová and her coworkers, employing the elegantly simple procedure of primulin-induced fluorescence staining to reveal further the patterns of growth and cytokinesis of various yeasts (Fig. 3; Streiblová and Beran, 1963). The cell wall of primulin-stained cells fluoresces with different intensities

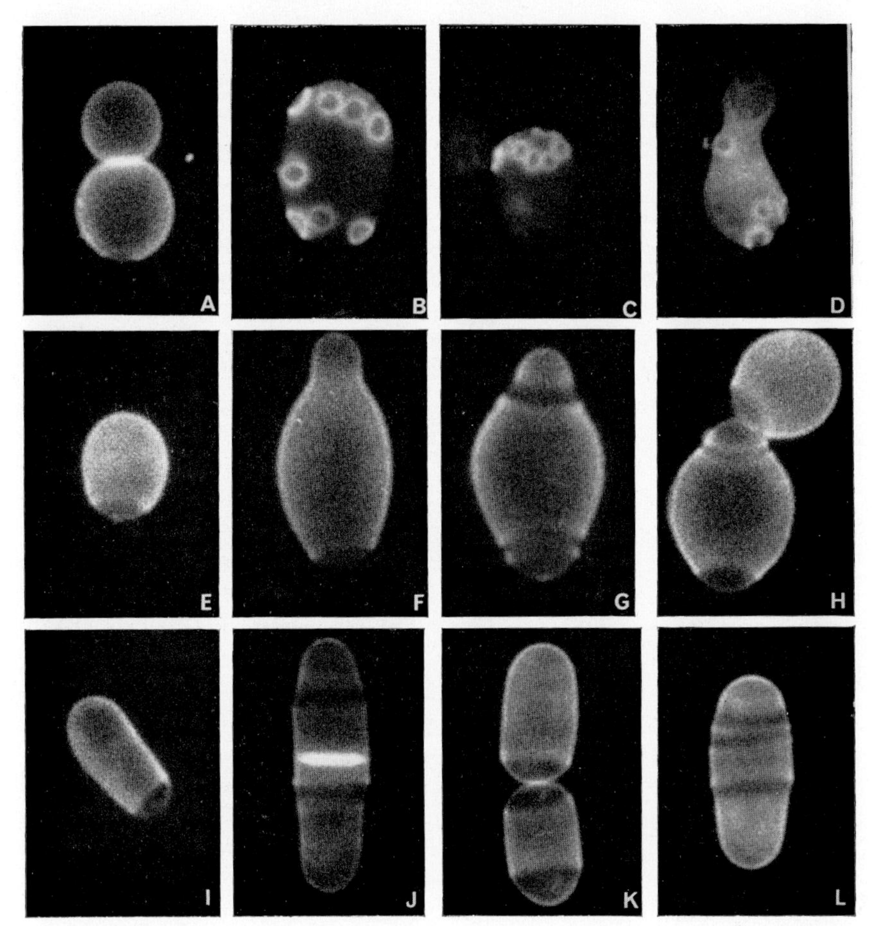


FIG. 3. Features of the cell walls of different yeasts related to their cell division processes as revealed by fluorescent staining with primulin. Upper row: a multipolar budding yeast. Middle row: a bipolar budding yeast. Lower row: a fission yeast. Reproduced by courtesy of E. Streiblová. A, Mother cell and bud of *Saccharomyces cerevisiae*, showing fluorescent band in the intercell region; B, distribution of bud scars on a diploid cell of *Sacch. cerevisiae*; C, distribution of bud scars on a haploid cell of *Sacch. cerevisiae*; D, diploid (large) and haploid (small) bud scars on a zygote of *Sacch. cerevisiae*; E, detached bud of *Saccharomycodes ludwigii*, showing its birth scar on the lower pole; F, budding cell of *Saccharomycodes ludwigii* with a daughter bud on the top and its birth scar at the bottom; G, cell of *Saccharomycodes ludwigii* with one division scar on the upper pole and two division scars on the lower pole; H, *Saccharomycodes ludwigii* in the process of separating from its mother cell, a scar from a previous budding is also evident on the upper pole; I, daughter cell of *Schizosaccharomyces pombe* (the lower dark pole is the "scar plug" which was formed from the septum wall in the mother cell); J, cell of *Schizosacch. pombe* prior to cleavage, the brightly fluorescent septum wall has been formed eccentrically, there are also two dark bands representing scars from two previous divisions; K, daughter cells of *Schizosacch. pombe* immediately after septum cleavage, the lower cell carries the birth scar of the original mother cell; L, cell of *Schizosacch. pombe* with three division scars.

in certain regions, indicating differences in wall texture (Streiblová, 1966; Beran, 1968). These differences in texture probably reflect changes in the fibrillar structure of the cell wall that take place as a result of wall growth and cell division. These changes in texture are of a permanent nature, and thus serve as markers to follow the patterns of cell-wall extension.

Based on the patterns of cell-wall development during growth and division, Streiblová (1969) recognized four types of cell-wall architecture corresponding to multipolar budding yeasts, bipolar budding (apiculate) yeasts, fission yeasts, and the type exhibited by *Endomyces magnusii*.

*a. Multipolar budding*. *Saccharomyces cerevisiae* exemplifies this type of budding. In contrast with bipolar budding yeasts, the buds of *Sacch. cerevisiae* never arise on the site of a previous budding (bud scar). Although budding occurs preferentially on the cell poles, the rest of the cell surface is potentially suitable for budding, especially after the budding pole has been covered by bud scars. Further features of this type of budding process are discussed in Sections III.A.2 and 3 (p. 467).

Electron micrographs of thin sections show the bud cell wall continuous with the parental cell wall (Agar and Douglas, 1955; McClary and Bowers, 1965; Sentandreu and Northcote, 1969b). However, the thin sections do not reveal a structural difference in fibril organization which probably exists in the connecting region between the two cells, and which is visualized by primulin staining as a strongly fluorescent band (Streiblová and Beran, 1965; see Fig. 3A). The intense fluorescence of this annular region is probably the result of the parallel circular re-orientation of wall microfibrils caused by the forceful extrusion of the protoplasm at the time of budding, according to the mechanism advanced by Nickerson and Falcone (1959).

The bud wall expands by acquisition of newly synthesized polymers and contains a negligible amount of preformed polymers from the parental wall (Chung *et al.*, 1965). Later, a septum grows centripetally in the neck region, cutting off the bud from the parent cell (Sentandreu and Northcote, 1969b; Marchant and Smith, 1967). Upon separation of the cells, characteristic scars are left on the parent cell (bud scar) and the daughter cell (birth scar). These scars are morphologically distinct, and can be recognized under the light microscope (Barton, 1950), fluorescence microscope (Streiblová and Beran, 1965) or electron microscope (Bartholomew and Mittwer, 1953; Agar and Douglas, 1955; Houwink and Kreger, 1953). Both scars appear as raised annuli, but the bud scar forms a pronounced crater-like depression whereas the birth scar, which is larger in diameter, has a convex character and is much less conspicuous. The rim of the bud scar has also a high affinity for the fluorescent dye primulin; in electron micrographs of chemically extracted walls

(Houwink and Kreger, 1953) it appears as a ring of closely packed, circularly arranged microfibrils. This microfibrillar ring is probably formed at the time of bud extrusion, as mentioned above.

Bud scars of *Sacch. cerevisiae* also have a distinct chemical composition. The floor of the crater, which was originally the parental half of the septum wall, exhibits cytochemical properties different from the rest

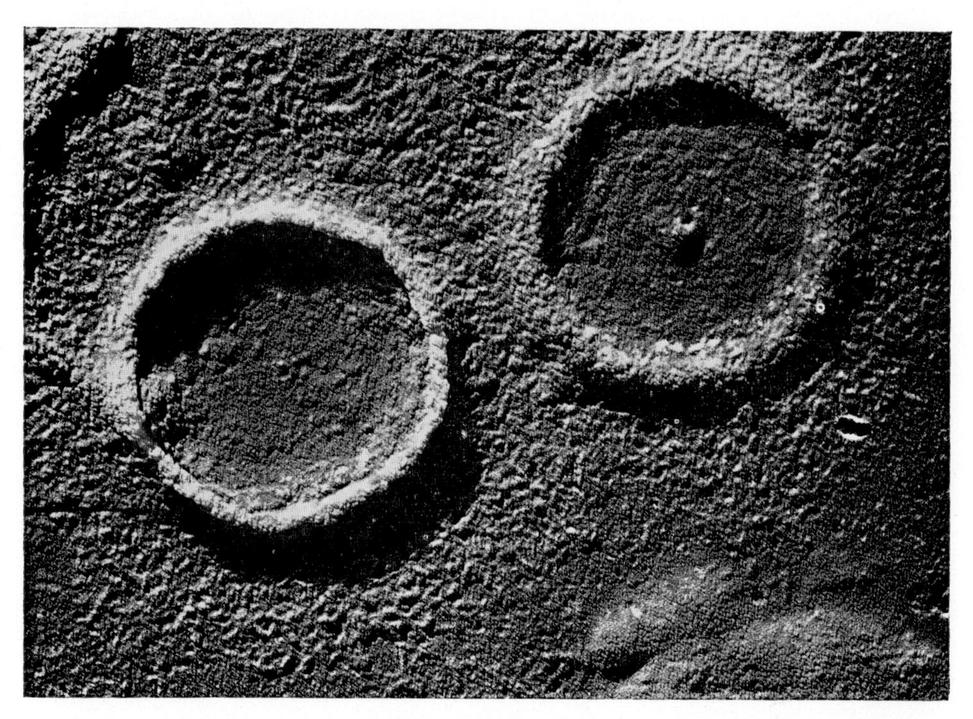

FIG. 4. Isolated bud scars of *Saccharomyces cerevisiae*, prepared by digestion of cell walls with a $\beta$-1,3-glucanase from *Cytophaga johnsoni*, with 2-mercaptoethanol for 8 days, pH 7·5. Reproduced by courtesy of J. S. D. Bacon. Magnification, × 40,000.

of the cell wall (Mundkur, 1960). Chemical studies of cell wall residues consisting principally of bud scars (Fig. 4) showed that most of the chitin in the cell wall of *Sacch. cerevisiae* is located in the bud-scar region, though its precise location has not been defined (Bacon *et al.*, 1965). This intriguing chemical differentiation invites questions about its morphogenetic significance and the possibility that septum-wall formation may be mediated by a specialized enzymic machinery different from that concerned with overall cell-wall synthesis.

Immunofluorescent and autoradiographic techniques have been applied to the elucidation of the pattern of cell-wall growth in developing buds of *Sacch. cerevisiae*, but contradictory conclusions were reached. By autoradiography, Johnson and Gibson (1966b) studied the incorporation of $H^3$-glucose into insoluble cell-wall components. Most of the silver grains over the buds were found at the distal ends, indicating that wall extension was principally by tip growth. Tip growth is the well recognized mode of elongation of tubular cells of fungi, including hyphae (see Fig. 5B) and sporangiophores (see Castle, 1958); it also occurs in long slender yeasts like *Pichia farinosa* (see below). The participation of tip growth

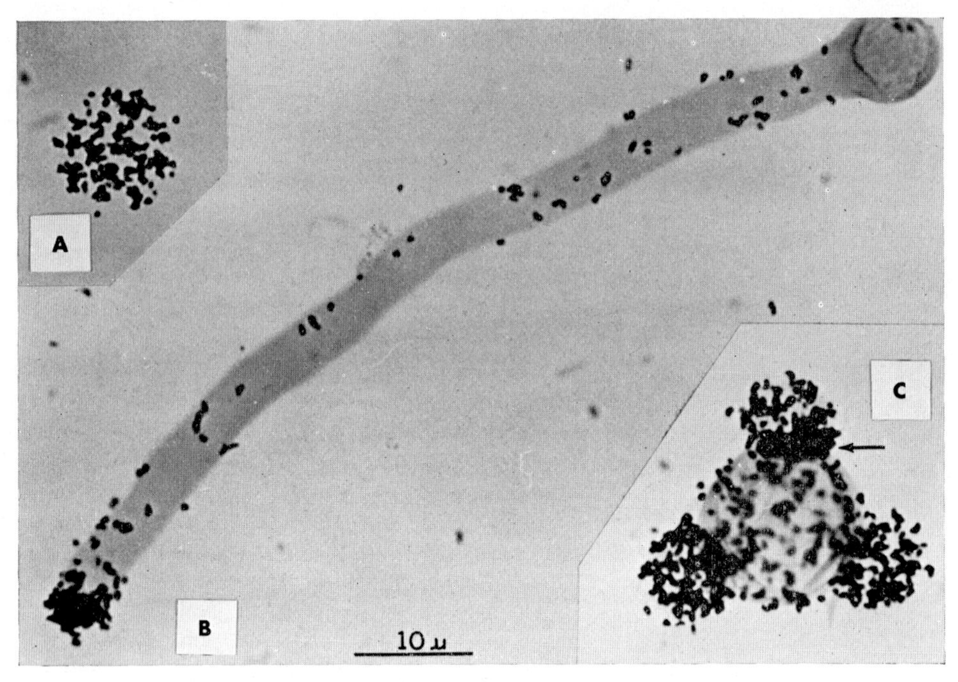


FIG. 5. Patterns of cell-wall synthesis in *Mucor rouxii* revealed by autoradiography (Bartnicki-Garcia, 1969). A, germinating sporangiospore; B, elongating hypha; C, budding yeast cell.

in the formation of ellipsoidal or nearly spherical yeast cells is less obvious. However, according to the morphometric data adduced by Johnson and Gibson (1966b), tip growth is compatible with the generation of the ellipsoidal shape of *Sacch. cerevisiae*. Additional indirect evidence in support of a tip-growth mechanism stems from the finding that the inhibitor 2-deoxyglucose causes localized lysis of the cells of *Sacch. cerevisiae* (as well as *Schizosacch. pombe* and *Pichia farinosa*) in the suspected region of wall growth, i.e. the distal end of a bud of

*Sacch. cerevisiae* (Johnson, 1968). A model of cell-wall growth was proposed by Johnson (1968) in which the glucan layer is postulated to grow by the concerted action of synthetic and hydrolytic enzymes. Accordingly, 2-deoxyglucose selectively nullifies the synthetic process, while the unimpaired lytic action weakens the cell wall to the bursting point and thus betrays the location of the wall growth area.

Chung *et al.* (1965) used fluorescent antibodies (prepared against whole cells of *Sacch. cerevisiae*) to label newly formed cell surfaces, and concluded that the cell wall was synthesized at an annular band located at the base of the bud. Although this conclusion is diametrically opposed to that obtained from autoradiographic studies there may be room for reconciliation. Each technique may have recognized different cell-wall components with presumably different patterns of synthesis. Thus, the immunofluorescent labelling probably detected the most antigenic wall components, i.e. mannan (Summers *et al.*, 1964), whereas in the autoradiographic study the insoluble cell-wall glucan was probably the polymer visualized (Johnson and Gibson, 1966b). Final judgement must, of course, await confirmatory studies in which incorporation patterns of unequivocally characterized polymers are traced. In further contrast to the above studies on *Sacch. cerevisiae*, the pattern of cell-wall synthesis for the spherical yeast form of *Mucor rouxii* was found to be largely random (Bartnicki-Garcia and Lippman, 1969). Tritiated N-acetyl-D-glucosamine was used to trace autoradiographically the formation of chitin and chitosan, two of the main cell-wall components of this organism. A more or less uniformly disperse pattern of incorporation was observed with only slightly higher activity near the base of the bud (Fig. 5C). The latter activity was thought to represent septum-wall synthesis rather than being an indication of basal growth of the bud itself. It was concluded that the wall of *Mucor rouxii* yeast cells grows by a rather uniform incorporation of material over its entire surface and thus acquires its spherical shape. It would be premature to draw any conclusions on the existence of fundamentally different modes of cell-wall extension in the yeast form of *Mucor* as compared with that of *Saccharomyces*. The yeast cells of *Mucor* show multipolar budding and have bud scars on the cell surface, but differ from those of *Sacch. cerevisiae* in being much larger, spherical and usually extruding two or more buds concurrently. Furthermore, these two organisms have totally different cell-wall compositions.

The mode of cell-wall extension in *Pichia farinosa*, an elongated multipolar budding yeast, was examined by Johnson and Gibson (1966a) who followed, autoradiographically, the incorporation of $H^3$-glucose into alkali-insoluble components (presumably cell-wall glucan). The distal end of a growing bud incorporated radioactivity far more rapidly than the middle or basal portions. The predominant growth of this yeast is by exponential extension of the wall at the distal end. Apparently the spindle-like shape of *Pi. farinosa* is largely the result of this marked tendency towards apical growth of the cell wall.

*b. Bipolar budding.* The apiculate morphology of bipolar budding yeasts seemingly derives from the characteristic features of bipolar budding. In these yeasts, and in contrast with multipolar budding yeasts, budding takes place repeatedly on the site of a previous scar (Streiblová and Beran, 1965). As shown by primulin-induced fluorescence staining of *Saccharomycodes ludwigii* (Fig. 3, p. 455), a succession of buddings leaves a series of annular ridges ("multiple scars") at both poles. Apparently bipolar budding, as opposed to multipolar budding, does not involve any marked circular reorientation of fibrils, and hence the scars are only faintly fluorescent. Also, the details of septum formation and septum cleavage of bipolar buds appear to be closer to those in the fission of *Schizosaccharomyces* than those in multipolar budding yeasts (Streiblová and Beran, 1965; compare Figs. 3, p. 455, 6 and 7, p. 461).

*c. Binary fission.* Unlike the controversies surrounding the mode of cell-wall extension in budding yeasts, all studies on *Schizosaccharomyces* agree that the cell wall extends by polar growth. These include light microscopy studies using the division scar as marker (Mitchison, 1957) immunofluorescence labelling (May, 1962), primulin-induced fluorescence staining (Streiblová *et al.*, 1966) and autoradiography (Johnson, 1965). Johnson concluded that *Schizosacch. pombe* has an exponentially increasing rate of tip growth, and also confirmed the observation that most cells grow mainly at one end (Mitchison, 1957). Only 20% of the cells were found to grow at both ends.

Evidently, polarized cell-wall growth is a primary determinant of the "sausage" shape (cylinder with hemispherical poles) of *Schizosacch. pombe.* The topography of individual cells in a culture of *Schizosacch. pombe* varies in accordance with the number, width and position of characteristic annular ridges (division scars). There are also subtle variations in wall texture readily manifested in primulin-stained cells (Streiblová *et al.*, 1966; see Fig. 3I–L). Cells that have undergone a few divisions exhibit a series of wall rings. These surface features permit reconstruction of the growth history of such cells (Fig. 7). During cell division of *Schizosacch. pombe*, a double septum wall is synthesized centripetally. The septum wall does not seem to be a direct continuation of the lateral walls but an independent wall which is somehow anchored against the lateral wall (Streiblová *et al.*, 1966). It also has different texture from that of lateral walls as revealed by its high affinity for primulin (Fig. 3J). Upon cleavage of the two daughter cells, the septum wall in each cell becomes distended, forming a hemispherical polar cap, and loses its fluorescence almost completely (see Fig. 3K). The rim of

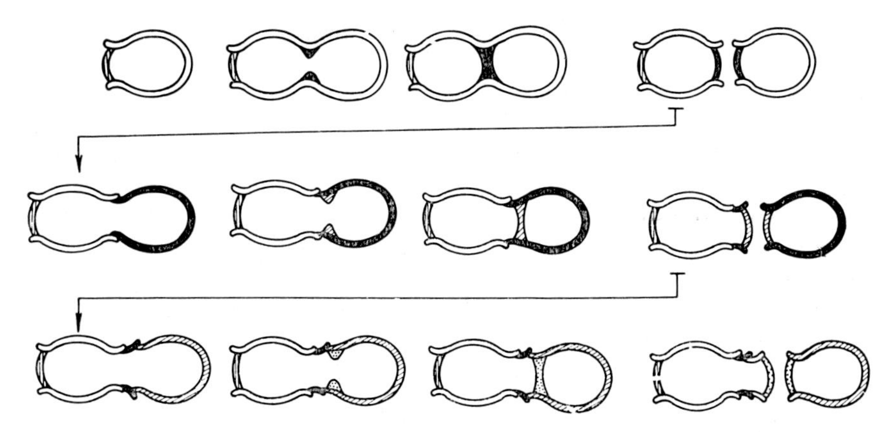

FIG. 6. Diagrams of cell-wall development in *Saccharomycodes ludwigii*. The sequence illustrates progressive steps in growth and of divisions leading to the formation of a cell with three division scars (Streiblová, 1966). Reproduced by courtesy of E. Streiblová.

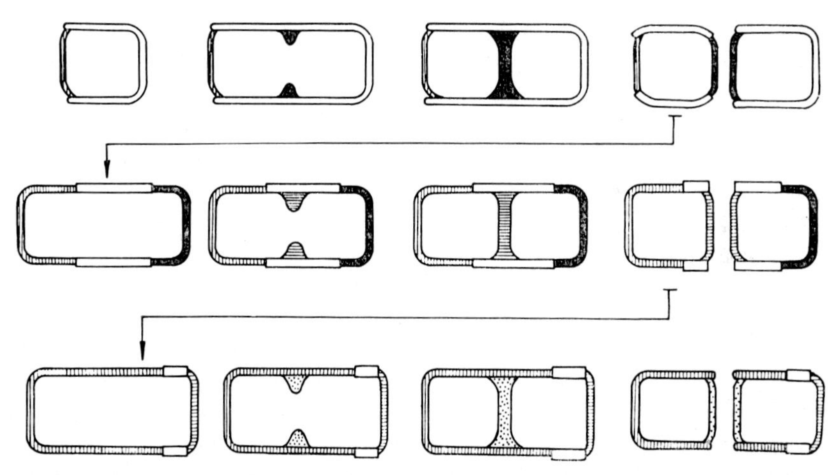

FIG. 7. Diagram of cell-wall development in *Schizosaccharomyces pombe*. The sequence illustrates progressive steps in growth and division leading to the formation of a cell with three division scars (Streiblová, 1966). Reproduced by courtesy of E. Streiblová.

the lateral wall, at the site of cell cleavage, becomes a fluorescent protruding annular ridge (division scar). Wall extension occurs primarily by insertion of new cell wall material into the polar cap.

Insertion is probably highly localized, since it does not seem to occur at the base of the polar cap. This leaves a belt of original septum wall, between the division scar and the growing end of the cell, recognized by its low fluorescence (Streiblová *et al.*, 1966).

### 3. *Cell Wall Regeneration by Protoplasts*

Upon dissolution of the glucan and mannan components of suitably sensitive yeast cells by treatment with complex mixtures of hydrolytic enzymes (e.g. snail-gut juice), osmotically-sensitive protoplasts (if the wall is entirely removed) or sphaeroplasts (if the wall is incompletely removed) are liberated. Studies made to test the ability of protoplasts, mainly of *Sacch. cerevisiae*, to regenerate cell walls have led to conflicting claims (see Villanueva, 1966). Protoplasts which have seemingly lost their capacity to regenerate a normal cell wall, retain their ability to synthesize and excrete most (perhaps all) of the components ordinarily present in the cell wall (*cf.* Lampen, 1968). Thus, protoplasts of *Sacch. cerevisiae* secrete into the medium the enzymes usually found in the cell walls, namely, melibiase (Friis and Ottolenghi, 1959), invertase (Sutton and Lampen, 1962), acid phosphatase (McLellan and Lampen, 1963), together with structural mannan and protein. Sentandreu and Northcote (1969a) showed that the mannan is secreted as a glycoprotein. In addition to the liberation of soluble wall components, a disorganized fibrillar meshwork composed of polymers of glucose and N-acetylglucosamine, but essentially devoid of protein, is deposited around the protoplast surface (Eddy and Williamson, 1959). Removal of the cell wall by hydrolytic enzymes seems to impair the ability of the cell to assemble cell-wall glycoproteins into an orderly three-dimensional network. The reason for this deficiency is still obscure. Eddy and Williamson (1959) attributed the inability of *Sacch. cerevisiae* protoplasts to regenerate normal cell walls to the absence of a specific protein template from the protoplast surface.

Against the idea of a permanent loss of structural information, are the claims that protoplasts can regenerate a normal cell wall under certain specified conditions. Thus, protoplasts of *Sacch. cerevisiae* prepared by digestion with snail enzymes were reported to regenerate normal cell walls if embedded in gelatin or agar (Nečas, 1961; Svoboda, 1966). Significantly, a normal cell wall was not regenerated immediately by the protoplast. The protoplast first grows hypertrophically, divides and then buds atypically before a cell with a normal cell wall is developed (Streiblová *et al.*, 1967). The disorganized cell wall formed in the early stages of regeneration (Nečas, 1965) is similar to the aberrant cell walls observed by Eddy and Williamson (1959) in liquid cultures of protoplasts. Substantial chemical and morphological changes also occur during cell wall regeneration of protoplasts of *Candida utilis* (García-Mendoza and Novaes-Ladieu, 1968; Uruburu *et al.*, 1968).

According to Ottolenghi (1967), the gelatin medium used in protoplast regeneration prevents the shedding of the protoplast outer membrane (possibly an inner layer of the cell wall) on which successful regeneration is thought to depend. In freeze-etched electron micrographs, yeast protoplasts sometimes show an innermost thin wall layer and remnants of a fibrillar middle-layer (Streiblová, 1968). Likewise, Bacon *et al.* (1969) demonstrated that protoplasts of baker's yeast prepared by prolonged treatment with snail digestive juice, retained some components of the original wall, including fibrils and chitinous bud scar residues. The bud scars were often seen attached to a thin membrane. Preliminary studies indicated that the undissolved components of the protoplast surface contained a variable mixture of chitin, $\beta$-1,3-glucan, mannan and protein. Following a more intensive digestion, bud-scar residues eventually disappeared (Bacon *et al.*, 1969). These observations question the total absence of cell-wall material from routine preparations of protoplasts, and cast some doubt on the ability of true protoplasts to synthesize a normal cell wall *de novo*, i.e. without the benefit of any surface primers or templates. On the other hand, Nečas *et al.* (1969) insist that their protoplasts are truly wall-free. They interpret the thin wall layers seen by Streiblová (1968) on *Sacch. cerevisiae* protoplasts as an artifact of freeze etching.

To conclude, although these studies have not yet afforded definitive answers on the existence of structural information in the cell wall, they permit us to state more precisely three possible alternatives, each one with a different morphogenetic implication: (i) Cell wall removal causes the loss of some irreplaceable template material. In such case, no *true* protoplast could possibly regenerate a normal cell wall. (ii) The loss of structural information, caused by complete removal of the cell wall, is not permanent but can be gradually regained. The recuperation may be attained through progressive spontaneous reassembly of cell wall polymers into envelopes having increasingly greater levels of structural organization until a normal cell wall is re-formed. If this were the case, the semi-solid media reportedly necessary for wall regeneration might serve as a non-specific barrier or anchoring substrate to prevent the diffusion or disintegration of the primordial wall (Nečas and Kopecká, 1969). (iii) The cell wall contains no intrinsic structural information to orient its own synthesis. The inability of protoplasts to regenerate a normal cell wall could be ascribed to undetected damage to the protoplast.

Of the three alternatives, the last seems least likely.

## III. Vegetative Development

### A. BIOCHEMICAL ASPECTS OF CELL DIVISION

#### 1. *Periodic Synthesis*

The morphogenetic sequence encompassed by one division cycle (Fig. 8) involves a precise co-ordination in time and space of a multitude of biochemical events which initiate, implement and terminate a progressive and seemingly irreversible sequence of structural changes. In

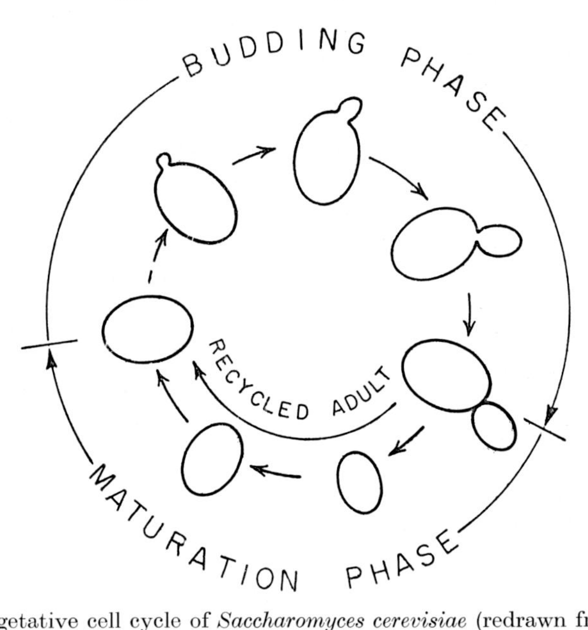


FIG. 8. Vegetative cell cycle of *Saccharomyces cerevisiae* (redrawn from Johnson and Gibson, 1966b).

each cycle, all cellular structures are duplicated. The regularity of shape and internal organization of a population of yeast cells in the midst of active vegetative growth (e.g. exponential growth) demands a harmonious co-ordination between the duplication of internal structures and that of the cell wall. When this co-ordination is altered or impaired, cell morphology may deviate drastically (see Section III.B, p. 473).

Biochemical investigations on the nature and control of these cyclic changes are preferably conducted in synchronized cultures in which the overall behaviour of large populations of cells can be assumed to represent the properties of a single cell. In certain studies, it has been possible to measure individual properties of single cells under the microscope. Thus, by interference microscopy, Mitchison (1957, 1958) measured the rate of dry mass increase during the division cycle of a budding yeast (*Sacch.*

*cerevisiae*) and a fission yeast (*Schizosacch. pombe.*) In both cases the dry mass increased as a linear function of time in each division cycle. In contrast, the cell volume of *Schizosacch. pombe* followed an almost exponential curve of increase for the first three-quarters of the division cycle. The cessation of volume increase in the last stages of the cycle coincided with nuclear division. In *Sacch. cerevisiae* a sigmoid curve was obtained for volume increase. These observations were extended with *Schizosacch. pombe* to the synthesis of major macromolecules (Mitchison and Lark, 1962; Mitchison and Walker, 1959; Mitchison and Wilbur, 1962). Cells were pulse-labelled with radio-active precursors whose incorporation into macromolecules was followed by grain counting under the microscope. Synthesis of RNA was followed by scanning ultraviolet micrographs of growing cells with a recording densitcmeter. It was found that the synthesis of protein, carbohydrate and RNA was approximately linear throughout the cell cycle (Mitchison, 1963). In contrast, the DNA content exhibited a "step pattern": there was a rapid increase during the initial one fourth to one third of the division cycle of *Sacch. cerevisiae* (at the time of bud emergence); thereafter the content remained constant until the next division (Williamson and Scopes, 1960; Gorman *et al.*, 1964; Williamson, 1965). Such findings correlate with the discrete and periodic nature of nuclear division (Williamson, 1966). In *Sacch. lactis*, replication of mitochondrial DNA was also found to be a periodic process (Smith *et al.*, 1968). Synthesis of the two types of DNA was not concomitant, however, mitochondrial DNA being synthesized before nuclear DNA. As an explanation for this observation, it was suggested that mitochondrial DNA requires for its synthesis a protein which is synthesized and coded by nuclear genes (Smith *et al.*, 1968).

The elegant technique of synchronous cultivation has also shown that a marked periodicity exists in the synthesis of induced enzymes. For instance, there are two periods of catalase synthesis in *Sacch. cerevisiae*, one occurring immediately before cell division and the other soon thereafter (Gifford and Pritchard, 1969). A similar periodic inducibility of α-glucosidase was ascribed to a variation in the accessibility of structural genes to transcription during the cell cycle (Tauro and Halvorson, 1966; Halvorson *et al.*, 1966). In another study in which the levels of aspartate carbamoyltransferase, ornithine carbamoyltransferase, alkaline phosphatase, invertase and maltase were determined in synchronous cultures of *Schizosacch. pombe*, it was concluded that both periodic and continuous syntheses of enzymes took place (Bostock *et al.*, 1966). The suspected existence of biochemical timing mechanisms in the development of yeast cells is further reinforced by the finding of cyclic variations in peptidase and proteinase activities in *Sacch. cerevisiae* (Sylvén *et al.*, 1959). Such activities reached their highest levels immediately prior to cell division. It was suggested that the activity of proteinases and peptidases may be a prerequisite for the replenishment of the amino-acid pools necessary for protein synthesis and hence cell growth.

One of the important contributions of synchronous yeast culture studies has been the demonstration of an increased generation of energy coincident with budding initiation. Meyenburg (1969) suggested that carbohydrate reserves are stored during the maturation phase and are then catabolized during bud formation to meet the demands of this process. Yeasts are by no means unique in this respect, since in general all cells mobilize energy reserves prior to division (Mazia, 1961). The disaccharide trehalose is one of the principal energy reserves of *Sacch. cerevisiae* (Brandt, 1941). It has been demonstrated that trehalose is synthesized optimally by yeast cells under conditions of non-proliferation, and is utilized prior to cell division (Panek, 1963). Since the enzymes responsible for trehalose synthesis are present in both growing and resting cells, competition for a common precursor by other synthetic systems may explain the submaximal synthesis of trehalose in growing cells. Further studies indicated that both trehalose and its cleaving enzyme, trehalase, co-exist in yeast cells; it was claimed, therefore, that trehalose synthesis and breakdown could be mediated by internal compartmentation (Souza and Panek, 1968). The phase of trehalose utilization would then be expected to be preceded by alterations in internal membrane organization so that enzyme and substrate could interact.

The foregoing observations on trehalose metabolism were made in batch cultures. By using synchronous cultures, the timing of these events was further clarified. Using a chemostat for the synchronous growth of *Sacch. cerevisiae*, Küenzi and Fiechter (1969) demonstrated unequivocally that, under conditions of glucose limitation, reserve carbohydrates (trehalose, glycogen) accumulate during the maturation phase. Immediately before bud extrusion, carbohydrate reserves are degraded and there ensues a period of structural polysaccharide synthesis. The specific activity of trehalase reached its peak during the phase of trehalose breakdown. The amount of reserve carbohydrate stored by *Sacch. cerevisiae*, under continuous cultivation, is inversely related to the rate of supply of fresh medium (Küenzi and Fiechter, 1969). A linear relationship was also observed between the content of trichloroacetic acid-soluble carbohydrate (mainly trehalose) and the growth rate of *Sacch. cerevisiae* cultivated in a chemostat under glucose limitation (McMurrough, 1966). Slow-growing cultures with a mean generation time of 14 h contained about ten times as much trichloroacetic acid-soluble carbohydrate as fast-growing cultures with a mean generation time of 1·5 h.

It seems, therefore, that during the maturation phase the cellular content of trehalose continuously increases provided that an adequate supply of fresh medium is maintained. The extent of this increase is related to the duration of the maturation phase, which in turn is regulated by the rate of supply of exogenous nutrients. In contrast, the dissimilation of trehalose is initiated and accomplished in a relatively short period of time. A smooth operation of the numerous endergonic biosynthetic reactions involved in budding can then ensue. As a result, the overall rate of the budding process is maintained almost constant (Meyenburg, 1969). An adequate store of energy is, of course, only one aspect of those requirements which must be fulfilled by the yeast prior to multiplication. It is possible, however, that the accumulation of a reserve of energy, sufficient to sustain a period of short-term independence from exogenous sources, might be a decisive factor in initiating yeast budding. Sensitive control over the mobilization of energy reserves would be greatly devalued if equally stringent regulations did not govern the overall cellular economy. It is therefore no surprise to find that the repression and derepression of at least seven enzymes other than trehalase are also correlated with the phase of the division cycle in *Sacch. cerevisiae* (Beck and Meyenburg, 1968).

In conclusion, the cell cycle can no longer be considered as a period of uniform steady growth (see review by Mitchison, 1969). Most of the enzymes examined are synthesized discontinuously at a particular time in the cycle characteristic of each enzyme. They in turn must bring about continuous changes in the chemical composition of the cell. To explain the discontinuous pattern of enzyme synthesis, Halvorson and co-workers (Halvorson *et al.*, 1966; Tauro *et al.*, 1966) proposed from observations on *Sacch. cerevisiae* that genes are transcribable only during a restricted portion of the cycle, and that they are transcribed in a sequence corresponding to their linear order in the chromosome. Oscillatory repression may be an additional mechanism responsible for the discontinuous synthesis of certain enzymes in the cell cycle of yeasts (Mitchison, 1969).

The foregoing studies point to the feasibility of eventually describing the yeast division cycle as a sequence of macromolecular events beginning with the progressive transcription of genetic information, its translation into enzymic and structural proteins, followed by the construction and differentiation of membranous structures, and terminating in the orderly construction of two cell walls (bud wall and septum wall).

### 2. *Bud Initiation*

The mechanism which times the onset of budding is poorly understood. The generation time of a strain of *Sacch. cerevisiae* growing in continuous culture under glucose limitation can be varied widely between 1·3 and 15 h, but the duration of the budding phase (from bud extrusion to abscission) remains approximately constant between 1·3 and 2·0 h (Meyenburg, 1968). This independence between the length of the budding phase and the length of the predivision phase implies the existence of control mechanisms of budding distinct from those regulating overall cell growth. Such a conclusion is in accord with the idea that cell division can be selectively uncoupled from cell growth, with profound effects on cellular morphology (see Section III.B.2, p. 474).

Initiation of cell division (budding) could depend on the accumulation of trigger compounds, and hence be regulated by the rate of their synthesis (Williamson and Scopes, 1961). Variations in the relative rates of accumulation of such hypothetical compounds would provide the observed flexible means of regulation. No such substance has been characterized to date, though it has been claimed that a sulphur-containing substance separated from cold trichloroacetic acid extracts of *Candida utilis* and *Chlorella pyrenoidosa* is capable of initiating budding (Vraná and Fencl, 1964, 1967). Although this may be taken as evidence in favour of the participation of specific trigger compounds, the case is by no means proven. The occurrence of specific trigger compounds in morphogenesis has been debated (Wright, 1966). Alternatively, the attainment of a particular "physiological state", a term which probably encompasses the sum total of most aspects of cellular metabolism, may send the cell into its budding phase. And, conceivably, the relative levels of metabolic pools may have a regulatory influence on the initiation of budding. Apart from purely biochemical considerations, low molecular-weight metabolites may have an indispensable role in that they provide the turgor pressure which is seemingly necessary for bud extrusion as well as cell expansion. Internal fluid pressure in an ellipsoidal cell is exerted maximally against the cell wall at the points of maximum curvature, i.e. the cell poles. According to Nickerson and Falcone (1959) budding is initiated by a localized increase in the activity of protein disulphide reductase acting against the gluco-mannan protein complexes of the cell wall. The ensuing splitting of —S—S— cross-links creates a deformable region in the cell-wall fabric where bud extrusion takes place forcibly. The exact manner and speed of bud extrusion remains controversial. Whereas Nickerson and Falcone (1959) maintain that the budding process is explosive in nature, with rupture of the cell wall and extrusion of a protoplast which is quickly covered with newly synthesized wall material, others contend that budding is a gradual bulging out of the parental cell with no appreciable decrease in wall thickness (Agar and Douglas, 1955; McClary and Bowers, 1965; Sentandreu and Northcote, 1969b). In any case, the proposed participation of protein disulphide reductase in the budding mechanism of *Candida albicans* (Nickerson and Falcone, 1956) was confirmed and expanded to other yeasts by subsequent workers. Robson and Stockley (1962) and Pitt (1969) detected by autoradiography and other cyto-chemical procedures, a preferential accumulation of –SH groups at the budding areas of *C. albicans*, *C. utilis* and *Eremothecium ashbyi*. Also, Brown and Hough (1966) confirmed the existence of protein disulphide reductase in cell-free extracts of *Sacch. cerevisiae* and, significantly, found that elongated cells contained considerably lower levels of protein disulphide reductase than did typically ellipsoidal cells—a finding in support of the idea that ellipsoidal budding yeast cells have a higher capacity to plasticize their cell walls compared to elongated cells.

As to the preferential occurrence of budding in the poles of an ellipsoidal cell, different hypotheses have been advanced. A ballistic explanation was proposed (Falcone and Nickerson, 1959) based on the higher probability of collision against the cell poles of intracellular particles (believed to be mitochondria) carrying protein disulphide reductase. An alternative and more elaborate cytological mechanism was adduced from Moor's (1967) freeze-etching electron micrographs of budding cells of *Sacch. cerevisiae*. Bud initiation was found to coincide with a proliferation of endoplasmic reticulum vesicles in the region between the nucleus and the budding site. The micrographs suggest a centrifugal displacement of these vesicles which Moor suggested carry protein disulphide reductase and discharge it upon fusing with the plasmalemma. Conceivably, these vesicles could also be the carriers of wall precursors (Marchant and Smith, 1967) and/or enzymes involved in the breakdown and growth of the cell wall. These vesicles have also been seen in electron micrographs of thin sections of cells fixed with glutaraldehyde-osmium tetroxide (Sentandreu and Northcote, 1969b). Assuming that these electron microscope observations may have identified the subcellular particles associated with budding, the fundamental question of what orients these, or other structures, to the budding sites remains unanswered. In the absence of an unequivocal explanation, other intriguing observations may be cited for they may eventually help to solve this riddle. For instance, Freifelder (1960) found that the position of the buds of *Sacch. cerevisiae* was correlated with the ploidy of the cell. Haploid cells produce their buds on the same pole as, and as near as possible to, the birth scar. On the other hand, polyploid cells bud at the pole opposite the birth scar. These observations demonstrate that the two cell poles of *Sacch. cerevisiae* are not necessarily equivalent in budding but dependent on the nuclear condition of the cell. How the latter can determine bud position is far from clear. Freifelder reported no correlation between nuclear position at interphase and bud position,

He also ruled out the possibility that bud position may have been governed by a genetic character.

Streiblová (1970) found that the bud scars of diploid cells of *Sacch. cerevisiae* are almost twice as large as those of haploid cells (Fig. 3B–D). Interestingly, she found that haploid scars were formed frequently in rows, rings and spirals, indicating, perhaps, the existence of some guiding mechanism, whereas in diploid cells the budding sequence occurred in the more scattered manner observed by Barton (1950), and described by Falcone and Nickerson (1959) as a progression of budding sites through permitted loci of maximal curvature.

### 3. *Termination of Growth and Ageing*

Our knowledge of the underlying biochemistry of growth termination is as meagre as that of bud initiation. Conceivably, the attainment of a particular physiological state terminates the growth phase, but its parameters have not been defined. It is known that in *Sacch. cerevisiae* the mass of a developing bud increases until it reaches about the same volume as the parent cell (Mitchison, 1958). Moreover, in steady-state cultures of *Sacch. cerevisiae*, such as can be maintained in a chemostat, it has been shown that the mean cell volume of populations is dependent on the specific growth rate of the cultures (McMurrough and Rose, 1967). Variations in cell volume from 68 $\mu m^3$ to 28 $\mu m^3$ were recorded for corresponding changes in mean generation time from 1·5 h to 14 h. According to Meyenburg (1968) the budding phase of *Sacch. cerevisiae* only varies from 1·3 h to 2·0 h over the aforementioned range of generation times, while the maturation phase or pre-budding period varies from about 0·2 h to 12 h. A steady-state yeast population growing with a mean generation time of 14 h should therefore be composed predominantly of single cells. Conversely, budding cells should outnumber single cells in yeast populations with a mean generation time of 1·5 h. This phenomenon is illustrated in Fig. 9. Both the mean cell volume and the percentage of budding cells in nitrogen-limited populations are shown to decrease with decreasing growth rate (i.e. increasing generation time). An identical effect was also observed with glucose-limited cultures. The chemical composition of chemostat cultures grown at high specific growth rates therefore principally reflects the properties of relatively large budding cells. As the growth rate of a chemostat culture is progressively lowered, the composition of the population as a whole will increasingly become that of single cells in their predivision phase. It is no surprise, therefore, to find that fast-growing yeast cultures contain more RNA and low molecular-weight components, and less reserve carbohydrate material than slow-growing cultures (McMurrough and Rose, 1967).

Growth of a developing bud, as judged by the expansion of volume, is terminated shortly before abscission from the parent cell. From the aforementioned observations, it seems that the volume of the bud is conditioned by the previous history of the parent cell. In this respect,

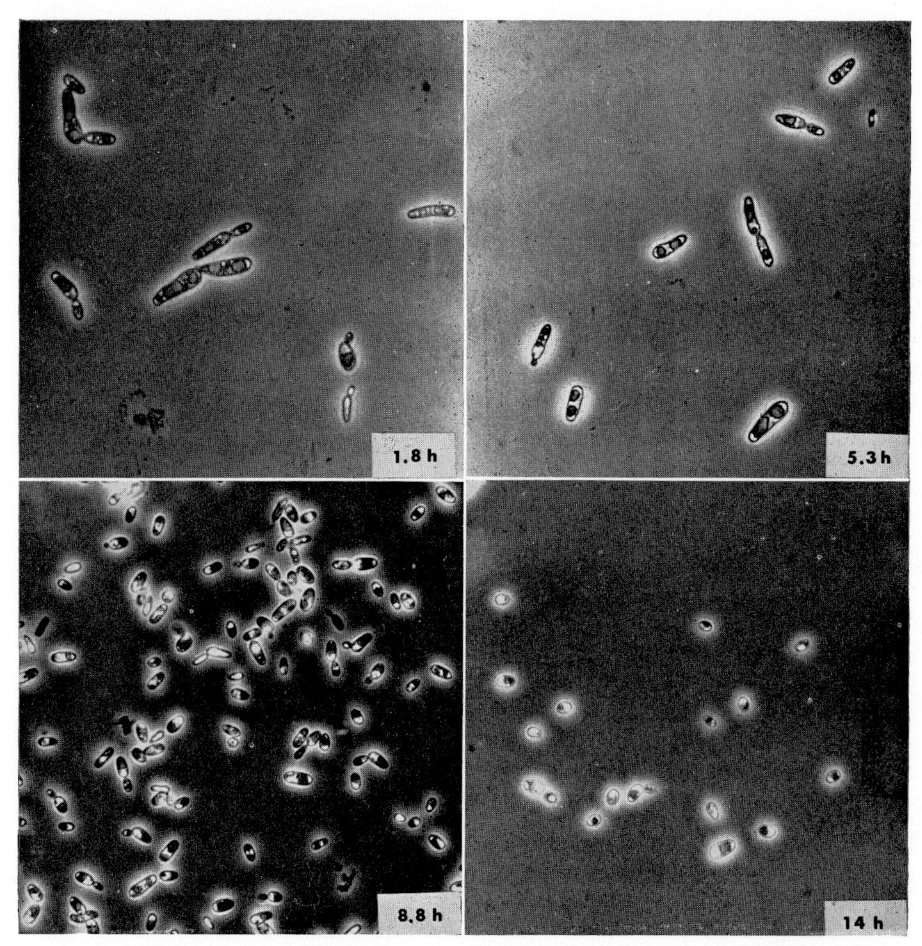


FIG. 9. Relation of growth rate to cellular morphology and budding frequency in *Saccharomyces cerevisiae*. Steady state populations growing in a chemostat under conditions of nitrogen ($NH_4^+$) limitation (McMurrough and Rose, 1967). Note decrease in mean volume and proportion of budding cells with increase in the generation times shown. Magnification, × 600.

the environment during the maturation phase of the parent probably has a profound bearing on the subsequent development of the bud. According to Mitchison (1958), the volume of a yeast cell is largely determined during the time of its formation by the budding process, but it has also been shown that the size of an adult yeast cell increases slightly with age (Mortimer and Johnston, 1959; Beran, 1968). The size gain is proportional to the number of bud scars, and is probably accounted for by the increase in cell-surface area contributed by each bud scar.

In many yeasts (e.g. *Sacch. cerevisiae*) budding does not occur on a previous scar, and the life span of an individual cell is thus limited to a finite number of cell divisions. It was calculated that up to 100 bud scars would be required to obliterate the surface of a yeast cell (Bartholomew and Mittwer, 1953; Beran *et al.*, 1966). In fact, only about 20 bud scars have been counted on *Sacch. cerevisiae* cells (Barton, 1950). Old cells, which have divided many times, contribute very little to the average dimensions or physiological properties of a growing yeast population. The number of bud scars on an individual cell, and hence the age of that cell, is most readily ascertained by using primulin dye. A recent analysis of yeast cultures using this technique revealed an almost ideal distribution of age throughout the population, as would be predicted from consideration of the kinetics of budding (Beran, *et al.*, 1966; Beran, 1968). Old cells carrying seven bud scars constituted only 0·5% of the total cell number.

4. *Bud Abscission*

The disconnection of a bud from its mother cell was explained by Barton (1950) in physical terms based on the idea that the independent increase in volume of the daughter cell results in a stretching of the birth scar. The shearing action between the two scars causes the mechanical connection to break. In agreement with this argument is the observed larger size of the birth scar than the bud scar. In the absence of criteria on which to evaluate the tensile strength of the cell wall and the shearing force generated by the expanding cell, no convincing argument can be raised against a purely physical explanation like this. However, one should seriously consider the possibility that physical forces may be only one part of the mechanism; it would not be unreasonable to assume that this process is greatly facilitated by the participation of cell wall-splitting enzymes.

Seemingly the abscission process of yeasts is quite sensitive to growth factor deficiencies. Thus, by cultivation in liquid media deficient in *meso*-inositol, *Sacch. carlsbergensis* manifested an impairment of the final steps of the division process (Ghosh *et al.*, 1960). As a consequence, daughter buds failed to separate from the parent cell and aggregates containing up to 50 cells were seen. A similar effect was demonstrated with *Sacch. cerevisiae* grown in inositol-deficient media (Challinor *et al.*, 1964; Power and Challinor, 1969) and in biotin-deficient media that had been supplemented with aspartate (Dunwell *et al.*, 1961). In all these instances, the formation of anatomically and functionally abnormal cell walls was accompanied by more than three-fold increases in the ratio of glucan to mannan in the walls. Further studies revealed a marked thickening, pronounced multiple layering and other distortions of the cell walls of *Sacch. cerevisiae* grown in biotin-deficient media containing aspartate (Dixon and Rose, 1964). The exact step(s) at which the bud-separation mechanism is blocked is not known. Dixon and Rose (1964) indicated that the failure of *Sacch. cerevisiae* to complete its cell division was in the inability to form a cross septum between mother and daughter cells. Whether the pronounced alteration in cell-wall composition is causally related to the absence of a septum wall, or is merely a co-incidental event has not been clarified.

## B. ENVIRONMENTAL CONTROL

### 1. *General*

The importance of the environment in yeast development is readily manifested. Spontaneous morphological changes occurring within a yeast colony growing over a solid substrate usually result from modifications in the environment caused by the yeast cells themselves. Recognition of the factor(s) affecting morphogenesis requires carefully controlled experiments in which the specific environmental condition under study can be critically examined. Observations made with batch cultures often leave an uncertainty as to the precise environmental conditions which affect morphogenesis, since growing cells inevitably cause progressive and extensive modifications of the initial compositions of the culture medium. Perhaps some of the contradictory findings on yeast morphogenesis reported in the literature can be ascribed to unsuspected alterations in the environment occurring after inoculation. To minimize this uncertainty, it is desirable to employ, whenever feasible, the technique of continuous cultivation. By this method a relatively constant environment can be maintained, and thus it is possible to discriminate between the intrinsic variability of a cell population in a given milieu and the variability imposed on the population by the fluctuating environment of a batch culture.

Identification of the exogenous factors controlling morphogenesis may be of great value in recognizing metabolic reactions that might be implicated in morphological differentiation. Regrettably, this ideal is often quite elusive in actual practice. For, in most cases, unrelated environmental factors can be shown to affect the same morphogenetic pathways, and it becomes exceedingly difficult to suggest any common biochemical *modus operandi*. If any generalization is to be made, it is that the exogenous control of form development is quite intricate and probably depends on the sum total of various mutually interdependent factors. A case in point is the environmental control of dimorphism of *Mucor rouxii* (see p. 477).

2. *Mycelial-Yeast Dimorphism*

Under certain conditions, some yeasts will grow as elongated cells, or even as a true mycelium. Conversely, some mycelial fungi can grow chiefly or exclusively as budding yeast cells. The ability of dimorphic fungi to develop into either yeast cells or mycelial forms depends on diverse environmental factors such as temperature, oxygen, carbon dioxide, quality and quantity of carbon and nitrogen sources, or the presence of thiols and other compounds. For general reviews on different aspects of dimorphism, see Scherr and Weaver (1953), Howard (1962), Mariat (1964) and Romano (1966).

*a. Candida albicans.* This is a well-studied example of a yeast with a pronounced tendency to develop filamentous cells (Fig. 10): the biochemistry of its morphogenesis has been the subject of extensive investigations by Nickerson and his coworkers. If *C. albicans* is grown on solid medium, with glucose as carbon source (e.g. Sabouraud medium), typical yeast colonies are formed. Days later, however, extensive filamentation may be observed around the periphery of the colonies (Fig. 10A). The filamentation consists of highly elongated cells and distinct hyphal forms. These filaments are produced in reponse to alterations in the composition of the medium caused by the metabolic activities of the yeast itself. If a less readily utilizable carbon source, such as glycogen, is employed in the growth medium, filamentation occurs from the beginning of cultivation (Nickerson and Mankowski, 1953). In both these cases the filamentation of *C. albicans* can be prevented by addition of —SH compounds (cysteine, glutathione, thioglycollate) to the medium.

It follows from the concepts advanced by Nickerson (1948) that the dimorphism of yeasts like *C. albicans* results from a variable interplay between two fundamental cellular processes: cell growth (elongation) and cell division (budding). Each process is subject to different control mechanisms, thus resulting in a flexible co-ordination or coupling. When cell division is optimally coupled to cell growth, regularly budding ellipsoidal, or spherical, cells are developed. Selective impairment of the division process (a lesion in a single enzymic locus should suffice) would uncouple cell division from cell growth, causing the cells to elongate with little or no budding. The stimulation of budding by —SH compounds was interpreted (Nickerson, 1948) as one further instance of the widespread requirement of —SH groups in the division process of living cells (Brachet, 1961). The dimorphic behavior of *C. albicans* was rationalized accordingly. When this organism grows in a glucose-rich medium,

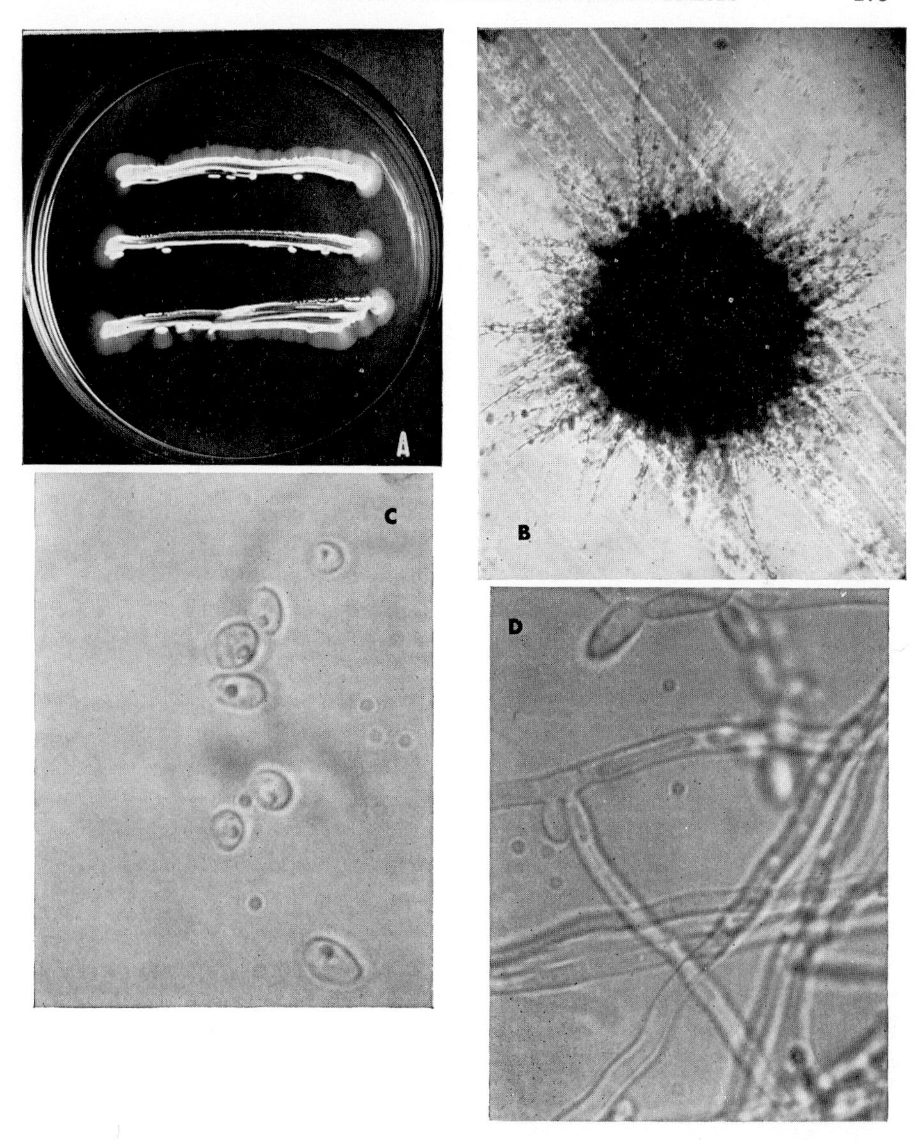


FIG. 10. Dimorphism of *Candida albicans*. A, wild strain 582 grown on glucose-glycine-yeast extract solid medium for 7 days, and showing the "Langeron" effect, i.e. filamentation at the free borders; B, colony of a filamentous mutant (Strain 806); C, yeast cells of wild strain 582); D, filamentous cells of wild strain 582. Reproduced by courtesy of W. J. Nickerson.

sufficient reducing power is generated to maintain the level of —SH necessary for budding. Should glucose become depleted or should only a slowly assimilated carbon source be available, the low level of intracellular

—SH thus attained would not support the requirements of cell division; consequently, elongated cells would be produced. If, however, an exogenous supply of —SH compounds is made available to the fungus, budding is resumed and the organism displays its yeast morphology.

This proposal was largely substantiated by the observation that a filamentous mutant of *C. albicans* (Fig. 10B) could also be induced to grow yeast cells by addition of —SH compounds to the growth medium (Nickerson and Chung, 1954). Comparative biochemical studies of the filamentous mutant and the normal budding strain of *C. albicans* led to the conclusion that reducing power was channelled into a cell division reaction *via* a metal-requiring flavoprotein (Nickerson, 1954). In the filamentous mutant, there was apparently some defect in the coupling between the flavoprotein and the specific hydrogen acceptor of cell division, causing the reducing power to "spill over" and become available for other reductions (e.g. dyes, oxygen).

The terminal hydrogen acceptor in the cell division process of *C. albicans* was found to be in the —S—S— links of the cell-wall gluco-mannan-protein complexes (Falcone and Nickerson, 1956; Nickerson *et al.*, 1961). The reduction, and hence the splitting, of these disulphide cross-links was catalysed by a protein disulphide reductase present in particles sedimenting in the mitochondrial cell fraction (Nickerson and Falcone, 1956). Significantly, the normal budding strain of *C. albicans* showed a strong protein disulphide reductase activity compared with the slight activity displayed by the filamentous mutant. This protein disulphide reductase is presumably a flavoprotein enzyme (FP). The splitting of disulphide bridges probably ruptures covalent connections between macromolecular complexes, and increases the plasticity of the cell-wall fabric. This plasticizing action, localized in certain areas of the cell wall, creates a condition of deformability essential for bud extrusion, and hence for yeast development, of *C. albicans*. The biochemical process may be represented as follows:

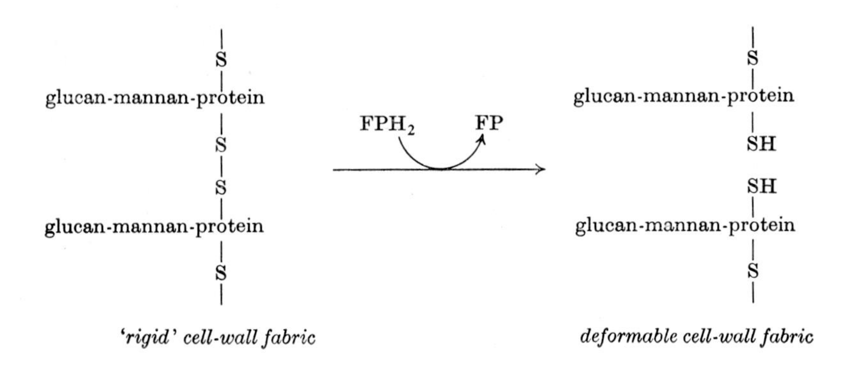

Two other observations provided additional support for the role of sulphydryl-sulphide balance in the dimorphism of *C. albicans*. First, addition of selenite or tellurite to the culture medium caused the filamentous mutant to grow yeast-like. These two more metallic elements of the sulphur family were incorporated into cell-wall proteins. Since selenhydryl or tellurhydryl groups are more stable than sulphydryl groups, it is reasonable to believe that they offset the deficiency of protein disulphide reductase present in the filamentous mutant (Nickerson *et al.*, 1956). Second, when the filamentous mutant of *C. albicans* was grown in sulphur-deficient medium, its cellular content of sulphur diminished and its development was mainly in the yeast form. Under these growth conditions, relatively few —S—S— cross-links would be formed in the cell wall, and the low level of protein disulphide reductase of the filamentous mutant would then seem sufficient to produce the degree of cell-wall plasticity compatible with budding (Falcone and Nickerson, 1959).

*b. Mucor rouxii.* This and other species of *Mucor* are typical mycelial fungi endowed with the potential capacity to develop into populations composed mainly, or exclusively, of budding spherical yeast cells (see review by Bartnicki-Garcia, 1963; Fig. 1A, p. 443). Vegetative morphogenesis of *Mucor rouxii* depends on a variety of environmental factors such as oxygen, carbon dioxide, hexoses, heavy metals, dicarboxylic acids, uncharacterized factors present in complex nutrients and others (Bartnicki-Garcia and Nickerson, 1962b, c; Bartnicki-Garcia, 1968b; Haidle and Storck, 1966; Elmer and Nickerson, 1970a, b). By suitably changing the concentration of any one of these factors, it is possible to induce either the development of yeast cells or hyphae. It would seem misleading, however, to assign to any one of these factors *the* causal role in dimorphic development, for the effect of each factor is conditioned by the concentration of the others. For example, in anaerobic cultures the control of morphogenesis of *Mucor rouxii* by carbon dioxide depends on hexose concentration (Bartnicki-Garcia, 1968b). A high $pCO_2$ value generally favours the development of yeast cells, but if the hexose concentration is below 0·1% no yeast development occurs even under an atmosphere of pure carbon dioxide. Conversely, if the glucose concentration is above 8%, exogenous carbon dioxide is no longer a requirement for yeast development. This interaction between hexose and carbon dioxide concentration is further conditioned by the concentrations of other factors in the medium derived from peptone and yeast extract (Bartnicki-Garcia, 1968b). Similarly, Elmer and Nickerson (1970b) found that a strain of *Mucor rouxii*, NRRL 1894, required supplementation with complex nutrients (peptone, yeast extract) to develop yeast cells under a carbon dioxide atmosphere. The activity was attributed to dialysable factors.

To explain these complex interactions of environmental factors one can only suggest, in the most speculative fashion, a mechanism in which yeast morphogenesis depends on the intracellular concentration of a certain metabolite "Y" (Bartnicki-Garcia and Nickerson, 1962c). If "Y" is assumed to be a common metabolic intermediate (like pyruvate or oxaloacetate) it might be easier to understand why such a diversity of nutritional factors exerts a controlling action on the same morphogenetic pattern. Conceivably, the attainment of a critical intracellular level of "Y" might influence the operation, synthesis, activity or localization of enzymes involved in cell-wall metabolism; as a consequence, a morphological differentiated cell wall would be produced.

Alternative mechanisms, which need not be mutually exclusive, have been proposed to explain the molecular basis of yeast morphogenesis in *Mucor rouxii* (Bartnicki-Garcia, 1963). One hypothesis suggests that the presence of an excess of mannose polymers in the cell wall of the yeast form (Bartnicki-Garcia and Nickerson, 1962a) somehow interferes with the orderly assembly of cell wall components into their "normal" cylindrical configuration. As a consequence, the cell assumes a spherical yeast shape dictated by isotropic physical forces. A second hypothesis, supported by autoradiographic evidence (Bartnicki-Garcia and Lippman, 1969; Fig. 5, p. 458), relates cellular form to the variable ability of the fungus to polarize its machinery of cell-wall growth. Accordingly, hyphae are formed as a result of the confinement of wall growth to a small region of the cell surface (hyphal apex). A subcellular corpuscle found in association with the apical cell wall of *Mucor rouxii* germ tubes was suggested as the seat of the polarizing factor(s) (Bartnicki-Garcia *et al.*, 1968). Seemingly, environmental factors which interfere with the establishment of polarized wall synthesis cause the cell wall to grow uniformly over its entire periphery, giving rise to spherical yeast cells.

c. *Elongation of yeast cells.* In this section we turn our attention to examples of typical yeasts which manifest only a limited capacity for cell elongation. This incipient tendency to elongate, by even the most typical yeasts, such as *Sacch. cerevisiae,* may be regarded as evidence of the seemingly universal capacity of fungi to develop tubular cells. Numerous factors, including temperature, nutrition age, pH value and the presence of certain chemicals in the medium, are known to affect elongation of yeast cells (Scherr and Weaver, 1953). Fusel oils have been implicated as morphogenetic inducers in brewing practice for some time; recently attention has been focused on the elongation of yeast cells in continuous brewing systems. The phenomenon was found to be associated with wort composition, substrate feed rate and available oxygen level (Geiger, 1961; Grant, 1964), but the biochemical cause was obscure. The problem was considerably clarified when Brown and Hough (1965a) found that the principal reason for yeast elongation in wort was a limitation of assimilable nitrogen—a condition common under continuous cultivation with forced aeration. Elongation also occurs in continuous culture with defined media, when the concentration of nitrogen supplied as ammonium sulphate, methionine or asparagine becomes growth limiting (Brown and Hough, 1965b). Significantly, under this condition inclusion of sodium thioglycollate, or sodium selenate in the medium effected a reversion of the elongated cells to ellipsoidal cells. In analogy with earlier observations on *Candida albicans* by Nickerson and his coworkers (see p. 474), Brown and Hough (1965b) concluded that the morphological effect in *Sacch. cerevisiae* was also due to an alteration in the cellular balance of sulphydryl-disulphide. This conclusion was supported by a subsequent report showing that the levels of protein disulphide reductase in the ellipsoidal cells were two or three times higher than those of elongated cells (Brown and Hough, 1966). The elongating effect caused by nitrogen limitation has been confirmed in chemostat cultures (McMurrough and Rose, 1967) with another strain of *Sacch. cerevisiae* which also showed no tendency to elongate in batch culture (Fig. 2, p. 449). Pronounced elongation was observed in continuous culture, particularly at high growth rates, with growth-limiting concentrations of ammonium sulphate. Elongation of the cells was permanent; when these cells were transferred to batch cultivation with excess nutrients, ellipsoidal cells budded off. Since the mean volumes of elongated and ellipsoidal cells were similar, at a particular growth rate, it was concluded that elongation was not effected primarily by a disruption of the cell division process (McMurrough and Rose, 1967).

Tryptophan has been shown to have a pronounced elongating effect on growing cells of *Hansenula schneggii* (Sundhagul and Hedrick, 1966). With tryptophan as the nitrogen source, up to 90% of the cells were elongated, reaching a length of 25 μm. Other amino acids (cystine, histidine, phenylalanine, tryosine and threonine) caused only a small degree of elongation involving 15% of the cells. Elongated cells transferred to a medium containing ammonium sulphate as the nitrogen source gave rise to a culture consisting almost entirely of ellipsoidal cells. The effect of tryptophan was simulated by kynurenine, a tryptophan catabolite. There were some differences in cell-wall composition between elongated and ellipsoidal cells; elongated cells contained much more mannan, more lipid and less protein than ellipsoidal cells. It was suggested that tryptophan may be responsible for partially rediverting fructose 6-phosphate into mannan synthesis (Sundhagul and Hedrick, 1966). In various dimorphic fungi, e.g. *Sporotrichum schenckii* (Drouhet and Mariat, 1952), *Ustilago sphaerogena* (Spoerl *et al.*, 1957), *Ustilago cynodontis* (Talou and Tavlitzki, 1969), *Candida albicans* (Mardon *et al.*,

1969), yeast morphogenesis has also been shown to be greatly affected by the nitrogen source. This suggests that nitrogen metabolites may have a controlling effect on the biochemical pathways related to form development. However, no simple correlation emanates from a comparison of morphological effects *vs.* nitrogen sources; in fact, any one compound may have opposite morphogenetic effects in different organisms. To cite but one example, methionine promotes the yeast growth of *Ustilago sphaerogena* (Spoerl *et al.*, 1957) but it causes pronounced pseudohyphae formation in an atypical strain of *C. albicans* (Mardon *et al.*, 1969). These divergent responses may reflect intrinsic metabolic differences between these two organisms, or may simply result from other factors in the environment which modify the response to methionine.

The concept of cellular form being attained through the operation of enzymes acting directly on the cell wall, as already indicated for protein disulphide reductase, has received some additional support. Shimoda and Yanagishima (1968) found that the application of exo-$\beta$-glucanase to *Sacch. cerevisiae* caused cell expansion. Interestingly, this response to glucanase was directly related to the ability of yeast strains to elongate under the influence of auxin (indole acetic acid). Moreover, isolated cell walls of auxin-sensitive cells were more susceptible to $\beta$-1,3-glucanase digestion than those isolated from refractory strains. These results tend to parallel findings made on cells of higher plants, where indole acetic acid has been found to induce the synthesis of $\beta$-glucanases (Masuda, 1968; Davies and Maclachlan, 1968). Since the exogenous application of $\beta$-1,3-glucanase caused elongation of the oat coleoptile too, it was proposed that a common mechanism may exist for the auxin-induced elongation of yeasts and higher plants (Shimoda and Yanagishima, 1968). It should be noted, however, that such a mechanism for the regulation of cell elongation in higher plants, by means of endogenous polysaccharidases, has not been accepted by other workers (Ruesink, 1969).

### 3. "*Triangular*" *Morphogenesis*

In the work of SentheShanmuganathan and Nickerson (1962a, b) on *Trigonopsis variabilis*, several lines of evidence, gathered from nutritional studies and chemical analyses, were consolidated into a working hypothesis correlating phospholipid metabolism with cellular morphology. This yeast has the property of developing "triangular" or ellipsoidal cells, depending on environmental conditions (Fig. 11). Formation of triangular cells was greatly stimulated by the availability of a suitable methyl donor in the growth medium (e.g. methionine, choline). Methionine was most effective, causing over 95% of the cells to be triangular as compared with a control culture containing ammonium sulphate, which was almost exclusively composed of ellipsoidal cells. Choline produced an effect similar to methionine, especially when it was supplied together with inositol, another phospholipid precursor. Triangular cells did not appear in the midst of exponential growth, but at the onset of the stationary growth phase; it was suggested that this timing coincided with the period of active phospholipid synthesis resulting from the exhaustion of the nitrogen supply. The possibility that the delayed development of triangular cells was due to a population-selection phenomenon was

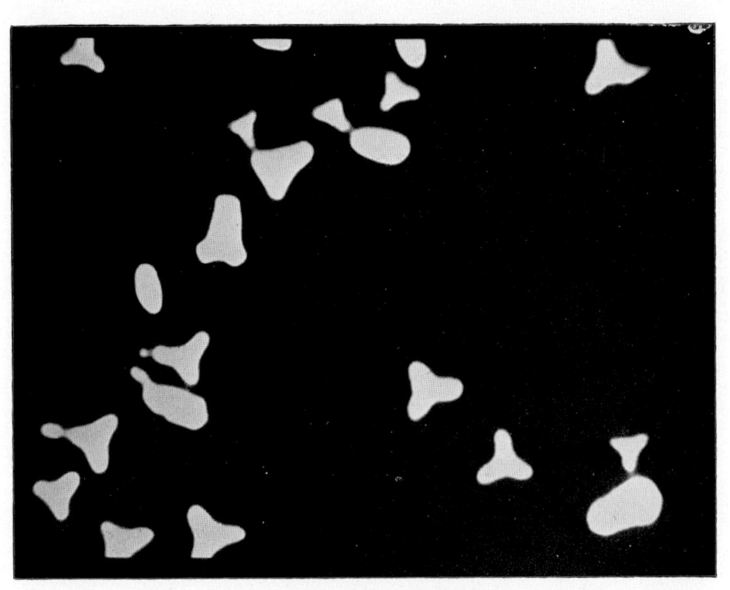

FIG. 11. *Trigonopsis variabilis* growing predominantly in the triangular form in medium with DL-methionine (SentheShanmuganathan and Nickerson, 1962a). Reproduced by courtesy of W. J. Nickerson.

eliminated. Chemical analysis revealed that triangular cells contained twice as much lipid as ellipsoidal cells. Significantly, the bound lipid of triangular cells was rich in choline whilst that of ellipsoidal cells contained much less choline, and higher amounts of ethanolamine and serine. Isolated cell walls of the triangular form, which retain the original outline of the cell (Fig. 12), were shown to have a somewhat higher total lipid content, and twice as much phospholipid, than the walls of the ellipsoidal form.

Seemingly, the stimulation of phospholipid metabolism by methyl-donor substances leads to the synthesis of a cell wall with an increased phospholipid content, but apparently devoid of fibrillar polysaccharides (SentheShanmuganathan and Nickerson, 1962b). Conceivably, these modifications in chemical composition force the cell wall to adopt, in an as yet unexplained manner, a triangular lozenge shape rather than the usual ellipsoidal shape of yeast cells.

The determinant role of methyl-donor substances in triangular cell morphogenesis was questioned by Šašek and Becker (1969) who found

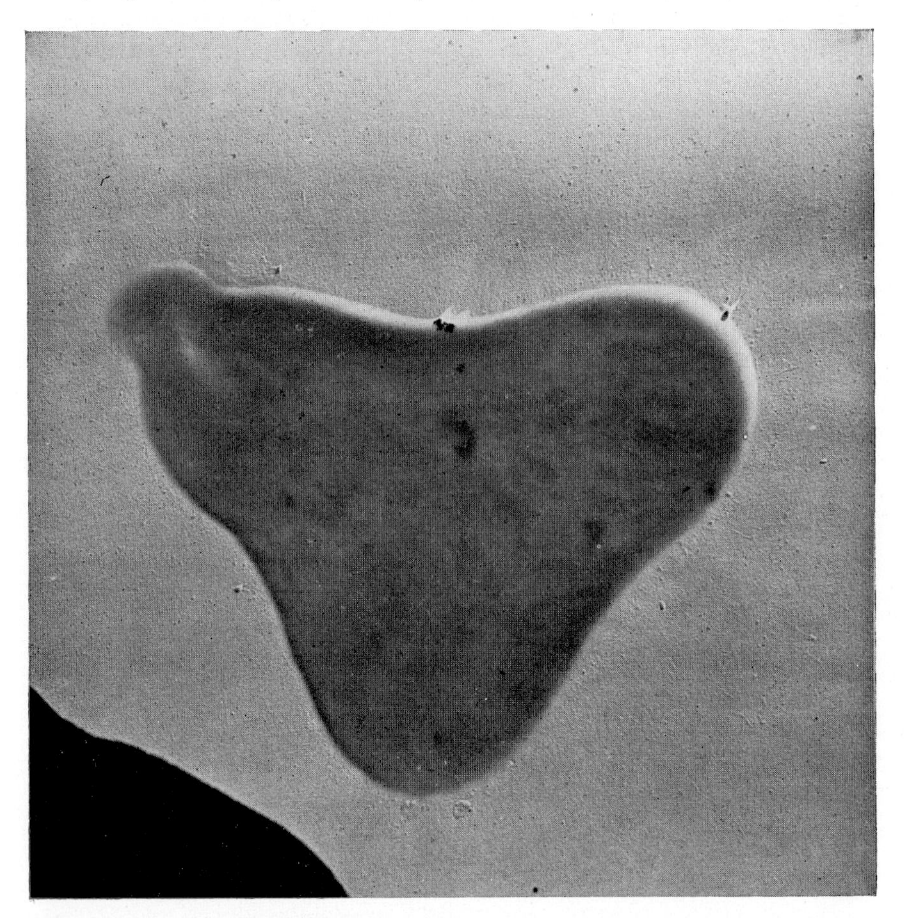

FIG. 12. Isolated cell wall of the triangular form of *Trigonopsis variabilis*. Note the cell wall retains the shape of the whole cell (SentheShanmuganathan and Nickerson, 1962b). Reproduced by courtesy of W. J. Nickerson. Magnification, × 21,750.

proline, alanine and hydroxyproline were as effective as methionine in promoting triangular cell morphogenesis. They also found that triangular cell development principally occurred during the early log phase, and not during the stationary growth phase. Interestingly, plots of percentage triangular cells in a culture against time of incubation revealed a peak of maximum formation of triangular cells during the early log phase, irrespective of the nitrogen source. These observations indicate that endogenous conditions conducive to triangular cell morphogenesis are highly transient. Šašek and Becker employed the same strain of *Trig. variabilis* used by Nickerson and SentheShanmuganathan; barring mutation, the reported discrepancy in the time of appearance of triangular cells (early log *vs.* stationary phase) probably reflects the inevitable minor differences in exogenous conditions of two separate laboratories. Since it is possible that an extracellular methyl-donor may not always be required for the synthesis of form-determining phospholipids, the findings of Šašek and Becker are not necessarily in conflict with the idea advanced by SentheShanmuganathan and Nickerson (1962a, b) that cell-wall phospholipids participate in triangular cell morphogenesis. A comparison of wall composition between triangular cells of *Trig. variabilis* grown under different conditions might be most revealing.

## IV. Sexual Sporulation

Yeast sporulation and its biochemical aspects have already been discussed by Fowell (1969) in Volume I of this treatise. Here we shall simply outline the role of cell wall phenomena in the various stages of sexual morphogenesis of yeasts, findings on *Hansenula wingei* serving as main examples (see the review by Crandall and Brock, 1968).

The participation of the cell wall in conjugation is vividly seen in the electron micrographs (Fig. 13) of Conti and Brock (1965). Mating cells of *H. wingei* recognize each other by cell-to-cell contact. This is a macromolecular recognition mediated by complementary glycoproteins present in the outer surface of the cell wall of opposite mating types. These glycoproteins interact specifically in a manner analogous to an antigen–antibody reaction causing the walls of mating cells to adhere to each other (Brock, 1959; Taylor, 1965).

Following adhesion, a conjugation tube is formed by fusion of the cell walls over the area of mutual contact (Fig. 13A–C). Conjugation requires precise controls so that fusion and lysis of the cell walls occur in correct sequence and in localized regions. Mutual induction of specific cell-wall glucanases may be one of the key phenomena in this mechanism (Brock, 1961). Recently, Yanagishima (1969) reported the isolation of two different sexual hormones from a pair of mating strains of *Sacch. cerevisiae* (hormones are apparently not involved in the mating of *H. wingei*). These two substances were claimed to be of steroidal nature and each was found to cause expansion of its complementary mating cell. This specific action resembled the expansion caused by auxin or by application of $\beta$-1,3-glucanase, and it was therefore proposed that these

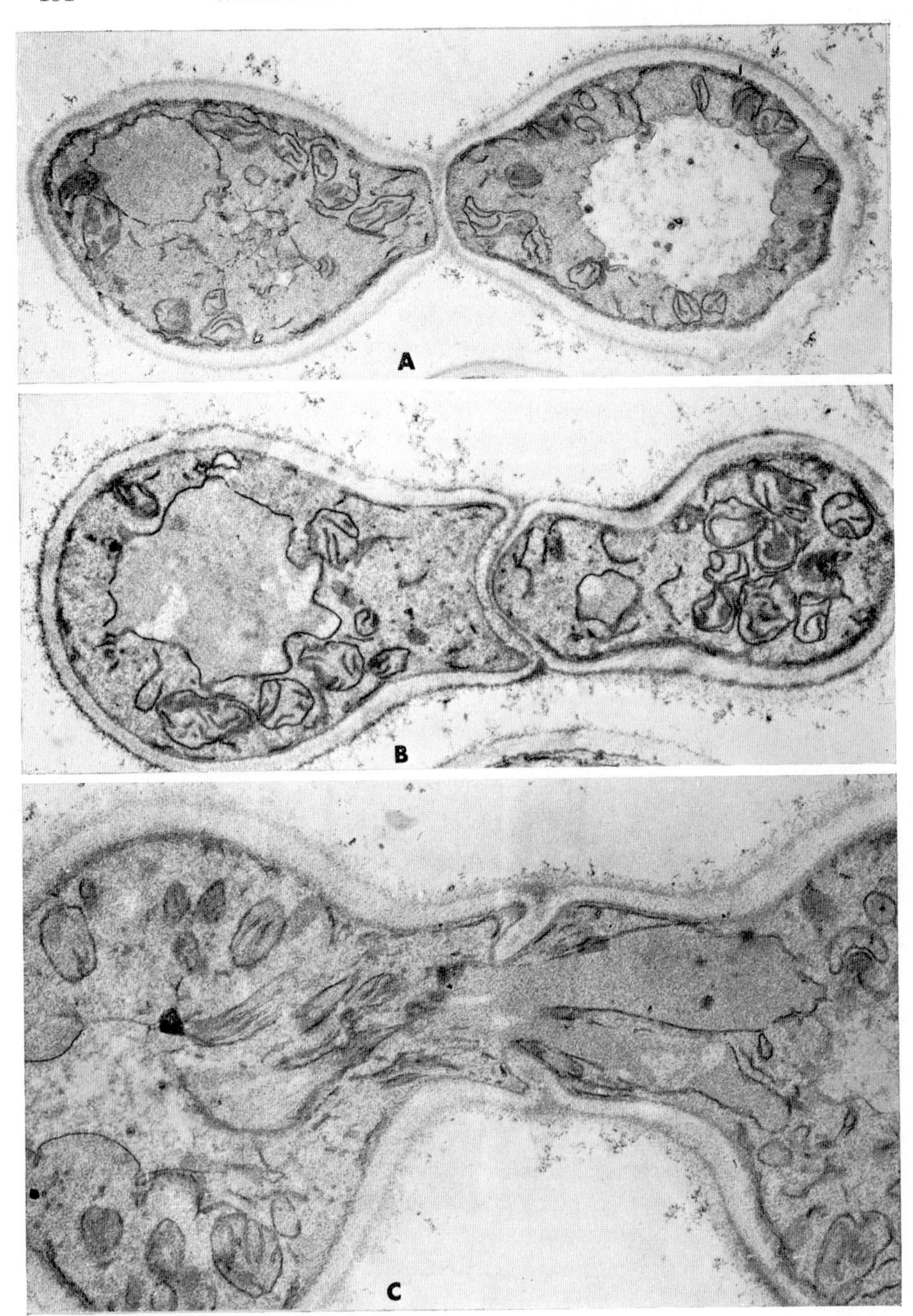


FIG. 13. Conjugation in *Hansenula wingei*. A and B, show cells in the initial phases of wall fusion, and formation of the conjugation tube. C, The cross walls in the completed conjugation tube have been partially dissolved to allow cytoplasmic fusion (Conti and Brock, 1965). Reproduced by courtesy of S. F. Conti.

hormones operate by activating cell-wall softening enzymes in the complementary mating type. Another related finding was the identification of a gene in *Schizosacch. pombe* which participates in the dissolution

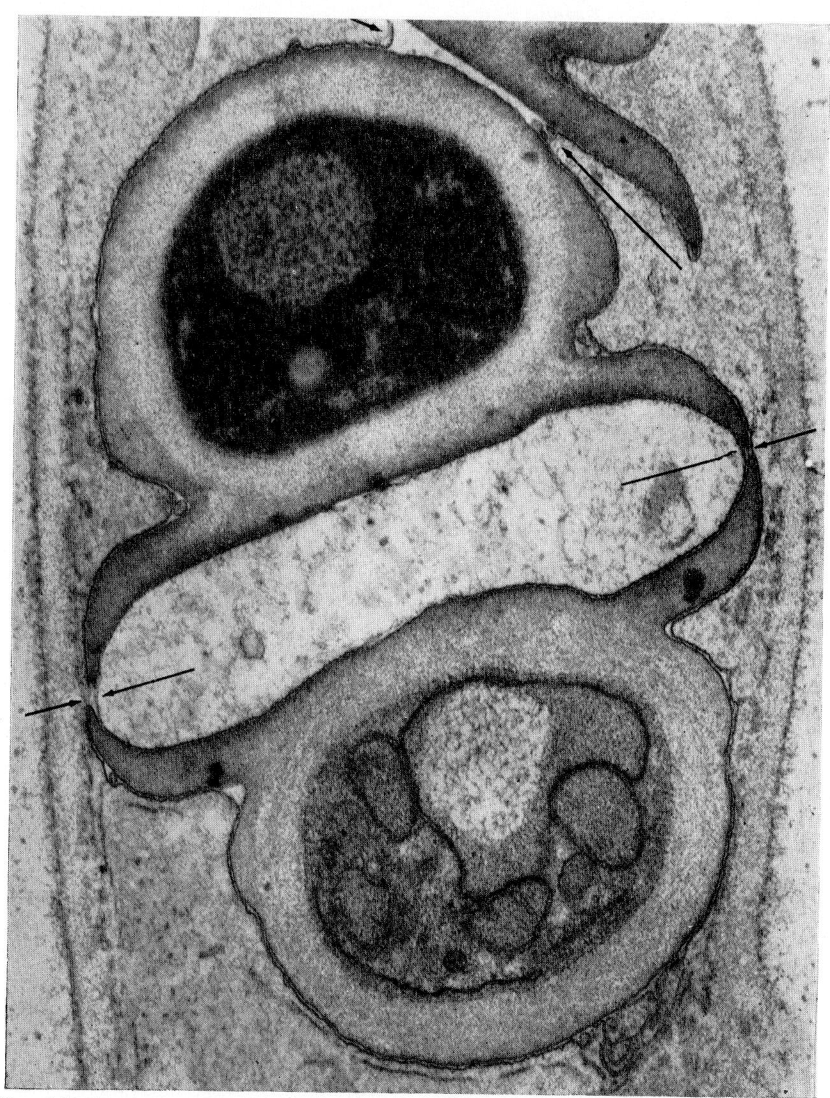

FIG. 14. Ascospores of *Hansenula anomala* inside their ascus (Bandoni *et al.*, 1967). Reproduced by courtesy of R. J. Bandoni. Magnification, × 43,000.

of cell walls during cell fusion (Bresch *et al.*, 1968). Further evidence of the essential role of the cell wall in yeast conjugation stems from the observation that protoplasts prepared from mating strains of *Schizosacch.*

*pombe* were unable to conjugate although sporulation could take place if the cell walls were removed after mating (Holter and Ottolenghi, 1960).

After nuclear fusion, diploid buds are formed on the conjugation tube of the zygotes of *H. wingei* or *Sacch. cerevisiae*. This directionality of the diploid bud may have some bearing on the problem of budding site determination in vegetative yeast cells (see Section III.A.2, p. 469). Finally, during sporogenesis, spore walls are synthesized intracellularly

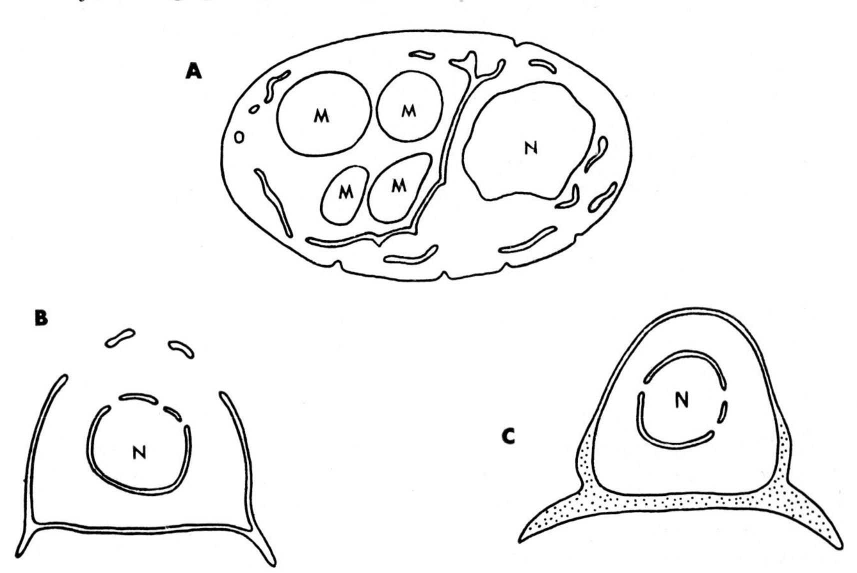


FIG. 15. Semi-diagrammatic representation of the process of ascospore wall formation in *Hansenula anomala*. A, Early stage in the fusion of endoplasmic reticulum vesicles to delimit an ascospore; N = nucleus, M = mitochondria. B, The coalescing membranes of the endoplasmic reticulum have already assumed a shape similar to the mature spore. C, Deposition of the first wall layer inside the cisternal space (Bandoni *et al.*, 1967). Reproduced by courtesy of R. J. Bandoni.

with characteristic ornamentations. The elaborate cytological sequence of ascosporogenesis may be illustrated by the work of Bandoni *et al.* (1967) on *H. anomala* (Figs. 14 and 15). The cytoplasm of the multinucleate ascus cell is cleaved by a system of coalescing membranes of the endoplasmic reticulum. Eventually, a continuous envelope of two "unit" membranes surrounds each nucleus and a corresponding portion of cytoplasm. This double-membrane system assumes the elegant hat shape of the mature ascospore wall (Fig. 15). The space between the membranes serves as a mould, so to speak, in which the ascospore wall is cast. According to this cytological scheme, the shape of the ascospore wall would be primarily determined by the differentiation of the internal membranes. The factors and forces which cause the membrane systems to undergo such precisely co-ordinated three-dimensional organization must remain for now a profound mystery.

## V. Acknowledgements

We are indebted to Eva Streiblová and W. J. Nickerson for reading the manuscript and offering valuable criticism.

S.B-G. wishes to thank the National Institutes of Health, U.S.A. for supporting in part all experimental work from his laboratory mentioned herein (Research grants AI-06205 and AI-05540). He also expresses his gratitude to Prof. Bengt Lindberg of Institutionen för Organisk Kemi, Stockholms Universitet, Sweden for his hospitality during sabbatical leave in his Institute, where part of this article was written.

### References

Agar, H. D. and Douglas, H. C. (1955). *J. Bact.* **70**, 427–434.

Bacon, J. S. D., Milne, B. D., Taylor, I. F. and Webley, D. M. (1965). *Biochem. J.* **95**, 28C–30C.

Bacon, J. S. D., Davidson, E. D., Jones, D. and Taylor, I. F. (1966). *Biochem. J.* **101**, 36C–38C.

Bacon, J. S. D., Jones, D. and Ottolenghi, P. (1969). *J. Bact.* **99**, 885–887.

Bandoni, R. J., Bisalputra, A. A. and Bisalputra, T. (1967). *Can. J. Bot.* **45**, 361–366.

Bartholomew, J. W. and Mittwer, T. (1953). *J. Bact.* **65**, 272–275.

Bartnicki-Garcia, S. (1963). *Bact. Rev.* **27**, 293–304.

Bartnicki-Garcia, S. (1968a). *A. Rev. Microbiol.* **22**, 87–108.

Bartnicki-Garcia, S. (1968b). *J. Bact.* **96**, 1586–1594.

Bartnicki-Garcia, S. (1969). *Phytopathology* **59**, 1065–1071.

Bartnicki-Garcia, S. (1970). *In* "Phytochemical Phylogeny" (J. B. Harborne, ed.), pp. 81–103. Academic Press, London.

Bartnicki-Garcia, S. and Lippman, E. (1969). *Science, N.Y.* **165**, 302–304.

Bartnicki-Garcia, S. and Nickerson, W. J. (1962a). *Biochim. biophys. Acta* **58**, 102–119.

Bartnicki-Garcia, S. and Nickerson, W. J. (1962b). *J. Bact.* **84**, 829–840.

Bartnicki-Garcia, S. and Nickerson, W. J. (1962c). *J. Bact.* **84**, 841–858.

Bartnicki-Garcia, S., Nelson, N. and Cota-Robles, E. (1968). *J. Bact.* **95**, 2399–2402.

Barton, A. A. (1950). *J. gen. Microbiol.* **4**, 84–86.

Beck, C. and Meyenburg, H. K. (1968). *J. Bact.* **96**, 479–486.

Beran, K. (1968). *Advanc. microbial Physiol.* **2**, 143–171.

Beran, K., Malek, I., Streiblová, E. and Lieblová, Y. (1966). *In* "Symposium on Microbial Physiology and Continuous Culture" (E. O. Powell *et al.*, eds.), pp. 57–69. H.M.I.O., London.

Bianchi, D. E. (1968). *Antonie van Leeuwenhoek* **33**, 324–332.

Bostock. C. J., Donachie, W. D., Master, M. and Mitchison, J. M. (1966). *Nature, Lond.* **210**, 808–810.

Brachet, J. (1961). *In* "Growth in Living Systems" (M. X. Zarrow, ed.), pp. 241–275. Basic Books, New York.

Brandt, K. M. (1941). *Biochem. Z.* **309**, 190–201.
Braun, V. and Schwarz, U. (1969). *J. gen. Microbiol.* **57**, iii.
Bresch, C., Müller, G. and Egel ,R. (1968). *Molec. gen. Genetics* **102**, 301–306.
Brock, T. D. (1959). *Science, N.Y.* **129**, 960–961.
Brock, T. D. (1961). *J. gen. Microbiol.* **26**, 487–497.
Brown, C. M. and Hough, J. S. (1965a). *Eur. Brew. Conv. Proc.* Stockholm, p. 223–237. Elsevier, Amsterdam.
Brown, C. M. and Hough, J. S. (1965b). *Nature, Lond.* **206**, 676–678.
Brown, C. M. and Hough, J. S. (1966). *Nature, Lond.* **211**, 201.
Brown, R. G. and Nickerson, W. J. (1965). *Bact. Proc.* p. 26.
Castle, E. S. (1958). *J. gen. Physiol.* **41**, 913–926.
Challinor, S. W., Power, D. M. and Tonge, R. J. (1964). *Nature, Lond.* **203**, 250–251.
Chattaway, F. W., Holmes, M. R. and Barlow A. J. E. (1968). *J. gen. Micrcbiol.* **51**, 367–376.
Chung, K. L., Hawirko, R. Z. and Isaac, P. K. (1965). *Can. J. Microbiol.* **11**, 953–956.
Conti, S. F. and Brock, T. D. (1965). *J. Bact.* **90**, 524–533.
Crandall, M. A. and Brock, T. D. (1968). *Bact. Revs* **32**, 139–163.
Crook, E. M. and Johnston, I. R. (1962). *Biochem. J.* **83**, 325–331.
Davies, E. and Maclachlan (1968). *Archs Biochem. Biophys.* **128**, 595–600.
Davies, R. and Elvin, P. A. (1964). *Biochem. J.* **93**, 8P.
Deshusses, J., Berthoud, S. and Posternak, T. (1969). *Biochim. biophys. Acta* **176**, 803–812.
Dixon, B. and Rose, A. H. (1964). *J. gen. Microbiol.* **35**, 411–419.
Domer, J. E., Hamilton, J. G. and Harkin, J. C. (1967). *J. Bact.* **94**, 466–474.
Drouhet, E, and Mariat, F. (1952). *Ann. Inst. Pasteur* **83**, 506–514.
Dunwell, J. L., Ahmad, F. and Rose, A. H. (1961). *Biochim. biophys. Acta* **51**, 604–607.
Dyke, K. G. H. (1964). *Biochim. biophys. Acta* **82**, 374–384.
Eddy, A. A. (1958). *Proc. R. Soc., Lond., Ser. B* **149**, 425–440.
Eddy, A. A. and Williamson, D. H. (1959). *Nature, Lond.* **183**, 1101–1104.
Elmer, G. W. and Nickerson, W. J. (1970a). *J. Bact.* **101**, 492–594.
Elmer, G. W. and Nickerson, W. J. (1970b). *J. Bact.* **101**, 595–602.
Elorza, M. V. and Sentandreu, R. (1969). *Biochem. Biophys. res. Commun.* **36**, 741–742.
Falcone, G. and Nickerson, W. J. (1956). *Science, N.Y.* **124**, 272–273.
Falcone, G. and Nickerson, W. J. (1959). *In* "Biochemistry of Morphogenesis" (W. J. Nickerson, ed.), pp. 65–70. Pergamon Press, London.
Farkaš, V., Svoboda, A. and Bauer, Š. (1969). *J. Bact.* **98**, 744–748.
Fowell, R. R. (1969). *In* "The Yeasts" (A. H. Rose and J. S. Harrison, eds.) Vol. 1, pp. 303–383. Academic Press, London.
Freifelder, D. (1960). *J. Bact.* **80**, 567–568.
Friis, J. and Ottolenghi, P. (1959). *C.r. Trav. Lab. Carlsberg. Sér. physiol.* **31**, 259–271.
Garcia-Mendoza, C. and Novaes-Ledieu, M. (1968). *Nature, Lond.* **220**, 1035.
Garzuly-Janke, R. (1940). *Zentbl. Bakt. ParasitKde* (*Abt. II*) **102**, 361–365.
Geiger, K. H. (1961). *Master Brewers Assn A. Proc.* pp. 8–14.
Ghosh, A., Charalampous, F., Sison, Y. and Borer, R. (1960). *J. biol. Chem.* **235**, 2522–2528.
Gifford, G. D. and Pritchard, G. C. (1969). *J. gen. Microbiol.* **46**, 143–149.
Gorman, J., Tauro, P., La Berge, M. and Halvorson, H. O. (1964). *Biochem. Biophys. res. Commun.* **15**, 43–49.

Grant, K. L. (1964). *Proc. A.M. Amer. Soc. Brew. Chem.* pp. 217–223.
Haidle, C. W. and Storck, R. (1966). *J. Bact.* **92**, 1236–1244.
Halvorson, H. O., Bock, R. M., Tauro, P., Epstein, R. and La Berge, M. (1966). *In* "Cell Synchrony" (I.L. Cameron and G. M. Padilla, eds.), pp. 102–116. Academic Press, New York.
Hassid, W. Z. (1969). *Science, N.Y.* **165**, 137–144.
Holter, H. and Ottolenghi, P. (1960). *C.r. Trav. Lab. Carlsberg. Sér. physiol.* **31**, 409–422.
Houwink, A. L. and Kreger, D. R. (1953). *Antonie van Leeuwenhoek* **19**, 1–24.
Howard, D. H. (1962). *Mycopathol. Mycol. appl.* **18**, 127–139.
Hurst, H. (1952). *J. expl Biol.* **29**, 30–53.
Johnson, B. F. (1965). *Expl Cell Res.* **39**, 613–624.
Johnson, B. F. (1968). *J. Bact.* **95**, 1169–1172.
Johnson, B. F. and Gibson, E. J. (1966a). *Expl Cell Res.* **41**, 297–306.
Johnson, B. F. and Gibson, E. J. (1966b). *Expl Cell Res.* **41**, 580–591.
Kanetsuna, F., Carbonell, L. M., Moreno, R. F. and Rodriguez, J. (1969). *J. Bact.* **97**, 1036–1041.
Kessler, G. and Nickerson, W. J. (1959). *J. biol. Chem.* **234**, 2281–2285.
Korn, E. D. and Northcote, D. H. (1960). *Biochem. J.* **75**, 12–17.
Kreger, D. R. (1954). *Biochim. biophys. Acta* **13**, 1–9.
Kreger-Van Rij, N. J. W. (1969). *In* "The Yeasts" (A. H. Rose and J. S. Harrison, eds.) Vol. I, pp. 5–73. Academic Press, London.
Küenzi, M. T. and Fiechter, A. (1969). *Arch. Mikrobiol.* **64**, 396–407.
Lampen, J. O. (1968). *Antonie van Leeuwenhoek* **34**, 1–18.
Lamport, D. T. A. (1970). *A. Rev. Plant Physiol.* **21**, 235–270.
Manners, D. J. and Patterson, J. C. (1966). *Biochem. J.* **98**, 19C.
Marchant, R. and Smith, D. G. (1967). *Arch. Mikrobiol.* **58**, 248–256.
Mardon, D., Balish, E. and Phillips, A. W. (1969). *J. Bact.* **100**, 701–707.
Mariat, F. (1964). *Symp. Soc. gen. Microbiol.* **14**, 85–111.
Masuda, Y. (1968). *Planta* **83**, 171–184.
May, J. W. (1962). *Expl Cell Res.* **27**, 170–172.
Mazia, D. (1961). *A. Rev. Biochem.* **30**, 669–688.
McClary, D. O. and Bowers, W. D. (1965). *Can. J. Microbiol.* **11**, 447–452.
McLellan, W. L., Jr. and Lampen, J. O. (1963). *Biochim. biophys. Acta* **67**, 324–326.
McMurrough, I. (1966). Ph.D. Thesis: University of Newcastle-upon-Tyne.
McMurrough, I. and Bartnicki-Garcia, S. (1969). *Bact. Proc.* p. 37.
McMurrough, I. and Rose, A. H. (1967). *Biochem. J.* **105**, 189–203.
Meyenburg, H. K. (1968). *Path. Microbiol., Basel* **31**, 117–127.
Meyenburg, H. K. (1969). *Arch. Mikrobiol.* **66**, 289–303.
Millbank, J. W. and Macrae, R. M. (1964). *Nature, Lond.* **201**, 1347.
Miller, M. W. and Phaff, H. J. (1958). *Antonie van Leeuwenhoek* **24**, 225–238.
Misaki, A., Johnson, J., Kirkwood, S., Scaletti, J. W. and Smith, F. (1968). *Carbohydrate Res.* **6**, 150–164.
Mitchison, J. M. (1957). *Expl Cell Res.* **13**, 244–262.
Mitchison, J. M. (1958). *Expl Cell Res.* **15**, 214–221.
Mitchison, J. M. (1963). *In* "Cell Growth and Cell Division". Soc. Cell Biol. Symp. (J. C. Harris, ed.) Vol. II, p. 151. Academic Press, New York.
Mitchison, J. M. (1969). *Science, N.Y.* **165**, 657–663.
Mitchison, J. M. and Lark K. G. (1962). *Expl Cell Res.* **28**, 452.
Mitchison, J. M. and Walker, P. M. B. (1959). *Expl Cell Res.* **16**, 49–58.
Mitchison, J. M. and Wilbur, K. M. (1962). *Expl Cell Res.* **26**, 144–157.
Moor, H. (1967). *Arch. Mikrobiol.* **57**, 135–146.

Moor, H. and Mühlethaler, K. (1963). *J. Cell Biol.* **17**, 609–628.
Mortimer, R. K. and Johnston, J. R. (1959). *Nature, Lond.* **183**, 1751–1752.
Mundkur, B. (1960). *Expl Cell Res.* **20**, 28–42.
Nečas, O. (1961). *Nature, Lond.* **192**, 580.
Nečas, O. (1965). *Folia Biol.* **11**, 97–102.
Nečas, O. and Kopecká, M. (1969). *Antonie van Leeuwenhoek* **35**, B7–B8.
Nečas, O., Kopecká, M. and Brichta, J. (1969). *Expl Cell Res.* **58**, 411–419.
Nečas, O., Svobodá, A. and Kopecká, M. (1968). *Expl Cell Res.* **53**, 291–293.
Nickerson, W. J. (1948). *Nature, Lond.* **162**, 241–245.
Nickerson, W. J. (1954). *J. gen. Physiol.* **37**, 483–494.
Nickerson, W. J. (1963). *Bact. Rev.* **27**, 305–324.
Nickerson, W. J. and Chung, C. W. (1954). *Am. J. Bot.* **41**, 114–120.
Nickerson, W. J. and Falcone, G. (1956). *Science, N.Y.* **124**, 722–723.
Nickerson, W. J. and Falcone, G. (1959). *In* "Sulfur in Proteins" (R. Benesch, ed.), pp. 409–424. Academic Press, New York.
Nickerson, W. J. and Mankowksi, Z. (1953). *Am. J. Bot.* **40**, 584–592.
Nickerson, W. J., Taber, W. A. and Falcone, G. (1956). *Can. J. Microbiol.* **2**, 575–584.
Nickerson, W. J., Falcone, G. and Kessler, G. (1961). *In* "Macromolecular Complexes" (M. V. Edds, ed.), pp. 205–228. Ronald Press, New York.
Northcote, D. H. (1963). *Pure appl. Chem.* **7**, 669–675.
Northcote, D. H. and Horne, R. W. (1952). *Biochem. J.* **51**, 232–236.
Ottolenghi, P. (1967). *C.r. Trav. Carlsberg. Sér physiol.* **35**, 363–368.
Panek, A. D. (1963). *Archs Bioch. Biophys.* **100**, 422–425.
Phaff, H. J. (1963). *A. Rev. Microbiol.* **17**, 15–30.
Pitt, D. (1969). *J. gen. Microbiol.* **59**, 257–262.
Power, D. M. and Challinor, S. W. (1969). *J. gen. Microbiol.* **55**, 169–176.
Reuvers, T., Tacoronte, E., Garcia-Mendoza, C. and Novaes-Ledicu, M. (1969). *Can. J. Microbiol.* **15**, 989–993.
Robson, J. E. and Stockley, M. H. (1962). *J. gen. Microbiol.* **28**, 57–68.
Romano, A. H. (1966). *In* "The Fungi" (G. C. Ainsworth and A. S. Sussman, eds.) Vol. II, pp. 181–209. Academic Press, New York.
Rost, K. and Venner, H. (1968). *Z. Allg. Mikrobiol.* **8**, 205–208.
Ruesink, A. W. (1969). *Planta* **89**, 95–107.
Šašek, V. and Becker, G. E. (1969). *J. Bact.* **99**, 891–892.
Scherr, G. H. and Weaver, R. H. (1953). *Bact. Rev.* **17**, 51–92.
Schopfer, W. H., Posternak, T. and Wustenfeld, D. (1962). *Arch. Mikrobiol.* **44**, 113–151.
Sentandreu, R. and Northcote, D. H. (1969a). *Biochem. J.* **115**, 231–240.
Sentandreu, R. and Northcote, D. H. (1969b). *J. gen. Microbiol.* **55**, 393–398.
SentheShanmuganathan, S. and Nickerson, W. J. (1962a). *J. gen. Microbiol.* **27**, 437–450.
SentheShanmuganathan, S. and Nickerson, W. J., (1962b). *J. gen. Microbiol.* **27**, 451–464.
Shifrine, M. and Phaff, H. J., (1958). *Antonie von Leeuwenhoek* **24**, 274–280.
Shimoda, C. and Yanagishima, N. (1968). *Physiologia Plantarum.* **21**, 1163–1169.
Smith, D., Tauro, P., Sweizer, E. and Halvorson, H. O. (1968). *Proc. natl Acad. Sci. U.S.A.* **60**, 936–942.
Souza, N. O. and Panek, A. D. (1968). *Archs Biochem. Biophys.* **125**, 22–28.
Spoerl, E., Sarachek, A. and Smith, S. B. (1957). *Am. J. Bot.* **44**, 252–258.
Streiblová, E. (1966). Thesis: Czechoslovak Academy of Sciences, Prague.
Streiblová, E. (1968). *J. Bact.* **95**, 700–707.

Streiblová, E. (1969). *Proc. 2nd Symp. Yeasts*, Bratislava, 1966, pp. 179–186. Publishing House of the Slovak Academy of Sciences, Praha.
Streiblová, E. (1970). *Can. J. Microbiol.* **16**, 827–831.
Streiblová, E. and Beran, K. (1963). *Expl Cell Res.* **30**, 603–605.
Streiblová, E. and Beran, K. (1965). *Folia Microbiol., Praha* **10**, 352–356.
Streiblová, E., Málek, I. and Beran, K. (1966). *J. Bact.* **91**, 428–435.
Streiblová, E., Svoboda, A. and Nečas, O. (1967). *In* "Symposium on Yeast Protoplasts" (R. Müller, ed.), p. 91. Akademie Verlag, Berlin.
Summers, D. F., Grollman, A. P. and Hansenclever, H. F. (1964). *J. Immunol.* **92**, 491–499.
Sundhagul, M. and Hedrick, L. R. (1966). *J. Bact.* **92**, 241–249.
Sutton, D. D. and Lampen, J. O. (1962). *Biochim. biophys. Acta* **56**, 303–312.
Svoboda, A. (1966). *Exptl Cell Res.* **44**, 640–642.
Svoboda, A. and Nečas, O. (1968). *Folia Biol.* **14**, 390–397.
Sylvén, B., Tobias, C. A., Malmgren, J., Ottoson, R. and Thorell, B. (1959). *Expl Cell Res.* **16**, 75–87.
Talou, B. and Tavlitzki, J. (1969). *C. r. hebd. Séanc. Acad. Sci., Paris* **268**, 2419–2422.
Tauro, P. and Halvorson, H. O. (1966). *J. Bact.* **92**, 652–661.
Taylor, N. W. (1965). *Archs Biochem. Biophys.* **111**, 181–186.
Uruburu, F., Elorza, V. and Villanueva, J. R. (1968). *J. gen. Microbiol.* **51**, 195–198.
Villanueva, J. R. (1966). *In* "The Fungi" (G. C. Ainsworth and A. S. Sussman, eds.) Vol. 2, pp. 3–62. Academic Press, New York.
Vraná, D. and Fencl, Z. (1964). *Folia Microbiol., Praha* **9**, 156–163.
Vraná, D. and Fencl, Z. (1967). *Folia Microbiol., Praha* **12**, 432–440.
Wang, M. C. and Bartnicki-Garcia, S. (1966). *Biochem. biophys. res. Commun.* **24**, 832–837.
Williamson, D. H. (1965). *J. Cell Biol.* **25**, 517–528.
Williamson, D. H. (1966). *In* "Cell Synchrony" (I. L. Cameron and G. M. Padilla, eds.), pp. 81–101. Academic Press, New York.
Williamson, D. H. and Scopes, A. W. (1960). *Expl Cell Res.* **20**, 338–349.
Williamson, D. H. and Scopes, A. W. (1961). *Symp. Soc. gen. Microbiol.* **14**, 217–242.
Wright, B. E. (1966). *Science, N.Y.* **153**, 830–837.
Yanagishima, N. (1969). *Planta* **87**, 110–118.

*Chapter 12*

# Carotenoid Pigments of Yeasts*

K. L. SIMPSON, C. O. CHICHESTER

*Department of Food and Resource Chemistry, University of Rhode Island, Kingston, Rhode Island, U.S.A.*

AND

H. J. PHAFF

*Department of Food Science and Technology, University of California, Davis, California, U.S.A.*

I. INTRODUCTION . . . 493
II. CULTURAL CONDITIONS FOR PIGMENT PRODUCTION . . . 494
A. Carbon Sources . . . 495
B. Nitrogen Sources . . . 496
C. Light . . . 497
D. Aeration . . . 498
E. Temperature . . . 498
F. Carotenoid Synthesis as a Function of Time of Growth . . . 499
III. EXTRACTION OF PIGMENTS . . . 499
IV. CHROMATOGRAPHY AND IDENTIFICATION . . . 501
V. BIOSYNTHESIS OF CAROTENOIDS . . . 504
A. Formation of Carotene Precursors . . . 504
B. Types of Carotenoid Pigments and Transformation of the $C_{40}$ Polyenes and Carotenoids . . . 504
C. Specific Carotenoid Inhibitors . . . 511
D. Stereochemistry of Carotenoid Biosynthesis . . . 512
VI. CONCLUSION . . . 513
REFERENCES . . . 513

## I. Introduction

The carotenoids are a group of pigments, yellow to red in colour, which are widely distributed in the Plant and Animal Kingdoms. They belong to a larger group of compounds known as terpenoids which possess repeating branched five-carbon units. The colours of these compounds

*Contribution number 1386 of the Rhode Island Agricultural Experiment Station.

vary, depending on the length of the chromophore and the type of oxygen-containing groups attached. β-Carotene (Fig. 1) is an example of a carotenoid found in yeast and in most carotene-forming systems.

Pigment formation is a characteristic of various species of *Rhodotorula*, *Cryptococcus* and *Sporobolomyces*. Other yeasts, though normally colourless, can be made to form pigments. For example, Cutts and Rainbow (1950) and Chamberlain *et al.* (1952) were able to induce

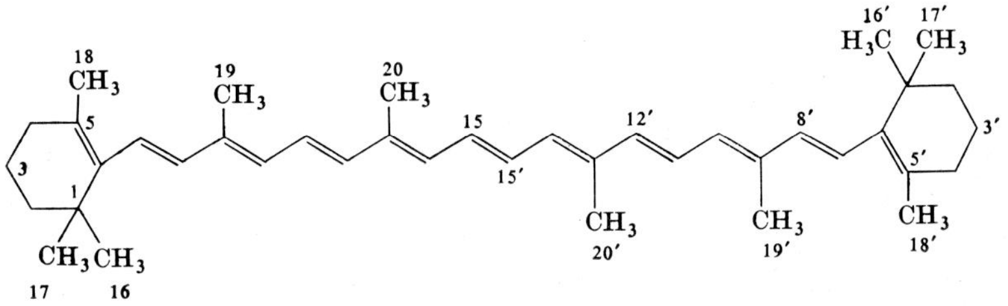

FIG. 1. Structure of β-carotene. The primed numbers always are assigned to that portion of the molecule which, in several related carotenoids, carries an open ring, e.g. γ-carotene or torularhodin (*cf*. Fig. 5, p. 510).

pigment formation in *Schizosaccharomyces pombe*, *Saccharomycodes ludwigii*, *Saccharomyces carlsbergensis* and *Sacch. cerevisiae* by culturing the yeasts in media containing 100–500 μg of DL-methionine per ml, but containing suboptimal amounts of D-biotin. In these species the pink pigments do not belong to the class of carotenoid pigments, which are usually associated with pigmented yeasts. Such and other non-carotenoid pigments will not be considered in the present review. Colourless polyenes structurally related to the carotenoids are known and are usually included in discussions of carotenoids. Phytoene (see Fig. 4) is an example of such a compound.

The general chemistry and biochemistry of carotenoids have been reviewed by a number of authors (Goodwin, 1965; Weedon, 1965; Davies, 1965; Porter and Anderson, 1967). Bacterial carotenoids have been reviewed by Jensen (1965), and those produced by fungi by Goodwin (1952a) and, more recently, by Valadon (1966).

## II. Cultural Conditions for Pigment Production

A number of authors have suggested media for producing high concentrations of microbial carotenoids, ease of extraction of the pigments, or maximum biomass. Generally, the media consist of glucose or some other sugar, yeast extract or more specific nitrogen sources, and salts. A simple medium giving good growth and pigment formation consists of glucose (5%) and yeast extract (0·5%) (Simpson *et al.*, 1964a). Defined media consisting of a carbon source, inorganic salts, organic and/or inorganic nitrogen sources and, in some instances, growth factors have been suggested (Bonner *et al.*, 1946; Tavlitzki, 1951; Wickerham, 1951; Wittmann, 1957; Peterson *et al.*, 1958; Nakagawa and Tatsumi, 1960a; Villoutreix, 1960b; Benigni *et al.*, 1963; Vecher *et al.*, 1965; Kakutani, 1966).

### A. CARBON SOURCES

In *Rhodotorula sanniei* (now considered synonymous with *Rh. rubra*), glycerol was shown to be effective in promoting carotenogenesis (Fromageot and Tchang, 1938), but it was less effective than glucose in *Rhodotorula* sp. no. 100, a strain which produces mainly β-carotene (Nakagawa and Tatsumi, 1960a) and in *Sporobolomyces roseus* (Bobkova, 1965c). In the latter case, Bobkova found that the best carbon sources for growth were maltose and sucrose, but the highest total carotenoid content was obtained with raffinose. In all cases, the medium contained also 0·2% sodium succinate.

Nakagawa and Tatsumi (1960c) observed that phenol, resorcinol and kojic acid stimulated β-carotene production by *Rhodotorula* sp. no. 100 in appropriate concentrations. The effect of kojic acid was especially striking. For another strain of *Rh. rubra*, grown with asparagine as the nitrogen source, glycerol was by far the best carbon source for highest yields of torularhodin, but sucrose gave the highest yields of torulene, β-carotene, γ-carotene and the best total carotene yield under otherwise identical growing conditions (Wittmann, 1957). Although she obtained the best cell yields on glucose, this carbon source was inferior to fructose, glycerol and sucrose in obtaining high yields of torularhodin and β-carotene, although glycerol was inferior to glucose in torulene production.

Petroleum hydrocarbons have been used as carbon sources in species of *Rhodotorula* (Nikolaev *et al.*, 1966; Vaskivnyuk and Kvasnikov, 1968). Strains of *Rh. glutinis*, *Rh. aurantiaca* and *Rh. mucilaginosa* were isolated from oil-containing soils. In mineral media with liquid paraffins as carbon source, the last species required vitamins for good growth. *Rhodotorula glutinis* gave the best cell yields and it accumulated about 360 μg of total carotenoids g dry weight (Vaskivnyuk and Kvasnikov, 1968).

Benigni *et al.* (1963) showed that, if the sugars of molasses were replaced by 10–20% carrot homogenate, production of carotenes by *Rhodotorula* was increased substantially. Similarly, Vecher *et al.* (1965) found that a strain of *Rh. gracilis* produced the highest carotenoid content when grown in depigmentized carrot juice without added mineral salts. Under these conditions 97% of the pigments consisted of β-carotene. In media with other vegetable extracts the total carotenoid content, as well as the percentage of β-carotene, was much lower.

### B. NITROGEN SOURCES

*Sporobolomyces roseus* (Bobkova, 1965c) has been shown to produce good yields of cell material and carotenoids on a glucose-containing medium with glycine, valine, proline, asparagine, glutamic acid and arginine as single nitrogen sources. The pH value did not change significantly during growth. Due to poor growth, carotenes were not formed with β-alanine. Of the amino acids tested, glutamic acid gave the highest pigment yield, and leucine, which supported only moderate growth, the lowest. Ammonium ion, as might be expected, supported carotene production and good growth. Although *Sp. roseus* can assimilate nitrate as sole nitrogen source, this organism gave poor growth on ammonium and potassium nitrates. With ammonium nitrate very low yields of carotenoid pigments were obtained. In spite of inferior growth with potassium nitrate (final pH value of the medium, 7·2), this nitrogen source gave a very significant increase in the carotene content on a dry weight basis. Interestingly, when the yeast was grown on potassium nitrate, torularhodin formed 34% of the total pigment content, whereas this pigment constituted only 3–6% of the total carotenoids when ammonium sulphate or asparagine was used as the nitrogen source.

Nakagawa and Tatsumi (1960a) also studied the effectiveness of various nitrogen sources and found that valine, leucine, and asparagine gave the best yields of carotenoids for their Rh-100 strain of *Rhodotorula*. The best C:N ratio was 50. Leucine and glutamic acid, when used as the sole nitrogen source for *Rh. gracilis*, gave good growth and carotene production (Vecher *et al.*, 1967). Lysine was ineffective in this study, probably because this nitrogen source is assimilated by only a few species of yeast. In a study with another strain of *Rh. gracilis*, Dyr and Praus (1955) found that this nitrate-assimilating species effectively synthesized carotene from ammonium nitrate, whereas glycine, valine, asparagine and leucine showed no stimulatory effect.

*Rhodotorula rubra*, a species not able to utilize nitrate, was investigated by Wittmann (1957). She compared asparagine, alanine, glutamic acid, urea, glycine histidine and ammonium nitrate as nitrogen sources with glucose as the source of carbon. Although ammonium nitrate gave the highest yield of total carotenoids per unit dry weight of cells, the cell yield was only moderate. Yields of individual carotenoids varied considerably with the nitrogen source. For example, the highest concentration of torularhodin was obtained with histidine. Maximal yields of torulene and β-carotene occurred on ammonium nitrate, while γ-carotene production was most abundant on asparagine.

It seems certain that few of the authors reporting the effects of nitrogen sources on carotenoid synthesis were familiar with the ability of the yeasts under study to assimilate specific nitrogen sources. Variation in effectiveness of certain nitrogen sources reported above might be due, at least in part, to the different strains and species used by various authors.

### C. LIGHT

Lederer (1933) observed that light stimulated development of carotenoid pigments in *Rh. rubra*, and Praus (1952) reported that light accelerated the biosynthesis of carotenoids during the growth phase of *Rh. gracilis*. Light was further shown to alter the ratio of α- and β-carotene to torulene from 2·29 in the light to 1·67 in the dark. Bobkova (1965b) noted a similar shift to torularhodin from β-carotene on increased intensity of illumination for *Rh. glutinis*, *Sp. pararoseus* and *Sp. roseus*. Bobkova further found that weak light stimulated carotene formation only in *Sp. pararoseus*, and that high intensities of illumination inhibited the formation of carotenoids in *Rh. glutinis*, *Sp. pararoseus* and *Sp. roseus*, the last species being the most sensitive.

In bacteria, it has been reported in numerous instances that carotenoid pigments offer protection against photodynamic effects (Griffiths *et al.*, 1955; Dworkin, 1958). In yeast, Maxwell *et al.* (1666) and Maxwell and Chichester (1967) compared the wild-type of *Rh. glutinis* strain 48-23T with a mutant obtained by ultraviolet irradiation which contained little or no carotenoids. They showed that the carotenoid-containing organism was protected from photodynamic death by monochromatic light at 632·8 nm from a gas laser even in the presence of the photosensitizer toluidine blue. The mutant was much more sensitive to photodestruction. At high intensities of radiation, using full-spectrum light above 300 nm, the same organism was not afforded any protection by carotenoid pigments. Chichester and Maxwell (1969) showed that the portion of the spectrum responsible for lethality lay between 300 and 400 nm, with an apparent maximum response around 300 nm. The irradiated cells suffered damage to their membranes, as indicated by increased leakage. Overall respiration of the cells decreased at a rate corresponding to the loss in viability. Such random damage to cells indicates that, whatever the damaging effect is, it must be quite non-specific. Their evidence would indicate that the sensitizing agent could be a free radical, since the sulphydryl-containing compounds cysteine and glutathione were capable of offering protection of a certain degree against irradiation damage. Little or no protection was offered by fat-soluble, free radical-trapping agents. This indicates that the primary lethality site must be in the water phase, thus supporting the argument that, at least in this organism, the carotenoids do not normally serve as protective compounds.

### D. AERATION

Bobkova (1965a, b) found that increasing the aeration of cultures of *Sp. pararoseus* and *Sp. roseus* beyond 0·4 g $O_2$/l. h had very little effect on the formation of carotenoids. *Sporobolomyces pararoseus* doubled its carotene content with increasing aeration from 0·1 to 0·4 g $O_2$/l. h under Bobkova's experimental conditions. Since the carotenogenic yeasts are all obligate aerobes, a minimal amount of oxygen is obviously required for growth.

### E. TEMPERATURE

Most yeasts do not change their pigment ratios with a change in cultivation temperature, although there may be a decrease in the level of the carotenoids (Fromageot and Tchang, 1938; Nakayama *et al.*, 1954; Bobkova, 1965b). However, Phaff *et al.* (1952) isolated two strains of *Rh. glutinis* (48-23T and 48-23W) which were pink to deep pink at 25° and yellow at 5°; another strain, which was named *Rh. peneaus* (now considered as a strain of *Cryptococcus laurentii* var. *flavescens*), was canary yellow at 25° and cream-coloured at 5°. Colour changes were also noted by Iguti (1957) in cultures of *Torula* (*Rh.*) *rubra* grown at similar temperatures. Nakayama *et al.* (1954) found that γ- and β-carotene comprised 92–96% of the total pigments of *Rh. glutinis* (str. 48-23T) at 5° but only 43–47% at 25°. The remainder of the pigments was made up of torularhodin and torulene. The total pigment concentration in *Rh. peneaus* decreased from about 9 mg/100 g yeast dry matter at 25° to about 1 mg/100 g at 5°. β-Carotene represented 93% of the total carotenes of *Rh. peneaus* at 25° but only 66% at 5°. Lycopene synthesis was increased from 3% to 13% of the total pigments when this yeast was cultured at the lower temperature.

Simpson *et al.* (1964b) reported that the percentage of γ-carotene remained nearly unchanged at 5° and 25° in *Rh. glutinis* 48-23T. However, at 25° the percentage of β-carotene in the total carotenoids was lower by approximately the percentage increase in the level of torulene and torularhodin. When the yeast cultures grown on agar media at 5°

were brought to 25°, only the cells on the surface became red over a period of 24–48 h. When the cells cultured at 5° were aerated at 25° in buffer solutions, no change in colour was observed (Simpson, 1963).

### F. CAROTENOID SYNTHESIS AS A FUNCTION OF TIME OF GROWTH

Bobkova (1965c) investigated the production of carotenoids in relation to the biomass formed and the use of nutrients in *Sporobolomyces roseus* as a function of time. The medium contained glucose, sodium succinate and ammonium sulphate. Sodium succinate was included in the medium to prevent strong acidification due to ammonium sulphate. By the fourth day the glucose and ammonium nitrogen had disappeared from the media and maximum cell growth had occurred. Carotenoid production greatly lagged behind the formation of cells and did not reach a maximum until the twenty-fifth day, after which time pigmentation decreased. It was also found that $\beta$-carotene and torulene were most abundant in younger cultures and torularhodin in older cultures. Similar results were obtained by Vecher and Kulikova (1968) for *Rh. gracilis* (now considered to be *Rh. glutinis*).

Simpson *et al.* (1964b) who studied a yeast subsequently identified as *Rh. pallida* (62-506) also observed profound changes in the ratios of carotenoids produced during growth. $\gamma$-Carotene, the major pigment in this yeast, was detected very early in the growth cycle, reached a maximum at four days, after which time its concentration decreased to less than 25% of the maximum after nine days. During the same period, the concentration of torulene and torularhodin increased continually, the proportion of the latter increasing from a minute fraction to approximately 50% of the total pigments.

## III. Extraction of Pigments

The forces required to disrupt most animal cells are mild compared to those required to disintegrate yeasts and bacteria. Due to difficulties inherent in physical methods during the early days of carotenoid studies, both chemical and enzymic methods were then used to disintegrate microbial cell walls. While the carotenoids of most pigmented fungi can be extracted with solvents, many workers have experienced difficulties in the extraction of carotenoids from yeast. Under certain cultural conditions some yeasts produce a heavy mucilaginous slime which may prevent complete and, in some cases, even partial extraction. Workers in the field have used one of three approaches for pigment extraction: (i) culture of the yeast on a medium that allows solvent extraction;

(ii) destruction of the slime layer and/or cell wall; (iii) physical disintegration of the cells.

Bonner *et al.* (1946) first treated yeast cells with strong alkali to destroy the cell walls. The cell mass was then extracted with methanolic potassium hydroxide and benzene and mechanically shaken for 1·0–1·5 h. After centrifugation the benzene supernatant was pipetted off and the residue was re-extracted twice with fresh benzene. As Mrak *et al.* (1949) pointed out, the method works well with some species but is unsuitable for others where very little pigment is found in the benzene layer. The cell wall is composed of polysaccharides (Phaff, 1971) and these are not hydrolysed by the potassium hydroxide, even though the capsule dissolves. Nakayama *et al.* (1954) noted that the method does not give a representative yield of pigments. Torularhodin, in particular, was poorly extracted by this method. This probably explains why Bonner *et al.* (1946) found very small amounts of acidic pigments in *Rh. rubra*. Other workers have encountered difficulties in extracting certain species of *Rhodotorula* with alcoholic potassium hydroxide (Mackinney, 1940; Lodder and Kreger-van Rij, 1952; Peterson *et al.*, 1954). Mrak *et al.* (1949) treated the yeast cells with 6 *N*-HCl and heated the suspension at 80–100° for varying periods of time, depending on the yeast culture. As these authors pointed out, the duration and temperature of the acid treatment are important since overheating destroys the pigments and underheating does not allow complete extraction. The method was refined by Nakayama *et al.* (1954), who used 1·2 *N*-HCl; they observed flocculation of the yeast as an indication of the boiling time.

Peterson *et al.* (1954) found that yeasts cultured in yeast nitrogen base (Difco) with 2% glucose (Wickerham, 1951) could be extracted with acetone without prior acid treatment. These authors compared their method with one in which the cells were boiled in 0·5 *N*-HCl, and found less destruction of pigments. Peterson *et al.* (1958), however, noted later that, although the method gave complete extraction of some yeasts, the carotenoids from others, such as some strains of *Rh. mucilaginosa*, *Rh. glutinis*, *Rh. aurantiaca*, *Rh. minuta* and *Rh. rubra*, were only partially extracted. Acetone did not extract pigments from *Rh. pallida*. In order to get complete pigment recovery, cells of some strains were first extracted with acetone, then boiled for 15 min in 0·5 *N*-HCl and re-extracted with acetone. Wittmann (1957) also was able to extract the carotenoids directly with acetone from cultures of *Rh. rubra* grown on a synthetic medium. With a synthetic growth medium Villoutreix (1960b) reported complete extraction of the carotenoid from *Rh. mucilaginosa* with acetone and an acetone:methanol (3:1) mixture. Other yeasts could be extracted by this method only after a prior treatment with benzene.

More recently, workers have used physical methods of disrupting yeast cells. With improved techniques it is possible to extract yeast cells cultured under less exacting conditions. While some cells may not be broken, complete extraction is probably attained from those cells that are disintegrated. Hughes and Cunningham (1963) have reviewed a number of physical methods available for the disruption of micro-organisms.

Simpson *et al.* (1963, 1964a) constructed a modified French press and found that two passages through the press were usually sufficient for complete extraction of the pigment. Scharf and Simpson (1968) found that a French press made by the American Instrument Co., Silver Springs, Md., U.S.A. gave similar breakage of cells. The maximum volume of cell suspension is limited to about 40 ml. A colloid mill (Simpson *et al.*, 1964a) has been used successfully for volumes of cells greater than 0·5 l and less than about 3 l (*cf.* Garver and Epstein, 1959). A thick yeast suspension is added to the mill, together with glass beads (average diameter 120 μm) in a ratio of 1·0:0·6. The mill is cooled with iced water and run under an atmosphere of nitrogen (Scharf and Simpson, 1968). Usually 1 h is required for 80–90% disintegration of yeast cells. Breakage of cells is probably caused by shear forces between the rotor, the stator and the glass beads. Without the glass beads little breakage occurs.

Maxwell *et al.* (1966) broke yeast cells in a small stainless-steel blendor cup with 120 μm glass beads and acetone for approximately 15 min. Small pieces of dry ice were continually added during blending for the purpose of cooling and to prevent oxidation. Bonaly and Villoutreix (1965) and Bae *et al.* (1970) successfully disintegrated yeast cells in acetone with glass beads under an atmosphere of nitrogen with a Braun MSK cell disintegrator. The disintegrator (Bae *et al.*, 1970) was run at 4000 rev./min for 2 min and was cooled with liquid carbon dioxide. Complete extraction of the cells was achieved.

## IV. Chromatography and Identification

Davies (1965) has reviewed many of the techniques involved in the isolation and identification of plant carotenoids. Only an outline will be given here, and the reader is referred to the general work by Davies or specific papers cited.

The carotenoids are usually obtained in an acetone extract from the cells, and the combined extracts are added to petroleum ether or diethyl ether and the pigments are transferred to this solvent by the addition of water. Water is used to wash the pigment solution free of acetone. This extract can be dried over anhydrous sodium sulphate and added to a magnesium oxide-Hyflo-Super-Cel column to remove torularhodin (Tefft *et al.*, 1970). Torularhodin, which contains a carboxyl group, is adsorbed tightly to the column and the other, less polar, pigments can be removed with acetone. These pigments are washed into petroleum ether or diethyl ether, and contaminating yeast lipids are saponified in methanolic potassium hydroxide. Alternatively, the crude extract can be saponified directly and torularhodin removed as the salt. The potassium salt of torularhodin is not very soluble and tends to form a solid at the interface of the methanolic potassium hydroxide and petroleum ether. The pigment solution in petroleum ether is washed free of alkali and dried with sodium sulphate prior to chromatography. The majority of the pigments can be separated on either de-activated alumina or on a mixture of magnesium oxide and Hyflo-Super-Cel.

Usually rechromatography of some of the bands is necessary depending on the amount of the pigment, the size of the column, and the amount and nature of interfering substances. The order of elution of the hydrocarbons is (Davies, 1955): phytoene, phytofluene, β-carotene, β-zeacarotene, ζ-carotene, γ-carotene, neurosporene, lycopene, torulene.

Wittmann (1957) used circular paper chromatography to separate, in an effective way, torularhodin, torulene, γ-carotene and β-carotene. Details on the techniques for the separation and purification of most of these compounds can be found in papers by Simpson *et al.* (1964b), Nusbaum-Cassuto *et al.* (1967) and Bonaly and Malenge (1968). Oxygenated carotenoids from yeast, other than torularhodin, have been reported recently, and techniques for their separation and identification have been given: for example, 3′,4′-dehydro-17′-oxo-γ-carotene (Bonaly and Villoutreix, 1965); 3′,4′-dehydro-17′-hydroxy-γ-carotene (Bonaly and Malenge, 1968); plectaniaxanthin (Bae *et al.*, 1970). For identification of the last compound, Bae and coworkers used nuclear magnetic resonance and high resolution mass spectroscopy.

The numbering system used here is that shown in Fig. 1 (p. 494) where 16′, 17′ are the methyl groups on carbon 1′, while 18′ is the methyl group on carbon 5′. This system of numbering is the same as that used by Winterstein *et al.* (1960) [*cf. J. Am. chem. Soc.* **82**, 5583 (1960), IUPAC 16th Conference, New York, 1951, p. 110]. Rüegg *et al.* (1959) refer to the "torularhodin" aldehyde and alcohol carbon as 18′. Bonaly and Villoutreix (1965) isolated and identified this aldehyde and called it 3′,4′-dehydro-17′-oxo-γ-carotene, but later Bonaly and Malenge (1968) used the 18′ carbon to describe these aldehyde and hydroxyl compounds. Isler *et al.* (1959) also termed the carboxyl carbon of torularhodin 18′.

The absorption data for carotenoids reported in yeast are given in Table I.

TABLE I. *Absorption Data for Yeast Carotenoids*

| Carotenoid | Maxima in petroleum ether (nm) | | | Extinction coefficient $E^{1\%}_{1\,cm}$ [a] | References |
|---|---|---|---|---|---|
| Phytoene | 275 | 286 | 296 | 1250 | Rabourn and Quackenbush (1953); Davies (1965) |
| Phytofluene | 332 | 348 | 367 | 1350 | Koe and Zechmeister (1952); Davies (1965) |
| ζ-Carotene | 378 | 400 | 424 | 2270 | Nash *et al.* (1948); Davies (1965); Davis *et al.* (1966) |
| Neurosporene | 416 | 440 | 470 | 2990 | Nakayama (1958); Davis *et al.* (1966) |
| Lycopene | 446 | 472 | 505 | 3450 | Isler and Schudel (1963) |
| β-Zeacarotene | 406 | 428 | 454 | 2520 | Rüegg *et al.* (1961) |
| Torulene | 460 | 484 | 518 | 3240 | Rüegg *et al.* (1961) |
| γ-Carotene | 435 | 460 | 491 | 2760 | Goodwin (1956) |
| | | | | 3100 | Isler and Schudel (1963) |
| β-Carotene | 424 | 450 | 477 | 2590 | Goodwin (1952b) |
| α-Carotene | 420 | 444 | 474 | 2800 | Rüegg *et al.* (1961a) |
| 17′ Hydroxytorulene | 462 | 488 | 519 | | Rüegg *et al.* (1958) |
| 17′ Oxotorulene | 480 | 514 | 548 | | Winterstein *et al.* (1960) |
| | | 508 | 540 | 2865 | Rüegg *et al.* (1959) |
| Torularhodin | 490[b] | 515[b] | 545[b] | 1932[b] | Simpson *et al.* (1964b) |
| Plectaniaxanthin | 445 | 471 | 502 | | Bae *et al.* (1970) |

[a] The extinction coefficients are based on the maxima given in the centre column. In some instances literature values vary significantly.

[b] In chloroform.

## V. Biosynthesis of Carotenoids

### A. FORMATION OF CAROTENE PRECURSORS

The biosynthesis of the carotenoids falls into three stages. In the first, the basic five-carbon terpenoid precursors are produced. The second stage includes formation of $C_{40}$ compounds, and the third is the alteration of the $C_{40}$ chains in carotenogenic systems.

Our knowledge of the terpenoid precursors has come from workers using micro-organisms, higher plants and animal cells (Rogers *et al.*, 1966a, b). While different enzyme systems are used for the final stages of the synthesis of sterols and carotenoids, they have common pathways through farnesyl pyrophosphate. At least in plant systems, spacially separated isoenzymes involved in the phosphorylation of mevalonic acid to mevalonic acid pyrophosphate have been demonstrated (Rogers *et al.*, 1966a, b). Workers using colourless as well as pigmented yeasts have contributed also to our understanding of the formation of the precursors to the terpenoids. Fig. 2 shows a general scheme for the formation of the carotenoid precursors.

The incorporation of specifically labelled mevalonic acid in *Rhodotorula* species is consistent with the view that carotene formation in yeast follows a pathway similar to carotenogenesis in other systems (Scharf and Simpson, 1968).

### B. TYPES OF CAROTENOID PIGMENTS AND TRANSFORMATION OF THE $C_{40}$ POLYENES AND CAROTENOIDS

Zopf (1890) first described fat-soluble pigments in red yeasts. Chapman (1916) reported that the pigments in yeast were different from "carotene" prepared from carrots. Lederer (1933) extracted the pigments from *Rh. rubra* and found four pigments: an acidic pigment (later named torularhodin); torulin (torulene); β-carotene; an unstable pigment. Lederer (1933) suggested that torulene was 3,3′-dimethoxy-γ-carotene. Torulene was later synthesized by Rüegg *et al.* (1961b) and found to be 3′,4′-dehydro-γ-carotene (Fig. 5). Torularhodin was at first thought to be an apocarotene ($C_{47}$), but was synthesized by Rüegg *et al.* (1958) and shown to be 3′,4′-dehydro-17′-γ-carotenoic acid (Fig. 5). Fromageot and Tchang (1938) found the three pigments reported by Lederer to be present in *Rh. sanniei* (*Rh. mucilaginosa*, and more recently placed in synonomy with *Rh. rubra*). These authors noted, in addition, trace amounts of γ-carotene and lycopene.

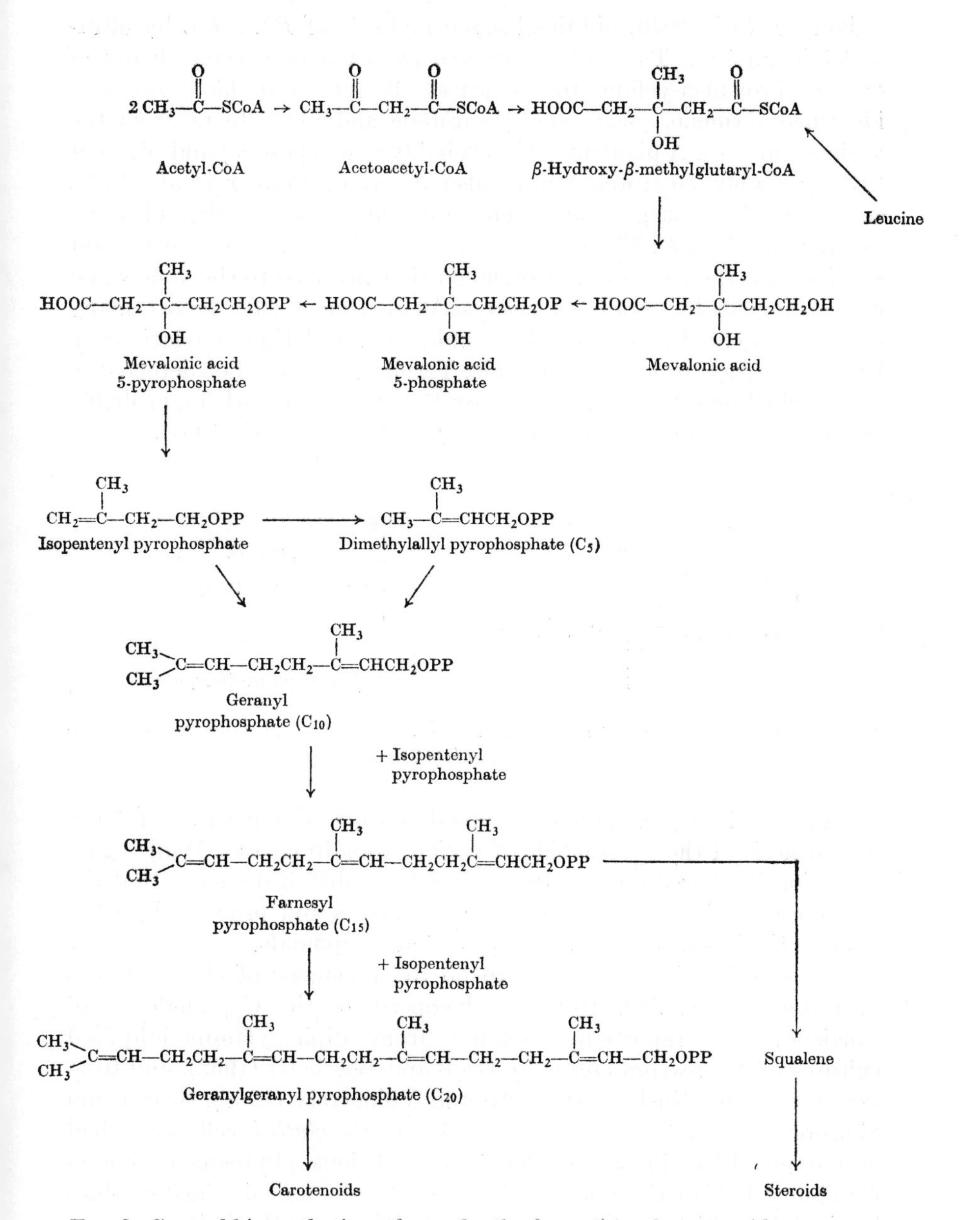


FIG. 2. General biosynthetic pathway for the formation of carotenoid precursors.

Bonner *et al.* (1946) obtained seven mutants of *Rh. rubra* by ultraviolet irradiation. The cultures ranged from red to various shades of orange, brownish-yellow to colourless. Bonner and his coworkers identified torulene, γ-carotene, β-carotene and phytofluene from the various cultures. A pigment "A" (probably neurosporene) and pigment "B" (probably ζ-carotene) were also reported. Bonner *et al.* (1946) suggested the first general scheme for the transformation of yeast carotenoids (Fig. 3). These workers concluded, on the basis of mutant studies, that the colourless carotenoids are converted to the yellow and red carotenoids by parallel pathways. At the time of their report, many of the carotenoid structures were unknown, and thus a relationship based on structure was not possible. Torularhodin was not detected in any marked quantity in *Rh. rubra* by Bonner and coworkers, probably because of the method of extraction (*cf.* Nakayama *et al.*, 1954).

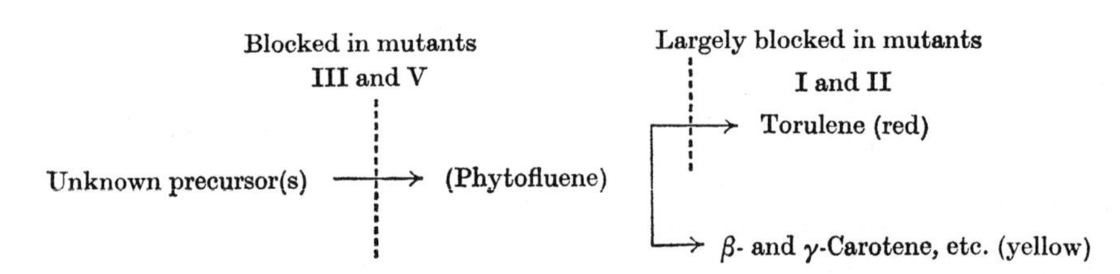


FIG. 3. An early scheme for the transformation of yeast carotenoids according to Bonner *et al.* (1946).

Porter and Lincoln (1950) proposed a sequential pathway for the biosynthesis of the carotenoids of various tomato crosses. While again a relationship based on structure was not possible, it did show that the carotenes are formed from common precursors and proceed from the more to the less saturated and oxygenated compounds.

Phytoene has been assumed to be the precursor of the more unsaturated carotenoids. However, lycopersene (the $C_{40}$ analogue of squalene) was reportedly isolated from diphenylamine-inhibited cultures of *Neurospora crassa* by Grob and Boschetti (1962), and from carrot roots by Nusbaum-Cassuto and Villoutreix (1965). Scharf and Simpson (1968) extracted large quantities of *Rh. glutinis* cells which had been exposed to β-ionone. Under these conditions, phytoene accounted for almost half of the polyenes isolated. A spot was obtained on thin-layer plates which could not be separated from synthetic lycopersene. However, the same substance could be isolated from the yeast extract used for the culture medium, and it could not be isolated from cultures grown on media not containing yeast extract. No significant carbon was incorporated into the lycopersene zone from $^{14}C$-labelled mevalonate, acetate or glucose by *Rh. glutinis* and a commercial baker's yeast. It would now seem that lycopersene, if it exists naturally, is not a direct precursor of the carotenoids in yeast or in other systems (*cf.* Goodwin, 1965).

Villoutreix (1960b) obtained a number of stable mutants from *Rh. mucilaginosa* by repeated ultraviolet irradiation. Torularhodin, torulene, $\gamma$-carotene and $\beta$-carotene were the principal pigments of the parent strain, whereas phytoene and phytofluene were absent. In mutants which contained cyclic pigments, there was a decrease in the levels of torularhodin and torulene, accompanied by an increase in $\beta$-carotene. Neurosporene, $\zeta$-carotene, pigment X ($\beta$-zeacarotene), phytofluene, phytoene, lycopene and spirilloxanthin, though not isolated from the parent strain, were detected, often in substantial amounts, in several of the mutants. One strain had apparently lost its ability to form cyclic carotenes, and was the only mutant which produced lycopene (86% of the total pigment content) and spirilloxanthin. Based on the levels of carotenoids in the various mutants and on the use of 2-hydroxybiphenyl as a reversible inhibitor of carotenoid synthesis, the following conversions were thought to be improbable in his strain of *Rh. mucilaginosa*: phytoene $\rightarrow$ phytofluene; phytofluene $\rightarrow$ $\zeta$-carotene; $\zeta$-carotene $\rightarrow$ neurosporene; neurosporene $\rightarrow$ lycopene; lycopene $\rightarrow$ $\gamma$-carotene $\rightarrow$ $\beta$-carotene. Villoutreix (1960b) concluded that his results supported the concept that the carotenoids in yeast are for the most part not mutually related, but are formed by independent pathways. Two possible relationships were observed, however, derived from pigment levels in two mutants. In one of these, a practically complete disappearance of torulene and a very large increase in the concentration of $\gamma$-carotene suggested that the latter is a precursor of torulene. In another mutant, the disappearance of torularhodin and a corresponding increase of $\beta$-carotene showed that $\beta$-carotene is a precursor of torularhodin. In a later report, Kayser and Villoutreix (1961) confirmed that, in an X-ray-produced mutant, production of torulene and torularhodin was decreased and synthesis of $\gamma$- and $\beta$-carotenes increased. This supports the hypothesis that $\gamma$-carotene is a precursor of torulene but does not rule out the possibility that $\beta$-carotene, whose level increased tenfold, could be a precursor of torulene as well.

Nakayama *et al.* (1954) examined the pigments from a number of *Rhodotorula* and *Cryptococcus* species. They showed that, depending on the growth temperature, the concentrations of either the red or yellow pigments can be altered. These results suggested a similar parallel pathway, as proposed by Bonner *et al.* (1946). Simpson *et al.* (1964b) re-examined the *Rh. glutinis* strain 48-23T which was reported by Nakayama *et al.* (1954) to decrease its concentration of pink carotenoids on being cultured at 5°. Total carotenoid concentration, on a dry weight basis, was found to be nearly equal at room temperature and at 5°. The level of γ-carotene remained fairly constant; however, there was a decrease in the percentage β-carotene content nearly equal to the gain in the levels of torulene and torularhodin on culturing the yeast at the higher temperature. These results suggest that γ-carotene lies at a branch point in the sequence, and that intermediates can be channelled through it either to β-carotene or to the red pigments, depending on the growth temperature. It was also shown with yeast 62-506 (*Rh. pallida*) that, as the level of γ-carotene dropped, the concentrations of torulene and then torularhodin increased. It was concluded from these results that γ-carotene is converted to torulene, which in turn is converted to torularhodin. Similar results were obtained by Bobkova (1965c) with *Sporobolomyces roseus*, and by Vecher and Kulikova (1968) for *Rh. gracilis* (*Rh. glutinis*).

Simpson *et al.* (1964b) exposed *Rh. glutinis* to the vapours of either methyl heptenone or β-ionone. Whereas phytoene, phytofluene and ζ-carotene were not isolated in the untreated yeast, these polyenes became the main carotenes in the treated culture. There was also an increased synthesis of neurosporene and β-zeacarotene upon treatment with these compounds, whereas synthesis of β-carotene, torularhodin and torulene was decreased. γ-Carotene either remained at a constant or lower level. These results showed that pathways are available for the formation of all of the $C_{40}$ intermediates of β-carotene and torulene, and that the Porter-Lincoln type of sequence holds for yeast carotenoids as well as for other carotene-forming systems.

The inhibitor 2-hydroxybiphenyl has been shown to block the formation of cyclic carotenoids and to cause an accumulation of lycopene (Villoutreix, 1960a). Nusbaum-Cassuto *et al.* (1967) treated a mutant with 2-hydroxybiphenyl, which inhibited pigmentation of the mutant and caused an accumulation of phytoene. After the inhibitor was removed, the cells under non-growing conditions rapidly synthesized pigments. An analysis of the pigments showed that the phytoene level dropped, and after a lag lycopene was formed. The levels of phytoene, ζ-carotene and neurosporene were shown to rise initially and then to fall. Bonaly and Malenge (1968) were able to show a similar relationship with *Rh. aurantiaca* (Fig. 4). In this organism, the end-products torularhodin and β-carotene accumulate while the levels of cyclic and acyclic carotenes rise and fall with the initial drop in phytoene.

It would thus seem that the evidence is now substantial that carotenoids in yeast are formed in a sequential manner from more saturated precursors. The point of cyclization, however, is not clear from these results. Simpson *et al.* (1964b) found increased levels of β-zeacarotene in

*Rh. glutinis* 48-23T treated with inhibitors, which suggests, since lycopene was not found in this yeast, that this compound is the precursor to β-carotene. However, Bonaly and Malenge (1968) found both lycopene and β-zeacarotene in their strains. They noted that the amount of

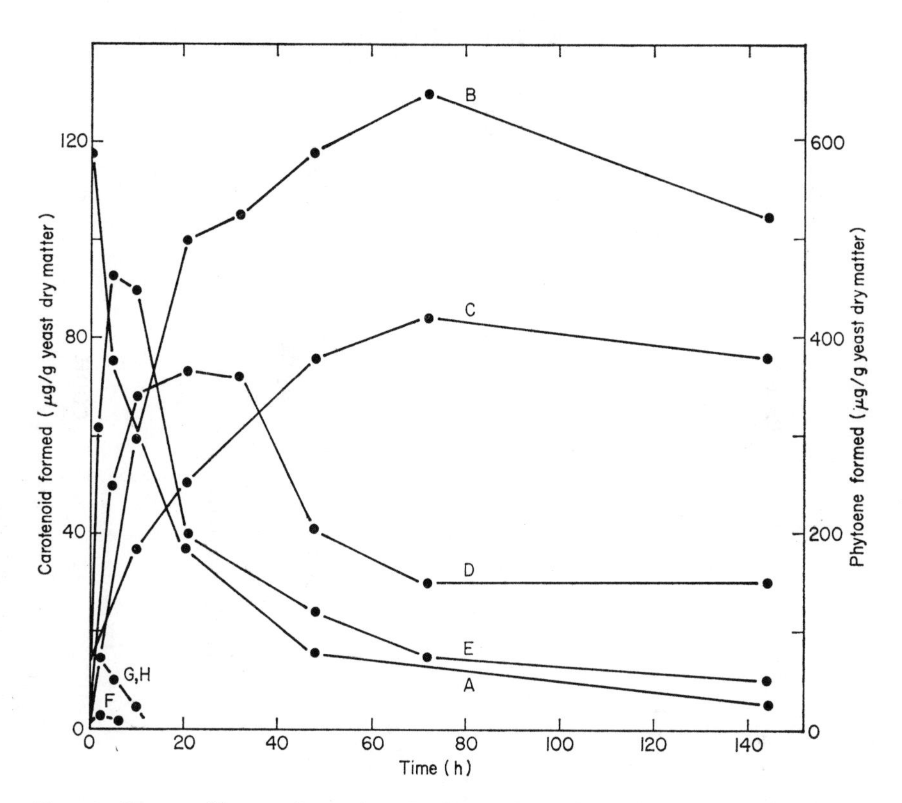


FIG. 4. Carotenoid transformation. Redrawn from data of Bonaly and Malenge (1968). Cells of *Rhodotorula aurantiaca* were grown in the presence of the inhibitor 2-hydroxybiphenyl. Cells were then washed free of inhibitor, placed in phosphate buffer and the development of coloured carotenoids was followed as a function of time. Curve A, phytoene (right-hand scale); B, torularhodin; C, β-carotene; D, torulene; E, γ-carotene; F, β-zeacarotene; G, lycopene; H, neurosporene (left-hand scale).

lycopene varied with β-zeacarotene content, which suggests that both β-zeacarotene and lycopene may be precursors of γ-carotene in *Rhodotorula*. This may depend in part on the organism and on conditions of growth.

The logical intermediates between torulene and torularhodin would be the 17′ hydroxy- and oxo-torulenes. Rüegg *et al.* (1959) prepared both

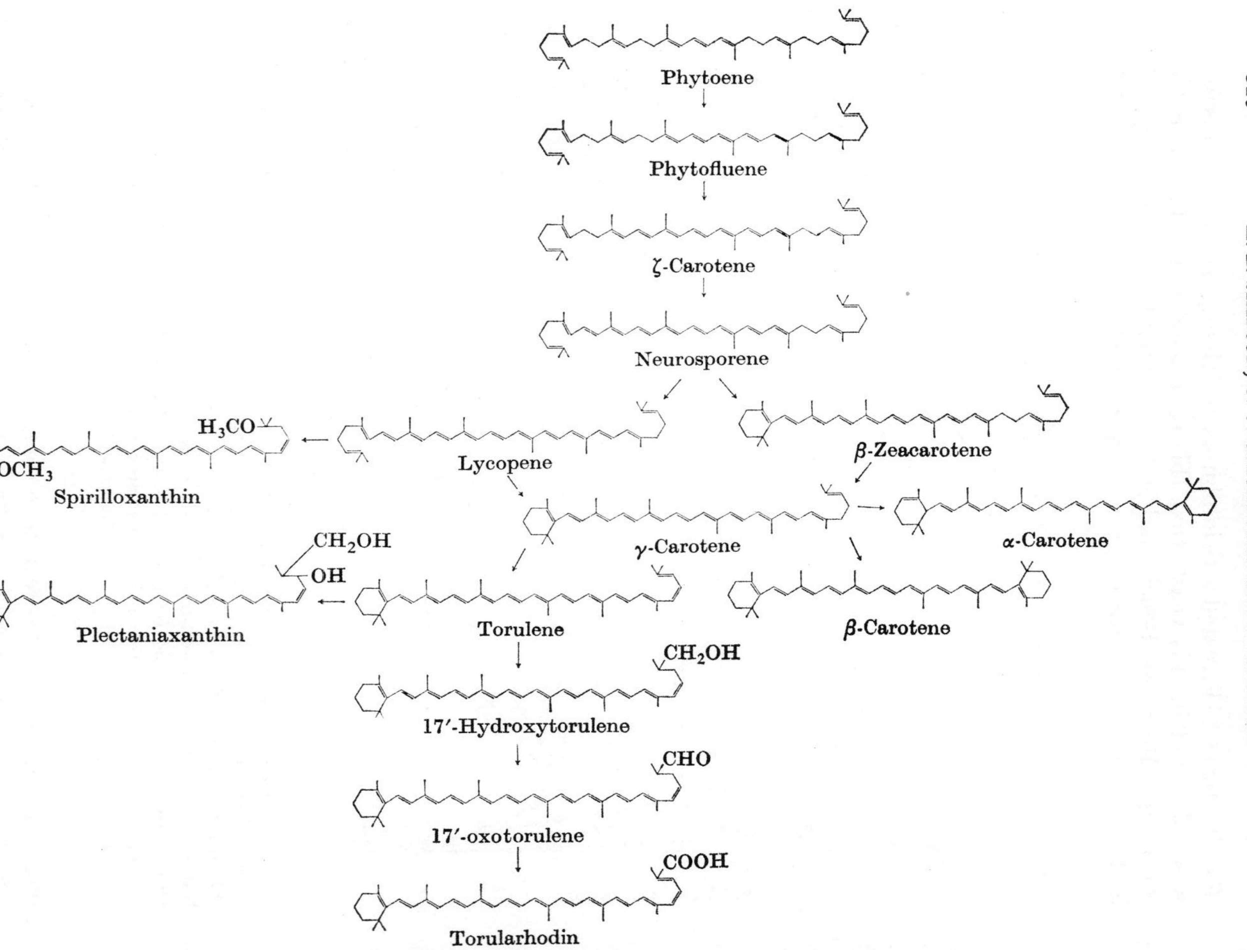


FIG. 5. Proposed pathways for the transformation of carotenoids in yeast.

of these oxygenated compounds synthetically. Winterstein *et al.* (1960) isolated 3′,4′-dehydro-17′-oxo-γ-carotene from an unnamed micro-organism and proposed this aldehyde as a precursor to torularhodin. Bonaly and Villoutreix (1956) used a mutant of *Rh. mucilaginosa* which had lost the ability to synthesize torularhodin. This mutant accumulated increased quantities of the postulated aldehydic intermediate, and they isolated and identified 3′,4′-dehydro-17-oxo-γ-carotene. Later Bonaly and Malenge (1968) also isolated 3′,4′-dehydro-17-hydroxy-γ-carotene from this strain. In the wild type, and in *Rh. aurantiaca*, these compounds were present in only trace amounts.

Recently, Bae *et al.* (1970) reported the isolation and identification of plectaniaxanthin (Arpin and Jensen, 1967) from *Cryptococcus laurentii*. These authors reported mass spectral evidence for several other oxygenated compounds which do not correspond to any compounds so far identified in yeast. Thus it would appear that, with more advanced techniques, other carotenoids which have previously been overlooked will be identified in yeast.

Fig. 5 represents proposed pathways for the formation of carotenoids in yeast.

### C. SPECIFIC CAROTENOID INHIBITORS

A number of chemical agents have been used in the study of carotenoid formation in yeast. Some of the results obtained by the use of inhibitors are contradictory, and in many cases the inhibitor affects cell growth as well as carotenoid formation.

Uehleke and Decker (1962) found that 250 mg of β-ionone/l inhibited the formation of carotenoids in *Rh. rubra*. When added to mature cultures, β-ionone destroyed torularhodin and β-carotene. Simpson *et al.* (1964b) and Scharf and Simpson (1968) added β-ionone to the cotton plug of the culture flask. Under these conditions β-ionone caused an increase in the more saturated polyenes in place of torulene, torularhodin and β-carotene. β-Ionone does not show the specificity of action in yeast, as it does in *Phycomyces* (Reyes *et al.*, 1964).

Diphenylamine has been shown in a number of micro-organisms to inhibit the conversion of carotenes to the more highly unsaturated ones. Šlechta *et al.* (1958) and Protiva *et al.* (1959) found that, at low concentrations ($10^{-3}$ M), formation of carotenoids by *Rh. gracilis* was inhibited, whereas fat synthesis and growth were not inhibited. Protiva *et al.* (1959) further observed that synthesis of phytoene was strongly increased, the β-carotene content was nearly doubled, and phytofluene and α-carotene were slightly increased by treatment with diphenylamine. Schneider *et al.* (1959), Villoutreix (1960a, b) and Nakagawa and Tatsumi (1960b, c) found a decrease in carotenoid formation by this inhibitor, with some effect shown by the type of medium used. Praus and Dyr (1959) showed that the effect of diphenylamine is reversible, and that riboflavin at low levels counteracts its effect in *Rh. gracilis*.

An effect similar to that of diphenylamine was obtained by Kakutani (1967a, b) with *Sporobolomyces shibatanus* by using several antioxidants and chelating agents. Villoutreix and Narbonne (1958) and Villoutreix (1960a) investigated the effect of a number of inhibitors on *Rh. mucilaginosa*, and found that, among the diphenyl derivatives tested, only those which were asymmetrically substituted inhibited the formation of carotenoids.

## D. STEREOCHEMISTRY OF CAROTENOID BIOSYNTHESIS

The use of various isotopes has yielded information on the mechanisms by which yeast pigments are formed. In particular, the oxidation of the 17′ methyl group is nearly unique to yeast, although Carlile and Friend (1956) isolated torularhodin from *Pyronema confluens*, a pigmented higher ascomycete.

Yamamoto *et al.* (1962) and Shneour (1962) had shown that the hydroxy and oxo-groups in the carotenoids are derived from gaseous oxygen. Simpson *et al.* (1963) cultured *Rh. rubra* in an $^{18}O_2$-enriched atmosphere, and found that one of the atoms in the carboxyl group of torularhodin came from the atmosphere. These results are consistent with the hypothesis that the formation of the hydroxy torulene is through the direct participation of atmospheric oxygen, and that oxidation of oxotorulene to torularhodin involves a pick-up of the other oxygen from the medium.

Information on the stereochemistry of torularhodin synthesis has come from the use of specifically labelled mevalonate substrates. Tefft *et al.* (1970) decarboxylated the torularhodin obtained from cultures grown on [2-$^{14}$C]mevalonic acid. If all of the reactions leading to the carboxyl group are free from randomization, either all or none of the $^{14}$C from the 16′ or 17′ position should be lost. Similarly, if [2-$^{3}$H]-mevalonic acid were used, all or none of the $^{3}$H should be lost from the 16′, 17′ positions upon conversion of torulene to torularhodin. These authors found that, in both experiments, the C-2 of mevalonic acid was used for the oxidation of torulene. The data proved that the hydroxylating enzyme was specific for only one methyl group (Fig. 5). These results require that the conversion of isopentenyl pyrophosphate to dimethylallyl pyrophosphate also be free from randomization. These results would rule out a classical carbonium ion mechanism for this reaction.

## VI. Conclusion

It is fairly obvious that the overall pathways for carotenogenesis in yeasts are now delineated. Subsidiary pathways for the synthesis of the oxygen-containing compounds still require additional work. Particularly needed is work which would describe the pathways for formation of the highly oxygenated compounds. An important area on which as yet little information is available in yeasts and other micro-organisms relates to the function of the pigments. A third area worth consideration is the compartmentalization of the pigments and of biosynthetic pathways in the cell, since it is becoming more evident that, in many organisms, different carotenoids are synthesized and stored in different parts of the cell.

### References

Arpin, N. and Jensen, S. L. (1967). *Phytochem.* **6**, 995–1005.
Bae, M., Lee, T. H., Yokoyama, H., Boettger, H. and Chichester, C. O. (1970). *Phytochem.* in press.
Benigni, F., Cieri, L. and Dreassi, M. (1963). *Ric. Sci. Rend. Sez. B.* **3**, 389–394.
Bobkova, T. S. (1965a). *Mikrobiologiya* **34**, 229–233.
Bobkova, T. S. (1965b). *Prikl. Biokhim. Mikrobiol.* **1**, 234–238.
Bobkova, T. S. (1965c). *Prikl. Biokhim. Mikrobiol.* **1**, 320–325.
Bonaly, R. and Malenge, J. P. (1968). *Biochim. biophys. Acta* **164**, 306–316.
Bonaly, R. and Villoutreix, J. (1965). *C.r. hebd. Séanc. Acad. Sci., Paris* **261**, 4904–4906.
Bonner, J., Sandoval, A., Tang, Y. W. and Zechmeister, L. (1946). *Archs Biochem.* **10**, 113–123.
Carlile, M. J. and Friend, J. (1956). *Nature, Lond.* **178**, 369–370.
Chamberlain, N., Cutts, N. S. and Rainbow, C. (1952). *J. gen. Microbiol.* **7**, 54–60.
Chapman, A. C. (1916). *Biochem. J.* **10**, 548–550.
Chichester, C. O. and Maxwell, W. A. (1969). *In* "Life Sciences and Space Research" (W. Vishniac and F. G. Favorite, eds.), VII(COSPAR), pp. 11–18. North-Holland Publ. Co., Amsterdam.
Cutts, N. S. and Rainbow, C. (1950). *Nature, Lond.* **166**, 1117.
Davies, B. H. (1965). *In* "Chemistry and Biochemistry of Plant Pigments" (T. W. Goodwin, ed.), pp. 489–532. Academic Press, New York.
Davis, J. B., Jackman, L. M., Siddons, P. T. and Weedon, B. L. (1966). *J. chem. Soc.* 2154–2165.
Dworkin, M. (1958). *J. gen. Physiol.* **41**, 1099–1112.
Dyr, J. and Praus, R. (1955). *Chem. Listy.* **49**, 1699–1702.
Fromageot, C. and Tchang, J. L. (1938). *Arch. Mikrobiol.* **9**, 424–434.
Garver, J. C. and Epstein, R. L. (1959). *Appl. Microbiol.* **7**, 318–319.
Goodwin, T. W. (1952a). *Bot. Rev.* **18**, 291–316.
Goodwin, T. W. (1952b). *Biochem. J.* **50**, 550–558.
Goodwin, T. W. (1956). *Biochem. J.* **62**, 346–352.
Goodwin, T. W. (1965). *In* "Chemistry and Biochemistry of Plant Pigments" (T. W. Goodwin, ed.), pp. 143–173. Academic Press, New York.
Griffiths, M., Sistrom, W. R., Cohen-Bazire, G. and Stanier, R. Y. (1955). *Nature, Lond.* **176**, 1211–1215.
Grob, E. C. and Boschetti, A. (1962). *Chimia* **16**, 15–16.

Hughes, D. E. and Cunningham, V. R. (1963). *Biochem. Soc. Symp. No.* **23**, "Separation of Subcellular Components", pp. 8–19. Cambridge University Press, Cambridge, England.
Iguti, S. (1957). *Bot. Mag., Tokyo* **70**, 190–197.
Isler, O. and Schudel, P. (1963). *In* "Carotine und Carotenoide" (Wissenschaftliche Veröffentlichungen der Deutschen Gesellschaft für Ernährung) Vol. **9**, p. 54. Steinkopff, Darmstadt.
Isler, O., Guex, W., Rüegg, R. Ryser, G. Saucy, G., Schwieter, U., Walter, M. and Winterstein, A. (1959). *Helv. chim. Acta* **42**, 864–871.
Jensen, S. L. (1965). *A. Rev. Microbiol.* **19**, 163–182.
Kakutani, Y. (1966). *J. Biochem., Japan* **59**, 135–138.
Kakutani, Y. (1967a). *J. Biochem., Japan* **62**, 179–183.
Kakutani, Y. (1967b). *J. Biochem., Japan* **61**, 193–198.
Kayser, F. and Villoutreix, J. (1961). *C. r. Séanc. Soc. Biol.* **155**, 1094–1095.
Koe, B. K. and Zechmeister, L. (1952). *Archs Biochem. Biophys.* **41**, 236–238.
Lederer, E. (1933). *C. r. hebd. Séanc. Acad. Sci., Paris* **197**, 1964–1695.
Lodder, J. and Kreger-van Rij, N. J. W. (1952). "The Yeasts. A Taxonomic Study". Interscience Publishers, Inc., New York.
Mackinney, G. (1940). *A. Rev. Biochem.* **9**, 459–490.
Maxwell, W. A. and Chichester, C. O. (1967). 164th Mtg Am. Chem. Soc., Sept. 1967, No. Q74. Chicago.
Maxwell, W. A., McMillan, J. D. and Chichester, C. O. (1966). *Photochem. Photobiol.* **5**, 567–577.
Mrak, E. M., Phaff, H. J. and Mackinney, G. (1949). *J. Bact.* **57**, 409–411.
Nakagawa, M. and Tatsumi, C. (1960a). *J. agric. Chem. Soc. Japan* **34**, 195–198.
Nakagawa, M. and Tatsumi, C. (1960b). *J. agric. Chem. Soc. Japan* **34**, 199–202.
Nakagawa, M. and Tatsumi, C. (1960c). *J. agric. Chem. Soc. Japan* **34**, 310–314.
Nakayama, T., Mackinney, G. and Phaff, H. J. (1954). *Antonie van Leeuwenhoek* **20**, 217–228.
Nakayama, T. O. M. (1958). *Archs Biochem. Biophys.* **75**, 352–355.
Nash, H. A., Quackenbush, F. W. and Porter, J. W. (1948). *J. Am. chem. Soc.* **70**, 3613–3615.
Nikolaev, P. I., Paleeva, M. A., Pomortseva, N. V., Samokhvalov, G. I., Vakulova, L. A. and Lopatik, M. D. (1966). *Izobret., Prom. Obraztsy Tovarnye Znaki* **43**, 47.
Nusbaum-Cassuto, E. and Villoutreix, J. (1965). *C. r. hebd. Séanc. Acad. Sci., Paris* **260**, 1013–1015.
Nusbaum-Cassuto, E. and Villoutreix, J. and Malenge, J. P. (1967). *Biochim. biophys. Acta* **136**, 459–472.
Peterson, W. J., Bell, T. A., Etchells, J. L. and Smart Jr., W. W. G. (1954). *J. Bact.* **67**, 708–713.
Peterson, W. J., Evans, W. R., Lecce, E., Bell, T. A. and Etchells, J. L. (1958). *J. Bact.* **75**, 586–591.
Phaff, H. J. (1971). *In* "The Yeasts" (A. H. Rose and J. S. Harrison, eds.) Vol. 2, pp. 135–210. Academic Press, London.
Phaff, H. J., Mrak, E. M. and Williams, O. B. (1952). *Mycologia* **44**, 431–451.
Porter, J. W. and Anderson, D. G. (1967). *A. Rev. Plant Physiol.* **18**, 197–228.
Porter, J. W. and Lincoln, R. E. (1950). *Archs Biochem.* **27**, 390–403.
Praus, R. (1952). *Chem. Listy* **46**, 643–645.
Praus, R. and Dyr, J. (1959). *Collection Czechoslovak Chem. Communs.* **24**, 116–121.
Protiva, J., Praus, R. and Dyr, J. (1959). *Folio microbiol., Praha* **4**, 332–335.
Rabourn, W. J. and Quackenbush, F. W. (1953). *Archs Biochem. Biophys.* **44**, 159–164.

Reyes, P., Chichester, C. O. and Nakayama, T. O. M. (1964). *Biochem. biophys. Acta* **90**, 578–592.

Rogers, L. J., Shah, S. P. J. and Goodwin, T. W. (1966a). *Biochem. J.* **100**, 14c–17c.

Rogers, L. J., Shah, S. P. J. and Goodwin, T. W. (1966b). *Biochem. J.* **99**, 381–388.

Rüegg, R., Guex, W., Montavon, M. Schwieter, U., Saucy, G. and Isler, O. (1958). *Chimia* **12**, 327.

Rüegg, R., Montavon, M., Ryser, G., Saucy, G., Schweiter, U. and Isler, O. (1959). *Helv. chim. Acta* **42**, 854–864.

Rüegg, R., Schwieter, U., Ryser, G., Schudel, P. and Isler, O. (1961a). *Helv. chim. Acta* **44**, 985–993.

Rüegg, R., Schwieter, U., Ryser, G., Schudel, P. and Isler, O. (1961b). *Helv. chim. Acta* **44**, 994–999.

Scharf, S. S. and Simpson, K. L. (1968). *Biochem. J.* **106**, 311–315.

Schneider, G., Matkovics, B. and Zsolt, J. (1959). *Acta Phys. Chem.* **5**, 55–58.

Shneour, E. A. (1962). *Biochim. biophys. Acta* **62**, 534–540.

Simpson, K. L. (1963). Ph.D. Thesis: University of California, Davis, U.S.A.

Simpson, K. L., Nakayama, T. O. M. and Chichester, C. O. (1964a). *Biochem. J.* **92**, 508–510.

Simpson, K. L., Nakayama, T. O. M. and Chichester, C. O. (1964b). *J. Bact.* **88**, 1688–1694.

Simpson, K. L., Wilson, A. W., Burton, E., Nakayama, T. O. M. and Chichester, C. O. (1963). *J. Bact.* **86**, 1126–1127.

Šlechta, L., Gabriel, O. and Hoffmann-Ostenhof, O. (1958). *Nature, Lond.* **181**, 268–269.

Tavlitzki, J. (1951). *Rev. Can. Biol.* **10**, 48–59.

Tefft, R. E., Goodwin, T. W. and Simpson, K. L. (1970). *Biochem. J.* **117**, 921–927.

Uehleke, H. and Decker, K. (1962). *Hoppe-Seyler's Z. physiol. Chem.* **327**, 225–233.

Valadon, L. R. G. (1966). *Ad. Frontiers Plant Sci.* **15**, 183–204.

Vaskivniuk, V. T. and Kvasnykov, E. I. (1968). *Mikrobiologiya* **37**, 876–878.

Vecher, A. S. and Kulikova, A. N. (1968). *Mikrobiologiya* **37**, 676–680.

Vecher, A. S., Gurinovich, E. S., Prokazov, G. F., Lukashik, A. N. and Astapovich, N. I. (1965). *Izv. vyssh. Uchebn. Zavedenii, Pishch. Technol.* **2**, 45–49.

Vecher, A. S., Karabava, G. Ya. and Kulikova, A. N. (1967). *Vesti Acad. Nauk Belarus. SSR, Biol Nauk* **3**, 87–90.

Villoutreix, J. (1960a). *Biochim. biophys. Acta* **40**, 434–441.

Villoutreix, J. (1960b). *Biochim. biophys. Acta* **40**, 442–457.

Villoutreix, J. and Narbonne, F. (1958). *Bull. Soc. Pharm. Nancy* **36**, 12–15.

Weedon, B. C. L. (1965). *In* "Chemistry and Biochemistry of Plant Pigments" (T. W. Goodwin, ed.), pp. 75–126. Academic Press, New York.

Wickerham, L. J. (1951). "Taxonomy of Yeasts". *U.S.D.A. Tech. Bull.* No. 1029, pp. 1–56.

Winterstein, A., Studer, A. and Rüegg, R. (1960). *Chem. Ber.* **93**, 2951–2965.

Wittmann, H. (1957). *Arch. Mikrobiol.* **25**, 373–391.

Yamamoto, H. Y., Chichester, C. O. and Nakayama, T. O. M. (1962). *Archs Biochem. Biophys.* **96**, 645–649.

Zopf, W. (1890). "Die Pilze in morphologischer, physiologischer, biologischer und systematischer Beziehung". Breslau.

# Author Index

Italic numbers indicate pages on which a reference is listed


## A

Abadie, F., 10, 11, 14, *60*, *64*
Abd-El-Al, A. T., 56, *60*, 136, 170, 188, 200, *202*
Abdullah, M., 427, *437*
Abelson, J., 393, 394, 395, *411*
Abercrombie, M. J., 142, 176, 183, 184, 186, *202*
Abranches, P., 98, *118*
Adamas, A., 320, 321, *328*
Adams, A., 321, *327*
Adams, A. M., 128, 131, *132*
Adams, B. G., 226, *264*
Afanasjewa, T. P., 188, *209*
Agar, H. D., 38, *60*, 254, 264, 456, 468, *487*
Agarwal, K. L., 392, *409*
Ahearn, D. G., 24, 25, 56, *60*, 123, 130, *132*, 176, *202*, *203*
Ahlborg, K., 7, *67*
Ahmad, F., 22, *60*, *62*, 149, *204*, 239, *264*, 472, *488*
Ahmed, A., 348, *409*
Akhtar, M., 246, 247, 248, 249, *264*
Aksl'rod, V. D., 321, *327*, *332*
Alberts, A. W., 241, *267*
Albo, P. W., 251, *264*
Aldous, J. G., 47, *60*
Algranati, I. D., 160, *203*, 431, *437*
Allen, C. M., 251, *264*
Allen, F. W., 316, *321*
Allen, N. E., 325, *327*
Allende, C. C., 391, *409*, *410*
Allende, J. E., 391, *409*, *410*
Allmann, D. W., 238, 242, *266*
Altmann, H., 19, *63*
Alworth, W., 251, *264*
Anderson, F. B., 189, 191, *203*
Anderson, D. G., 494, *514*
Anderson, L. G., 27, *62*, 213, 216, *265*
Anderson, R. F., 175, 178, *203*, *207*
Andrean, B. A. G., 406, *414*
Andreasen, A. A., 26, *60*, 229, *264*
Andrews, J., 55, *60*
Andrieu, S., 159, *203*
Anfinsen, C. B., 388, *409*
Ankel, E., 184, *203*
Ankel, H., 184, 185, 186, *203*, *207*
Anow, P. M., 402, *413*
Anton, J., 125, *133*
Apgar, J., 310, 319, 320, 321, *327*, *329*, 392, 393, *412*
Appelby, C. A., 324, *327*
Archibald, A. R., 424, *437*
Aref, H., 129, 130, *132*
Aris, R., 82, *118*
Armstrong, A., 321, *327*
Armstrong, W. McD., 51, *60*
Arnow, P., 158, *206*
Arpin, N., 511, *513*
Arst, H. N., 372, *409*
Artamonova, V. A., 322, *327*
Asahi, T., 365, 372, *409*, *417*
Asano, K., 321, *327*
Aschner, M., 175, *203*, *206*
Ascione, R., 320, 321, *328*
Asensio, C., 276, *306*
Ashby, D. R., 246, *267*
Astapovich, N. I., 495, *515*
Astrachan, L., 396, *417*
Atkin, L., 6, 14, 18, 19, 20, 22, 23, 24, 25, 30, 45, *60*, *70*
Atkinson, D. E., 276, 299, 302, *303*, *305*, *306*
Atzpodien, W., 278, 288, 289, *303*, *304*
Aubel-Sadron, G., 318, 319, 323, *327*
Aubert, J. P., 182, 183, *203*
Auerbach, C., 314, *331*
Augustsson, A. M., 47, *63*
Auret, B. J., 24, *66*
Avers, C. J., 324, *327*

Avigad, G., 294, *303*
Axelrod, A. E., 21, 26, *60*, *68*
Axelrod, V. D., 320, *327*, 392, *409*
Axelson, E., 7, *71*
Ayuso, M. S., 396, *409*
Äyräpää, T., 13, 36, 37, 520, *6*, *61*, *68*
Azam, F., 43, *60*, 86, *116*, 300, *303*
Azoulay, E., 254, *267*, 279, 292, *306*

## B

Babcock, G. E., 209, *210*
Bacilă, M., 290, *304*
Bacon, E. E., 171, 172, 187, *203*
Bacon, J. S. D., 140, 141, 142, 143, 144, 145, 163, 164, 171, 172, 176, 187, 190, 194, 198, 199, 210, *203*, *205*, *210*, 420, *437*, 451, 452, 457, 463, *487*
Baddiley, J., 223, 225, 233, *265*, *269*, *270*
Bae, M., 501, 502, 503, 511, *513*
Baev, A. A., 321, *327*, *330*, *332*
Baich, A., 357, *409*
Baiev, A. A., 392, 395, *409*, *414*
Bak, A. L., 313, *332*
Bakerspigel, A., 38, *60*
Bakes, J., 395, *409*
Baldwin, I. L., 19, *63*
Balish, E., 479, 480, *489*
Ballou, C. E., 147, 148, 149, 150, 151, 153, 154, 157, 159, 201, 209, *203*, *205*, *206*, *208*, 225, *267*, 420, *438*
Bamann, E., 6, *73*
Bandoni, R. J., 176, 184, *202*, *208*, 485, 486, *487*
Bandurski, R. S., 365, 372, *409*, *416*, *417*
Baraud, J., 222, 223, 225, 227, 251, 258, *264*, *265*
Barbaresi, G., 215, 217, *265*
Barker, S. A., 182, *207*
Barlow, A. J. E., 164, *203*, 452, *488*
Bärlund, H., 37, *61*
Barnett, J. A., 291, *303*
Baronowsky, P., 26, *60*
Barron, E. J., 217, 222, *265*
Barroso-Lopes, C., 175, 186, *203*
Barry, V. C., 137, *203*
Bartholemew, J. W., 456, 472, *487*
Bartley, W., 9, 10, *68*, *72*, 262, *268*, 301, *304*, *305*, 315, *331*
Bartnicki-Garcia, S., 137, *203*, 443, 444, 445, 446, 447, 448, 451, 453, 458, 459, 477, 478, *487*, *489*, *491*
Barton, A. A., 456, 470, 472, *487*
Barton, D. H. R., 226, 248, *265*
Barton-Wright, E. C., 12, 13, 30, 31, *60*, *63*
Bathgate, G. N., 427, 429, 433, *437*
Battley, E. H., 107, *116*, 122, *133*
Bauchop, T., 106, 107, *116*
Bauer, H., 254, *265*
Bauer, S., 8, *60*, *62*, 150, 151, 166, 175, 177, 180, 181, 197, 210, *204*, *206*, *208*, 431, *437*, 445, *488*
Baum, H. M., 131, *133*
Baxter, R. M., 125, 130, 132, *133*, 213, 215, 218, 220, 221, 222, 229, *267*
Beamer, P. R., 129, 130, *133*
Beary, M. E., 19, 53, *61*
Bechet, J., 46, *63*, 354, 355, 356, *409*, *415*
Beck, G., 318, 319, 323, *327*, 467, *481*
Becker, G. E., 482, *490*
Beeman, W. W., 393, *413*
Befort, N., 395, *409*
Behrens, N. H., 160, *203*, 262, *265*
Belaich, J-P., 43, *60*
Bell, D. J., 138, 144, *203*
Bell, T. A., 495, 500, *514*
Bellomy, B. R., 19, *63*
Belozersky, A. N., 311, *327*
Bender, H., 427, *437*
Benedict, R. G., 178, *203*
Benend, W., 225, *270*
Berends, W., 401, *413*
Benham, R. W., 230, *265*
Benigni, F., 495, *513*
Beran, K., 76, *116*, 199, *203*, 454, 456, 460, 462, 472, *487*, *491*
Berg, F., 216, *265*
Berg, P., 320, 322, 324, *327*, *329*, 389, 391, *409*, *414*, *417*
Bergel'son, L. D., 217, 218, *265*, *270*
Bergmann, F. H., 322, *327*
Bergquist, P. L., 321, 322, *327*, *328*, *331*, 392, *409*
Berke, H. L., 297, *303*
Bernardi, G., 312, 325, *328*
Bernhardt, W., 281, *305*, 361, *409*, *414*
Bernhauer, K., 27, *60*
Bernofsky, C., 301, *306*
Berthelot, M., 8, *60*

Berthoud, S., 446, 447, *488*
Betz, A., 300, *304*
Bevan, E. A., 327, *332*
Bhandary, U. L. R., 324, *329*
Bhattacharjee, J. K., 359, 360, *409*
Bhattacharjee, S. S., 150, 151, 152, 155, 160, 201, 202, *203*, *204*
Bhattacharyya, S. N., 23, *63*
Bialy, J. J., 9, *72*
Bianchi, D. E., 452, *487*
Biely, P., 8, *60*, 431, *437*
Bigger, L. C., 47, *61*
Biggs, D. R., 301, *305*
Biguet, J., 159, *203*
Billheimer, F. E., 324, *327*
Binkley, W. W., 28, 29, *60*
Birch, G. G., 420, *437*
Birch-Andersen, A., 188, 194, *203*
Bishop, C. T., 141, 144, 149, 150, 168, *203*, 209
Bisalputra, A. A., 485, 486, *487*
Bisalputra, T., 485, 486, *487*
Björk, G. R., 376, 408, *409*, *416*
Björklund, A., 24, 25, 29, *71*
Black, S., 282, *303*, 364, 365, 366, *409*
Blandamer, A., 182, 183, *203*
Blank, F., 141, 144, 149, 150, 168, *203*, *209*
Blew, D., 392, *411*
Blix, R., 57, *62*
Bloch, K., 26, *60*, *65*, 132, *133*, 217, 229, 230, 242, 246, 249, 251, 257, *264*, *265*, *267*, *268*, *270*
Bloemers, H. P. J., 55, *60*
Bloomfield, D., 242, *265*
Blumenthal, H. J., 9, *60*, 282, *303*
Bobkova, T. S., 495, 496, 497, 498, 499, 508, *513*
Bock, R. M., 322, *332*, 395, 398, 399, 400, 401, *410*, *411*, *413*, *416*, 465, 467, *489*
Boer, P., 36, 56, *60*, 173, *203*
Boettger, H., 501, 502, 503, 511, *513*
Bohonas, J. N., 401, *417*
Boiteux, A., 300, *305*
Boll, M., 292, *303*, 378, 379, 402, *410*, *412*
Boman, H., 320, *329*
Booij, H. L., 41, 42, 43, 52, 56, *62*, *65*, *72*, 84, *118*, 221, 253, 260, 261, *265*, *267*
Bonaly, R., 501, 502, 508, 509, 511, *513*
Bouner, J., 495, 500, 506, 507, *513*
Bordes, A. M., 45, 46, 47, *72*
Borei, H., 48, *69*
Borek, E., 320, *328*, 407, 408, *409*, *415*
Borer, R., 472, *488*
Borges, W-D., 20, 22, 23, 24, 25, *73*
Borowskio, Z., 402, *413*
Borst, P., 324, *328*, *332*
Borst Pauwels, G. W. F. H., 53, *70*, 317, *328*
Borstel, R. C., von. 341, 354, 361, *409*
Bos, P., 180, *208*
Boschetti, A., 506, *513*
Bostock, C. J., 315, *328*, 398, 399, *409*, 465, *487*
Botsford, J. L., 365, 370, 371, *409*
Boulton, A. A., 36, 57, *60*, 255, 256, *265*
Bourgeois, C., 354, 356, 360, *409*
Bours, J., 252, *265*
Bowen, H. J. M., 18, *61*
Bowers, W. D., 362, *409*, 456, 468, *489*
Boxer, G. E., 288, *303*
Boyd, W. L., 21, *65*
Brachet, J., 474, *487*
Brady, T. G., 50, *61*, 297, *303*
Brahms, J., 320, *330*
Brandt, K. M., 7, 26, 27, 61, 466, *488*
Brandts, J. F., 93, *116*
Braun, V., 452, *488*
Breen, J., 18, *61*
Breivik, O. N., 225, 257, *265*
Brenner, S., 393, 394, 395, *411*
Bresch, C., 485, *488*
Bretthauer, R. K., 399, 400, *413*
Brichta, J., 210, *210*, 463, *490*
Brightwell, R., 263, *265*
Brock, T. D., 483, 484, *488*
Brodie, A. F., 283, *303*
Brooks, W. A., 246, *264*
Broquist, H. P., 358, 359, *409*, *412*
Brown, B. I., 432, *437*
Brown, C. M., 132, *133*, 229, *265*, 451, 469, 478, 479, *488*
Brown, D. H., 293, *303*, 432, *437*
Brown, G. B., 20, 21, *66*, 347, *412*
Brown, R. G., 448, *488*
Bruce, M., 49, 50, 51, *69*
Brüggemann, J., 365, 373, *415*
Brukner, B., 28, *61*
Brunner, A., 379, *409*, *410*

Bruns, F. H., 277, *305*
Bryce, W. A. B., 425, 429, *438*
Bryne, M. J., 436, 437, *438*
Bücher, T., 277, *304*
Büchi, H., 392, *409*
Bucovaz, E. T., 389, 390, *412*
Buday, A., 19, *62*
Buecher, E. J., Jr., 143, 164, 165, 166, *203*
Buening, G., 401, *410*
Buestein, H. G., 391, *410*
Buettner, M., 375, *416*
Bulder, C. J. E. A., 107, *116*
Bull, A. T., 137, *203*
Burger, M., 171, 172, 187, *203*
Burk, D., 20, *73*
Burkholder, P. R., 20, 22, 24, 25, *61*
Burnett, J. K., 263, *265*
Burns, D. J. W., 322, *327*, 292, *409*
Burns, J. A., 110, 112, 113, 115, *116*, *117*
Burns, R. O., 380, *411*
Burns, V. W., 343, *410*
Burrell, R. C., 225, *270*
Burton, E., 136, *208*, 501, 512, *515*
Burton, K., 277, *304*
Burton, K. A., 178, 202, *209*
Busch, H., 406, *414*
Bussey, H., 378, 380, 381, *410*
Button, D. K., 85, 90, 91, *116*
Byrne, W. J., 15, 16, *66*

C

Cabeca-Silva, C., 98, *118*
Cabet, D., 343, *410*
Cabib, E., 160, *203*, 422, 431, 435, *437*, *438*, *439*
Cadmus, M. C., 175, 178, 179, *203*, *207*, *208*
Calfee, R. K., 19, *66*
Callow, R. K., 226, *265*
Campbell, A., 79, *116*
Campbell, J. D., 130, *133*
Campbell, J. J. R., 322, *332*
Cantoni, G. L., 319, 321, 322, *328*, *329*, *330*, 389, 390, 391, *410*, *413*
Cantor, C. R., 321, *328*
Carbonell, L. M., 142, *205*, 448, 450, 451, *489*
Carlile, M. J., 512, *513*
Carlson, A. S., 176, *205*
Carminatti, H., 160, *203*
Carmu-Sousa, L., do. 175, 186, *203*
Carnevali, F., 312, 325, *328*
Carroll, W. R., 282, *305*
Caruthers, M. H., 392, *409*
Cartledge, T. G., 263, *265*
Carton, E., 50, 51, *61*
Casey, J., 315, 326, *331*
Cassagne, C., 222, 223, 251, *264*
Castelli, A., 215, 217, *265*
Castle, E. S., 458, *488*
Catalina, L., 289, *306*
Cathcart, W. H., 131, *133*
Catley, B. J., 427, *437*
Cawley, T. N., 155, 169, *203*
Ceci, L. N., 360, *413*
Cennamo, C., 378, 379, 402, *410*, *412*
Cerbón, J., 53, *61*, 261, *265*
Cerna, J., 322, *328*
Chakravorty, M., 290, *304*
Challinor, S. W., 23, *61*, *68*, 216, 243, *265*, 451, 472, 488, *490*
Chamberlain, N., 494, 513
Chambers, R. W., 320, 322, *328*, *331*, 394, *415*
Champagnat, A., 34, *61*
Chance, B., 108, *116*, 288, 300, *304*
Chang, S. H., 321, *331*, 392, *415*
Chang, W-S., 20, *61*
Chang, Y. Y., 238, *269*
Changeux, J. P., 377, 379, *410*, *414*
Chantrenne, H., 57, *61*
Chao, F. H., 319, *321*
Chapman, A. C., 504, *513*
Chapman, C., 301, *304*
Chappell, J. B., 262, *265*
Charalampous, F., 472, *488*
Chargaff, E., 310, 311, 313, *332*, *333*
Chattaway, F. W., 164, *203*, 452, 488
Chayen, S., 11, *67*
Cheldelin, V. H., 9, 17, 22, *62*, *69*, *72*
Chen, A. W-C., 56, *61*
Chen, S. L., 9, *61*, 107, *116*
Cherayil, J. D., 322, *332*, *395*, *416*
Cherest, H., 365, 368, 369, 372, 373, *410*, *415*
Chester, V. E., 293, 296, *304*, 419, 421, 436, 437, *438*
Chesters, C. G. C., 137, *203*
Chevreul, M. E., 211, *265*
Chiang, N., 212, 214, *268*

Chichester, C. O., 136, *208*, 495, 497, 498, 499, 501, 502, 503, 507, 508, 511, 512, *513*, *514*, *515*
Chimenes, A. M., 315, *328*
Christensen, B. E., 9, *72*
Christian, W., 279, *307*
Christiansen, V., 6, *71*
Christman, J. F., 21, *65*
Chugueo, I. I., 320, *327*
Chung, C. W., 476, *490*
Chung, K. L., 456, 459, *488*
Cieri, L., 495, *513*
Cifonelli, J. A., 146, 167, *203*
Cirillo, V. P., 39, 40, 41, 44, *61*, 84, *116*, 125, *133*, 276, *304*
Claisse, L. M., 297, *306*
Clark, D. J., 102, 103, 104, *117*
Clarke, A. E., 136, *203*
Clark-Walker, G. D., 301, *305*, 402, *410*, *413*
Clavilier, L., 340, 352, *410*, *413*
Cleland, W. W., 277, *307*
Clinton, R. H., 122, 125, *133*
Cohen, A., 185, *203*
Cohen, D., 326, *332*
Cohen, G. N., 45, *63*, 336, *410*
Cohen-Bazire, G., 497, *513*
Cohn, M., 279, *304*
Coker, L., 362, *414*
Coleman, R., 236, *265*
Collander, R., 37, 47, 48, 52, *61*, *68*
Colowick, S. P., 277, 293, *304*
Combs, T. J., 215, *265*
Contardi, A., 252, *265*
Conti, S. F., 483, 484, *488*
Contois, D. E., 82, *116*
Conway, E. J., 18, 19, 36, 47, 48, 50, 51, 53, *61*, *62*
Cook, A. H., 7, *61*
Cooke, W. B., 124, *133*
Cooney, D. G., 120, *133*
Cooper, E. J., 59, *66*
Cooper, F. P., 141, 144, 150, *209*
Cooper, J. M., 385, 386, *410*
Cordes, E. H., 336, 340, *413*
Cori, C. F., 293, *304*, 420, 432, *438*, *439*
Cori, G. T., 293, *304*, 432, *438*
Corneo, G., 312, 315, 324, *328*
Corner, T. R., 260, *265*
Corrivaux, D., 365, 366, *415*
Cota-Robles, E., 478, *487*
Cotman, C., 262, *266*
Cotter, R. I., 400, *410*
Coulon, A., 251, *269*
Courtois, C., 57, *61*
Cowie, D. B., 45, 46, *61*, *63*
Cox, B. S., 314, *328*
Crabtree, H. G., 301, *304*
Craig, B. M., 215, 227, 228, *265*
Cramer, F., 320, *331*, 393, *410*
Crandall, M. A., 483, *488*
Crane, F. K., 423, *437*
Cremona, T., 283, *305*
Cresson, E. L., 20, 21, *73*
Crestfield, A. M., 316, *328*
Crick, F. H. C., 395, *410*
Crocomo, O. J., 17, 54, *61*
Croes, A. F., 316, *328*
Croft, J. H., 325, *332*
Crompton, D. W. T., 274, *307*
Crook, E. M., 446, *488*
Cruess, W. V., 129, 130, *132*
Cruz, F. S., 254, *256*
Csáky, T. Z., 11, *72*
Čtvrtnicek, O., 195, 196, *205*
Cunningham, V. R., 501, *514*
Curtis, N. S., 30, 31, *63*
Cury, A., 254, *265*
Custers, M. Th. J., 8, 10, *65*
Cutts, N. S., 25, *61*, 494, *513*
Cwiakala, C. E., 213, *268*

## D

Dainko, J. L., 38, 39, 59, *70*, *74*, 187, 188, *208*
Danforth, W. F., 125, *134*
Daniel, V., 394, *410*
Daniels, N. W. R., 23, *61*, 216, 243, *265*
Danielsson, H., 243, 245, *265*
Danishefsky, I., 182, 183, *203*
Dankert, M., 223, *270*
Darling, S., 188, 194, *203*
Darrow, R. A., 277, *304*
Dastidar, S. G., 315, *328*
Daubert, B. F., 26, *60*
Davidson, E. D., 163, 194, *203*, 452, *487*
Davies, B. H., 495, 501, 502, 503, *513*
Davies, E., 480, *488*
Davies, R., 47, *61*, 172, 188, 210, *203*, *205*, *210*, 451, *488*
Davis, B. D., 343, *417*

Davis, J. B., 503, *513*
Davis, R. H., 343, *410*
Davis, R. M., 343, *417*
Dawson, E. C., 42, *72*
Dawson, E. R., 29, *62*
Dawson, P. S. S., 76, *116*, 215, 227, 228, 253, *265*
Dean, A. C. R., 87, *116*
Decker, K., 432, *438*, 511, *515*
Dee, E., 51, *62*
Deierkauf, F. A., 42, 43, *72*, 221, 260, 261, *265*
Deinema, M. H., 174, 199, *207*, *208*, 231, 232, 233, *265*, *269*
De Deken, R. H., 301, *304*, 353, 354, 355, *410*
DeLaFuente, G., 4, 6, 7, 8, 39, 40, 41, 45, 55, *62*, *64*, *70*, 171, *203*, 276, 290, 293, 294, 300, *304*, *306*
Dela Peña, R., 294, *304*
Delavier-Klutchko, C., 373, 375, *410*
Delisle, A. L., 14, 15, 58, *62*
Demis, C., 18, *69*
Demis, D. J., 5, 55, *62*
Densky, H., 19, *62*
Denton, J. F., 213, *265*
De Pahn, E. M., 297, *306*
Deshusses, J., 446, 447, *488*
Deutsch, J., 326, *332*
Devillers-Mire, A., 379, *410*
Devlin, T. M., 288, *303*
De Witt, W. G., 19, *63*
De Wulf, H., 437, *438*
Diamond, R. J., 260, *265*
Di Carlo, F. J., 14, *62*
Dickens, F., 9, *62*, 298, *304*
Dieckmann, M., 321, 322, *327*
Dillon, T., 137, *203*
Di Menna, M. E., 120, 123, *133*
Di Salvo, A. F., 213, *265*
Dische, Z., 182, *207*
Dittmer, J. C., 251, *264*
Dittmer, K., 20, 21, *62*, *66*
Dixon, B., 22, *62*, 473, *488*
Doctor, B. P., 321, *328*, 393, *410*
Doepman, H., 393, *410*
Domanski, R. E., 192, *203*
Domer, J. E., 448, *488*
Domnas, A. J., 55, *69*
Donachie, W. D., 315, *328*, 398, 399, *409*, 465, *487*
Dorfman, B., 340, *410*
Dorfman, B-Z., 340, 341, *410*
Dorn, F. L., 131, *133*
Dorp, D. A., van 24, 56, *73*
Doty, P., 311, *330*, *331*
Douglas, H. C., 23, 38, *60*, *69*, 254, *264*, 456, 468, *487*
Douglas, L. J., 223, *265*
Downey, M., 36, 48, *61*
Doy, C. H., 382, 384, 385, 386, *410*
Doyle, D., 147, *207*
Dreassi, M., 495, *513*
Drekter, L., 20, 21, *69*
Drews, B., 14, *62*
Dreyfuss, J., 372, 373, *414*, *417*
Drouhet, E., 182, 183, *203*, 479, *488*
Duell, E. A., 188, *203*, 301, *306*
Duerre, J. A., 371, *410*
Duetting, D., 321, 322, *333*
Duggan, P. F., 47, 51, 53, *61*, 297, *303*
Dulaney, E. L., 225, 226, *265*
Dunn, D. B., 407, *413*
Dunphy, P. J., 223, *265*
Duntze, W., 288, 289, *303*, 385, 386, *410*, *413*
Dunwell, J. L., 22, *62*, 149, *204*, 472, *488*
Duphil, M., 346, *412*
Durr, I. F., 243, *266*
Dütting, D., 392, *417*
Dutton, H. J., 232, *269*
Duval, J., 318, 319, 321, *327*, *328*, 394, 396, *410*, *417*
Duvnjak, Z., 254, *267*
Dworkin, M., 497, *513*
Dworschack, R. G., 171, *204*, *209*
Dworsky, P., 23, *62*
Dyke, K. G. H., 452, *488*
Dyr, J., 496, 511, 512, *513*, *514*


## E


Eakin, R. E., 20, 22, 23, *73*
Eastcott, E. V., 22, *62*
Eaton, N. R., 293, 296, *304*, 339, 341, 343, *414*, *416*, 435, 436, *438*
Ebel, J. P., 318, 319, 320, 321, 323, *327*, *328*, *330*, 392, 394, 395, 396, *409*, *410*, *411*, *417*
Eckstein, H., 316, *328*, 398, 402, *410*
Eddy, A. A., 36, 47, 57, *60*, *62*, 142, 152, 163, 164, 169, 170, 173, 187, 194, *204*, 227, 255, *265*, 450, 462, *488*

Edwards, T. E., 138, 139, 146, *207*, 428, *438*
Egambardiev, N. B., 87, *116*
Egel, R., 485, *488*
Ehninger, D. J., 371, *415*
Ehresman, C., 395, *409*
Eichler, F., 365, 372, 373, *410*
Elander, M., 7, 8, *62*, 216, *265*
Elmer, G. W., 477, *488*
Elorza, M-V., 454, *488*
Elorza, V., 198, *208*, 448, 463, *491*
Elsden, S. R., 106, 107, *116*
Elsworth, R., 80, *116*
Elvehjem, C. A., 19, *62*
Elvin, P. A., 188, *203*, 451, *488*
Emerson, R., 120, *133*
Emerson, R. L., 401, *417*
Enari, T-M., 19, *62*
Enebo, L., 13, 27, 31, *62*, *69*, 213, 216, *265*
Ephrussi, B., 262, *266*, 301, *305*, 310, 325, *328*
Epstein, R., 398, *411*, 465, 467, *489*
Epstein, R. L., 136, *204*, 317, 323, *332*, 398, *416*, 501, *513*
Enns, L. H., 49, *69*
Ercoli, A., 252, *265*
Erkama, J., 19, *62*
Ernster, L., 288, *305*
Eschenbecher, F., 20, 22, 23, 24, 25, 26, 32, *73*
Esposito, R. E., 315, *328*
Etchells, J. L., 495, 500, *514*
Euler, H., von, 5, 11, 57, 59, *62*
Evans, E. E., 182, *204*, *207*
Evans, G. H., 34, *62*
Evans, J. M., 138, *207*
Evans, W. C., 34, *62*
Evans, W. R., 495, 500, *514*
Eveleigh, D. E., 156, 157, 201, *204*
Everett, G. A., 310, 319, 320, 321, *327*, *329*, *330*, 392, 393, *412*, *413*
Evison, L. M., 122, 125, 126, *133*, *134*
Eyring, H., 91, 92, 93, 94, 97, *116*, *117*

## F

Fabre, F., 314, 315, *330*
Fabritius, E., 37, 47, 52, *61*
Falcone, G., 165, 167, 168, 188, *204*, *207*, 445, 446, 451, 456, 468, 469, 470, 476, 477, *488*, *490*
Farias, G., 191, 192, *207*
Farkaš, V., 8, *62*, 197, 210, *204*, *210*, 431, *437*, 445, *488*
Farmer, V. C., 140, 141, 142, 143, 144, 145, 164, 176, 190, 198, 199, *203*, *205*, 420, *437*
Farrell, J., 124, 126, 127, *133*, 228, 261, *266*
Faulkner, R. D., 321, 322, *331*, *332*, 392, 395, *415*, *416*
Fauman, M., 326, *328*
Faures, M., 312, 325, 326, *328*, *329*
Favre, A., 393, *414*
Feeney, J., 223, *265*
Feingold, D. S., 184, 185, *203*
Feinstein, M., 18, 20, 23, *60*
Feldbruegge, D. H., 245, *267*
Feldman, H., 392, *417*
Feldman, M. Y., 320, *327*
Feldmann, H., 321, 322, *333*
Fell, J. W., 120, *133*
Fels, G., 17, *62*
Fencl, Z., 48, *62*, 80, *116*, 468, *491*
Fennessey, P., 223, *270*
Ferguson, J. J., 243, *266*
Fernández, M. J., 282, *306*
Fersco, J. R., 321, *327*
Fiechter, A., 466, *489*
Fieser, L. F., 225, *266*
Fieser, M., 225, *266*
Filosa, J., 34, *61*
Fimognari, G. M., 250, *266*
Fink, G. R., 348, 351, *410*
Fink, H., 10, 14, 23, 24, 27, 32, 33, 57, *62*
Finkelstein, A., 36, *69*
Fischer, E., 4, 5, 8, *62*
Fischer, E. H., 427, *438*
Fisher, C. R., 339, 341, *410*
Fisher, E. H., 293, *305*
Fittler, F., 321, *328*, 395, *410*
Fjellstedt, T., 360, *415*
Flavin, M., 365, 367, 368, 369, 373, 374, 375, *410*, *411*, *412*, *414*
Fleming, I. D., 424, *437*
Flower, D., 20, 21, *69*
Foda, M. S., 176, *204*
Fodor, I., 321, *327*
Fogel, S., 315, *328*
Folkes, J. P., 47, *61*
Follett, E. A. C., 189, *209*
Forbusch, I. A., 9, *72*

Forrest, W. W., 106, 107, *116*
Forsander, O., 24, *63*
Forss, K., 33, *63*
Foulkes, E. C., 48, 50, *63*
Fowell, R. R., 483, *488*
Frederick, E. W., 403, 404, *411*
Fredrickson, A. G., 82, *117*, *118*
Freed, S., 320, *329*
Freese, E., 313, *328*
Freifelder, D., 469, *488*
Fresco, J. R., 320, *328*, *330*
Freundlich, M., 377, 380, *411*
Frey, C. N., 6, 14, 22, 24, 25, 30, *60*, *70*
Frey, S. W., 19, *63*
Friedkin, M., 347, *411*
Friend, J., 512, *513*
Fries, N., 25, *63*, *68*, 122, 130, *133*
Friïs, J., 8, 55, *63*, 171, *204*, 462, *488*
Frolova, L-Y., 322, *327*, 394, *412*
Fromageot, C., 495, 498, 504, *513*
Fry, W., 45, *63*
Fry-Wyssling, A., 193, *206*
Fuhr, B. W., 23, *64*
Fuhrmann, G. F., 52, 53, *63*
Fuhs, G. W., 10, 34, *63*
Fusimura, Y., 322, *328*
Fukazawa, Y., 159, *208*
Fukuhara, H., 316, 318, 323, 325, 326, *328*, *329*, *331*, 397, 405, 406, *411*, *416*
Fukui, K., 197, *204*
Fukumoto, J., 191, *207*
Fulco, A. J., 251, *266*
Fuller, W., 393, *410*
Fulmer, E. I., 18, 19, *63*, 122, 131, *134*
Furuichi, Y., 394, *417*
Furuya, A., 191, *204*

## G

Gabriel, O., 511, *515*
Gadebusch, H. H., 192, *204*
Gaffney, H. M., 53, *61*
Gale, E. F., 47, *61*
Gancedo, C., 90, *116*, 234, *266*, 278, 279, 281, 289, 291, 301, 303, *304*
Gancedo, J. M., 90, *116*, 234, *266*, 279, 281, 288, 289, 291, 301, *303*, *304*
Gans, M., 343, *410*
Garcia Acha, I., 192, *207*
Garcia Lopez, M. D., 193, *204*
Garcia Mendoza, C., 166, 192, 193, *204*, *207*, 255, 256, 261, *266*, 447, 448, 463, *488*, *490*
Gardner, P. E., 141, 144, 149, 150, 168, *203*
Garfinkel, D., 110, *116*
Garg, N. K., 22, *60*
Garner, H. R., 365, 368, 369, 373, 375, *417*
Garver, J. C., 85, 90, 91, *116*, 136, *204*, 501, *513*
Garzuly-Janke, R., 149, 155, *204*, 445, *488*
Gascón, S., 172, 173, 187, 192, 196, *204*, *206*, *207*
Gaye, P., 320, *331*
Gaylor, J. L., 247, 248, 250, *268*, *269*
Gazith, J., 277, *306*
Gebicki, J. M., 320, *329*
Geddes, W. F., 30, *65*
Geige, R., 394, 396, *411*
Geiger, K. H., 478, *488*
Genevois, L., 45, *63*, 222, 223, 251, *264*
Genghof, D. S., 20, 21, *66*
Genin, C., 326, *329*
Gerbach, G., 252, *266*
Gerrits, J. P., 106, *116*
Gesswagner, D., 19, *63*
Getz, G. S., 315, 325, 326, *327*, *331*, *332*
Geyer, Fenzl, M., 253, *266*
Ghadially, R. C., 23, *64*
Ghosh, A., 23, *63*, 472, *488*
Ghosh, H. P., 392, 393, *415*
Giaja, J., 186, *204*
Gibbons, N. E., 125, 130, *133*
Gibbs, M. H., 243, *269*
Gibson, E. J., 458, 459, 464, *489*
Gibson, F., 382, *411*
Gifford, G. D., 465, *488*
Gilbert, W., 396, *411*
Gillam, I., 392, *411*
Gillespie, D., 405, *411*
Gilliland, R. B., 55, *60*, *63*
Gilmore, R. A., 393, *411*
Gilmour, C. M., 9, *72*
Giner, A., 278, 303, *304*
Giovanelli, J., 373, 375, *411*
Gits, J. J., 46, 47, *63*
Godkin, W. J., 131, *133*
Goebel, W., 384, 385, 386, 387, *413*
Goedde, H. W., 277, *305*

Goldfine, H., 26, *60*
Goldman, P., 236, *266*
Goldstein, J., 320, *329*
Goldthwait, D. A., 397, *417*
Goldwasser, E., 320, *329*
Golovkina, L. S., 217, *270*
Gonzales, G., 191, 192, *207*
Gooding, R. H., 277, *306*
Goodman, H. M., 293, 394, 395, *411*
Goodman, J., 15, 53, 54, *63*
Goodwin, T. W., 494, 502, 503, 504, 507, 512, *513*, *515*
Goodyear, G. H., 22, *73*
Gorin, S. A. J., 137, 147, 148, 149, 150, 151, 152, 155, 156, 157, 160, 175, 176, 177, 181, 182, 185, 186, 201, 202, *203*, *204*, *208*
Gorman, J., 7, *63*, 317, *329*, 397, *411*, 465, *488*
Gorman, J. A., 351, *411*
Görts, C. P., 301, *304*
Gorts, C. P. M., 10, *63*
Gosse, C., 324, *329*
Gottschalk, A., 4, 7, 8, 9, *63*
Gough, G. A. C., 226, *270*
Grandchamp, C., 324, *329*
Grant, D. W., 127, *133*
Grant, K. L., 478, *489*
Gratzer, W. B., 319, 320, *329*, *330*, 400, *410*
Gray, M. W., 320, *329*
Gray, P. J., 19, *62*
Gray, P. P., 18, 19, 20, 23, *60*
Greaves, J. D., 19, *63*
Greaves, J. E., 19, *63*
Green, D. E., 238, 242, *266*, 279, *304*
Green, S. R., 30, *63*
Greenberg, D. M., 234, *266*
Greene, M. L., 251, *266*
Greenwood, C. T., 425, 429, *438*
Gregg, C. T., 9, *72*
Grenson, M., 46, 47, *63*, *64*, 343, 344, 345, 346, 354, 355, *409*, *412*
Griffiths, M., 497, *513*
Grisolia, S., 277, 279, *304*, *306*
Grivell, A. R., 347, *411*
Grob, E. C., 506, *513*
Grollman, A. P., 158, *208*, 459, *491*
Groodt, A., de, 347, *411*
Gros, F., 396, *411*
Gross, P. R., 318, 319, *330*
Gross, S. R., 336, *411*
Grossman, L. I., 312, 315, 324, *328*, *331*
Gruber, M., 24, 56, *73*
Grunberg-Manago, M., 336, 402, 403, 405, *411*
Grunberger, D., 395, *414*
Guarneri, J. J., 215, *265*
Guerineau, M., 311, 324, *329*
Guex, W., 502, 503, *514*
Guilliermond, A., 122, 123, 128, *133*
Gunja, Z. H., 293, *304*, 427, 428, 429, 430, 432, *438*
Gunsalus, I. C., 25, *65*, 302, *304*
Günther, Th., 16, *63*
Güntner, H., 252, *266*
Gupta, N. K., 392, 393, *409*, *411*
Gurinovich, E. S., 495, *515*
Gurr, M. I., 243, *266*
Guthenberg, H., 13, 31, *69*

H

Hadjipetrou, L. P., 106, *116*
Haar, F. de, 393, *410*
Haeckel, R., 277, 279, *304*
Hagen, P. O., 126, *133*
Hägglund, E., 32, 47, 48, *63*
Hagopian, H., 321, *327*
Haidle, C. W., 477, *489*
Haley, A. B., 230, *267*
Hall, R H., 320, 321, *328*, *331*, 395, 407, *410*, *411*
Halvorson, H. O., 7, 44, 45, 46, *63*, *69*, 84, *117*, 311, 315, 317, 322, 323, 326, *329*, *331*, *332*, 388, 393, 397, 398, 399, 400, 405, *411*, *413*, *415*, *416*, 465, 467, *488*, *489*, *490*, *491*
Halvorson, H. P., 348, 351, *416*
Hamilton, D. M., 176, *205*
Hamilton, J. G., 448, *488*
Hampel, A., 322, *332*, 395, *416*
Hanahan, D. J., 214, 217, 218, 220, 222, 253, *265*, *266*
Hansen, R. G., 432, *438*
Hansford, R. G., 262, *265*
Hanson, A. M., 19, *63*
Harden, A., 425, *438*
Harkin, J. C., 448, *488*
Harold, F. M., 292, *304*
Harriman, P., 394, 396, *411*
Harris, D. L., 48, *69*
Harris, G., 12, 30, 31, 44, 45, 55, *60*, *63*, *64*

Harrison, D. M., 226, 248, *265*
Harrison, J. S., 26, 29, 57, 59, *62*, *64*, *72*, 214, 253, *266*, 420, 427, 435, *439*
Hartelius, V., 12, 13, 14, 22, *64*, *67*
Hartlief, R., 396, 397, *411*
Hartman, L., 213, *266*
Hartwell, L. H., 124, 127, *133*, 322, *329*, 369, 391, 392, 401, *411*, *412*, *414*
Hasegawa, S., 142, 145, 191, *205*
Hasenclever, H. F., 141, 144, 150, 158, 159, *205*, *208*, *209*, 459, *491*
Haskell, B. E., 216, 218, 220, *266*
Haškovec, C., 5, 44, *64*, *65*
Haslam, J. M., 263, *266*, *268*, *270*
Haslbrunner, E., 311, 324, *331*
Hassid, W. Z., 137, *205*, 453, *489*
Hatanaka, H., 250, *267*
Hathaway, J. A., 276, 299, *305*, *306*
Hauber, J., 284, 302, *306*
Havelková, M., 193, 195, *205*, *206*
Havinga, E., 317, *328*
Hawirko, R. Z., 456, 459, *488*
Hawke, J. C., 213, *266*
Haworth, W. N., 145, 146, *205*
Hawthorne, D. C., 343, 348, 392, *412*, *414*
Hay, J., 408, *415*
Hayashi, H., 322, *329*, 394, *412*
Hayatsu, H., 395, *416*
Hayatu, H., 322, *332*
Hayes, A. D., 52, 53, 54, *69*
Haynes, R. H., 314, *331*
Heath, R. L., 145, 146, *205*
Hecht, L., 18, 54, *70*
Hedrick, L. R., 448, 451, 479, *491*
Hehre, E. J., 176, *205*
Heide, S., 216, *266*
Heidelberger, M., 182, *207*
Heinemeyer., C, 391, *415*
Heinrich, J., 394, 396, *410*, *411*
Heinrikson, R. L., 320, *329*
Held, H. R., 31, *66*
Hellström, H., 57, *62*
Hemming, F. W., 223, 250, *266*
Henipfling, W. P., 104, *116*
Henley, D., 320, 321, *328*
Henley, D. D., 320, *330*
Hensley, D. E., 209, *210*
Heppel, L. A., 36, *64*
Herbert, D., 80, 85, 100, 108, *116*, 279, *304*
Heredia, C., 388, 396, *409*, *411*
Heredia, C. F., 4, 39, 40, 41, *64*, 276, 293, 300, *304*
Hereford, L. M., 378, 380, 381, *413*
Herman, H. P., 348, 351, *416*
Hernandez, E., 106, 107, *116*
Herrera, T., 59, *64*
Hers, H. G., 433, 434, 437, *438*
Heslot, H., 340, 341, 343, 347, 350, 352, *411*, *412*, *414*, *417*
Hess, B., 277, 279, 300, *304*, *305*
Hess, H., 225, *270*
Hestrin, S., 6, 7, 45, *65*
Hiatt, H., 396, *411*
Hierholzer, G., 292, *305*, 361, *412*
Higashi, Y., 223, 251, *266*
Higgins, L. W., 144, 190, 201, *208*
Hilby, W., 320, *329*
Hilz, H., 316, *328*, 372, 398, 402, *410*, *412*
Himes, H. W., 25, *66*
Hinselwood, C. N., 87, 92, *116*
Hirschmann, H., 217, *266*
Hirst, E. L., 145, 146, *205*
Hobson, P. N., 427, *438*
Hochberg, M., 25, *66*
Hodnett, J. L., 406, *414*
Hoette, I., 162, *207*
Höfer, M., 9, 44, *64*, *65*, 276, *305*
Hoffee, P., 283, *305*
Hoffmann, H. P., 324, *327*
Hoffmann-Ostenhof, O., 23, 27, *62*, *64*, *70*, 253, *266*, 511, *515*
Hoffmann-Walbeck, H. P., 28, *70*
Hofmann, K., 20, 21, 26, *60*, *64*, *68*
Holden, M., 189, *205*
Holderby, J. M., 33, *64*
Holler, B. W., 406, *417*
Holley, R. W., 310, 319, 320, 321, *327*, *329*, 392, 393, *412*
Holmes, H., 360, *413*
Holmes, M. R., 164, *203*, 452, *488*
Holoday, D., 22, *73*
Holter, H., 38, *64*, 187, *203*, 486, *489*
Holy, A., 395, *414*
Holzer, E., 278, *305*
Holzer, H., 53, *64*, 277, 278, 281, 288, 292, 300, *303*, *304*, *305*, 317, *332*, 361, 378, 379, 402, *409*, *410*, *412*, *414*, *417*
Holzman, M., 9, 40, *70*
Hommes, F. A., 300, *305*
Hooghwinkel, G. J. M., 56, *65*, 253, *267*

Hooper, D. C., 52, 54, *64*, 69
Hopkins, R. H., 4, 5, 31, 32, 55, *64*, 171, *205*
Horecker, B. L., 283, 290, *304*, *305*
Horens, J., 11, *66*
Horikoshi, K., 190, *205*
Horitsu, K., 177, *204*
Horne, R. W., 36, *67*, 152, 164, 166, 168, *207*, 227, *268*, 446, 447, *490*
Horschak, R., 18, *64*
Horwood, M., 5, *64*
Hoskinson, R. M., 392, *415*
Hottinguer, H., 325, *328*
Hottinguer-Margerie, H., 314, 315, *330*
Hough, J. S., 56, *66*, 103, 104, *118*, 451, 469, 478, 479, *488*
Hounsell, J., *330*
Houwink, A. L., 141, 162, 164, 196, *205*, 447, 456, 457, *489*
Howard, D. H., 474, *489*
Hu, A. S. L., 351, *411*
Huang, M., 301, *305*
Hübscher, G., 236, *265*
Hudson, C. S., 7, 27, *71*, 421, *439*
Hudson, J. R., 32, *64*
Hue, L., 434, *438*
Huff, J. W., 243, *269*
Hughes, D. E., 501, *514*
Hughes, D. T. D., 25, *67*
Hughes, L. P., 33, *64*
Hunsley, J. R., 277, *306*
Hunt, P. F., 248, 249, *264*
Hunter, J. R., 100, *117*
Hunter, K., 213, 215, 228, 229, 238, 255, 257, 260, *266*
Hurst, H., 452, *489*
Hurwitz, J., 403, 404, *411*
Hutchings, B. L., 21, *70*
Hutchinson, H. T., 322, *329*
Hwang, Y. L., 359, 360, *412*

## I

Ibsen, K., 301, *305*
Ichishima, E., 252, *268*
Ierusalimsky, D. D., 87, *116*
Ierusalimsky, N. D., 82, 87, *117*
Iguti, S., 498, *514*
Ikeda, Y., 191, *204*
Illingworth, B., 293, *303*
Imahori, K., 322, *329*, 389, 390, 394, *412*, *414*
Imai, K., 401, *416*
Imamura, A., 325, *332*
Imura, N., 322, *331*
Indge, K. J., 260, 263, *266*
Ingraham, J. L., 58, *73*, 120, 124, 132, *133*, *134*
Ingram, M., 11, 25, *64*, 129, 131, *133*
Ingram, V. H., 321, *332*
Ingram, V. M., 321, *327*
Inoue, S., 188, *203*
Inoue, T., 58, *64*
Inskeep, G. C., 33, *64*
Ipata, P. L., 322, *329*
Isaac, P. K., 456, 459, *488*
Isbell, H. S., 4, *64*
Isherwood, F. A., 145, 146, *205*
Ishikura, H., 321, *328*
Islam, M. F., 172, *205*
Isler, O., 502, 503, 504, 509, *514*, *515*
Itaka, 316
IUPAC-IUB Commission on Biochemical Nomanclature, 217, *266*
Iwashima, A., 403, *412*

## J

Jackman, L. M., 503, *513*
Jackson, J. F., 324, *329*, 347, *411*
Jacob, F., 10, *64*, 396, *412*
Jaenicke, L., 22, *64*, 369, 371, *415*
Jahnke, L., 239, *267*
Jakob, H., 312, 315, 324, 325, *330*
James, A. T., 243, *266*
James, H. L., 389, 390, *412*
Jannasch, H. W., 80, *117*
Janosko, N., 385, 386, *413*
Jayaraman, J., 262, *266*, 312, 324, *332*
Jayko, M. E., 214, 218, *266*
Jeanes, A., 155, 178, 183, 184, *205*
Jeanloz, R., 428, 429, *438*
Jefferies, J. P., 226, *269*
Jennings, D. H., 51, 52, 53, 54, *64*, *69*
Jensen, S. L., 494, 511, *513*, *514*
Jeuniaux, C., 163, *205*
Johnson, B. F., 197, *205*, 458, 459, 460, 464, *489*
Johnson, F. H., 92, 93, 94, 97, *117*
Johnson, J., 420, *438*, 447, *489*
Johnson, J., Jr., 139, 140, 143, *206*
Johnson, J. D., 182, *204*

Johnson, M. J., 10, 19, 20, 34, *64*, *68*, 89, 90, 91, 106, 107, *116*, *117*, 292, 289, *305*
Johnston, I. R., 446, *488*
Johnston, J. A., 23, *64*
Johnston, J. R., 189, *205*, 472, *490*
Joiris, C. R., 47, *64*
Jollow, D., 213, 221, 222, 229, 262, *266*, *267*
Jollow, D. J., 26, *68*, 230, *268*
Joneau, M., 222, 223, 251, *264*
Jones, A. S., 316, *329*
Jones, D., 140, 141, 142, 145, 163, 164, 176, 190, 191, 194, 198, 199, 210, *203*, *205*, *210*, 452, 463, *487*
Jones, D. S., 322, *332*, 395, *416*
Jones, E. E., 358, 359, *412*
Jones, E. W., 340, *412*
Jones, G. H., 147, 153, 154, 157, *205*, 420, *438*
Jones, I. C., 425, 429, *438*
Jones, J. K. N., 142, 147, 176, 183, 184, 186, *202*, *205*
Jones, M., 13, 14, 31, *64*, *68*
Joslyn, M. A., 24, *67*, 137, *205*
Julien, J., 319, *330*
Jung, C., 53, *64*
Just, F., 4, 14, 23, 24, 27, 32, 33, 34, *62*, *64*

## K

Kaback, H. R., 261, *268*
Kacser, H., 115, *117*
Kahane, E., 214, *267*
Kahanpää, H., 12, *71*
Kakar, S. N., 378, 379, 380, *412*
Kakntani, Y., 495, 512, *514*
Kalb, H. W., 48, *66*
Kallio, R. E., 34, 35, *66*, *73*
Kalousek, F., 322, *328*
Kamiya, H., 224, *268*
Kaneshiro, T., 251, *266*
Kanetsuna, F., 142, *205*, 448, 450, 451, *489*
Kaplan, A., 18, *67*
Kaplan, J. G., 55, 57, *64*, *65*, 171, *205*, 346, *412*, *413*
Kaplan, M. M., 374, *412*
Kaplan, N. O., 22, *66*
Karabava, G. Ya, 496, *515*
Karassevitch, Y., 365, 366, *412*, *415*
Karau, W., 321, *333*, 392, *417*
Karkkainen, V., 16, 56, *68*
Katagiri, H., 401, *416*
Kates, M., 132, *133*, 213, 215, 218, 220, 221, 222, 224, 229, *267*, *269*
Katsuki, H., 249, 250, 257, *267*
Kattner, W., 16, *63*
Katz, A. M., 48, *69*
Kauppila, O., 58, *71*
Kauppinen, V., 19, *62*, *65*
Kawabata, Y., 34, *73*
Kawade, Y., 394, *414*
Kawaguchi, A., 250, *267*
Kawakita, S., 159, *208*
Kawasaki, C., 24, 25, *65*
Kawata, M., 320, 322, *329*, *330*, 392, *414*
Kayser, F., 507, *514*
Kazarinova, L., 321, *327*
Kaziro, Y., 24, *54*, 394, *417*
Kearney, E. B., 9, *70*
Keith, A. D., 230, *267*
Kellerman, G. M., 213, 222, 229, *266*
Kellog, D. A., 321, *328*
Kempfle, M., 277, *306*
Kempner, E. S., 323, *329*
Kennedy, E. P., 238, *269*
Kent, A. M., 14, *62*
Keppler, D., 432, *438*
Keränen, A. J. A., 14, 17, 20, 22, 29, 48, 58, *65*, *71*, 231, *269*
Kernan, R. P., 51, *61*
Kerr, D. S., 369, 374, 375, *412*
Kerr, J. D., 223, *265*
Kerr, S. E., 347, *412*
Kerridge, D., 323, *329*, 401, *412*
Kessel, J. F., 182, *204*
Kessler, G., 165, 167, 168, *205*, *207*, 227, *267*, 445, 447, 450, 452, 476, *489*, *490*
Khin Maung, 293, *304*, 427, 428, 429, 430, 432, *438*
Khmel, I. A., 82, *117*
Khorana, H. G., 321, 322, 324, *329*, *331*, *332*, 392, 393, 395, *409*, *411*, *415*, *416*
Kidby, D. K., 172, 210, *205*, *210*
Kijima, S., 389, 390, *412*
Kemril, R. J., 178, *205*
Kinoshita, S. H., 353, *417*
Kirkwood, S., 139, 140, 142, 143, 145, 191, *205*, *206*, 420, *438*
Kiselev, L. L., 322, *327*

Kishi, T., 24, 25, *65*
Kisselev, L. L., 394, *412*
Kitazume, Y., 317, *329*
Kitunen, M., 22, 29, *71*
Kjeldgaard, N. O., 80, *117*
Kjellin-Straby, B., 320, *329*, *331*
Kjellin-Straby, K., 376, 408, *412*, *415*
Kjølberg, O., 293, *305*, 428, 433, *438*
Kleeff, B. H. A. van, 19, *65*
Klein, H. P., 216, 239, 242, *267*, *268*, *270*
Kleinschmidt, A., 312, 324, *329*, *331*
Kleinzeller, A., 4, 17, 39, 54, *65*, 86, *117*
Kleppe, K., 392, *409*
Klinistra, P. D., 34, *73*
Klink, F., 396, *415*
Kloet, S. R. de, 316, 323, *329*, 406, 407, *417*, *414*
Klungsoyr, S., 91, *72*
Klyne, W., 225, *267*
Kluyver, A. J., 8, 10, *65*
Knights, B. A., 36, *66*, 164, *206*, 213, 220, 222, 225, 227, 228, 255, 256, 257, *267*
Knudsen, R. C., 376, *412*
Kobayashi, T., 293, *305*
Kobel, M., 282, *305*
Koch, E. M., 420, *438*
Koch, F. C., 420, *438*
Koch, R. B., 30, *65*
Kocourek, J., 150, 151, 153, 159, 201, *205*
Kodicek, E., 246, *267*
Koe, B. K., 503, *514*
Koeber, H-J., 30, *68*
Koenigsberger, R., 108, *117*
Kögl, F., 20, 21, *65*
Kokke, R., 19, 56, *65*, 253, *267*
Kolattukudy, P. E., 251, *267*
Kollar, K., 222, *267*
König, J., 8, *70*
Koningsberger, V. V., 55, *60*, 396, 397, *411*
Konishi, S., 365, 366, 367, *417*
Konttinen, K., 38, 47, 48, *71*
Kooiman, P., 176, *205*
Kopecká, M., 195, 196, 197, 198, 210, *205*, *206*, *210*, 454, 463, *490*
Korn, E. D., 36, *65*, 152, 163, 164, 168, *205*, 450, *489*
Kornberg, A., 286, *305*, 324, *329*
Kornberg, H. L., 289, 290, *305*
Kornberg, R. D., 324, *329*
Kornberg, S. R., 16, *65*
Kortstee, G. J. J., 253, *276*
Kotyk, A., 4, 5, 17, 39, 40, 41, 43, 44, 54, *60*, *64*, *65*, 84, 86, *116*, *117*, 276, 300, *303*, *305*
Kouttinen, K., 291, *306*
Kováč, L., 17, 54, *65*, 222, *267*
Kowal, J., 283, *305*
Kramer, K., 161, *207*
Krebs, E. G., 277, *305*
Krebs, H. A., 10, *65*
Krebs, Jos., 10, 14, *62*
Kredich, N. M., 373, *412*
Kreger, D. R., 141, 142, 155, 162, 163, 164, 196, *205*, 446, 447, 456, 457, *489*
Kreger-van Rij., N. J. W., 176, *206*, 444, *489*, 500, *514*
Kremer, E. H., 4, *70*
Krishna, G., 245, *267*
Kritchevsky, G., 212, *268*
Kroger, H., 320, *329*
Krohn, V., 32, *65*
Kroon, A. M., 324, *332*
Krutilina, A. I., 321, *327*, *330*, *332*, 392, *409*
Kudo, Y., 322, *329*
Küenzi, M. T., 466, *489*
Kuhn, N. J., 236, *267*
Kulikova, A. N., 496, 498, 508, *515*
Kumagawa, M., 214, *267*
Kumar, H., *392*, *409*
Kung, H., 392, *413*
Kung, H. K., 321, *330*
Kurland, C. G., 396, *411*
Kuroda, A., 191, *205*
Kuroiwa, Y., 58, *64*
Kuronen, T., 19, *71*
Kurz, W., 136, *206*
Kvasnykov, E. I., 495, *515*
Kylä-Sivrola, A-E., 16, 56, *68*

## L

Laberge, M., 317, *329*, 397, 398, *411*, 465, 467, *489*
Lachowicz, T-M., 368, *416*
Lacroute, F., 343, 344, 345, 346, 354, 355, 384, *412*, *414*
Lagerkvist, U., 389, 390, 391, 394, *412*, *413*
Lagoda, A. A., 175, *203*
Lai, C. Y., 283, *305*

Lainé, B., 34, *61*
Lake, J. A., 393, *413*
Lamanna, C., 102, *117*
Lamb, A. J., 402, *413*
Lamonthezie, N., 311, *329*
Lampen, J. O., 5, 6, 55, 56, *65*, *66*, *67*, *72*, 158, 171, 172, 173, 179, 192, 196, *205*, *206*, *208*, 402, *413*, 451, 462, *489*, *491*
Lamport, D. T. A., 453, *489*
Landy, A., 393, 394, 395, *411*
Lane, B. G., 320, *329*
Lark, K. G., 465, *489*
Larkin, J. M., 120, 123, *133*
Larsson-Raznikiewicz, M., 277, *305*
La Rue, T. A., 12, 14, *65*
Laskin, A. I., 402, *413*
Lasnitzki, A., 18, *65*, 215, *267*
Latuasan, H. E., 401, *413*
Lauterborn, W., 277, 279, *304*
Law, J. H., 251, *266*
Lawrence, N. L., 120, 123, 130, *133*
Lazowska, J., 343, *413*
Lebeault, J. M., 254, *267*
Lebovitz, P., 322, *329*
Lechner, R., 10, 33, *62*, *65*
Lederer, E., 248, *267*, 497, 504, *514*
Lee, C-P., 288, *304*
Lee, E. Y. C., 293, *305*, 427, *437*, *438*
Lee, J. M., 20, 21, *66*
Lee, L. W., 391, *413*
Lee, T. C., 151, 59, *65*
Lee, T. H., 501, 502, 503, 511, *513*
Lee, Y. V., 148, *206*, 225, *267*
Leece, E., 498, 500, *514*
Leeuw, A., de., 385, 386, *413*
Lehmann, J., 427, *437*
Leibowitz, J., 6, 7, 45, *65*, 175, *203*
Leissner, E., 7, *67*
Leloir, L. F., 262, *265*, 422, 430, *438*
Lengyel, P., 336, 388, 390, 391, 395, *413*
Lennarz, W. J., 26, *60*, *65*, 161, *207*, 223, 229, 230, *267*, *269*
Leonian, L. H., 21, *66*
Leonian, R. H., 20, 22, 23, 25, *65*
Lesh, J. B., 18, *63*
Leslie, L., 405, *413*
Letters, R., 155, 169, *203*, 213, 214, 217, 218, 219, 221, 222, 228, 236, 253, *267*
Leupold, U., 341, *413*
Lewin, L. M., 23, *65*, 218, 220, 222, 231, 243, *269*
Lewis, D., 314, *333*
Lewis, K. F., 9, *60*, 282, *303*
Lewis, M. J., 14, 15, 54, 58, 59, *65*, *70*
Li, L., 321, *327*, *330*, *332*, 392, *409*
Lichstein, H. C., 21, 22, 25, *65*
Liddle, A. M., 424, *437*
Lieblová, J., 76, *116*, 472, *487*
Light, R., 26, *60*
Light, R. F., 225, 257, *265*
Light, R. J., 26, *65*, 229, 230, *267*
Lilly, V. G., 20, 21, 22, 23, 25, *65*, *66*
Lincoln, R. E., 506, *514*
Lindahl, T., 320, 321, *327*, *328*, *330*
Lindberg, M., 246, *268*
Lindebren, C., 315, *331*, 359, 360, *412*
Lindegren, C. C., 5, 16, 27, *66*, 315, *331*
Lindegren, G., 359, 360, 378, 382, *412*, *415*
Lindell, T. J., 277, *305*
Lindner, P., 11, *62*, 216, *267*
Lindquist, 152
Lindstedt, G., 146, *206*
Lingens, F., 384, 385, 386, 387, *413*
Linko, M., 7, 16, 56, *71*, *72*
Linnahalme, T., 58, 59, *71*, *72*
Linnane, A. W., 26, *68*, 213, 221, 222, 229, 230, 262, 263, *266*, *267*, *268*, *270*, 301, *305*, 318, 319, *331*, 402, *410*, *413*
Lipe, R. S., 82, 102, 103, 104, *117*
Lipmann, F., 22, *66*, 216, *267*, 283, *303*, 365, 372, *412*, *415*
Lippman, E., 458, 459, 478, *487*
Lis, A. W., 320, *330*
Littaru, G. P., 215, 217, *265*
Littauer, U. Z., 394, *410*
Littlefield, J. W., 407, *413*
Littman, M. L., 184, *206*
Ljungdahl, L., 13, 31, *66*, *69*
Lloyd, D., 263, *265*
Lochmann, E. R., 323, *330*
Lock, M. V., 142, 176, 183, 186, *202*
Lockwood, L. B., 171, *209*
Lodder, J., 176, *206*, 500, *514*
Loebel, J. E., 321, *328*
Loef, H. W., 317, *328*
Loening, U. E., 318, *330*
Loew, O., 227, *268*
Lohrman, R., 395, *416*
Longley, R. P., 36, *66*, 174, *206*, 213, 220, 222, 225, 227, 228, 255, 256, 257, *267*
Longton, J., 170, *204*

Lopatik, M. D., 495, *514*
Losada, M., 279, 282, 289, 291, *305*, *306*
Lovimann, R., 322, *332*
Luber, K., 224, *268*
Lue, F., 346, *413*
Lukashik, A. N., 495, *515*
Lukins, H. B., 221, 262, *267*
Lund, A., 12, *67*, 122, 123, 130, 131, *133*
Lundahl, P., 376, 408, *416*
Lundin, H., 27, *62*, 213, 216, *265*, *269*
Luzzati, M., 340, 343, 352, *410*, *413*
Lynian, C. M., 22, *73*
Lynen, F., 22, 48, *64*, *66*, 108, *117*, 236, 239, 241, *267*, *270*, 295, *305*, 365, 373, *415*
Lynes, K. J., 31, *66*

M

Maaløe, O., 80, *117*, 124, *133*
Maas, W. K., 355, *417*
Maas-Förster, M., 16, *66*, 216, *267*
McBurney, C. H., 22, *73*
McClary, D. O., 372, *409*, 456, 468, *489*
McClure, F. T., 45, *61*
McCready, R. M., 137, *205*
Macfarlane, M. G., 221, *267*
McFarlane, W. D., 31, *66*
McGann, C., 297, *303*
McHargue, J. S., 19, *66*
Machida, H., 252, *270*
MacKechnie, C., 315, 323, *331*, 393, 405, *415*
McKenna, E. J., 34, *66*
Mackenzie, D. W. R., 24, *66*
MacKenzie, S. L., 176, *204*
Mackinney, G., 498, 500, 506, 507, *514*
Maclachlan, 480, *488*
Mackler, B., 312, 324, *332*
McLaughlin, C. S., 127, *133*, 269, 391, 392, 401, *412*, *414*
McLellan, W. L., Jr., 56, *66*, 158, 171, 179, 192, *206*, 462, *489*
McManus, D. K., 14, 16, *62*, *70*
McMillan, J. D., 497, 501, *514*
McMurrough, I., 101, *117*, 227, *268*, 448, 449, 450, 451, 452, 453, 466, 470, 471, 479, *489*
McPhie, P., 319, *330*, 400, *410*
McQuillan, A. M., 325, *327*
Macrae, A., 251, *264*
Macrae, R. M., 187, *206*, 451, *489*
McVeigh, I., 20, 22, 24, 25, *61*
Madeira-Lopes, A., 92, 96, 97, 98, 99, 100, *118*
Maddox, I. S., 56, *66*
Madison, J. T., 310, 319, 320, 321, *329*, *330*, 392, 393, *412*, *413*
Madyasthia, P. B., 226, *267*
Magaña-Schwencke, N., 47, *66*, *70*
Magasanik, B., 338, 340, *412*, *413*
Magee, P. T., 378, 379, 380, 381, *413*, *415*
Mager, J., 175, *203*, *206*
Magrath, D. I., 320, *330*
Maguigan, W. H., 226, *267*
Magus, R. J., 156, 201, *204*
Mahler, H. R., 262, *266*, 213, 324, 325, *330*, *332*, 336, 340, *413*
Maitra, V., 403, 404, *411*
Majerus, P. W., 241, *267*
Makman, M. H., 321, 322, *329*, *330*, 389, 390, *413*
Málek, I., 76, *116*, 460, 462, 472, *487*, *491*
Malenge, J. P., 502, 508, 509, 511, *513*, *514*
Malm, M., 48, 54, *66*
Malmgren, H., 317, *332*, 398, *416*
Malmgren, J., 465, *491*
Malström, B. G., 277, *305*
Mangold, H. K., 217, *267*
Mankowski, Z., 474, *490*
Mann, P. F. E., 427, *439*
Manners, D. J., 139, 140, 141, 144, *206*, 293, *304*, *305*, 420, 424, 425, 426, 427, 428, 429, 430, 432, 433, 436, *437*, *438*, 447, *489*
Manney, T. R., 385, 386, 387, *410*, *413*
Maragoudakis, M. E., 359, 360, *413*
Marchant, R., 254, 255, 263, *267*, 456, 469, *489*
Marcker, K. A., 393, *416*
Marcot-Queiroz, J., 319, *330*
Marcovitch, H., 325, *330*
Marcus, L., 399, 400, *413*, 479, 480, *489*
Mariani, E., 28, *66*
Mariat, F., 474, 479, *488*, *489*
Marinetti, G. V., 217, *267*
Marker, K. A., 326, *331*
Markham, E., 15, 16, *66*
Markovitz, A. J., 35, *66*

Marmur, J., 310, 311, 312, 315, 324, *328*, *330*, *331*
Marquis, R. E., 260, *265*
Marquisee, M., 310, 319, 321, *329*, 392, 393, *412*
Marr, A. G., 76, 79, 102, 103, 104, *117*, 132, *133*, 230, *269*
Marsh, J. B., 212, *267*
Maruo, B., 293, *305*
Masler, L., 150, 151, 166, 175, 177, 180, 181, *206*, *208*
Massart, L., 11, *66*
Masşchelein, C. A., 165, 166, 173, *206*
Masson, A. J., 140, 141, 144, *206*, 420, 425, *438*
Masters, M., 315, *328*, 398, 399, *409*, 465, *487*
Masuda, Y., 200, *208*, 480, *489*
Masuyma, K., 58, *64*
Matile, Ph., 57, *66*, 198, 199, *206*, 254, 255, 258, 261, 263, 264, *268*
Matkovics, B., 511, *515*
Matsuoka, T., 197, *204*
Maurice, A., 225, 227, 258, *265*
Maw, G. A., 16, 17, 46, 54, *66*, 371, *414*
Maxwell, I. H., 320, *333*, 392, *417*
Maxwell, W. A., 497, 501, *513*, *514*
May, J. W., 460, *489*
Maynard Smith, J., 110, *117*
Mayo, V. S., 406, *414*
Mazia, D., 466, *489*
Mazlen, A. S., 343, *414*
Mazur, P., 131, *133*
Mazurek, M., 201, *204*
Mead, J. F., 251, *270*
Medina, R., 9, *70*
Medrano, L., 282, 291, *305*, *306*
Meek, G. A., 9, 10, *68*, 262, *268*, 301, *305*
Megnet, R., 197, *206*, 341, *414*
Mehl, J. W., 182, *204*
Mehrotra, B. D., 325, *330*
Meier, R., 5, 55, *62*
Mela, L., 288, *304*
Melchers, F., 321, *333*, 392, *417*
Melnick, D., 25, *66*
Melville, D. B., 20, 21, *62*, *66*
Menard, L. N., 17, 54, *61*
Mendershausen, P. B., 148, 161, *208*
Menna, M. E., di, 213, *266*
Menzinsky, G., 4, 29, *66*, *67*
Mercer, G. A., 424, *437*
Merdinger, E., 213, *268*
Merkenschlager, M., 136, *206*
Merrill, S. H., 310, 319, 320, 321, *327*, *329*, 392, 393, *412*
Merritt, N. R., 126, 141, *133*
Messenguy, F., 356, 358, *414*
Messmer, I., 346, *412*
Metcalfe, G., 11, *67*
Metzenberg, R. L., 173, *206*
Meuris, P., 384, 385, *414*
Meyenburg, H. K., von, 76, 102, 106, 107, 108, *117*, 466, 467, 468, 470, *487*, *489*
Meyer, F., 132, *133*, 217, 242, *268*
Meyer, K. H., 428, *438*
Meyer, S. A., 313, *330*
Meyerhof, O., 18, *67*
Meyers, S. P., 24, 25, 26, *60*
Michaljaničová, D., 41, *65*
Mickelson, M. N., 29, *69*
Middlehoven, W. J., 354, 355, 356, *414*
Miettinen, J. K., 15, *67*, *70*
Miikulainen, P., 15, *68*
Mill, P. J., 153, 154, 155, *206*
Millbank, J. W., 187, 189, *203*, *206*, 451, *489*
Miller, J. J., 56, *61*, 128, *132*
Miller, M. W., 122, 123, *164*, 155, 165, 191, *206*, 446, *489*
Miller, R. E., 192, *203*
Millin, D. J., 45, *63*
Mills, A. K., 15, *66*
Milne, B. D., 163, *203*, 451, 452, 457, *487*
Milner, L. S., 261, *268*
Minckler, S., 315, *331*
Mirsabekov, A. D., 392, *409*
Mirzabekov, A. D., 321, *327*, *330*, *332*, 395, *414*
Misaki, A., 139, 104, 143, *206*, 420, *438*, 447, *489*
Mitchell, P., 41, *67*
Mitchell, W. O., 158, 159, *205*
Mitchison, J. M., 315, 318, 319, *328*, *330*, 398, 399, *409*, *414*, 460, 464, 465, 467, 470, 471, *487*, *489*
Mitz, M., 26, *60*
Mittwer, T., 456, 472, *487*
Miura, K., 322, *328*, 392, 394, *412*, *414*
Miyazaki, M., 320, 322, *329*, *330*, 392, *414*, *416*
Miyazaki, T., 183, *206*
Mizutani, T., 392, *416*
Mommaerts, W. F. H. M., 320, *330*

Monier, R., 319, 321, *330*, *331*
Monod, J., 80, 82, 83, 103, *117*, 377, 396, *412*, *414*
Montavon, M., 502, 503, 509, *515*
Montencourt, B. S., 173, 196, *206*
Montreuil, J., 30, 31, *67*
Moor, H., 57, *66*, 193, 198, 199, *206*, 264, 255, 258, 261, 263, *268*, 454, 469, *489*, *490*
Moore, C., 312, 315, 324, *328*
Moore, D. P., 373, *416*
Moore, D. T., 18, *61*
Moore, J. T., 052, *268*
Moore, K., 376, *411*
Moreno, R. E., 142, *205*, 448, 450, 451, *489*
Morita, R. Y., 127, *133*
Mortimer, R. K., 189, *205*, 230, *268*, 315, *328*, 343, 348, 365, 366, 367, 392, 393, *412*, *414*, *415*, 472, *490*
Morpurgo, G., 325, *328*
Morris, E. O., 17, 18, 19, 20, *67*
Morris, J. G., 35, *67*
Morton, R. K., 324, *327*
Moser, H., 82, *117*
Moses, W., 18, 20, 23, 24, *60*, *67*
Mosher, W. A., 22, *73*
Moss, J. A., de, 385, 386, *414*, *417*
Mossel, D. A. A., 252, *265*
Motai, H., 252, *268*
Mothes, K., 11, *67*
Motta, R., 343, *410*
Mounolou, J. C., 312, 315, 324, 325, 326, *330*
Mousset, M., 46, *63*
Moustacchi, E., 311, 312, 314, 315, 324, 325, *330*
Moyer, D., 20, 22, 24, 25, *61*
Mrak, E. M., 122, 123, 127, 128, *134*, 498, 500, *514*
Mudd, S. H., 373, 375, 376, *411*, *414*
Muhammed, A., 314, *330*
Mühlethaler, K., 57, *66*, 193, 198, *206*, 258, 261, *268*, 454, *490*
Mulder, C., 25, *67*
Mulder, E., 56, *65*, 253, *267*
Müller, G., 485, *488*
Mullet, S., 30, 31, *67*
Munch-Petersen, A., 423, *438*
Mundkur, B., 199, *206*, 457, *490*
Munns, D. J., 17, 18, 19, 20, 22, 23, 29, 31, *67*, *73*, 122, 131, *134*



Muntz, J. A., 435, *438*
Murakami, M., 392, *416*
Muramatsu, M., 406, *414*
Murgier, M., 43, *60*
Musfeld, W., 4, *67*
Myrbäck, K., 5, 6, 7, 8, 26, 27, 55, 59, *62*, *67*, 216, *265*

## N

Nabel, K., 162, *206*
Nagai, H., 325, *330*
Nagai, S., 325, *330*, 365, 368, 374, *414*
Nagasaki, S., 158, *206*
Nageli, C., 227, *268*
Nagy, M., 340, 341, 343, 350, 352, *412*, *414*, *417*
Naiki, N., 365, 372, *414*
Najjar, V. A., 423, *438*
Nakagawa, N., 9, *67*, 495, 496, 511, *514*
Nakagaway, Y., 276, 277, *305*
Nakayama, T. O. M., 136, *208*, 498, 500, 506, 507, *514*
Nakazawa, K., 320, *330*, 392, *414*
Nannii, G., *269*
Napias, C., 225, 227, 258, *265*
Narbonne, F., 512, *515*
Nash, C. H., 127, 130, *133*
Nash, H. A., 503, *514*
Nashed, N., 325, *331*
Nasim, A., 314, *330*, *331*
Nečas, O., 194, 195, 196, 197, 198, 210, *205*, *206*, *208*, *210*, 445, 454, 462, 463, *490*, *491*
Neciullah, N., 48, *66*
Nelson, F. E., 252, *268*
Nelson, N., 478, *487*
Nelson, V. E., 18, 19, *63*
Netter, P., 326, *332*
Neuberg, C., 282, *305*
Neumann, N. D. P., 6, *67*
Neumann, N. P., 158, 172, 173, 196, *206*
Nevalainen, P., 26, *72*
Nichols, R. A., 56, *60*
Nicholson, W. H., 147, *205*
Nickerson, W. J., 19, *67*, 165, 166, 167, 168, 188, *204*, *205*, *207*, *208*, 227, *267*, 282, *305*, 443, 445, 446, 447, 448, 450, 451, 452, 456, 468, 469, 470, 474, 476, 477, 478, 480, 481, 482, 483, *487*, *488*, *489*, *490*

Nicolaieff, A., 312, 325, *328*
Nielsen, H., 216, *268*
Nielsen, N., 12, 13, 14, 16, 22, 27, *67*
Nielson, L. D., 293, *305*, 427, *438*
Nieudorp, P. J., 19, *65*, 180, *208*
Nikolaev, P. I., 495, *514*
Nilson, E. H., 102, 103, 104, *117*
Nilsson, N. G., 16, 27, *67*, 216, *268*
Nilsson, R., 216, *265*
Ninio, J., 393, *414*
Nishihara, T., 24, 25, *65*
Nishimura, S., 322, *332*, 392, *416*
Noltmann, E. A., 276, 277, *305*
Nordin, J. H., 142, 145, 191, *205*, 432, *438*
Norkans, B., 120, *133*
Norris, A. T., 26, *50*, 391, *414*
Norris, F. W., 31, *66*
Northcote, D. H., 36, *67*, 136, 138, 144, 146, 152, 155, 161, 162, 163, 164, 166, 168, 169, 196, 197, *203*, *205*, *207*, 227, *268*, 425, 426, 428, *438*, 445, 446, 447, 450, 453, 456, 462, 468, 469, *489*, *490*
Novaes-Ledieu, M., 166, *207*, 261, *266*, 448, 463, *488*, *490*
Novelli, G. D., 389, *416*
Novick, A., 80, *117*
Nozawa, M., 296, *306*
Nurminen, T., 9, 24, 29, 36, 56, 57, 59, *67*, *68*, *71*, 170, 188, *207*, 252, 256, 258, 259, 261, *268*, *269*
Nusbaum-Cassuto, E., 502, 506, 508, *514*
Nykänen, L., 58, *71*
Nyman, B., 25, *68*
Nyns, E. J., 212, 214, *268*


## O


Ochoa, A. G., 192, *204*, *207*
Oda, T., 224, *268*
Oeser, A., 317, *332*
Ofengand, E. J., 322, *327*
Ogur, M., 310, 315, *331*, 360, 361, 362, *409*, *414*, *416*
Ogur, S., 360, 361, 362, *414*, *415*, *416*
Ogura, Y., 277, *305*
Ohnishi, T., 288, *305*
Ohta, T., 277, *305*, 389, 390, 394, *412*, *414*
Ohtsika, E., 322, *332*
Ohtsuka, E., 392, 395, *409*, *416*
Okabe, K., 325, *332*
Okada, H., 84, *117*
Okada, K., 58, *64*
Okada, T., 34, *73*
Okamoto, T., 394, *414*
O'Kane, D. J., 6, *71*
Olbrich, H., 28, 29, *68*
Oliviera-Baptista, A., 96, *117*
Olley, J. N., 220, *266*
Olson, B. H., 19, 20, *68*
Olson, J. A., 246, *268*
Onihara, T., 34, *73*
Onishi, H., 27, *68*, 124, *133*
Ono, M., 191, *207*
Oota, Y., 318, *331*
Oro, J., 251, *269*
Ørskov, S. L., 4, 37, *68*
Ortenblad, B., 7, 8, 27, *67*
Osawa, S., 316, 318, 320, *331*, 407, *414*
Oser, B. L., 25, *66*
Ota, Y., 252, *268*
Otaka, E., 317, 320, *331*
Otaka, Y., 319, *331*
Ottolenghi, P., 8, 23, 55, 59, *63*, *68*, 171, 172, 187, 193, 210, *204*, *205*, *207*, *210*, 462, 463, 486, *487*, *488*, *489*, *490*
Ottoson, R., 317, *332*, 398, *416*, 465, *491*
Oura, E., 5, 6, 7, 9, 16, 24, 25, 26, 29, 36, 37, 38, 45, 47, 48, 52, 56, 57, 59, *67*, *68*, *71*, *72*, 188, *207*, 256, 258, 259, 260, 261, *268*, *269*, 291, *306*
Ourisson, G., 225, 246, *268*
Owades, J. L., 225, 257, *265*


## P


Paduch, V., 316, *328*, 398, 402, *410*
Painter, P. R., 76, 79, *117*
Palacián, E., 279, 289, *306*
Paleeva, M. A., 495, *514*
Palmer, E. T., 5, 6, *73*
Palmqvist, U., 13, *68*
Panek, A., 8, *68*, 422, *438*
Panek, A. D., 294, *306*, 466, *490*
Panten, K., 361, *414*
Paoletti, C., 324, *329*
Pardee, A. B., 36, *68*, 367, 372, *414*, *417*
Parks, L. W., 26, *68*, 124, 132, *134*, 226, 248, *264*, *267*, *268*, *332*, 365, 370, 371, 375, 376, *409*, *414*, *415*, *416*

Parnas, J. K., 293, *305*
Parrish, F. W., 143, *207*
Parry, J. M., 314, *328*
Parsons, D. G., 316, *329*
Parvez, M. A., 247, 248, 249, *264*
Passarge, W. E., 320, *330*
Pasteur, L., 275, *305*
Patrick, M. H., 314, *331*
Patterson, J. C., 139, 144, *206*, 420, *438*, 447, *489*
Paul, R. M., 324, *327*
Pavloff, M., 45, *63*
Peat, S., 138, 139, 145, 146, 147, *205*, *207*, 427, 428, *438*
Pedersen, T. A., 314, *268*
Peetrissant, G., 320, *331*
Pelshenke, P., 30, *68*
Peltonen, R. J., 58, *71*
Pennock, J. F., 223, *265*
Penswick, J. R., 310, 319, 321, *329*, 392, 393, *412*
Peppler, H. J., 59, *64*
Perlin, A., 436, *439*
Perlin, A. S., 143, 147, 148, 149, 175, *204*, *207*
Perrodin, G., 315, 325, 326, *330*, *332*
Perry, M. B., 142, 176, 183, 184, 186, *202*
Perry, R. P., 406, *415*
Peters, I. I., 252, *268*
Peterson, C. S., 120, 123, 130, *133*
Peterson, W. H., 20, 21, 59, *61*, *64*, *70*, 495, 500, *514*
Petrochild, E., 326, *332*
Pettijohn, O. G., 171, *209*
Peynaud, E., 32, *69*
Pfäffli, S., 5, 7, 8, 16, 26, 27, *72*, 420, 423, *439*
Pfennig, N., 80, *117*
Phaff, H. J., 14, 15, 56, 58, *60*, *62*, *65*, *70*, 120, 122, 123, 127, 128, *133*, *134*, 136, 143, 144, 155, 165, 170, 171, 176, 188, 190, 200, 201, 202, *202*, *204*, *206*, *207*, *208*, 255, 259, 261, *268*, 313, *330*, 450, *490*, 498, 500, 506, 507, *514*
Philipps, G. R., 393, *415*
Phillips, A. W., 7, 31, *61*, *68*, 479, 480, *489*
Phillips, J. H., 320, *331*, 376, 408, *412*, *415*
Picard, S., 224, *268*
Piediscalzi, N., 360, *415*
Pierard, A., 343, 344, 345, 346, 354, 355, *412*
Pierce, J. S., 13, 14, 31, *64*, *68*
Pierson, D. J., 357, *409*
Pigg, J., 365, 371, *415*
Pigman, W. W., 4, *64*
Pilgrim, F. J., 21, *68*
Pillinger, D. J., 408, *415*
Piperno, G., 312, 325, *328*
Pirschle, K., 10, *68*
Pirt, S. J., 103, 104, *117*
Pisano, M. A., 215, *265*
Pitt, D., 469, *490*
Pittard, J., 382, *411*
Pittsley, J. E., 178, 183, 184, *205*
Planta, R. J., 322, 323, *331*, 405, 406, *415*
Plinston, C. A., 322, *328*, 392, *409*
Poirier, L., 343, 347, *411*, *414*
Polakis, E. S., 9, 10, *68*, 262, *268*, 301, *305*, 315, *331*
Polissar, M. J., 92, 93, 94, 97, *117*
Pomeranz, Y., 30, *68*
Pomortseva, N.V., 495, 514
Pomper, S., 11, 14, *70*, 347, 371, *415*
Ponsinet, G., 225, 246, *268*
Popkova, G. A., 217, *265*
Porter, J. W., 245, *267*, 494, 503, 506, *514*
Posternak, T., 446, 447, *488*, *490*
Postgate, J. R., 100, *117*
Pourquie, J., 340, 341, 343, *415*
Powell, E. O., 82, 83, *117*
Power, D. M., 14, 23, *61*, *64*, *68*, 451, 472, 488, *490*
Pragnell, M. J., 13, *64*
Praus, R., 496, 497, 511, 512, *513*, *514*
Preiss, J. W., 55, *68*, 171, 172, 199, *207*
Prestidge, L. S., 372, *414*
Preston, B. W., 318, 319, *331*
Prévost, G., 343, *410*
Pricer, W., 286, *305*
Prinsen, Geerligs, H. C., 19, *68*
Pritchard, G. C., 465, *488*
Prokazov, G. F., 495, *515*
Prokazova, N. V., 217, *270*
Prokozova, N. V., 217, *265*
Protiva, J., 511, *514*
Proudlock, J. W., 26, *68*, 230, 263, *266*, *268*
Pugh, E. L., 241, *268*
Pulver, R., 18, 49, *68*

Pye, E. K., 300, *306*
Pyke, M., 29, *68*

## Q

Quackenbush, F. W., 503, *514*
Quayle, J. R., 10, 34, *68*
Questiaux, L. M., 55, *69*
Quetsch, M. F., 125, *134*

## R

Rabinowitz, M., 311, 312, 315, 324, 325, 326, *328*, *331*, *332*, 403, *411*
Rabourn, W. J., 503, *514*
Racker, E., 262, *268*, 277, *306*
Rahimtula, A. D., 246, *264*
Rahn, O., 131, *133*
Rainbow, C., 494, *513*
Rajbhandary, U. L., 321, *331*, 392, 393, 395, *415*, *416*
Rake, A. V., 321, *331*
Ramaih, A., 276, *306*
Rainbow, C., 25, *61*, *68*
Raj Bhandary, V. L., 392, *409*
Ramkrishna, D., 82, *117*
Ramos, F., 354, 356, *415*
Raskunas, S. R., 321, *328*
Rasmussen, R. K., 242, *268*
Rath, F., 225, *270*
Raut, C., 325, *331*
Rautanen, N., 11, 15, 16, 56, *68*, *72*
Ravel, J. M., 391, *413*, *415*
Rebers, P. A., 182, *207*
Reed, D. J., 9, *72*
Reese, E. T., 143, *207*
Reeves, R. H., 322, *331*
Regord, M. Th., 214, *267*
Reichert, E., 22, *66*
Reilly, C., 49, 51, *68*
Reindel, F., 224, *268*
Reiner, M., 9, 40, *70*
Reinfurth, E., 282, *305*
Renvall, S., 6, 24, *67*, *70*
Resnick, M. A., 230, *268*
Resnick, M. R., 230, *267*
Retel, J., 322, 323, *331*, 405, 406, *415*
Rether, B., 394, *410*
Reutter, W., 432, *438*
Reuvers, T., 166, *207*, 447, *490*
Reyes, P., 511, *515*
Rhoades, H. E., 6, *68*
Ribéreau-Gayon, J., 32, *69*
Richard, H. H., 322, *329*
Richard, O., 9, *73*
Richards, H. H., 319, 321, *328*
Richards, O. W., 19, *69*, 122, *134*
Richter, D., 396, *415*
Richtmyer, N. K., 7, 27, *71*, 421, *439*
Ridgway, G. J., 23, *69*
Riemersma, J. C., 49, 51, *69*
Rilling, H. C., 245, *269*
Ris, H., 399, 400, *417*
Rischbiet, P., 8, *69*
Risebrough, R. W., 396, *411*
Robbins, P. W., 223, *270*, 365, 372, *415*
Roberts, C., 7, *73*
Roberts, D. G., 316
Roberts, E. R., 11, *69*
Roberts, R. H., 4, *64*
Roberts, R. N., 23, *64*
Robertson, J. J., 44, *69*
Robichon-Szulmajster, de H., 45, 46, 47, *72*, 325, *332*, 347, 365, 366, 367, 368, 369, 372, 373, 378, 379, 380, *409*, *410*, *412*, *413*, *415*, *416*
Robinow, C. F., 198, 199, *206*, 254, 255, 263, *268*
Robins, M. J., 320, *331*
Robinson, M. P., 243, *266*
Robson, J. E., 469, *490*
Robyt, J., 427, *437*
Rocca, E., 9, *70*
Roche, B., 254, *267*, 279, 292, *306*
Rodriguez, J., 142, *205*, 448, 450, 451, *489*
Rodwell, V. W., 250, *266*, 279, *306*
Roehm, R. R., 23, *73*
Roelofsen, P. A., 162, 165, *205*, *207*
Rogers, D., 29, *69*
Rogers, L. J., 504, *515*
Rogers, P. J., 318, 319, *331*
Rogosa, M., 8, 24, 25, *69*
Rogovin, S. P., 178, *207*
Rohdewald, M., 6, *73*
Rohrman, E., 22, *73*
Roine, P., 11, 13, *69*
Roman, H., 339, 340, *415*
Romano, A. H., 474, *490*
Ronkainen, P., 57, 68, *72*
Roodyn, D. B., 262, *268*
Rose, 260

Rose, A. H., 22, 36, *60*, *62*, *66*, *69*, 101, *117*, 120, 122, 124, 125, 126, 127, 132, *133*, *134*, 149, 174, *204*, *206*, 213, 215, 220, 222, 225, 227, 228, 229, 238, 239, 255, 256, 257, 260, 261, *264*, *265*, *266*, *267*, *268*, 448, 449, 450, 451, 452, 470, 471, 472, 473, 479, *488*, *489*
Rose, I. A., 301, *306*
Rose, Z. B., 301, *306*
Rosen, G., 310, *331*
Rosen, F., 20, 21, *69*
Rosen, O. M., 283, *305*
Rosen, S., 283, *305*
Rosenbaum, E., 15, *69*
Rosenberg, T., 84, *118*
Roshanmanesh, A., 360, 361, 362, *414*
Rosset, R., 319, 321, *330*, *331*
Rossi, C., 284, 302, *306*
Rossiter, R. J., 236, *268*
Rost, K., 187, 188, 194, *207*, 453, *490*
Roth, F. J., Jr., 24, 25, *60*, 123, 130, *132*
Roth, J. F., 176, *203*
Rothfield, L., 36, *69*
Rothman, L. B., 435, *439*
Rothstein, A., 5, 16, 18, 39, 41, 49, 50, 51, 52, 53, 54, 55, *60*, *62*, *63*, *64*, *69*, *72*, 84, *118*, 260, *270*, 297, *303*
Rotsch, A., 30, *68*
Rouser, G., 212, *268*
Roush, A. H., 55, *69*
Rowe, K. L., 427, *438*
Rozijn, T. H., 314, *331*, *332*
Rozijn, Th. H., 264, *268*, *269*
Rublin, S. H., 20, 21, *69*
Rudert, F., 7, *69*
Rüdiger, M., 369, 371, *415*
Rudney, H., 243, *266*, *268*
Rüegg, R., 502, 503, 504, 509, 511, *514*, *515*
Ruesink, A. W., 480, *490*
Ruinen, J., 199, *207*, 231, 232, *268*, *269*
Ruiz-Amil, M., 282, 289, 291, *305*, *306*
Runner, C. M., 324, *332*
Runnström, J., 48, 54, *69*
Rupert, C. S., 314, *333*
Russ, G., 222, *267*
Ruttenberg, G. J. C. M., 324, *328*, *332*
Rutter, W. J., 277, 278, *306*
Ryan, H., 50, 51, *61*
Rychlik, I., 322, *328*
Rymo, L., 391, 394, 412, *413*
Ryser, G., 502, 503, 504, 509, *514*, *515*

## S

Safranski, M. J., 209, *210*
Sagara, Y., 197, *204*
Saito, T., 159, *208*
Sakaguchi, K., 190, *205*
Salas, J., 300, 303, *306*
Salas, M., 276, 278, 300, *306*
Salas, M. L., 276, 277, 278, 300, 303, *304*, *306*
Salazar, J., 385, 386, *413*
Salkowski, E., 137, 145, *207*
Salton, M. R. J., 189, *207*
Samokhvalov, G. I., 495, *514*
Samson, F. E., 48, *69*
Samuel, L. W., 30, *69*
Sanadi, D. R., 311, 312, 315, 324, *328*
Sandegren, E., 13, 31, *66*, *69*
Sandeman, W., 251, *269*
Sandoval, A., 495, 500, 506, 507, *513*
Šandula, J., 166, *206*
Sangavi, P., 312, 324, *331*
Sarachek, A., 479, 480, *490*
Sarett, H. P., 22, *69*
Sarin, P. S., 321, *331*
Sarkar, P. K., 319, 320, *331*
Sašek, V., 482, *490*
Satomura, Y., 191, *207*
Sattler, L., 5, *70*, *74*
Satyanarayana, T., 478, 382, *415*
Saucy, G., 502, 503, 509, *514*, 515
Saunders, D. H., 22, *73*
Savioja, T., 15, *70*
Scaletti, J. V., 139, 140, 143, *206*, 420, *438*, 447, *489*
Schanderl, H., 11, 34, *70*
Schapiro, L., 324, *331*
Scharf, S. S., 501, 504, 506, 511, *515*
Scharff, T. G., 4, *70*
Schatz, G., 311, 324, *331*
Scheer, C., van der, 199, *270*
Scher, B., 283, *305*
Scher, M., 161, *207*, 223, *269*
Scheraga, H. A., 321, *332*
Scherr, G. H., 474, 478, *490*
Scheverbrandt, G., 26, *60*
Schildkraut, C. L., 311, *331*
Schimpfessel, L., 279, *306*
Schlegel, H. G., 104, *117*
Schlenk, F., 38, 39, 59, *70*, *74*, 187, 188, *208*, 369, 376, *415*

Schlimme, E., 393, *410*
Schlossmann, K., 136, *206*, 365, 373, *415*
Schmidt, G., 18, 54, *70*
Schmidt, M., 162, *207*
Schnabel, W., 33, *64*
Schnable, L. D., 158, *206*
Schneider, F., 28, *70*
Schneider, G., 511, *515*
Schneider, S., 281, *305*
Scholfield, C. R., 232, *269*
Schönherr, O. Th., 53, *70*
Schopper, W. H., 447, *490*
Schudel, P., 503, 504, *514*, *515*
Schulman, L. H., 322, *331*, 394, *415*
Schultz, A. S., 6, 11, 14, 16, 22, 24, 25, 30, 45, *60*, *62*, *70*
Schultz, G., 278, *305*
Schultz, J., 241, *270*
Schulze, I. T., 277, *306*
Schulze, K. L., 16, *70*, 82, 102, 103, 104, *117*
Schuster, C. W., 302, *304*
Schuster, E., 104, *117*
Schuster, K., 31, 32, *70*
Schutzbach, J. S., 185, 186, *207*
Schvam, H., 322, *331*
Schwaier, R., 325, *331*
Schwartz, U., 452, *488*
Schweizer, E., 241, *269*, 315, 323, 326, *331*, *332*, 393, 405, *415*
Schwemmin, D., 45, *63*
Schwencke, J., 47, *66*, *70*, 191, 192, *207*
Schwers, W., 251, *269*
Schwieter, U., 502, 503, 504, 509, *514*, *515*
Scopes, A. W., 315, *333*, 465, 468, *491*
Scott, J. F., 321, *331*
Scriban, R., 30, 31, *67*
Segretain, G., 182, 183, *203*
Seidel, H., 320, *331*, 393, *410*
Sekiya, T., 322, *331*
Senez, J. C., 43, *60*, 106, *117*
Sentandreu, R., 155, 161, 162, 164, 168, 169, 197, 199, *207*, 454, 456, 462, 468, 469, *488*, *490*
Senthe Shanmuganathan, S., 165, 166, *208*, 452, 480, 481, 482, 483, *490*
Seraidarian, K., 347, *412*
Sgaramella, V., 392, *409*
Shafai, T., 23, *65*, 218, 220, 222, 231, 243, *269*
Shah, S. P. J., 504, *515*
Shapiro, S. K., 369, 371, 376, *415*, *416*
Sharp, C. W., 262, *266*
Shaw, D. C., 320, *330*
Shaw, M. K., 132, *134*
Shaw, N., 225, 233, *269*, *270*
Shaw, W. H. C., 226, *269*
Shchennikov, V. A., 218, *270*
Sheifinger, C., 360, *416*
Sherman, F., 124, *134*, 393, *411*
Sherwood, F. F., 18, 19, *63*, 122, 131, *134*
Shieh, T. R., 55, *69*
Shifrine, M., 165, *208*, 230, *269*, 450, *490*
Shimada, I., 394, *414*
Shimoda, C., 200, *208*, *209*, 480, *490*
Shimura, K., 365, 366, 367, *417*
Shive, W., 391, *413*, *415*
Shneour, E. A., 512, *515*
Shoppee, C. W., 225, *269*
Shorland, F. B., 213, *266*
Shortman, K., 316, 318, *331*, 405, *416*
Shull, G. M., 21, *70*
Shuurmans-Stekhoven, F. M. A. H., 324, *332*
Siddons, P. T., 503, *513*
Siegel, M. R., 323, *331*, 401, 402, *415*
Siewert, G., 251, *266*
Sih, C. J., 246, *269*
Šikl, D., 150, 151, 166, 175, 177, 180, 181, *206*, *208*
Sillevis, 264
Silven, B., 398, *416*
Silver, J. M., 339, 340, 341, *416*
Simon, H., 9, *70*
Simpf, K., 225, 226, *265*
Simpson, K. L., 136, *208*, 495, 498, 499, 501, 502, 503, 504, 506, 507, 508, 511, 512, *515*
Simpson, W. L., 325, *331*
Sinclair, J. H., 311, 312, 324, *331*
Sinclair, N. A., 120, 121, 123, 125, 126, 127, 130, 131, *133*, *134*
Singer, T. P., 9, *70*, 284, 302, *306*
Sire, J., 45, 46, 47, *72*
Sisler, H. D., 323, *331*, 401, 402, *416*
Sison, Y., 472, *488*
Sistrom, W. R., 497, *513*
Sivak, A., 23, *70*
Skeggs, H. R., 20, 21, *73*
Skraup, Z. H., 8, *70*
Slator, A., 4, 8, *70*, 131, *134*
Slaughter, C., 367, 374, 375, *411*

Slechta, L., 511, *515*
Slodki, M. E., 175, 176, 178, 179, 180, 184, 202, 209, *203*, *208*, 210
Sloneker, J. H., 178, *205*
Slonimski, P. P., 262, *266*, 301, *305*, 310, 312, 315, 324, 325, 326, *328*, *329*, *330*, *332*, 340, 352, 368, 384, *410*, *413*, *414*, *415*, *416*
Slooff, W. Ch., 180, *208*
Sly, W., 45, 46, 47, *72*
Smart, W. W. G., Jr., 500, *514*
Smith, A. E., 326, *331*, 393, *416*
Smith, D., 315, 326, *332*, 465, *490*
Smith, D. G., 254, 255, 263, *267*, 465, 469, *489*
Smith, E. E., 433, *439*
Smith, F., 30, *65*, 139, 140, 143, 146, 167, *203*, *206*, 420, *438*, 447, *489*
Smith, J. D., 311, *332*, 393, 394, 395, *411*
Smith, K. C., 316, *328*
Smith, R. H., 22, 23, *70*
Smith, S. B., 479, 480, *490*
Smitt, W. W. S., *269*
Smythe, C. V., 48, *70*
Snell, B. K., 214, 221, 222, *267*
Snell, E. E., 20, 22, 23, 25, *70*, *73*, 216, 218, 220, *266*
Snow, R., 314, *332*
Snyder, H. E., 56, *70*, 171, *208*
Sobotka, H., 9, 40, *70*
Sofer, S. S., 245, *269*
Söll, D., 322, *332*, 336, 388, 390, 391, 395, *413*, *416*
Sols, A., 4, 6, 7, 8, 39, 40, 41, 45, 55, *62*, *64*, *70*, 86, 90, *116*, *118*, 171, *203*, 234, *266*, 276, 277, 278, 279, 281, 289, 290, 291, 293, 299, 300, 301, 303, *304*, *306*
Somers, J. M., 327, *332*
Somogyi, M., 45, *70*
Sonderhoff, R., 243, *269*
Sorms, F., 395, *414*
Sorsoli, W. A., 375, *416*
Soška, J., 197, *208*
Sošková, L., 197, *208*
Sottocasa, G., 288, *305*
Souter, G. A., 51, *64*
Souza, N. O., 294, *306*, 466, *490*
Spence, K. D., 365, 371, 376, *415*, *416*
Spencer, J. F. T., 12, 14, *65*, 137, 149, 151, 152, 155, 156, 157, 160, 174, 175, 176, 177, 181, 182, 185, 186, 201, 202, *204*, *207*, *208*, 231, 232, 233, *269*, *270*
Sperber, E., 24, 48, 54, *69*, *70*
Spiegelman, S., 296, *306*, 405, *411*
Spirin, A. S., 311, *327*
Spoerl, E., 479, 480, *490*
Sprinson, D. B., 251, *269*
Sprössler, B., 384, 385, *413*
Srinivasan, P. R., 407, *409*
Stalmans, W., 437, *438*
Stanacev, N. Z., 224, 238, *269*
Stanier, R. Y., 346, *416*, 497, *513*
Stanley, P. E., 80, *118*
Stanley, S. O., 120, *134*
Stanley, W. M., 225, *270*
Stannard, J. N., 296, *306*
Staple, E., 243, 254, *269*
Stapley, E. O., 225, 226, *265*
Stark, I. E., 45, *70*
Starkey, R. L., 213, *269*
Starr, P. R., 26, *68*, 124, 132, *134*
Stearn, A. E., 91, *117*
Steibelt, W., 6, *73*
Stein, W. D., 39, *70*, 261, *269*
Steiner, A., 420, *439*
Steinschneider, A., 326, *332*
Stellwaggen, E., 277, *305*
Stenderup, A., 313, *332*
Stephanopoulos, D., 54, 58, *65*, *70*
Stern, I., 9, *72*
Steveninck, J., van, 84, *118*
Stevens, A., 405, *416*
Stevens, B. J., 311, 312, 324, *331*
Stewart, J. E., 34, *73*
Stewart, J. W., 393, *411*
Stewart, L. C., 7, 27, *71*, 421, *439*
Stewart, P. S., 230, *270*
Stewart, T. S., 148, 149, 150, 151, 153, 154, 209, *208*
Steyn-Parvé, E. P., 36, 56, *60*, *70*, *72*, 171, 173, *203*, *208*
Stier, T. J. B., 26, *60*, 229, *264*, 296, *306*
Stille, B., 131, *134*
Stochley, M. A., 469, *490*
Stodola, F. H., 174, *208*, 231, 232, 233, 234, *269*, *270*
Stokes, J. L., 120, 121, 123, 124, 125, 126, 127, 130, 131, *133*, *134*
Stone, B. A., 136, *203*
Stone, I., 30, *63*
Stone, W. E., 5, *71*

Stoodley, R. J., 142, 176, 183, 186, *202*
Stoppani, A. O. M., 297, *306*
Storck, R., 313, 325, *332*, 477, *489*
Stouthamer, A. H., 106, 107, *116*, *118*
Straka, R. P., 120, 123, 125, *134*
Strassman, M., 358, 359, 360, *409*, *413* *416*
Streiblová, E., 76, *116*, 164, 193, 194, 199, 210, *208*, 454, 456, 460, 461, 462, 463, 470, 472, *487*, *490*, *491*,
Strickland, K. P., 236, 238, *269*
Strijkert, P. J., 323, *329*
Strominger, J. L., 223, 251, *266*
Stuart, A., 321, 324, *331*, 392, *415*
Studer, A., 502, 503, 511, *515*
Stulberg, M. P., 389, *416*
Suassuna, E. N., 254, *265*
Subik, J., 222, *267*
Subrahmanyan, V., 279, *304*
Sueoka, N., 322, *333*
Sugimori, T., 401, *416*
Sugimura, T., 325, *332*
Summers, D. F., 158, *208*, 459, *491*
Sumner, J. B., 6, *71*
Sunayama, H., 159, *208*
Sundhagne, M., 448, 451, 479, *491*
Sundman, J., 33, *71*
Suomalainen, H., 5, 6, 7, 8, 9, 12, 14, 16, 17, 19, 22, 24, 25, 26, 27, 29, 33, 36, 37, 38, 45, 47, 48, 52, 55, 56, 57, 58, 59, *67*, *68*, *71*, *72*, 170, 188, *207*, 231, 252, 256, 258, 259, 260, 261, *269*, 291, *306*, 420, 423, *439*
Surdin, Y., 45, 46, 47, *72*, 365, 366, 367, 368, *415*, *416*
Surdin-Kerjan, Y., 365, 366, 368, *416*
Suto, K., 214, *267*
Sutton, D. D., 55, *72*, 172, *208*, 451, 462, *491*
Suyter, M., 234, *270*
Suzuki, S., 159, *208*
Suzuki, T., 27, *68*
Suzuoki, Z., 24, *72*
Svensson, I., 376, 391, 408, *409*, *416*
Svihla, G., 187, 188, *208*
Svoboda, A., 194, 195, 196, 197, 210, *204*, *206*, *208*, *210*, 445, 454, 462, *488*, *490*, *491*
Sweeley, C. C., 161, *207*, 223, 251, *266*, *269*
Sweizer, E., 465, *490*
Swift, H., 325, *332*
Swift, H. H., 315, 326, *331*
Sword, R. W., 243, *266*
Sylvén, B., 317, *332*, 465, *491*
Szilard, L., 80, *117*
Szörenyi, E., 18, *65*, 215, *267*

## T

Taber, R. L., 323, *332*, 407, *416*
Taber, W. A., 477, *490*
Tacoronte, E., 166, *207*, 447, *490*
Tacreiter, W., 55, *65*
Tahara, M., 379, *417*
Takahashi, J., 34, *73*
Takeishi, K., 322, *331*, *332*, 392, 393, *416*
Takemura, S., 320, 322, *329*, *330*, 392, *414*, *416*
Talbot, G., 369, *410*
Talou, B., 443, 479, *491*,
Tang, Y. W., 495, 500, 506, 507, *513*
Tamaki, H., 314, *332*
Tanaka, H., 143, 144, 190, 201, *208*
Tanaka, K., 321, *328*
Tanner, F. W., 129, 130, *133*
Tanner, M. J. A., 321, *332*
Tanner, W., 161, *208*, 225, 238, 261, *269*
Tatsumi, C., 9, *67*, 495, 496, 511, *514*
Tauro, P., 315, 317, 323, 326, *329*, *332*, 397, 398, 399, *411*, *416*, 465, 467, *488*, *489*, *490*, *491*,
Tavlitzki, J., 262, *266*, 301, *305*, 443, 479, 491, 495, *515*
Tavormina, P. A., 243, *269*
Tawada, N., 191, *205*
Taylor, I. F., 140, 141, 163, 189, 194, 198, 199, *203*, *209*, 451, 452, 457, *487*
Taylor, N. W., 483, *491*
Taylor, P. M., 433, *439*
Tchang, J. L., 495, 498, 504, *513*
Tchen, T. T., 243, 245, *265*
Tecce, G., 312, 325, *328*
Tefft, R. E., 502, 512, *515*
Teissier, G., 82, *118*
Telling, R. C., 80, *116*
Tempest, D. W., 101, *118*
Tener, G. M., 320, 321, *331*, *333*, 392, *411*, *417*
Teulings, F. A. G., 106, *116*
Tewari, J. J., 312, 324, *332*

Thannhauser, S. J., 18, 54, *70*
Thedford, R., 320, *331*
Theilade, J., 188, 194, *203*
Theriault, R. J., 182, *204*
Thierfelder, H., 4, 5, 8, *62*
Thijsse, G. J. E., 34, *72*
Thomas, H., 243, *269*
Thompson, C. C., 44, 45, 55, *63*, *74*
Thomson, J. F., 373, *416*
Thorell, B., 317, *332*, 398, *416*, 465, *491*
Thorne, R. S. W., 11, 12, 13, 19, *60*, *72*
Thorpe, S. R., 251, *269*
Thoss, G., 24, 26, 32, *73*
Thuriaux, P., 354, 356, *415*
Thurston, C. F., 357, *416*
Tigerstrom, M., von, 392, *411*
Tingle, M., 348, 351, *416*
Tinoco, I., 321, *328*
Tipper, E. J., 30, *72*
Titchener, E. B., 318, 319, *331*
Tobias, C. A., 317, *332*, 398, *416*, 465, *491*
Toivonen, T., 5, 9, *72*
Tokumaru, Y., 191, *205*
Tollens, B., 5, 8, *69*, *71*
Tomkins, G. M., 373, *412*
Tomlinson, G. A., 322, *332*
Tonge, R. J., 23, *61*, 471, *488*
Tonio, G. de M., 314, *332*
Tonino, G. J. M., 56, *72*, 171, *208*, 264, *268*, *269*, 314, *331*
Tonnis, B., 20, 21, *65*
Topham, R. W., 247, 248, *269*
Torii, K., 365, 372, *416*
Tornabene, T. G., 251, *269*
Tornqvist, E., 216, *269*
Torraca, G., 28, *66*
Torrôntegui, G., de, 279, 289, *306*
Toth, G., 137, *209*
Towne, J. C., 279, *306*
Tracey, M. V., 189, *205*
Tran van Ky, P., 159, *203*
Travassos, L. R., 254, *265*
Tresguerres, E. F., 279, *306*
Trevelyan, W. E., 26, 57, 59, *64*, *72*, 214, 220, 225, 253, *266*, *270*, 419, 420, 421, 427, 435, *439*
Tristram, H., 357, *416*
Truesdail, J. H., 22, *83*
Trupin, J. S., 358, *409*
Tsuchiya, H. M., 82, *117*, *118*
Tsuchiya, T., 159, *208*
Tsukada, Y., 401, *416*
Tucci, A. F., 359, 360, *416*
Tulloch, A. P., 232, *270*
Tully, E., 277, *303*
Tuppy, H., 311, 324, *331*
Turpeinen, O., 37, 47, 52, *61*
Turula, P., 224, *268*
Turvey, J. R., 138, 146, 147, *207*
Tustanoff, E. R., 9, 10, *72*

U

Uchida, K., 319, *331*
Udaka, S., 353, *417*
Uden, N., van, 81, 82, 83, 86, 87, 88, 89, 91, 92, 96, 97, 98, 99, 100, 103, 104, 105, 106, 108, 109, *117*, *118*
Uehleke, H., 511, *515*
Uessler, H., 384, 385, 386, 387, *413*
Ukita, C., 321, *333*
Ukita, T., 322, *331*, *332*, 392, 393, 394, *416*, *417*
Ullmann, S., 33, 34, *64*
Ullrich, J., 277, *306*
Umbarger, E., 336, 343, 377, *417*
Umbarger, H. E., 377, 378, 380, 381, 382, *410*, *411*, *415*
Umbreit, W. W., 25, *65*
Underkofler, L. A., 18, *63*
Uruburu, F., 193, 198, *204*, *208*, 448, 463, *491*
Usden, V. R., 225, *270*
Ushakov, A. N., 217, *265*, *270*
Utter, M. F., 188, *203*, 301, *306*

V

Vagelos, P. R., 236, 241, *266*, *267*
Valenlova, L. A., 495, *514*
Valadon, L. R. G., 494, *515*
Van Bruggen, E. F. J., 324, *323*, *332*
Van Deenen, L. L. M., 56, *65*, 253, *267*, *270*
Van den Bosch, H., 56, *65*, 253, *267*, *270*
Van den Elzen, H. M., 253, *270*
Van der Linden, A., 34, *72*
Vandesando, J. H., 392, *409*

Van Steveninck, J., 40, 41, 42, 43, 52, *72*, 260, 261, *270*
Vaver, V. A., 213, *265*, *270*
Vecher, A. S., 495, 496, 499, 508, *515*
Veiga, L. A., 290, *304*
Veldman, H., 24, 56, *73*
Venkstern, T. V., 321, *327*, *330*, *332*, 392, *409*
Venner, H., 187, 188, 194, *207*, 453, *490*
Verhue, W., 433, *439*
Vernet, C., 34, *61*
Verzár, F., 18, 49, *68*
Vesonder, R. F., 233, 234, *269*
Vickery, J. R., 252, *270*
Viehhauser, G., 326, *333*
Vigneaud, V., du. 20, 21, *62*, *66*, *73*
Vihervaara, K., 24, 25, 29, *71*
Villa, V. D., 313, 325, *332*
Villaneuva, J. R., 187, 192, 193, 198, 199, *204*, *207*, *208*, 255, 256, *266*, 448, 462, 463, *491*
Villar-Palasi, C., 276, *306*
Villoutreix, J., 495, 500, 501, 502, 506, 507, 508, 511, 512, *513*, *514*, *515*
Vincent, W. S., 317, 323, *332*, *333*, 396, 407, *416*, *417*
Viñuela, E., 276, 278, 300, *306*
Virtanen, A. I., 11, *72*
Vischer, E., 310, 311, *332*
Vishniac, W. J., 104, *116*
Vito, P. C., de, 373, *417*
Vitols, E., 262, *270*
Vogel, H., 28, *72*
Vogt, E., 32, *72*
Vold, B. S., 392, *417*
Volkin, E., 396, *417*
Votsch, W., 312, 324, *332*
Vournakis, J. N., 321, *332*
Vraná, D., 468, *491*
Vyas, S., 355, *417*

W

Wagner, E., 224, 225, 253, *266*, *270*
Wagner, E. K., 321, *332*
Wagner, R. P., 378, 379, 380, *412*
Wainwright, T., 365, 372, *417*
Waite, M., 22, *72*
Wakil, S. J., 22, *72*, 241, *268*
Waldenström, J., 389, 390, 391, 394, *412*, *413*
Waldner, H., 193, *206*
Walker, E., 226, *267*
Walker, G. J., 427, *439*
Walker, P. M. B., 465, *489*
Wallace, P. G., 221, 262, *267*
Wallenfels, K., 427, *437*
Walter, M., 502, *514*
Walton, G. M., 302, *303*
Wang, C. H., 9, *72*
Wang, M. C., 453, *491*
Wang, S. F., 391, *415*
Warburg, O., 279, 301, *306*, *307*
Ward, G. E., 171, *209*
Ward, P. F. V., 274, *307*
Warren, W. A., 397, *417*
Warrington, R. C., 392, *411*
Wartenberg, H., 18, *64*
Wase, D. A. J., 103, 104, *118*
Watanabe, Y., 365, *366*, 367, *417*
Watkinson, I. A., 246, *264*
Watson, J. D., 396, *411*
Watson, K., 263, *270*
Watson, P. R., 155, 178, 183, 184, *205*
Wawzonek, S., 34, *73*
Weaver, R. H., 474, 478, *490*
Webb, A. D., 58, *73*
Webb, N. L., 393, *410*
Weber, H., 392, *409*
Webley, D. M., 142, 145, 163, 164, 176, 189, 190, 191, *203*, *205*, *209*, 451, 452, 457, *487*
Weedon, B. C. L., 494, *515*
Weedon, B. L., 503, *513*
Wegman, J., 385, 386, *417*
Wehr, C. T., *332*
Weickmann, A., 224, *268*
Weidenhagen, R., 7, 55, *73*
Weigert, W., 27, *64*
Weil, J. H., 320, *330*, 394, 395, 396, *409*, *411*
Weil, J. M., 394, *410*
Weil, J. W., 394, 396, *411*
Weiler, P. G., 361, *417*
Weinfurtner, F., 20, 22, 23, 24, 25, 26, 32, *73*
Weinhold, G., 14, *62*
Weinhouse, S., 9, *60*, 282, *303*, 358, *416*
Weinstein, D. B., 212, *267*
Weinstock, H. H., Jr, 22, *73*
Weiss, G. B., 322, *331*
Wells, W. W., 241, *270*
Welsh, K., 233, *270*
Werkman, C. H., 10, *73*, 216, *270*

Werner, H., 252, *270*
Wertheimer, E., 39, *73*
Wessels, J. G. H., 141, *209*
Westenbrink, H. G. K., 24, 56, *73*
Westhead, E. W., 277, *307*
Westphal, H., 317, *332*
Whalley, H. C. S., de, 28, *73*
Wheeldon, L. W., 26, *68*, 230, *268*
Whelan, W. J., 138, 139, 146, *207*, 425, 427, 428, 433, *437*, *438*, *439*
Whiffen, A. J., 401, *417*
Whipple, M. B., 372, *414*
White, A. G. C., 10, *73*, 216, *270*
White, D., 242, *270*
White, J., 10, 17, 18, 19, 20, 22, 23, 28, 29, 30, 31, *67*, *72*, *73*, 122, 131, *134*
Whitehead, E., 340, 341, 343, 350, 352, *412*, *417*
Whiteley, H. R., 379, *417*
Whitlock, H. W., 246, *269*
Whittle, K. J., 223, *265*
Wiamie, J. M., 15, 16, 27, 46, *63*, *73*, 343, 344, 345, 346, 354, 355, 356, 358, *409*, *412*, *414*, *415*
Wiaux, A. L., 212, 214, *268*
Wickerham, L. J., 10, *73*, 171, 178, 179, 180, 184, 202, *204*, *209*, 232, 233, 234, *269*, *270*, 495, 500, *515*
Widdowson, D. A., 226, 248, *265*
Wiebers, J. L., 365, 368, 369, 373, 375, *417*
Wieland, H., 225, 226, *270*
Wieland, O., 234, *270*
Wiemken, A., 263, 264, *268*
Wiken, T., 9, *73*
Wilbrandt, W., 84, *117*
Wilbur, K. M., 465, *489*
Wilde, P. F., 230, *270*
Wiles, A. E., 24, *73*
Wiley, A. J., 33, *64*
Wilkes, B. G., 5, 6, *73*
Wilkie, D., 262, *268*, 314, 325, 326, *332*, *333*
Wilkins, P. O., 40, 41, *61*, *73*
Williams, L. G., 343, *417*
Williams, O. B., 498, *514*
Williams, P. O., 125, *133*
Williams, R. J., 20, 22, 23, *73*
Williams, W. L., 22, *60*
Williamson, D. H., 36, *62*, 76, *118*, 187, 194, *204*, 255, *265*, 311, 312, 315, 324, 326, *330*, *333*, 462, 465, 468, *488*, *491*



Willstaedt, E., 55, *67*
Willstätter, R., 6, *73*
Wilson, A. W., 136, *208*, 501, 512, *515*
Wilson, D. C., 120, 123, 130, *133*
Wilson, L. G., 365, 372, *409*, *417*
Wilson, T. G. G., 11, *69*
Wilton, D. C., 246, *264*
Wimmer, E., 320, *333*, 392, *417*
Winderman, S., 7, *63*
Winge, Ö., 7, *73*
Winnick, T., 20, 21, *64*, *68*
Winstein, W. A., 322, *333*
Wintersberger, E., 326, *333*
Winterstein, A., 502, 503, 511, *514*, *515*
Winzler, R. J., 20, *73*
Witt, I., 360, *417*
Wittmann, H., 495, 496, 500, 502, *515*
Wolfrom, M. L., 28, 29, *60*
Wood, T. H., 97, *118*
Wood, W. A., 136, *209*
Woodbine, M., 216, *270*
Woodhead, J. S., 142, *204*
Woof, J. B., 14, *68*
Woolley, D. W., 23, *73*
Wormser, E. H., 367, *417*
Wratten, C. C., 277, *307*
Wright, A., 223, *270*, 424, 436, *437*, *438*
Wright, B. E., 468, *491*
Wright, L. D., 20, 21, *73*
Wright, N. G., 364, 365, 366, *409*
Wulff, D. L., 314, *333*
Wustenfeld, D., 447, *490*
Wüster, K. H., 277, 279, *304*
Wyman, J., 377, *414*

## Y

Yall, I., 376, *412*
Yamada, K., 34, *73*, 252, *268*, *270*
Yamada, T., 392, *409*
Yamamoto, A., 212, *268*
Yamamoto, H. Y., 512, *515*
Yamomato, T., 323, *333*
Yamomato, Y., 58, *64*
Yamane, T., 322, *333*
Yamataka, A., 16, *74*
Yanagishima, N., 200, *208*, *209*, 480, 483, *491*
Yang, J. T., 319, 320, *331*
Yanguishima, N., 325, *330*
Yaniv, M., 393, *414*
Yarus, M., 391, *417*

Ycas, M., 317, *329*, *333*, 396, *417*
Yokoyama, H., 501, 502, 503, 511, *513*
Yoshida, A., 16, 27, *74*
Yoshida, F., 252, *268*
Yoshida, M., 321, *333*
Yoshida, H., 394, 396, *417*
Yoshida, M., 394, *417*
Yoshida, N., 197, *204*
Yotsuyanagi, Y., 262, *266*, *270*, 301, *305*, 324, 326, *329*, *333*
Young, W. J., 425, *438*
Yphantis, D. A., 38, 39, 59, *74*
Yu, R. J., 141, 144, 150, *209*
Yuan, C., 242, *270*

Z

Zabin, I., 251, *270*
Zachau, H. G., 321, 322, *333*, 392, 394, 396, *411*, *417*
Zahn, R. K., 312, *329*
Zaitseva, G. N., 82, *117*
Zamecnick, P. C., 321, *331*
Zamenhof, S., 310, 311, 313, *332*, *333*
Zamir, A., 310, 319, 320, 321, *329*, 392, 393, *412*
Zechmeister, L., 137, *209*, 495, 500, 503, 506, 507, *513*, *514*
Zelles, L., 56, *65*, 253, *267*
Zemek, J., 8, *62*
Zemplén, G., 8, *74*
Zerahn, K., 19, *67*
Zerban, F. W., 5, *74*
Zimmermann, F. K., 325, *331*, 379, 380, 406, *412*, *417*
Zink, M., 361, *409*
Zofesik, W., 224, 225, *270*
Zopf, W., 504, *515*
Zsolt, J., 511, *515*
Zvjagilskaja, R. A., 188, *209*

# Subject Index

## A


Ability of yeasts to grow at 37°, 124
Acetaldehyde, active, 286
Acetaldehyde excretion by yeasts, 58
Acetic acid, action of *Saccharomyces cerevisiae* on, 10
Acetic acid uptake of by yeasts, 47
Acetohydroxy acid synthetase, 380
Acetolysis of yeast wall mannan, 148
Acetylated phosphogalactans produced by yeasts, 180
Acetylated polysaccharides produced by *Cryptococcus* spp., 184
Acetyl-CoA carboxylase in yeasts, 239
N-Acetylglucosamine as a link between mannan and protein in the yeast wall, 164
Acetylornithinase, 355
Acidic heteropolysaccharides, biosynthesis of, 184
  production of by yeasts, 182
Acid phosphatase in yeast walls, 171
Acid phosphatase, location of in yeast cell, 56, 462
Aconitase, 360, 361
Aconitic acid in molasses, 28
Actinomycetes as a source of glucanases, 192
Active dried yeast, permeability of, 59
Active transport of sugars into yeasts, 84
Active uptake of sugars by yeasts, 41
Acyl carrier protein in fatty acid synthesis, 241
Acyl-CoA dehydrogenase, 295, 296
Adenine as a source of nitrogen, 14
Adenine in deoxyribonucleic acids, 310
Adenine, transport of into *Candida utilis*, 55
Adenosine monophosphate pyrophosphorylase, 347
Adenosine triphosphatase activity of yeast plasma membranes, 261
Adenosine triphosphatase activity of yeasts, 57
Adenosine triphosphatase in yeast walls, 170
Adenosine triphosphate as energy carrier for yeast metabolism, 272
Adenosine triphosphate content in yeast, 302
Adenosine triphosphate pool in yeasts, effect of growth rate on, 102
  role of in cation uptake by yeasts, 51
Adenosine triphosphate sulphurylase, 372, 373
Adenosine triphosphate yields by yeasts, 106
S-Adenosylmethionine in sterol biosynthesis, 248
S-Adenosylmethionine, metabolic roles of, 376
Adenylate kinase, 302
Adenylosuccinase, 339
Advantages of *Saccharomyces cerevisiae* as an experimental organism, 2
Aeration, effect on pigment formation, 498
Aerobic conditions, substrates used under, 9
Aerobic metabolism, 286–290
  of yeasts, 9
Agar gels, use of in cell-wall regeneration, 196
Ageing of yeast cells, 470–472
Alanine aminotransferase, 366
$\beta$-Alanine, ability of to replace pantothenate requirement of yeasts, 22
  utilization of by yeasts, 12

Alcohol dehydrogenase, 277, 279, 281, 282, 398
Alcoholic fermentation, 274–281
  energy balance, 281
Aldehyde dehydrogenase, 282
Aldolase, 277, 278
Aldonic acids in sulphite waste liquor, 33
Alkali extraction of yeast walls, 168
Alkali metals, effect of on yeasts, 18
Alkanes, utilization of by yeasts, 34
Alkenes, utilization of by yeasts, 34
Alkenyl ethers in yeast lipids, 218
Alkylation in biosynthesis of yeast sterols, 248
Allantoin as a source of nitrogen, 5, 14
Allosteric control of phosphofructokinase, 300, 303
Allulose, action of yeast on, 5
Amido nitrogen, utilization of by yeasts, 11
Amino acid excretion by yeasts, 58
Amino acids, aromatic, biosynthesis, 383, 385
  as energy source for yeasts, 292
  biosynthesis of, 348–387
  excretion of, 14
  in molasses, 28
  in yeast wall proteins, 165
  uptake of by yeasts, 45
  utilization of by yeasts, 11
Aminoacetyl-t-RNA synthetases, 388–392, 394
β-Aminobutyric acid, utilization of by yeasts, 12
Aminopeptidase production by yeasts, 56
Ammonia, permeability of in yeast, 36
Ammonium ion uptake in yeasts, 47
Ammonium ions, uptake of by yeasts, 51
Ammonium salts, utilization of by yeasts, 10
Amylase, 293
α-Amylase, 426, 428, 431, 432
  in yeast walls, 171
β-Amylase, 427, 428, 429, 431, 432
Amylo-1,6-glycosidase in yeasts, 294, 427, 434
Amylopectin, 427
  6-glucanohydrolase, 427
Anaerobic conditions, factors required under for yeast growth, 26
Anaerobic growth, effect of on lipid composition of yeasts, 222
  effect of on lipid synthesis, 229
Anaerobic growth of yeast, metabolic pathways of glucose metabolism, 283–285
Anaerobic metabolism, 273–285
Anaerobiosis, need for nicotinic acid under conditions of, 24
Analysis of plasma membranes in yeast, 255
Anaplerotic pathways, 289, 290
Antarctic, psychrophilic yeasts found in, 120
Anthranilate synthetase, 386
Antibiotics, effect on yeasts, 402
Anticodon, participation in recognition of t-RNA, 394
Antigen, mannan as in yeast walls, 158
Arabinomannans produced extracellularly by yeasts, 181
Arabinose, counter transport of, 40
  transport of, 44
Arctic, psychrophilic yeasts found in, 120
Arginase, 356
Arginine biosynthesis, 353–356
Arginine permease in yeasts, 46
Argininosuccinase, 398
Arginyl-t-RNA synthetase, 389
Aromatic hydrocarbons, utilization of by yeasts, 34
Arrhenius plots of maximum specific growth rates of respiration-deficient *Saccharomyces cerevisiae*, 92
Arrhenius relations in yeast growth, 91
Arsenate ions, uptake of, 53
Aryl-β-glucosidase in yeast walls, 171
Asci, yeast, susceptibility to enzymic attack, 188
Ascosterol in yeasts, 225
Ash content of molasses, 29
*Ashbya gossypii*, action of enzymes on glucan from walls of, 144
Asparagine, utilization of by yeasts, 11
Aspartate aminotransferase, 366
Aspartate semialdehyde dehydrogenase, 366, 387
Aspartate transcarbamylase, 346, 398

Aspartic acid, utilization of by yeasts, 11
Aspartokinase, 366, 398
*Aureobasidium pullulans*, production of extracellular lipids by, 232
proteolytic activity of, 56
starch production by, 176
Auxin, 483
Auxin-responsive mutants of yeast, 200

B

*Bacillus circulans*, as a source of, enzymes for wall lysis, 190
as a source of glucanase, 143
as a source of mannanase, 158
as a source of phosphomannanase, 179
Bacteria, polyprenols, in, 223
Baker's yeast, branching enzyme in, 433
carotenoid synthesis, 507
decarboxylation of $\alpha$-keto acids by, 37
glucose-transferring enzyme in, 431
glycogen in, 424
lipids in, 295, 452
reserve carbohydrate content, 292
structure of glycogen in, 428, 429
trehalose in, 421, 423
trehalose phosphate phosphatase in, 422
t-RNA in, 392
Balanced growth of yeasts, 79
Barium in yeasts, 17
Barium ions, uptake of, 52
Betaine in molasses, 28
Binary fission, 460–462
Biochemical basis of cardinal temperatures for growth, 124
Biochemistry of capsular polysaccharides, 174
Biochemistry of yeast cell walls, 137
Biocytin, as a replacement for biotin in yeasts, 21
Biogenesis of mitochondria in yeasts, 263
Biomass, production of yeast, 81
Biosynthesis of lipids in yeasts, 234
Biosynthesis of side chain in yeast sterols, 248
Biosynthesis of sterols in yeasts, 243
Biosynthesis of the yeast cell envelope, 135
Biosynthesis of yeast wall mannans, 160
Biotin content of molasses, 28
Biotin deficiency, effect of on lipid composition of yeasts, 231
Biotin, metabolic role of in yeasts, 22
need of by yeasts, 20
sulphoxide, use of by yeasts, 20
Birth scars, 455, 456
Bivalent cations, uptake of by yeasts, 52
Blackstrap molasses for yeast growth, 28
*Blastomyces dermatitidis*, lipid content of, 213
Boron, effect of on yeasts, 19
Branched nature of yeast cell-wall glucan, 138
Branched nature of yeast wall mannan, 148
Branching enzyme, 431, 432, 433, 434
Branching, extent of in yeast cell-wall glucan, 141
*Brettanomyces anomalus*, mannan from walls of, 152
*Brettanomyces bruxellensis*, action of on malt wort, 31
Brewer's yeast, 423, 432, 436
lipids in, 295
phosphatase in, 422
reserve carbohydrate content, 292
structure of glycogen in, 429, 430
trehalose in, 421
t-RNA in, 392
Bud abscission, 472, 473
Bud extrusion in *Candida albicans*, 476
Bud initiation, 467–470
Bud scars, 455, 456, 457
composition of, 199
yeast, chitin in, 163
Budding, bipolar, 460
multipolar, 456–460
*Bullera alba*, production of starch by, 176
Butane-1,3-diol in yeast lipids, 218
Butanol, effect of on yeast plasma membrane, 59
Butyric acid, uptake of by yeasts, 47

## C


Caesium, effect of on yeasts, 18
Calcium ions, uptake of, 52
Calcium, need for by yeasts, 18
Canavanine resistance in yeasts, 47
*Candida albicans*, bud formation, 469
*Candida albicans* cell walls, 445, 450, 451, 452
*Candida albicans*, chitin in walls of, 164
  deoxyribonucleic acids in, 315
  filamentous cells, 474–477, 479, 480
  mannan as an antigen in walls of, 158
  mannan from walls of, 149
  mannan-protein complexes in walls of, 166
  mycelial forms, 327
  structure of wall glucan in, 141
*Candida atmosphaerica*, mannan in walls of, 153
*Candida bogoriensis*, extracellular polysaccharides of, 186
  phospholipids in, 219
  production of extracellular lipids by, 233
  production of extracellular polysaccharides by, 185
*Candida bovina*, utilization of choline by, 254
*Candida buffonii*, extracellular polysaccharides of, 186
*Candida curvata*, production of starch by, 176
*Candida cylindricaceae*, lipase production by, 252
*Candida diffluens*, production of heteropolysaccharides by, 186
*Candida foliarum*, heteropolysaccharide production by, 186
*Candida frigida*, cardinal growth temperatures of, 123
*Candida gelida*, cardinal growth temperatures of, 123
*Candida*, glucan and mannan in cell walls of, 446
  taxonomy in relation to deoxyribonucleic acid, 313
  uptake of hydrocarbons, 292
  uptake of sugars by species of, 44
*Candida guilliermondii*, effect of iron deficiency on, 19
*Candida humicola*, lipase production by, 252
  production of heteropolysaccharides by, 186
  starch production by, 176
*Candida javanica*, extracellular polysaccharides of, 186
*Candida lipolytica*, cardiolipin in, 221
  effect of growth temperature on lipid content of, 215
  effect of growth temperature on synthesis of unsaturated lipids by, 229
  effect of temperature on lipid synthesis in, 132
  galactomannan in walls of, 156
  growth of on hydrocarbons, 33
  lipid content of, 213
  lipolysis by, 252
  phospholipids in, 19
  proteolytic activity of, 56
*Candida macedoniensis*, cardinal growth temperatures of, 122
*Candida parapsilosis*, cardinal growth temperatures of, 122
  mannan from walls of, 150
  structure of wall glucan in, 141
*Candida pulcherrima*, fat production by, 27
  sphaeroplast production from, 189
*Candida reukaufii*, phospholipids in, 219
*Candida scottii*, cardinal growth temperatures of, 123
  cardiolipin in, 221
  lipid content of, 213
  production of heteropolysaccharides by, 186
*Candida slooffii*, cardinal growth temperatures of, 122
*Candida* spp., excretion of lipids by, 174
  psychrophilic, cardinal growth temperatures of, 120
  use of glycerol as a carbon source by, 234
*Candida stellatoidea*, mannan from walls of, 150, 154
*Candida tropicalis*, growth of on hydrocarbons, 33
  mannan from walls of, 150
*Candida utilis*, 355
  aerobic metabolism of sugars in, 9
  amino acid pools in, 45

bud formation, 468, 469
cardinal growth temperatures of, 122
cell-wall regeneration by, 195
cerebrosides in, 224
chitin production by regenerating sphaeroplasts of, 261
composition of plasma membrane from, 256
effect of growth temperature on synthesis of unsaturated lipids by, 229
effect of sodium chloride on lipid content of, 215
fatty acid synthesis in, 242
glucosamine accumulation by, 261
growth of on sulphite waste liquor, 33
kinetics of exponential growth of on glycerol, 90
oxygen-limited growth of, 91
production of sphaeroplasts from, 187
protoplasts, 448, 463
sulphur sources for, 16
synthesis of long-chain fatty acids in, 251
total lipid content of, 215
transport of nucleic acid bases into, 55
t-RNA in, 392
uptake of proteins by, 38
utilization of nitrates by, 11
utilization of pentoses by, 10
utilization of xylose by, 290
Cane molasses as a source of yeast nutrients, 28
Capsular polysaccharide of *Cryptococcus laurentii*, structure of, 183
Capsular polysaccharide of *Cryptococcus neoformans*, structure of, 183
Capsular polysaccharides, chemistry of, 174
Carbamyl phosphate synthetases, 343, 344, 346, 355
Carbohydrate-protein complexes in walls of yeasts, 164
Carbohydrate reserves, content in yeast, 419
Carbohydrates, reserve, 292
reserve, in yeasts, 26
Carbon dioxide concentration, effect of on lipid content of yeasts, 215
Carbon sources for carotenoid formation, 495, 496
Carbon sources for yeasts, 4
Cardinal growth temperatures of yeasts, 120
Cardiolipin in intracellular membranes, 228
Cardiolipin in yeast lipids, 220
α-Carotene, 503, 510, 511
β-Carotene, 494, 497, 498, 499, 502, 503, 504, 506, 507, 508, 509, 510, 511
γ-Carotene, 498, 499, 502, 503, 506, 507, 508, 509, 510
ζ-Carotene, 502, 503, 506, 507, 508, 510
Carotene precursors, formation of, 504
Carotenoid biosynthesis, 504–512
specific inhibitors of, 511, 512
stereochemistry of, 512
Carotenoid pigments, 493–513
extraction from yeast, 499–501
types of, 504–511
Carotenoid synthesis as function of time of growth, 499
Carotenoids, absorption data, 503
cultural conditions for production of, 494–499
cyclic, 508
identification, 501–503
Carriers in sugar uptake by yeasts, 38
Catabolism of lipids in yeasts, 252
Catalase, 465
in yeast walls, 171
production by yeasts, 57
Catechol, utilization of by yeasts, 34
Cations and yeast plasma membranes, 260
Cations, monovalent, uptake of by yeasts, 48
Cell composition, yeast, effect of temperature on, 132
Cell division, biochemical aspects, 464–473
Cell envelope, yeast, structure and biosynthesis of, 135
Cell membrane of yeast, 254
Cell size of yeasts, variation of with growth rate, 85
Cell surface, enzymes located at the, 55
Cell volume, effect of temperature on in *Candida utilis*, 132
Cell wall as a barrier in solute uptake by yeasts, 36
Cell-wall biochemistry, morphogenetic aspects, 444–463

Cell-wall biogenesis, 453–463
Cell-wall composition, 444–452
as a basis for differentiating yeasts, 200
Cell-wall construction in growth, 454–462
Cell-wall proteins, amino acids in, 165
Cell-wall regeneration, 194
by yeast sphaeroplasts, 261
Cell-wall softening enzymes, 485
Cell-wall synthesis by yeast plasma membranes, 261
Cell walls, regeneration of from sphaeroplasts, 193
yeast enzymic lysis of, 186
yeast preparation of, 136
protein content of, 164
Cellobiose, fermentation of by yeasts, 8
location of in yeast cell, 55
Cells, disruption of yeast, 501
Cellulases in snail juice, 189
Cerebrosides, extracellular, produced by yeasts, 232
nature of, 223
Chain length, importance of in utilization of hydrocarbons by yeasts, 34
Chain length of fatty acids, regulation of in yeasts, 241
Chain-lengthening reaction in yeast fatty-acid synthesis, 239
Chemical structure of yeast wall mannans, 145
Chemistry of capsular polysaccharides, 174
Chemostat, use of with yeasts, 80
Chitin in yeast walls, 162
Chitin, location of in yeast wall, 199
Chitin production by regenerating sphaeroplasts of *Candida utilis*, 261
Chitinase in snail juice, 189
Choline, utilization of by yeasts, 253
Chorismate biosynthesis, 384–386
Chorismate mutase, 387
Chromium in yeasts, 17
*Citeromyces matritensis*, mannan from walls of, 152
Citrate, effect of on glyceride synthesis in yeast, 242
Citric acid cycle, 286–289, 296
Cobalt in yeasts, 17
Cobalt ions, uptake of, 52
Codecarboxylase, thiamin and, 24
Codon-anticodon interaction in amino-acid transfer, 394, 395
Codons specifying for amino acids, 388
Combined sterols in yeasts, 226
Compartmentalization, 408
chemical in yeast cells, 284
mitochondrial, of citric-acid cycle, 299
Competitive inhibition of yeast growth, 87
Composition, cellular, effect of temperature on, 132
Composition of yeast cell walls, 137
Composition of yeast lipids, 216
Composition of yeast membranes, 254
Composition of yeast vacuoles, 263
Concurrent exponential death in yeasts, 96
Conjugation tube, formation of, 483
Continuous cultures of yeasts, 80
Copper in yeasts, 17
Copper ions, uptake of, 52
Counter transport of sugars, 40
Coupling in lipid synthesis in yeasts, 243
Crabtree effect, 301
Cryptic catalase in yeasts, 57
*Cryptococcus*, 313, 507
pigment formation in, 494
*Cryptococcus* sp., biochemical basis for maximum growth temperature of, 126
biochemical basis of optimum temperature for growth of, 126
excretion of lipids by, 174
production of extracellular acidic heteropolysaccharides by, 182
psychrophilic, 120
use of inositol as carbon source by, 253
*Cryptococcus albidus*, chitin fibrils in walls of, 164
structure of wall glucan of, 142
*Cryptococcus laurentii*, extracellular polysaccharide production by, 175
*Cryptococcus laurentii* var. *flavescens*, 498
*Cryptococcus neoformans*, production of starch by, 175

*Cryptococcus terreus*, chitin fibrils in walls of, 164
structure of wall glucan in, 142
*Cryptococcus terricolus*, lipid content of, 214
Culture age, effect of on lipid content of *Candida utilis*, 215
Cyclization of squalene in yeasts, 245
Cycloheximide, action on RNA synthesis, 406, 407
blocking protein synthesis, 316
use of to prevent cell-wall regeneration, 197
Cyclopropane acids, incorporation of into yeasts, 230
β-Cystathionase, 375
Cystathionine β-synthetase, 375
Cystathionine γ-synthetase, 374, 375
Cysteine, acceleration of yeast sphaeroplast formation by, 187
Cysteine biosynthesis, 362–376
Cysteine-homocysteine transulphuration pathway, 373–375
Cysteine synthetase, 373
Cysteine utilization by yeasts, 16
Cysteinyl-t-RNA synthetase, 389, 390
Cytochrome oxidase, 361
biosynthesis, 367–368
*Cytophaga johnsonii* as a source of enzymes for wall lysis, 189
Cytoplasmic inheritance of mitochondrial characters, 310, 325
Cytoplasmic membrane of yeast, 254
Cytosine as a source of nitrogen, 14
Cytosine in deoxyribonucleic acids, 310

D

Death, concurrent exponential in yeasts, 96
*Debaryomyces globosus*, thermal death times of, 129
*Debaryomyces hansenii*, action of enzymes on glucan from walls of, 144
cardinal growth temperatures of, 122
*Debaryomyces subglobosus*, kinetics of growth of under conditions of phenol limitation, 105
Debranching enzyme in *Saccharomyces cerevisiae*, 293
Debranching enzymes, action on glycogen, 427
Decarboxylation of α-keto acids by yeasts, 37
Decarboxylation of α-ketoglutarate in yeast, 38
Deficiencies, vitamin, effect of on lipid composition of yeasts, 231
Dehydroergosterol in yeasts, 225
Dehydrosqualene in sterol biosynthesis in yeasts, 244
*Dematium pullulans*, starch production by, 176
3-Deoxy-D-arabinoheptulosonate 7-phosphate aldolases, 385, 386
Deoxyglucose, effects of on cell-wall regeneration, 197
Deoxyribonuclease, 313
Deoxyribonucleic acid, circular, 312
heterogeneity in yeast, 311
mitochondrial, 312, 324–326
nuclear, composition and structure, 312
structure, 311
Deoxyribonucleic acid content of yeast, effect of growth rate on, 101
Deoxyribonucleic acid polymerase, 398, 402, 403
Deoxyribonucleic acid replication, 315
Deoxyribonucleic acid synthesis, 402, 403
Deoxyribonucleic acids, 310–316
effect of mutagens and radiation in yeasts, 313, 314
in relation to taxonomy of yeasts, 313
isolation from yeast, 310, 311
Deoxyribonucleic acids and chromosomes, 314, 315
Dependence of specific growth rate on concentration of limiting nutrient, 82
Dependence of yeast specific growth rate on temperature, 91
Desaturation of fatty acids in yeasts, 242
Desthiobiotin in yeasts, 20
Detergents, effect of on yeast plasma membrane, 59
α-Dextrin 6-glucanohydrolase, 427
Diacetyl production by yeasts, 58
Diacylglycerols in yeast lipids, 218

Diacylglycerols, reactions leading to synthesis of in yeasts, 235
Diaminopelargonic acid, effect of on biotin requirement of yeasts, 20
Differentiation of yeasts, cell-wall composition as a basis for, 200
Diffusion transfer into yeasts, 89
Digitonides, use of in extracting sterols, 225
Dihydric alcohols, diesters of in *Lipomyces starkeyi*, 217
$N_2$-Dimethylguanine methylating enzyme, 408
Dimorphism of mycelial yeasts, 474, 480
Diols in yeast lipids, 217
Dipeptides, utilization of by yeasts, 13
Diphenylamine, effect on carotenoid formation, 511, 512
Diphosphatidylglycerol in yeast lipids, 220
Direct fermentation of maltose by yeasts, 6
Disaccharides, direct and indirect fermentation of, 290
  uptake of, 44
Distribution of lipids in yeast cells, 227
Disulphide reductase in yeast wall biosynthesis, 167
Dolichols in yeast lipids, 223
Dolichols, yeast, biosynthesis of, 250
Dough as a source of yeast nutrients, 30
Dried yeast cells, permeability of, 59
Drying of yeast, effect on cell-wall permeability, 425
Dulcitol as a storage product in yeast, 27

## E

Effect of growth conditions on lipid composition of yeasts, 228
Effect of subzero temperatures on yeasts, 130
Efficiency of oxidative phosphorylation in yeast, 108
Elements, mineral, need for by yeasts, 17
Embden-Meyerhof pathway in yeasts, 9
Endogenous metabolism, 296
Endoglucanases, action of on yeast cell-wall glucans, 144
*Endomyces*, mycelial growth form, 446
*Endomyces decipiens*, structure of wall glucan of, 142
*Endomyces magnusii*, 456
  sphaeroplast production from, 188, 192
*Endomyces* spp., lipid production by, 216
*Endomycopsis*, mycelial growth form, 446
*Endomycopsis capsularis*, chitin in, 162
*Endomycopsis fibuligera*, α-amylase in walls of, 171
  chitin in, 162
  structure of mannan from walls of, 157
Energetics of yeast growth, 75
Energy and mass relations in an open continuous culture system, 81
Energy charge, 302, 303
Energy metabolism, regulation of, 297–303
Energy reserves, mobilization of, 292
Energy-rich pyrophosphate bonds, 272
Energy sources, other than glucose, for yeast, 290–292
Enolase, 277, 279
Enoyl hydrase, 295
Envelope, yeast cell, structure and biosynthesis of, 135
Enzymes in yeast walls, 170
Enzymes located at the yeast cell surface, 55
Enzymes of fatty-acid synthesis in yeasts, 239
Enzymic analysis of yeast extracellular phosphomannans, 179
Enzymic lysis of yeast cells, 186
Enzymic studies on yeast cell-wall glucans, 143
Enzymic studies on yeast wall mannans, 157
Epicholestanol as a yeast growth factor, 26
Epiprotein activity, 356
Episterol in yeasts, 225
*Eremascus fertilis*, chitin in, 162
Ergosterol, ability of to stimulate growth of yeasts at high temperatures, 124
  biosynthesis of in yeasts, 246
Ergosterol as a yeast growth factor, 26

Ergosterol biosynthesis in yeasts, regulation of, 250
Ergosterol in yeasts, 225
Ergosterol palmitate in yeasts, 226
Esterified sterols in yeasts, 226
Ethanol excretion by yeasts, 58
Ethanol production, effect of temperature on, 131
Ethanol, utilization by yeasts, 291
Ethylenediamine extraction of yeast walls, 168
Excretion of compounds from yeasts, 57
Excretion of nitrogen-containing compounds, 14
Excretion of succinic acid by yeasts, 50
Exergonic and endogonic reactions, 272
Exo-β-glucanase, 480
Expansion growth of the yeast wall, 200
Experimental organism, *Saccharomyces cerevisiae* as an, 1
Exponential death, concurrent, in yeasts, 96
Exponential growth of yeasts, 79
Extensibility of yeast plasma membranes, 259
Extracellular lipases produced by yeasts, 252
Extracellular lipids synthesized by yeasts, 231
Extraction of lipids, methods for, 212

F

*Fabospora fragilis*, glucanase production by, 170
  inulinase in walls of, 171
Facilitated diffusion, 276
Facilitated diffusion of sugars into yeasts, 84
Facilitated diffusion of sugars in yeasts, 41
Factors required under anaerobic conditions, 26
Facultative psychrophiles, yeast, 121
Fat in yeasts in relation to phosphate availability, 16
Fat reserve in yeasts, 26
Fat yeasts, nature of, 212
Fatty-acid desaturase mutants of *Saccharomyces cerevisiae*, 230
Fatty acid oxidation by yeasts, 254
Fatty acid synthesis in yeast, 238
Fatty acid synthetase in yeasts, 241
Fatty acids, biotin-sparing action of, 20, 239
  oxidation of, 295
  uptake of by yeasts, 47
Fecosterol, in yeasts, 225
Fermentation, definition, 273
  direct and indirect of disaccharides, 290
  effect of temperature on yeast, 131
Fermentation of hexoses, 4
Fermentation of sugars by yeasts, 4
Fermentative activity, effect of temperature on, 127
Ferrous ions, uptake of, 52
Filamentous yeasts, chitin in, 162
Fingerprinting of yeast wall polysaccharides, 201
Fission yeasts, cell-wall composition, 446
Flours as sources of yeast nutrients, 30
Fluoride ions, uptake of, 54
Fluorine, effect of on yeasts, 17
Folic acid and yeast growth, 26
Free energy of glucose oxidation reactions, 272
Free fatty acids in yeast lipids, 222
Freezing, effect of on yeasts, 130
β-Fructofuranosidase activity in yeasts, 172
Fructose, fermentation of, 4
Fructosediphosphatase, 303, 434
β-Fructosidase, production by yeasts, 55, 291
  hydrolysis of sucrose by, 291
Fucopyranosyl residues in yeast polysaccharides, 185
Fucose, transport of, 44
Function of yeast membranes, 254
Furanose sugars, fermentation of, 4
*Fusarium* spp., lytic activity of towards yeast, 191
Fusel oil production by yeasts, 58

G

Galactomannan, extracellular of *Lipomyces starkeyi*, structure of, 181
Glucomannan-protein fractions in yeast walls, 165
Galactomannans in yeast walls, 155

Galactomannans produced extracellularly by yeasts, 180
Galactopyranosyl residues in yeast polysaccharides, 185
Galactose, fermentation of, 5
inducible transport system, 290
Galactose transport in yeasts, 41
Galactosides, fermentation of by yeasts, 8
Gangliosides, nature of, 223
Gastric juice of the snail, use of in sphaeroplast production, 186
Gay-Lussac equation, 275
Gelatin medium, use of in cell-wall regeneration, 194
Genes, correspondence between proteins and, 397–399
Gentiobiose fermentation by yeasts, 8
*Geotrichum lactis*, sphaeroplast formation from, 192
Glucamylase in yeast walls, 171
Glucamylase, location of in yeast cell, 55
Glucan, 445–450
content in yeast cell walls, 446
location of in yeast wall, 199
yeast cell wall, preparation of, 138
structure of, 140
Glucan fibril production during cell-wall regeneration, 195
β-1,3-Glucanase, 483
β-Glucanase production by yeasts, 56
Glucanases, action of on yeast cell-wall glucans, 143
effect of on yeast asci, 188
extracellular, production by yeasts, 170
Glucanases in snail juice, 189
Glucanases produced by *Bacillus circulans*, 190
Glucans, production of by yeasts, 175
wall, chemical nature of, 137
Glucosamine accumulation in *Candida utilis*, 261
Glucosamine in yeast invertase, 6
Glucosamine in yeast walls, 168
Glucosamine transport in yeasts, 125
Glucose concentrations in continuous culture of *Saccharomyces cerevisiae*, 88
Glucose, effect of on invertase synthesis, 172
fermentation of, 4
inhibitory effect of on aerobic metabolism, 9
Glucose in medium, effect of on lipid content of *Candida utilis*, 215
Glucose limitation, effect of on lipid content of yeast walls, 227
Glucose oxidation, effect of temperature on in yeasts, 126
Glucose 6-phosphate dehydrogenase, 283
Glucose phosphate isomerase, 276, 277
Glucose uptake during fermentaton, 276
Glucosidase, 397, 398, 427
α-Glucosidase, location of in yeast cell, 55
α-Glucosides, action of yeasts on, 6
transport of, 45
β-Glucosides, fermentation of by yeasts, 8
Glucosidic linkages, α-(1 → 4)-D-, biosynthesis of, 430–432
α-(1 → 6)-D-, biosynthesis of, 432, 433
Glutamate biosynthesis, 361, 362
Glutamate dehydrogenase, 292, 361, 362, 402
Glutamic-α-ketoadipic transaminase, 360
Glutamic-oxaloacetic transaminase, 360
Glutamine, utilization of by yeasts, 11
Glutamyl-t-RNA synthetase, 391
Glutathione utilization by yeasts, 16
Glutose, action of yeast on, 5
Glyceraldehyde 3-phosphate dehydrogenase, 277, 278, 280
Glyceride synthesis in yeasts, regulation of, 242
Glycerides, biosynthesis of in yeasts, 234
Glycerol as a carbon source for yeasts, 234
Glycerol formation, effect of temperature on, 131
Glycerol formation in yeast, 281–282, 284
Glycerol kinase in yeasts, 235
Glycerol, kinetics of growth of *Candida utilis* on, 90
utilization by yeasts, 291

Glycerol 3-phosphate dehydrogenase, 282, 289, 398
Glycerophosphate oxidase, 289
α-Glycerophosphate shuttle, 288
Glycero-3-phosphoric acid, effect of on glyceride synthesis in yeast, 242
Glycero-3-phosphoric acid in lipid synthesis by yeasts, 234
Glycine, utilization of by yeasts, 12
Glycogen, 293–294, 423–437
  characterization methods, 425–428
  in endogenous metabolism, 296
  methods of isolation, 424, 425
  pools in yeast, 296, 297, 435
  structure in baker's yeast, 428, 429
  structure in brewer's yeast, 429, 430
Glycogen formation by yeasts, 26
Glycogen 6-glucanohydrolase, 427
Glycogen synthetase, 431, 432, 433, 434, 435
Glycogen-uridine diphosphate glucosyl transferase, 430, 431
Glycolipids, excretion of by yeasts, 174
Glycolipids in yeast lipids, 225
Glycolysis, energy gain in, 280, 281
  oscillations in, 300
Glycolytic enzymes, 276–280
Glycolytic pathway, 301, 434
  multi-site control of, 300
Glycosidases, inducible, 290
  surface, 291
Glycosides as energy supply for yeast, 290, 291
Glycylleucine, utilization of by yeasts, 13
Grape must as a source of yeast nutrients, 32
Growth conditions, effect of on lipid composition of yeasts, 228
  effect of on yeast lipid content, 214
Growth, effect of temperature on yeast, 120
  exponential, of yeasts, 79
Growth factors for yeasts, 20
Growth of yeasts, influence of temperature on, 119
Growth rate, effect of on lipid composition of yeasts, 228
  effect of on lipid content of *Saccharomyces cerevisiae*, 215
Growth temperature, effect of on lipid composition of yeasts, 228
  effect of on lipid content of yeasts, 215
Growth, termination of, 470–472
Guanine as a source of nitrogen, 14
Guanine in deoxyribonucleic acids, 310
Guanine, transport of into *Candida utilis*, 55
Guanosine diphosphomannose in mannan biosynthesis in yeasts, 160

## H

*Hanseniaspora uvarum*, protein content of walls of, 165
*Hanseniaspora valbyensis*, effect of pyridoxine deficiency on lipid content of, 216
  glucanase production by, 170
  mannan from walls of, 151
  phosphatidylcholine in, 218
*Hansenula angusta*, serology of, 160
*Hansenula anomala*, action of enzymes on glucan from walls of, 143
  ascospores, 485, 486
  glucanase production by, 170
  lipid production by, 27
  phospholipids in, 219
  sphaeroplast production from, 192
  lipid production by, 27
  uptake of biotin by, 20
*Hansenula capsulata*, phosphomannan production by, 178
*Hansenula ciferrii*, action of glucanases on glucans from walls of, 144
  phytosphingosine production by, 174
  production of extracellular sphingolipids by, 232
*Hansenula*, glucan and mannan in cell walls of, 446
*Hansenula holstii*, extracellular polysaccharide production by, 175
*Hansenula minuta*, phospholipids in, 219
*Hansenula saturnus*, phospholipids in, 219
*Hansenula schneggii*, cell walls of, 448, 451
  elongation of cells, 451, 479
*Hansenula* spp., production of extracellular phosphomannans by, 177

*Hansenula suaveolens*, cardinal growth temperatures of, 122
phospholipids in, 219
*Hansenula valbyensis*, cardinal growth temperatures of, 122
*Hansenula wingei*, phospholipids in, 219
sexual morphogenesis in, 483, 484
Heat, effect of on yeast survival, 127
Heavy metals, effect of on yeasts, 19
*Helix pomatia*, as a source of enzymes for formation of sphaeroplasts, 186
use of to prepare yeast sphaeroplasts, 255
Heteropolysaccharides, acidic, production of by yeasts, 182
Hexokinase, 276, 277, 281, 290, 423, 433, 398, 434
Hexose monophosphate oxidative pathway, 282–283
Hexoses, constitutive transport system, 290
fermentation of, 4
Higher alcohol formation, effect of temperature on, 131
Histidine as a source of nitrogen, 14
Histidine biosynthesis, 342, 348–352
Histidinol dehydrogenase, 398
*Histoplasma capsulatum*, yeast form, 448
Homocysteine biosynthesis, 368–369
Homocysteine methylation process, 369–372
Homocysteine synthetase, 369, 373, 374, 375
Homocysteine transmethylase, 371
Homoserine biosynthesis, 364–366
Homoserine dehydrogenase, 366, 398
Homoserine kinase, 367
Homoserine-*O*-transacetylase, 368, 369
Homoserine-*O*-transacetylase-deficient mutants, 375
Hydrazine, permeability of in yeast, 37
Hydrocarbons, as yeast nutrients, 33
in yeast lipids, 222
utilization by yeasts, 10, 292
yeast, biosynthesis of, 251
Hydrogen ions, uptake of, 49
Hydrolytic enzymes in yeast walls, 170
3-Hydroxyacyl-CoA dehydrogenase, 295
Hydroxy fatty-acid esters of amino alcohols in yeasts, 223
Hydroxymethylglutaryl-CoA in sterol biosynthesis in yeasts, 243
17′-Hydroxytorulene, 502, 503, 509, 510, 511, 512
Hypoxanthine, transport of into *Candida utilis*, 55

## I

Immunochemistry of yeast wall mannans, 158
Indoleglycerol phosphate synthetase, 386
Induction of invertase synthesis in yeast, 171
Industrial sources of yeast nutrients, 28
Inhibited Michaelis-Menten transfer in yeasts, 87
Inhibitors, effect of on specific growth rate of yeasts, 87
Inhibitors of sugar uptake in yeasts, 41
Inorganic anions, uptake of, 53
Inorganic nitrogen, sources of for yeasts, 10
Inosine in yeast t-RNA, 320
Inositol as an effector in yeast lipid synthesis, 243
Inositol as a source of carbon for yeasts, 253
Inositol-containing sphingolipid in yeasts, 225
Inositol deficiency, effect of on lipid composition of *Saccharomyces cerevisiae*, 231
effect of on lipid content of yeasts, 216
Inositol requirement in yeasts, 22
Insoluble cell-wall glucan, nature of, 140
Intracellular lipids in yeasts, composition of, 216
Inulinase in yeast walls, 171
Inulinase production by yeasts, 55
Inulin impermeability in yeasts, 36
Invaginations in the yeast plasma membrane, 38
Invertase, 291
action of, 5

chemical nature of, 5
glycoprotein nature of, 172
internal in yeasts, 173
in yeast walls, 171
location of in yeast cell, 55
location of in yeast wall, 199
polysaccharides associated with in yeast, 167
yeasts, location of, 5
Iodine, effect of on yeasts, 19
Iodoacetate, effect of on sugar transport, 41
β-Ionone, as inhibitor of carotenoid formation, 511
Iron, effect of on yeasts, 19
Isoamylase, 293, 427, 428, 429, 432
Isocitratase, 290
Isocitrate dehydrogenase, 286
feedback control of, 299
Isoleucine biosynthesis, 376–382
Isoleucine-valine, combined pathway, 379–382
Isoleucyl-t-RNA synthetase, 391, 392
Isopentenyl pyrophosphate in sterol biosynthesis in yeasts, 244


K


β-Ketoacyl-CoA thiolase, 295
Killer character in yeasts, 327
Kinetics of sugar uptake by yeasts, 38
Kinetics of yeast growth, 75
*Kloeckera africana*, mannan from walls of, 151
*Kloeckera spiculata*, cardinal growth temperatures of, 122
mannan from walls of, 151
sphaeroplast formation from, 192
*Kloeckera brevis*, mannan from walls of, 151, 153
serology of, 159
*Kloeckera magna*, mannan from walls of, 151


L


*Lactobacillus arabinosus*, specificity of biotin requirement of, 21
*Lactobacillus casei*, specificity of biotin requirement of, 21
Lactose fermentation by yeasts, 8
Lactose, transport by inducible system, 290
Lanosterol, biosynthesis of, 245
Lanosterol in yeasts, 225
Layering in the yeast wall, 198
Lead in yeasts, 17
Leucine biosynthesis, 376–382
Leucyl-t-RNA synthetase, 389
Light, effect on pigment formation, 497, 498
Lignosulphonic acids in sulphite waste liquor, 33
Limit dextrin, 293
Limit dextrinase, 427
Limiting nutrient, dependence of specific growth rate on, 82
Linoleic acid, synthesis of in yeasts, 242
Lipase in snail juice, 189
Lipases in yeasts, 252
Lipid composition, effect of temperature on, 132
Lipid composition of yeasts, effect of growth conditions on, 228
Lipid content of yeast plasma membranes, 255
Lipid content of yeasts, effect of growth conditions on, 214
Lipid droplets in yeasts, 227
Lipid intermediates in mannan biosynthesis in yeasts, 161
Lipid metabolism in yeasts, 234
Lipid reserves, production of by yeasts, 27
Lipid solubility as a factor in solute permeability, 37
Lipids, 295, 296
catabolism of in yeasts, 252
extracellular, synthesized by yeasts, 231
in cell walls, 452
in yeast walls, 167, 173
yeast, 211
Lipolytic enzymes in yeasts, 252
*Lipomyces*, lipids in, 295
*Lipomyces lipofer*, phospholipids in, 219
*Lipomyces* spp., production of starch by, 176
*Lipomyces starkeyi*, extracellular galactomannans produced by, 180
lipid content of, 213
neutral lipids of, 217
phospholipids in, 219

Lithium, effect of on yeasts, 17
Localization of components in the yeast wall, 198
Location of invertase in yeast cells, 172
*Lodderomyces elongisporus*, action of enzymes on glucan from walls of, 144
Logarithmic growth of yeasts, 80
Long-chain fatty acids, biosynthesis of in yeasts, 251
Lycopene, 498, 502, 503, 507, 509, 510
Lycopersene, 506
Lyosome, the yeast vacuole as a, 264
Lysine biosynthesis, 358–361
Lysine permease in *Saccharomyces cerevisiae*, 46
Lysis, enzymic, of yeast cells, 186
Lysis of yeast walls by microbial enzymes, 189
Lysophosphoglycerides in yeast lipids, 221
Lysyl-t-RNA synthetase, 389

M

Magnesium content of cells, effect of growth rate on, 101
Magnesium ions, uptake of, 52
Magnesium, need for by yeasts, 18
Maintenance analogues in yeast growth kinetics, 105
Maintenance analysis in yeast growth kinetics, 108
Maintenance rate analogue, 106
Maintenance relations in yeast growth kinetics, 100
Maintenance yield factor, 103
Malate dehydrogenase, 288, 289
Malate synthase, 290
Malonyl-CoA pathway of fatty-acid synthesis in yeasts, 238
Malt worts as sources of yeast nutrients, 30
Maltase, 398
  location of in yeast cell, 55
Maltose, action of yeast on, 10
  action of yeasts on, 6
  transport by inducible system, 290
  uptake of by yeasts, 44
Maltose permease in yeasts, 45
Maltotriose fermentation by yeasts, 7
Maltotriose, transport of, 45
Manganese, effect of on yeasts, 19
Manganese ions, uptake of, 52
Mannan, 445–450
  content in yeast cell walls, 446
  wall, chemical structure of, 146
    containing phosphate, 152
Mannanases, action of on yeast wall mannans, 157
Mannanases in snail juice, 189
Mannan-protein complexes in yeast walls, 166
Mannan-protein on the outside of yeast wall, 199
Mannans in yeast capsules, 151
Mannans of yeast walls, 145
Mannans, production of by yeasts, 177
  wall, fingerprinting of, 201
  yeast wall, action of enzymes on, 157
    biosynthesis of, 160
    immunochemistry of, 158
Mannose, fermentation of, 4
Mannose in yeast invertase, 6
Mannosidase, action of on yeast wall mannan, 154
Mannosyl peptides in yeast walls, 161
Maximum growth temperature, biochemical basis of, 126
Maximum temperature for growth, kinetic aspects of, 91
Maximum temperature for yeast growth, 120
Mechanism of cation uptake by yeasts, 51
Melibiase, 462
  location of in yeast cell, 55
Melibiase in yeast walls, 171
Melibiose fermentation by yeasts, 8
Melibiose, hydrolysis, 291
Membrane-bound lipase in *Saccharomyces cerevisiae*, 252
Membrane damage as a factor in effect of temperature on yeasts, 127
Membranes, yeast, 211
  yeast, composition, structure and function of, 254
Mercaptoethylamine, acceleration of yeast sphaeroplast formation by, 188
Mercury ions, uptake of, 52
Mercury toxicity for yeasts, 17
Metabolic integration, 302, 303

Metabolic role of inositol in yeasts, 23
Metabolic wheels, turnover in, 284
Metabolism, endogenous, 296
Metabolism of lipids in yeasts, 234
Metabolism of yeasts, influence of temperature on, 119
Metal contents of molasses, 29
Metaphosphate granules in yeasts, 27
Metaphosphate in yeasts, 15
Methionine biosynthesis, 362–376
Methionine uptake in yeasts, 46
Methionine utilization by yeasts, 16
Methionyl-t-RNA synthetase, 389, 391, 392, 393
Methylamine, permeability of in yeast, 36
Methylase, 408
Methylated sterols in yeasts, 225
Methylation in sterol biosynthesis in yeasts, 246
Methylation of yeast cell-wall glucan, 138
Methylation of yeast wall mannan, 146
Methylglucoside, fermentation of by yeasts, 7
Methyl-4a-zymosterol in yeasts, 225
Mevaldyl-CoA in sterol biosynthesis in yeasts, 244
Mevalonate, synthesis of in yeasts, 243
Michaelis constant in yeast growth kinetics, 82
Michaelis-Menten transfer in yeast growth kinetics, 83
Microbial enzymes, lysis of yeast walls by, 189
*Micromonospora* spp., as a source of glucanases, 192
Mineral content of malt worts, 32
Minerals, need for by yeasts, 17
Minimum temperature for growth, biochemical basis of, 125
Minimum temperature for yeast growth, 120
Miscellaneous heteropolysaccharides produced by yeasts, 185
Mitochondria, 286, 288, 289, 301, 311, 312, 402, 469
  cristate, 362
  t-RNA in, 393
  yeast, phospholipids in, 228
Mitochondrial characters, cytoplasmic inheritance of, 310, 326
Mitochondrial membranes, yeast, 262
Mitotic recombination, 315
Mixtures of amino acids, utilization of by yeasts, 12
Mixtures of sugars, fermentation of by yeasts, 8
Model for kinetics of yeast growth, 109
Molasses as a source of yeast nutrients, 28
Molecular nitrogen, possible utilization of by yeasts, 11
Molybdenum in yeasts, 17
*Monilia candida*, thermal death times of, 129
Monoacylglycerols in yeast lipids, 218
Monod growth equation, application to yeasts, 82
Monogalactosyldiacylglycerols in yeast lipids, 225
Monosaccharide uptake by yeasts, 38
Monosaccharides, other than glucose, as energy sources for yeast, 290, 291
Monovalent cations, uptake of by yeasts, 48
Morphogenesis, environmental control of, 473–483
  examples in fungi, 443
Morphology of yeast cell wall, 198
*Mucor rouxii*, cell-wall synthesis, 458, 459
  vegetative morphogenesis, 477, 478
Multi-site control of glycolytic pathway, 300
Must, grape, as a source of yeast nutrients, 32
Mutagens, effects on yeast, 314
Mutants, cytoplasmic *petites*, 324
  temperature-sensitive, 391, 392
Mycoglycolipid in yeast lipids, 224
*Myxococcus fulvus* as a source of enzymes for wall lysis, 189

## N

*Nadsonia elongata*, action of enzymes on glucan from walls of, 144
  cardinal growth temperatures of, 122
  cell-wall regeneration by, 195
  cell walls, 452
  galactomannan from walls of, 155

*Nadsonia fulvescens*, chitin in, 163
galactomannan from walls of, 155
Native enzymes, steady-state concentration of in yeasts, 94
Natural sources of yeast nutrients, 28
Neuberg's forms of fermentation, 282
*Neurospora* sp., invertase in, 173
Neurosporene, 502, 503, 506, 507, 508, 509, 510
Neutral polysaccharides produced by yeasts, 186
Nickel in yeasts, 17
Nickelous ions, effect of on solute transport through yeast plasma membranes, 260
effect of on sugar uptake, 41
uptake of, 52
Nicotinic acid as a yeast growth factor, 24
Nicotinic acid content of molasses, 29
Nitrates, utilization of by yeasts, 10
Nitrogen compounds as energy source for yeasts, 292
Nitrogen sources for carotenoid formation, 496, 497
Nitrogen sources for yeasts, 10
Nitrogen, urea as a source of, 14
Nomenclature of yeast lipids, 217
Non-metabolizable sugars, uptake of by yeasts, 40
Nuclei, yeast, preparation of, 264
Nucleic acid biosyntheiss, 402–409
Nucleic acids, methylation of, 407–409
mitochondrial, 324–326
Nucleotides, excretion of, 14, 58
uptake of by yeasts, 55
Nutrition of yeasts, 3

## O

Obligately psychrophilic yeasts, 120
Oleic acid, synthesis of in yeasts, 242
Oligopeptides, excretion of, 14
Oligosaccharidases in yeast walls, 170
Oligosaccharides, action of yeast on, 5
*Oospora suaveolens*, sphaeroplast formation from, 192
Optical isomers of glucose, fermentation of, 4
Optimum growth temperature, biochemical basis of, 125
Optimum temperature for growth, kinetic aspects of, 91
Optimum temperature for yeast fermentation, 131
Optimum temperatures for yeast growth, 120
Organic acids, uptake of by yeasts, 47
utilization by yeasts, 291
Organic dry matter in sulphite waste liquor, 33
Ornithine transacetylase, 355
Ornithine transaminase, 356, 398
Ornithine transcarbamylase, 355, 356, 398
Orotidine 5′-phosphate decarboxylase, 298
Orthophosphate ions, uptake of, 53
Orthophosphate, utilization of by yeasts, 15
Osmium, toxicity for yeasts, 17
Osmophilic yeasts, production of extracellular lipids by, 233
Osmotic lysis of yeast spheroplasts, 260
Outer layers of the yeast wall, 199
Outer metabolic region of the yeast cell, 36
Outflow of compounds from yeasts, 57
Oxaloacetate cycle, 288
Oxidation of alkanes by yeasts, 34
Oxidation of glycerol by yeasts, 253
Oxidative metabolism of yeasts, 9
Oxidative phosphorylation, 288
efficiency of in yeast, 108
2,3-Oxidosqualene in sterol biosynthesis in yeasts, 246
17′-Oxotorulene, 502, 503, 509, 510, 511, 512
Oxybiotin, as a replacement for biotin in yeasts, 21
Oxygen-limited growth of *Candida utilis*, 91
Oxygen tension, effect of on lipid synthesis, 229

## P

Palladium, toxicity for yeasts, 17
Palmityl-CoA, effect of on fatty-acid synthesis in yeast, 242

Pantothenate content of molasses, 29
Pantothenate deficiency, effect of on lipid content of *Saccharomyces cerevisiae*, 216
Pantothenic acid requirement in yeasts, 22
Papain treatment of yeast walls, 169
*Para*-aminobenzoic acid as a yeast growth factor, 25
*Paracoccidioides brasiliensis*, structure of wall glucan of, 142
yeast form, 448, 450, 451
Pasteur effect, 101, 297–301
Pentose phosphate cycle, 282, 283, 290, 291
Pentoses, fermentation of, 9
Pentosylmannans in yeast walls, 155
Pentosylmannans, production of extracellular by yeasts, 181
Peptidase, 398
Peptides, utilization of by yeasts, 13
Permeases, amino acid, in yeasts, 45
Petite formation by yeasts growing at high temperatures, 124
Petroleum, utilization of by yeasts, 34
*o*-Phenanthroline, effect of on cation uptake by yeasts, 49
Phenol limitation, growth of *Debaryomyces subglobosus* under, 105
Phenylalanine biosynthesis, 382–387
Phenylalanyl-t-RNA synthetase, 389
Phosphatase, acid, 398, 462
alkaline, 398
Phosphatase in brewer's yeast, 422
Phosphate deficiency, effect of on lipid contents of yeasts, 216
Phosphate deficiency in yeasts, effects of, 16
Phosphate groups, anhydride-bound, 302
Phosphate in yeast wall mannan, 152
Phosphate ions, uptake of, 53
Phosphate reserves in yeasts, 26
Phosphate starvation, effect of on synthesis of acid phosphtase in yeasts, 56
Phosphatidic acid in lipid metabolism in yeasts, 236
Phosphatidic acid in yeast lipids, 222
Phosphatidylcholine, biosynthesis of in yeasts, 236
Phosphatidylcholine in intracellular membranes, 228
Phosphatidylcholine in yeast lipids, 218
Phosphatidylethanolamine, biosynthesis of in yeasts, 236
Phosphatidylethanolamine in yeasts, 218
Phosphatidylglycerol, biosynthesis of in yeasts, 238
role in solute transport through yeast plasma membranes, 260
Phosphatidylglycerol in yeast lipids, 221
Phosphatidylglycerol phosphate in sugar transport, 42
Phosphatidylglycerol phosphate in yeast lipids, 221
Phosphatidylinositol, biosynthesis of in yeasts, 238
role of in solute transport in yeast, 261
Phosphatidylinositol in yeast lipids, 220
Phosphatidylserine, biosynthesis of in yeasts, 236
Phosphtidylserine in yeast lipids, 220
Phosphofructokinase, 276, 277, 281, 303, 434, 435
feedback inhibition of, 300, 303
Phosphogalactans, extracellular, structure of, 180
Phosphogalactans produced extracellularly by yeasts, 180
Phosphoglucoisomerase, 434
Phosphoglucomutase, 290, 423, 434
Phosphogluconate dehydrogenase, 283
6-Phosphogluconolactone, 283
Phosphoglycerate kinase, 277, 279, 280
Phosphoglycerate mutase, 277, 279
Phosphoglycopeptide in yeast walls, 155
*O*-Phosphohomoserine mutaphosphatase, 367
Phospholipase activity of yeasts, 56
Phospholipases, activation of in yeast, 214
Phospholipases produced by yeasts, 252
Phospholipids, effect of growth temperature on synthesis of, 229
reactions leading to synthesis of in yeasts, 235
Phospholipids in yeast lipids, 218

Phosphomannan in yeast walls, 152
Phosphomannanase produced by *Bacillus circulans*, 179
Phosphomannans, extracellular, structure of, 178
production of by yeasts, 177
$N'$-(5'-Phosphoribosyl)-adenosine triphosphate pyrophosphorylase, 350, 352, 398
Phosphorus, sources of for yeasts, 15
Phosphorylase, 293, 431, 432, 434
Phosphorylase-amylo-1,6-glucosidase, 436
Phosphorylation of sugars during transport, 42
pH value, effect of on cation uptake by yeasts, 50
effect of on yeast invertase, 6
Phyllosphere, extracellular lipids synthesized by yeasts from the, 231
Physical nature of chitin in the yeast wall, 163
Phytoene, 502, 503, 506, 507, 508, 510, 511
Phytofluene, 502, 503, 506, 507, 510, 511
Phytosphingosine production by *Hansenula ciferrii*, 174
Phytosphingosines in yeast lipids, 224
*Pichia bispora*, serology of, 160
*Pichia farinosa*, 458, 459, 460
*Pichia membranefaciens*, cardinal growth temperatures of, 122
*Pichia pastoris*, mannan from walls of, 152
structure of mannan from walls of, 157
*Pichia polymorpha*, fibrils produced during cell-wall regeneration, 198
*Pichia* spp., production of extracellular phosphomannans by, 177
Pinitol, use of by yeasts, 23
Pinocytosis, possible operation of in yeast, 38
*Pityrosporum ovale*, lipid requirements of, 230
Plasma membrane as a solute barrier in yeasts, 36
composition of, 36
Plasma membrane of yeast, 254
Plasmalemma, 254, 453
Plasmalogens in yeasts, 217
Plectaniaxanthin, 502, 503, 510, 511
Ploidy in yeasts, 315
Polygalacturonase in snail juice, 189
Polynucleotide synthesis, enzymic systems for, 402–405
Polyol fatty-acid esters produced extracellularly by yeasts, 232
Polyols, utilization by yeasts, 291
Polypeptides, utilization of by yeasts, 13
Polyphosphates, content in yeast, 16, 27, 292
Polyprenol kinases, 251
Polyprenols in yeast lipids, 223
Polyprenols, yeast, biosynthesis of, 250
Polysaccharides, endogenous, 480
hydrolysis of, 291
of cell walls, 445–450
Potassium, need for in yeasts, 18
Potassium uptake in yeasts, 47, 48
Preparation of yeast mitochondrial membranes, 262
Preparation of yeast plasma membranes, 258
Preparation of yeast vacuoles, 263
Prephenate dehydrase, 387
Prephenate dehydrogenase, 387
Prephenate, in biosynthesis of phenylalanine, 387
in biosynthesis of tyrosine, 387
PR factor, nature of, 192
Priming reaction in yeast fatty-acid synthesis, 239
Project K, nature of, 2
Project Y, nature of, 2
Proline, assimilation of from wort, 13
Proline biosynthesis, 356–358
Proline transport in *Saccharomyces cerevisiae*, 47
Protease, 398
Protease production by yeasts, 56
Protein and carbohydrate complexes in walls, 164
Protein biosynthesis, 388–402
inhibition of, 401
Protein content of yeast, effect of growth rate on, 101
Protein content of yeast walls, 164
Protein denaturation in yeast, reversible, 93
Protein disulphide reductase, 469, 480

Proteins, correspondence between genes and, 397, 399
Proteins, in cell walls, 450
 uptake of by yeasts, 38
Proteolytic activity in yeasts, 56
Proton magnetic resonance spectra of yeast walls, 201
Protoplast membrane, *see* Plasma membrane
Protoplasts, 310
 cell-wall regeneration, 462, 463
 yeast, *see* Sphaeroplast
*Prototheca* spp., trehalase in walls of, 171
*Prototheca wickerhamii*, trehalase in, 173
Pseudokeratin in yeast cell walls, 167
Pseudouridine in yeast t-RNA, 320, 321
Psychrophilic yeasts, ecology of, 119
 thermolability of, 130
Pullulan, 427
Pullulan production by yeasts, 176
Pullulanase, 427
*Pullularia pullulans*, glucan production by, 176
 lipid content of, 213
Purine biosynthesis, 349
 gene-enzyme relationships, 341
Purine derivatives, biosynthesis of, 338–347
Purines as a source of nitrogen, 14
Purines, utilization by yeasts, 347
Pyranose sugars, fermentation of, 4
Pyridoxal as a yeast growth factor, 25
Pyridoxamine as a yeast growth factor, 25
Pyridoxine as a yeast growth factor, 25
Pyridoxine deficiency, effect of on lipid content of *Hanseniaspora valbyensis*, 216
Pyrimidine biosynthesis, 349
Pyrimidine derivatives, biosynthesis of, 338–347
Pyrimidine moiety of thiamin, effect of on yeasts, 24
Pyrimidines as a source of nitrogen, 14
Pyrimidines, utilization by yeasts, 347
Pyrophosphate bonds, energy rich, 272
Pyrophosphorylase, 434
 t-RNA-CCA-, 396
Pyruvate carboxylase, 289
Pyruvate decarboxylase, 277, 279, 282, 289
Pyruvate utilization by yeasts, 291
Pyruvate kinase, 277, 279, 280, 431
Pyruvic acid, uptake of by yeasts, 47

## R

Raffinose, fermentation of by yeasts, 8
Rate, growth, effect of on lipid content of *Saccharomyces cerevisiae*, 215
Rate of glucose uptake by yeasts, 4
Reconstitution of active dried yeast, 59
Refinery cane molasses for yeast growth, 28
Regeneration of cell walls from sphaeroplasts, 193
Regulation of chain length of fatty acids in yeasts, 241
Regulation of glyceride synthesis in yeasts, 242
Regulation of sterol biosynthesis in yeasts, 250
R-enzyme, 427
Repeating unit in yeast wall mannan, 147
Requirements of yeasts for growth factors, 20
Reserve carbohydrates in yeasts, 26
Residues, cell wall, presence of during sphaeroplast formation, 210
Respiration-deficient mutant of *Saccharomyces cerevisiae*, kinetics of growth of, 87
Respiration, inhibition by fermentation, 301
Resting yeast cells, 297
Reversible heat denaturation of yeast enzymes, 93
Rhamnopyranosyl residues in yeast polysaccharides, 185
Rhamnose uptake by yeasts, 44
*Rhodotorula*, 313
 accumulative transport of monosaccharides in, 276
 cell-wall composition, 446
 lipids in, 295
 pigment formation in, 494
 uptake of hydrocarbons by, 292
 uptake of sugars by species of, 44

*Rhodotorula aurantiaca*, 495, 500, 508, 509, 511
*Rhodotorula glutinis*, 495, 497, 498, 499, 500, 506, 507, 509
cardinal growth temperatures of, 122
cell-wall regeneration by, 195
chitin in, 163
lipid content of, 213
phospholipids in, 219
production of extracellular lipids by, 232
pyruvate uptake by, 291
*Rhodotorula gracilis*, 496, 499, 508, 511, 512
cardinal growth temperatures of, 122
lipid production by, 27, 216
sugar fermentation by, 9
*Rhodotorula graminis*, lipid content of, 213
phospholipids in, 219
production of extracellular lipids by, 232
*Rhodotorula infirmo-miniata*, cardinal growth temperatures of, 123
*Rhodotorula macerans*, production of starch by, 176
*Rhodotorula minuta*, 500
*Rhodotorula mucilaginosa*, 495, 500, 504, 511, 512
*Rhodotorula pallida*, 499, 500, 508
*Rhodotorula rubra*, 495, 496, 500, 504, 506, 511
*Rhodotorula* spp., excretion of lipids by, 174
possible utilization of molecular nitrogen by, 11
production of extracellular mannans by, 177
production of extracellular polysaccharides by, 175
Rhodozonic acid, use of by yeasts, 23
Riboflavin as a yeast growth factor, 25
Ribonuclease, 310
Ribonucleic acid content, effect of temperature on in *Candida utilis*, 132
Ribonucleic acid content of yeast, effect of growth rate on, 101
Ribonucleic acid polymerase, 403–405
Ribonucleic acid synthesis, control of, 322, 323
synthesis of, 403–405
Ribonucleic acids, 316–323
bulk, isolation from yeast, 316, 317
fractionation of, 318, 406
free ribosomal, 400, 401
messenger, 317, 318, 396–399
polycistronic, 407
mitochondrial, 326
rapidly-labelled, 397
ribosomal, 318, 319
transfer, 319–322, 392–396
role in protein synthesis, 393–396
Ribose uptake by yeasts, 44
Ribosomes, 399–401
binding of aminoacyl-t-RNA to, 395, 396
mitochondrial and cytoplasmic, 318, 319
sedimentation values of, 400
Rubidium, effect of on yeasts, 18

## S

*Saccharomyces carlsbergensis*, biosynthesis of mannan by, 160
bud formation, 472
cardinal growth temperatures of, 122
glucosidase in, 397
lipid content of walls of, 227
m- and t-RNA in, 323
mannan from walls of, 149
melibiase in walls of, 171
phospholipids in, 219
protoplasts of, 396
r-RNA biosynthesis in, 406, 407
structure of cell-wall glucan of, 142
*Saccharomyces*, cell-wall composition, 446, 447
glucan and mannan in cell walls of, 446
phosphorylation in, 288
t-RNA from, 322, 323
*Saccharomyces cerevisiae*, 315, 343, 345, 347, 348, 351, 353, 354, 355, 358, 359, 363, 367, 369, 370, 371, 372, 375, 376, 377, 379, 381, 382, 383, 385, 391, 402, 404, 408, 421, 428, 429, 436, 445, 448, 449, 450, 452, 453, 455, 456, 457, 458, 459, 462, 463, 465, 466, 469, 470, 472, 473, 483, 486, 494
acid phosphatase in walls of, 171

adenine synthesis in, 338, 339, 340
aerobic and anerobic states, 327
aerobic metabolism of sugars in, 9
amino-acid transport in, 45
aminopeptidase production by, 56
anaerobic adenosine triphosphate yield by, 107
as an experimental organism, 1
biosynthesis of sterols in, 243
carbohydrate composition, 437
cardinal growth temperatures of, 122
catalase in walls of, 171
cell wall, 419
cell-wall regeneration by, 195
composition of plasma membrane in, 36, 256
debranching enzyme of, 293
density profile of deoxyribonucleic acids in, 313
deoxyribonucleic acid in, 315
effect of freezing on, 131
effect of growth temperature on lipid content of, 215
effect of growth temperature on synthesis of unsaturated lipids by, 229
effect of sodium chloride on lipid content of, 216
effect of temperature on glucose oxidation by, 126
effect of temperature on sporulation by, 128
effect of temperature on sterol synthesis in, 132
effect of temperature on viable count of, 97
elongation of cells, 478–480
enzyme synthesis in, 398, 399
fatty acid desaturase mutants of, 230
fatty-acid synthesis in, 238
for study of developmental biology, 336
gene-enzyme relationships, 340, 344
glucose uptake, 276
glycogen in, 293
glycolytic enzymes, 277
lipid content of, 213
lipid content of walls of, 227
lipid synthesis in, 236
mannan biosynthesis in, 161
m-RNA in, 318
nucleic acid content of in relation to growth rate, 101
phosphatidylcholine in, 218
phosphoglycopeptide in walls of, 155
phospholipase activation in, 214
phospholipases in, 252
phospholipid involvement in sugar transport in, 42
phospholipids in, 219
phosphomannan in walls of, 153
production of sphaeroplasts from, 187
protein content of walls of, 165
regulation of glyceride synthesis in, 242
relation of growth rate to budding frequency, 471
respiration-deficient, Arrhenius plots of maximum specific growth rates of, 92
serology of, 159
specificity of biotin requirement of, 21
structure of wall glucan in, 141
survival curves of exposed to temperature, 96
synchronized cells of, 403
temperature-sensitive mutants, 401
thermal death times of, 129
transport-limited growth of, 86
t-RNA in, 392
vegetative cell cycle of, 464, 467
*Saccharomyces chevalieri*, cell-wall regeneration by, 195
*Saccharomyces diastaticus*, action of on malt worts, 31
glucamylase in walls of, 171
starch fermentation by, 55
*Saccharomyces dobzhanskii*, enzyme synthesis in, 398
*Saccharomyces dobzhanskii* × *Saccharomyces fragilis*, enzyme synthesis in, 398, 399
r-RNA in, 400
*Saccharomyces ellipsoideus*, effect of temperature on sporulation by, 128
response of to auxin, 200
thermal death times of, 129
*Saccharomyces fragilis*, 401, 402
cardinal growth temperatures of, 122
inulinase production by, 55

*Saccharomyces fragilis*—cont.
protoplasts, 453
ribosomes in, 400
sphaeroplast formation by, 188
*Saccharomyces intermedeus*, cardinal growth temperatures of, 122
effect of temperature on sporulation of, 128
*Saccharomyces lactis*, 315, 348, 351
amino-acid content of protein in walls of, 166
cell-wall regeneration by, 195
free r-RNA in, 400
phosphomannan in walls of, 153
replication of mitochondrial DNA, 465
serology of, 159
*Saccharomyces lodderi*, action of mannanase on mannan from walls of, 157
*Saccharomyces ludwigii*, cardinal growth temperatures of, 122
hydrocarbons in lipids of, 223
*Saccharomyces marxianus*, cardinal growth temperatures of, 123
*Saccharomyces mellis*, cardinal growth temperatures of, 123
*Saccharomyces octosporus*, cardinal growth temperatures of, 123
*Saccharomyces odessa*, thermal death times of, 129
*Saccharomyces oviformis*, hydrocarbons in lipids of, 223
lipid content of walls of, 227
*Saccharomyces pastorianus*, 402
cardinal growth temperatures of, 123
effect of temperature on sporulation by, 128
RNA in, 323
*Saccharomyces rouxii*, mannan from walls of, 149
structure of mannan from wall of, 157
*Saccharomyces* spp., effect of temperature on sporulation by, 127
utilization of amino acids by, 11
*Saccharomyces turbidans*, cardinal growth temperatures of, 123
effect of temperature on sporulation by, 128
thermal death times of, 129
*Saccharomyces validus*, cardinal growth temperatures of, 123
effect of temperature on sporulation by, 128
*Saccharomycodes ludwigii*, budding, 455, 460, 461
cell-wall regeneration by, 195
pigment formation in, 494
sphaeroplast formation by, 188
*Saccharomycopsis guttulata*, action of enzymes on glucan from walls of, 143
cardinal growth temperatures of, 122
cell walls, 450
chitin in walls of, 164
Saccharopine dehydrogenase, 398
Saccharopine reductase, 398
*Sarcina lutea*, biosynthesis of hydrocarbons in, 251
Sauterne yeasts, sugar fermentation by, 9
Scars, division, 460
*Schizosaccharomyces octosporus*, absence of chitin in, 162
structure of wall glucan of, 141
*Schizosaccharomyces pombe*, 315, 343, 347, 350, 352, 407, 455, 458, 460, 465, 485, 486, 494
cell-wall composition, 446, 447
cell-wall development, 461
cell-wall regeneration by, 195
enzyme synthesis in, 398, 399
gene-enzyme relationships, 339, 340
m-RNA in, 318
production of sphaeroplasts from, 187
*Schizosaccharomyces* spp., galactomannans in walls of, 156
*Sclerotinia libertiana* as a source of glucanases, 191
Selective fermentation by yeasts, 8
Selenate, effect of on utilization of sulphate by yeasts, 17
Sensitivity coefficient in yeast growth kinetics, 114
Separation, chromatographic, of yeast lipids, 217
Serology of *Cryptococcus neoformans*, 182
Serology of yeast mannans, 159
Seryl-t-RNA synthetase, 389, 390
Side chain in yeast sterols, biosynthesis of, 248
Silver content of yeasts, 17

Simulation experiments with model for yeast growth, 113
Snail juice, enzymic composition of, 189
Snails, garden, as a source of enzymes for preparation of sphaeroplasts, 186
Sodium chloride, effect of on lipid content of yeasts, 215
Sodium, effect of on yeasts, 18
Sodium ion uptake in yeasts, 47
Solute transport through yeast plasma membranes, 260
Solute uptake by yeasts, 3, 36
Solvents used for extracting yeast lipids, 214
Sophorosides of hydroxy fatty acids, production of extracellularly by yeasts, 233
Sorbose transport in yeasts, 41, 125
Sources of carbon for yeasts, 4
Sources of nitrogen for yeasts, 10
Sources of phosphorus for yeasts, 15
Sources of sulphur for yeasts, 16
Species variation in lipid content of yeasts, 212
Specific diffusion coefficient in kinetics of yeast growth, 89
Specific growth rate, dependence of on temperature, 91
Specific growth rate of yeasts, 79
Specific inactivation rates in yeast growth kinetics, 98
Specificity of cation carriers in yeast, 50
Specific maintenance rate and yield factor, 100
Specific maintenance rate of yeasts, 102
Specific transfer rate of nutrient, in kinetics of yeast growth, 83
Sphaeroplast production from yeasts, 186
Sphaeroplasts, 462
Sphaeroplast, yeast, osmotic lysis of, 255
Sphingolipids, biosynthesis of in yeasts, 251
  extracellular, produced by yeasts, 232
Sphingolipids in yeast lipids, 223
Sphingomyelins, nature of, 223
Sphingosine production by yeasts, 174
Sphingosines, biosynthesis of in yeasts, 251
Spirilloxanthin, 507, 510
Spores, yeast, thermostability of, 130
*Sporobolomyces*, cell-wall composition, 446
  pigment formation in, 494
  production of extracellular phosphogalactans by, 180
*Sporobolomyces pararoseus*, 497, 498
*Sporobolomyces roseus*, 495, 496, 497, 498, 508
  chitin in, 163
*Sporobolomyces shibatanus*, 512
*Sporobolomyces* spp., production of extracellular polysaccharides by, 175
  sphaeroplast formation by, 188
Sporulation, hormonal control of, 483
  sexual in yeast, 483–487
Sporulation in yeasts, effect of temperature on, 127
Squalene, biosynthesis of in yeasts, 244
Squalene in yeast lipids, 222
Starch-like polysaccharides, production of by yeasts, 175
Steady-state concentrations of native enzymes in yeasts, 94
Stereospecificity in polyprenol biosynthesis, 250
Sterol biosynthesis, 250
Sterol composition of yeast plasma membranes, 257
Sterol glycosides in yeast lipids, 225
Sterols, biosynthesis of in yeasts, 243
  requirement for under anaerobic conditions, 230
Sterols in yeast lipids, 225
Sterol synthesis, effect of temperature on in *Saccharomyces cerevisiae*, 132
Stratification in the yeast wall, 198
*Streptomyces* spp. as a source of enzymes for wall lysis, 189
Strontium ions, uptake of, 52
Structure of the yeast cell envelope, 135
Structure of yeast membranes, 254
Substituted acids, production of extracellularly by yeasts, 233
Substrate constant, nature of, 82
Substrates used under aerobic conditions, 9

Subzero temperatures, effect of on yeasts, 130
Succinate dehydrogenase, 288
Succinate formation, reductive pathway of, 302
Succinic acid, excretion of by yeasts, 50
Sucrase, 398
Sucrose, action of yeasts on, 5
hydrolysis by β-fructosidase, 291
Sugar composition of malt worts, 31
Sugar content of grape must, 32
Sugar content of sulphite waste liquor, 32
Sugar fermentation by yeasts, 4
Sugar metabolism, schematic outline, 274
Sugar transport as a factor in determining the minimum temperature for growth, 125
Sugar uptake by yeasts, kinetics of, 84
Sugars, uptake of, 39
Sulphate assimilation, 16, 363
pathway of, 372–373
Sulphate ions, uptake of, 54
Sulphate permease, 372
Sulphatides in yeast lipids, 225
Sulphide, utilization of by *Candida utilis*, 16
Sulphite ions, uptake of, 54
Sulphite reductase, 373
Sulphite, utilization by yeasts, 16
Sulphite waste liquor as a source of yeast nutrients, 32
Sulphur content of yeast cell walls, 167
Sulphur, sources of for yeasts, 16
Sulphydrylase, 374
Super-attenuating yeasts, action of on malt worts, 31
Surface area in relation to cell mass of yeasts, 4
Survival curves of *Saccharomyces cerevisiae* exposed to various temperatures, 96
Survival of yeasts, effect of temperature on, 127
Susceptibility of yeasts to sphaeroplast formation, 187
Synchronously-dividing cells of *Saccharomyces cerevisiae*, 403
Synchronously dividing culture, enzyme synthesis in, 397, 398
Synchronously dividing culture of *Schizosaccharomyces pombe*, 399
Synchronously dividing yeast cells, 315, 317, 326
Synthesis of fatty acids in yeasts, 238
Synthesis, periodic, 464–467
γ-Synthetase, 368
Systematics, yeast cell-wall composition and, 201

## T

Taurine, utilization of by *Candida utilis*, 16
Taxonomy, yeast, cell-wall composition and, 200
Temperature adaptation in yeasts, 124
Temperature, dependence of specific growth rate on, 91
effect on pigment formation, 498
growth, effect of on lipid composition of yeasts, 228
effect of on lipid content of yeasts, 215
influence of on growth and metabolism of yeasts, 119
survival curves of *Saccharomyces cerevisiae* exposed to, 96
Temperature functions in yeast growth kinetics, 93
Terminal reaction in yeast fatty-acid synthesis, 239
Tetraethenoid sterols in yeasts, 225
Tetrahydrofolate-serine transhydroxymethylase, 371
Tetrasaccharides, hydrolysis of, 291
Thallium, effect of on yeasts, 17
Thermal death of yeasts, 98
Thermal death times of yeasts, 127
Thermolability of yeast enzymes, 126
Thermosensitive sites in yeasts, 97
Thiamin, as a yeast growth factor, 23
metabolic role of in yeasts, 24
Thiazole moiety of thiamin, effect of on yeast, 24
Thickness of the yeast wall, 198
Thiohemiacetal, 280
Thiokinase, 295
Thiosulphate ions, uptake of, 54
Thiosulphate utilization by yeasts, 16
Threonine biosynthesis, 362–376

Threonine deaminases, 376, 377–379, 387, 398
Threonyl-t-RNA synthetase, 368
Thymidine incorporation by yeast, 315
Thymidine kinase, 347
Thymidylate synthetase, 347
Thymine, incorporation by yeast, 315
  in deoxyribonucleic acids, 310
Tin in yeasts, 17
*Torula monosa*, thermal death times of, 129
Torularhodin, 496, 497, 498, 499, 500, 502, 503, 504, 507, 508, 509, 510, 511, 512
*Torula*, t-RNA from, 322
Torulene, 498, 499, 502, 503, 504, 506, 507, 508, 509, 510, 511, 512
*Torulopsis apicola*, production of extracellular lipids by, 233
  structure of mannan from walls of, 157
*Torulopsis candida*, cardinal growth temperatures of, 123
  purine transport into, 55
*Torulopsis colliculosa*, growth of on hydrocarbons, 33
*Torulopsis gropengiesseri*, galactomannan in walls of, 156
*Torulopsis ingeniosa*, glucan production by, 177
*Torulopsis lactis-condensi*, galactomannan in walls of, 156
*Torulopsis magnoliae*, galactomannan in walls of, 156
*Torulopsis molischiana*, cardinal growth temperatures of, 123
*Torulopsis nodaensis*, galactomannan in walls of, 157
*Torulopsis pintolopesii*, utilization of choline by, 254
*Torulopsis rotundata*, production of starch by, 175
*Torulopsis* spp., excretion of lipids by, 174
*Torulopsis* sp., production of extracellular lipids by, 231
Toxic elements for yeasts, 17
Transaminases, 360
Transport-limited growth of yeasts, 86
Transport of hexoses, allosteric feedback inhibition of, 300
Transport, solute, as a factor in determining the minimum temperature for growth, 125
Transport system, constitutive for hexoses, 290
  inducible for galactose, 290
Trehalase, 291, 294, 466
Trehalase in yeast walls, 171
Trehalose, 291, 294, 295, 297, 420–423, 466
  biosynthesis, 421–423
  in endogenous metabolism, 296
Trehalose as a storage product in yeasts, 27
Trehalose fermentation by yeasts, 7
Trehalose phosphate phosphatase, 422
Trehalose phosphate synthetase, 294, 421, 431
*Tremella mesenterica*, production of acetylated polysaccharides by, 184
*Tremella spp.*, production of starch by, 176
Triacylglycerols, biosynthesis of in yeasts, 234
  composition of yeast, 217
  reactions leading to synthesis of in yeasts, 235
Triangular yeast, 446, 480
  morphogenesis of, 480–483
*Trichoderma viride* as a source of glucanase, 145
*Trichoderma viride*, glucanase production by, 191
*Trichosporon cutaneum*, production of extracellular pentosylmannans by, 181
  starch production by, 176
*Trichosporon fermentans*, galactomannan in walls of, 155
*Trichosporon hellenicum*, galactomannan in walls of, 155
*Trichosporon inkin*, production of extracellular xylomannans by, 181
*Trichosporon penicillatum*, galactomannan in walls of, 155
*Trichosporon sericeum*, production of extracellular xylomannans by, 181
*Trichosporon* spp., galactomannans produced extracellularly by, 180

*Trichosporon undulatum*, production of extracellular xylomannans by, 181
Triglycerides, *see* Triacylglycerols
*Trigonopsis variabilis*, 446, 452, 480–483
amino-acid content of protein in walls of, 166
protein content of walls of, 165
Triosephosphate isomerase, 277, 278
Trisaccharides, hydrolysis, 291
Tryptophan biosynthesis, 382–387
Tryptophan synthetase, 398
Tyrosine biosynthesis, 382–387
Tyrosyl-t-RNA synthetase, 389, 390
Tween 80 as a yeast growth factor, 26
Tween 80, requirement for under anaerobic conditions, 230

U

Ultraviolet radiation, effect on deoxyribonucleic acid duplication, 315
effect on mutation frequency, 436
effect on yeast, 314
forming mutants of *Rhodotorula mucilaginosa*, 507
forming mutants of *Rhodotorula rubra*, 506
sensitivity of mitochondrial DNA to, 325
Unfermentable sugars for yeasts, 5
Unsaturated fatty acids as yeast growth factors, 26
Unsaturated fatty acids, requirement for under anaerobic conditions, 230
synthesis of in yeasts, 242
Unsaturated lipids, distribution of in yeast cell, 228
effect of growth temperature on synthesis of, 228
Unsaturation in yeast sterols, biosynthesis and, 247
Uptake of amino acids by yeasts, 45
Uptake of bivalent cations by yeast, 52
Uptake of inorganic anions by yeasts, 53
Uptake of monovalent cations by yeasts, 48
Uptake of nucleotides by yeasts, 55
Uptake of organic acids by yeasts, 47
Uptake of solutes by yeasts, 3, 36
Uptake of sugars by yeasts, 39
Uranyl ion inhibition on yeasts, 5
Uranyl ions, effect of on solute transport through the yeast plasma membrane, 260
effect of on sugar uptake, 41
effect of on uptake of bivalent cations by yeasts, 52
Urea as a source of nitrogen, 14
Uric acid, effect of temperature on accumulation of by *Candida utilis*, 125
transport of into *Candida utilis*, 55
Uridine diphosphate-glucose epimerase, 290
Uridine diphosphate glucose-glucose 6-phosphate transglucosylase, 421
Uridine diphosphoglucose pyrophosphorylase, 423, 433

V

Vacuolar membranes, yeast, 263
Valine biosynthesis, 376–382
pH Value, effect of on lipid content of yeasts, 215
Valyl-t-RNA synthetase, 389, 390
Vanadium in yeasts, 17
Vegetative development of yeast, 464–483
Viable count of *Saccharomyces cerevisiae*, effect of temperature on, 97
Viscous cultures due to glucan production, 177
Vitamin $B_6$ as a yeast growth factor, 25
Vitamin content of malt worts, 31
Vitamin deficiencies, effect of on lipid composition of yeasts, 231
Vitamin deficiency, effect of on lipid contents of yeasts, 216
Vitamin excretion by yeasts, 59
Volutin granules in yeasts, 15, 27

W

Wall, cell, as a barrier in solute uptake by yeasts, 36

Walls, yeast, enzymes in, 170
yeast, lipid contents of, 227
Weak bases, permeability of, 36
Wheat flour as a source of yeast nutrients, 30
*Willia anomala*, thermal death times of, 129
Wobble hypothesis, relation to genetic code degeneracy, 395
Wort, utilization of amino acids in, 12

## X

Xanthine, effect of temperature on accumulation of by *Candida utilis*, 125
transport of into *Candida utilis*, 55
X-ray diffraction studies on yeast cell-wall glucans, 142
X-ray mutants, 507
Xylomannans produced extracellularly by yeasts, 181
Xylose fermentation by yeasts, 9
Xylose, utilization by *Candida utilis*, 290

## Y

Yeast cell envelope, structure and biosynthesis of, 135
Yeast cellulose, 137
Yeast cell-wall glucan, structure of, 140
Yeast cell wall, morphology of, 198
Yeast cerebrosides, 224
Yeast growth, kinetics and energetics of, 75
Yeast gum, nature of, 145
Yeast lipids, 211
composition of, 216
methods of estimating, 212
Yeast mannan containing both α- and β-linkages, 151
Yeast membranes, composition, structure and function of, 254
Yeast nutrition, 3
Yeast polyose, 137
Yeasts, similarities to heterotrophic bacteria, 75
Yield analogues in yeast growth kinetics, 105
Yield analysis in yeast growth kinetics, 108
Yield factor and specific maintenance rate, 100
Yield factor in yeast growth kinetics, 83
Yield factor, variation of with specific growth rate of yeasts, 85
Yield relations in yeast growth kinetics, 100
Yields of adenosine triphosphate in yeasts, 106

## Z

β-Zeacarotene, 502, 503, 507, 508, 509, 510
Zinc in yeasts, 17
Zinc ions, uptake of, 52
*Zygosaccharomyces acidifaciens*, glycolytic process in, 282
Zymohexoses, fermentation of, 4
Zymosterol, biosynthesis of in yeasts, 246
Zymosterol in yeasts, 225